Life

The Science of Biology

NINTH EDITION

NINTH EDITION *Life* The Science of Biology

DAVID
SADAVA
The Claremont Colleges
Claremont, California

DAVID M.
HILLIS
University of Texas
Austin, Texas

H. CRAIG
HELLER
Stanford University
Stanford, California

MAY R.
BERENBAUM
University of Illinois
Urbana-Champaign, Illinois

About the Cover

The cover of *Life* captures many themes that echo throughout the book. The photograph shows a lesser long-nosed bat pollinating a saguaro cactus. This cactus has evolved large flowers that produce copious quantities of nectar. The nectar attracts many species that pollinate the cactus, including bats. The ability of bats to hover as they feed on the nectar of the cactus is an excellent example of adaptation of body form and physiology. These themes of adaptation, evolution, nutrition, reproduction, species interactions, and integrated form and function are ideas that are repeated throughout the chapters of *Life*. Photograph copyright © Dr. Merlin D. Tuttle/Photo Researchers, Inc.

The Frontispiece

Blue wildebeest and Burchell's zebra migrate together through Serengeti National Park, Tanzania. Copyright © Art Wolfe, www.artwolfe.com.

LIFE: The Science of Biology, Ninth Edition

Address editorial correspondence to:
Sinauer Associates Inc., 23 Plumtree Road, Sunderland, MA 01375 U.S.A.
www.sinauer.com
publish@sinauer.com

Address orders to:
MPS / W. H. Freeman & Co., Order Dept., 16365 James Madison Highway,
U.S. Route 15, Gordonsville, VA 22942 U.S.A.
Examination copy information: 1-800-446-8923
Orders: 1-888-330-8477

Planet Friendly Publishing
✔ Made in the United States
✔ Printed on Recycled Paper
Text: 10%   Cover: 10%
GREEN EDITION   Learn more: www.greenedition.org

© **Mixed Sources**
Product group from well-managed forests, controlled sources and recycled wood or fibre
www.fsc.org Cert no. SW-COC-002985
© 1996 Forest Stewardship Council
FSC

Library of Congress Cataloging-in-Publication Data
Life, the science of biology / David Sadava .. [et al.]. — 9th ed.
    p. cm.
    Includes index.
    ISBN 978-1-4292-1962-4 (hardcover) — 978-1-4292-4645-3 (pbk. : v. 1) —
ISBN 978-1-4292-4644-6 (pbk. : v. 2) — ISBN 978-1-4292-4647-7 (pbk. : v. 3)
1. Biology. I. Sadava, David E.
    QH308.2.L565 2011
    570—dc22   2009036693

Printed in U.S.A.
First Printing October 2009
The Courier Companies, Inc.

To Bill Purves and Gordon Orians,
extraordinary colleagues, biologists, and teachers,
and the original authors of LIFE

The Authors

CRAIG HELLER   DAVID HILLIS   MAY BERENBAUM   DAVID SADAVA

DAVID SADAVA is the Pritzker Family Foundation Professor of Biology, Emeritus, at the Keck Science Center of Claremont McKenna, Pitzer, and Scripps, three of The Claremont Colleges. In addition, he is Adjunct Professor of Cancer Cell Biology at the City of Hope Medical Center. Twice winner of the Huntoon Award for superior teaching, Dr. Sadava taught courses on introductory biology, biotechnology, biochemistry, cell biology, molecular biology, plant biology, and cancer biology. In addition to *Life: The Science of Biology*, he is the author or coauthor of books on cell biology and on plants, genes, and crop biotechnology. His research has resulted in many papers coauthored with his students, on topics ranging from plant biochemistry to pharmacology of narcotic analgesics to human genetic diseases. For the past 15 years, he has investigated multi-drug resistance in human small-cell lung carcinoma cells with a view to understanding and overcoming this clinical challenge. At the City of Hope, his current work focuses on new anti-cancer agents from plants.

DAVID HILLIS is the Alfred W. Roark Centennial Professor in Integrative Biology and the Director of the Center for Computational Biology and Bioinformatics at the University of Texas at Austin, where he also has directed the School of Biological Sciences. Dr. Hillis has taught courses in introductory biology, genetics, evolution, systematics, and biodiversity. He has been elected into the membership of the National Academy of Sciences and the American Academy of Arts and Sciences, awarded a John D. and Catherine T. MacArthur Fellowship, and has served as President of the Society for the Study of Evolution and of the Society of Systematic Biologists. His research interests span much of evolutionary biology, including experimental studies of evolving viruses, empirical studies of natural molecular evolution, applications of phylogenetics, analyses of biodiversity, and evolutionary modeling. He is particularly interested in teaching and research about the practical applications of evolutionary biology.

CRAIG HELLER is the Lorry I. Lokey/BusinessWire Professor in Biological Sciences and Human Biology at Stanford University. He earned his Ph.D. from the Department of Biology at Yale University in 1970. Dr. Heller has taught in the core biology courses at Stanford since 1972 and served as Director of the Program in Human Biology, Chairman of the Biological Sciences Department, and Associate Dean of Research. Dr. Heller is a fellow of the American Association for the Advancement of Science and a recipient of the Walter J. Gores Award for excellence in teaching. His research is on the neurobiology of sleep and circadian rhythms, mammalian hibernation, the regulation of body temperature, the physiology of human performance, and the neurobiology of learning. Dr. Heller has done research on a huge variety of animals and physiological problems ranging from sleeping kangaroo rats, diving seals, hibernating bears, photoperiodic hamsters, and exercising athletes. Some of his recent work on the effects of temperature on human performance is featured in the opener to Chapter 40, "Physiology, Homeostasis, and Temperature Regulation."

MAY BERENBAUM is the Swanlund Professor and Head of the Department of Entomology at the University of Illinois at Urbana-Champaign. She has taught courses in introductory animal biology, entomology, insect ecology, and chemical ecology, and has received awards at the regional and national level for distinguished teaching from the Entomological Society of America. A fellow of the National Academy of Sciences, the American Academy of Arts and Sciences, and the American Philosophical Society, she served as President of the American Institute for Biological Sciences in 2009. Her research addresses insect–plant coevolution, from molecular mechanisms of detoxification to impacts of herbivory on community structure. Concerned with the practical application of ecological and evolutionary principles, she has examined impacts of genetic engineering, global climate change, and invasive species on natural and agricultural ecosystems. Devoted to fostering science literacy, she has published numerous articles and five books on insects for the general public.

Contents in Brief

Investigating Life/Tools for Investigating Life

INVESTIGATING LIFE

TOOLS FOR INVESTIGATING LIFE

Preface

Biology is a dynamic, exciting, and important subject. It is dynamic because it is constantly changing, with new discoveries about the living world being made every day. (Although it is impossible to pinpoint an exact number, approximately 1 million new research articles in biology are published each year.) The subject is exciting because life in all of its forms has always fascinated people. As active scientists who have spent our careers teaching and doing research in a wide variety of fields, we know this first hand.

Biology has always been important in peoples' daily lives, if only through the effects of achievements in medicine and agriculture. Today more than ever the science of biology is at the forefront of human concerns as we face challenges raised both by recent advances in genome science and by the rapidly changing environment.

Life's new edition brings a fresh approach to the study of biology while retaining the features that have made the book successful in the past. A new coauthor, the distinguished entomologist May R. Berenbaum (University of Illinois at Urbana-Champaign) has joined our team, and the role of evolutionary biologist David Hillis (University of Texas at Austin) is greatly expanded in this edition. The authors hail from large, medium-sized, and small institutions. Our multiple perspectives and areas of expertise, as well as input from many colleagues and students who used previous editions, have informed our approach to this new edition.

Enduring Features

We remain committed to blending the presentation of core ideas with an emphasis on introducing students to the *process of scientific inquiry*. Having pioneered the idea of depicting seminal experiments in specially designed figures, we continue to develop this here, with 79 **INVESTIGATING LIFE** figures. Each of these figures sets the experiment in perspective and relates it to the accompanying text. As in previous editions, these figures employ a structure: Hypothesis, Method, Results, and Conclusion. They often include questions for further research that ask students to conceive an experiment that would explore a related question. Each *Investigating Life* figure has a reference to BioPortal (*yourBioPortal.com*), where citations to the original work as well as additional discussion and references to follow-up research can be found.

A related feature is the **TOOLS FOR INVESTIGATING LIFE** figures, which depict laboratory and field methods used in biology. These, too, have been expanded to provide more useful context for their importance.

Over a decade ago—in *Life's* Fifth Edition—the authors and publishers pioneered the much-praised use of **BALLOON CAPTIONS** in our figures. We recognized then, and it is even truer today, that many students are visual learners. The balloon captions bring explanations of intricate, complex processes directly into the illustration, allowing students to integrate information without repeatedly going back and forth between the figure, its legend, and the text.

Life is the only introductory textbook for biology majors to begin each chapter with a story. These **OPENING STORIES** provide historical, medical, or social context and are intended to intrigue students while helping them see how the chapter's biological subject relates to the world around them. In the new edition, all of the opening stories (some 70 percent of which are new) are revisited in the body of the chapter to drive home their relevance.

We continue to refine our well-received *chapter organization*. The chapter-opening story ends with a brief **IN THIS CHAPTER** preview of the major subjects to follow. A **CHAPTER OUTLINE** asks questions to emphasize scientific inquiry, each of which is answered in a major section of the chapter. A **RECAP** at the end of each section asks the student to pause and answer questions to review and test their mastery of the previous material. The end-of-chapter summary continues this inquiry framework and highlights key figures, bolded terms, and activities and animated tutorials available in BioPortal.

New Features

Probably the most important new feature of this edition is *new authorship*. Like the biological world, the authorship team of *Life* continues to evolve. While two of us (Craig Heller and David Sadava) continue as coauthors, David Hillis has a greatly expanded role, with full responsibility for the units on evolution and diversity. New coauthor May Berenbaum has rewritten the chapters on ecology. The perspectives of these two acclaimed experts have invigorated the entire book (as well as their coauthors).

Even with the enduring features (see above), this edition has a different look and feel from its predecessor. A fresh *new design* is more open and, we hope, more accessible to students. The extensively *revised art program* has a contemporary style and color palette. The information flow of the figures is easier to follow, with numbered balloons as a guide for students. There are new conceptual figures, including a striking visual timeline for the evolution of life on Earth (Figure 25.12) and a single overview figure that summarizes the information in the genome (Figure 17.4).

In response to instructors who asked for more real-world data, we have incorporated a feature introduced online in the Eighth Edition, **WORKING WITH DATA**. There are now 36 of these exercises, most of which relate to an *Investigating Life* figure. Each is referenced at the end of the relevant chapter and is available online via BioPortal (*yourBioPortal.com*). In these exercises, we describe in detail the context and approach of the research paper that forms the basis of the figure. We then ask the student to examine the data, to make calculations, and to draw conclusions.

We are proud that this edition is a *greener Life,* with the goal of reducing our environmental impact. This is the first introductory biology text to be printed on paper earning the Forest Stewardship Council label, the "gold standard" in green paper products, and it is manufactured from wood harvested from sustainable forests. And, of course, we also offer *Life* as an eBook.

The Ten Parts

We have reorganized the book into ten parts. **Part One, The Science of Life and Its Chemical Basis**, sets the stage for the book: the opening chapter focuses on biology as an exciting science. We begin with a startling observation: the recent, dramatic decline of amphibian species throughout the world. We then show how biologists have formed hypotheses for the causes of this environmental problem and are testing them by carefully designed experiments, with a view not only to understanding the decline, but reversing it. This leads to an outline of the basic principles of biology that are the foundation for the rest of the book: the unity of life at the cellular level and how evolution unites the living world. This is followed by chapters on the basic chemical building blocks that underlie life. We have added a new chapter on nucleic acids and the origin of life, introducing the concepts of genes and gene expression early and expanding our coverage of the major ideas on how life began and evolved at its earliest stages.

In **Part Two, Cells**, we describe the view of life as seen through cells, its structural units. In response to comments by users of our previous edition, we have moved the chapter on cell signaling and communication from the genetics section to this part of the book, with a change in emphasis from genes to cells. There is an updated discussion of ideas on the origin of cells and organelles, as well as expanded treatment of water transport across membranes.

Part Three, Cells and Energy, presents an integrated view of biochemistry. For this edition, we have worked to clarify such challenging concepts as energy transfer, allosteric enzymes, and biochemical pathways. There is extensive revision of the discussions of alternate pathways of photosynthetic carbon fixation, as well as a greater emphasis on applications throughout these chapters.

Part Four, Genes and Heredity, is extensively revised and reorganized to improve clarity, link related concepts, and provide updates from recent research results. Separate chapters on prokaryotic genetics and molecular medicine have been removed and their material woven into relevant chapters. For example, our chapter on cell reproduction now includes a discussion of how the basic mechanisms of cell division are altered in cancer cells. The chapter on transmission genetics now includes coverage of this phenomenon in prokaryotes. New chapters on gene expression and gene regulation compare prokaryotic and eukaryotic mechanisms and include a discussion of epigenetics. A new chapter on mutation describes updated applications of medical genetics.

In **Part Five, Genomes**, we reinforce the concepts of the previous part, beginning with a new chapter on genomes—how they are analyzed and what they tell us about the biology of prokaryotes and eukaryotes, including humans. This leads to a chapter describing how our knowledge of molecular biology and genetics underpins biotechnology (the application of this knowledge to practical problems). We discuss some of the latest uses of biotechnology, including environmental cleanup. Part Five finishes with two chapters on development that explore the themes of molecular biology and evolution, linking these two parts of the book.

Part Six, The Patterns and Processes of Evolution, emphasizes the importance of evolutionary biology as a basis for comparing and understanding all aspects of biology. These chapters have been extensively reorganized and revised, as well as updated with the latest thinking of biologists in this rapidly changing field. This part now begins with the evidence and mechanisms of evolution, moves into a discussion of phylogenetic trees, then covers speciation and molecular evolution, and concludes with the evolutionary history of life on Earth. An integrated timeline of evolutionary history shows the timing of major events of biological evolution, the movements of the continents, floral and faunal reconstructions of major time periods, and depicts some of the fossils that form the basis of the reconstructions.

In **Part Seven, The Evolution of Diversity**, we describe the latest views on biodiversity and evolutionary relationships. Each chapter has been revised to make it easier for the reader to appreciate the major changes that have evolved within the various groups of organisms. We emphasize understanding the big picture of organismal diversity, as opposed to memorizing a taxonomic hierarchy and names (although these are certainly important). Throughout the book, the tree of life is emphasized as a way of understanding and organizing biological information. A *Tree of Life Appendix* allows students to place any group of organisms mentioned in the text of our book into the context of the rest of life. The web-based version of this appendix provides links to photos, keys, species lists, distribution maps, and other information to help students explore biodiversity of specific groups in greater detail.

After modest revisions in the past two editions, **Part Eight, Flowering Plants: Form and Function**, has been extensively reorganized and updated with the help of Sue Wessler, to include both classical and more recent approaches to plant physiology. Our emphasis is not only on the basic findings that led to the elucidation of mechanisms for plant growth and reproduction, but also on the use of genetics of model organisms. There is expanded coverage of the cell signaling events that regulate gene expression in plants, integrating concepts introduced earlier in the book. New material on how plants respond to their environment is included, along with links to both the book's earlier descriptions of plant diversity and later discussions of ecology.

Part Nine, Animals: Form and Function, continues to provide a solid foundation in physiology through comprehensive coverage of basic principles of function of each organ system and then emphasis on mechanisms of control and integration. An important reorganization has been moving the chapter on immunology from earlier in the book, where its emphasis was on molecular genetics, to this part, where it is more closely allied to the information systems of the body. In addition, we have added a number of new experiments and made considerable effort to clarify the sometimes complex phenomena shown in the illustrations.

Part Ten, Ecology, has been significantly revised by our new coauthor, May Berenbaum. A new chapter of biological interactions has been added (a topic formerly covered in the community ecology chapter). Full of interesting anecdotes and discussions of field studies not previously described in biology texts, this new ecology unit offers practical insights into how ecologists acquire, interpret, and apply real data. This brings the book full circle, drawing upon and reinforcing prior topics of energy, evolution, phylogenetics, Earth history, and animal and plant physiology.

Exceptional Value Formats

We again provide *Life* both as the full book and as a cluster of *paperbacks*. Thus, instructors who want to use less than the whole book can choose from these split volumes, each with the book's front matter, appendices, glossary, and index.

Volume I, The Cell and Heredity, includes: Part One, The Science of Life and Its Chemical Basis (Chapters 1–4); Part Two, Cells (Chapters 5–7); Part Three, Cells and Energy (Chapters 8–10); Part Four, Genes and Heredity (Chapters 11–16); and Part Five, Genomes (Chapters 17–20).

Volume II, Evolution, Diversity, and Ecology, includes: Chapter 1, Studying Life; Part Six, The Patterns and Processes of Evolution (Chapters 21–25); Part Seven, The Evolution of Diversity (Chapters 26–33); and Part Ten, Ecology (Chapters 54–59).

Volume III, Plants and Animals, includes: Chapter 1, Studying Life; Part Eight, Flowering Plants: Form and Function (Chapters 34–39); and Part Nine, Animals: Form and Function (Chapters 40–53).

Responding to student concerns, we offer two options of the entire book at a *significantly reduced cost*. After it was so well received in the previous edition, we again provide *Life* as a *looseleaf version*. This shrink-wrapped, unbound, 3-hole punched version fits into a 3-ring binder. Students take only what they need to class and can easily integrate any instructor handouts or other resources.

Life was the first comprehensive biology text to offer the entire book as a truly robust *eBook*. For this edition, we continue to offer a flexible, interactive ebook that gives students a new way to read the text and learn the material. The ebook integrates the student media resources (animations, quizzes, activities, etc.) and offers instructors a powerful way to customize the textbook with their own text, images, Web links, documents, and more.

Media and Supplements for the Ninth Edition

The wide range of media and supplements that accompany *Life*, Ninth Edition have all been created with the dual goal of helping students learn the material presented in the textbook more efficiently and helping instructors teach their courses more effectively. Students in majors introductory biology are faced with learning a tremendous number of new concepts, facts, and terms, and the more different ways they can study this material, the more efficiently they can master it.

All of the *Life* media and supplemental resources have been developed specifically for this textbook. This provides strong consistency between text and media, which in turn helps students learn more efficiently. For example, the animated tutorials and activities found in BioPortal were built using textbook art, so that the manner in which structures are illustrated, the colors used to identify objects, and the terms and abbreviations used are all consistent.

For the Ninth Edition, a new set of Interactive Tutorials gives students a new way to explore many key topics across the textbook. These new modules allow the student to learn by doing, including solving problem scenarios, working with experimental techniques, and exploring model systems. All new copies of the Ninth Edition include access to the robust new version of BioPortal, which brings together all of *Life's* student and instructor resources, powerful assessment tools, and new integration with Prep-U adaptive quizzing.

The rich collection of visual resources in the Instructor's Media Library provides instructors with a wide range of options for enhancing lectures, course websites, and assignments. Highlights include: layered art PowerPoint® presentations that break down complex figures into detailed, step-by-step presentations; a collection of approximately 200 video segments that can help capture the attention and imagination of students; and PowerPoint slides of textbook art with editable labels and leaders that allow easy customization of the figures.

For a detailed description of all the media and supplements available for the Ninth Edition, please turn to "*Life's* Media and Supplements," on page xvii.

Many People to Thank

"If I have seen farther, it is by standing on the shoulders of giants." The great scientist Isaac Newton wrote these words over 330 years ago and, while we certainly don't put ourselves in his lofty place in science, the words apply to us as coauthors of this text. This is the first edition that does not bear the names of Bill Purves and Gordon Orians. As they enjoy their "retirements," we are humbled by their examples as biologists, educators, and writers.

One of the wisest pieces of advice ever given to a textbook author is to "be passionate about your subject, but don't put your ego on the page." Considering all the people who looked over our shoulders throughout the process of creating this book, this advice could not be more apt. We are indebted to many people who gave invaluable help to make this book what it is. First and foremost are our colleagues, biologists from over 100 institutions. Some were users of the previous edition, who suggested many improvements. Others reviewed our chapter drafts in detail, including advice on how to improve the illustrations. Still others acted as accuracy reviewers when the book was almost completed. All of these biologists are listed in the Reviewer credits.

Of special note is Sue Wessler, a distinguished plant biologist and textbook author from the University of Georgia. Sue looked critically at Part Eight, Flowering Plants: Form and Function, wrote three of the chapters (34–36), and was important in the revision of the other three (37–39). The new approach to plant biology in this edition owes a lot to her.

The pace of change in biology and the complexities of preparing a book as broad as this one necessitated having two developmental editors. James Funston coordinated Parts 1–5, and Carol Pritchard-Martinez coordinated Parts 6–10. We benefitted from the wide experience, knowledge, and wisdom of both of them. As the chapter drafts progressed, we were fortunate to have experienced biologist Laura Green lending her critical eye as in-house editor. Elizabeth Morales, our artist, was on her third edition with us. As we have noted, she extensively revised almost all of the prior art and translated our crude sketches into beautiful new art. We hope you agree that our art program re-

mains superbly clear and elegant. Our copy editors, Norma Roche, Liz Pierson, and Jane Murfett, went far beyond what such people usually do. Their knowledge and encyclopedic recall of our book's chapters made our prose sharper and more accurate. Diane Kelly, Susan McGlew, and Shannon Howard effectively coordinated the hundreds of reviews that we described above. David McIntyre was a terrific photo editor, finding over 550 new photographs, including many new ones of his own, that enrich the book's content and visual statement. Jefferson Johnson is responsible for the design elements that make this edition of *Life* not just clear and easy to learn from, but beautiful as well. Christopher Small headed the production department—Joanne Delphia, Joan Gemme, Janice Holabird, and Jefferson Johnson—who contributed in innumerable ways to bringing *Life* to its final form. Jason Dirks once again coordinated the creation of our array of media and supplements, including our superb new Web resources. Carol Wigg, for the ninth time in nine editions, oversaw the editorial process; her influence pervades the entire book.

W. H. Freeman continues to bring *Life* to a wider audience. Associate Director of Marketing Debbie Clare, the Regional Specialists, Regional Managers, and experienced sales force are effective ambassadors and skillful transmitters of the features and unique strengths of our book. We depend on their expertise and energy to keep us in touch with how *Life* is perceived by its users. And thanks also to the Freeman media group for eBook and BioPortal production.

Finally, we are indebted to Andy Sinauer. Like ours, his name is on the cover of the book, and he truly cares deeply about what goes into it. Combining decades of professionalism, high standards, and kindness to all who work with him, he is truly our mentor and friend.

DAVID SADAVA

DAVID HILLIS

CRAIG HELLER

MAY BERENBAUM

Reviewers for the Ninth Edition

Between-Edition Reviewers

David D. Ackerly, University of California, Berkeley

Amy Bickham Baird, University of Leiden

Jeremy Brown, University of California, Berkeley

John M. Burke, University of Georgia

Ruth E. Buskirk, University of Texas, Austin

Richard E. Duhrkopf, Baylor University

Casey W. Dunn, Brown University

Erika J. Edwards, Brown University

Kevin Folta, University of Florida

Lynda J. Goff, University of California, Santa Cruz

Tracy A. Heath, University of Kansas

Shannon Hedtke, University of Texas, Austin

Richard H. Heineman, University of Texas, Austin

Albert Herrera, University of Southern California

David S. Hibbett, Clark University

Norman A. Johnson, University of Massachusetts

Walter S. Judd, University of Florida

Laura A. Katz, Smith College

Emily Moriarty Lemmon, Florida State University

Sheila McCormick, University of California, Berkeley

Robert McCurdy, Independence Creek Nature Preserve

Jacalyn Newman, University of Pittsburgh

Juliet F. Noor, Duke University

Theresa O'Halloran, University of Texas, Austin

K. Sata Sathasivan, University of Texas, Austin

H. Bradley Shaffer, University of California, Davis

Rebecca Symula, Yale University

Christopher D. Todd, University of Saskatchewan

Elizabeth Willott, University of Arizona

Kenneth Wilson, University of Saskatchewan

Manuscript Reviewers

Tamarah Adair, Baylor University

William Adams, University of Colorado, Boulder

Gladys Alexandre, University of Tennessee, Knoxville

Shivanthi Anandan, Drexel University

Brian Bagatto, University of Akron

Lisa Baird, University of San Diego

Stewart H. Berlocher, University of Illinois, Urbana-Champaign

William Bischoff, University of Toledo

Meredith M. Blackwell, Louisiana State University

David Bos, Purdue University

Jonathan Bossenbroek, University of Toledo

Nicole Bournias-Vardiabasis, California State University, San Bernardino

Nancy Boury, Iowa State University

Sunny K. Boyd, University of Notre Dame

Judith L. Bronstein, University of Arizona

W. Randy Brooks, Florida Atlantic University

James J. Bull, University of Texas, Austin

Darlene Campbell, Cornell University

Domenic Castignetti, Loyola University, Chicago

David T. Champlin, University of Southern Maine

Shu-Mei Chang, University of Georgia

Samantha K. Chapman, Villanova University

Patricia Christie, MIT

Wes Colgan, Pikes Peak Community College

John Cooper, Washington University

Ronald Cooper, University of California, Los Angeles

Elizabeth Cowles, Eastern Connecticut State University

Jerry Coyne, University of Chicago

William Crampton, University of Central Florida

Michael Dalbey, University of California, Santa Cruz

Anne Danielson-Francois, University of Michigan, Dearborn

Grayson S. Davis, Valparaiso University

Kevin Dixon, Florida State University

Zaldy Doyungan, Texas A&M University, Corpus Christi

Ernest F. Dubrul, University of Toledo

Roland Dute, Auburn University

Scott Edwards, Harvard University

William Eldred, Boston University

David Eldridge, Baylor University

Joanne Ellzey, University of Texas, El Paso

Susan H. Erster, State University of New York, Stony Book

Brent E. Ewers, University of Wyoming

Kevin Folta, University of Florida

Brandon Foster, Wake Technical Community College

Richard B. Gardiner, University of Western Ontario

Douglas Gayou, University of Missouri, Columbia

John R. Geiser, Western Michigan University

Arundhati Ghosh, University of Pittsburgh

Alice Gibb, Northern Arizona University

Scott Gilbert, Swarthmore College

Matthew R. Gilg, University of North Florida

Elizabeth Godrick, Boston University

Lynda J. Goff, University of California, Santa Cruz

Elizabeth Blinstrup Good, University of Illinois, Urbana-Champaign

John Nicholas Griffis, University of Southern Mississippi

Cameron Gundersen, University of California, Los Angeles

Kenneth Halanych, Auburn University

E. William Hamilton, Washington and Lee University

Monika Havelka, University of Toronto at Mississauga

Tyson Hedrick, University of North Carolina, Chapel Hill

Susan Hengeveld, Indiana University, Bloomington

Albert Herrera, University of Southern California

Kendra Hill, South Dakota State University

Richard W. Hill, Michigan State University

Erec B. Hillis, University of California, Berkeley

Jonathan D. Hillis, Carleton College

William Huddleston, University of Calgary

Dianne B. Jennings, Virginia Commonwealth University

Norman A. Johnson, University of Massachusetts, Amherst

William H. Karasov, University of Wisconsin, Madison

Susan Keen, University of California, Davis

Cornelis Klok, Arizona State University, Tempe

Olga Ruiz Kopp, Utah Valley University

William Kroll, Loyola University, Chicago

Allen Kurta, Eastern Michigan University

Rebecca Lamb, Ohio State University

Brenda Leady, University of Toledo

Hugh Lefcort, Gonzaga University

Sean C. Lema, University of North Carolina, Wilmington

Nathan Lents, John Jay College, City University of New York

Rachel A. Levin, Amherst College

Donald Levin, University of Texas, Austin

Bernard Lohr, University of Maryland, Baltimore County

Barbara Lom, Davidson College

David J. Longstreth, Louisiana State University

Catherine Loudon, University of California, Irvine

Francois Lutzoni, Duke University

Charles H. Mallery, University of Miami

Kathi Malueg, University of Colorado, Colorado Springs

Richard McCarty, Johns Hopkins University

Sheila McCormick, University of California, Berkeley

Francis Monette, Boston University

Leonie Moyle, Indiana University, Bloomington

Jennifer C. Nauen, University of Delaware

Jacalyn Newman, University of Pittsburgh

Alexey Nikitin, Grand Valley State University

Shawn E. Nordell, Saint Louis University

Tricia Paramore, Hutchinson Community College

Nancy J. Pelaez, Purdue University

Robert T. Pennock, Michigan State University

Roger Persell, Hunter College

Debra Pires, University of California, Los Angeles

Crima Pogge, City College of San Francisco

Jaimie S. Powell, Portland State University

Susan Richardson, Florida Atlantic University

David M. Rizzo, University of California, Davis

Benjamin Rowley, University of Central Arkansas

Brian Rude, Mississippi State University

Ann Rushing, Baylor University

Christina Russin, Northwestern University

Udo Savalli, Arizona State University, West

Frieder Schoeck, McGill University

Paul J. Schulte, University of Nevada, Las Vegas

Stephen Secor, University of Alabama

Vijayasaradhi Setaluri, University of Wisconsin, Madison

H. Bradley Shaffer, University of California, Davis

Robin Sherman, Nova Southeastern University

Richard Shingles, Johns Hopkins University

James Shinkle, Trinity University

Richard M. Showman, University of South Carolina

Felisa A. Smith, University of New Mexico

Ann Berry Somers, University of North Carolina, Greensboro

Ursula Stochaj, McGill University

Ken Sweat, Arizona State University, West

Robin Taylor, Ohio State University

William Taylor, University of Toledo

Mark Thogerson, Grand Valley State University

Sharon Thoma, University of Wisconsin, Madison

Lars Tomanek, California Polytechnic State University

James Traniello, Boston University

Jeffrey Travis, State University of New York, Albany

Terry Trier, Grand Valley State University

John True, State University of New York, Stony Brook

Elizabeth Van Volkenburgh, University of Washington

John Vaughan, St. Petersburg College

Sara Via, University of Maryland

Suzanne Wakim, Butte College (Glenn Community College District)

Randall Walikonis, University of Connecticut

Cindy White, University of Northern Colorado

Elizabeth Willott, University of Arizona

Mark Wilson, Humboldt State University

Stuart Wooley, California State University, Stanislaus

Lan Xu, South Dakota State University

Heping Zhou, Seton Hall University

Accuracy Reviewers

John Alcock, Arizona State University

Gladys Alexandre, University of Tennessee, Knoxville

Lawrence A. Alice, Western Kentucky University

David R. Angelini, American University

Fabia U. Battistuzzi, Arizona State University

Arlene Billock, University of Louisiana, Lafayette

Mary A. Bisson, State University of New York, Buffalo

Meredith M. Blackwell, Louisiana State University

Nancy Boury, Iowa State University

Eldon J. Braun, University of Arizona

Daniel R. Brooks, University of Toronto

Jennifer L. Campbell, North Carolina State University

Peter C. Chabora, Queens College, CUNY

Patricia Christie, MIT

Ethan Clotfelter, Amherst College

Robert Connour, Owens Community College

Peter C. Daniel, Hofstra University

D. Michael Denbow, Virginia Polytechnic Institute

Laura DiCaprio, Ohio State University

Zaldy Doyungan, Texas A&M University, Corpus Christi

Moon Draper, University of Texas, Austin

Richard E. Duhrkopf, Baylor University

Susan A. Dunford, University of Cincinnati

Brent E. Ewers, University of Wyoming

James S. Ferraro, Southern Illinois University

Rachel D. Fink, Mount Holyoke College

John R. Geiser, Western Michigan University

Elizabeth Blinstrup Good, University of Illinois, Urbana-Champaign

Melina E. Hale, University of Chicago

Patricia M. Halpin, University of California, Los Angeles

Jean C. Hardwick, Ithaca College

Monika Havelka, University of Toronto at Mississauga

Frank Healy, Trinity University

Marshal Hedin, San Diego State University

Albert Herrera, University of Southern California

David S. Hibbett, Clark University

James F. Holden, University of Massachusetts, Amherst

Margaret L. Horton, University of North Carolina, Greensboro

Helen Hull-Sanders, Canisius College

C. Darrin Hulsey, University of Tennessee, Knoxville

Timothy Y. James, University of Michigan

Dianne B. Jennings, Virginia Commonwealth University

Norman A. Johnson, University of Massachusetts, Amherst

Susan Jorstad, University of Arizona

Ellen S. Lamb, University of North Carolina, Greensboro

Dennis V. Lavrov, Iowa State University

Hugh Lefort, Gonzaga University

Rachel A. Levin, Amherst College

Bernard Lohr, University of Maryland, Baltimore County

Sharon E. Lynn, College of Wooster

Sarah Mathews, Harvard University

Susan L. Meacham, University of Nevada, Las Vegas

Mona C. Mehdy, University of Texas, Austin

Bradley G. Mehrtens, University of Illinois, Urbana-Champaign

James D. Metzger, Ohio State University

Thomas W. Moon, University of Ottowa

Thomas M. Niesen, San Francisco State University

Theresa O'Halloran, University of Texas, Austin

Thomas L. Pannabecker, University of Arizona

Nancy J. Pelaez, Purdue University

Nicola J. R. Plowes, Arizona State University

Gregory S. Pryor, Francis Marion University

Laurel B. Roberts, University of Pittsburgh

Anjana Sharma, Western Carolina University

Richard M. Showman, University of South Carolina

John B. Skillman, California State University, San Bernadino

John J. Stachowicz, University of California, Davis

Brook O. Swanson, Gonzaga University

Robin A. J. Taylor, Ohio State University

William Taylor, University of Toledo

Steven M. Theg, University of California, Davis

Mark Thogerson, Grand Valley State University

Christopher D. Todd, University of Saskatchewan

Jeffrey Travis, State University of New York, Albany

Joseph S. Walsh, Northwestern University

Andrea Ward, Adelphi University

Barry Williams, Michigan State University

Kenneth Wilson, University of Saskatchewan

Carol L. Wymer, Morehead State University

LIFE's Media and Supplements

BIO P⊛RTAL featuring Prep-U
yourBioPortal.com

BioPortal is the new gateway to all of *Life's* state-of-the-art on-line resources for students and instructors. BioPortal includes the breakthrough quizzing engine, Prep-U; a fully interactive eBook; and additional premium learning media. The textbook is tightly integrated with BioPortal via in-text references that connect the printed text and media resources. The result is a powerful, easily-managed online course environment. Bio-Portal includes the following features and resources:

Life, Ninth Edition eBook

- Integration of all activities, animated tutorials, and other media resources.
- Quick, intuitive navigation to any section or subsection, as well as any printed book page number.
- In-text links to all glossary entries.
- Easy text highlighting.
- A bookmarking feature that allows for quick reference to any page.
- A powerful Notes feature that allows students to add notes to any page.
- A full glossary and index.
- Full-text search, including an additional option to search the glossary and index.
- Automatic saving of all notes, highlighting, and bookmarks.

Additional eBook features for instructors:

- Content Customization: Instructors can easily add pages of their own content and/or hide chapters or sections that they do not cover in their course.
- Instructor Notes: Instructors can choose to create an annotated version of the eBook with their own notes on any page. When students in the course log in, they see the instructor's personalized version of the eBook. Instructor notes can include text, Web links, images, links to all Bio-Portal content, and more.

✔PrepU
Smarter than the average quiz

Built by educators, Prep-U focuses student study time exactly where it should be, through the use of personalized, adaptive quizzes that move students toward a better grasp of the material—and better grades. For *Life*, Ninth Edition, Prep-U is fully integrated into BioPortal, making it easy for instructors to take advantage of this powerful quizzing engine in their course. Features include:

- Adaptive quizzing
- Automatic results reporting into the BioPortal gradebook

- Misconception index
- Comparison to national data

Student Resources

Diagnostic Quizzing. The diagnostic quiz for each chapter of *Life* assesses student understanding of that chapter, and generates a Personalized Study Plan to effectively focus student study time. The plan includes links to specific textbook sections, animated tutorials, and activities.

Interactive Summaries. For each chapter, these dynamic summaries combine a review of important concepts with links to all of the key figures from the chapter as well as all of the relevant animated tutorials, activities, and key terms.

Animated Tutorials. Over 100 in-depth animated tutorials, in a new format for the Ninth Edition, present complex topics in a clear, easy-to-follow format that combines a detailed animation with an introduction, conclusion, and quiz.

Activities. Over 120 interactive activities help students learn important facts and concepts through a wide range of exercises, such as labeling steps in processes or parts of structures, building diagrams, and identifying different types of organisms.

NEW! Interactive Tutorials. New for the Ninth Edition, these tutorial modules help students master key concepts through hands-on activities that allow them to learn through action. With these tutorials, students can solve problem scenarios by applying concepts from the text, by working with experimental techniques, and by using interactive models to discover how biological mechanisms work. Each tutorial includes a self-assessment quiz that can be assigned.

Interactive Quizzes. Each question includes an image from the textbook, thorough feedback on both correct and incorrect answer choices, references to textbook pages, and links to eBook pages, for quick review.

BioNews from Scientific American. BioNews makes it easy for instructors to bring the dynamic nature of the biological sciences and up-to-the minute currency into their course. Accessible from within BioPortal, BioNews is a continuously updated feed of current news, podcasts, magazine articles, science blog entries, "strange but true" stories, and more.

NEW! BioNavigator. This unique visual resource is an innovative way to access the wide variety of *Life* media resources. Starting from the whole-Earth view, instructors and students can zoom to any level of biological inquiry, encountering links to a wealth of animations, activities, and tutorials on the full range of topics along the way.

Working with Data. Built around some of the original experiments depicted in the Investigating Life figures, these exercises help build quantitative skills and encourage student interest in how scientists do research, by looking at real experimental data and answering questions based on those data.

Flashcards. For each chapter of the book, there is a set of flashcards that allows the student to review all the key terminology from the chapter. Students can review the terms in study mode, and then quiz themselves on a list of terms.

Experiment Links. For each Investigating Life figure in the textbook, BioPortal includes an overview of the experiment featured in the figure and related research or applications that followed, a link to the original paper, and links to additional information related to the experiment.

Key Terms. The key terminology introduced in each chapter is listed, with definitions and audio pronunciations from the glossary.

Suggested Readings. For each chapter of the book, a list of suggested readings is provided as a resource for further study.

Glossary. The language of biology is often difficult for students taking introductory biology to master, so BioPortal includes a full glossary that features audio pronunciations of all terms.

Statistics Primer. This brief introduction to the use of statistics in biological research explains why statistics are integral to biology, and how some of the most common statistical methods and techniques are used by biologists in their work.

Math for Life. A collection of mathematical shortcuts and references to help students with the quantitative skills they need in the laboratory.

Survival Skills. A guide to more effective study habits. Topics include time management, note-taking, effective highlighting, and exam preparation.

Instructor Resources

Assessment

- Diagnostic Quizzing provides instant class comprehension feedback to instructors, along with targeted lecture resources for those areas requiring the most attention.

- Question banks include questions ranked according to Bloom's taxonomy.

- Question filtering: Allows instructors to select questions based on Bloom's category and/or textbook section.

- Easy-to-use customized assessment tools allow instructors to quickly create quizzes and many other types of assignments using any combination of the questions and resources provided along with their own materials.

- Comprehensive question banks include questions from the test bank, study guide, textbook self-quizzes, and diagnostic quizzes.

Media Resources (see Instructor's Media Library below for details)

- Videos

- PowerPoint® Presentations (Textbook Figures, Lectures, Layered Art)

- Supplemental Photos

- Clicker Questions

- Instructor's Manual

- Lecture Notes

Course Management

- Complete course customization capabilities

- Custom resources/document posting

- Robust Gradebook

- Communication Tools: Announcements, Calendar, Course Email, Discussion Boards

Note: The printed textbook, the eBook, BioPortal, and Prep-U can all be purchased individually as stand-alone items, in addition to being available in a package with the printed textbook.

Student Supplements

Study Guide (ISBN 978-1-4292-3569-3)

Jacalyn Newman, *University of Pittsburgh;* Edward M. Dzialowski, *University of North Texas;* Betty McGuire, *Cornell University;* Lindsay Goodloe, *Cornell University;* and Nancy Guild, *University of Colorado*

For each chapter of the textbook, the *Life* Study Guide offers a variety of study and review tools. The contents of each chapter are broken down into both a detailed review of the Important Concepts covered and a boiled-down Big Picture snapshot. New for the Ninth Edition, Diagram Exercises help students synthesize what they have learned in the chapter through exercises such as ordering concepts, drawing graphs, linking steps in processes, and labeling diagrams. In addition, Common Problem Areas and Study Strategies are highlighted. A set of study questions (both multiple-choice and short-answer) allows students to test their comprehension. All questions include answers and explanations.

Lecture Notebook (ISBN 978-1-4292-3583-9)

This invaluable printed resource consists of all the artwork from the textbook (more than 1,000 images with labels) presented in the order in which they appear in the text, with ample space for note-taking. Because the Notebook has already done the drawing, students can focus more of their attention on the concepts. They will absorb the material more efficiently during class, and their notes will be clearer, more accurate, and more useful when they study from them later.

Companion Website www.thelifewire.com

(Also available as a CD, which can be optionally packaged with the textbook.)

For those students who do not have access to BioPortal, the *Life,* Ninth Edition Companion Website is available free of charge (no access code required). The site features a variety of resources, including animations, flashcards, activities, study ideas, help with math and statistics, and more.

CatchUp Math & Stats

Michael Harris, Gordon Taylor, and Jacquelyn Taylor (ISBN 978-1-4292-0557-3)

This primer will help your students quickly brush up on the quantitative skills they need to succeed in biology. Presented in brief, accessible units, the book covers topics such as working with powers, logarithms, using and understanding graphs, calculating standard deviation, preparing a dilution series, choosing the right statistical test, analyzing enzyme kinetics, and many more.

Student Handbook for Writing in Biology, Third Edition

Karen Knisely, *Bucknell University* (ISBN 978-1-4292-3491-7)

This book provides practical advice to students who are learning to write according to the conventions in biology. Using the standards of journal publication as a model, the author provides, in a user-friendly format, specific instructions on: using biology databases to locate references; paraphrasing for improved comprehension; preparing lab reports, scientific papers, posters; preparing oral presentations in PowerPoint®, and more.

Bioethics and the New Embryology: Springboards for Debate

Scott F. Gilbert, Anna Tyler, and Emily Zackin (ISBN 978-0-7167-7345-0)

Our ability to alter the course of human development ranks among the most significant changes in modern science and has brought embryology into the public domain. The question that must be asked is: Even if we can do such things, should we?

BioStats Basics: A Student Handbook

James L. Gould and Grant F. Gould (ISBN 978-0-7167-3416-1)

BioStats Basics provides introductory-level biology students with a practical, accessible introduction to statistical research. Engaging and informal, the book avoids excessive theoretical and mathematical detail, and instead focuses on how core statistical methods are put to work in biology.

Instructor Media & Supplements

Instructor's Media Library

The *Life*, Ninth Edition Instructor's Media Library (available both online via BioPortal and on disc) includes a wide range of electronic resources to help instructors plan their course, present engaging lectures, and effectively assess student comprehension. The Media Library includes the following resources:

Textbook Figures and Tables. Every image and table from the textbook is provided in both JPEG (high- and low-resolution) and PDF formats. Each figure is provided both with and without balloon captions, and large, complex figures are provided in both a whole and split version.

Unlabeled Figures. Every figure is provided in an unlabeled format, useful for student quizzing and custom presentation development.

Supplemental Photos. The supplemental photograph collection contains over 1,500 photographs (in addition to those in the text), giving instructors a wealth of additional imagery to draw upon.

Animations. Over 100 detailed animations, revised and enlarged for the Ninth Edition, all created from the textbook's art program, and viewable in either narrated or step-through mode.

Videos. A collection of over 200 video segments that covers topics across the entire textbook and helps demonstrate the complexity and beauty of life. Includes the Cell Visualization Videos.

PowerPoint® Resources. For each chapter of the textbook, several different PowerPoint presentations are available. These give instructors the flexibility to build presentations in the manner that best suits their needs. Included are:

- Textbook Figures and Tables
- Lecture Presentation
- Figures with Editable Labels
- Layered Art Figures
- Supplemental Photos
- Videos
- Animations

Clicker Questions. A set of questions written specifically to be used with classroom personal response systems, such as the iClicker system, is provided for each chapter. These questions are designed to reinforce concepts, gauge student comprehension, and engage students in active participation.

Chapter Outlines, Lecture Notes, and the complete **Test File** are all available in Microsoft Word® format for easy use in lecture and exam preparation.

Intuitive Browser Interface provides a quick and easy way to preview and access all of the content on the Instructor's Media Library.

Instructor's Resource Kit

The *Life*, Ninth Edition Instructor's Resource Kit includes a wealth of information to help instructors in the planning and teaching of their course. The Kit includes:

Instructor's Manual, featuring (by chapter):

- A "What's New" guide to the Ninth Edition
- Brief chapter overview
- Chapter outline
- Key terms section with all of the boldface terms from the text

Lecture Notes. Detailed notes for each chapter, which can serve as the basis for lectures, including references to figures and media resources.

Media Guide. A visual guide to the extensive media resources available with the Ninth Edition of *Life*. The guide includes thumbnails and descriptions of every video, animation, lecture PowerPoint®, and supplemental photo in the Media Library, all organized by chapter.

Overhead Transparencies

This set includes over 1,000 transparencies—including all of the four-color line art and all of the tables from the text—along with convenient binders. All figures have been formatted and color-enhanced for clear projection in a wide range of conditions. Labels and images have been resized for improved readability.

Test File

Catherine Ueckert, *Northern Arizona University;* Norman Johnson, *University of Massachusetts;* Paul Nolan, *The Citadel;* Nicola Plowes, *Arizona State University*

The Test File offers more than 5,000 questions, covering the full range of topics presented in the textbook. All questions are referenced to textbook sections and page numbers, and are ranked according to Bloom's taxonomy. Each chapter includes a wide range of multiple choice and fill-in-the-blank questions. In addition, each chapter features a set of diagram questions that involve the student in working with illustrations of structures, graphs, steps in processes, and more. The electronic versions of the Test File (within BioPortal, the Instructor's Media Library, and the Computerized Test Bank CD) also include all of the textbook end-of-chapter Self-Quiz questions, all of the BioPortal Diagnostic Quiz questions, and all of the Study Guide multiple-choice questions.

Computerized Test Bank

The entire printed Test File, plus the textbook end-of-chapter Self-Quizzes, the BioPortal Diagnostic Quizzes, and the Study Guide multiple-choice questions are all included in Wimba's easy-to-use Diploma® software. Designed for both novice and advanced users, Diploma enables instructors to quickly and easily create or edit questions, create quizzes or exams with a "drag-and-drop" feature, publish to online courses, and print paper-based assignments.

Course Management System Support

As a service for *Life* adopters using WebCT, Blackboard, or ANGEL for their courses, full electronic course packs are available.

www.whfreeman.com/facultylounge/ majorsbio
NEW! The new Faculty Lounge for Majors Biology is the first publisher-provided website for the majors biology community that lets instructors freely communicate and share peer-reviewed lecture and teaching resources. It is continually updated and vetted by majors biology instructors—there is always something new to see. The Faculty Lounge offers convenient access to peer-recommended and vetted resources, including the following categories: Images, News, Videos, Labs, Lecture Resources, and Educational Research.

In addition, the site includes special areas for resources for lab coordinators, resources and updates from the *Scientific Teaching* series of books, and information on biology teaching workshops.

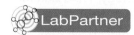

Developed for educators by educators, iclicker is a hassle-free radio-frequency classroom response system that makes it easy for instructors to ask questions, record responses, take attendance, and direct students through lectures as active participants. For more information, visit www.iclicker.com.

www.whfreeman.com/labpartner

NEW! LabPartner is a site designed to facilitate the creation of customized lab manuals. Its database contains a wide selection of experiments published by W. H. Freeman and Hayden-McNeil Publishing. Instructors can preview, choose, and re-order labs, interleave their original experiments, add carbonless graph paper and a pocket folder, and customize the cover both inside and out. LabPartner offers a variety of binding types: paperback, spiral, or loose-leaf. Manuals are printed on-demand once W. H. Freeman receives an order from a campus bookstore or school.

The Scientific Teaching Book Series is a collection of practical guides, intended for all science, technology, engineering and mathematics (STEM) faculty who teach undergraduate and graduate students in these disciplines. The purpose of these books is to help faculty become more successful in all aspects of teaching and learning science, including classroom instruction, mentoring students, and professional development. Authored by well-known science educators, the Series provides concise descriptions of best practices and how to implement them in the classroom, the laboratory, or the department. For readers interested in the research results on which these best practices are based, the books also provide a gateway to the key educational literature.

Scientific Teaching

Jo Handelsman, Sarah Miller, and Christine Pfund, *University of Wisconsin-Madison* (ISBN 978-1-4292-0188-9)

NEW! Transformations: Approaches to College Science Teaching

A Collection of Articles from CBE Life Sciences Education
Deborah Allen, *University of Delaware;* Kimberly Tanner, *San Francisco State University* (ISBN 978-1-4292-5335-2)

Contents

PART TWO

CELLS

5 Cells: The Working Units of Life 76

6 Cell Membranes 105

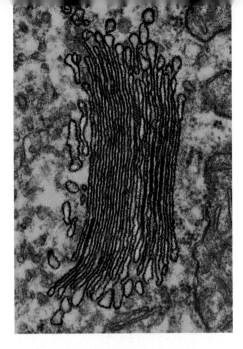

7 Cell Signaling and Communication 128

PART THREE

CELLS AND ENERGY

8 Energy, Enzymes, and Metabolism 148

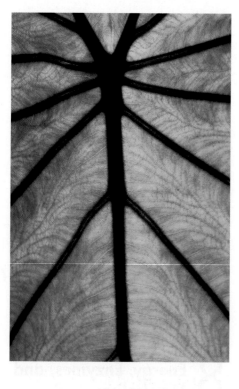

PART FOUR
GENES AND HEREDITY

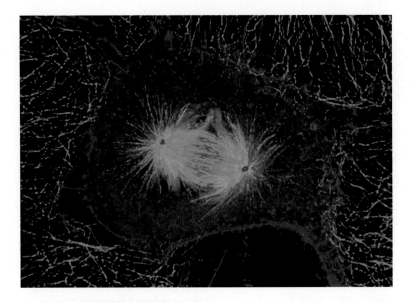

PART FIVE
GENOMES

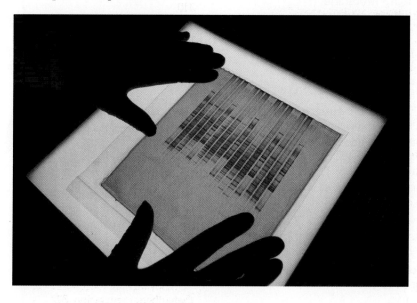

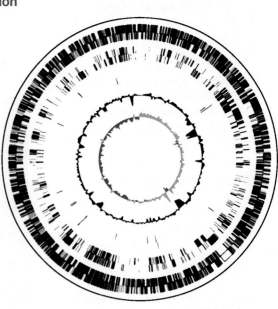

PART SIX
THE PATTERNS AND PROCESSES OF EVOLUTION

25 The History of Life on Earth 518

PART SEVEN
THE EVOLUTION OF DIVERSITY

26 Bacteria and Archaea: The Prokaryotic Domains 536

PART EIGHT

FLOWERING PLANTS: FORM AND FUNCTION

43 Animal Reproduction 899

44 Animal Development 922

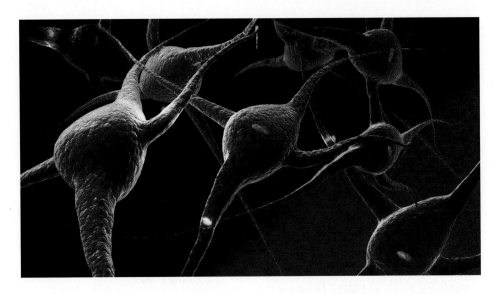

53 Animal Behavior 1113

PART TEN
ECOLOGY

Life
The Science of Biology
NINTH EDITION

1 Studying Life

Why are frogs croaking?

Amphibians—frogs, toads, and salamanders—have been around for a long time. They watched the dinosaurs come and go. But today amphibian populations around the world are in dramatic decline, with more than a third of the world's amphibian species threatened with extinction. Why?

Biologists work to answer this question by making observations and doing experiments. A number of factors may be involved, and one possible cause may be the effects of agricultural pesticides and herbicides. Several studies have shown that many of these chemicals tested at realistic concentrations do not kill amphibians. But Tyrone Hayes, a biologist at the University of California at Berkeley, probed deeper.

Hayes focused on atrazine, the most widely used herbicide in the world and a common contaminant in fresh water. More than 70 million pounds of atrazine are applied to farmland in the United States every year, and it is used in at least 20 countries. Atrazine is usually applied in the spring, when many amphibians are breeding and thousands of tadpoles swim in the ditches, ponds, and streams that receive runoff from farms.

In his laboratory, Hayes and his associates raised frog tadpoles in water containing no atrazine and in water with concentrations ranging from 0.01 parts per billion (ppb) up to 25 ppb. The U.S. Environmental Protection Agency considers environmental levels of atrazine of 10 to 20 ppb of no concern; the level it considers safe in drinking water is 3 ppb. Rainwater in Iowa has been measured to contain 40 ppb. In Switzerland, where the use of atrazine is illegal, the chemical has been measured at approximately 1 ppb in rainwater.

In the Hayes laboratory, concentrations as low as 0.1 ppb had a dramatic effect on tadpole development: it feminized the males. In some of the adult males that developed from these larvae, the vocal structures used in mating calls were smaller than normal, female sex organs developed, and eggs were found growing in the testes. In other studies, normal adult male frogs exposed to 25 ppb had a tenfold reduction in testosterone levels and did not produce sperm. You can imagine the disastrous effects these developmental and hormonal changes could have on the capacity of frogs to breed and reproduce.

But Hayes's experiments were performed in the laboratory, with a species of frog bred for laboratory use. Would his results be the same in nature? To find out, he and his students traveled from Utah to Iowa, sampling water and collecting frogs. They analyzed the water

Frogs Are Having Serious Problems An alarming number of species of frogs, such as this tiny leaf frog (*Agalychnis calcarifer*) from Ecuador, are in danger of becoming extinct. The numerous possible reasons for the decline in global amphibian populations have been a subject of widespread scientific investigation.

A Biologist at Work Tyrone Hayes grew up near the great Congaree Swamp in South Carolina collecting turtles, snakes, frogs, and toads. Now a professor of biology at the University of California at Berkeley, he has more than 3,000 frogs in his laboratory and studies hormonal control of their development.

for atrazine and examined the frogs. In the only site where atrazine was undetectable in the water, the frogs were normal; in all the other sites, male frogs had abnormalities of the sex organs.

Like other biologists, Hayes made observations. He then made predictions based on those observations, and designed and carried out experiments to test his predictions. Some of the conclusions from his experiments, described at the end of this chapter, could have profound implications not only for amphibians but also for other animals, including humans.

IN THIS CHAPTER we identify and examine the most common features of living organisms and put those features into the context of the major principles that underlie all biology. Next we offer a brief outline of how life evolved and how the different organisms on Earth are related. We then turn to the subjects of biological inquiry and the scientific method. Finally we consider how knowledge discovered by biologists influences public policy.

1.1 What Is Biology?

Biology is the scientific study of living things. Biologists define "living things" as all the diverse organisms descended from a single-celled ancestor that evolved almost 4 billion years ago. Because of their common ancestry, living organisms share many characteristics that are not found in the nonliving world. Living organisms:

- consist of one or more cells
- contain genetic information
- use genetic information to reproduce themselves
- are genetically related and have evolved
- can convert molecules obtained from their environment into new biological molecules
- can extract energy from the environment and use it to do biological work
- can regulate their internal environment

This simple list, however, belies the incredible complexity and diversity of life. Some forms of life may not display all of these characteristics all of the time. For example, the seed of a desert plant may go for many years without extracting energy from the environment, converting molecules, regulating its internal environment, or reproducing; yet the seed is alive.

And what about viruses? Viruses do not consist of cells, and they cannot carry out physiological functions on their own; they must parasitize host cells to do those jobs for them. Yet viruses contain genetic information, and they certainly mutate and evolve (as we know, because evolving flu viruses require constant changes in the vaccines we create to combat them). The existence of viruses depends on cells, and it is highly probable that viruses evolved from cellular life forms. So, are viruses alive? What do you think?

This book explores the characteristics of life, how these characteristics vary among organisms, how they evolved, and how they work together to enable organisms to survive and reproduce. *Evolution* is a central theme of biology and therefore of this book. Through differential survival and reproduction, living systems evolve and become adapted to Earth's many environments. The processes of evolution have generated the enormous diversity that we see today as life on Earth.

Cells are the basic unit of life

We lay the chemical foundation for our study of life in the next three chapters, after which we will turn to cells and the processes by which they live, reproduce, age, and die. Some organisms are *unicellular,* consisting of a single cell that carries out

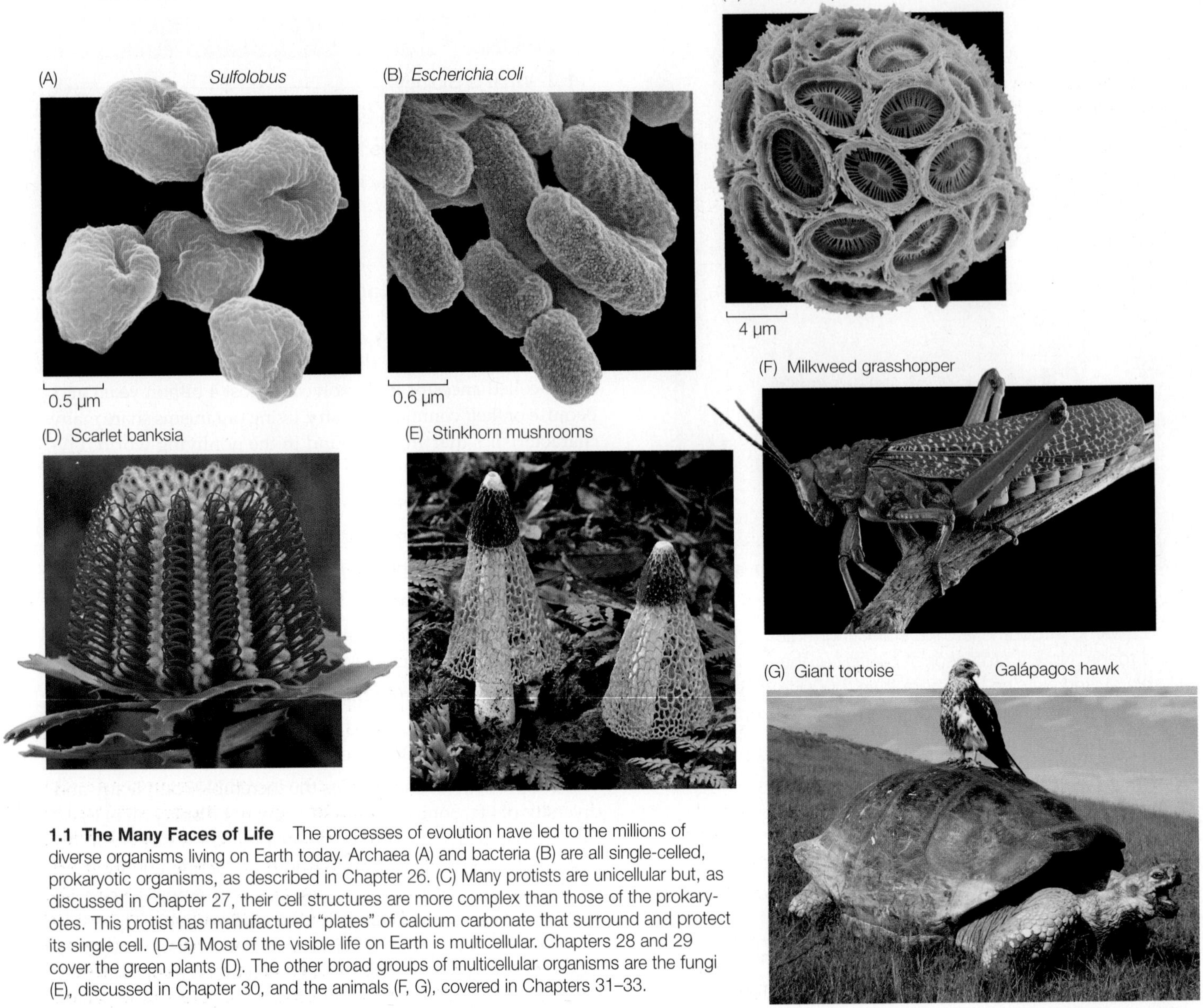

(A) *Sulfolobus*

0.5 µm

(B) *Escherichia coli*

0.6 µm

(C) Coccolithophore

4 µm

(D) Scarlet banksia

(E) Stinkhorn mushrooms

(F) Milkweed grasshopper

(G) Giant tortoise   Galápagos hawk

1.1 The Many Faces of Life   The processes of evolution have led to the millions of diverse organisms living on Earth today. Archaea (A) and bacteria (B) are all single-celled, prokaryotic organisms, as described in Chapter 26. (C) Many protists are unicellular but, as discussed in Chapter 27, their cell structures are more complex than those of the prokaryotes. This protist has manufactured "plates" of calcium carbonate that surround and protect its single cell. (D–G) Most of the visible life on Earth is multicellular. Chapters 28 and 29 cover the green plants (D). The other broad groups of multicellular organisms are the fungi (E), discussed in Chapter 30, and the animals (F, G), covered in Chapters 31–33.

all the functions of life (**Figure 1.1A–C**). Others are *multicellular*, made up of many cells that are specialized for different functions (**Figure 1.1D–G**). Viruses are *acellular*, although they depend on cellular organisms.

The discovery of cells was made possible by the invention of the microscope in the 1590s by the Dutch spectacle makers Hans and Zaccharias Janssen (father and son). In the mid- to late 1600s, Antony van Leeuwenhoek of Holland and Robert Hooke of England both made improvements on the Janssens' technology and used it to study living organisms. Van Leeuwenhoek discovered that drops of pond water teemed with single-celled organisms, and he made many other discoveries as he progressively improved his microscopes over a long lifetime of research. Hooke put pieces of plants under his microscope and observed that they were made up of repeated units he called *cells* (**Figure 1.2**). In 1676, Hooke wrote that van Leeuwenhoek had observed "a vast number of small animals in his Excrements which were most abounding when he was troubled with a Loosenesse and very few or none when he was well." This simple observation

represents the discovery of bacteria—and makes one wonder why scientists do some of the things they do.

More than a hundred years passed before studies of cells advanced significantly. As they were dining together one evening in 1838, Matthias Schleiden, a German biologist, and Theodor Schwann, from Belgium, discussed their work on plant and animal tissues, respectively. They were struck by the similarities in their observations and came to the conclusion that the basic structural elements of plants and animals were essentially the same. They formulated their conclusion as the **cell theory**, which states that:

• Cells are the basic structural and physiological units of all living organisms.

• Cells are both distinct entities and building blocks of more complex organisms.

But Schleiden and Schwann also believed (wrongly) that cells emerged by the self-assembly of nonliving materials, much as crystals form in a solution of salt. This conclusion was in ac-

1.2 Cells Are the Building Blocks of Life The development of microscopes revealed the microbial world to seventeenth-century scientists such as Robert Hooke, who proposed the concept of cells based on his observations. (A) Hooke drew the cells of a slice of plant tissue (cork) as he saw them under his optical microscope. (B) A modern optical, or "light," microscope reveals the intricacies of cells in a leaf. (C) Transmission electron microscopes (TEMs) allow scientists to see even smaller objects. TEMs do not visualize color; here color has been added to a black-and-white micrograph of cells in a duckweed stem.

(A)

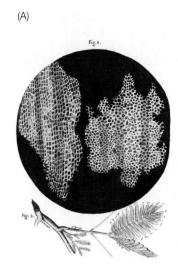

(B)

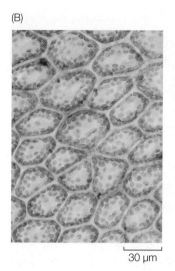

30 µm

(C)

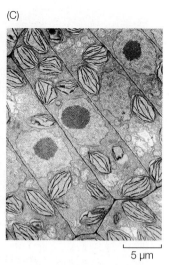

5 µm

cordance with the prevailing view of the day, which was that life can arise from non-life by spontaneous generation—mice from dirty clothes, maggots from dead meat, or insects from pond water.

The debate continued until 1859, when the French Academy of Sciences sponsored a contest for the best experiment to prove or disprove spontaneous generation. The prize was won by the great French scientist Louis Pasteur, who demonstrated that sterile broth directly exposed to the dirt and dust in air developed a culture of microorganisms, but a similar container of broth not directly exposed to air remained sterile (see Figure 4.7). Pasteur's experiment did not prove that it was microorganisms in the air that caused the broth to become infected, but it did uphold the conclusion that life must be present in order for new life to be generated.

Today scientists accept the fact that all cells come from preexisting cells and that the functional properties of organisms derive from the properties of their cells. Since cells of all kinds share both essential mechanisms and a common ancestry that goes back billions of years, modern cell theory has additional elements:

- All cells come from preexisting cells.
- All cells are similar in chemical composition.
- Most of the chemical reactions of life occur in aqueous solution within cells.
- Complete sets of genetic information are replicated and passed on during cell division.
- Viruses lack cellular structure but remain dependent on cellular organisms.

At the same time Schleiden and Schwann were building the foundation for the cell theory, Charles Darwin was beginning to understand how organisms undergo evolutionary change.

All of life shares a common evolutionary history

Evolution—change in the genetic makeup of biological populations through time—is the major unifying principle of biol-

ogy. Charles Darwin compiled factual evidence for evolution in his 1859 book *On the Origin of Species*. Since then, biologists have gathered massive amounts of data supporting Darwin's theory that all living organisms are descended from a common ancestor. Darwin also proposed one of the most important processes that produce evolutionary change. He argued that differential survival and reproduction among individuals in a population, which he termed **natural selection**, could account for much of the evolution of life.

Although Darwin proposed that living organisms are descended from common ancestors and are therefore related to one another, he did not have the advantage of understanding the mechanisms of genetic inheritance. Even so, he observed that offspring resembled their parents; therefore, he surmised, such mechanisms had to exist. That simple fact is the basis for the concept of a **species**. Although the precise definition of a species is complicated, in its most widespread usage it refers to a group of organisms that can produce viable and fertile offspring with one another.

But offspring do differ from their parents. Any population of a plant or animal species displays variation, and if you select breeding pairs on the basis of some particular trait, that trait is more likely to be present in their offspring than in the general population. Darwin himself bred pigeons, and was well aware of how pigeon fanciers selected breeding pairs to produce offspring with unusual feather patterns, beak shapes, or body sizes (see Figure 21.2). He realized that if humans could select for specific traits in domesticated animals, the same process could operate in nature; hence the term *natural selection* as opposed to artificial (human-imposed) selection.

How would natural selection function? Darwin postulated that different probabilities of survival and reproductive success would do the job. He reasoned that the reproductive capacity of plants and animals, if unchecked, would result in unlimited growth of populations, but we do not observe such growth in nature; in most species, only a small percentage of offspring survive to reproduce. Thus any trait that confers even a small increase in the probability that its possessor will survive and reproduce would be spread in the population.

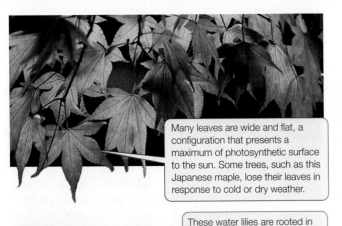

Many leaves are wide and flat, a configuration that presents a maximum of photosynthetic surface to the sun. Some trees, such as this Japanese maple, lose their leaves in response to cold or dry weather.

The leaves of many evergreen conifers, such as spruce trees, are waxy-coated needles that resist water loss and are not shed on a yearly basis.

These water lilies are rooted in the pond bottom; their large leaves are flat "pads" that float on the surface.

The leaves of pitcher plants form a vessel that holds water. The plant receives extra nutrients from the decomposing bodies of insects that drown in the pitcher.

The ability to climb can be advantageous to a plant, enabling it to reach above other plants to obtain more sunlight. Some of the leaves of this climbing cucumber are tightly furled tendrils that wrap around a stake.

1.3 Adaptations to the Environment The leaves of all plants are specialized for photosynthesis—the sunlight-powered transformation of water and carbon dioxide into larger structural molecules called carbohydrates. The leaves of different plants, however, display many different adaptations to their individual environments.

Because organisms with certain traits survive and reproduce best under specific sets of conditions, natural selection leads to **adaptations**: structural, physiological, or behavioral traits that enhance an organism's chances of survival and reproduction in its environment (**Figure 1.3**). In addition to natural selection, evolutionary processes such as sexual selection (selection due to mate choice) and genetic drift (the random fluctuation of gene frequencies in a population due to chance events) contribute to the rise of diverse adaptations. These processes operating over evolutionary history have led to the remarkable array of life on Earth.

If all cells come from preexisting cells, and if all the diverse species of organisms on Earth are related by descent with modification from a common ancestor, then what is the source of information that is passed from parent to daughter cells and from parental organisms to their offspring?

Biological information is contained in a genetic language common to all organisms

Cells are the basic building blocks of organisms, but even a single cell is complex, with many internal structures and many functions that depend on information. The information required for a cell to function and interact with other cells—the "blueprint" for existence—is contained in the cell's **genome**, the sum total of all the DNA molecules it contains. **DNA** (deoxyribonucleic acid) molecules are long sequences of four different subunits called **nucleotides**. The sequence of the nucleotides contains genetic information. **Genes** are specific segments of DNA encoding the information the cell uses to make **proteins** (**Figure 1.4**). Protein molecules govern the chemical reactions within cells and form much of an organism's structure.

By analogy with a book, the nucleotides of DNA are like the letters of an alphabet. Protein molecules are the sentences. Combinations of proteins that form structures and control biochemical processes are the paragraphs. The structures and processes that are organized into different systems with specific tasks (such as digestion or transport) are the chapters of the book, and the complete book is the organism. If you were to write out your own genome using four letters to represent the four nucleotides, you would write more than 3 billion letters. Using the size type you are reading now, your genome would fill about a thousand books the size of this one. The mechanisms of evolution, including natural selection, are the authors and editors of all the books in the library of life.

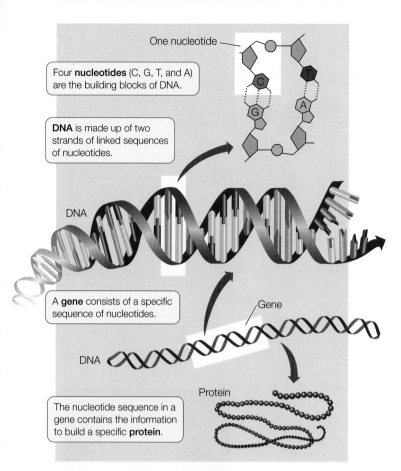

One nucleotide

Four **nucleotides** (C, G, T, and A) are the building blocks of DNA.

DNA is made up of two strands of linked sequences of nucleotides.

DNA

A **gene** consists of a specific sequence of nucleotides.

Gene

DNA

Protein

The nucleotide sequence in a gene contains the information to build a specific **protein**.

1.4 DNA Is Life's Blueprint The instructions for life are contained in the sequences of nucleotides in DNA molecules. Specific DNA nucleotide sequences comprise genes. The average length of a single human gene is 16,000 nucleotides. The information in each gene provides the cell with the information it needs to manufacture molecules of a specific protein.

of biochemical reactions that occur inside cells. Some of these reactions break down nutrient molecules into smaller chemical units, and in the process some of the energy contained in the chemical bonds of the nutrients is captured by high-energy molecules that can be used to do different kinds of cellular work.

One obvious kind of work cells do is mechanical—moving molecules from one cellular location to another, moving whole cells or tissues, or even moving the organism itself, as muscles do (**Figure 1.5A**). The most basic cellular work is the building, or *synthesis*, of new complex molecules and structures from smaller chemical units. For example, we are all familiar with the fact that carbohydrates eaten today may be deposited in the body as fat tomorrow (**Figure 1.5B**). Still another kind of work is the electrical work that is the essence of information processing in nervous systems. The sum total of all the chemical transformations and other work done in all the cells of an organism is its **metabolism**, or **metabolic rate**.

The myriad of biochemical reactions that go on in cells are integrally linked in that the products of one are the raw materials of the next. These complex networks of reactions must be integrated and precisely controlled; when they are not, the result is disease.

Living organisms regulate their internal environment

Multicellular organisms have an *internal environment* that is not cellular. That is, their individual cells are bathed in extracellular fluids, from which they receive nutrients and into which they excrete waste products of metabolism. The cells of multicellu-

All the cells of a multicellular organism contain the same genome, yet different cells have different functions and form different structures—contractile proteins form in muscle cells, hemoglobin in red blood cells, digestive enzymes in gut cells, and so on. Therefore, different types of cells in an organism must express different parts of the genome. How cells control gene expression in ways that enable a complex organism to develop and function is a major focus of current biological research.

The genome of an organism consists of thousands of genes. If the nucleotide sequence of a gene is altered, it is likely that the protein that gene encodes will be altered. Alterations of the genome are called *mutations*. Mutations occur spontaneously; they can also be induced by outside factors, including chemicals and radiation. Most mutations are either harmful or have no effect, but occasionally a mutation improves the functioning of the organism under the environmental conditions it encounters. Such beneficial mutations are the raw material of evolution and lead to adaptations.

Cells use nutrients to supply energy and to build new structures

Living organisms acquire *nutrients* from the environment. Nutrients supply the organism with energy and raw materials for carrying out biochemical reactions. Life depends on thousands

(A)

(B)

1.5 Energy Can Be Used Immediately or Stored (A) Animal cells break down and release the energy contained in the chemical bonds of food molecules to do mechanical work—in this kangaroo's case, to jump. (B) The cells of this Arctic ground squirrel have broken down the complex carbohydrates in plants and converted their molecules into fats, which are stored in the animal's body to provide an energy supply for the cold months.

lar organisms are specialized, or *differentiated*, to contribute in some way to the maintenance of the internal environment. With the evolution of specialization, differentiated cells lost many of the functions carried out by single-celled organisms, and must depend on the internal environment for essential services.

To accomplish their specialized tasks, assemblages of differentiated cells are organized into *tissues.* For example, a single muscle cell cannot generate much force, but when many cells combine to form the tissue of a working muscle, considerable force and movement can be generated (see Figure 1.5B). Different tissue types are organized to form *organs* that accomplish specific functions. For example, the heart, brain, and stomach are each constructed of several types of tissues. Organs whose functions are interrelated can be grouped into *organ systems*; the stomach, intestine, and esophagus, for example, are parts of the digestive system. The functions of cells, tissues, organs, and organ systems are all integral to the multicellular *organism.* We cover the biology of organisms in Parts Eight and Nine of this book.

Living organisms interact with one another

The internal hierarchy of the individual organism is matched by the external hierarchy of the biological world (**Figure 1.6**). Organisms do not live in isolation. A group of individuals of the same species that interact with one another is a *population,* and populations of all the species that live and interact in the same area are called a *community.* Communities together with their abiotic environment constitute an *ecosystem.*

Individuals in a population interact in many different ways. Animals eat plants and other animals (usually members of another species) and compete with other species for food and other resources. Some animals will prevent other individuals of their own species from exploiting a resource, whether it be food, nesting sites, or mates. Animals may also *cooperate* with members of their species, forming social units such as a termite colony or a flock of birds. Such interactions have resulted in the evolution of social behaviors such as communication.

Plants also interact with their external environment, which includes other plants, animals, and microorganisms. All terrestrial plants depend on complex partnerships with fungi, bacteria, and animals. Some of these partnerships are necessary to obtain nutrients, some to produce fertile seeds, and still others to disperse seeds. Plants compete with each other

1.6 Biology Is Studied at Many Levels of Organization
Life's properties emerge when DNA and other molecules are organized in cells. Energy flows through all the biological levels shown here.

—————— **yourBioPortal.com** ——————
GO TO Web Activity 1.1 • The Hierarchy of Life

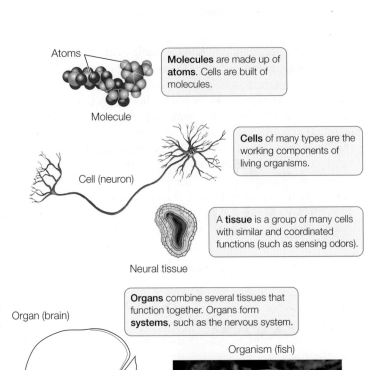

Atoms

Molecule

> **Molecules** are made up of **atoms**. Cells are built of molecules.

Cell (neuron)

> **Cells** of many types are the working components of living organisms.

Neural tissue

> A **tissue** is a group of many cells with similar and coordinated functions (such as sensing odors).

Organ (brain)

> **Organs** combine several tissues that function together. Organs form **systems**, such as the nervous system.

Organism (fish)

> An **organism** is a recognizable, self-contained individual. Complex multicellular organisms are made up of organs and organ systems.

Population (school of fish)

> A **population** is a group of many organisms of the same species.

> **Communities** consist of populations of many different species.

Community (coral reef)

> Biological communities in the same geographical location form **ecosystems**. Ecosystems exchange energy and create Earth's **biosphere**.

Biosphere

for light and water, and they have ongoing evolutionary interactions with the animals that eat them, evolving anti-predation adaptations or ways to attract the animals that assist in their reproduction. The interactions of populations of different plant and animal species in a community are major evolutionary forces that produce specialized adaptations.

Communities interacting over a broad geographic area with distinguishing physical features form ecosystems; examples might include an Arctic tundra, a coral reef, or a tropical rainforest. The ways in which species interact with one another and with their environment in communities and in ecosystems is the subject of *ecology* and of Part Ten of this book.

Discoveries in biology can be generalized

Because all life is related by descent from a common ancestor, shares a genetic code, and consists of similar building blocks—cells—knowledge gained from investigations of one type of organism can, with care, be generalized to other organisms. Biologists use **model systems** for research, knowing that they can extend their findings to other organisms, including humans. For example, our basic understanding of the chemical reactions in cells came from research on bacteria but is applicable to all cells, including those of humans. Similarly, the biochemistry of photosynthesis—the process by which plants use sunlight to produce biological molecules—was largely worked out from experiments on *Chlorella*, a unicellular green alga (see Figure 10.13). Much of what we know about the genes that control plant development is the result of work on *Arabidopsis thaliana,* a relative of the mustard plant. Knowledge about how animals develop has come from work on sea urchins, frogs, chickens, roundworms, and fruit flies. And recently, the discovery of a major gene controlling human skin color came from work on zebrafish. Being able to generalize from model systems is a powerful tool in biology.

1.1 RECAP

Living organisms are made of (or depend on) cells, are related by common descent and evolve, contain genetic information and use it to reproduce, extract energy from their environment and use it to do biological work, synthesize complex molecules to construct biological structures, regulate their internal environment, and interact with one another.

- Describe the relationship between evolution by natural selection and the genetic code. **See pp. 6–7**

- Why can the results of biological research on one species often be generalized to very different species? **See p. 9**

Now that you have an overview of the major features of life that you will explore in depth in this book, you can ask how and when life first emerged. In the next section we will summarize briefly the history of life from the earliest simple life forms to the complex and diverse organisms that inhabit our planet today.

1.2 How Is All Life on Earth Related?

What do biologists mean when they say that all organisms are *genetically related*? They mean that species on Earth share a *common ancestor*. If two species are similar, as dogs and wolves are, then they probably have a common ancestor in the fairly recent past. The common ancestor of two species that are more different—say, a dog and a deer—probably lived in the more distant past. And if two organisms are very different—such as a dog and a clam—then we must go back to the *very* distant past to find their common ancestor. How can we tell how far back in time the common ancestor of any two organisms lived? In other words, how do we discover the evolutionary relationships among organisms?

For many years, biologists have investigated the history of life by studying the *fossil record*—the preserved remains of organisms that lived in the distant past (**Figure 1.7**). Geologists supplied knowledge about the ages of fossils and the nature of the environments in which they lived. Biologists then inferred the evolutionary relationships among living and fossil organisms by comparing their anatomical similarities and differences. Frequently big gaps existed in the fossil record, forcing biologists to predict the nature of the "missing links" between two lineages of organisms. As the fossil record became more complete, those missing links were filled in.

Molecular methods for comparing genomes, described in Chapter 24, are enabling biologists to more accurately establish the degrees of relationship between living organisms and to use that information to interpret the fossil record. Molecular information can occasionally be gleaned from fossil specimens, such as recently deciphered genetic material from fossil bones of Ne-

1.7 Fossils Give Us a View of Past Life This fossil, formed some 150 million years ago, is that of an *Archaeopteryx*, the earliest known representative of the birds. Birds evolved from the same group of reptiles as the modern crocodiles.

anderthals that led to the conclusion that even though Neanderthals and modern humans coexisted, they did not interbreed.

In general, the greater the differences between the genomes of two species, the more distant their common ancestor. Using molecular techniques, biologists are exploring fundamental questions about life. What were the earliest forms of life? How did simple organisms give rise to the great diversity of organisms alive today? Can we reconstruct a family tree of life?

Life arose from non-life via chemical evolution

Geologists estimate that Earth formed between 4.6 and 4.5 billion years ago. At first, the planet was not a very hospitable place. It was some 600 million years or more before the earliest life evolved. If we picture the history of Earth as a 30-day month, life first appeared somewhere toward the end of the first week (**Figure 1.8**).

When we consider how life might have arisen from nonliving matter, we must take into account the properties of the young Earth's atmosphere, oceans, and climate, all of which were very different than they are today. Biologists postulate that complex biological molecules first arose through the random physical association of chemicals in that environment. Experiments simulating the conditions on early Earth have confirmed that the generation of complex molecules under such conditions is possible, even probable. The critical step for the evolution of life, however, had to be the appearance of molecules that could reproduce themselves and also serve as templates for the synthesis of large molecules with complex but stable shapes. The variation of the shapes of these large, stable molecules (described in Chapters 3 and 4) enabled them to participate in increasing numbers and kinds of chemical reactions with other molecules.

Cellular structure evolved in the common ancestor of life

The second critical step in the origin of life was the enclosure of complex biological molecules by *membranes* that contained them in a compact internal environment separate from the surrounding external environment. Fatlike molecules played a critical role because they are not soluble in water and they form membranous films. When agitated, these films can form spherical *vesicles*, which could have enveloped assemblages of biological molecules. The creation of an internal environment that concentrated the reactants and products of chemical reactions opened up the possibility that those reactions could be integrated and controlled. As described in Section 4.4, scientists postulate that this natural process of membrane formation resulted in the first cells with the ability to replicate themselves—the evolution of the first cellular organisms.

For more than 2 billion years after cells originated, all organisms consisted of only one cell. These first unicellular organisms were (and are, as multitudes of their descendants exist in similar form today) **prokaryotes**. Prokaryotic cells consist of DNA and other biochemicals enclosed in a membrane.

These early prokaryotes were confined to the oceans, where there was an abundance of complex molecules they could use as raw materials and sources of energy. The ocean shielded them from the damaging effects of ultraviolet light, which was intense at that time because there was little or no oxygen (O_2) in the atmosphere, and hence no protective ozone (O_3) layer.

Photosynthesis changed the course of evolution

To fuel their cellular metabolism, the earliest prokaryotes took in molecules directly from their environment and broke these small molecules down to release and use the energy contained in their chemical bonds. Many modern species of prokaryotes still function this way, and very successfully. During the early eons of life on Earth, there was no oxygen in the atmosphere. In fact, oxygen was toxic to the life forms that existed then.

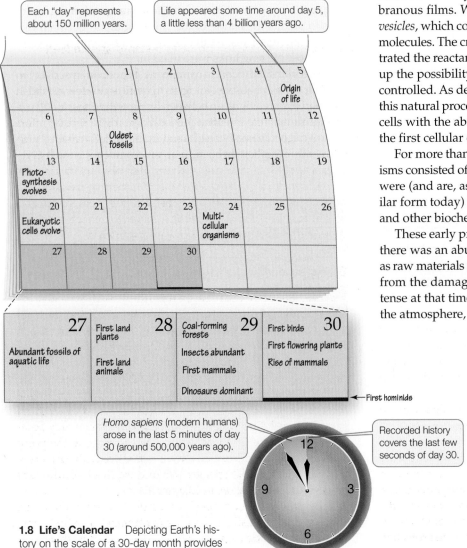

1.8 Life's Calendar Depicting Earth's history on the scale of a 30-day month provides a sense of the immensity of evolutionary time.

350 μm

1.9 Photosynthetic Organisms Changed Earth's Atmosphere
These strands are composed of many cells of cyanobacteria. This modern species (*Oscillatoria tenuis*) may be very similar to the early photosynthetic prokaryotes responsible for the buildup of oxygen in Earth's atmosphere.

About 2.7 billion years ago, the evolution of **photosynthesis** changed the nature of life on Earth. The chemical reactions of photosynthesis transform the energy of sunlight into a form of biological energy that can power the synthesis of large molecules (see Chapter 10). These large molecules are the building blocks of cells, and they can be broken down to provide metabolic energy. Photosynthesis is the basis of much of life on Earth today because its energy-capturing processes provide food for other organisms.

Early photosynthetic cells were probably similar to present-day prokaryotes called *cyanobacteria* (**Figure 1.9**). Over time, photosynthetic prokaryotes became so abundant that vast quantities of O_2, which is a by-product of photosynthesis, slowly began to accumulate in the atmosphere. Oxygen was poisonous to many of the prokaryotes that lived at that time. Those organisms that did tolerate oxygen, however, were able to proliferate as the presence of oxygen opened up vast new avenues of evolution. *Aerobic metabolism* (energy production based on the conversion of O_2) is more efficient than *anaerobic* (non-O_2-using) *metabolism*, and today it is used by the majority of Earth's organisms. Aerobic metabolism allowed cells to grow larger.

Oxygen in the atmosphere also made it possible for life to move onto land. For most of life's history, ultraviolet (UV) radiation falling on Earth's surface was too intense to allow life to exist outside the shielding water. But the accumulation of photosynthetically generated oxygen in the atmosphere for more than 2 billion years gradually produced a layer of ozone in the upper atmosphere. By about 500 million years ago, the ozone layer was sufficiently dense and absorbed enough UV radiation to make it possible for organisms to leave the protection of the water and live on land.

Eukaryotic cells evolved from prokaryotes

Another important step in the history of life was the evolution of cells with discrete intracellular compartments, called **organelles**, which were capable of taking on specialized cellular functions. This event happened about 3 weeks into our calendar of Earth's history (see Figure 1.8). One of these organelles, the dense-appearing *nucleus* (Latin *nux*, "nut" or "core"), came to contain the cell's genetic information and gives these cells their name: **eukaryotes** (Greek *eu*, "true"; *karyon*, "kernel" or "core"). The eukaryotic cell is completely distinct from the cells of prokaryotes (*pro*, "before"), which lack nuclei and other internal compartments.

Some organelles are hypothesized to have originated by **endosymbiosis** when cells ingested smaller cells. The *mitochondria* that generate a cell's energy probably evolved from engulfed prokaryotic organisms. And *chloroplasts*—organelles specialized to conduct photosynthesis—could have originated when photosynthetic prokaryotes were ingested by larger eukaryotes. If the larger cell failed to break down this intended food object, a partnership could have evolved in which the ingested prokaryote provided the products of photosynthesis and the host cell provided a good environment for its smaller partner.

Multicellularity arose and cells became specialized

Until just over a billion years ago, all the organisms that existed—whether prokaryotic or eukaryotic—were unicellular. An important evolutionary step occurred when some eukaryotes failed to separate after cell division, remaining attached to each other. The permanent association of cells made it possible for some cells to specialize in certain functions, such as reproduction, while other cells specialized in other functions, such as absorbing nutrients and distributing them to neighboring cells. This **cellular specialization** enabled multicellular eukaryotes to increase in size and become more efficient at gathering resources and adapting to specific environments.

Biologists can trace the evolutionary tree of life

If all the species of organisms on Earth today are the descendants of a single kind of unicellular organism that lived almost 4 billion years ago, how have they become so different? A simplified answer is that as long as individuals within a population mate with one another, structural and functional changes can evolve within that population, but the population will remain one species. However, if something happens to isolate some members of a population from the others, the structural and functional differences between the two groups may accumulate over time. The two groups may diverge to the point where their members can no longer reproduce with each other and are thus distinct species. We discuss this evolutionary process, called *speciation*, in Chapter 23.

Biologists give each species a distinctive scientific name formed from two Latinized names (a **binomial**). The first name identifies the species' *genus*—a group of species that share a recent common ancestor. The second is the name of the species. For example, the scientific name for the human species is *Homo sapiens*: *Homo* is our genus and *sapiens* our species. *Homo* is Latin for "man"; *sapiens* is from the Latin for word for "wise" or "rational."

Tens of millions of species exist on Earth today. Many times that number lived in the past but are now extinct. Many millions of speciation events created this vast diversity, and the unfolding of these events can be diagrammed as an evolutionary "tree" whose branches describe the order in which populations split and eventually evolved into new species, as described in Chapter 22. Much of biology is based on comparisons among species, and these comparisons are useful precisely because we can place species in an evolutionary context relative to one another. Our ability to do this has been greatly enhanced in recent decades by our ability to sequence and compare the genomes of different species.

Genome sequencing and other molecular techniques have allowed *systematists*—scientists who study the evolution and classification of life's diverse organisms—to augment evolutionary knowledge based on the fossil record with a vast array of molecular evidence. The result is the ongoing compilation of *phylogenetic trees* that document and diagram evolutionary relationships as part of an overarching tree of life, the broadest categories of which are shown in **Figure 1.10**. (The tree is expanded in this book's Appendix; you can also explore the tree interactively at http://tolweb.org/tree.)

Although many details remain to be clarified, the broad outlines of the tree of life have been determined. Its branching patterns are based on a rich array of evidence from fossils, structures, metabolic processes, behavior, and molecular analyses of genomes. Molecular data in particular have been used to separate the tree into three major **domains**: Archaea, Bacteria, and Eukarya. The organisms of each domain have been evolving separately from those in the other domains for more than a billion years.

Organisms in the domains **Archaea** and **Bacteria** are single-celled prokaryotes. However, members of these two groups differ so fundamentally in their metabolic processes that they are believed to have separated into distinct evolutionary lineages very early. Species belonging to the third domain—**Eukarya**—have eukaryotic cells whose mitochondria and chloroplasts may have originated from the ingestion of prokaryotic cells, as described on page 11.

The three major groups of multicellular eukaryotes—plants, fungi, and animals—each evolved from a different group of the eukaryotes generally referred to as *protists*. The chloroplast-containing, photosynthetic protist that gave rise to plants was completely distinct from the protist that was ancestral to both animals and fungi, as can be seen from the branching pattern of Figure 1.10. Although most protists are unicellular (and thus sometimes called *microbial eukaryotes*), multicellularity has evolved in several protist lineages.

The tree of life is predictive

There are far more species alive on Earth than biologists have discovered and described to date. In fact, most species on Earth

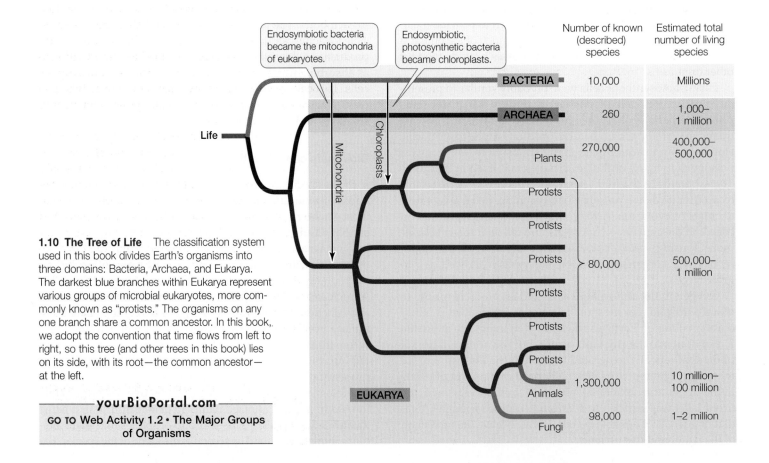

1.10 The Tree of Life The classification system used in this book divides Earth's organisms into three domains: Bacteria, Archaea, and Eukarya. The darkest blue branches within Eukarya represent various groups of microbial eukaryotes, more commonly known as "protists." The organisms on any one branch share a common ancestor. In this book, we adopt the convention that time flows from left to right, so this tree (and other trees in this book) lies on its side, with its root—the common ancestor—at the left.

yourBioPortal.com
GO TO Web Activity 1.2 • The Major Groups of Organisms

have yet to be discovered by humans (see Section 32.4 for a discussion of how we know this). When we encounter a new species, its placement on the tree of life immediately tells us a great deal about its biology. In addition, understanding relationships among species allows biologists to make predictions about species that have not yet been studied, based on our knowledge of those that have.

For example, until phylogenetic methods were developed, it took years of investigation to isolate and identify most newly encountered human pathogens, and even longer to discover how these pathogens moved into human populations. Today, pathogens that cause diseases such as the flu are identified quickly on the basis of their evolutionary relationships. Placement in an evolutionary tree also gives us clues about the disease's biology, possible effective treatments, and the origin of the pathogen (see Chapters 21 and 22).

1.2 RECAP

The first cellular life on Earth was prokaryotic and arose about 4 billion years ago. The complexity of the organisms that exist today is the result of several important evolutionary events, including the evolution of photosynthesis, eukaryotic cells, and multicellularity. The genetic relationships of all organisms can be shown as a branching tree of life.

● Discuss the evolutionary significance of photosynthesis. **See pp. 10–11**

● What do the domains of life represent? What are the major groups of eukaryotes? **See p. 12 and Figure 1.10**

In February of 1676, Robert Hooke received a letter from the physicist Sir Isaac Newton in which Newton famously remarked, "If I have seen a little further, it is by standing on the shoulders of giants." We all stand on the shoulders of giants, building on the research of earlier scientists. By the end of this course, you will know more about evolution than Darwin ever could have, and you will know infinitely more about cells than Schleiden and Schwann did. Let's look at the methods biologists use to expand our knowledge of life.

1.3 How Do Biologists Investigate Life?

Regardless of the many different tools and methods used in research, all scientific investigations are based on *observation* and *experimentation*. In both, scientists are guided by the *scientific method,* one of the most powerful tools of modern science.

Observation is an important skill

Biologists have always observed the world around them, but today our ability to observe is greatly enhanced by technologies such as electron microscopes, DNA chips, magnetic resonance imaging, and global positioning satellites. These technologies have improved our ability to observe at all levels, from the distribution of molecules in the body to the distribution of fish in the oceans. For example, not too long ago marine biologists were only able to observe the movement of fish in the ocean by putting physical tags on the fish, releasing them, and hoping that a fisherman would catch that fish and send back the tag—and even that would reveal only where the fish ended up. Today we can attach electronic recording devices to fish that continuously record not only where the fish is, but also how deep it swims and the temperature and salinity of the water around it (**Figure 1.11**). The tags download this information to a satellite, which relays it back to researchers. Suddenly we are acquiring a great deal of knowledge about the distribution of life in the oceans—information that is relevant to studies of climate change.

Technologies that enable us to *quantify* observations are very important in science. For example, for hundreds of years species were classified by generally qualitative descriptions of the physical differences between them. There was no way of objectively calculating evolutionary distances between organisms, and biologists had to depend on the fossil record for insight. Today our ability to rapidly analyze DNA sequences enables quantitative estimates of evolutionary distances, as described in Parts Five and Six of this book. The ability to gather quantitative observations adds greatly to the biologist's ability to make strong conclusions.

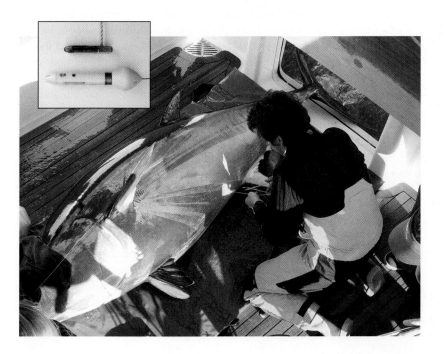

1.11 Tuna Tracking Marine biologist Barbara Block attaches computerized data recording tags (inset) to a live bluefin tuna before returning it to the ocean. Such tags make it possible to track an individual tuna wherever it travels in the world's oceans.

The scientific method combines observation and logic

Observations lead to questions, and scientists make additional observations and do experiments to answer those questions. The conceptual approach that underlies most modern scientific investigations is the **scientific method**. This powerful tool, also called the *hypothesis–prediction (H–P) method*, has five steps: (1) making *observations*; (2) asking *questions*; (3) forming *hypotheses*, or tentative answers to the questions; (4) making *predictions* based on the hypotheses; and (5) *testing* the predictions by making additional observations or conducting experiments (**Figure 1.12**).

After posing a question, a scientist uses *inductive logic* to propose a tentative answer. Inductive logic involves taking observations or facts and creating a new proposition that is compatible with those observations or facts. Such a tentative proposition is called a **hypothesis**. In formulating a hypothesis, scientists put together the facts they already know to formulate one or more possible answers to the question. For example, at the opening of

this chapter you learned that scientists have observed the rapid decline of amphibian populations worldwide and are asking why. Some scientists have hypothesized that a fungal disease is a cause; other scientists have hypothesized that increased exposure to ultraviolet radiation is a cause. Tyrone Hayes hypothesized that exposure to agricultural chemicals could be a cause. He knew that the most widely used chemical herbicide is atrazine; that it is mostly applied in the spring, when amphibians are breeding; and that atrazine is a common contaminant in the waters in which amphibians live as they develop into adults.

The next step in the scientific method is to apply a different form of logic—*deductive logic*—to make predictions based on the hypothesis. Deductive logic starts with a statement believed to be true and then goes on to predict what facts would also have to be true to be compatible with that statement. Based on his hypothesis, Tyrone Hayes predicted that frog tadpoles exposed to atrazine would show adverse effects of the chemical once they reached adulthood.

Good experiments have the potential to falsify hypotheses

Once predictions are made from a hypothesis, experiments can be designed to test those predictions. The most informative experiments are those that have the ability to show that the prediction is wrong. If the prediction is wrong, the hypothesis must be questioned, modified, or rejected.

There are two general types of experiments, both of which compare data from different groups or samples. A *controlled* experiment manipulates one or more of the factors being tested; *comparative* experiments compare unmanipulated data gathered from different sources. As described at the opening of this chapter, Tyrone Hayes and his colleagues conducted both types of experiment to test the prediction that the herbicide atrazine, a contaminant in freshwater ponds and streams throughout the world, affects the development of frogs.

In a **controlled experiment**, we start with groups or samples that are as similar as possible. We predict on the basis of our hypothesis that some critical factor, or **variable**, has an effect on the phenomenon we are investigating. We devise some method to manipulate *only that variable* in an "experimental" group and compare the resulting data with data from an unmanipulated "control" group. If the predicted difference occurs, we then apply statistical tests to ascertain the probability that the manipulation created the difference (as opposed to the difference being the result of random chance). **Figure 1.13** describes one of the many controlled experiments performed by the Hayes laboratory to quantify the effects of atrazine on male frogs.

The basis of controlled experiments is that one variable is manipulated while all others are held constant. The variable that is manipulated is called the *independent variable,* and the response that is measured is the *dependent variable.* A good controlled experiment is not easy to design because biological variables are so interrelated that it is difficult to alter just one.

A **comparative experiment** starts with the prediction that there will be a difference between samples or groups based on the hypothesis. In comparative experiments, however, we can-

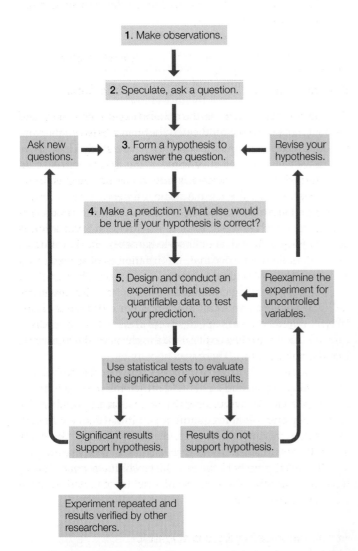

1.12 The Scientific Method The process of observation, speculation, hypothesis, prediction, and experimentation is the cornerstone of modern science. Answers gleaned through experimentation lead to new questions, more hypotheses, further experiments, and expanding knowledge.

INVESTIGATING LIFE

1.13 Controlled Experiments Manipulate a Variable

The Hayes laboratory created controlled environments that differed only in the concentrations of atrazine in the water. Eggs from leopard frogs (*Rana pipiens*) raised specifically for laboratory use were allowed to hatch and the tadpoles were separated into experimental tanks containing water with different concentrations of atrazine.

HYPOTHESIS Exposure to atrazine during larval development causes abnormalities in the reproductive system of male frogs.

METHOD

1. Establish 9 tanks in which all attributes are held constant except the water's atrazine concentrations. Establish 3 atrazine conditions (3 replicate tanks per condition): 0 ppb (control condition), 0.1 ppb, and 25 ppb.
2. Place *Rana pipiens* tadpoles from laboratory-reared eggs in the 9 tanks (30 tadpoles per replicate).
3. When tadpoles have transitioned into adults, sacrifice the animals and evaluate their reproductive tissues.
4. Test for correlation of degree of atrazine exposure with the presence of abnormalities in the reproductive systems of male frogs.

RESULTS

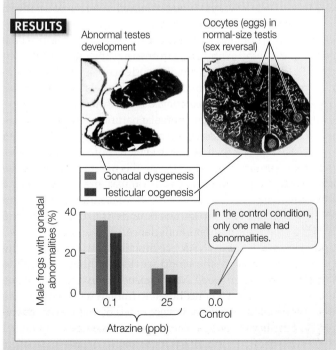

Abnormal testes development

Oocytes (eggs) in normal-size testis (sex reversal)

Gonadal dysgenesis
Testicular oogenesis

In the control condition, only one male had abnormalities.

Male frogs with gonadal abnormalities (%)

Atrazine (ppb)

CONCLUSION Exposure to atrazine at concentrations as low as 0.1 ppb induces abnormalities in the male reproductive systems of frogs. The effect is not proportional to the level of exposure.

Go to **yourBioPortal.com** for original citations, discussions, and relevant links for all INVESTIGATING LIFE figures.

not control the variables; often we cannot even identify all the variables that are present. We are simply gathering and comparing data from different sample groups.

When his controlled experiments indicated that atrazine indeed affects reproductive development in frogs, Hayes and his colleagues performed a comparative experiment. They collected frogs and water samples from eight widely separated sites across the United States and compared the incidence of abnormal frogs from environments with very different levels of atrazine (**Figure 1.14**). Of course, the sample sites differed in many ways besides the level of atrazine present.

The results of experiments frequently reveal that the situation is more complex than the hypothesis anticipated, thus raising new questions. In the Hayes experiments, for example, there was no clear direct relationship between the *amount* of atrazine present and the percentage of abnormal frogs: there were fewer abnormal frogs at the highest concentrations of atrazine than at lower concentrations. There are no "final answers" in science. Investigations consistently reveal more complexity than we expect. The scientific method is a tool to identify, assess, and understand that complexity.

— **yourBioPortal.com** —

GO TO Animated Tutorial 1.1 • The Scientific Method

Statistical methods are essential scientific tools

Whether we do comparative or controlled experiments, at the end we have to decide whether there is a difference between the samples, individuals, groups, or populations in the study. How do we decide whether a measured difference is enough to support or falsify a hypothesis? In other words, how do we decide in an unbiased, objective way that the measured difference is significant?

Significance can be measured with statistical methods. Scientists use statistics because they recognize that variation is always present in any set of measurements. Statistical tests calculate the probability that the differences observed in an experiment could be due to random variation. The results of statistical tests are therefore probabilities. A statistical test starts with a **null hypothesis**—the premise that no difference exists. When quantified observations, or **data**, are collected, statistical methods are applied to those data to calculate the likelihood that the null hypothesis is correct.

More specifically, statistical methods tell us the probability of obtaining the same results by chance even if the null hypothesis were true. We need to eliminate, insofar as possible, the chance that any differences showing up in the data are merely the result of random variation in the samples tested. Scientists generally conclude that the differences they measure are significant if statistical tests show that the *probability of error* (that is, the probability that the same results can be obtained by mere chance) is 5 percent or lower.

Not all forms of inquiry are scientific

Science is a unique human endeavor that is bounded by certain standards of practice. Other areas of scholarship share with science the practice of making observations and asking ques-

INVESTIGATING LIFE

1.14 Comparative Experiments Look for Differences among Groups

To see whether the presence of atrazine correlates with reproductive system abnormalities in male frogs, the Hayes lab collected frogs and water samples from different locations around the U.S. The analysis that followed was "blind," meaning that the frogs and water samples were coded so that experimenters working with each specimen did not know which site the specimen came from.

HYPOTHESIS Presence of the herbicide atrazine in environmental water correlates with reproductive system abnormalities in frog populations.

METHOD
1. Based on commercial sales of atrazine, select 4 sites (sites 1–4) less likely and 4 sites (sites 5–8) more likely to be contaminated with atrazine.
2. Visit all sites in the spring (i.e., when frogs have transitioned from tadpoles into adults); collect frogs and water samples.
3. In the laboratory, sacrifice frogs and examine their reproductive tissues, documenting abnormalities.
4. Analyze the water samples for atrazine concentration (the sample for site 7 was not tested).
5. Quantify and correlate the incidence of reproductive abnormalities with environmental atrazine concentrations.

RESULTS

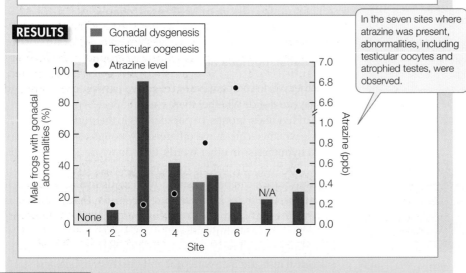

In the seven sites where atrazine was present, abnormalities, including testicular oocytes and atrophied testes, were observed.

CONCLUSION Reproductive abnormalities exist in frogs from environments in which aqueous atrazine concentration is 0.2 ppb or above. The incidence of abnormalities does not appear to be proportional to atrazine concentration at the time of transition to adulthood.

FURTHER INVESTIGATION: The highest proportion of abnormal frogs was found at site 3, located on a wildlife reserve in Wyoming. What kind of data and observations would you need to suggest possible explanations for this extremely high incidence?

Go to **yourBioPortal.com** for original citations, discussions, and relevant links for all INVESTIGATING LIFE figures.

tions, but scientists are distinguished by what they do with their observations and how they answer their questions. Data, subjected to appropriate statistical analysis, are critical in the testing of hypotheses. The scientific method is the most powerful way humans have devised for learning about the world and how it works.

Scientific explanations for natural processes are objective and reliable because the hypotheses proposed *must be testable* and *must have the potential of being rejected* by direct observations and experiments. Scientists must clearly describe the methods they use to test hypotheses so that other scientists can repeat their results. Not all experiments are repeated, but surprising or controversial results are always subjected to independent verification. Scientists worldwide share this process of testing and rejecting hypotheses, contributing to a common body of scientific knowledge.

If you understand the methods of science, you can distinguish science from non-science. Art, music, and literature all contribute to the quality of human life, but they are not science. They do not use the scientific method to establish what is fact. Religion is not science, although religions have historically purported to explain natural events ranging from unusual weather patterns to crop failures to human diseases. Most such phenomena that at one time were mysterious can now be explained in terms of scientific principles.

The power of science derives from the uncompromising objectivity and absolute dependence on evidence that comes from *reproducible and quantifiable observations*. A religious or spiritual explanation of a natural phenomenon may be coherent and satisfying for the person holding that view, but it is not testable, and therefore it is not science. To invoke a supernatural explanation (such as a "creator" or "intelligent designer" with no known bounds) is to depart from the world of science.

Science describes the facts about how the world works, not how it "ought to be." Many scientific advances that have contributed to human welfare have also raised major ethical issues. Recent developments in genetics and developmental biology, for example, enable us to select the sex of our children, to use stem cells to repair our bodies, and to modify the human genome. Although scientific knowledge allows us to do these things, science cannot tell us whether or not we should do them, or, if we choose to do so, how we should regulate them.

To make wise decisions about public policy, we need to employ the best possible ethical reasoning in deciding which outcomes we should strive for.

1.3 RECAP

The scientific method of inquiry starts with the formulation of hypotheses based on observations and data. Comparative and controlled experiments are carried out to test hypotheses.

- Explain the relationship between a hypothesis and an experiment. **See p. 14 and Figure 1.12**

- What is controlled in a controlled experiment? **See p. 14 and Figure 1.13**

- What features characterize questions that can be answered only by using a comparative approach? **See pp. 14–15 and Figure 1.14**

- Do you understand why arguments must be supported by quantifiable and reproducible data in order to be considered scientific? **See pp. 15–16**

The vast scientific knowledge accumulated over centuries of human civilization allows us to understand and manipulate aspects of the natural world in ways that no other species can. These abilities present us with challenges, opportunities, and above all, responsibilities.

1.4 How Does Biology Influence Public Policy?

Agriculture and medicine are two important human activities that depend on biological knowledge. Our ancestors unknowingly applied the principles of evolutionary biology when they domesticated plants and animals, and people have speculated about the causes of diseases and searched for methods to combat them since ancient times. Long before the microbial causes of diseases were known, people recognized that infections could be passed from one person to another, and the isolation of infected persons has been practiced as long as written records have been available.

Today, thanks to the deciphering of genomes and our new-found ability to manipulate them, vast new possibilities exist for controlling human diseases and increasing agricultural productivity, but these capabilities raise ethical and policy issues. How much and in what ways should we tinker with the genes of humans and other species? Does it matter whether the genomes of our crop plants and domesticated animals are changed by traditional methods of controlled breeding and crossbreeding or by the biotechnology of gene transfer? What rules should govern the release of genetically modified organisms into the environment? Science alone cannot provide all the answers, but wise policy decisions must be based on accurate scientific information.

Biologists are increasingly called on to advise government agencies concerning the laws, rules, and regulations by which society deals with the increasing number of challenges that have at least a partial biological basis. As an example of the value of scientific knowledge for the assessment and formulation of public policy, let's return to the tracking study of bluefin tuna introduced in Section 1.3. Prior to this study, both scientists and fishermen knew that bluefins had a western breeding ground in the Gulf of Mexico and an eastern breeding ground in the Mediterranean Sea (**Figure 1.15**). Overfishing had led to declining numbers of fish in the western-breeding populations, to the point of these populations being endangered.

1.15 Bluefin Tuna Do Not Recognize Boundaries It was assumed that tuna from western-breeding populations and those from eastern-breeding populations also fed on their respective sides of the Atlantic, so separate fishing quotas were established on either side of 45° W longitude (dashed line) to allow the endangered western population to recover. However, tracking data shows that the two populations *do not* remain separate after spawning, so in fact the established policy does not protect the western population.

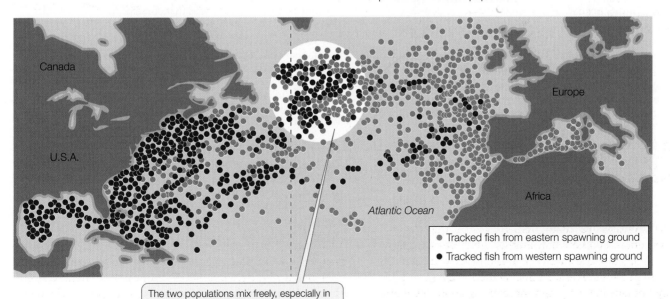

Canada

U.S.A.

Europe

Africa

Atlantic Ocean

- Tracked fish from eastern spawning ground
- Tracked fish from western spawning ground

The two populations mix freely, especially in the heavily fished waters of the North Atlantic.

Initially it was assumed by scientists, fishermen, and policy makers alike that the eastern and western populations had geographically separate feeding grounds as well as separate breeding grounds. Acting on this assumption, an international commission drew a line down the middle of the Atlantic Ocean and established strict fishing quotas on the western side of the line, with the intent of allowing the western population to recover. New tracking data, however, revealed that in fact the eastern and western bluefin populations mix freely on their feeding grounds across the entire North Atlantic—a swath of ocean that includes the most heavily fished waters in the world. Tuna caught on the eastern side of the line could just as likely be from the western breeding population as the eastern; thus the established policy was not achieving its intended goal.

Policy makers take more things into consideration than scientific knowledge and recommendations. For example, studies on the effects of atrazine on amphibians have led one U.S. group, the Natural Resources Defense Council, to take legal action to have atrazine banned on the basis of the Endangered Species Act. The U.S. Environmental Protection Agency, however, must also consider the potential loss to agriculture that such a ban would create and has continued to approve atrazine's use as long as environmental levels do not exceed 30 to 40 ppb—which is 300 to 400 times the levels shown to induce abnormalities in the Hayes studies. Scientific conclusions do not always prevail in the political world.

Another reason for studying biology is to understand the effects of the vastly increased human population on its environment. Our use of natural resources is putting stress on the ability of Earth's ecosystems to continue to produce the goods and services on which our society depends. Human activities are changing global climates, causing the extinctions of a large number of species like the amphibians featured in this chapter, and spreading new diseases while facilitating the resurgence of old ones. The rapid spread of flu viruses has been facilitated by modern modes of transportation, and the recent resurgence of tuberculosis is the result of the evolution of bacteria that are resistant to antibiotics. Biological knowledge is vital for determining the causes of these changes and for devising wise policies to deal with them.

Beyond issues of policy and pragmatism lies the human "need to know." Human beings are fascinated by the richness and diversity of life, and most people want to know more about organisms and how they interact. Human curiosity might even be seen as an adaptive trait—it is possible that such a trait could have been selected for if individuals who were motivated to learn about their surroundings were likely to have survived and reproduced better, on average, than their less curious relatives. Far from ending the process, new discoveries and greater knowledge typically engender questions no one thought to ask before. There are vast numbers of questions for which we do not yet have answers, and the most important motivator of most scientists is curiosity.

CHAPTER SUMMARY

1.1 What Is Biology?

- **Biology** is the scientific study of living organisms, including their characteristics, functions, and interactions. Cells are the basic structural and physiological units of life. The **cell theory** states that all life consists of cells and that all cells come from preexisting cells.

- All living organisms are related to one another through descent with modification. **Evolution** by **natural selection** is responsible for the diversity of **adaptations** found in living organisms.

- The instructions for a cell are contained in its **genome**, which consists of **DNA** molecules made up of sequences of **nucleotides**. Specific segments of DNA called **genes** contain the information the cell uses to make **proteins**. Review Figure 1.4

- Living organisms regulate their internal environment. They also interact with other organisms of the same and different species. Biologists study life at all these levels of organization. **Review Figure 1.6, WEB ACTIVITY 1.1**

- Biological knowledge obtained from a **model system** may be generalized to other species.

1.2 How Is All Life on Earth Related?

- Biologists use fossils, anatomical similarities and differences, and molecular comparisons of genomes to reconstruct the history of life. **Review Figure 1.8**

- Life first arose by chemical evolution. Cells arose early in the evolution of life.

- **Photosynthesis** was an important evolutionary step because it changed Earth's atmosphere and provided a means of capturing energy from sunlight.

- The earliest organisms were **prokaryotes**. Organisms called **eukaryotes**, with more complex cells, arose later. Eukaryotic cells have discrete intracellular compartments, called **organelles**, including a nucleus that contains the cell's genetic material.

- The genetic relationships of **species** can be represented as an evolutionary tree. Species are grouped into three **domains**: **Archaea**, **Bacteria**, and **Eukarya**. Archaea and Bacteria are domains of unicellular prokaryotes. Eukarya contains diverse groups of protists (most but not all of which are unicellular) and the multicellular plants, fungi, and animals. **Review Figure 1.10, WEB ACTIVITY 1.2**

1.3 How Do Biologists Investigate Life?

- The **scientific method** used in most biological investigations involves five steps: making observations, asking questions, forming hypotheses, making predictions, and testing those predictions. **Review Figure 1.12**

- **Hypotheses** are tentative answers to questions. Predictions made on the basis of a hypothesis are tested with additional observations and two kinds of **experiments**: **comparative** and **controlled experiments**. Review Figures 1.13 and 1.14, **ANIMATED TUTORIAL 1.1**

- Statistical methods are applied to **data** to establish whether or not the differences observed are significant or whether they could be the result of chance. These methods start with the **null hypothesis** that there are no differences.

- Science can tell us how the world works, but it cannot tell us what we should or should not do.

1.4 How Does Biology Influence Public Policy?

- Biologists are often called on to advise government agencies on the solution of important problems that have a biological component.

FOR DISCUSSION

1. Even if we knew the sequences of all of the genes of a single-celled organism and could cause those genes to be expressed in a test tube, it would still be incredibly difficult to create a functioning organism. Why do you think this is so? In light of this fact, what do you think of the statement that the genome contains all of the information for a species?

2. Why is it so important in science that we design and perform tests capable of falsifying a hypothesis?

3. What features characterize questions that can be answered only by using a comparative approach?

4. Cite an example of how you apply aspects of the scientific method to solve problems in your daily life.

ADDITIONAL INVESTIGATION

1. The abnormalities of frogs in Tyrone Hayes's studies were associated with the presence of a herbicide in the environment. That herbicide did not kill the frogs, but it feminized the males. How would you investigate whether this effect could lead to decreased reproductive capacity for the frog populations in nature?

2. Just as all cells come from preexisting cells, all mitochondria—the cell organelles that convert energy in food to a form of energy that can do biological work—come from preexisting mitochondria. Cells do not synthesize mitochondria from the genetic information in their nuclei. What investigations would you carry out to understand the nature of mitochondria?

WORKING WITH DATA (GO TO yourBioPortal.com)

Feminization of Frogs Analogous to the experiment shown in Figure 1.13, this exercise asks you to graph data about the size of the laryngeal (throat) muscles required to produce male mating calls in the frog *Xenopus laevis*. After plotting data from frogs exposed to different levels of the herbicide atrazine during their development, you will formulate conclusions about the effects of the herbicide on this physical attribute and speculate about what these effects might mean.

34 The Plant Body

The doomsday vault

Carved into a sandstone mountain, surrounded by permafrost on the Arctic island of Spitsbergen, Norway, the Svalbard vault is almost the size of a soccer field. It is stable enough to withstand a major earthquake, even though earthquakes are very unlikely at that location. It is high enough in elevation to remain above sea level even if all the polar ice caps were melted by global warming. A cooling system run on energy from local coal holds its interior at −18°C, and even if this system fails, insulation and the cold weather outside ensure that it will be weeks before the interior temperature rises. It is surrounded by a technologically advanced security system.

The 120 nations that participated in establishing this facility agree that the vault at Svalbard needs to be very secure. Does it contain gold bars? No, it contains seeds, the carriers of plant embryos. As photosynthesizers, plants are the keys to the biosphere. They have also been the mainstay of human survival. Of the approximately 300,000 species of plants, humans depend directly on only a few dozen. You can name some of them—wheat, rice, and corn for food; cotton for fiber; forest trees for paper. Mil-

lennia ago, humans selected certain plants growing in the wild for their own uses and began to cultivate them. After many generations of artificial selection, the plants looked very distinct from their wild relatives. In addition, because these plants were grown in different environments and for different purposes, different genetic strains of plants came into being.

By the twentieth century, there were thousands of genetically distinct varieties of crop plants—an amazing 100,000 varieties of rice alone. A crop plant variety, also called a *cultivar*, is a member of a species that has been artificially selected for one or more of its useful traits. Although only a small number of these varieties are in use at a given time, plant biologists realized that the genetic diversity of the remaining species should not be allowed to disappear. Because plant seeds are generally quite hardy

The Svalbard Global Seed Vault The vault is located 390 feet inside a sandstone mountain on the Norwegian island of Spitsbergen, about 700 miles from the North Pole. The location was chosen for secure, long-term storage of seeds because earthquakes are unlikely there and the ground is permeated with permafrost.

Seed vaults   Airlock doors   Office and handling area   Sleeve to protect tunnel from erosion and climactic changes   Tunnel entrance   Bridge

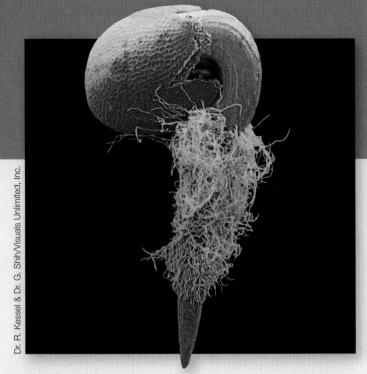

Plant in Storage   Every seed contains an embryo with the means to create all parts of the plant body.

and relatively easy to store for long periods, seed banks were established. One of the largest is the National Seed Bank in Colorado, USA, where over half a million plant varieties are stored as seeds. When a plant breeder wants to develop a new genetic strain of corn—for example, one that is naturally resistant to a fungal disease—samples of seed can be withdrawn from the seed bank and used for cross-breeding.

The newly established seed vault in Norway is not so much a bank as a safe deposit box. Seed banks from all over the world are depositing samples of seeds for safe-keeping in the vault in the event of a disaster that might destroy the seed banks. Such destruction does happen; for example, two valuable seed banks in Iraq and Afghanistan were destroyed during recent wars. As the existence of the Svalbard vault clearly demonstrates, seeds, and the plants that they form, are vital to humanity.

IN THIS CHAPTER we will examine plant structure at the levels of cells, tissues, organs, and tissue systems. We will see how that structure arises from clusters of undifferentiated cells, called meristems, that permit continuous growth throughout a plant's lifetime. The chapter concludes with a look at how humans have altered plant form through crop domestication.

34.1 What Is the Basic Body Plan of Plants?

Plants live by harvesting energy from sunlight and by collecting water and mineral nutrients from the soil. These resources, however, are incredibly sparse in the environment, so plants face the challenge of collecting them from huge areas, both above and below ground. Another challenge plants face is their inability to move. Plants cannot relocate themselves from, say, a dry, shady location to one that is wet and sunny.

The plant body plan allows plants to respond to these challenges. Stems, leaves, and roots enable plants anchored to one spot to capture scarce resources effectively, both above and below the ground. More important, to compensate for their inability to move, plants can grow throughout their lifetimes. Thus, while plants cannot move to a new water source or a new sunny clearing, they can respond to environmental cues by redirecting their growth to exploit opportunities that arise in their immediate environment.

In Chapters 28 and 29 we saw how modern plants arose from aquatic ancestors, giving rise to simple land plants and then vascular plants. Despite their obvious differences in size and form, all vascular plants have essentially the same simple structural organization. This chapter describes the basic architecture of the largest group of vascular plants, the angiosperms (flowering plants), and shows how so much diversity can literally grow out of such a simple basic form.

As we saw in Figure 29.1, angiosperms first appeared about 140 million years ago, radiated explosively over a period of about 60 million years, and became the dominant plant life on this planet. There are over 250,000 angiosperm species today. Flowers, the angiosperms' devices for sexual reproduction and their main distinguishing feature, consist of modified leaves and stems and will be considered in detail in Section 38.1. In this chapter we'll focus on the three kinds of vegetative (nonsexual) organs angiosperms possess: roots, stems, and leaves. Each of these vegetative organs can be understood in terms of its structure. By *structure* we mean both its overall form, called its *morphology*, and its internal component cells and tissues and their arrangement, known as its *anatomy*.

Plant organs are organized into two systems (**Figure 34.1**):

• The **root system** anchors the plant in place, absorbs water and dissolved minerals, and stores the products of photosynthesis from the shoot system. The extreme branching of plant roots and their high surface area-to-volume ratios allow them to absorb water and mineral nutrients from the soil efficiently.

34.1 Vegetative Organs and Systems The basic plant body plan, with root and shoot systems, and the principal vegetative organs are similar in eudicots and monocots, although there are also some differences between the two clades.

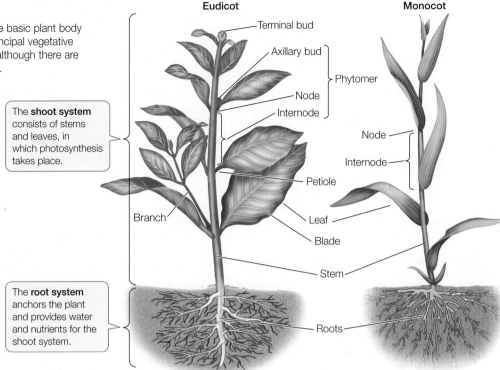

Eudicot

- Terminal bud
- Axillary bud
- Node
- Internode
- Phytomer
- Branch
- Petiole
- Leaf
- Blade
- Stem

The **shoot system** consists of stems and leaves, in which photosynthesis takes place.

Monocot

- Node
- Internode
- Roots

The **root system** anchors the plant and provides water and nutrients for the shoot system.

- The **shoot system** of a plant consists of the stems, leaves, and flowers. Broadly speaking, the **leaves** are the chief organs of photosynthesis. The **stems** hold and display the leaves to the sun and provide connections for the transport of materials between roots and leaves.

As we saw in Section 29.3, most angiosperms belong to one of two major clades. *Monocots* are generally narrow-leaved flowering plants such as grasses, lilies, orchids, and palms. *Eudicots* are broad-leaved flowering plants such as soybeans, roses, sunflowers, and maples. These two clades, which account for 97 percent of flowering plant species, differ in several important basic characteristics (see Figure 34.1). Let's take a closer look at how the root and shoot systems are elaborated in eudicots and monocots.

The root system anchors the plant and takes up water and dissolved minerals

Water and minerals enter most plants through the root system, which is located in the soil. Because light does not penetrate the soil, roots typically lack the capacity for photosynthesis. Although hidden from view, the root system is often larger than the visible shoot system. For example, the root system of a 4-month-old winter rye plant (*Secale cereale*) was found to be 130 times larger than the shoot system, with almost 13 million branches that had a cumulative length of over 500 km!

The root system of angiosperms originates in an embryonic root called the *radicle*. From this common starting point, the root systems of monocots and eudicots develop differently. Following seed germination, the radicle of most eudicots develops as a primary root (called the **taproot**), which extends downward by tip growth and outward by initiating **lateral roots**. The taproot and the lateral roots form a **taproot system**, which can take a variety of forms. For example, the taproot itself often functions as a nutrient storage organ, as in carrots (*Daucus carota*), sugar beets (*Beta vulgaris*), and sweet potato (*Ipomoea batatas*) (**Figure 34.2A**).

In contrast, the primary root of monocots (and some eudicots) is short-lived. Because they originate from the stem at

(A) Taproots

(B) Fibrous root system

(C) Prop roots

34.2 Root Systems of Eudicots and Monocots (A) The taproot systems of eudicots, such as carrots, sugar beets, and sweet potato, contrast with (B) the fibrous root system of a leek and (C) the adventitious prop roots of corn.

ground level or below, the roots of a typical monocot are called **adventitious** ("arriving from outside") **roots**, and they form a **fibrous root system** composed of numerous thin roots that are all roughly equal in diameter (**Figure 34.2B**). Many fibrous root systems have a large surface area for the absorption of water and minerals. A fibrous root system clings to soil very well. The fibrous root systems of grasses, for example, may protect steep hillsides where runoff from rain would otherwise cause erosion.

In some monocots—corn, banyan trees, and some palms, for example—adventitious roots function as props to help support the shoot. **Prop roots** are critical to these plants, which, unlike most eudicot tree species, are unable to support aboveground growth through the thickening of their stems (**Figure 34.2C**).

The stem supports leaves and flowers

The central function of stems is to elevate and support the photosynthetic organs (leaves) as well as the reproductive organs (flowers).

Unlike roots, stems bear buds of various types. A *bud* is an undeveloped shoot that may or may not develop further to produce additional branches or leaves. Shoots are composed of repeating modules called **phytomers** (see Figure 34.1). A phytomer includes one or more leaves, which are attached to the stem at a **node**; an **internode** (the interval of stem between two nodes); and one or more **axillary buds**, which form in the angle (*axil*) where each leaf meets the stem. The axillary buds are distinguished from the bud at the end of a stem or branch, which

is called a **terminal bud**. If it becomes active, an axillary bud can develop into a new *branch*, or extension of the shoot system. The arrangement of leaves along the stem (called the *phyllotaxy*) is often characteristic of the plant species.

Various modifications of stems are seen in nature. The *tuber* of a potato, for example—the part of the plant eaten by humans—is not a root, but rather an underground stem. The "eyes" of a potato are depressions containing axillary buds—in other words, a sprouting potato is just a branching stem (**Figure 34.3A**). Many desert plants have enlarged, water-retaining stems (**Figure 34.3B**). The *runners* of strawberry plants are horizontal stems from which roots grow at frequent intervals (**Figure 34.3C**). If the links between the rooted portions are broken, independent plants can develop on each side of the break—a form of vegetative (asexual) reproduction (see Section 38.3).

Although most young stems are green and capable of photosynthesis, leaves are the principal sites of photosynthesis in most plants. There are, however, exceptions: for example, photosynthesis occurs primarily in the stem of the barrel cactus featured in Figure 34.3B.

Leaves are the primary sites of photosynthesis

In gymnosperms and in most flowering plants, the leaves are responsible for most of the plant's photosynthesis. Leaves are marvelously adapted for gathering light. Typically, the **blade** of a leaf is a thin, flat structure attached to the stem by a stalk called a **petiole** (see Figure 34.1). In many plants, the leaf blade is held by its petiole at an angle almost perpendicular to the rays of the sun. This orientation, with the leaf surface facing the sun, maximizes the amount of light available for photosynthesis. Some leaves track the sun over the course of the day, moving so that they constantly face it.

In some plant species, leaves are highly modified for special functions. For example, some modified leaves serve as storage depots for energy-rich molecules, as in the bulbs of onions. In other species, such as succulents, the leaves store water. The protective spines of cacti are modified leaves (see Figure 34.3B). Other plants, such as peas, have modified portions of leaves called *tendrils* that support the plant by wrapping around other structures or plants.

(A) Tuber (modified stem)   Branches

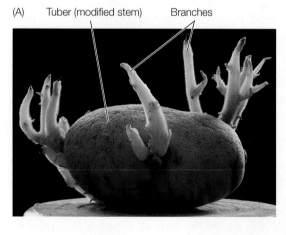

(B)

(C)

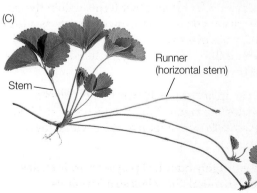

Stem

Runner
(horizontal stem)

"Barrel"
(enlarged stem)

Spines
(modified leaves)

34.3 Modified Stems (A) A potato is a modified stem called a tuber; the sprouts that grow from its eyes are shoots, not roots. (B) The stem of this barrel cactus is enlarged to store water. Its highly modified leaves serve as thorny spines. Most of this plant's photosynthesis occurs in the stem. (C) The runners of beach strawberry are horizontal stems that produce roots and shoots at intervals. Rooted portions of the plant can live independently if the runner is cut.

34.1 RECAP

The basic vegetative plant body plan consists of a root system and a shoot system. Stems and leaves, which are part of the shoot system, may be highly modified.

- How do plants explore their environment for resources even though they cannot move? **See p. 720**

- How would you distinguish between a piece of a root and a piece of a shoot? **See pp. 721–722**

- What are the differences between the root systems of eudicots and monocots? **See p. 721 and Figure 34.2**

Our closer examination of the plant body begins with its fundamental building blocks: its cells. Although plant cells share many features with animal cells, including nucleus, mitochondria, plasma membrane, and Golgi apparatuses (see Section 5.3), it is their distinguishing features that we will consider in this chapter.

34.2 How Does the Cell Wall Support Plant Growth and Form?

Plant cells have the essential organelles that are shared by all eukaryotes (see Figure 5.7), but certain additional structures and organelles distinguish them from many other eukaryotic cells:

- *Chloroplasts* or other plastids
- A central *vacuole*
- Rigid, cellulose-containing *cell walls*

As mentioned earlier, plant form is dictated in part by the need to collect energy for photosynthesis, which takes place in the chloroplasts (see Section 10.2). Less obvious is the importance of vacuoles and cell walls in determining plant form.

Cell walls and vacuoles help determine plant form

Mature plant cells usually contain a single **central vacuole**, which may account for a staggering 90 percent of its volume (see Figure 5.16). The vacuole is a watery sac containing a high concentration of solutes, including enzymes, amino acids, and sugars produced by photosynthesis. Many of these solutes are pumped into the vacuole by transporter proteins located in the **tonoplast**, the vacuolar membrane. This active accumulation of solutes provides the osmotic force for water uptake into the vacuole (as we will see in Section 35.1). As the vacuole expands, it exerts turgor pressure on the cell wall (see Figure 6.10). Turgor pressure not only keeps plants upright, but also is essential for plant growth.

The structure of cell walls allows plants to grow

Cell walls are a feature of bacteria, fungi, algae, and plants. They serve to regulate cell volume, determine cell shape, and protect the cell contents. Plant cell walls have unique features that derive from their chemical composition. Furthermore, most of the

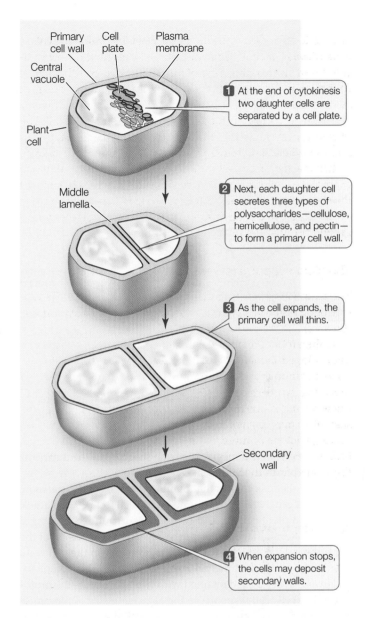

Primary cell wall Cell plate Plasma membrane
Central vacuole
Plant cell

1 At the end of cytokinesis two daughter cells are separated by a cell plate.

Middle lamella

2 Next, each daughter cell secretes three types of polysaccharides—cellulose, hemicellulose, and pectin—to form a primary cell wall.

3 As the cell expands, the primary cell wall thins.

Secondary wall

4 When expansion stops, the cells may deposit secondary walls.

34.4 Plant Cell Wall Formation Plant cell walls form as the final step in cell division.

carbon in terrestrial ecosystems is sequestered in the molecules that make up plant cell walls. As such, it is worth taking a closer look at their formation and structure.

The cytokinesis of a plant cell is completed when the two daughter cells are separated by a cell plate (**Figure 34.4**; see also Figure 11.13). A gluelike substance that forms within the cell plate constitutes the **middle lamella**, which persists as a thin layer between the walls of the two daughter cells. Each daughter cell then secretes three types of polysaccharides to form a **primary cell wall** (**Figure 34.5**):

- *Cellulose* is made up of linear polymers of thousands of glucose molecules (see Figure 3.16) that are organized into bundles of **microfibrils**, which form a lattice within the cell wall.

- *Hemicelluloses* are highly branched polysaccharide chains that extensively cross-link the cellulose microfibrils.

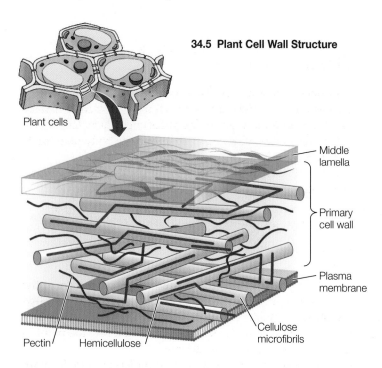

Plant cells

34.5 Plant Cell Wall Structure

Middle lamella

Primary cell wall

Plasma membrane

Cellulose microfibrils

Pectin   Hemicellulose

thought to assist in cell wall loosening by disrupting the non-covalent bonds that link the hemicelluloses and pectins to the cellulose microfibrils. To prevent the cell wall from becoming too thin (so that it does not blow out like an overinflated balloon), new cell wall components are synthesized and integrated.

When cell expansion stops, some types of plant cells deposit one or more additional cellulosic layers to form a thick **secondary cell wall** internal to the primary cell wall (see Figure 34.4). Secondary cell walls provide the mechanical support that allows some plants to produce large stems. Like the primary wall, the secondary wall contains layers of ordered cellulose microfibrils. However, rather than being embedded in pectins, the microfibrils are embedded in a remarkable substance called **lignin**. When secondary walls become lignified, the primary wall and even the middle lamella are also lignified. Lignin is a complex, carbon-containing polymer that forms a hydrophobic matrix that is strong, waterproof, and resistant to digestion by animals. After cellulose, lignin is the most abundant biological polymer on Earth, accounting for 20–35 percent of the dry weight of wood.

Scientists have just begun to dissect the complexity and dynamics of plant cell walls. Their basic components—celluloses, hemicelluloses, pectins, and lignins—are classes of molecules that can be built from several components and modified in a variety of ways. Thus the composition of plant cell walls varies among different types of plant cells. In addition, the composition of the wall of a single plant cell may not be uniform. For example, it is possible that directional growth reflects the deposition of cell wall components that are more easily loosened at one end of the cell. One measure of how much remains unknown is the finding that the genome of the tiny plant *Arabidopsis thaliana* contains more than a thousand genes related to cell wall biosynthesis, the functions of only a small fraction of which are currently known.

- *Pectins* are heterogeneous polysaccharides that are more soluble than the other components. (Pectin is responsible for the gel properties of fruit jams and jellies.)

This secretion and deposition of polysaccharides continues as the cell expands to its final size.

One of the major ways that plants grow is by cell expansion. Some cells can increase in volume by 100,000 to 1,000,000 times! How can a plant cell expand when it is surrounded by a rigid cell wall? Recall that osmotic pressure leads to expansion of the central vacuole, which exerts turgor pressure on the cell wall. The living contents of the plant cell—that is, the plasma membrane and everything contained within it—constitute the **protoplast**. The cell wall responds to the increasing size of the protoplast by loosening the linkages between cellulose microfibrils. A class of proteins that reside in the cell wall, called *expansins*, are

Building a plant body requires cooperation between groups of cells. Although they may appear to be isolated by their cell walls, plant cells interact in two ways to build and maintain a complex organism. First, in most areas, the cell wall is permeable to water and mineral ions and allows small molecules to

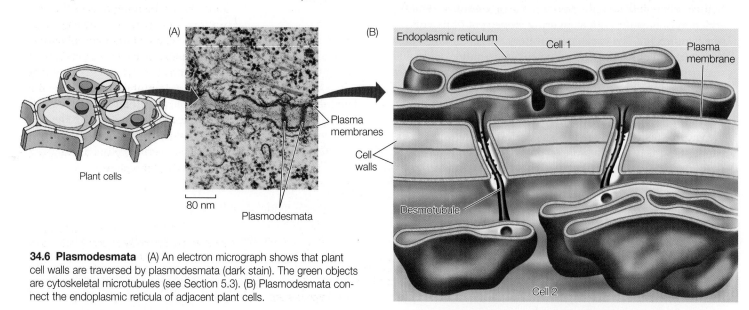

(A)

Plant cells

80 nm

Plasma membranes

Plasmodesmata

(B)

Endoplasmic reticulum

Cell 1

Plasma membrane

Cell walls

Desmotubule

Cell 2

34.6 Plasmodesmata   (A) An electron micrograph shows that plant cell walls are traversed by plasmodesmata (dark stain). The green objects are cytoskeletal microtubules (see Section 5.3). (B) Plasmodesmata connect the endoplasmic reticula of adjacent plant cells.

reach the plasma membrane. Second, the endoplasmic reticula (ER) of adjacent cells are connected by cytoplasm-filled canals called **plasmodesmata** that pass through the primary wall, allowing direct communication between plant cells (**Figure 34.6**). A single plant cell may be connected to its neighbors by up to a thousand plasmodesmata, which permit the movement of proteins and even RNAs from cell to cell. Some of these plasmodesmata are formed during cytokinesis when the cell plate is deposited.

Evolution has given some plant viruses a clever way to use this intercellular highway to their advantage. Tobacco mosaic virus (TMV), for example, encodes a protein called movement protein, or MP, that helps the virus spread throughout the plant. Without MP, the RNA genome of TMV cannot move from cell to cell. However, in some unknown way, the MP–RNA complex is able to move easily from cell to cell via plasmodesmata.

34.2 RECAP

Plants synthesize a primary cell wall during cell division and cell expansion. In some types of plant cells, a secondary cell wall, reinforced with lignin, forms within the primary cell wall when cell expansion stops, providing additional structural support.

- What are the components of plant cell walls in which most of the carbon in terrestrial ecosystems is sequestered? See **pp. 723–724**

- How do plant cell walls accommodate an expanding protoplast? See **p. 724 and Figure 34.5**

- Describe two features that allow plant cells to interact with one another. See **pp. 724–725 and Figure 34.6**

That there are dramatic differences between plant and animal body plans should not be surprising, since the multicellular forms of plants and animals evolved independently from entirely distinct protist ancestors (see Figure 1.10). In the next two sections we will look more closely at the unique characteristics of the plant body by following its development from a zygote into an adult.

34.3 How Do Plant Tissues and Organs Originate?

How does a single plant cell (a zygote) divide and grow into an organism like a redwood tree, which may grow continuously for over a thousand years to a height of over 100 meters? While still in the seed, a plant establishes the basic body plan for its mature form.

Two patterns that contribute to the plant body plan are established in the embryo:

- The *basal–apical* axis: the arrangement of cells and tissues along the main axis from root to shoot

- The *radial* axis: the concentric arrangement of the tissue systems

In addition, two clusters of undifferentiated cells form at the tips of the embryonic root and shoot. These clusters, called **meristems** (from the Greek *merizein*, "to divide"), will orchestrate all postembryonic development and allow the plant to form organs throughout its lifetime.

Both axes and meristems are best understood in developmental terms. We focus here on embryogenesis (embryo formation) in the model eudicot *Arabidopsis thaliana*, in which the process has been most intensively studied.

The first step in the formation of a plant embryo is a mitotic division of the zygote that gives rise to two daughter cells (**Figure 34.7, step 1**). These two cells face different fates (see Section 19.4). An asymmetrical (uneven) distribution of cytoplasm within the zygote causes one daughter cell to produce the embryo proper and the other daughter cell to produce a supporting structure, the **suspensor** (**Figure 34.7, step 2**). This asymmetrical division of the zygote establishes polarity as well as the basal–apical axis of the new plant. A long, thin suspensor and a more spherical or globular embryo are distinguishable after just four mitotic divisions. The suspensor soon ceases to elongate.

In eudicots, the initially globular embryo develops into the characteristic *heart stage* as the **cotyledons** ("seed leaves") start to grow (**Figure 34.7, step 3**). Further elongation of the cotyledons and of the main axis of the embryo gives rise to the *torpedo stage*, during which some of the internal tissues begin to differentiate (**Figure 34.7, step 4**). Between the cotyledons is the **shoot apical meristem**; at the other end of the axis is the **root apical meristem**. Each of these regions contains undifferentiated cells that will continue to divide to give rise to the organs developing over the life of the plant.

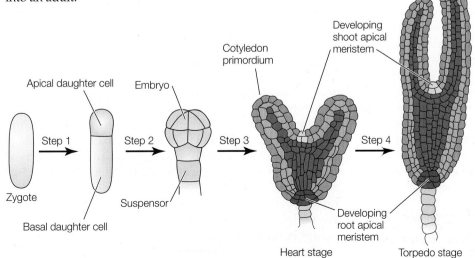

34.7 Plant Embryogenesis The basic body plan of the model eudicot (*Arabidopsis thaliana*) is established in several steps. By the heart stage, the three tissue systems are established: the dermal (gold), ground (light green), and vascular (blue) tissue systems.

By the end of embryogenesis, the radial axis of the plant has also been established. The embryonic plant contains three tissue systems, arranged concentrically, that will give rise to the tissues of the adult plant body.

The plant body is constructed from three tissue systems

A *tissue* is an organized group of cells that have features in common and that work together as a structural and functional unit. In plants, tissues, in turn, are grouped into **tissue systems**. Despite their structural diversity, all vascular plants are constructed from three tissue systems: *dermal, vascular,* and *ground.* These three tissue systems are established during embryogenesis and ultimately extend throughout the plant body in a concentric arrangement (**Figure 34.8**). Each tissue system has distinct functions and is composed of different mixtures of cell types.

DERMAL TISSUE SYSTEM The **dermal tissue system** forms the *epidermis,* or outer covering, of a plant, which usually consists of a single cell layer. The stems and roots of woody plants develop a dermal tissue called *periderm.*

During plant development, the epidermis must grow to cover the expanding plant body. The cells of the epidermis are small and round and usually have a small central vacuole or none at all. Once cell division ceases in the epidermis of an organ, the epidermal cells expand. Some epidermal cells differentiate to form one of three specialized structures:

- *Stomatal guard cells,* which form stomata (pores) for gas exchange in leaves
- *Trichomes,* or leaf hairs, which provide protection against insects and damaging solar radiation
- *Root hairs,* which greatly increase root surface area, thus providing more surface for the uptake of water and mineral nutrients.

Aboveground epidermal cells secrete a protective extracellular **cuticle** made of *cutin* (a polymer composed of long chains of fatty acids), a complex mixture of waxes, and cell wall polysaccharides. The cuticle limits water loss, reflects potentially damaging solar radiation, and serves as a barrier against pathogens.

GROUND TISSUE SYSTEM Virtually all the tissue lying between dermal tissue and vascular tissue in both shoots and roots is part of the **ground tissue system**, which therefore makes up most of the plant body. Ground tissue functions primarily in storage, support, and photosynthesis. To fulfill these diverse functions, ground tissues incorporate three cell types that are classified according to their cell wall structure: *parenchyma, collenchyma,* and *schlerenchyma.*

The most common cell type in plants is the **parenchyma** cell (**Figure 34.9A**). Parenchyma cells have large vacuoles and thin walls consisting only of a primary wall and the shared middle

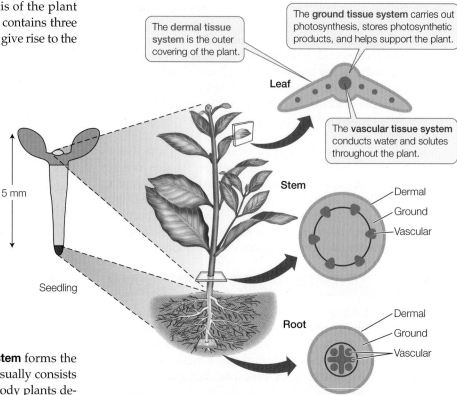

The **dermal tissue system** is the outer covering of the plant.

The **ground tissue system** carries out photosynthesis, stores photosynthetic products, and helps support the plant.

The **vascular tissue system** conducts water and solutes throughout the plant.

Leaf

Stem

Seedling

5 mm

Stem
— Dermal
— Ground
— Vascular

Root
— Dermal
— Ground
— Vascular

34.8 Three Tissue Systems Extend Throughout the Plant Body The arrangement shown here is typical of eudicots, but the three tissue systems are continuous in the bodies of all vascular plants.

lamella. They play important roles in photosynthesis (in leaves) and in the storage of, for example, protein (in fruits) and starch (in roots). Many retain the capacity to divide and hence may give rise to new cells, as when a wound results in cell proliferation.

Collenchyma cells resemble parenchyma cells that have been modified to provide flexible support. Their primary walls are characteristically thick at the corners of the cells (**Figure 34.9B**). Collenchyma cells are generally elongated. In these cells, the primary wall thickens in part due to the deposition of pectins, but no secondary wall forms. Collenchyma provides support to leaf petioles, nonwoody stems, and growing organs. Tissue made of collenchyma cells is flexible, permitting stems and petioles to sway in the wind without snapping. The familiar "strings" in celery consist primarily of collenchyma cells.

Sclerenchyma cells have thickened secondary walls that perform their major function: support. Many sclerenchyma cells undergo programmed cell death (apoptosis; see Section 11.6) after lignifying their cell walls, and thus perform their supporting function when dead. There are two types of sclerenchyma cells: elongated **fibers** and variously shaped **sclereids**. Fibers provide relatively rigid support to wood and other parts of the plant, in which they are often organized into bundles (**Figure 34.9C**). The bark of trees owes much of its mechanical strength to long fibers. Sclereids may pack together densely, as in a nut's shell or in some seed coats (**Figure 34.9D**). Isolated clumps of sclereids, called *stone cells,* in pears and some other fruits give them their characteristic gritty texture.

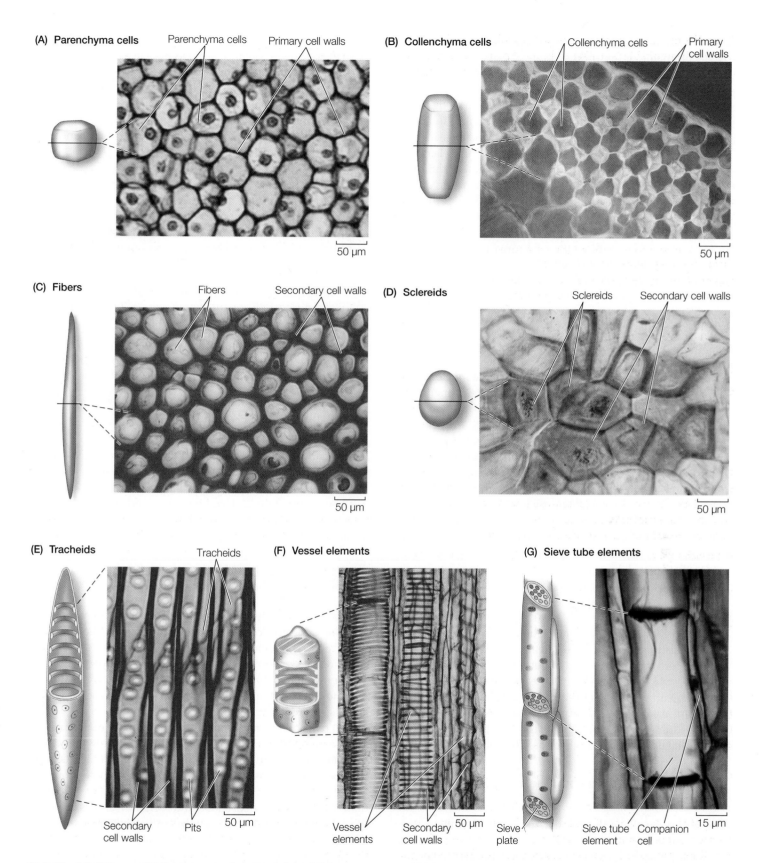

(A) Parenchyma cells
Parenchyma cells
Primary cell walls
50 μm

(B) Collenchyma cells
Collenchyma cells
Primary cell walls
50 μm

(C) Fibers
Fibers
Secondary cell walls
50 μm

(D) Sclereids
Sclereids
Secondary cell walls
50 μm

(E) Tracheids
Tracheids
Secondary cell walls
Pits
50 μm

(F) Vessel elements
Vessel elements
Secondary cell walls
50 μm

(G) Sieve tube elements
Sieve plate
Sieve tube element
Companion cell
15 μm

34.9 Plant Cell Types (A) Parenchyma cells in the petiole of *Coleus*. Note the thin, uniform cell walls. (B) Collenchyma cells make up the five outer cell layers of this spinach leaf vein. Their walls are thick at the corners of the cells and thin elsewhere. (C) Sclerenchyma: fibers in a sunflower stem (*Helianthus*). The thick secondary walls are stained red. (D) Sclerenchyma: sclereids. The extremely thick secondary walls of sclereids are laid down in layers. They provide support and a hard texture to structures such as nuts and seeds. (E, F) Tracheary elements: (E) Tracheids in pinewood. The thick secondary walls are stained dark red. (F) Vessel elements in the stem of a squash. The secondary walls are stained red; note the different patterns of thickening, including rings and spirals. (G) Sieve tube elements and companion cells in the stem of a cucumber.

VASCULAR TISSUE SYSTEM   The **vascular tissue system** is the plant's plumbing or transport system—the distinguishing feature of vascular plants. Its two constituent tissues, the xylem and phloem, distribute materials throughout the plant. The **xylem** distributes water and mineral ions taken up by the roots to all the cells of the stems and leaves. **Phloem**, as a result of its cellular complexity, can perform a variety of functions, including transport, support, and storage. All the living cells of the plant body require a source of energy and chemical building blocks. The phloem meets these needs by transporting carbohydrates from sites of production (called *sources*, primarily leaves) to sites of utilization or storage (called *sinks*, such as growing tissue, storage tubers, and developing flowers).

Let's take a closer look at the structure of the diverse cell types that make up these vascular tissues. In Chapter 35 we will see how they transport water and materials throughout the plant body.

Cells of the xylem transport water and dissolved minerals

Xylem contains conducting cells called **tracheary elements**, which undergo apoptosis before assuming their function of transporting water and dissolved minerals. There are two types of tracheary elements: *tracheids* and *vessel elements*. **Tracheids** are spindle-shaped cells, found in gymnosperms and other vascular plants, that are evolutionarily more ancient than vessel elements (**Figure 34.9E**). When the protoplast disintegrates upon cell death, water and minerals can move with little resistance from one tracheid to its neighbors by way of *pits*, interruptions in the secondary wall that leave the primary wall unobstructed.

Flowering plants evolved a water-conducting system made up of *vessels*, formed from individual cells, called **vessel elements**, that are laid down end-to-end. Vessel elements have pits in their cell walls, as do tracheids, but are generally larger in diameter than tracheids. Vessel elements secrete lignin into their secondary walls, then partially break down their end walls before undergoing apoptosis. The result is a continuous hollow tube consisting of many vessel elements, providing an open pipeline for water conduction (**Figure 34.9F**). In the course of angiosperm evolution, vessel elements have become shorter, and their end walls have become less and less obliquely oriented and less obstructed, presumably increasing the efficiency of water transport through them. The xylem of many angiosperms includes tracheids as well as vessel elements.

Cells of the phloem transport the products of photosynthesis

The transport cells of the phloem, unlike those of the mature xylem, are living cells. In flowering plants, the characteristic cells of the phloem are **sieve tube elements** (**Figure 34.9G**). Like vessel elements, these cells meet end-to-end. They form long *sieve tubes*, which transport carbohydrates and many other materials from their sources to tissues that consume or store them. In plants with mature leaves, for example, products of photosynthesis move from leaves to root tissues.

Unlike vessel elements, which break down their end walls, sieve tube elements contain plasmodesmata in their end walls that enlarge to form pores. The result is end walls that look like sieves, called *sieve plates*. Although the sieve tube elements remain alive, some components of the protoplast break down. They are closely connected to *companion cells* that retain all their organelles and function as a "life support system" for the sieve tube elements.

34.3 RECAP

Plant embryos have an embryonic root and shoot containing three concentric tissue systems: dermal, ground, and vascular. These tissue systems carry out different functions through their unique combinations of specialized cell types.

- What distinguishes the three tissue systems in terms of their location and functions? See pp. 726–728 and Figure 34.8
- What structural differences make tissues made of collenchyma cells more flexible than those consisting primarily of sclerenchyma cells? See p. 726
- Outline the differences between tracheids and vessel elements. See p. 728 and Figure 34.9

By the end of embryogenesis, the plant embryo is encased in a seed and is ready to germinate. We will return to seeds and their germination in the chapters that follow. For now, let's consider the beginning of life as faced by plants and see how the cells and tissues we have just described allow the embryo to build an adult plant body.

34.4 How Do Meristems Build a Continuously Growing Plant?

As noted at the beginning of this chapter, plants and animals develop and function differently. While animals use their mobility to forage for food, plants are sessile and must collect scarce resources from above and below ground by growing. Plants thus grow in two directions—toward sunlight, and toward water and dissolved minerals in the soil—through the growth of shoots and roots, respectively.

All parts of the animal body grow as an individual develops from embryo to adult, but in most animals this growth is **determinate**—that is, the growth of the individual and all its parts ceases when the adult state is reached. Determinate growth is also characteristic of some plant organs, such as leaves, flowers, and fruits. The growth of shoots and roots, however, is a lifelong process. Such open-ended growth is called **indeterminate**.

Plants increase in size through primary and secondary growth

Plants increase their surface area above and below ground by growing. All plants experience **primary growth**, which is characterized by the lengthening of roots and shoots and by the proliferation of new roots and shoots through branching. In addition, many gymnosperms and eudicots, especially trees, experience **secondary growth**, by which they increase in girth.

Primary and secondary growth lead to distinctive traits in the plant body. Primary growth develops what is called the **primary plant body**, while secondary growth develops the **secondary plant body**. All seed plants have a primary plant body, which consists of all the *nonwoody* parts of the plant. Many herbaceous plants—monocots in particular—consist entirely of a primary plant body. *Woody* plants, such as trees and shrubs, have, in addition to the primary plant body, a secondary plant body consisting of wood and bark. As the tissues of the secondary plant body are laid down, the stems and roots thicken. The secondary plant body continues to grow and thicken throughout the life of the plant. The primary plant body also continues to grow, lengthening and branching the shoot and root systems and forming new leaves.

A hierarchy of meristems generates the plant body

Meristems, as we have seen, are localized regions of undifferentiated cells that are the source of all new organs in the adult plant. Even before seed germination, the plant embryo has two meristems: a shoot apical meristem at the end of the embryonic shoot, and a root apical meristem near the end of the embryonic root (see Figure 34.7).

Meristematic cells are small and closely packed, with very small central vacuoles and a very thin primary cell wall. Meristematic cells are undifferentiated and forever young, retaining the ability to produce new cells indefinitely. The cells that perpetuate the meristems, called **initials**, are comparable to animal stem cells (discussed in Section 19.2). When an initial divides, one daughter cell develops into another meristem cell the size of its parent, while the other daughter cell differentiates into a more specialized cell.

While the plant embryo experiences primary growth through the activities of the root and shoot apical meristems, growth of the adult plant reflects the activity of additional meristem types. Our discussion of postembryonic plant growth begins with a closer look at how the adult plant grows throughout its lifetime and the critical role of meristems in that growth.

Two types of meristems contribute to the growth and development of the adult plant (**Figure 34.10**):

- **Apical meristems** orchestrate primary growth, giving rise to the primary plant body. This growth is characterized by cell division followed by cell enlargement (vertical elongation).

- **Lateral meristems** orchestrate secondary growth. Two lateral meristems, *vascular cambium* and *cork cambium*, contribute to the secondary plant body.

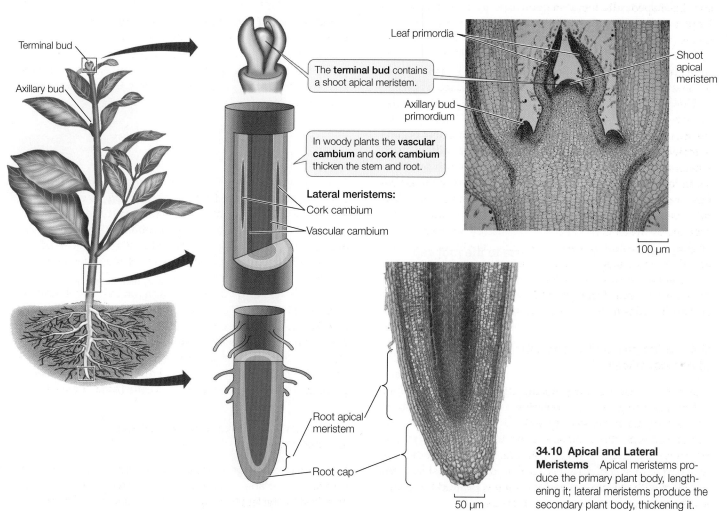

Terminal bud

Axillary bud

The **terminal bud** contains a shoot apical meristem.

In woody plants the **vascular cambium** and **cork cambium** thicken the stem and root.

Lateral meristems:
Cork cambium
Vascular cambium

Leaf primordia

Shoot apical meristem

Axillary bud primordium

100 µm

Root apical meristem

Root cap

50 µm

34.10 Apical and Lateral Meristems Apical meristems produce the primary plant body, lengthening it; lateral meristems produce the secondary plant body, thickening it.

Indeterminate primary growth originates in apical meristems

Because apical meristems can perpetuate themselves indefinitely, a shoot or root can continue to lengthen and grow indefinitely; in other words, growth of the shoot or root is indeterminate. Primary growth leads to elongation of shoots and roots and formation of organs (see Figure 34.10). All plant organs arise ultimately from cell divisions in apical meristems, followed by cell expansion and differentiation. Several types of apical meristems play roles in organ formation:

- *Shoot apical meristems* supply the cells that extend stems and branches, allowing more leaves to form and photosynthesize. Apical meristems that form leaves are called *vegetative meristems*. Flowers are formed by apical meristems that become *inflorescence meristems* (see Section 38.2 for more on floral development).

- *Root apical meristems* supply the cells that extend roots, enabling the plant to penetrate and explore the soil for water and minerals.

Apical meristems in both the shoot and the root give rise to a set of cylindrical **primary meristems**, which produce the tissues of the primary plant body. From the outside to the inside of the root or shoot, which are both cylindrical organs, the primary meristems are the **protoderm**, the **ground meristem**, and the **procambium**. These meristems, in turn, give rise to the three tissue systems:

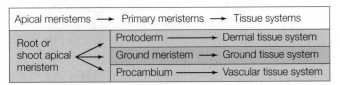

Because meristems can continue to produce new organs throughout the lifetime of the plant, the plant body is much more variable in form than the animal body, which produces each organ only once. To see how meristems function, let's look more closely at how the root apical meristem produces the root system.

The root apical meristem gives rise to the root cap and the root primary meristems

The root apical meristem produces all the cells that contribute to growth in the length of a root (**Figure 34.11A**). Some of the daughter cells from the apical (tip) end of the root apical meristem contribute to a **root cap**, which protects the delicate growing region of the root as it pushes through the soil. The root cap secretes a mucopolysaccharide (slime) that acts as a lubricant. Even so, the cells of the root cap are often damaged or scraped away and must therefore be replaced constantly. The root cap

is also the structure that detects the pull of gravity and thus controls the downward growth of roots.

In the middle of the root apical meristem is a *quiescent center*, in which cell divisions are rare. The quiescent center can become more active when needed—following injury, for example. The daughter cells produced above the quiescent center (that is, away from the root cap) become the three cylindrical primary meristems: the protoderm, the ground meristem, and the procambium.

The apical and primary meristems constitute the **zone of cell division**, the source of all the cells of the root's primary tissues. Just above this zone is the **zone of cell elongation**, where the newly formed cells are elongating and thus pushing the root farther into the soil. Above that zone is the **zone of maturation**, where the cells are differentiating, taking on specialized forms and functions. These three zones grade imperceptibly into one another; there is no abrupt line of demarcation.

The products of the root's primary meristems become root tissues

The products of the three primary meristems (the protoderm, ground meristem, and the procambium) are the tissue systems of the mature root (**Figure 34.12**).

The protoderm gives rise to the **epidermis**, an outer layer of cells that is adapted for protection of the root and absorption of mineral ions and water. Many of the epidermal cells become

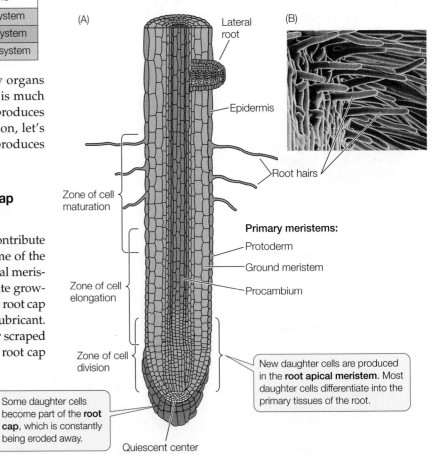

34.11 Tissues and Regions of the Root Tip
(A) Extensive cell division creates the complex structure of the root. (B) Root hairs, seen with a scanning electron microscope.

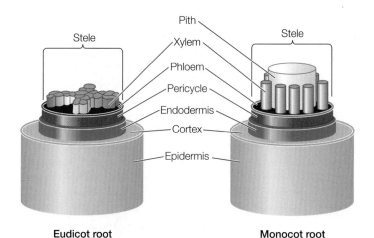

34.12 Products of the Root's Primary Meristems The protoderm gives rise to the outermost layer (epidermis). The ground meristem produces the cortex, the innermost layer of which is the endodermis. The primary vascular tissues of the root are found in the stele, which is the product of the procambium. The arrangement of tissues in the stele differs in the roots of eudicots and monocots.

---

yourBioPortal.com
GO TO Web Activity 34.1 • Eudicot Root
AND Web Activity 34.2 • Monocot Root

---

long, delicate **root hairs**, which vastly increase the surface area of the root (**Figure 34.11B**). Root hairs grow out among the soil particles, probing nooks and crannies and taking up water and minerals.

Internal to the epidermis, the ground meristem gives rise to a region of ground tissue that is many cells thick, called the **cortex**. The cells of the cortex are relatively unspecialized and often serve as storage depots. The innermost layer of the cortex is the **endodermis**. Unlike those of other cortical cells, the cell walls of the endodermal cells contain *suberin*, a waterproof substance. Strategic placement of suberin in only certain parts of the cell wall enables the cylindrical ring of endodermal cells to control the movement of water and dissolved mineral ions into the vascular tissue system.

Moving inward past the endodermis, we enter the vascular cylinder, or **stele**, produced by the procambium. The stele consists of three tissues: pericycle, xylem, and phloem. The **pericycle** consists of one or more layers of relatively undifferentiated cells. It has three important functions:

- It is the tissue within which lateral roots arise (**Figure 34.13A**).
- It can contribute to secondary growth by giving rise to lateral meristems that thicken the root.
- Its cells contain membrane transport proteins that export nutrient ions into the cells of the xylem.

At the very center of the root of a eudicot lies the xylem. Seen in cross section, it typically has the shape of a star with a variable number of points (**Figure 34.13B**). Between the points are bundles of phloem. In monocots, a region of parenchyma cells, called the **pith**, typically lies in the center of the root, surrounded by xylem and phloem (**Figure 34.13C**). Pith, which often stores carbohydrate reserves, is also found in the stems of both eudicots and monocots.

The products of the stem's primary meristems become stem tissues

Recall that shoots are composed of repeating modules called phytomers, each consisting of a node with its attached leaf or leaves, the internode between nodes, and axillary buds in the angle between each leaf and the stem (see Figure 34.1). Shoots grow by adding new phytomers. Those new phytomers originate from cells in shoot apical meristems, which are formed at the tips of stems and in axillary buds.

The shoot apical meristem, like the root apical meristem, forms three primary meristems: protoderm, ground meristem, and procambium. These primary meristems, in turn, give rise to the three shoot tissue systems. The shoot apical meristem

34.13 Root Anatomy (A) Cross section through the tip of a lateral root in a willow tree. Cells in the pericycle divide and the products differentiate, forming the tissues of a lateral root. (B, C) Cross sections of the stele of (B) a representative eudicot (the buttercup, *Ranunculus*) and (C) a representative monocot (corn, *Zea mays*), showing the arrangement of the primary root tissues.

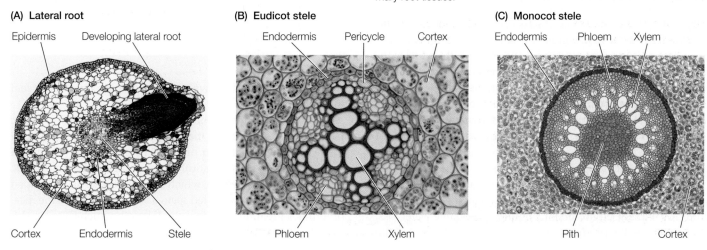

(A) Lateral root   **(B) Eudicot stele**   **(C) Monocot stele**

repetitively lays down the beginnings of leaves and axillary buds. Leaves arise from bulges called *leaf primordia*, which form as cells divide on the sides of the shoot apical meristem (see Figure 34.10). *Bud primordia* form at the bases of the leaf primordial and where they may become new apical meristems and initiate new shoots. The growing stem has no protective structure analogous to the root cap, but the leaf primordia can act as a protective covering for the shoot apical meristem.

The plumbing of stems differs from that of roots. In a root, the vascular tissue lies deep in the interior, with the xylem at or near the center (see Figure 34.13B and C). The vascular tissue of a young stem, however, is divided into discrete **vascular bundles** (**Figure 34.14**). Each vascular bundle contains both xylem and phloem. In eudicots, the vascular bundles generally form a cylinder, but in monocots, they are seemingly scattered throughout the stem.

In addition to the vascular tissues, the stem contains other important storage and supportive tissues. In eudicots, the pith lies to the inside of the ring of vascular bundles and also extends between them, forming regions called *pith rays*. To the outside lies the cortex, which may contain supportive collenchyma cells with thickened walls. The pith and cortex constitute the ground tissue system of the stem. The outermost cell layer of the young stem is the epidermis.

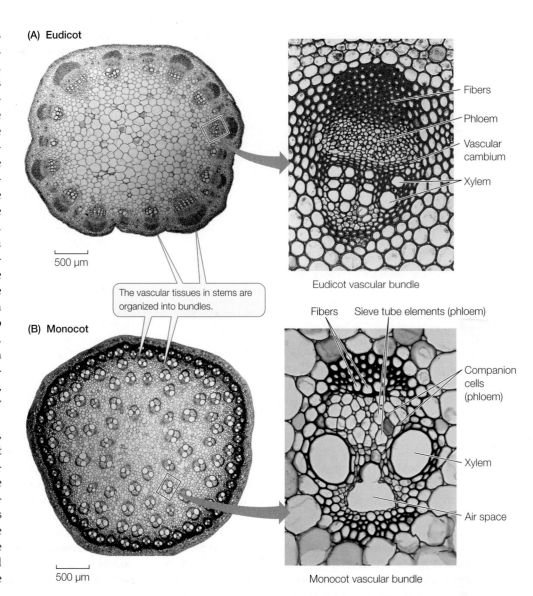

The vascular tissues in stems are organized into bundles.

34.14 Vascular Bundles in Stems (A) In herbaceous eudicot stems, the vascular bundles are arranged in a cylinder, with pith in the center and the cortex outside the cylinder. (B) A scattered arrangement of vascular bundles is typical of monocot stems.

─── **yourBioPortal.com** ───
GO TO Web Activity 34.3 • Eudicot Stem
AND Web Activity 34.4 • Monocot Stem

Leaves are determinate organs produced by shoot apical meristems

For most of its life a plant produces leaves from apical meristems. Apical meristems that produce leaves are called **vegetative meristems**. As shown in Figure 34.10, leaves originate from the edges of the apical meristem as initial cells that differentiate into leaf primordia. A highly simplified way to think of the development of the leaf from the leaf primordia is to imagine leaves as flattened stems. However, there are two important differences. First, unlike the indeterminate growth of the stem, the growth of a leaf is determinate. Second, while the tissues of the stem are arranged in a radial pattern, the leaf, as a flat organ, has a distinct top side and bottom side.

Leaf anatomy is beautifully adapted to carry out photosynthesis and to support that process by exchanging the gases O_2 and CO_2 with the environment, limiting evaporative water loss, and exporting the products of photosynthesis to the rest of the plant. **Figure 34.15A** shows a section of a typical eudicot leaf in three dimensions.

Most eudicot leaves have two zones of photosynthetic parenchyma tissue called **mesophyll** (which means "middle of the leaf"). The upper layer or layers of mesophyll, which consist of elongated cells, constitute a zone called *palisade mesophyll*. The lower layer or layers, which consist of irregularly shaped cells, constitute a zone called *spongy mesophyll*. Within the mesophyll is a great deal of air space through which CO_2 can diffuse to photosynthesizing cells.

(A)

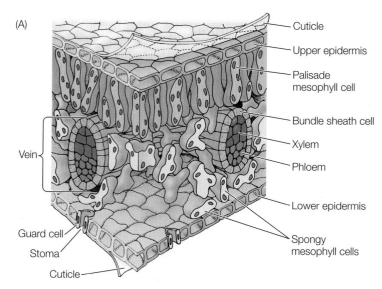

Cuticle

Upper epidermis

Palisade mesophyll cell

Bundle sheath cell

Xylem

Phloem

Vein

Lower epidermis

Guard cell

Stoma

Spongy mesophyll cells

Cuticle

(B)

(C)

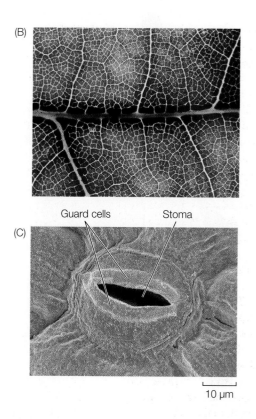

Guard cells   Stoma

10 μm

34.15 The Eudicot Leaf (A) This three-dimensional diagram shows a section of a eudicot leaf. (B) The network of fine veins in this maple leaf carries water to the mesophyll cells and carries photosynthetic products away from them. (C) Carbon dioxide enters the leaf through stomata like this one on the epidermis of a eudicot leaf.

── yourBioPortal.com ──
GO TO Web Activity 34.5 • Eudicot Leaf

Vascular tissue branches extensively throughout the leaf, forming a network of *veins* (**Figure 34.15B**). Veins extend to within a few cell diameters of all the cells of the leaf, ensuring that the mesophyll cells are well supplied with water and minerals. The products of photosynthesis are loaded into the veins for export to the rest of the plant.

Covering virtually the entire leaf on both its upper and lower surfaces is a layer of nonphotosynthetic cells, the epidermis. The epidermal cells have an overlying waxy cuticle that is impermeable to water. Although this impermeability prevents excessive water loss, it also poses a problem: while the epidermis keeps water in the leaf, it also keeps out CO_2—the other raw material of photosynthesis.

The problem of balancing water retention and carbon dioxide availability is solved by an elegant regulatory system that will be discussed in more detail in Section 35.3. Stomatal *guard cells* are modified epidermal cells that can change their shape, thereby opening or closing pores called *stomata* (singular *stoma*), which serve as passageways between the environment and the leaf's interior (**Figure 34.15C**). When the stomata are open, carbon dioxide can enter and oxygen can leave, but water can also be lost.

Many eudicot stems and roots undergo secondary growth

As we have seen, the roots and stems of some eudicots develop a secondary plant body, the tissues of which we commonly refer to as *wood* and *bark*. These tissues are derived by secondary growth from the two lateral meristems, the vascular cambium and the cork cambium.

The **vascular cambium** is a cylindrical tissue consisting predominantly of elongated cells that divide frequently. It supplies the cells of the secondary xylem and secondary phloem, which eventually become wood and bark. The **cork cambium** produces mainly waxy-walled protective cells. It supplies some of the cells that become bark.

Each year, deciduous trees lose their leaves, leaving bare branches and twigs in winter. These twigs illustrate both primary and secondary growth (**Figure 34.16**). The apical meristems of the twigs are enclosed in buds protected by *bud scales*. When the buds begin to grow in spring, the scales fall away, leaving scars, which show us where the bud was and allow us to identify each year's growth. The dormant twig shown in Figure 34.16 is the product of primary and secondary growth. Only the buds consist entirely of primary tissues.

The vascular cambium is initially a single layer of cells lying between the primary xylem and the primary phloem within the vascular bundles (see Figure 34.16). The root or stem increases in diameter when the cells of the vascular cambium divide, producing secondary xylem cells toward the inside of the root or stem and producing secondary phloem cells toward the outside (**Figure 34.17**). In the stem, cells in the pith rays between the vascular bundles also divide, forming a continuous cylinder of vascular cambium running the length of the stem. This cylinder, in turn, gives rise to complete cylinders of secondary xylem (**wood**) and secondary phloem, which contributes to the bark. It also produces vascular rays for lateral transport, a structure not found in primary xylem and phloem. The principal cell products of the vascular cambium are vessel elements, tracheids, and supportive fibers in the secondary xylem, and sieve tube elements, companion cells, fibers, and parenchyma cells in the secondary phloem.

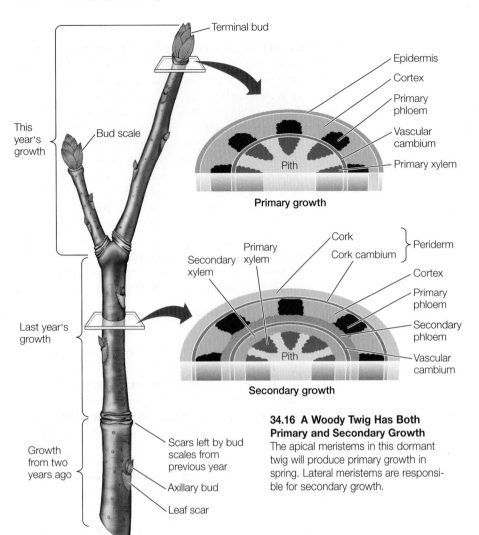

34.16 A Woody Twig Has Both Primary and Secondary Growth The apical meristems in this dormant twig will produce primary growth in spring. Lateral meristems are responsible for secondary growth.

As secondary growth of stems or roots continues, the expanding vascular tissue stretches and breaks the epidermis and the outer layers of the cortex, which ultimately flake away. Tissue derived from the secondary phloem then becomes the outermost part of the stem. Before the dermal tissues are broken away, cells lying near the surface of the secondary phloem begin to divide, forming a cork cambium. This meristematic tissue produces layers of *cork,* a protective tissue composed of cells with thick walls waterproofed with suberin. The cork soon becomes the outermost tissue of the stem or root (see Figure 34.16). Without the activity of the cork cambium, the sloughing off of the outer primary tissues would expose the plant to potential damage, such as excessive water loss or invasion by microorganisms. Sometimes the cork cambium produces cells to the inside as well as to the outside; these cells constitute a tissue known as the *phelloderm.*

The cork cambium, cork, and phelloderm constitute a secondary dermal tissue called **periderm**. As the vascular cambium continues to produce secondary vascular tissue, these corky layers are lost, but the continuous formation of new cork cambia in the underlying secondary phloem gives rise to new corky layers. The periderm and the secondary phloem—that is, all the tissues external to the vascular cambium—constitute the **bark**.

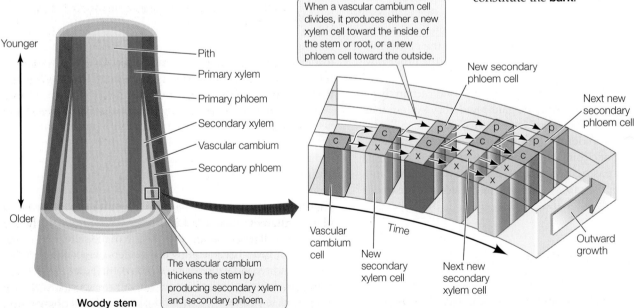

34.17 Vascular Cambium Thickens Stems and Roots Stems and roots grow thicker because a thin layer of cells, the vascular cambium, remains meristematic. These highly diagrammatic images emphasize the pattern of deposition of secondary xylem and phloem by the vascular cambium.

When a vascular cambium cell divides, it produces either a new xylem cell toward the inside of the stem or root, or a new phloem cell toward the outside.

The vascular cambium thickens the stem by producing secondary xylem and secondary phloem.

yourBioPortal.com
GO TO Animated Tutorial 34.1 • Secondary Growth: The Vascular Cambium

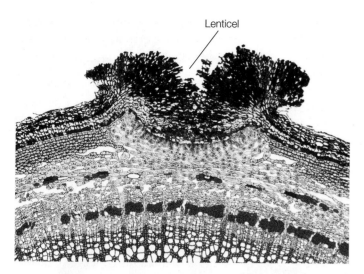

Lenticel

34.18 Lenticels Allow Gas Exchange through the Periderm The region of periderm that appears broken open is a lenticel in a year-old elderberry (*Sambucus*) twig; note the spongy tissue that constitutes the lenticel.

When periderm forms on stems or roots, the underlying tissues still need to release carbon dioxide and take up oxygen for cellular respiration. *Lenticels* are spongy regions in the periderm that allow such gas exchange (**Figure 34.18**).

Cross sections of most trunks (mature stems) of trees in temperate-zone forests show annual rings of wood (**Figure 34.19**), which result from seasonal environmental conditions. In spring, when water is relatively plentiful, the tracheids or vessel elements produced by the vascular cambium tend to be large in diameter and thin-walled. Such wood is well adapted for transporting water and minerals. As water becomes less available during the summer, narrower cells with thicker walls are pro-

duced, making this summer wood darker and perhaps more dense than the wood formed in spring. Thus each growing season is usually recorded in a tree trunk by a clearly visible annual ring. Trees in the moist tropics do not undergo seasonal growth, so they do not lay down such obvious regular rings. Variations in temperature or water supply can lead to the formation of more than one "annual" ring in a single year, but commonly each year brings a new annual ring and a new batch of leaves.

Only eudicots and other non-monocot angiosperms, along with many gymnosperms, have a vascular cambium and a cork cambium and thus undergo secondary growth. The few monocots that form thickened stems—palms, for example—do so without secondary growth. Palms have a very wide apical meristem that produces a wide stem, and dead leaf bases add to the diameter of the stem. All monocots grow in essentially this way, as do other angiosperms that lack secondary growth.

34.4 RECAP

Meristems are localized regions of cell division that are the source of all new organs in the adult plant. Apical meristems are responsible for primary growth, which is associated with the lengthening and branching of shoots and roots. Lateral meristems increase plant girth and form wood and bark in many eudicots.

- Explain how an apical meristem can be maintained for years while continuing to form leaves. **See p. 729 and Figure 34.10**

- What cells are derived from the root apical meristem and what is the general process of root growth? **See pp. 730-731 and Figure 34.11**

- How does the vascular cambium give rise to thicker stems and roots? **See p. 733 and Figure 34.16**

The building of the plant body by meristems allows a plant to respond to its environment by redirecting its growth. Thus individual plants of the same species can vary greatly in form. What underlies this variation, and how have humans used it to our advantage?

34.5 How Has Domestication Altered Plant Form?

We have seen in this chapter that a very simple plant body plan—with roots, stems, leaves, meristems, and relatively few tissue and cell types—underlies the remarkable diversity of the flowering plants that cover our planet. However, while a difference in plant form between members of different species is expected, members of the same species can be remarkably diverse in form as well. From a genetic perspective, this observation suggests that minor differences in gene content or gene regulation can underlie dramatic differences in plant form. (Nevertheless, different plant species do differ greatly sometimes in gene content and genome organization.)

Secondary xylem (one year's growth) Bark Cork cambium Cork

Pith Spring wood Summer wood Secondary phloem

34.19 Annual Rings Rings of secondary xylem are the most noticeable feature of this cross section from a tree trunk.

Let's return to the Doomsday Vault. We saw at the opening of this chapter that the vault will be used as a backup for seed banks around the world. Many of these seed banks concentrate on seed from a particular crop species, such as the Maize Stock Center at the University of Illinois (corn) or the Genetic Resources Center of the International Rice Research Institute in the Philippines (rice). In addition to containing large collections of cultivated varieties, these seed banks also contain seeds from populations of their wild relatives.

Why maintain collections of seed from both cultivated crops and their wild relatives? Despite sometimes vast morphological differences, crops and their wild relatives are still members of the same species. As such, when they are crossbred, they can produce viable progeny. These progeny will carry new combinations of their parents' traits.

It is hard to believe that modern corn was domesticated from the wild grass teosinte, which still grows in the hills of Mexico (**Figure 34.20**). One of the most conspicuous differences is that teosinte, like other wild grasses, is highly branched, with many shoots, while domesticated corn has a single shoot. This morphological difference is due in large part to the activity of a single gene called *teosinte branched 1* (*tb1*). The protein product of *tb1* regulates the growth of axillary buds (see Figure 34.1). The allele of *tb1* in domesticated corn represses branching, while the allele in teosinte permits branching.

Even harder to believe is that a single species, *Brassica oleracea* (wild mustard), is the ancestor of so many familiar and morphologically diverse crops such as kale, broccoli, Brussels sprouts, and cabbage (see Figure 21.4). An understanding of how the basic body plan of plants arises makes it possible to appreciate how each of these crops was domesticated. Starting with morphologically diverse populations of the wild ancestor, humans selected and planted the seed from variants with the trait they found desirable. Many generations of such artificial selection produced the crops that fill the produce section of the supermarket or the stands of the farmers' market.

Just as they were for ancient farmers, the genomes of plants are priceless resources today. The genetic variation in crop plants and their wild relatives can be used to improve our crop plants or adapt them to changing conditions. The improvement of crop plants is a work in progress that is being carried out in plant breeding programs worldwide. In fact, these programs are more important than ever. Increased human activity is dramatically changing our planet and leading to the extinction of more and more plant species. When seen in this light, the

34.20 Modern Corn Was Domesticated from the Wild Grass Teosinte   Beginning more than 8,000 years ago in Mexico, farmers favored plants with minimal branching. Reducing the number of branches results in fewer ears per plant, but allows each ear to grow larger and produce more seeds.

Doomsday Vault is an insurance policy for our crop plants against the loss of our most valuable resource, the genetic diversity underlying plant form and growth.

34.5 RECAP

Crop domestication involves artificial selection of certain desirable traits found in wild plant populations. By understanding the basic body plan of plants, one can more easily understand the morphological relationship between a crop plant and its wild relatives.

- Why is seed from wild relatives of crop plants valuable? See p. 736 and Figure 34.20

CHAPTER SUMMARY

34.1 What Is the Basic Body Plan of Plants?

- The vegetative organs of flowering plants are roots, which form a **root system**, and stems and leaves, which form a **shoot system**. Review Figure 34.1
- The two major clades of flowering plants, eudicots and monocots, differ from each other in a number of structural respects.

Most eudicots have a **taproot system**, and most monocots have a **fibrous root system**. Review Figure 34.2

- **Stems** bear undeveloped shoots called buds. **Axillary buds** can develop into new branches. A **terminal bud** is found at the end of a shoot.
- **Leaves** are the primary sites of photosynthesis. The leaf **blade** is attached to the stem by a **petiole**.

34.2 How Does the Cell Wall Support Plant Growth and Form?

- Plant cells differ from other eukaryotic cells in having chloroplasts or other plastids, large **central vacuoles**, and cellulose-containing cell walls.

- Once cytokinesis of a plant cell is complete, each daughter plant cell produces a **primary cell wall**. The walls of the two cells are separated by a **middle lamella**. Review Figure 34.4

- The primary cell wall is made up of bundles of cellulose **microfibrils** cross-linked by hemicellulose and pectin. **Review Figure 34.5**

- The primary cell wall is rigid but dynamic. By loosening the linkages between microfibrils, the cell wall can expand in volume by up to a million times.

- Some cells produce a thick **secondary cell wall**. **Lignin** in the secondary cell wall offers exceptional structural support.

- **Plasmodesmata** connect adjacent plant cells and allow direct communication between them. **Review Figure 34.6**

34.3 How Do Plant Tissues and Organs Originate?

- During embryogenesis, the basal–apical axis and the radial axis of the plant body are established. **Review Figure 34.7**

- The **shoot apical meristem** and the **root apical meristem** are also established during embryogenesis. These clusters of undifferentiated cells will orchestrate all postembryonic development.

- Three **tissue systems**, arranged concentrically, extend throughout the plant body: the vascular tissue, dermal tissue, and ground tissue systems. **Review Figure 34.8**

- The **dermal tissue system** protects the plant body surface. Dermal cells form the epidermis and, in woody plants, the periderm.

- The **ground tissue system** contains cells of three types. Some **parenchyma** cells carry out photosynthesis; others store starch. **Collenchyma** cells provide flexible support. **Sclerenchyma** cells include **fibers** and **sclereids** that provide strength and mechanical support. **Review Figure 34.9**

- The **vascular tissue system** includes **xylem**, which conducts water and minerals absorbed by the roots, and **phloem**, which conducts the products of photosynthesis throughout the plant body.

- **Tracheary elements** include **tracheids** and **vessel elements**, which are the conducting cells of the xylem. **Sieve tube elements** are the conducting cells of the phloem.

34.4 How Do Meristems Build a Continuously Growing Plant?

- All seed plants possess a **primary plant body** consisting of non-woody tissues. Woody plants also possess a **secondary plant body** consisting of wood and bark. **Apical meristems** generate the primary plant body, and **lateral meristems** generate the secondary plant body. **Review Figure 34.10**

- Apical meristems are responsible for **primary growth** (lengthening of roots and shoots). Apical meristems at the tips of stems and roots give rise to three **primary meristems** (**protoderm**, **ground meristem**, and **procambium**), which in turn produce the three tissue systems of the primary plant body

- The root apical meristem gives rise to the **root cap** and to three primary meristems. Root tips have overlapping **zones of cell division**, **elongation**, and **maturation**. **Review Figure 34.11**

- The vascular tissue of roots is contained within the **stele**. It is arranged differently in eudicot and monocot roots. **Review Figures 34.12 and 34.13, WEB ACTIVITIES 34.1 and 34.2**

- In nonwoody stems, the vascular tissue is divided into **vascular bundles**, each containing both xylem and phloem. **Review Figure 34.14, WEB ACTIVITIES 34.3 and 34.4**

- Eudicot leaves have two zones of photosynthetic **mesophyll** that are supplied by **veins** with water and minerals. Veins also carry the products of photosynthesis to other parts of the plant body. A waxy **cuticle** limits water loss from the leaf. **Guard cells** control openings (stomata) in the leaf that allow CO_2 to enter, but also allow some water to escape. **Review Figure 34.15, WEB ACTIVITY 34.5**

- Two lateral meristems, the **vascular cambium** and **cork cambium**, are responsible for secondary growth. The vascular cambium produces secondary xylem (wood) and secondary phloem. The cork cambium produces a protective tissue called **cork**. **Review Figures 34.16 and 34.17, ANIMATED TUTORIAL 34.1**

34.5 How Has Domestication Altered Plant Form?

- The plant body plan is simple, yet it can be changed dramatically by minor differences in genes, as evidenced by the natural diversity of wild plants.

- Crop domestication involves artificial selection of certain desirable traits found in wild populations. **Review Figure 34.20**

SELF-QUIZ

1. Which of the following is a difference between monocots and eudicots?
   a. Only eudicots have phytomers.
   b. Only monocots have shoot and root apical meristems.
   c. Monocot stems do not undergo secondary growth.
   d. The vascular bundles of monocot stems are commonly arranged as a cylinder.
   e. Eudicot embryos commonly have one cotyledon.

2. Roots
   a. always form a fibrous root system that holds the soil.
   b. possess a root cap at their tip.
   c. form branches from axillary buds.
   d. are commonly photosynthetic.
   e. do not show secondary growth.

3. The primary plant cell wall
   a. lies immediately inside the plasma membrane.
   b. is an impermeable barrier between cells.
   c. is always waterproofed with either lignin or suberin.
   d. always consists of a primary wall and a secondary wall, separated by a middle lamella.
   e. contains cellulose and other polysaccharides.

4. Which statement about parenchyma cells is *not* true?
   a. They are alive when they perform their functions.
   b. They typically lack a secondary wall.
   c. They often function as storage depots.
   d. They are the most common cell type in the plant body.
   e. They are found only in stems and roots.

5. Tracheids and vessel elements
   a. must die to become functional.
   b. are important constituents of all seed plants.
   c. have no secondary cell wall.
   d. are always accompanied by companion cells.
   e. are found only in the secondary plant body.

6. Which statement about meristems is *not* true?
   a. They are formed during embryogenesis.
   b. They have secondary cell walls.
   c. Their cells have small central vacuoles.
   d. They are clusters of undifferentiated cells.
   e. They retain the ability to produce new cells indefinitely.

7. The pericycle
   a. is the innermost layer of the cortex.
   b. is the tissue within which lateral roots arise.
   c. consists of highly differentiated cells.
   d. forms a star-shaped structure at the very center of the root.
   e. is waterproofed by suberin.

8. Which of these statements is true of secondary growth, but *not* primary growth?
   a. It occurs in eudicots and monocots.
   b. It involves the proliferation of roots and shoots through branching.
   c. It derives from the vascular cambium and the cork cambium.
   d. It occurs in palms.
   e. It derives from the shoot apical meristem

9. Periderm
   a. contains lenticels that allow for gas exchange.
   b. is produced during primary growth.
   c. is permanent; it lasts as long as the plant does.
   d. is the innermost part of the plant.
   e. contains vascular bundles.

10. Which statement about leaf anatomy is *not* true?
    a. Opening of stomata is controlled by guard cells.
    b. The cuticle is secreted by the epidermis.
    c. The veins contain xylem and phloem.
    d. The cells of the mesophyll are packed together, minimizing air space.
    e. The spines of cacti are actually modified leaves.

FOR DISCUSSION

1. When a young oak was 5 m tall, a thoughtless person carved his initials in its trunk at a height of 1.5 m above the ground. Today that tree is 10 m tall. How high above the ground are those initials? Explain your answer in terms of plant growth.

2. Distinguish between the primary cell wall and the secondary cell wall. When do secondary walls form? What cell types lack secondary walls?

3. Distinguish between sclerenchyma cells and collenchyma cells in terms of structure and function.

4. Distinguish between primary and secondary growth. Do all angiosperms undergo secondary growth? Explain.

5. What anatomical features make it possible for a plant to retain water? Describe the plant tissues involved and how and when they form.

6. The Doomsday Vault contains the seeds of both domesticated and wild plants. Why is it important to preserve collections of seeds of both domesticated and wild plants? What kinds of situations would necessitate the withdrawal of seeds from the Doomsday vault?

7. Take a walk through a farmer's market or the produce section of a supermarket. Use your knowledge of plant growth and form to figure out what desirable trait was selected to produce some of your favorite vegetables.

ADDITIONAL INVESTIGATION

Of the approximately 20,000 genes in the sequenced genome of *Arabidopsis thaliana*, over 1,000 are involved in cell wall biosynthesis. Based on the composition, growth, and functions of cell walls, what types of proteins would you predict some of these genes encode?

35 Transport in Plants

Engineering water-conserving crops

Everyone knows that plants need water to grow. However, it may come as a surprise that the cultivation of crop plants consumes far more water than all other human activities combined. Worldwide demand for water is increasing at the same time that supplies are declining. This situation makes it imperative that we understand how plants use water so that we can breed plants that use it more efficiently.

The question of just how much water plants use while they grow was addressed in 1690 by John Woodward, a professor at Cambridge University. He reported that a plant that gained just 1 g in weight used 76,000 g of water over 77 days. He proposed that most of the water taken up by plants was "drawn off and conveyed through the pores of the leaves and exhaled into the atmosphere."

We know now, of course, that much of the mass plants acquire as they grow is due to net fixation of atmospheric CO_2 into carbohydrates through photosynthesis. But Woodward nevertheless articulated a crucial insight: plants need to take up a lot of water to grow. Plant biologists have a name for the ratio of net photosynthetic carbon fixation to water uptake: *water-use efficiency*.

Droughts and a dwindling water supply are challenging farmers all over the world. One of the least water-efficient of all crop plants is, unfortunately, one of our most important: rice. Rice plants use up to 3 times more water per unit of growth than other crops such as wheat and maize (corn). The precariousness of heavily water-dependent rice farming was dramatically demonstrated in eastern India between 1997 and 2003, when drought reduced rice production by over 5 million tons—some farmers lost up to 50 percent of their crop.

Clearly, a strain of rice that needs less water would not only make the world supply of rice less vulnerable to drought but also help conserve water for other uses. A team of molecular biologists, plant physiologists, and crop scientists led by Andrew Pereira at Virginia Polytechnic began their quest for such a strain of rice by studying an entirely different plant—the model organism *Arabidopsis thaliana* (thale cress). They searched for genetic variants of *Arabidopsis* that had improved

Thirsty Rice Cultivation of rice, the most important food crop in Asia, requires large quantities of water.

A Need for Improved Water-Use Efficiency in Plants
Rice that could use water more efficiently would be less vulnerable to drought and might help maintain or even increase crop yields.

water-use efficiency. One variant they chose to study was particularly hard to pull out of the ground because of its extensive root system (indicating more capacity for water uptake) and had thick leaves with abundant photosynthetic tissue (indicating prolific photosynthesis). Molecular and physiological characterization of this *Arabidopsis* strain showed that its improved water usage was linked to a mutation in a single gene that codes for a transcription factor. When this gene (called *HARDY*) was isolated and put into rice plants using recombinant DNA technology, the rice plants also were more efficient, and indeed tolerated dry soil much better than their normal counterparts. While the *HARDY* gene may or may not lead to crops with higher water-use efficiency, many laboratories around the world are using *Arabidopsis* to isolate genes that can be used to improve water usage and other important characteristics of crop plants.

IN THIS CHAPTER we will consider the uptake of water and minerals from the soil and the transport of these materials up the plant in the xylem. We will also look at the control of evaporative water loss from leaves, and the translocation (movement from one location to another) of dissolved substances in the phloem.

35.1 How Do Plants Take Up Water and Solutes?

Terrestrial plants must obtain both water and mineral nutrients from the soil, usually through their roots. The roots, in turn, obtain carbohydrates and other important materials from the leaves (**Figure 35.1**). Water is required for carbohydrate production by photosynthesis in leaves (see Section 10.1), for transporting solutes between plant organs, for cooling the plant, and for developing the internal pressure that supports the plant body.

As our opening story conveys, plants lose large quantities of water to evaporation. To balance this loss, an equally large amount of water must be absorbed through the roots, continue up the stem, and be transported into the leaves. The minerals that a plant needs are transported along with the water. Several steps in water and mineral transport will be considered in this chapter. In this section we will focus on the first part of the journey—the uptake of water and minerals into the roots and their transport into the xylem.

Water potential differences govern the direction of water movement

The process of water uptake by plants requires water to move through at least one, and usually many cell membranes. Accordingly, we will begin our discussion of water transport by examining the rules that govern the movement of water across membranes. As described in Section 6.3, the movement of water through a membrane in accordance with the laws of diffusion is called *osmosis*.

The overall tendency of a solution to take up water from pure water, across a membrane, is called its **water potential** and is represented as ψ, the Greek letter *psi* (pronounced "sigh"). The water potential of a solution is measured as the sum of its (negative) solute potential (ψ_s) and its (usually positive) pressure potential (ψ_p):

$$\psi = \psi_s + \psi_p$$

Whenever water moves by osmosis, the following important rule applies: *water always moves across a selectively permeable membrane toward the region of lower (more negative) water potential.*

We can measure solute potential, pressure potential, and water potential in *megapascals* (MPa), a unit of pressure. Atmospheric pressure, "one atmosphere," is about 0.1 MPa, or 14.7 pounds per square inch; a typical pressure in an automobile tire is about 0.2 MPa.

We explore the meaning of the water potential equation in **Figure 35.2**, which assigns values to ψ_s and ψ_p and illustrates how changes in the values of these parameters alter the water potential (ψ) and determine the direction of water movement between two compartments (for example, the inside and outside of a plant cell) separated by a semipermeable membrane.

The **solute potential** (ψ_s, also called the *osmotic potential*) of a solution is a measure of the effect of dissolved solutes on the osmotic behavior of the solution. The addition of solutes removes free water from the solution because the solute molecules bind water molecules to their surfaces. This is reflected in a more negative value for ψ_s (–0.4 MPa in our example; see Figure 35.2A), which lowers the water potential ($\psi = -0.4$ MPa) and leads to the movement of water through the membrane to the region of lower ψ. Equilibrium is reached when there is no difference in ψ on either side of the membrane (see Figure 35.2B). Now let's see how these same forces determine the direction of water flow through plant cells.

Mature plant cells usually contain a large central vacuole filled with solutes, which are often pumped into the vacuole by transporter proteins. The active accumulation of solutes provides the osmotic force for water uptake into the vacuole. Plant cells are surrounded by a relatively rigid cell wall that resists the expansion of the underlying protoplast. The pressure exerted by the cell wall is equivalent to the positive pressure exerted by the piston on the water column in the idealized example shown in Figure 35.2C. When the pressure potential equals the solute potential, there is no net movement of water through the membrane.

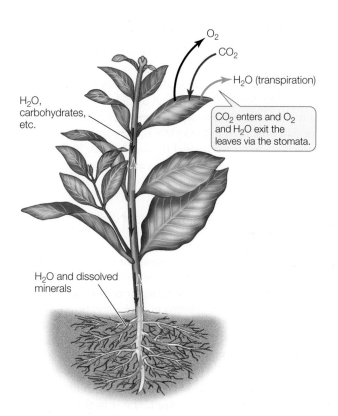

35.1 The Pathways of Water and Solutes in a Plant
Water travels from the soil to the atmosphere, with only a small fraction used within the plant.

35.2 Water Potential, Solute Potential, and Pressure Potential
As can be seen in these idealized examples, water flows towards regions of lower water potential (ψ), which is the sum of the solute potential (ψ_s) and the pressure potential (ψ_p). For pure water under no applied pressure, all three of these parameters are equal to zero.

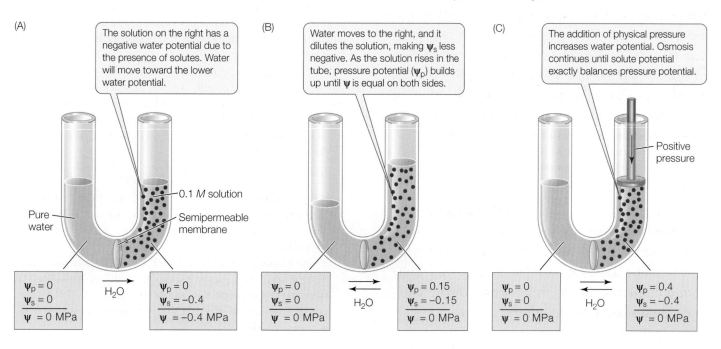

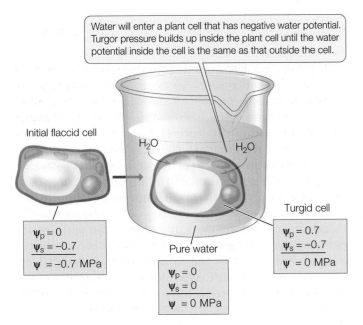

Water will enter a plant cell that has negative water potential. Turgor pressure builds up inside the plant cell until the water potential inside the cell is the same as that outside the cell.

Initial flaccid cell

H_2O H_2O

Turgid cell

Pure water

$\psi_p = 0$
$\psi_s = -0.7$
$\psi = -0.7$ MPa

$\psi_p = 0$
$\psi_s = 0$
$\psi = 0$ MPa

$\psi_p = 0.7$
$\psi_s = -0.7$
$\psi = 0$ MPa

35.3 Turgor   Turgor pressure builds up inside the cell as the cell wall resists further expansion of the cell.

When the wall of a plant cell is exerting no pressure on the underlying protoplast, the cell is said to be *flaccid* (**Figure 35.3**, left). In this situation,

$$\psi_s = -0.7 \text{ MPa and } \psi_p = 0 \text{ MPa}$$

$$\text{So, } \psi = \psi_s + \psi_p = -0.7 \text{ MPa}$$

When a flaccid cell is placed in pure water, water initially moves into the cell due to its negative solute potential (Figure 35.3, right). However, the cell can't expand because it is contained by the cell wall; thus, as water enters, the cell's internal pressure increases and resists the further entry of water.

$$\psi_s = -0.7 \text{ MPa and } \psi_p = 0.7 \text{ MPa}$$

$$\text{So, } \psi = 0$$

This opposing pressure is called **turgor pressure** in plants and is equivalent to the pressure potential (ψ_p) exerted by the piston in Figure 35.2C. Water will enter plant cells by osmosis until the pressure potential exactly balances the solute potential. At this point, the cell is **turgid**; that is, it has a significant positive pressure potential. The physical structure of many plants is maintained by the (positive) pressure potential of their cells; if the pressure potential drops (for example, if the plant does not have enough water), the plant *wilts* (**Figure 35.4**).

Within living plant tissues, the movement of water from cell to cell follows a gradient of water potential. Over long distances, in unobstructed tubes such as xylem vessels and phloem sieve tubes, the flow of water and dissolved solutes is driven by a *gradient of pressure potential*, not a gradient of water potential. The movement of a solution from a region of higher pressure potential to a region of lower pressure potential is called **bulk flow**. We'll see that bulk flow in the xylem is between regions of differing *negative* pressure potential (tension), while bulk flow in the phloem is between regions of differing *positive* (turgor) pressure potential.

Aquaporins facilitate the movement of water across membranes

The large quantities of water lost to evaporation from the leaves must be balanced by water taken up by the roots. Yet only a trickle of water can pass through the hydrophobic environment created by the phospholipid bilayers of cell membranes. How do plants turn this trickle into a gusher? The answer is that water diffuses through cell membranes mainly through channels called **aquaporins** (see Figure 6.13), which are located in both the plasma membrane and the tonoplast (vacuolar membrane) of plant cells. Aquaporins allow water to move rapidly from environment to cell and from cell to cell. The abundance of aquaporins in a plant cell varies with environmental conditions, depending on a cell's need to obtain and retain water. The permeability of some aquaporins also can be regulated. Alterations in aquaporin abundance and permeability change the *rate* of osmosis across the membrane. Note, however, that water movement through aquaporins is always passive, so the *direction* of water movement is unchanged.

Uptake of mineral ions requires membrane transport proteins

Although water molecules can cross membranes through aquaporins, mineral ions generally cannot. The ions, which carry electric charges, are blocked by the hydrophobic interior of the

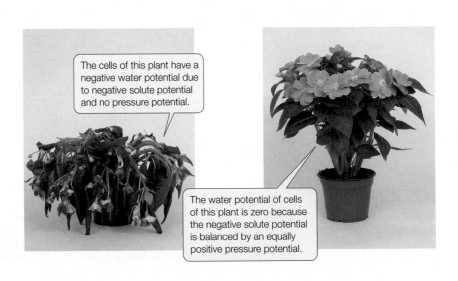

The cells of this plant have a negative water potential due to negative solute potential and no pressure potential.

The water potential of cells of this plant is zero because the negative solute potential is balanced by an equally positive pressure potential.

35.4 A Wilted Plant   A plant wilts when the pressure potential of its cells is zero.

membrane, and they are too large to pass through aquaporins. Instead, mineral ions generally cross membranes through transport proteins, including ion channels and carrier proteins (see Sections 6.3 and 6.4).

We have just seen that water moves through a water-permeable membrane in response to a water potential gradient. Other molecules and ions also follow their own concentration gradients, as permitted by the characteristics of the membrane. When the concentration of charged ions in the soil is greater than that in the plant, transport proteins can move them into the plant by facilitated diffusion, which is a passive process. The concentrations of most ions in the soil solution, however, are lower than those required inside the plant. In these cases, the plant must actively take up ions *against* their concentration gradients—a process that requires energy.

Electric charge differences also play a role in the uptake of mineral ions. For example, a negatively charged ion that moves into a negatively charged compartment is moving against an *electrical gradient*, and this requires energy. Concentration and electrical gradients combine to form an *electrochemical gradient*. Uptake against an electrochemical gradient involves *active transport*, which requires specific transport proteins and is fueled by ATP generated by cellular respiration.

Unlike animals, plants do not have a sodium–potassium pump (see Section 6.4) for active transport. Rather, plants have a **proton pump**, which uses energy obtained from ATP to move protons out of the cell against a proton concentration gradient (**Figure 35.5, step 1**). Because protons (H^+) are positively charged, their accumulation outside the cell has two results:

- An electrical gradient is created such that the region outside the cell becomes more positively charged relative to the region inside.

- A proton concentration gradient develops, with more protons outside the cell than inside.

Each of these results has consequences for the movement of other ions. Because the inside of the cell is now more negative than the outside, cations (positively charged ions) such as potassium (K^+) move into the cell by facilitated diffusion through their specific membrane channels (**Figure 35.5, step 2**). In addition, the proton concentration gradient can be harnessed to drive secondary active transport, in which anions (negatively charged ions) such as chloride (Cl^-) are moved into the cell against an electrochemical gradient by a symport protein that couples their movement with that of H^+ (**Figure 35.5, step 3**).

In sum, there is vigorous traffic of water molecules and mineral ions across plant cell membranes. This traffic involves specific membrane channels and transport proteins, and both active and passive processes. Now we will step back and see how these membrane transport processes participate in the journey of water and nutrients from the soil to the xylem.

Water and ions pass to the xylem by way of the apoplast and symplast

The journey from the soil through the roots to the xylem occurs primarily by one of two pathways, the fast lane (called the *apoplast*) and the slow(er) lane (called the *symplast*) (**Figure 35.6**):

- The **apoplast** (Greek *apo*, "away from"; *plast*, "living material") consists of the cell walls, which lie outside the plasma membranes, and the intercellular spaces (spaces between cells) that are common in many plant tissues. The apoplast is a continuous meshwork through which water and dissolved substances can flow without ever having to cross a membrane. Movement of materials through the apoplast is thus unregulated and rapid—until it reaches the *Casparian strips* of the endodermis.

- The **symplast** (Greek *sym*, "together with") passes through the continuous cytoplasm of the living cells connected by plasmodesmata. The selectively permeable plasma membranes of the root cells control access to the symplast, so movement of water and dissolved substances into the symplast is tightly regulated.

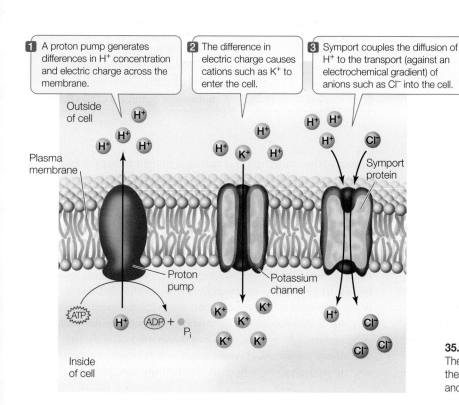

1 A proton pump generates differences in H^+ concentration and electric charge across the membrane.

2 The difference in electric charge causes cations such as K^+ to enter the cell.

3 Symport couples the diffusion of H^+ to the transport (against an electrochemical gradient) of anions such as Cl^- into the cell.

Outside of cell

Plasma membrane

Proton pump

Potassium channel

Symport protein

ATP

ADP + P$_i$

Inside of cell

35.5 The Proton Pump in Transport of K$^+$ and Cl$^-$
The active transport of hydrogen ions (H^+) out of the cell by the proton pump (1) drives the movement of both cations (2) and anions (3) into the cell.

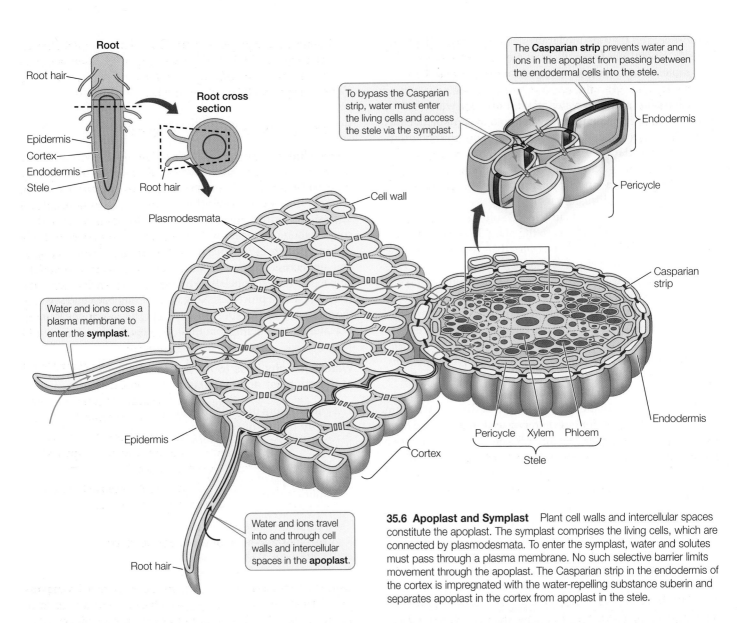

The **Casparian strip** prevents water and ions in the apoplast from passing between the endodermal cells into the stele.

To bypass the Casparian strip, water must enter the living cells and access the stele via the symplast.

Endodermis

Pericycle

Root

Root hair

Root cross section

Epidermis

Cortex

Endodermis

Stele

Root hair

Cell wall

Plasmodesmata

Casparian strip

Water and ions cross a plasma membrane to enter the **symplast**.

Endodermis

Epidermis

Pericycle   Xylem   Phloem

Cortex

Stele

Root hair

Water and ions travel into and through cell walls and intercellular spaces in the **apoplast**.

35.6 Apoplast and Symplast   Plant cell walls and intercellular spaces constitute the apoplast. The symplast comprises the living cells, which are connected by plasmodesmata. To enter the symplast, water and solutes must pass through a plasma membrane. No such selective barrier limits movement through the apoplast. The Casparian strip in the endodermis of the cortex is impregnated with the water-repelling substance suberin and separates apoplast in the cortex from apoplast in the stele.

yourBioPortal.com

GO TO Web Activity 35.1 • Apoplast and Symplast of the Root

Water and minerals that pass from the soil solution through the apoplast can travel as far as the endodermis, the innermost layer of the root cortex. The endodermis is distinguished from the rest of the ground tissue by the presence of the **Casparian strip**. This waxy, suberin-impregnated region of the endodermal cell wall forms a water-repelling (hydrophobic) belt around each endodermal cell where it is in contact with other endodermal cells. The Casparian strip acts as a seal that prevents water and ions from moving between the cells (see Figure 35.6).

The Casparian strip of the endodermis completely separates the apoplast of the cortex from the apoplast of the stele. Accordingly, the only way water and ions can enter the stele is by way of the symplast—that is, by entering and passing through the cytoplasm of the endodermal cells. Water and ions already in the symplast can enter the endodermal cells through plasmadesmata, but those in the apoplast must cross the plasma membranes of the endodermal cells (this is possible because the Casparian strip does not obstruct the inner or outer faces of endodermal cells). Thus transport proteins in the plasma membranes of the epidermal and cortical cells (for mineral ions traveling through the symplast) and endodermal cells (for those traveling through the apoplast) determine which mineral ions pass into the stele, and at what rates.

Once they have passed the endodermal barrier, water and minerals remain in the symplast until they reach parenchyma cells in the pericycle or xylem. These cells then actively export mineral ions into the apoplast of the stele. As mineral ions are transported into the solution in the cell walls of the stele, the water potential in the apoplast becomes more negative; consequently, water moves out of the cells and into the apoplast by osmosis. In other words, ions are transported actively, and water follows passively. The end result is that water and minerals end up in the xylem, where they constitute the *xylem sap*.

35.1 RECAP

Differences in water potential govern the osmotic flow of water from the soil into the plant stele; this is a passive process. Uptake of minerals from the soil occurs along an electrochemical gradient and is therefore an active process requiring energy and membrane transport proteins. Water and minerals can move into the root through either the apoplast or the symplast, but must enter and leave the symplast to reach the xylem.

- What distinguishes water potential, solute potential, and pressure potential? **See pp. 740–741 and Figure 35.2**

- Explain why the cell wall is so important in determining the direction of water movement. **See pp. 741–742**

- What are aquaporins? Why are they needed? **See p. 742**

- Describe the differences between the apoplast and the symplast. **See p. 743 and Figure 35.6**

So far we've described the movement of water and minerals into plant roots and their entry into the root xylem. How does the xylem sap move once it is in the xylem?

35.2 How Are Water and Minerals Transported in the Xylem?

Water has arrived in the xylem—it is all uphill from there! Before we consider the ascent to the leaves, let's revisit the cells that make up the xylem, the xylem vessels (see Figure 34.9E and F). In this section you will learn that the properties of xylem vessels make it possible for water and solutes to be transported efficiently over long distances. Recall that xylem vessels are dead and lack all cell contents. When fused end to end, the xylem vessels form a long tubular "straw" of lignified cell walls that provide both structural support and the rigidity needed to maintain a gradient of pressure.

Consider the magnitude of what xylem accomplishes in transporting a *large amount of water over a great distance* within the plant. A single maple tree 15 meters tall has been estimated to have some 177,000 leaves, with a total leaf surface area of 675 square meters—half again the area of a basketball court. During a summer day, that tree loses 220 liters of water *per hour* to the atmosphere by evaporation from the leaves. So to prevent wilting, the xylem needs to transport 220 liters of water from the roots to the leaves every hour. (By comparison, a 50-gallon drum holds 189 liters.) The tallest leaves can be quite far from the root. The tallest gymnosperm, the coast redwood *Sequoia sempervirens*, and the tallest angiosperm, the Australian *Eucalyptus regnans*, are more than 110 meters tall. Any successful explanation of water transport in the xylem must account for the transport of water to these great heights.

Scientists have proposed various models to explain the ascent of xylem sap. We begin by reviewing some illuminating experiments that ruled out early models, and then turn to evidence in support of the current model.

Xylem sap is not pumped by living cells

Some of the earliest attempts to explain the rise of sap in the xylem were based on the hypothesis that a pumping action by living cells in the stem might push the sap upward. However, in 1893 the German botanist Eduard Strasburger conducted and published experiments that definitively ruled out such models.

Strasburger worked with trees about 20 meters tall. He sawed through the trunk of each tree at its base and plunged the cut end into a solution of a poison, such as copper sulfate. The solution rose through the trunk, as was evident from the progressive death of the bark higher and higher up. When the solution reached the leaves, the leaves died too, at which point the movement of the solution stopped (as shown by the liquid level in the bucket, which stopped dropping).

This simple experiment established three important points:

- Living, "pumping" cells were not responsible for the upward movement of the solution, because the solution itself killed all living cells with which it came in contact.

- The leaves played a crucial role in transport. As long as they were alive, the solution continued to move upward; when the leaves died, movement ceased.

- The roots did not cause the movement, because the trunk had been completely separated from the roots.

Root pressure alone does not account for xylem transport

A second hypothesis about xylem transport involved **root pressure**—pressure exerted by the root tissues that would force liquid up the xylem. The basis for root pressure is a higher solute concentration, and accordingly a more negative water potential, in the xylem sap than in the soil solution. This water potential draws water into the stele; once there, the water has nowhere to go but up, so it rises in the xylem vessels.

Root pressure certainly exists—for example, it is responsible for the phenomenon of *guttation*, in which liquid water is forced out through special openings at the margins of leaves. Guttation occurs only when atmospheric humidity is high and soil water is plentiful, conditions that occur most commonly at night. Root pressure is also the source of the sap that oozes from the cut stumps of some plants when their tops are cut off.

Root pressure, however, cannot account for the ascent of sap in trees. Root pressure seldom exceeds 0.1–0.2 MPa (1–2 atmospheres). If root pressure were driving sap up the xylem, we would observe a positive pressure potential in the xylem at all times. In fact, as we are about to see, the xylem sap in most trees is under *tension*—has a *negative* pressure potential—when it is ascending. Furthermore, as Strasburger had already shown, materials can be transported upward in the xylem even when the roots have been removed. If the roots are not pushing the xylem sap upward, what causes it to rise?

The transpiration–cohesion–tension mechanism accounts for xylem transport

The current model of xylem transport relies on an alternative to pushing: pulling. The evaporative loss of water from the leaves indirectly generates a pulling force—**tension**—on the water in the apoplast of the leaves, which pulls the xylem sap upward. Hydrogen bonding between water molecules makes the sap in the xylem cohesive enough to withstand the tension and rise by bulk flow. Let's see how this process works (**Figure 35.7**).

The concentration of water vapor in the atmosphere is lower than that in the leaf. Because of this difference, water vapor diffuses from the intercellular spaces of the leaf, through the *stomata* (which we will consider in more detail later) to the outside air,

in a process called **transpiration**. Within the leaf blade, water evaporates from the moist walls of the mesophyll cells and enters the intercellular spaces. As water evaporates from the aqueous film coating each cell, the film shrinks back into tiny spaces in the cell walls, increasing the curvature of the water surface and thus increasing its surface tension. This increased tension (negative pressure potential) in the surface film draws more water into the cell walls, replacing that which was lost. The resulting tension in the mesophyll draws water from the xylem of the nearest vein into the apoplast surrounding the mesophyll cells. The removal of water from the veins, in turn, establishes tension on the entire column of water contained within the xylem, so that the column is drawn upward all the way from the roots.

Water can be pulled upward through tiny tubes because of the remarkable **cohesion** of water—the tendency of water molecules to stick to one another by hydrogen bonding (see Section 2.4 and Figure 35.7, step 6). The narrower the tube, the greater the tension the water column can withstand without breaking. The integrity of the column is also maintained by the *adhesion* of water to the xylem walls.

In summary, the key elements of water transport in the xylem are:

- *Transpiration* of water molecules from the leaves by evaporation
- *Tension* in the xylem sap resulting from transpiration from the leaves
- *Cohesion* of water molecules in the xylem sap, from the leaves to the roots

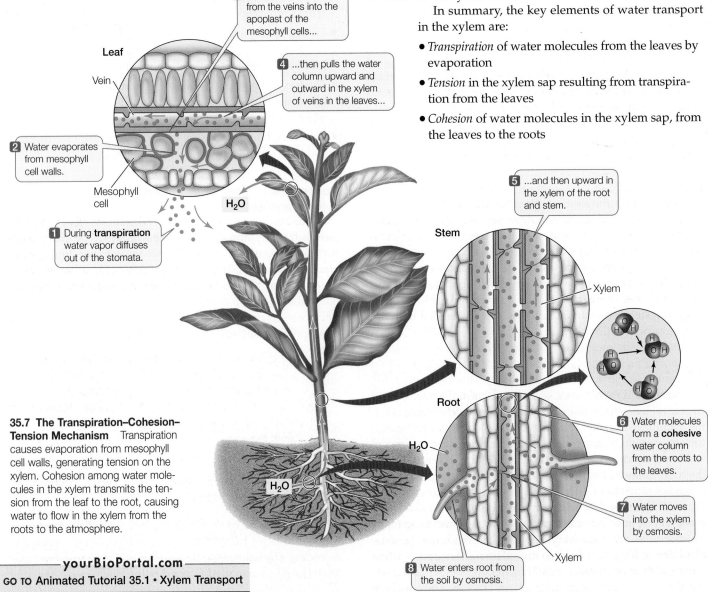

3 **Tension** pulls water from the veins into the apoplast of the mesophyll cells...

4 ...then pulls the water column upward and outward in the xylem of veins in the leaves...

2 Water evaporates from mesophyll cell walls.

1 During **transpiration** water vapor diffuses out of the stomata.

5 ...and then upward in the xylem of the root and stem.

6 Water molecules form a **cohesive** water column from the roots to the leaves.

7 Water moves into the xylem by osmosis.

8 Water enters root from the soil by osmosis.

Leaf
Vein
Mesophyll cell
Stem
Xylem
Root
Xylem
H_2O

35.7 The Transpiration–Cohesion–Tension Mechanism Transpiration causes evaporation from mesophyll cell walls, generating tension on the xylem. Cohesion among water molecules in the xylem transmits the tension from the leaf to the root, causing water to flow in the xylem from the roots to the atmosphere.

yourBioPortal.com
GO TO Animated Tutorial 35.1 • Xylem Transport

The water transport process we have described, called the **transpiration–cohesion–tension mechanism**, *requires no work (that is, no expenditure of energy) on the part of the plant.* At each step between soil and atmosphere, water moves passively; first toward a region with lower water potential, and then to a region of lower pressure potential. Dry air has the most negative water potential (–95 MPa at 50% relative humidity), and the soil solution has the least negative water potential (between –0.01 and –3 MPa). Xylem sap has a water potential more negative than that of cells in the cortex of the root, but less negative than that of mesophyll cells in the leaf.

In the tallest trees, such as a 110-meter *Sequoia*, the difference in pressure potential between the top and the bottom of the column may be as great as 3 MPa. Compare this to root pressure and the pressure in a typical automobile tire, which seldom exceed 0.2 MPa. The cohesion of water in the xylem is great enough to withstand the huge tensions that develop in the tallest trees.

Mineral ions contained in the xylem sap rise passively with water as it ascends from root to leaf. In this way the nutritional needs of the shoot are met. Some of the mineral elements brought to the leaves are subsequently redistributed to other parts of the plant by way of the phloem, but the initial delivery from the roots is through the xylem.

In addition to promoting the transport of minerals, transpiration has an added benefit of cooling a plant's leaves. The evaporation of water from mesophyll cells consumes heat, thereby decreasing the leaf temperature. A farmer can hold a leaf between thumb and forefinger to estimate its temperature; if the leaf doesn't feel cool, that means that transpiration is not occurring and it must be time to water.

The cooling effect of evaporation (also evident in the cooling of our skin when we sweat) may also be important in enabling plants to live in hot environments. However, while transpiration may lead to the cooling of the leaf, this effect is a consequence of the need to transpire, not a reason for it.

A pressure chamber measures tension in the xylem sap

The transpiration–cohesion–tension model holds true only if the column of sap in the xylem is under tension (has a negative pressure potential). The most elegant demonstrations of this tension, and of its adequacy to account for the ascent of xylem sap in tall trees, were performed by the biologist Per Scholander, who measured tension in stems with an instrument called a **pressure chamber** (**Figure 35.8**).

Scholander used the pressure chamber to study dozens of plant species from diverse habitats, growing under a variety of conditions. The rate at which xylem sap ascends is not the same at all times. No flow of xylem sap takes place at night, when there is little or no transpiration. By day, when the sap is ascending, the rate of ascent depends on several factors. These include temperature, light intensity, and wind velocity, all of which affect the transpiration rate, and hence the rate of sap flow. In addition, Scholander found that the xylem sap in developing vines was not under tension until leaves formed. Once leaves developed, the tension increased and transport in the xylem began.

TOOLS FOR INVESTIGATING LIFE

35.8 Measuring the Pressure of Xylem Sap with a Pressure Chamber Xylem sap pulls away from a cut stem because the pressure in the intact xylem is lower than that of the atmosphere. The negative pressure potential originally present in the plant can be measured in a pressure chamber in which the pressure can be raised. The cut surface remains outside the chamber. As gas pressure increases, the xylem sap is pushed back to the cut surface. When the sap first becomes visible again at the cut surface, the pressure in the chamber is recorded. This pressure is equal in magnitude but opposite in sign to the tension (negative pressure potential) originally present in the xylem.

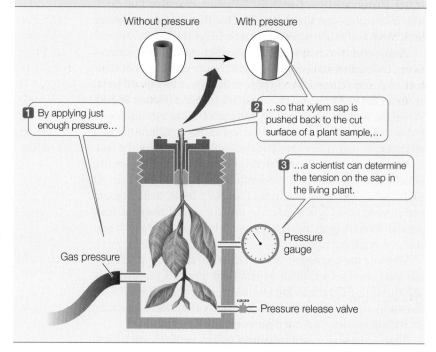

Without pressure   With pressure

1 By applying just enough pressure…

2 …so that xylem sap is pushed back to the cut surface of a plant sample,…

3 …a scientist can determine the tension on the sap in the living plant.

Gas pressure

Pressure gauge

Pressure release valve

35.2 RECAP

The transpiration–cohesion–tension mechanism explains the ascent of xylem sap. Transpiration draws water out of leaves, resulting in tension that pulls water from the xylem. Because of cohesion between water molecules, water is pulled passively through the xylem vessels in continuous columns, always toward a region with lower pressure potential.

- What are the roles of transpiration, cohesion, and tension in xylem transport? **See p. 746 and Figure 35.7**

- What properties of the water molecule contribute to cohesion and tension? **See p. 746**

- Describe how mineral ions get from the roots to the leaves. **See p. 747**

Although transpiration provides the driving force for the transport of water and minerals in the xylem, it also results in the loss of tremendous quantities of water from the plant. How plants control this loss is the subject of the next section.

35.3 How Do Stomata Control the Loss of Water and the Uptake of CO_2?

The epidermis of leaves and stems minimizes transpirational water loss by secreting a waxy cuticle, which is impermeable to water. However, the cuticle is also impermeable to carbon dioxide. This poses a problem: how can the plant balance its need to retain water with its need to obtain CO_2 for photosynthesis?

An elegant compromise has evolved in plants in the form of pores called **stomata** (singular *stoma*) in the epidermis of their leaves. A pair of specialized epidermal cells, called **guard cells**, controls the opening and closing of each stoma (**Figure 35.9A**). When the stomata are open, CO_2 can enter the leaf by diffusion— but water vapor diffuses out of the leaf at the same time. Closed stomata prevent water loss, but also exclude CO_2 from the leaf.

Most plants open their stomata only when the light intensity is sufficient to maintain a moderate rate of photosynthesis. At night, when darkness precludes photosynthesis, their stomata are closed; no CO_2 is needed at this time, and water is conserved. Even during the day, the stomata close if water is being lost at too great a rate.

Stomata are ancient structures that are found in plant fossils that are over 400 million years old. For this reason, they are thought to predate the evolution of leaves. Stomata are found in all vascular plants and in many nonvascular plants, including mosses (but not liverworts; see Chapter 28).

The stoma and guard cells seen in Figure 35.9A are typical of eudicots. Monocots typically have specialized epidermal cells associated with their guard cells. However, the principle of operation, which we will now describe in more detail, is the same for both monocot and eudicot stomata.

The guard cells control the size of the stomatal opening

Light causes the stomata of most plants to open, admitting CO_2 for photosynthesis. Another cue for stomatal opening is the level of CO_2 in the intercellular spaces inside the leaf. A low level favors opening of the stomata, thus allowing the uptake of more CO_2.

Stomata can respond to changes in light and CO_2 in a matter of minutes. How can such an important biological process happen so rapidly? The answer is that the opening and closing of stomata is controlled by turgor pressure changes in the guard cells. Changes in turgor pressure are in turn driven by changes in K^+ concentration in the guard cells. Blue light, absorbed by a pigment in the guard cell plasma membrane, activates a proton pump, which actively transports H^+ out of the guard cells and into the apoplast of the surrounding epidermis. The resulting electrochemical gradient drives K^+ into the guard cells, where it accumulates (**Figure 35.9B**). The increased internal concentration of K^+ makes the water potential of the

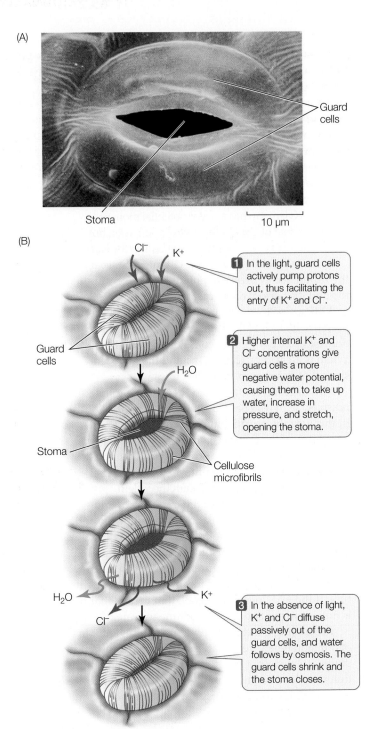

(A)

Guard cells

Stoma

10 µm

(B)

Cl^- K^+

Guard cells

Stoma

H_2O

Cellulose microfibrils

H_2O

Cl^-

K^+

1 In the light, guard cells actively pump protons out, thus facilitating the entry of K^+ and Cl^-.

2 Higher internal K^+ and Cl^- concentrations give guard cells a more negative water potential, causing them to take up water, increase in pressure, and stretch, opening the stoma.

3 In the absence of light, K^+ and Cl^- diffuse passively out of the guard cells, and water follows by osmosis. The guard cells shrink and the stoma closes.

35.9 Stomata (A) Scanning electron micrograph of an open stoma formed by two sausage-shaped guard cells. (B) Potassium ion concentrations affect the water potential of the guard cells, controlling the opening and closing of stomata. Negatively charged ions (e.g., Cl^-) that accompany K^+ maintain electrical balance and contribute to the changes in water potential that open and close the stomata.

guard cells more negative. Negatively charged chloride ions and organic ions also move into and out of the guard cells along with the potassium ions, maintaining electrical balance and contributing to the change in the solute potential of the guard cells. Water enters by osmosis (guard cell membranes are particularly rich in aquaporin protein channels), increasing the pressure potential of the guard cells. The arrangement of the cellulose microfibrils in their cell walls (see Figure 34.5) is such that the guard cells change shape in response to the increase in pressure potential, so that a gap—the stoma—appears between them.

The stoma closes in the absence of blue light. The proton pump becomes less active, potassium ions diffuse passively out of the guard cells, water follows by osmosis, the pressure potential decreases, and the guard cells sag together and seal off the stoma.

INVESTIGATING LIFE

35.10 Measuring Potassium Ion Concentration in Guard Cells

G. D. Humble and Klaus Raschke used the electron probe microanalyzer to examine individual stomata of the broad bean. In electron probe microanalysis, electron bombardment of the sample causes it to emit X rays. The wavelength and intensity of the lines in the X-ray spectrum can be analyzed to identify the elements present in the specimen and estimate their concentrations.

HYPOTHESIS Guard cells of open stomata contain more potassium ions than do those of closed stomata.

METHOD
1. Peel strips of epidermis from leaves of broad beans in the dark (closed stomata) and in the light (open stomata).
2. Examine the strips to locate stomata.
3. Scan across guard cells with the electron probe microanalyzer set to measure K$^+$ concentration.

RESULTS

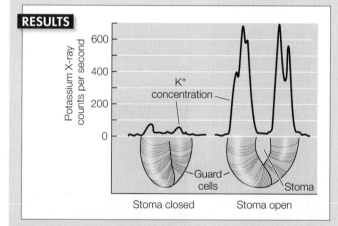

CONCLUSION K$^+$ concentration within the guard cells surrounding an open stoma was much greater than that in the guard cells surrounding a closed stoma.

FURTHER INVESTIGATION: What other ion or ions would you study in order to further explore the mechanism of stomatal opening?

Go to **yourBioPortal.com** for original citations, discussions, and relevant links for all INVESTIGATING LIFE figures.

Showing how much potassium moves into the guard cells to open a stoma was a difficult feat. A typical guard cell is very small, with a total volume of less than 0.03 nanoliters when the stoma is closed and almost 0.05 nanoliters when it is open. The scientists who solved the problem used an *electron probe microanalyzer*, an instrument normally used by metallurgists to study the fine structure of alloys (**Figure 35.10**).

Individual stomata are tiny and plants lose large amounts of water (a single corn plant can lose 2 quarts per day); even so, you may find it surprising that there can be up to 250,000 stomata per square inch of leaf surface! To survive, plants limit water loss by controlling stomata in two very different ways:

- By regulating stomatal opening and closing
- By controlling the total number of stomata

The opening and closing of stomata is regulated by several environmental and endogenous factors. For example, water stress is a common problem for plants, especially on hot, windy days. In response to these conditions, plants will close their stomata, even when the sun is shining. The water potential of the mesophyll cells is the cue for this protective response. If the mesophyll is too dehydrated—that is, if its water potential is too negative—its cells release a plant hormone called *abscisic acid*. Abscisic acid causes the guard cells to close the stomata and prevent further drying of the leaf. This response reduces the rate of photosynthesis, but it protects the plant. Stomata also close in most plants when CO$_2$ levels in the mesophyll spaces are high. Can you think of a reason why this makes good biological sense?

A plant can also reduce its total number of stomata when water is in short supply. Trees can do this by losing some of their leaves. Other plants reduce the number of stomata on the new leaves they produce. For example, if the model plant *Arabidopsis* is exposed to high CO$_2$ levels, the new leaves that form on the plant have fewer stomata than they would have had under normal conditions. Why do you think this might be advantageous?

35.3 RECAP

Leaf pores called stomata admit the CO$_2$ needed for photosynthesis and also permit the exit of water by transpiration. Stomata can be opened or closed by guard cells to regulate water loss.

- What is the role of K$^+$ ions in the functioning of guard cells? See p. 748 and Figure 35.9

- Describe how the external environment (including CO$_2$ level and light intensity) can affect stomatal function and number during the life of a plant. See p. 749

Stomata are normally open during daylight hours, allowing CO$_2$ to be fixed and converted to the products of photosynthesis. In the next section we'll see how these products are delivered to other parts of the plant, supporting plant growth.

35.4 How Are Substances Translocated in the Phloem?

Photosynthesis occurs primarily in the leaf (see Figure 10.1). The carbohydrate products of photosynthesis (mainly sucrose) diffuse to the nearest small vein (composed of xylem and phloem), where they are actively transported into sieve tube elements. The movement of carbohydrates and other solutes through the plant in the phloem is called **translocation**. Phloem content has several names, including *phloem sap*, *photosynthate*, and *assimilates*.

Substances in the phloem are translocated from *sources* to *sinks*.

- A **source** is an organ (such as a mature leaf or a storage root) that *produces* (by photosynthesis or by digestion of stored reserves) more sugars than it requires.

- A **sink** is an organ (such as a root, flower, developing fruit or tuber, or immature leaf) that *consumes* sugars for its own growth and storage needs.

Sugars (primarily sucrose), amino acids, some minerals, and a variety of other solutes are translocated between sources and sinks in the phloem. However, an organ that is a sink can turn into a source. For example, storage roots (such as sweet potatoes) are sinks when they accumulate carbohydrates but are sources when the stored reserves are needed to nourish other organs in the plant.

How do we know that such organic solutes are translocated in the phloem, rather than in the xylem? Over 300 years ago, the Italian scientist Marcello Malpighi performed a classic experiment in which he removed a ring of bark (containing the phloem) from the trunk of a tree—that is, he *girdled* the tree (**Figure 35.11**). Over time, the bark in the region above the girdle swelled. We now know that the swelling resulted from the accumulation of organic solutes that came from higher up the tree and could no longer continue downward because of the disruption of the phloem. Later, the bark below the girdle died because it no longer received sugars from the leaves. Eventually the roots, and then the entire tree, died.

Any explanation of the translocation of organic solutes must account for a few important observations:

- Translocation stops if the phloem tissue is killed by heating or other methods; thus the mechanism must be different from that of transport in the xylem. Recall that xylem is composed of dead cells.

- Translocation often proceeds in both directions—with some phloem transporting up the stem and parallel phloem transporting down the stem. The direction depends on the location of sources and sinks.

- Translocation is inhibited by compounds that inhibit respiration and thus limit the ATP supply in the source. Thus transport in the phloem, unlike the xylem, depends on the input of energy.

Let's first revisit the structure of the phloem to find clues to how it functions. Recall from Chapter 34 that the characteristic cells of the phloem are **sieve tube elements** (see Figure 34.9G). Like vessel elements, these cells meet end-to-end. However, unlike vessel elements, which break down their end walls as they mature, sieve tube elements contain plasmodesmata in the end walls. During sieve tube development, the diameter of these plasmodesmata increases 10- to 100-fold to form pores, enhancing the connection between neighboring cells. The result is end walls that look like sieves, called **sieve plates** (**Figure 35.12**).

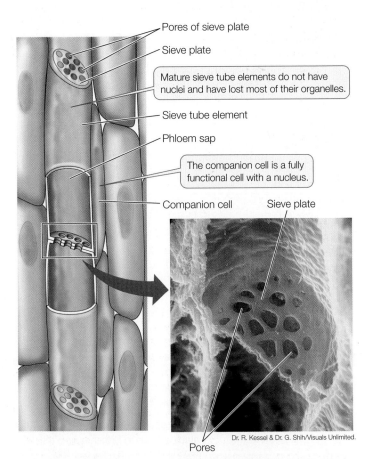

35.12 Sieve Tubes Individual sieve tube elements join together to form long tubes that transport carbohydrates and other nutrient molecules throughout the plant body in the phloem. Sieve plates form at the ends of each sieve tube element, and phloem sap passes through the pores in the sieve plate.

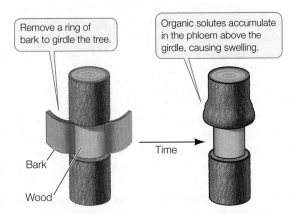

35.11 Girdling Blocks Translocation in the Phloem By girdling—removing a ring of bark containing the phloem—Malpighi blocked the translocation of organic solutes in a tree. Bark below the girdle died because it no longer received nutrients; eventually the entire tree died.

What happens next is truly remarkable and makes sieve tube elements among the most unusual cell types in nature. As the holes in the sieve plates expand, most of the cell contents are lost, including the nucleus, Golgi apparatus, and most of the ribosomes and cytoskeleton. Despite this, sieve tube elements live for an entire growing season in trees and for decades in some cases. How can they live for so long with no nucleus? The answer is that each sieve tube element has one or more **companion cells** (see Figure 35.12), produced as daughter cells along with the sieve tube element when a parent cell divides. Numerous plasmodesmata link a companion cell with its sieve tube element. Companion cells retain all their organelles and, through the activities of their nuclei, provide all the functions needed to maintain the sieve tube elements—they may be thought of as the "life support systems" of the sieve tube elements.

A mature sieve tube element is filled with phloem sap consisting of water, dissolved sugars, and other solutes. This solution moves through the sieve tube within the symplast, moving from cell to cell through enlarged plasmodesmata. However, because of the unique structural features of the living sieve tubes, phloem sap is able to move rapidly by bulk flow (like xylem sap).

Now that we have a better picture of the structure of the phloem, let's consider the experiments that led to our current understanding of phloem function.

To investigate translocation, plant physiologists needed to obtain samples of pure sieve tube sap from individual sieve tube elements. This difficult task was simplified when it was discovered that a common garden pest, the aphid, feeds on plants by drilling into a sieve tube with its specialized feeding organ, the *stylet*. The pressure within the sieve tube is higher than outside the plant, so the nutritious sieve tube sap is forced through the stylet and into the aphid's digestive tract. So great is the pressure that sugary liquid is forced through the insect's body and out the anus (**Figure 35.13**). This works because the phloem sap is under strongly positive pressure, unlike the negative pressure potential in the xylem. (We will discuss the forces underlying phloem movement a bit later in this section.)

Plant physiologists use aphids to collect phloem sap. When liquid appears on the aphid's abdomen, indicating that the insect has connected with a sieve tube, the physiologist quickly freezes the aphid and cuts its body away from the stylet, which remains in the sieve tube element and can exude phloem sap for hours. Chemical analysis of phloem sap collected from such a stylet reveals the contents of a single sieve tube element over time. Physiologists can also infer the rates at which different substances are translocated by measuring how long it takes for radioactive tracers administered to a leaf to appear at stylets at different distances from the leaf.

Recall that different substances can move in opposite directions in the phloem of a stem. Experiments with aphid stylets have shown that all the contents of any given sieve tube element move in the same direction. Thus, to account for bidirectional

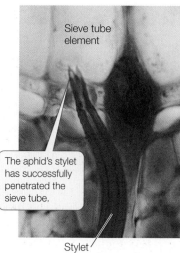

35.13 Aphids Collect Sap Aphids feed on sap drawn from a sieve tube, which they penetrate with a modified feeding organ, the stylet. Pressure inside the sieve tube forces sap through the aphid's digestive tract.

translocation in the phloem, different sieve tubes must be conducting sap in opposite directions. These and other experiments led to the general adoption of the *pressure flow model* as an explanation for translocation in the phloem.

The pressure flow model appears to account for translocation in the phloem

As briefly noted above, phloem sap flows under positive pressure through the sieve tubes, moving from one sieve tube element to the next by bulk flow through the sieve plates, without crossing a membrane. We need to understand how this pressure is generated in order to understand translocation in the phloem.

Two steps in translocation require metabolic energy:

- Transport of sucrose and other solutes from sources into the sieve tubes; called *loading*
- Removal of the solutes from the sieve tubes into sinks; called *unloading*

According to the **pressure flow model** of translocation in the phloem, sucrose is actively transported into sieve tube elements at a source, giving those cells a greater sucrose concentration than the surrounding cells (think lower solute potential—more negative ψ_s). Water therefore enters the sieve tube elements from xylem vessels by osmosis. The entry of this water causes a greater pressure potential (turgor pressure) at the source end of the sieve tube, so that the entire fluid content of the sieve tube is pushed toward the sink end of the tube—in other words, the sap moves by bulk flow in response to a pressure gradient (**Figure 35.14**). In the sink, the sucrose is unloaded both passively and by active transport and water moves back to xylem vessels. In this way the gradient of solute potential and pressure potential needed for movement of phloem sap is maintained.

The pressure flow model of translocation in the phloem is contrasted with the transpiration–cohesion–tension model of xylem transport in **Table 35.1**.

TABLE 35.1
Mechanisms of Sap Flow in Plant Vascular Tissues

	XYLEM	PHLOEM
Driving force for bulk flow	Transpiration from leaves	Active transport of sucrose at source
Site of bulk flow	Nonliving vessel elements and tracheids	Living sieve tube elements
Pressure potential in sap	Negative (pull from top; tension)	Positive (push from source; pressure)

The pressure flow model has been experimentally tested

Even though the pressure flow model was first proposed more than half a century ago, some of its features are still being debated. Other mechanisms have been proposed to account for translocation in sieve tubes, but some have been disproved, and others do not have as much support as the pressure flow model.

Two requirements must be met in order for the pressure flow model to be valid:

- The sieve plates must be unobstructed, so that bulk flow from one sieve tube element to the next is possible.

- There must be an effective method for loading sucrose and other solutes into the phloem in source tissues and unloading them in sink tissues.

Let's see whether these requirements are met.

ARE THE SIEVE PLATES OPEN? Early electron microscopic studies of phloem samples cut from plants seemed to contradict the pressure flow model. The pores in the sieve plates always appeared to be plugged with masses of a fibrous protein, suggesting that sieve tube sap could not flow freely.

However, the fibrous protein seen in early electron micrographs turned out to be a normal plant response to wounding. When phloem is cut, the contents of the sieve elements are forced out through the sieve plate pores. To prevent massive losses of phloem sap, a fibrous protein, called P-protein, is synthesized in companion cells, enters the sieve elements through plasmodesmata, moves to the sieve plate pores, and seals them off. This short-term solution gives the plant time to repair the wound

site by synthesizing a polysaccharide called *callose*, which is thought to strengthen the cell wall.

More recent studies using rapid freezing, which prevents the wounding response, show open pores (see Figure 35.13) in the sieve plates.

HOW ARE SIEVE TUBE ELEMENTS LOADED AND UNLOADED? If the pressure flow model is correct, there must be mechanisms for loading sugars and other solutes into the phloem in source regions and for unloading them in sink regions.

Two general routes can be taken by sugars and other solutes as they move from the mesophyll cells to the phloem: apoplastic and symplastic. The exact details vary widely among plant species. In many plants, sugars and other solutes follow the *apoplastic pathway*; they leave the mesophyll cells and enter the apoplast before they reach the sieve elements. Specific sugars and amino acids are then actively transported into cells of the phloem, and in this way reenter the symplast. Because the solutes cross at least one selectively permeable membrane in the apoplastic pathway, selective transport can be used to regulate which specific substances enter the phloem. In other plants, solutes follow a *symplastic pathway*; the solutes remain within the symplast all the way from the mesophyll cells to the seive tube cells. Because no membranes are crossed in the symplastic pathway, a mechanism that does not involve membrane transport is used to load sucrose into the phloem.

In the apoplastic pathway, sucrose is actively loaded into the companion cells and sieve tubes by sucrose–proton symport, a secondary active transport mechanism. A proton pump actively pumps protons out of the phloem cells, increasing the concentration of protons in the apoplast. The protons then diffuse back

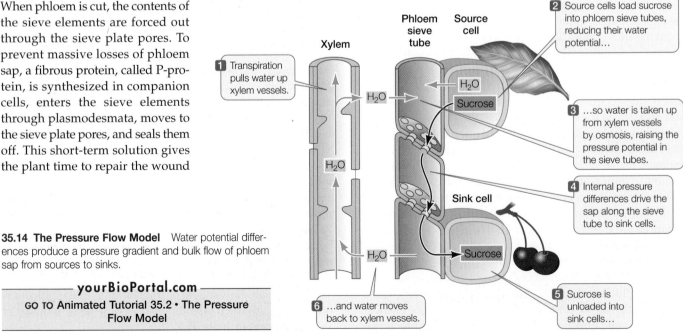

35.14 The Pressure Flow Model Water potential differences produce a pressure gradient and bulk flow of phloem sap from sources to sinks.

yourBioPortal.com
GO TO Animated Tutorial 35.2 • The Pressure Flow Model

into the phloem cells through sucrose–proton symport proteins, bringing sucrose with them.

In sink regions, the solutes are actively transported *out* of the sieve tube elements and into the surrounding tissues. This unloading serves two purposes: it helps maintain the gradient of solute potential and hence of pressure potential in the sieve tubes; and it helps build up high concentrations of sugars and starch in storage organs, such as developing fruits and seeds. Thus the second requirement of the pressure flow model is met, and the model is supported.

35.4 RECAP

Carbohydrates produced by photosynthesis are translocated from source to sink through the phloem by a pressure flow mechanism.

- Explain the difference between a source and a sink. See p. 750

- How does loading of sucrose at the source result in bulk flow toward the sink? See p. 751 and Figure 35.14

CHAPTER SUMMARY

35.1 How Do Plant Cells Take Up Water and Solutes?

- Water moves through biological membranes by osmosis, always moving toward regions with a more negative water potential. The **water potential** (ψ) of a cell or solution is the sum of the **solute potential** and the **pressure potential**. Review Figure 35.2

- **Turgid** plant cells have significant positive pressure potential because the rigid cell wall pushes back on the protoplast. This positive pressure (**turgor pressure**) maintains the physical structure of many plant cells; if the pressure potential drops, the plant *wilts*. Review Figures 35.3 and 35.4

- The movement of a solution due to a difference in pressure potential between two parts of a plant is called **bulk flow**.

- **Aquaporins** are channel proteins that facilitate movement of water molecules through biological membranes.

- Mineral uptake requires transport proteins. Some minerals enter the plant passively by facilitated diffusion; others enter by active transport. A **proton pump** provides energy for the active transport of many mineral ions across membranes in plants. Review Figure 35.5

- Water and minerals pass from the soil into the root by way of the **apoplast** and **symplast**, but must pass through the symplast to cross the endodermis and enter the xylem. The **Casparian strip** in the endodermis blocks movement of water and minerals through the apoplast. Review Figure 35.6, **WEB ACTIVITY 35.1**

35.2 How Are Water and Minerals Transported in the Xylem?

- **Root pressure** is responsible for guttation and for the oozing of sap from cut stumps, but it cannot account for the ascent of xylem sap in trees.

- Water transport in the xylem results from the combined effects of **transpiration**, **cohesion**, and **tension**—the **transpiration–**

cohesion–tension mechanism. Evaporation from the leaf produces tension in the mesophyll cells, which pulls a column of water—held together by cohesion—up through the xylem from the root. Review Figure 35.7, **ANIMATED TUTORIAL 35.1**

- Transport in the xylem is by bulk flow. It does not require the expenditure of energy. Dissolved minerals are carried passively in the water.

35.3 How Do Stomata Control the Loss of Water and the Uptake of CO_2?

- **Stomata** allow a compromise between water retention and carbon dioxide uptake.

- A pair of **guard cells** controls the size of the stomatal opening. A light-activated proton pump, moves protons out of the guard cells to the walls of surrounding epidermal cells, setting up an electrochemical gradient that drives the transport of potassium ions into the guard cells. Water follows osmotically, swelling the guard cells and opening the stomata. Review Figures 35.9 and 35.10

- When threatened by dehydration, mesophyll cells release abscisic acid, which causes guard cells to close the stomata, even in the light.

35.4 How Are Substances Translocated in the Phloem?

- Products of photosynthesis, as well as some minerals, are translocated through sieve tubes in the phloem by way of living **sieve tube elements**. Review Figure 35.12

- **Translocation** in the phloem can proceed in both directions in the stem, although it only goes one way in any given sieve tube. Translocation requires a supply of ATP.

- Translocation in the phloem is explained by the **pressure flow model**: the difference in solute concentration between **sources** and **sinks** creates a difference in (positive) pressure potential along the sieve tubes, resulting in bulk flow. Review Figure 35.14 and Table 35.1, **ANIMATED TUTORIAL 35.2**

SELF-QUIZ

1. Osmosis
   a. requires ATP.
   b. results in the bursting of plant cells placed in pure water.
   c. can cause a cell to become turgid.
   d. is independent of solute concentrations.
   e. continues until the pressure potential equals the water potential.

2. Water potential
   a. is the difference between the solute potential and the pressure potential.
   b. is analogous to the air pressure in an automobile tire.
   c. is the movement of water through a membrane.
   d. determines the direction of water movement between cells.
   e. is defined as 1.0 MPa for pure water under no applied pressure.

3. Which statement about aquaporins is *not* true?
   a. They are membrane transport proteins.
   b. Water movement through aquaporins is always active.
   c. The permeability of some aquaporins is subject to regulation.
   d. They vary in abundance depending on environmental conditions.
   e. They enable water to pass through the phospholipid bilayer without encountering a hydrophobic environment.

4. Which statement about proton pumping across the plasma membrane of plants is *not* true?
   a. It requires ATP.
   b. The region inside the membrane becomes positively charged with respect to the region outside.
   c. It enhances the movement of K^+ ions into the cell.
   d. It pushes protons out of the cell against a proton concentration gradient.
   e. It can drive the secondary active transport of negatively charged ions.

5. Which statement is *not* true?
   a. The symplast consists of the interconnected cytoplasm of living cells.
   b. Water can enter the stele without entering the symplast.
   c. The Casparian strips prevent water from moving between endodermal cells.
   d. The endodermis is a cell layer in the cortex.
   e. Water can move freely in the apoplast without entering cells.

6. In the xylem,
   a. the products of photosynthesis travel down the stem.
   b. living, pumping cells push the sap upward.
   c. the driving force is in the roots.
   d. the sap is often under tension.
   e. the sap must pass through sieve plates.

7. Which of the following is *not* part of the transpiration–cohesion–tension mechanism?
   a. Water evaporates from the walls of mesophyll cells.
   b. Removal of water from the xylem exerts a pull on the water column.
   c. Water is remarkably cohesive.
   d. The wider the tube, the greater the tension its water column can withstand.
   e. At each step, water moves to a region with a more strongly negative water potential.

8. Stomata
   a. control the opening of guard cells.
   b. release less water to the environment than do other parts of the epidermis.
   c. open when CO_2 levels inside the leaf are high.
   d. do not respond to light.
   e. close when water is being lost at too great a rate.

9. Which statement about phloem transport is *not* true?
   a. It takes place in sieve tubes.
   b. It depends on mechanisms for loading solutes into the phloem at sources.
   c. It stops if the phloem is killed by heat.
   d. A high pressure potential is maintained in the sieve tubes.
   e. At sinks, solutes are actively transported into sieve tube elements.

10. The fibrous protein in sieve tube elements
    a. may plug leaks when a plant is damaged.
    b. clogs the sieve plates at all times.
    c. never clogs the sieve plates.
    d. serves no known function.
    e. provides the driving force for transport in the phloem.

FOR DISCUSSION

1. Epidermal cells protect against excess water loss. How do they perform this function? What differences might you expect to find in the structure of the epidermis in stems, roots, and leaves?

2. Phloem transports material from sources to sinks. Give examples of each. How might the distribution of sources and sinks change in the course of a year?

3. What is the minimum number of plasma membranes a water molecule would have to cross in order to get from the soil solution to the atmosphere by way of the stele? To get from the soil solution to a mesophyll cell in a leaf?

4. Transpiration exerts a powerful pulling force on the water column in the xylem. When would you expect transpiration to proceed most rapidly? Why? Describe the source of the pulling force.

ADDITIONAL INVESTIGATION

In the story that opened this chapter we saw that a mutation in the *HARDY* gene resulted in *Arabidopsis* plants with a more extensive root system and thicker leaves than wild-type plants. When the *HARDY* gene was isolated, it was found to encode a transcription factor that stimulates expression of genes for increased water use efficiency. What type of mutation in the *HARDY* gene would cause *Arabidopsis* plants to use water more efficiently? How would you investigate the effect of the *HARDY* mutation on stomata? What results would you expect?

36 Plant Nutrition

When the land blew away

Two of the greatest disasters of recent times occurred because of the mismanagement of soil resources. One was in North America in the 1930s and the other is ongoing in Haiti.

Beginning in the early 1930s, a prolonged drought, combined with a culturally modified landscape, turned the central plains of North America into the Dust Bowl. The native vegetation in the Plains States in the nineteenth century was long grass in the east and short grass in the west. Cattlemen moved in to where their herds could eat a seemingly endless supply of food. But in fact the supply wasn't endless. As one area was overgrazed, the cattle were moved on to new areas, leaving damaged soil in their wake. Settlers followed, "busted the sod" with plows, planted crops, and disrupted vegetation cycles that had persisted for centuries.

In the early twentieth century, farmers in the Plains States began growing wheat, plowing both good and marginal soils. But wheat requires more water than did the native grasses, and rainfall was irregular throughout the 1920s. In 1932 rainfall was almost nonexistent and did not return at adequate levels until 1939.

Without water, crops failed—if they even started to grow. The U.S. plains are windy regions, and without plant roots there was nothing to hold the soil in place when the winds blew. The farms literally blew away. Farmers spent the last of their money on seeds, but dry year followed dry year. Destitute farmers migrated westward, along with others whose livelihoods had depended on the farmers. But they only encountered more difficulties, as these events took place during the Great Depression.

Unfortunately Dust Bowl conditions are not just a thing of the past. Today many countries around the world are struggling with land use issues. The consequences of poor land management, principally deforestation, are dramatically visible on the Caribbean island of Hispaniola. The island is shared by the countries of Haiti, which has mismanaged its soil resources, and the Dominican Republic, which hasn't. From an airplane window it is easy to see the 120-mile-long bor-

Dreams Disappeared in a Cloud of Dust
This photograph of a family displaced by the Dust Bowl was taken by Dorothea Lange in the winter of 1936. The family had traveled to northern California looking for work as migrant farm labor. In Lange's words, "I saw and approached the hungry and desperate mother...I did not ask her name or her history. ...She said that they had been living on frozen vegetables from the surrounding fields, and birds that the children killed. She had just sold the tires from her car to buy food."

A Border Marks Life and Death An aerial view of the border between Haiti (left) and the Dominican Republic (right) provides a dramatic illustration of the extent of deforestation in Haiti.

der between these two countries. The Dominican side is verdant and forested, while the Haitian side is devoid of plant life and most of its soil is gone.

Haiti's population, in addition to living in the poorest nation in the Western Hemisphere, has paid a huge price for the loss of its trees and soil. Because there are few plant roots to stabilize the soil in mountainous areas, rain washes the soil into the sea. Not only is this soil loss detrimental to agriculture, but the runoff hurts offshore reefs and the fisheries they support. To make matters worse, tropical storms all too often result in devastating landslides like the one that killed 2,600 Haitians in 2004.

As human activity increases, ecological disasters become all too common. Today, parts of sub-Saharan Africa's farmland are losing their topsoil as a result of poor land management, swelling populations, and a challenging climate. Crop failures, starvation, and large-scale human displacements are inevitable consequences.

IN THIS CHAPTER we consider the nutritional conditions that foster healthy and sustained plant growth. We identify nutrients that are essential to plants and how plants acquire them. Because most plant nutrients come from the soil, we discuss the formation of soils and the effects of plants on soils. We devote a section to the role played by fungi and bacteria in the uptake of phosphorus and nitrogen by plants, and we conclude with a look at carnivorous and parasitic plants.

36.1 How Do Plants Acquire Nutrients?

Every living thing—and plants are no exception—must obtain raw materials from its environment. These **nutrients** include the major ingredients of macromolecules: carbon, hydrogen, oxygen, and nitrogen. Plants are *autotrophs*, and obtain carbon from atmospheric carbon dioxide through the carbon-fixing reactions of photosynthesis (see Chapter 10). Hydrogen and oxygen come mainly from water, so these elements are plentiful with an adequate water supply. Nitrogen, as you will see later in this chapter, enters plants primarily through the activities of bacteria.

Living organisms need other **mineral nutrients** as well, which most plants obtain from the soil. For example, proteins contain sulfur (S), nucleic acids contain phosphorus (P), chlorophyll contains magnesium (Mg), cytochromes contain iron (Fe), and cellular signaling can involve calcium (Ca). Within the soil, these and other minerals dissolve in water as ions, forming a solution—called the **soil solution**—that contacts the roots of plants.

How does a stationary organism find nutrients?

Many organisms can move from place to place to find the nutrients they need. An organism that cannot move, termed a *sessile* organism, must obtain nutrients from sources that are somehow brought to it. With the exception of the carbon and oxygen in CO_2, a plant's supply of nutrients is strictly local, and a plant may use up the water and mineral nutrients in its local environment as it grows. How does a plant cope with the problem of scarce nutrient supplies?

As discussed in Chapter 34, plants differ fundamentally from animals in that they grow throughout their lifetimes. In fact, growth is a plant's version of movement. For example, roots obtain most of the mineral nutrients plants need. By growing through the soil, roots mine it for new sources of mineral nutrients and water. The growth of stems and leaves helps a plant secure light and carbon dioxide, which in turn allows the roots to continue their growth through the soil.

As it grows, a plant—or even a single root—must deal with a variable environment. Animal droppings create high local concentrations of nitrogen. A particle of calcium carbonate may make a tiny area of the soil alkaline, while dead organic matter may make a nearby area acidic. Such microenvironments encourage or discourage the proliferation of a root system and help direct its growth.

36.1 RECAP

Plants are autotrophs that obtain carbon by photosynthesis, and mineral nutrients and water from the soil.

- Why do plants need phosphorus? Why do they need nitrogen? **See p. 756**
- How does the ability to grow throughout their lifetime allow plants to seek out nutrients? **See p. 756**

We know that plants need nutrients to support their growth. Let's look in more detail at the specific mineral nutrients they need.

36.2 What Mineral Nutrients Do Plants Require?

Plants require many mineral nutrients (**Table 36.1**). Except for nitrogen, all mineral nutrients derive from rock and are usually taken up from the soil solution. A nutrient is called an **essential element** if its absence causes severe disruption of normal plant growth and reproduction. An essential element cannot be replaced by another element.

Essential elements fall roughly into two categories—*macronutrients* and *micronutrients* (see Table 36.1)—based on the amounts required by plants.

- A plant needs **macronutrients** in concentrations of at least 1 gram per kilogram of the plant's dry matter.
- A plant needs **micronutrients** in concentrations of less than 100 milligrams per kilogram of the plant's dry matter.

How do we know if a plant is getting enough of a particular nutrient?

Deficiency symptoms reveal inadequate nutrition

Before a plant that is deficient in an essential element dies, it usually displays characteristic **deficiency symptoms**. Table 36.1 lists the symptoms of some common mineral deficiencies, one of which is also shown in **Figure 36.1**. Such symptoms help horticulturists diagnose mineral nutrient deficiencies in plants. With proper diagnosis, the missing nutrient(s) can be provided in the form of a **fertilizer** (an added source of mineral nutrients).

We know that the elements listed in Table 36.1 are essential to the life of all plants. How did biologists discover which elements are essential?

TABLE 36.1
Mineral Elements Required by Plants

ELEMENT	ABSORBED FORM	MAJOR FUNCTIONS	DEFICIENCY SYMPTOMS
MACRONUTRIENTS			
Nitrogen (N)	NO_3^- and NH_4^+	In proteins, nucleic acids	Oldest leaves turn yellow and die prematurely; plant is stunted
Phosphorus (P)	$H_2PO_4^-$ and HPO_4^{2-}	In nucleic acids, ATP, phospholipids	Plant is dark green with purple veins and is stunted
Potassium (K)	K^+	Enzyme activation; water balance; ion balance; stomatal opening	Older leaves have dead edges
Sulfur (S)	SO_2^{4-}	In proteins and coenzymes	Young leaves are yellow to white with yellow veins
Calcium (Ca)	Ca^{2+}	Affects the cytoskeleton, membranes, and many enzymes; second messenger	Growing points die back; young leaves are yellow and crinkly
Magnesium (Mg)	Mg^{2+}	In chlorophyll; required by many enzymes; stabilizes ribosomes	Older leaves have yellow stripes between veins
MICRONUTRIENTS			
Iron (Fe)	Fe^{2+} and Fe^{3+}	In active site of many redox enzymes and electron carriers; chlorophyll synthesis	Young leaves are white or yellow
Chlorine (Cl)	Cl^-	Photosynthesis; ion balance	Leaf tips wilt; leaves turn yellow and die
Manganese (Mn)	Mn^{2+}	Activation of many enzymes	Younger leaves are pale with green veins
Boron (B)	$B(OH)_3$	Required for proper cell wall formation and expansion	Poor growth of leaves and roots
Zinc (Zn)	Zn^{2+}	Enzyme activation; auxin synthesis	Young leaves are abnormally small; older leaves have many dead spots
Copper (Cu)	Cu^{2+}	In active site of many redox enzymes and electron carriers	New leaves are dark green, may have dead spots
Nickel (Ni)	Ni^{2+}	Activation of the enzyme urease	Leaf tips die; deficiency is rare
Molybdenum (Mo)	MoO_4^{2-}	Nitrate reduction	Leaves turn yellow between veins; older leaves die

36.1 Iron Deficiency Symptoms In crop plants, mineral deficiencies can often be detected in leaves, as in these blueberry leaves. As is typical of iron deficiency, the younger leaves are yellow, whereas the older leaves look normal.

used in nineteenth-century experiments on plant nutrition were sometimes so impure that they provided micronutrients that investigators thought they had excluded. Furthermore, some micronutrients are required in such tiny amounts that a seed may contain enough to supply the embryo and the resultant second-generation plant throughout its lifetime. There might even be enough left over to pass on to third-generation plants. Because of such difficulties, nutrition experiments must be performed in tightly controlled laboratories with special air filters that exclude microscopic salt particles in the air, and must use only the purest available chemicals.

Iron was the first micronutrient to be clearly established as essential, in the 1840s. The last micronutrient to be listed as essential was nickel, in 1983 (the experiment is described in **Figure 36.3**).

yourBioPortal.com
GO TO Animated Tutorial 36.1 • Nitrogen and Iron Deficiencies

Hydroponic experiments identified essential elements

An element is considered essential to plants if a plant fails to complete its life cycle or grows abnormally when that element is absent or insufficient. The essential elements for plants were identified by growing plants **hydroponically**—that is, with their roots suspended in nutrient solutions instead of soil (**Figure 36.2**). Growing plants in this manner allows for greater control of nutrient availability than is possible in a complex medium like soil. In the first successful experiments of this type, performed a century and a half ago, plants grew seemingly normally in solutions containing only calcium nitrate [$Ca(NO_3)_2$], magnesium sulfate ($MgSO_4$), and potassium phosphate (KH_2PO_4). A solution missing any of these compounds could not support normal growth. Tests with other compounds that included various combinations of these elements soon established six macronutrients—calcium, nitrogen, magnesium, sulfur, potassium, and phosphorus—as essential elements.

Identifying essential *micronutrients* by this experimental approach proved to be more difficult. The chemicals

TOOLS FOR INVESTIGATING LIFE

36.2 Growing Plants Hydroponically
Hydroponics is used to grow plants without soil. It is a classic procedure for identifiying nutrients essential to plants. Nitrogen is used here as an example.

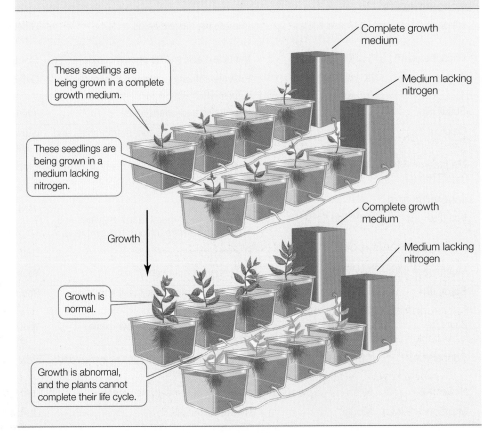

These seedlings are being grown in a complete growth medium.

These seedlings are being grown in a medium lacking nitrogen.

Complete growth medium

Medium lacking nitrogen

Growth

Complete growth medium

Medium lacking nitrogen

Growth is normal.

Growth is abnormal, and the plants cannot complete their life cycle.

INVESTIGATING LIFE

36.3 Is Nickel an Essential Element for Plant Growth?

Using highly purified salts in growth media, Patrick Brown and his colleagues tested whether barley can complete its life cycle in the absence of nickel. Other investigators showed that no other element could substitute for nickel.

HYPOTHESIS Nickel is an essential element for a plant to complete its life cycle.

METHOD
1. Grow barley plants for 3 generations in nutrient solutions containing 0, 0.6, and 1.0 μM $NiSO_4$.
2. Harvest seeds from 5–6 third-generation plants in each of the groups.
3. Determine the nickel concentration in seeds from each plant.
4. Germinate other seeds from the same plants on nickel-free medium and plot the success of germination against nickel concentration.

RESULTS There was a positive correlation between seed germination and seed nickel concentration. There was significantly less germination at the lowest nickel concentrations.

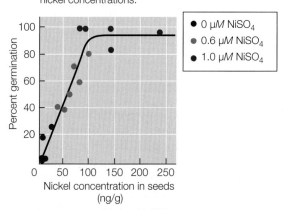

- 0 μM $NiSO_4$
- 0.6 μM $NiSO_4$
- 1.0 μM $NiSO_4$

x-axis: Nickel concentration in seeds (ng/g)
y-axis: Percent germination

CONCLUSION Barley seeds require nickel in order to germinate and thereby complete the life cycle.

Go to **yourBioPortal.com** for original citations, discussions, and relevant links for all INVESTIGATING LIFE figures.

36.2 RECAP

Mineral nutrients required by plants are classified as macronutrients and micronutrients, depending on the amount needed. Micronutrients are often needed in such minute amounts that only sophisticated chemical experiments can determine their essentiality.

- What are some specific mineral deficiency symptoms seen in plants? **See p. 757 and Table 36.1**

- Outline an experimental method for determining whether an element is essential to a plant. **See Figures 36.2 and 36.3**

As we have seen, all plants require nutrients for growth. Plants get nutrients in two ways. For plants growing in natural settings such as forests and fields, nutrients are derived from minerals in the soil. Crop plants, which often need large quantities of nutrients to support rapid growth, may be given nutrient supplements in the form of fertilizer. However, in either case, nutrition is not the only role the soil plays in the life of plants.

36.3 How Does Soil Structure Affect Plants?

Most terrestrial plants grow in soil. Soils provide:

- mechanical support
- mineral nutrients and water from the soil solution
- O_2 for root respiration

Soils also harbor many bacteria and other organisms; some of these are beneficial to plant life, but others are harmful. Some soils contain toxic levels of metal ions such as cadmium, chromium, and lead (see Chapter 39).

Soils are modified by natural phenomena, such as rain, temperature extremes, and the activities of plants and animals, and by the practices of humans—particularly agriculture. In this section, we examine the composition, structure, and formation of soils, as well as their role in plant nutrition.

Soils are complex in structure

Soils have living and nonliving components (**Figure 36.4**). The living components include plant roots as well as populations of bacteria, fungi, protists, and animals such as earthworms and insects. The nonliving portion of the soil includes rock fragments

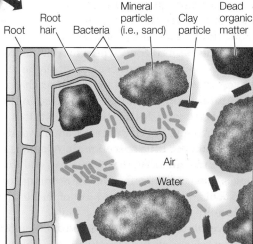

36.4 The Complexity of Soil Soils favorable for plant growth contain both clay and larger mineral particles, as well as water, air, and organic matter. Other organisms are also present.

Root · Root hair · Bacteria · Mineral particle (i.e., sand) · Clay particle · Dead organic matter · Air · Water

A horizon
Topsoil

B horizon
Subsoil

C horizon
Weathering
parent rock
(bedrock)

36.5 A Soil Profile The A, B, and C horizons can sometimes be seen in road cuts such as this one in Australia. The dark upper layer (the A horizon) is home to most of the living organisms in the soil.

ranging in size from large stones to sand to silt and finally to tiny particles of *clay* that are 2 μm or less in diameter. Soil also contains water and dissolved mineral nutrients, air spaces, and dead organic matter. The air spaces in soil contain O_2.

Although soils vary greatly, almost all of them have a *soil profile* consisting of several recognizable horizontal layers, called **horizons**, lying on top of one another. Soil scientists recognize three major *horizons*—termed A, B, and C—in the profile of a typical soil (**Figure 36.5**).

Topsoil is the **A horizon**, which contains most of the soil's living and dead organic matter. Successful agriculture depends on the presence of a suitable A horizon; the A horizon is what blew away from the U.S. plains during the Dust Bowl (as we saw at the beginning of this chapter).

Topsoils vary greatly in their proportions of sand, silt, and clay, and this influences their ability to support plant growth. For example, mineral nutrients tend to be **leached** from the upper soil horizons—dissolved in rain or irrigation water and carried to deeper horizons, where they are unavailable to plant roots. Because sand particles are relatively large and cannot hold water, dissolved minerals are readily leached from sandy soil. Clay binds more water than sand does, and the charged surfaces of clay particles bind mineral ions that plant roots ultimately take up. But clay particles are tiny and pack tightly together, leaving little space for air. A **loam** is a soil that is an optimal mixture of sand, silt, and clay, and thus has sufficient levels of air, water, and available nutrients for plants. Loams also contain organic matter. Most of the best topsoils for agriculture are loams.

Below the A horizon is the **B horizon**, or **subsoil**, which is the zone of infiltration and accumulation of materials leached from

above. Farther down, the **C horizon** is the **parent rock**, also called bedrock, that is breaking down to form soil. Some deep-growing roots extend into the B horizon to obtain water and nutrients, but roots rarely enter the C horizon.

Soils form through the weathering of rock

Rocks are broken down into soil particles (**weathered**) in two ways: First, there is *mechanical weathering*, which is the physical breakdown of materials by wetting, drying, and freezing. Second there is *chemical weathering*, the alteration of the chemistry of the materials in the rocks. Several types of chemical weathering occur, all of which influence the availability of mineral nutrients:

- *Oxidation* by atmospheric oxygen
- *Hydrolysis* (reaction with water)
- Reaction with *acids* (particularly carbonic acid)

The parent rock and the weathering it undergoes determine the basic structure and chemical composition of a soil. However, a key soil characteristic for plants is the *availability* of nutrients, which must be dissolved in the soil solution for uptake by the plant. Chemical weathering results in clay particles covered with negatively charged chemical groups, which bind positively charged mineral nutrients (**Figure 36.6**). How do roots obtain these mineral nutrients?

Soils are the source of plant nutrition

Negatively charged clay particles form ionic bonds (see Section 2.2) with the positively charged ions (cations) of many minerals that are important for plant nutrition, such as potassium (K^+), magnesium (Mg^{2+}), and calcium (Ca^{2+}). To become available to plants or other organisms, these cations must be detached from the clay particles.

Recall that the root surface is covered with root hair cells (see Figure 34.11). Protein transporters in the plasma membrane of these cells actively pump protons (H^+) out of the cell. In addition, cellular respiration in the roots releases CO_2, which dissolves in the soil water and reacts with it to form carbonic acid. This acid ionizes to form bicarbonate and free protons:

$$CO_2 + H_2O \rightleftharpoons H_2CO_3 \rightleftharpoons H^+ + HCO_3^-$$

Proton-pumping by the root and ionization of carbonic acid both act to increase the proton concentration in the soil surrounding the root. The protons bind more strongly to clay particles than do mineral cations; in essence, they trade places with the cations in a process called **ion exchange** (see Figure 36.6). Ion exchange releases important cations into the soil solution, where they are available to be taken up by the roots. The capacity of a soil to support plant growth, called *soil fertility*, is determined in part by its ability to provide nutrients in this manner.

There is no comparable mechanism for binding and releasing negatively charged ions. As a result, important anions such as nitrate (NO_3^-) and sulfate (SO_4^{2-})—direct sources of nitrogen and sulfur, respectively—may leach rapidly from the A horizon.

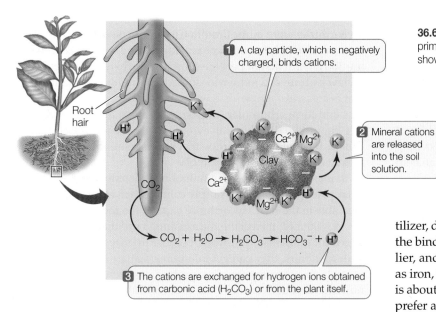

1 A clay particle, which is negatively charged, binds cations.

2 Mineral cations are released into the soil solution.

3 The cations are exchanged for hydrogen ions obtained from carbonic acid (H_2CO_3) or from the plant itself.

$$CO_2 + H_2O \rightarrow H_2CO_3 \rightarrow HCO_3^- + H^+$$

36.6 Ion Exchange Plants obtain mineral nutrients from the soil primarily in the form of positive ions; potassium (K^+) is the example shown here.

Fertilizers and lime are used in agriculture

Agricultural soils are often deficient in one or more essential elements. Irrigation and rainwater leach mineral nutrients from the soil, and the harvesting of crops removes the nutrients that the crops took up from the soil during their growth. Unless the soil is replenished, crop yields will decrease. Mineral nutrients may be replaced by adding **fertilizers**: inorganic fertilizers of various types; or organic fertilizers, such as compost or rotted manure.

INORGANIC AND ORGANIC FERTILIZERS The three elements most commonly added to agricultural soils are nitrogen (N), phosphorus (P), and potassium (K). Commercial inorganic fertilizers are characterized by their "N-P-K" percentages. A 5-10-10 fertilizer, for example, contains 5 percent nitrogen, 10 percent phosphate (P_2O_5), and 10 percent potash (K_2O) by weight (of the nutrient-containing compound, not as weights of the elements N, P, and K). Sulfur, in the form of ammonium sulfate, is also occasionally added to soils.

Organic fertilizers such as manure or crop residues can also be used to supply mineral nutrients. Organic fertilizers have both advantages and disadvantages over inorganic fertilizers. Among the advantages:

• Organic fertilizers release nutrients slowly, which results in less leaching than occurs with a one-time application of an inorganic fertilizer.

• They contain residues of plant or animal materials that improve the structure of the soil, providing spaces for air movement, root growth, and drainage.

However, the nutrients in organic fertilizers are not in a form that is immediately available for absorption, as are the nutrients in inorganic fertilizers. Furthermore, unlike organic fertilizers, inorganic fertilizers can be formulated to meet the specific requirements of a particular soil and a particular crop.

pH EFFECTS ON NUTRIENTS The availability of nutrient ions, whether they are naturally present in the soil or added as fertilizer, depends on soil pH. The proton concentration can affect the binding of nutrient cations to clay particles, as we saw earlier, and can also affect the solubility of other nutrients, such as iron, in the soil solution. The optimal soil pH for most crops is about 6.5, but so-called acid-loving crops such as blueberries prefer a pH closer to 4.

Rainfall and decomposition of organic substances lower the pH of soil, sometimes making it so acidic that plant growth is inhibited. Such acidification can be reversed by **liming**—the application of compounds commonly known as *lime*, such as calcium carbonate, calcium hydroxide, or magnesium carbonate. The addition of these compounds removes H^+ from the soil, and also increases the availability of calcium to plants.

Sometimes, on the other hand, a soil is not acidic enough for a crop. In this case, sulfur can be added in the form of elemental sulfur, which soil bacteria convert to sulfuric acid. Iron and some other elements are more available to plants at a slightly acidic pH. Because soil pH is so important for soil fertility, measuring pH is often the first step in deciding which amendments to add to soils for home gardens and agriculture.

SPRAY APPLICATION OF NUTRIENTS Spraying leaves with a nutrient solution is another effective way to deliver some essential elements to growing plants. Plants take up copper, iron, and manganese more readily from foliar (leaf) sprays than from the soil. Many foliar applications contain chemicals that partially dissolve the protective covering of leaf cells (the cuticle) to increase nutrient uptake.

Plants affect soil fertility and pH

The relationship between plants and soils is not a one-way affair—soils affect plants, but plants also affect soils. The soil that forms in a particular place depends not only on the underlying parent rock, mechanical weathering, and other such factors, but also on the particular plants that grow there. For example, dead plant matter provides most of the carbon-rich materials that break down to form **humus**—a dark-colored, organic soil component, each particle of which is too small to be recognizable with the naked eye. Soil bacteria and fungi produce humus by breaking down plant litter (such as fallen leaves and dead roots), animal feces, dead organisms, and other organic material. Humus is rich in mineral nutrients, especially nitrogen (from animal excrement). Humus also favors plant growth by trapping supplies of water and oxygen for absorption by roots.

Plants also affect the pH of the soil in which they grow. Roots maintain a balance of electric charges. If roots absorb more cations than anions, they excrete H^+, thus lowering the soil pH. If they absorb more anions than cations, they excrete OH^- or HCO_3^-, raising the soil pH. Roots can also actively change the pH in their immediate vicinity by exuding organic acids, such as citric acid and malic acid, that acidify the soil, making it easier to take up certain ions such as ferric iron (Fe^{3+}). Looking at the big picture, we see that successful plant growth can help create conditions that favor further plant growth.

36.3 RECAP

Land plants live anchored in the soil and obtain water and mineral nutrients from it. Plants and soil interact in many ways. Plants can affect many aspects of the soil in which they grow, including mineral content and availability, pH, and the amount of humus. Many of these effects are beneficial to future plant growth.

- Explain how mechanical and chemical weathering form soil from rock. **See p. 760**

- How is soil fertility enhanced by the process of ion exchange? **See p. 760 and Figure 36.6**

Thus far we have focused on the uptake of nutrients in the soil by plant roots. An understanding of how plants acquire nutrients from the soil would be incomplete, however, without taking into account the involvement of *soil microbes*, including fungi and bacteria. In the next section we will focus on the intimate interactions of plants with these organisms, which are essential to the success of most terrestrial plants.

36.4 How Do Fungi and Bacteria Increase Nutrient Uptake by Plant Roots?

One gram of soil contains 6,000–50,000 bacterial *species* and up to 200 *meters* of fungal hyphae (the long branching cells of fungi), although both are largely invisible to the naked eye. In Chapter 39 we describe the strategies plants use to prevent infection by harmful soil microbes. It may surprise you that plants actively encourage a few species of fungi and bacteria to infect their roots and even invade root cells. In this section we describe the resulting "intracellular trading posts," where products are exchanged to the mutual benefit of plants and a few very special soil microbes.

Mycorrhizae expand the root system of plants

The association of fungi with roots is so prevalent that it has its own name: **mycorrhizae** (singular, *mycorrhiza*) (from the Greek *mycos*, "fungus," and *rhiza*, "root"). Recall from Chapter 30 that a multicellular fungus is called a **mycelium** (plural, *mycelia*) and that it is composed of rapidly growing individual tubular filaments called **hyphae** (singular, *hypha*). Two types of

mycorrhizae were introduced in Chapter 30 (see Section 30.2). In *ectomycorrhizae*, fungal hyphae wrap around the root (see Figure 30.11A) but do not penetrate the cells. In this section we will review features of a more widespread and intimate association: that of *arbuscular mycorrhizae,* where the fungal hyphae enter the root and form arbuscular (treelike) structures inside root cells (see Figure 30.11B). This is an evolutionarily ancient association. What is it about this interaction that makes it so enduring? What benefit does each partner derive?

In most cases, roots alone cannot nutritionally support vascular plant growth—they simply cannot reach all the nutrients available in the soil. Mycorrhizae expand the root surface area 10- to 1000-fold, increasing the amount of soil that can be scavenged for nutrients. In addition, because hyphae are much finer than root hairs, they can get into pores that are inaccessible to roots. In this way, mycorrhizae probe a vast expanse of soil for nutrients and deliver them into root cortical cells.

The primary nutrient that the plant obtains from a mycorrhizal interaction is phosphorus. In exchange, the fungus obtains an energy source, largely in the form of simple sugars. In fact, up to 20 percent of the photosynthate (the product[s] of photosynthesis) of terrestrial plants is directed to and consumed by arbuscular mycorrhiza fungi. Such associations are excellent examples of *mutualism*, an interaction between two species in which both species benefit (further discussed in Chapter 56). They are also examples of *symbiosis*, in which two different species live in close contact for a significant portion of their life cycles.

The events in the formation of arbuscular mycorrhiza are shown in **Figure 36.7**. Plant roots produce compounds called **strigolactones** that stimulate rapid growth of fungal hyphae toward the root. (We will return to strigolactones at the end of this chapter.) In response, fungi produce signals that stimulate expression of plant symbiosis-related genes. The products of some of these genes give rise to the *prepenetration apparatus* (PPA), which guides the growth of the fungal hyphae into the root cortex. The sites of nutrient exchange between fungus and plant are the *arbuscules*, which form within root cortical cells. Despite the intimacy of this association, the plant and fungal cytoplasms never mix—they are separated by two membranes, the fungal plasma membrane and the *periarbuscular membrane* (PAM), which is continuous with the plant plasma membrane. We will return to this structure and the features it shares with bacteria-induced root nodules in the next section.

Soil bacteria are essential in getting nitrogen from air to plant cells

The essential mineral nutrient most commonly in short supply, in both natural and agricultural situations, is nitrogen. This is surprising because elemental nitrogen (N_2) makes up almost four-fifths of Earth's atmosphere. However, plants cannot use N_2 directly as a nutrient. The triple bond linking the two nitrogen atoms is extremely stable, and a great deal of energy is required to break it; thus N_2 is a highly unreactive substance. How, then, do plants obtain usable nitrogen for the synthesis of proteins and nucleic acids?

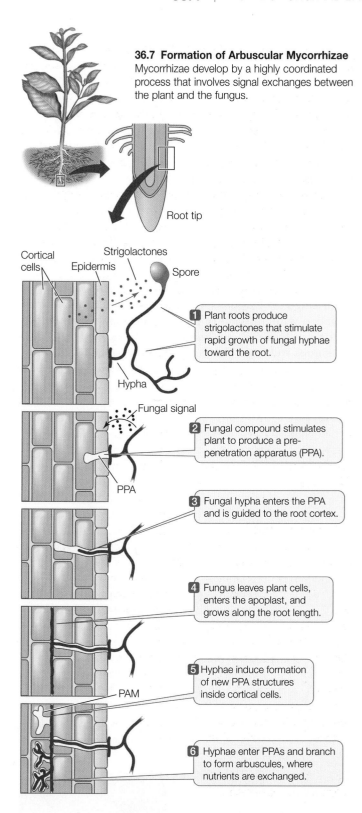

36.7 Formation of Arbuscular Mycorrhizae Mycorrhizae develop by a highly coordinated process that involves signal exchanges between the plant and the fungus.

Root tip

Cortical cells
Epidermis
Strigolactones
Spore

1 Plant roots produce strigolactones that stimulate rapid growth of fungal hyphae toward the root.

Hypha

Fungal signal

2 Fungal compound stimulates plant to produce a pre-penetration apparatus (PPA).

PPA

3 Fungal hypha enters the PPA and is guided to the root cortex.

4 Fungus leaves plant cells, enters the apoplast, and grows along the root length.

PAM

5 Hyphae induce formation of new PPA structures inside cortical cells.

6 Hyphae enter PPAs and branch to form arbuscules, where nutrients are exchanged.

A few species of bacteria have an enzyme that enables them to convert N_2 into a more reactive and biologically useful form by a process called **nitrogen fixation**. These prokaryotic organisms—*nitrogen fixers*—convert N_2 to ammonia (NH_3). Although there are relatively few species of nitrogen fixers, and their biomass is small compared to that of the organisms that depend on them, these talented prokaryotes are essential to the biosphere as we know it.

Nitrogen fixers make all other life possible

By far the greatest share of total world nitrogen fixation is performed biologically by **nitrogen-fixing bacteria**, which fix approximately 170 million metric tons of nitrogen per year. About 80 million metric tons is fixed industrially by humans. Smaller amounts of nitrogen, about 20 million metric tons per year, are fixed in the atmosphere by nonbiological means such as lightning, volcanic eruptions, and forest fires. Rain brings these atmospherically formed products to the ground.

Several groups of bacteria fix nitrogen. In the oceans, various photosynthetic bacteria, including cyanobacteria, fix nitrogen. In fresh water, cyanobacteria are the principal nitrogen fixers. On land, free-living soil bacteria make some contribution to nitrogen fixation, but they fix only what they need for their own use and release the fixed nitrogen only when they die.

Important groups of nitrogen-fixing bacteria live in close association with plant roots. The plant obtains fixed nitrogen from the bacterium, and the bacterium obtains energy sources from the plant. As with arbuscular mycorrhizae, the relationship nitrogen-fixing bacteria have with plants is both mutualistic and symbiotic.

We will look at nitrogen-fixing symbioses in more detail later. But first: how does biological nitrogen fixation work?

Nitrogenase catalyzes nitrogen fixation

Nitrogen fixation is the *reduction* of nitrogen gas (see Section 9.1). It proceeds by the stepwise addition of three pairs of hydrogen atoms to N_2 (**Figure 36.8**). In addition to N_2, these reactions require three things:

- A strong reducing agent to transfer hydrogen atoms (protons and electrons) to N_2 and to the intermediate products of the reaction
- A great deal of energy, which is supplied by ATP
- The enzyme **nitrogenase**, which catalyzes the reaction

Depending on the species of nitrogen fixer, either respiration or photosynthesis provides the necessary reducing agent and ATP.

Nitrogenase is strongly inhibited by oxygen, and many nitrogen fixers are anaerobes that live in environments with little or no O_2. But rhizobia are aerobic and fix nitrogen in aerobic plant roots. How can nitrogenase function under these circumstances?

Plants typically house nitrogen-fixing bacteria in special root structures called **nodules**. Within a nodule, O_2 is maintained at a low level that is sufficient to support respiration, but not so high as to inactivate nitrogenase. This is possible because the cytoplasm of nodule cells contains a plant-produced protein called **leghemoglobin**, which is an O_2 carrier. Leghemoglobin is a close relative of hemoglobin, the red, oxygen-carrying pigment of animals, and is thus an evolutionarily ancient molecule. Some plant nodules contain enough of it to be bright pink inside. Leghemoglobin, with its iron-containing heme groups, transports enough oxygen to the nitrogen-fixing bacteria to support their respiration, while keeping free oxygen concentrations low enough to protect nitrogenase.

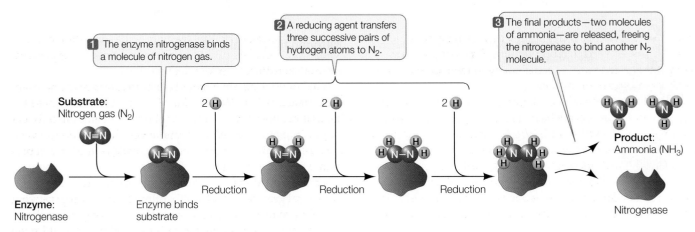

1 The enzyme nitrogenase binds a molecule of nitrogen gas.

2 A reducing agent transfers three successive pairs of hydrogen atoms to N₂.

3 The final products—two molecules of ammonia—are released, freeing the nitrogenase to bind another N₂ molecule.

Substrate: Nitrogen gas (N₂)

Enzyme: Nitrogenase

Enzyme binds substrate

Reduction Reduction Reduction

Product: Ammonia (NH₃)

Nitrogenase

36.8 Nitrogenase Fixes Nitrogen Throughout the chemical reactions of nitrogen fixation, the reactants are bound to the enzyme nitrogenase. A reducing agent transfers hydrogen atoms to nitrogen, and eventually the final product—ammonia—is released. This reaction requires a large input of energy: about 16 ATPs are consumed per reaction.

Some plants and bacteria work together to fix nitrogen

Bacteria of several different genera, collectively known as **rhizobia** (singular, *rhizobium*), fix nitrogen in close, mutualistic association with the roots of plants in the legume family. The legumes include peas, soybeans, clover, alfalfa, and many tropical shrubs and trees. The bacteria infect the plant's roots, and in response the roots develop nodules that house the bacteria.

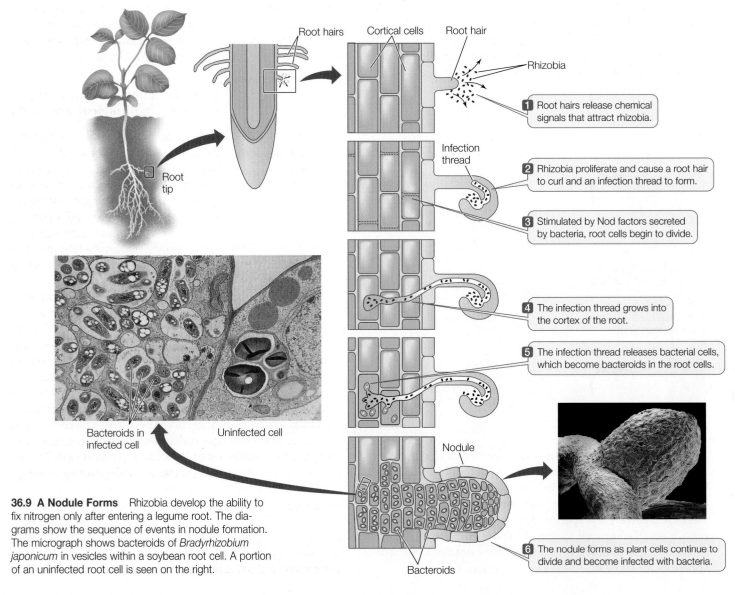

Root hairs Cortical cells Root hair

Rhizobia

1 Root hairs release chemical signals that attract rhizobia.

Root tip

Infection thread

2 Rhizobia proliferate and cause a root hair to curl and an infection thread to form.

3 Stimulated by Nod factors secreted by bacteria, root cells begin to divide.

4 The infection thread grows into the cortex of the root.

5 The infection thread releases bacterial cells, which become bacteroids in the root cells.

Bacteroids in infected cell

Uninfected cell

Nodule

6 The nodule forms as plant cells continue to divide and become infected with bacteria.

36.9 A Nodule Forms Rhizobia develop the ability to fix nitrogen only after entering a legume root. The diagrams show the sequence of events in nodule formation. The micrograph shows bacteroids of *Bradyrhizobium japonicum* in vesicles within a soybean root cell. A portion of an uninfected root cell is seen on the right.

Bacteroids

Because of their ability to form nitrogen-fixing associations, legumes are often used in crop rotations; for example, farmers might plant clover or alfalfa occasionally to increase the available nitrogen content of the soil.

The legume–rhizobium association is not the only nitrogen-fixing symbiosis. Some cyanobacteria fix nitrogen in association with fungi (in lichens) or with ferns, cycads, or nonvascular plants. Rice farmers can increase crop yields by growing the water fern *Azolla*, with its symbiotic nitrogen-fixing cyanobacterium, in the flooded fields where rice is grown. Another group of bacteria, the filamentous actinobacteria, fix nitrogen in association with woody species such as alder and mountain lilacs.

Legumes and rhizobia communicate using chemical signals

Neither free-living rhizobia nor uninfected legumes can fix nitrogen. Only when the two are closely associated in root nodules does the reaction take place. The establishment of this symbiosis between a rhizobium and a legume requires a complex series of steps, with active contributions by both the bacterium and the plant root (**Figure 36.9**). First the root releases flavonoids and other chemical signals that attract soil-living rhizobia to the vicinity of the root. Flavonoids trigger the transcription of bacterial *nod* genes, the products of which synthesize Nod (nodulation) factors. These factors, secreted by the bacteria, cause cells in the root cortex to divide, leading to the formation of a primary nodule meristem. This meristem gives rise to the plant tissue that constitutes the root nodule.

The bacteria enter the root via an infection thread and eventually reach cells in the interior of the root nodule. There the bacteria are released into the cytoplasm of the nodule cells, enclosed in plant-derived membrane vesicles. Inside the vesicles, the bacteria differentiate into **bacteroids**—the form of bacteria that can fix nitrogen.

The legume–rhizobium interaction is very specific. For example, only one species of rhizobium will form a nitrogen-fixing symbiosis with alfalfa; another rhizobium will only infect clover. The specificity of the interaction is determined in part by the specificity of the chemical signals exchanged by the plant and bacterium. The soil may not have the correct bacterium for a given legume crop, so farmers and gardeners often coat legume seeds with the appropriate rhizobium before planting.

There is increasing evidence that nodule formation depends on some of the same genes and mechanisms that allow mycorrhizae to develop. For example, both processes involve invagination of the plasma membrane to allow entry of the fungal hypha or rhizobia. The similarities of the structures formed during the development of mycorrhizae and nodules are especially striking considering that the symbioses involve members of two different *kingdoms* (fungi and bacteria) (**Figure 36.10**).

Biological nitrogen fixation does not always meet agricultural needs

Bacterial nitrogen fixation is not always sufficient to support the needs of agriculture. Traditional farmers used to plant dead fish along with corn; the decaying fish released nitrogen that the developing corn could use. Today farmers use inorganic nitrogen fertilizers produced through industrial nitrogen fixation to meet the food needs of a rapidly expanding population.

Most industrial nitrogen fixation is done by the *Haber process*, a chemical reduction that requires a great deal of energy. (Recall that biological nitrogen fixation consumes a lot of ATP—about 16 ATP per N fixed; see Figure 36.8.) At present in the United States, the manufacture of nitrogen-containing fertilizer takes more energy—primarily natural gas and hydroelectric—than does any other aspect of crop production. The rising cost and dwindling supply of energy

Mycorrhizal infection

Plant cell wall
Plant cell membrane
Periarbuscular membrane
Fungal cell wall
Fungal cell membrane

Vacuole

Vacuole

Nucleus

In both types of infection, a strand of plant cytoplasm forms in the next cell to be crossed, ahead of the hypha or infection thread.

A plant membrane separates the fungus and rhizobia from the plant cell cytoplasm.

Infection thread membrane

Rhizobial infection

Rhizobia

Preinfection thread

36.10 Intracellular Structures in Plant–Fungus and Plant–Rhizobium Symbioses Several steps in the development of mycorrhizae and nodules involve similar structures.

36.11 The Nitrogen Cycle
Nitrogen fixation, nitrification, nitrate reduction, and denitrification are components of an essential chemical cycle that converts atmospheric nitrogen gas into ammonium ions and nitrate ions—forms of nitrogen that can be taken up by plants—and returns N_2 to the atmosphere.

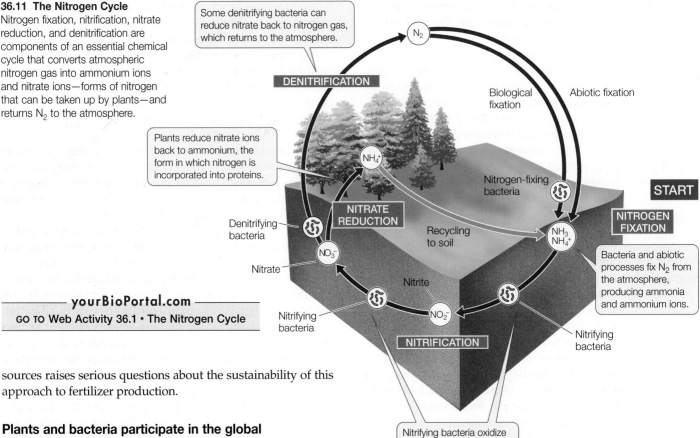

Some denitrifying bacteria can reduce nitrate back to nitrogen gas, which returns to the atmosphere.

DENITRIFICATION

Plants reduce nitrate ions back to ammonium, the form in which nitrogen is incorporated into proteins.

Biological fixation

Abiotic fixation

N_2

START

NITRATE REDUCTION

NITROGEN FIXATION

Nitrogen-fixing bacteria

Denitrifying bacteria

NH_4^+

NH_3
NH_4^+

Recycling to soil

Bacteria and abiotic processes fix N_2 from the atmosphere, producing ammonia and ammonium ions.

Nitrate

NO_3^-

Nitrite

Nitrifying bacteria

NO_2^-

NITRIFICATION

Nitrifying bacteria

Nitrifying bacteria oxidize ammonia to nitrate ions.

—— **yourBioPortal.com** ——
GO TO Web Activity 36.1 • The Nitrogen Cycle

sources raises serious questions about the sustainability of this approach to fertilizer production.

Plants and bacteria participate in the global nitrogen cycle

Nitrogen moves through the biosphere in a **global nitrogen cycle** (**Figure 36.11**), which includes four key steps:

1. *Fixation* of atmospheric N_2 to NH_3 and NH_4^+ by bacteria and by abiotic processes

2. *Nitrification* of these molecules to nitrate by bacteria

3. *Nitrate reduction* by plants

4. *Denitrification* of nitrate by bacteria back to N_2, which is then released to the atmosphere to begin another cycle

The nitrogen released into the soil as a result of nitrogen fixation is primarily in the form of ammonia (NH_3) and ammonium ions (NH_4^+). Although ammonia can be toxic to plants if it accumulates in tissues, ammonium ions can be taken up safely at low concentrations. Soil bacteria called **nitrifiers** oxidize ammonia to nitrate ions (NO_3^-)—another form that plants can take up—by the process of **nitrification**. Soil pH affects which form of nitrogen is taken up by plants: nitrate ions are taken up preferentially under more acidic conditions, ammonium ions under more basic ones. To use nitrate, a plant must first reduce it to ammonium in a process called **nitrate reduction**. This occurs in two enzyme-catalyzed steps. The first step, from nitrate (NO_3^-) to nitrite (NO_2^-), takes place in the cytoplasm; the second, from nitrite (NO_2^-) to ammonia (NH_4^+), in the plastids. The plant uses the ammonia to manufacture amino acids, from which the plant's proteins and all its other nitrogen-containing compounds are formed. Animals cannot reduce nitrogen, and they depend on plants to supply them with reduced nitrogenous compounds.

The nitrogen cycle is essential for life on Earth: nitrogen-containing compounds constitute 5–30 percent of a plant's total dry weight. The nitrogen content of animals is even higher, and all of it arrives there by way of the plant kingdom.

36.4 RECAP

Two mutualistic interactions with soil microbes are critical to the success of terrestrial plants. Fungi and plants form mycorrhizae, which greatly increase the soil area that roots can scavenge for nutrients. Bacteria in soils and root nodules fix inert, atmospheric nitrogen into forms that plants and ultimately animals can use. Denitrification returns nitrogen from dead organisms and animal waste back to the atmosphere, continuing the global nitrogen cycle.

● What is exchanged between plants and fungi in mycorrhizae? **See p. 762**

● What, besides nitrogenase, is required to reduce nitrogen gas to a form plants can use? **See p. 763 and Figure 36.8**

● How is the formation of a root nodule on a legume similar to the formation of an arbuscular mychorriza? **See p. 765 and Figure 36.10**

Let's turn now to some special mechanisms for obtaining nutrients that have evolved in plant species with unusual lifestyles.

36.5 How Do Carnivorous and Parasitic Plants Obtain a Balanced Diet?

Most plants obtain their mineral nutrients from the soil solution (with the help of fungi), but some use other sources. Carnivorous and parasitic plants are examples of such plants.

Carnivorous plants supplement their mineral nutrition

Some plants augment their nitrogen supply by capturing and digesting flies and other insects. There are about 500 of these **carnivorous plant** species, the best known of which are Venus flytraps (genus *Dionaea*; **Figure 36.12A**), sundews (genus *Drosera*; **Figure 36.12B**), and pitcher plants (genus *Sarracenia*).

Carnivorous plants are typically found in boggy habitats that are acidic and nutrient deficient. To obtain extra nitrogen, these plants capture animals, digest their proteins, and absorb the amino acids. Pitcher plants have pitcher-shaped leaves that collect small amounts of rainwater. Insects and even small rodents are lured into the pitchers by bright colors or attractive scents and are prevented from leaving by stiff, downward-pointing hairs. The animals eventually die and are digested by a combination of plant enzymes and bacteria in the water. Sundews have leaves covered with hairs that secrete a clear, sticky, sugary liquid. Insects become stuck to these hairs, and more hairs curve over to further entrap them. Enzymes secreted by the plant digest the insects. Venus flytraps have specialized leaves with two halves that fold together. When an insect touches trigger hairs on a leaf, its two halves quickly come together, their spiny margins interlocking and imprisoning the insect before it can escape. The leaf then secretes enzymes that digest its prey.

The closing of the Venus flytrap's leaf is one of the fastest movements in the plant world, requiring only 0.1 sec. To find out how this happens, Dr. Lakshminarayanan Mahadevan and colleagues painted fluorescent dots on the surface of the flytrap's leaf surface and used high-speed cameras to record the trap snapping shut when its trigger hairs were touched. They then used computer analysis of the recorded dot movements to generate a mathematical model to help explain the movement. The researchers found that the first step is the elongation of cells on the outer surface of the leaf. The expansion of only one side of the leaf

causes it to snap from a convex into a concave shape, much like a contact lens flipping inside out.

Carnivorous plants do not need to feed on insects, but doing so helps them grow faster in their natural habitats. They use the additional nitrogen from the insects to make more proteins, chlorophyll, and other nitrogen-containing compounds.

Parasitic plants take advantage of other plants

Approximately 1 percent of flowering plant species derive some or all of their water, nutrients, and sometimes even photosynthate from other plants. In these **parasitic plants**, absorptive organs called **haustoria** have evolved that invade the host and tap into the vascular tissues in the root or stem.

Parasitic plants are divided into two broad classes based on their nutritional interactions with their hosts. **Hemiparasites** can still photosynthesize, but derive water and mineral nutrients from the living bodies of other plants. Perhaps the most familiar hemiparasites are the several genera of mistletoes. Mistletoes are green and carry on some photosynthesis, but they parasitize other plants for water and mineral nutrients and may derive photosynthetic products from them as well. Dwarf mistletoe (*Arceuthobium americanum*) is a serious parasite in forests of the western United States, destroying more than 3 billion board feet of lumber per year.

Holoparasites are completely parasitic and do not perform photosynthesis. They are taxonomically and morphologically diverse. Some, such as members of the dodder family, are plantlike in appearance, with small leaf remnants and flowers (**Figure 36.13**). Some holoparasites do not have leaves or stems because they spend most of their life cycle underground and only break the surface to flower.

Several parasitic plant species lack many of the genes normally present in the chloroplast genome (which in turn is only a remnant of the genome in the original endosymbiont from which the chloroplast evolved; see Sections 5.5 and 27.1). These genes, which are needed for photosynthesis, have been lost because there is no evolutionary pressure to retain them. Thus, while the parasitic lifestyle can be viewed as a free ride, for some

(A) *Dionaea muscipula*

(B) *Drosera rotundifolia*

36.12 Carnivorous Plants Some plants have adapted to nitrogen-poor environments by becoming carnivorous. (A) The Venus flytrap obtains nitrogen from the bodies of insects trapped inside the plant when its hinges snap shut. (B) Sundews trap insects on sticky hairs. Secreted enzymes will digest the carcass externally.

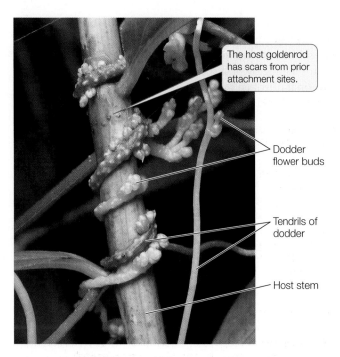

The host goldenrod has scars from prior attachment sites.

Dodder flower buds

Tendrils of dodder

Host stem

36.13 A Parasitic Plant Tendrils of dodder (genus *Cuscuta*) wrap around a goldenrod (genus *Solidago*). The parasitic dodder obtains water, sugars, and other nutrients through tiny, rootlike protuberances that penetrate the surface of the host plant.

plants it is also a one-way ticket, with no possibility of return to self-sufficiency.

The plant–parasite relationship is similar to plant–fungi and plant–bacteria associations

Plant–bacteria and plant–fungus relations both involve reciprocal signaling between the two species (see Figures 36.7 and 36.9). Parasitic plants also need to detect nearby plants so they can obtain their nutrients, but obviously this is to the disadvantage of the potential host plant. In one interesting case, a mechanism has evolved in a parasitic plant to recognize the chemical signals produced by plants to attract beneficial fungi.

The holoparasite *Striga* (witchweed) is a serious pest of cereal crops in Africa (see the opening story of Chapter 30). Earlier in this chapter, you learned that arbuscular fungi are attracted to plant roots by compounds called strigolactones. One of these same molecules was discovered over 60 years ago as an inducer of germination of some parasitic plants, including *Striga*. Scientists strongly suspect that this is no coincidence. The mycorrhizal interaction is ancient (over 400 million years old) and predates the evolution of parasitic plants. For this reason scientists hypothesize that a mechanism evolved in the ancestors of modern *Striga* to recognize a compound that was already produced by plants to attract soil microbes.

In *Striga* we thus find an example of "opportunistic evolution"—that is, the repurposing of preexisting processes rather than the invention of new processes from scratch. This is not the first time we have encountered this phenomenon in this chapter. Recall that the formation of nodules by rhizobia uses some of the same mechanisms utilized by arbuscular fungi to establish residence inside plant cells (see Figure 35.10), implying an evolutionary connection between the two symbioses.

36.5 RECAP

Carnivorous plants supplement their nutrition by extracting materials from animals. Rapid reflexes have evolved in some of these plants for trapping their prey. Parasitic plants, on the other hand, get some or all of their sustenance from other plants. Extreme holoparasites cannot function as autotrophs, having lost chloroplast genes coding for photosynthetic machinery. At least one parasitic plant responds to the same signaling molecule that the host plant uses to attract beneficial fungi.

- Why do carnivorous plants eat animals? **See p. 767**
- How do the needs of holoparasitic plants differ from those of carnivorous plants? **See p. 767**
- What characteristics are shared among plant–parasite, plant–fungus, and plant–bacteria associations?

CHAPTER SUMMARY

36.1 How Do Plants Acquire Nutrients?

- Plants are photosynthetic autotrophs that can produce all their organic molecules from carbon dioxide, water, and minerals, including a nitrogen source.
- **Mineral nutrients** are obtained from the **soil solution**.
- Root growth allows plants, which are sessile, to search for mineral resources.
- Microenvironments within the soil, such as acidic or alkaline areas, affect the direction of root growth.

36.2 What Mineral Nutrients Do Plants Require?

- Plants require 14 **essential elements**. Of these, six are **macronutrients** and eight are **micronutrients**. **Deficiency symptoms** suggest what essential element a plant lacks. **Review Table 36.1 and Figure 36.1**
- The essential elements were discovered by growing plants on **hydroponic solutions** that lacked individual elements. **Review Figures 36.2 and 36.3, ANIMATED TUTORIAL 36.1**

36.3 How Does Soil Structure Affect Plants?

- **Soils** contain water, air, and inorganic and organic substances. Soils have living (biotic) and nonliving (abiotic) components. **Review Figure 36.4**

- A soil typically consists of two or three horizontal zones called **horizons**. **Topsoil** forms the uppermost or **A horizon**. Topsoil tends to lose mineral nutrients through **leaching**. **Loams** are excellent agricultural topsoils, with a good balance of sand, silt, clay, and organic matter. **Review Figure 36.5**
- Soils form by mechanical and chemical **weathering** of rock. Chemical weathering imparts mineral nutrients to **clay** particles. Plant litter and other organic matter decomposes to form **humus**. Plants obtain some mineral nutrients through **ion exchange** between the soil solution and the surface of clay particles. **Review Figure 36.6**
- Farmers use **fertilizers** to make up for deficiencies in soil mineral nutrient content. **Liming** can reverse acidification.
- Plants can influence the characteristics, including the pH, of the soil in which they grow.

36.4 How Do Fungi and Bacteria Increase Nutrient Uptake by Plant Roots?

- **Mycorrhizae** are symbiotic root–fungus associations that greatly increase a plant's absorption of water and minerals, especially phosphorus. They occur in 80 percent of plant species.
- The fungal **mycelia** invade root cortex cells and form arbuscules, which are the sites of nutrient exchange between the fungus and plant. **Review Figure 36.7**
- In the earliest stages of mycorrhiza formation, the **hyphae** of arbuscular fungi grow toward **strigolactones**, compounds that are produced by the plant roots.
- Some **nitrogen-fixing bacteria** live free in the soil; others live symbiotically as **bacteroids** within plant roots. In **nitrogen fixation**, nitrogen gas (N_2) is reduced to ammonia (NH_3) or ammonium ions (NH_4^+) in a reaction catalyzed by **nitrogenase**. **Review Figure 36.8**
- Nitrogenase requires anaerobic conditions, but the bacteroids in root **nodules** require oxygen, which is maintained at the proper level by **leghemoglobin**.

- The formation of a root nodule requires interaction between the root system of a legume and a **rhizobium**. **Review Figure 36.9**
- Several steps in the formation of root nodules and arbuscules are similar and probably involve some of the same plant genes. **Review Figure 36.10**
- In agriculture, biological nitrogen fixation must usually be supplemented with commercial nitrogen fertilizers made by the Haber process.
- Plants and bacteria interact in the **global nitrogen cycle**, which involves a series of reductions and oxidations of nitrogen-containing molecules. **Review Figure 36.11, WEB ACTIVITY 36.1**
- **Nitrification** by bacteria converts ammonia to nitrate ions in the soil. **Nitrate reduction** is carried out by plant enzymes, enabling plants to form their own nitrogen compounds. **Denitrification** returns nitrogen from animal wastes and dead organisms to the atmosphere.

36.5 How Do Carnivorous and Parasitic Plants Obtain a Balanced Diet?

- **Carnivorous plants** are autotrophs that supplement a low nitrogen supply by feeding on insects or other small animals.
- **Parasitic plants** draw on other plants to meet their needs, which may include minerals, water, or the products of photosynthesis.
- **Hemiparasites**, such as mistletoes, can still photosynthesize. Extreme **holoparasites** cannot function as auxotrophs because they have lost chloroplast genes that code for components of the photosynthetic apparatus (which they no longer need).
- A strigolactone—a compound in the same category of compounds plants use to attract mycorrhizal fungi—also induces the germination of some parasitic plants, including *Striga*.
- Scientists hypothesize that a mechanism evolved in the ancestors of modern *Striga* to recognize a compound that was already produced by plants to attract arbuscular fungi.

SELF-QUIZ

1. Macronutrients
   a. are so called because they are more essential than micronutrients.
   b. include manganese, boron, and zinc, among others.
   c. function as catalysts.
   d. are required in concentrations of at least 1 gram per kilogram of plant dry matter.
   e. are obtained by the process of photosynthesis.

2. Which of the following is *not* an essential mineral element for plants?
   a. Potassium
   b. Magnesium
   c. Calcium
   d. Lead
   e. Phosphorus

3. Fertilizers
   a. are often characterized by their N-P-O percentages.
   b. are not required if crops are removed frequently enough.
   c. restore needed mineral nutrients to the soil.
   d. are needed to provide carbon, hydrogen, and oxygen to plants.
   e. are needed to destroy soil pests.

4. In a typical soil,
   a. the topsoil tends to lose mineral nutrients by leaching.
   b. there are four or more horizons.
   c. the C horizon consists primarily of loam.
   d. the dead and decaying organic matter gathers in the B horizon.
   e. more clay means more air space and thus more oxygen for roots.

5. Which of the following is *not* true for arbuscules?
   a. They are an ancient association between plants and fungi.
   b. They expand the effective root area of plants and allow more efficient water uptake.
   c. They are a significant source of fixed nitrogen for plants.
   d. They are a significant source of phosphorous for plants.
   e. Most land plants have them.

6. Nitrogen fixation is
   a. performed only by plants.
   b. the oxidation of nitrogen gas.
   c. catalyzed by the enzyme nitrogenase.
   d. a single-step chemical reaction.
   e. possible because N_2 is a highly reactive substance.

7. Nitrification is
   a. performed only by plants.
   b. the reduction of ammonium ions to nitrate ions.
   c. the reduction of nitrate ions to nitrogen gas.
   d. catalyzed by the enzyme nitrogenase.
   e. performed by certain bacteria in the soil.

8. Which of the following is an early step in the formation of *both* arbuscules and root nodules?
   a. Invasion of a plant root by a fungus
   b. Invasion of a plant root by a bacterium
   c. Strigolactones produced by the root are recognized by the microbe
   d. Root cells are invaded but there is no direct contact between plant and microbe cell contents
   e. Root cells are invaded and there is direct contact between plant and microbe cell contents

9. Which of the following is a parasite?
   a. Venus flytrap
   b. Pitcher plant
   c. Sundew
   d. Dodder
   e. Tobacco

10. All carnivorous plants
    a. are parasites.
    b. depend on animals as a source of carbon.
    c. are incapable of photosynthesis.
    d. depend on animals as their sole source of phosphorus.
    e. obtain supplemental nitrogen from animals.

FOR DISCUSSION

1. Methods for determining whether a particular element is essential have been known for more than a century. Since these methods are so well established, why was the essentiality of some elements discovered only recently?

2. If a Venus flytrap were deprived of soil sulfates and hence made unable to synthesize the amino acids cysteine and methionine, would it die from lack of protein? Explain.

3. Soils are dynamic systems. What changes might result when land is subjected to heavy irrigation for agriculture after being relatively dry for many years? What changes in the soil might result when a virgin deciduous forest is cut down and replaced by crops that are harvested each year? Even though the countries share the same island, why are hurricanes frequently accompanied by the loss of life in Haiti but not in the Dominican Republic?

4. We mentioned that important positively charged ions are held in the soil by clay particles, but other, equally important, negatively charged ions are leached deeper into the soil's B horizon. Why doesn't leaching cause an electrical imbalance in the soil? (Hint: think of the ionization of water.)

5. The biosphere of Earth as we know it depends on the existence of a few species of nitrogen-fixing prokaryotes. What do you think might happen if one of these species were to become extinct? If all of them were to disappear?

6. Holoparasitic plants have lost many of the morphological and genetic traits necessary for an autotrophic lifestyle. From an evolutionary point of view, how do you think this happened? (Hint: think about selection pressures.)

ADDITIONAL INVESTIGATION

Some mutant *Arabidopsis* plants that are very bushy (their shoots are more highly branched than wild-type plants) cannot make strigolactones because of a mutation in a gene necessary for strigolactone biosynthesis. If an investigator applies strigolactones to the plants, they grow normally. What does this experiment suggest about the role of strigolactones in plant growth? How does this add to the story of strigolactones as signals for arbuscules and parasitic plants?

WORKING WITH DATA (GO TO yourBioPortal.com)

Is Nickel an Essential Element for Plant Growth?   In this hands-on exercise, you will critically examine the experimental approach used by Patrick Brown and his colleagues to show that barley plants require nickel for their life cycle (Figure 36.3).

Analyzing data from the original paper, you will calculate the critical value, the tissue concentration below which growth is significantly reduced.

37 Regulation of Plant Growth

Saving millions of lives by regulation of plant development

There is no Nobel Prize for agriculture as there is for medicine, the other branch of applied biology. This is not because plant biology is unimportant; it is because agriculture was not mentioned in the will of the prize's benefactor, Alfred Nobel. But a Nobel prize was awarded in this field: plant geneticist Norman Borlaug received the Peace Prize for research on wheat that has been estimated to have saved a billion lives.

In their constant search for ways to help farmers produce more food for a growing population, biologists have developed crop plants whose physiology allows them to produce more grain per plant (higher yield). However, when a plant produces a lot of seeds, the sheer weight of the load may cause the stem to bend over, or even break. This makes harvesting the seeds impossible: think of how hard it would be to get enough food for your family if you had to pick up seeds on the ground, some of which had already sprouted.

In 1945, the U.S. Army temporarily occupied Japan, which it had defeated in World War II. During the war, Japan, an island nation, was blockaded and could not import food. How had they been able to grow enough grain to feed their people? The answer lay in the fields: the Japanese had bred genetic strains of rice and wheat with short, strong stems that could bear a high yield of grain without bending or breaking. This innovation made an impression on an agricultural advisor who happened to be among the first wave of U.S. occupiers, and seeds of the Japanese strains were sent back to the U.S.

A decade later, Borlaug, who was working in Mexico at the time, began genetic crosses of what were known as semi-dwarf wheat plants from Japan with varieties that had genes conferring rapid growth, adaptability to varying climates, and resistance to fungal diseases. The results were genetic strains of wheat that gave record yields, first in Mexico and then in India and Pakistan in the 1960s. At about the same time and using a similar strategy, scientists in the Philippines developed semi-dwarf rice with equally spectacular results. People who had lived on the edge of starvation now produced enough food. Countries that had been relying on food aid from

Norman Borlaug Seen here in a field of semi-dwarf wheat, plant geneticist Norman Borlaug carried out a program of genetic crosses that led to high-yielding varieties and saved millions from starvation.

Semi-Dwarf Rice   The short variety of rice can give higher yields of grain than its taller counterpart (right). The difference is that the latter can respond to the hormone gibberellin.

other countries were now growing so much grain that they could export the surplus. The development of these semi-dwarf grains began what was called the "Green Revolution."

It is only recently that plant biologists have discovered why semi-dwarf wheat and rice have short stems. In normal plants a hormone called gibberellin stimulates stem elongation. But in the semi-dwarf plants, a mutation affects the signal transduction mechanism for gibberellin so that the stem cells do not respond to it and growth is reduced. The lives of countless people have been saved by a disruption of hormone signaling.

IN THIS CHAPTER we will give a brief overview of the life of a flowering plant and its developmental stages. We will explore the nature of the environmental cues, photoreceptors, and hormones (including gibberellin) that regulate plant growth and development. We will also consider the multiple roles and interactions of these different elements.

37.1 How Does Plant Development Proceed?

As Chapter 34 describes, plants are sessile organisms that must seek out resources above and below the ground. Features that maximize the ability of plants to obtain the resources that they need to grow and reproduce include:

- *Meristems.* Plants have permanent collections of stem cells (undifferentiated, constantly dividing cells) that allow them to continue growing throughout their lifetimes (see Section 34.4).

- *Post-embryonic organ formation.* Unlike animals, plants can initiate development of new organs such as leaves and flowers throughout their lifetimes.

- *Differential growth.* Plants can allocate their resources so that they grow more of the organs that will benefit them most; for example, more leaves to harvest more sunlight or more roots to obtain more water and nutrients.

To use growth for maximal advantage, plants must continuously monitor their environment and redirect their growth as appropriate. Under normal circumstances a plant's environment is never completely stable. For example, the amount of light changes from day to night and from season to season. In addition, other plants are often vying for what light there is, and plants modulate their growth to compete with their neighbors for this precious resource. As you will see in this chapter, several mechanisms have evolved in plants to sense their environment and trigger appropriate growth responses.

The *development* of a plant—the series of progressive changes that take place throughout its life—is regulated in many ways. Key factors involved in regulating plant growth and development are:

- *Environmental cues,* such as day length

- *Receptors* that allow a plant to sense environmental cues, such as photoreceptors that absorb light, and chemoreceptors that signal the presence of pathogens (see Chapter 39)

- *Hormones*—chemical signals that mediate the effects of the environmental cues, including those sensed by receptors

- The plant's *genomes,* which encode regulatory proteins and enzymes that catalyze the biochemical reactions of development

We will explore these regulatory mechanisms in more detail later in this chapter. But first let's look at the initial steps of plant development—from seed to seedling—and the types of internal and external cues that guide them.

In early development, the seed germinates and forms a growing seedling

If all developmental activity is suspended in a seed, even when conditions appear to be suitable for its growth, the seed is said to be **dormant**. Cells in dormant seeds do not divide, expand, or differentiate. For the embryo to begin developing, seed dormancy must be broken by one of the mechanisms discussed later in this section.

As the seed begins to **germinate**—to develop into a seedling—it takes up water. The growing embryo then obtains chemical building blocks—carbohydrate, amino acid, and lipid monomers—for its development by digesting the polysaccharides, fats, and proteins stored in the seed. As we will see later, the embryos of some plant species secrete hormones that direct the mobilization of these reserves. Germination is completed when the **radicle** (embryonic root) emerges from the seed coat. The plant is then called a **seedling**.

If the seed germinates underground, the new seedling must elongate rapidly (in the right direction!) and cope with a period of life in darkness or dim light. A series of photoreceptors direct this stage of development and prepare the seedling for growth in the light.

Early shoot development varies among the flowering plants. **Figure 37.1** shows the shoot development patterns of monocots and eudicots.

Environment cues can initiate seed germination

The seeds of some plant species are capable of germinating as soon as they have matured. All they need for germination is water. But the seeds of many species are dormant at maturity. Seed storage and dormancy may last for weeks, months, years, or even centuries, as we saw in the opening story of Chapter 29: in 2005 a botanist was able to germinate a date palm seed recovered from a 2000-year-old storage bin at Masada in Israel. The mechanisms that maintain seed dormancy are numerous and diverse, but three principal strategies dominate:

- *Exclusion of water or oxygen* from the embryo by an impermeable seed coat
- *Mechanical restraint* of the embryo by a tough seed coat
- *Chemical inhibition* of germination

Seed dormancy must be broken before germination can begin. The dormancy of seeds with impermeable seed coats can be broken if the coat is abraded as the seed tumbles across the ground or through a creek bed or passes through the digestive tract of an animal. Cycles of freezing and thawing can also aid in making the seed coat permeable, as can soil microorganisms. Fire can end seed dormancy by melting waterproof wax in seed coats, allowing water to reach the embryo. Fire can also release mechanical restraint by cracking the seed coat. *Leaching*—the dissolving and diffusing away of water-soluble chemical inhibitors by prolonged exposure to water—is another way in which dormancy can be broken.

---

yourBioPortal.com

GO TO Web Activity 37.1 • Monocot Shoot Development
AND Web Activity 37.2 • Eudicot Shoot Development

37.1 Patterns of Early Shoot Development (A) In grasses and some other monocots, growing shoots are protected by a coleoptile until they reach the soil surface. (B) In most eudicots, the growing point of the shoot is protected within the cotyledons. (C) In some eudicots, the cotyledons remain in the soil, and the apex is protected by the first true leaves.

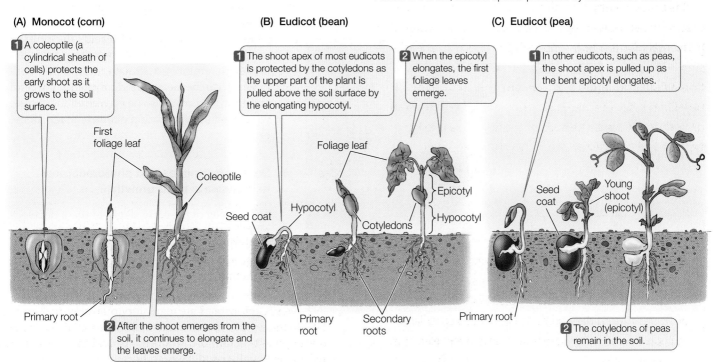

(A) Monocot (corn)

1 A coleoptile (a cylindrical sheath of cells) protects the early shoot as it grows to the soil surface.

First foliage leaf

Coleoptile

Primary root

2 After the shoot emerges from the soil, it continues to elongate and the leaves emerge.

(B) Eudicot (bean)

1 The shoot apex of most eudicots is protected by the cotyledons as the upper part of the plant is pulled above the soil surface by the elongating hypocotyl.

2 When the epicotyl elongates, the first foliage leaves emerge.

Foliage leaf

Epicotyl

Seed coat

Hypocotyl

Cotyledons

Hypocotyl

Primary root

Secondary roots

(C) Eudicot (pea)

1 In other eudicots, such as peas, the shoot apex is pulled up as the bent epicotyl elongates.

Seed coat

Young shoot (epicotyl)

Primary root

2 The cotyledons of peas remain in the soil.

Seed dormancy affords adaptive advantages

What are the potential advantages of seed dormancy? For many plant species, dormancy ensures survival during unfavorable conditions and results in germination when conditions are more favorable for growth. To avoid germination in the dry days of late summer, for example, some seeds require exposure to a long cold period (winter) before they will germinate. Other seeds will not germinate until a certain amount of time has passed, regardless of how they are treated. Among other things, this strategy prevents germination while the seeds are still attached to the parent plant. Plants whose seeds need fire to break dormancy avoid competition with other plants by germinating only where an area has been cleared by fire. Dormancy also helps seeds to survive long-distance dispersal, allowing plants to colonize new territory.

The dormancy of some seeds is broken by exposure to light. These seeds, which germinate only at or near the surface of the soil, are generally tiny and have few food reserves. Seedlings from such seeds would not have enough food to reach the light if they germinated deep in the soil. Conversely, the germination of some other seeds is inhibited by light; these seeds germinate only when deeply buried. Light-inhibited seeds are often large and well-stocked with nutrients.

Dormancy may also increase the likelihood of a seed germinating in a favorable ecological setting. Some cypress trees, for example, grow in standing water, and their seeds germinate only if germination inhibitors are leached away by water (**Figure 37.2**).

Seed germination begins with the uptake of water

Seeds begin to germinate when dormancy is broken and environmental conditions are satisfactory. The first step in germination is the uptake of water, called **imbibition** (from *imbibe*, "to

37.2 Breaking dormancy The seeds of bald cypress, a tree adapted to moist or wet environments, germinate only after being leached by water, which increases the chances that they will germinate in a location suitable for their growth.

drink in"). A dormant seed contains very little water; only 5 to 15 percent of a seed's weight is water compared to 80 to 95 percent for most other plant parts. Seeds also contain polar macromolecules, such as cellulose and starch, that attract and bind polar water molecules. Consequently a seed has a very negative water potential (see Section 35.1), and will take up water if the seed coat is permeable. The force exerted by imbibing seeds, which expand several-fold in volume, demonstrates the magnitude of seed water potential. Imbibing cocklebur seeds can exert a pressure of up to 1,000 atmospheres (approximately 100 kilopascals or 15,000 pounds per square inch)!

As a seed takes up water, it undergoes metabolic changes: enzymes are activated upon hydration, RNA and then proteins are synthesized, the rate of cellular respiration increases, and other metabolic pathways are activated. In many seeds, there is no initiation of the cell division cycle during the early stages of germination. Instead, growth results solely from the expansion of small preformed cells. DNA is synthesized only after the radicle begins to grow and ruptures the seed coat.

The embryo must mobilize its reserves

To fuel its metabolic activities, the embryo must use the reserves of energy and raw materials stored in the seed. Until the young plant is able to photosynthesize, it depends on these reserves, which are stored in the **cotyledons** (see Figure 37.1) or in the **endosperm** (the specialized nutritive tissue) of the seed. The principal reserve of energy and carbon in many seeds, such as wheat, is starch. Other seeds, such as sunflower, store fats or oils. Usually the seed holds amino acid reserves in the form of proteins, rather than as free amino acids.

Before they can be used to support growth of the embryo, starch, lipids, and proteins must be broken down by enzymes into monomers. Starch breakdown yields glucose for energy metabolism and for the synthesis of cellulose and other cell wall constituents. Digestion of stored proteins provides the amino acids the embryo needs to synthesize its own proteins. Lipids are broken down into glycerol and fatty acids, both of which can be metabolized for energy. Glycerol and fatty acids can also be converted to glucose, which permits fat-storing plants to make all the building blocks they need for growth.

Several hormones and photoreceptors help regulate plant growth

This survey of the early stages of plant development illustrates the many internal and external cues that influence plant growth. A plant's responses to these cues are initiated and maintained by two types of regulators: *hormones* and *photoreceptors*.

Hormones are regulatory molecules that act at very low concentrations at sites often distant from where they are produced. Unlike animals, which usually produce each hormone in specific cells within the body, plants produce hormones in many types of cells. Each plant hormone plays multiple regulatory roles, affecting several different aspects of plant development (**Table 37.1**). Interactions among these hormones can be complex. Several hormones regulate plant growth from seedling to adult. Other hormones are involved in the plant's defenses against herbivores and microorganisms (discussed in Chapter 39).

Photoreceptors, like hormones, are involved in many developmental processes in plants. However, unlike plant hormones, which are small molecules, plant photoreceptors consist of *pigments* (molecules that absorb light) associated with proteins. Light acts directly on photoreceptors, which in turn regulate developmental processes that need to be responsive to light, such as the many changes that occur as a young seedling germinates and emerges from the soil.

Signal transduction pathways are involved in all stages of plant development

Plants, like other organisms, make extensive use of *signal transduction pathways*, sequences of biochemical reactions by which a cell generates a response to a stimulus (see Chapter 7). Cell signaling in plant development generally involves a receptor (for a hormone or for light) and a signal transduction pathway, and concludes with a cellular response that is relevant to development. Protein kinase cascades often amplify responses to signals in plants, as they do in other organisms (see Figure 7.12). We will look at several plant signal transduction pathways in more detail in the remaining sections of this chapter.

No matter what cues regulate development, the plant's genome ultimately determines the limits of plant development. The genome encodes the master plan, but its interpretation depends on conditions in the environment. For several decades biologists focused on identifying the hormones and photoreceptors that control plant development, but recent advances in molecular genetics have now made it possible to explore the underlying processes that regulate development, such as signal transduction pathways.

Studies of *Arabidopsis thaliana* have increased our understanding of plant signal transduction

Many recent advances in understanding plant growth and development have come from work with *Arabidopsis thaliana*, a weed in the mustard family. This plant is used as a model organism by researchers because its body and seeds are tiny, its nuclear genome is unusually small for a flowering plant (125 million base pairs), and it flowers and forms many seeds (up to 10,000 per plant) within weeks of germination. Furthermore, its genomes (nuclear, plastid, and mitochondrial) are fully sequenced, so researchers have an accounting of all genes in the plant.

In Chapter 19, we describe how genetics can be used to identify the steps along a developmental pathway. You will recall the theme of these experiments: *if a mutation for a certain biochemical process disrupts a developmental event, then the biochemical process must be essential for that developmental event.* Similarly, genetics can be used to dissect pathways for receptor activation and signal transduction in

TABLE 37.1
Plant Growth Hormones

HORMONE	STRUCTURE	TYPICAL ACTIVITIES
Abscisic acid*		Maintains seed dormancy; closes stomata
Auxins (e.g., indole-3-acetic acid)		Promote stem elongation, adventitious root initiation, and fruit growth; inhibit axillary bud outgrowth, leaf abscission, and root elongation
Brassinosteroids		Promote stem and pollen tube elongation; promote vascular tissue differentiation
Cytokinins		Inhibit leaf senescence; promote cell division and axillary bud outgrowth; affect root growth
Ethylene		Promotes fruit ripening and leaf abscission; inhibits stem elongation and gravitropism
Gibberellins		Promote seed germination, stem growth, and fruit development; break winter dormancy; mobilize nutrient reserves in grass seeds

*See Chapter 38.

plants: if proper signaling does not occur in a mutant strain, the mutant gene must be involved in the signal transduction process. Mapping the mutant gene and characterizing its molecular phenotype is a starting point for understanding the signaling pathway.

One technique for identifying genes involved in a plant signal transduction pathway is illustrated in **Figure 37.3**. Called a **genetic screen**, the process involves creating a collection of mutants and identifying those individuals that are likely to have a defect in the pathway being studied. Genes can be randomly mutated in two ways:

- Insertion of a transposon (see Section 17.2)
- Point mutation by a chemical mutagen, usually ethyl methane sulfonate

In both cases, a large number of mutated plants are then examined for a specific phenotype, usually a characteristic that is easy to see or measure (e.g., height). The growth conditions and plant characteristics used for the screen are carefully chosen to maximize the chances that the selected plants will have a defect in the pathway of interest. Once mutant plants have been selected, their genotypes and phenotypes are compared to those of wild-type plants. *Arabidopsis* mutants with altered developmental patterns have provided a wealth of new information about the hormones present in plants and the mechanisms of hormone and photoreceptor action.

37.1 RECAP

Plant development is under the control of external cues in the environment as well as internal hormones. In both cases, signal transduction pathways regulate plant development. Genetic screens have been useful in describing signal transduction pathways in the model plant *Arabidopsis thaliana*. Seed dormancy often precedes seed germination.

- Describe how monocots and eudicots differ in early development. See p. 773 and Figure 37.1

- Under what circumstances is seed dormancy advantageous? See p. 774

- What fuels the metabolic activities of a young plant embryo before it is able to commence photosynthesis? See p. 774

- What is a genetic screen and how can it be used to analyze the regulation of plant development? See p. 776 and Figure 37.3

You have now seen the early stages of plant development and growth, and how the environment influences these processes. Plant hormones are central to the internal regulation of development, a subject to which we now turn. We will describe how hormones were discovered and what physiological effects they

TOOLS FOR INVESTIGATING LIFE

37.3 A Genetic Screen

Genetics of the model plant *Arabidopsis thaliana* can be used to identify the steps of a signal transduction pathway. If a mutant strain does not respond to a hormone (in this case, ethylene), the corresponding wild-type gene must be essential for the pathway (in this case, ethylene response). This method has been instrumental to scientists in understanding plant growth regulation.

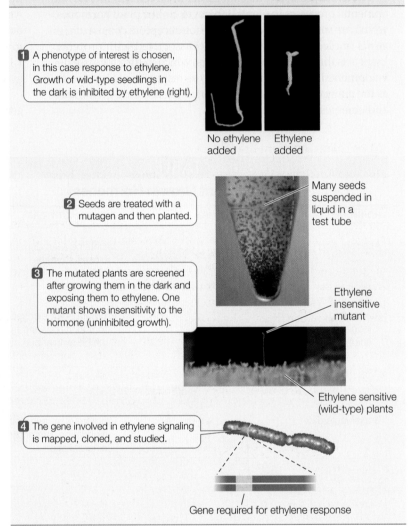

1 A phenotype of interest is chosen, in this case response to ethylene. Growth of wild-type seedlings in the dark is inhibited by ethylene (right).

No ethylene added   Ethylene added

2 Seeds are treated with a mutagen and then planted.

Many seeds suspended in liquid in a test tube

3 The mutated plants are screened after growing them in the dark and exposing them to ethylene. One mutant shows insensitivity to the hormone (uninhibited growth).

Ethylene insensitive mutant

Ethylene sensitive (wild-type) plants

4 The gene involved in ethylene signaling is mapped, cloned, and studied.

Gene required for ethylene response

have on plants. We will emphasize how genetic screens and other methods have led to a deeper molecular understanding of the action of plant hormones.

37.2 What Do Gibberellins Do?

In Asia, rice farmers have known about "foolish seedling disease" (*bakanae* in Japanese) for centuries. Seedlings affected by this disease grow more rapidly than their healthy neighbors, but this rapid growth gives rise to tall, spindly plants that die before producing rice grains, having expended most of their energy on vegetative growth. At first, the disease was attributed to an inherited defect in the plants themselves. But by the twentieth century it was clear that it is caused by infection with the ascomycete fungus *Gibberella fujikuroi*. How does the infection cause the disease?

The mystery of foolish seedling disease was solved in 1925 by Japanese biologist Eiichi Kurosawa. He hypothesized that the fungus must release a molecule that overstimulates plant growth. To isolate it, he grew the fungus in a liquid medium and then removed the fungus from the medium by filtration. He heated the filtered medium to kill any remaining fungus, but found that the resulting heat-treated filtrate was still capable of inducing rapid growth in rice seedlings. Medium that had never contained the fungus did not have this effect. This experiment established that *G. fujikuroi* produces a growth-promoting chemical substance, which Kurosawa called a **gibberellin**. Soon, gibberellin was isolated and its chemical nature described (see Table 37.1).

Gibberellins are plant hormones

Once externally applied gibberellin was shown to affect rice plant growth, a question arose: *does a plant make the same or similar molecules to regulate its own growth?* Biologists used a genetic approach to answer this question, by studying mutant strains of corn and tomato that were dwarfs—they had abnormally short stems. The stems hardly grew, even though other parts of the plants appeared normal. Because this was the exact opposite of the effect of *too much* gibberellin (in the fungus-infected rice plants), the biologists hypothesized that the dwarf mutants had *too little* gibberellin. This hypothesis was tested in two ways:

- Gibberellin was applied to the dwarf plants. As a result, they grew to normal height (**Figure 37.4**).
- The levels of gibberellin were measured in wild-type and dwarf plants, and the wild-type plants had much more gibberellin.

These experiments clearly showed that gibberellin is made by plants and acts to stimulate stem elongation. Numerous chemically related gibberellins exist, all belonging to a family of common plant metabolites called *diterpenoids*.

Gibberellins have many effects on plant growth and development

The functions of gibberellins can be inferred from the effects of experimentally decreasing gibberellins or blocking their action at various points in plant development. Such experiments reveal that gibberellins have multiple roles in regulating plant growth.

FRUIT GROWTH Gibberellins and other hormones regulate the growth of fruits. Grapevines that produce seedless grapes develop smaller fruit than varieties that produce seed-bearing grapes. Biologists wanting to explain this phenomenon removed seeds from immature seeded grapes and found that this prevented normal fruit growth, suggesting that the seeds are sources of a growth regulator. Biochemical studies showed that developing seeds produce gibberellins, which diffuse out into the immature fruit tissue. Spraying young seedless grapes with a gibberellin solution causes them to grow as large as seeded ones, and this is now a standard commercial practice (**Figure 37.5**).

MOBILIZATION OF SEED RESERVES Early in seed germination hydrolytic enzymes are produced to break down stored reserves of starch, proteins, and lipids. Just after imbibition in germinating seeds of barley and other cereals, the embryo secretes gibberellins. The hormones diffuse through the endosperm to a surrounding tissue called the **aleurone layer**, which lies underneath the seed coat. The gibberellins trigger a cascade of events in the aleurone layer, causing it to syn-

22 days after being sprayed with a dilute gibberellin solution, this plant reached the size of a nondwarf plant.

This untreated control plant remained a dwarf.

37.4 The Effect of Gibberellins on Dwarf Plants Both of the dwarf tomato plants in this photograph were the same size when the one on the right was treated with gibberellins.

37.5 Gibberellin and Fruit Growth Spraying developing seedless grapes with gibberellins (right) increases their size compared to untreated fruit (left).

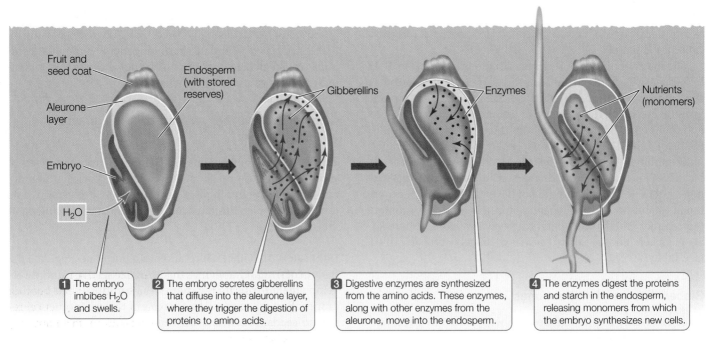

1. The embryo imbibes H_2O and swells.

2. The embryo secretes gibberellins that diffuse into the aleurone layer, where they trigger the digestion of proteins to amino acids.

3. Digestive enzymes are synthesized from the amino acids. These enzymes, along with other enzymes from the aleurone, move into the endosperm.

4. The enzymes digest the proteins and starch in the endosperm, releasing monomers from which the embryo synthesizes new cells.

37.6 Embryos Mobilize Their Reserves During seed germination in cereal grasses, gibberellins trigger a cascade of events that result in the conversion of starch and protein reserves into monomers that can be used by the developing embryo.

—— **yourBioPortal.com** ——
GO TO Web Activity 37.3 • Events of Seed Germination

thesize and secrete enzymes that digest proteins and starch stored in the endosperm (**Figure 37.6**). These observations have practical importance: in the beer brewing industry, gibberellins are used to enhance the "malting" (germination) of barley and the breakdown of its endosperm, producing sugar that is fermented to alcohol.

STEM ELONGATION The effects of gibberellins on wild-type plants are not as dramatic as those seen on dwarf plants. However, gibberellins are indeed active in wild-type plants, because inhibitors of gibberellin synthesis cause a reduction in stem elongation. Such inhibitors can be put to practical uses. For example, plants such as chrysanthemums that are grown in greenhouses tend to get too tall; but leggy plants, unfortunately, do not appeal to consumers. Flower growers thus spray such plants with gibberellin synthesis inhibitors to control their height. Some wheat crops are similarly sprayed to keep them short, so they do not fall over when they produce grain—this is essentially a chemically induced version of the semi-dwarf genetic varieties described in the opening story of this chapter. In some plants, such as cabbage, the normal growth habit is to be a squat, leafy head near the ground. When environmental signals are right, however, the plant "bolts," quickly producing a tall stem with flowers. This response is mediated by gibberellins.

Gibberellins act by initiating the breakdown of transcriptional repressors

The molecular mechanisms underlying gibberellin action have been worked out with the help of genetic screens. Biologists started by identifying mutant plants whose growth and devel-

opment are *insensitive* to gibberellins; that is, they are *not* affected by added gibberellins. Several such mutants have been found—both natural mutant strains and induced mutants selected from genetic screens—and they fall into two general categories:

- *Excessively tall plants.* These plants resemble wild-type plants given an excess of gibberellin, and get no taller when given extra gibberellin. They are also tall even when treated with inhibitors of gibberellin synthesis. Their gibberellin response is always "on," even in the absence of the hormone. It is presumed that the normal allele for the mutant gene codes for an *inhibitor* of the gibberellin signal transduction pathway. In wild-type plants, the pathway is "off" but in the mutant plants, the pathway is "on" and the plant grows tall.

- *Dwarf plants.* These plants resemble dwarf tomato or maize plants that are deficient in gibberellin synthesis, but they do not respond to added gibberellin. In these mutants the gibberellin response is always "off," regardless of the presence of the hormone.

The two types of mutations described above *affect the same protein,* which turns out to be a *repressor* of a transcription factor that stimulates the expression of growth-promoting genes. The repressor protein has two important domains, explaining how mutations in the same protein can have seemingly opposite effects:

- *One region of the repressor protein binds to the transcription complex to inhibit transcription.* This is the region mutated in the excessively tall plants: the growth-promoting genes are always "on" because the repressor does not bind to the transcription complex.

- *Another part of the repressor protein causes it to be removed from the transcription complex.* This is the region mutated in the

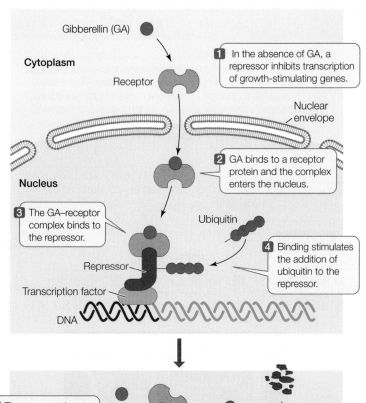

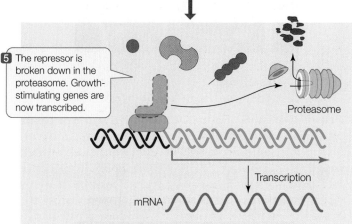

37.7 How Gibberellin Works Gibberellin acts to stimulate gene transcription by inactivating a repressor protein, a common mechanism for plant hormone action.

dwarf plants: the growth-promoting genes are always "off" because the repressor is always bound to the complex.

How does gibberellin act in this system? *Gibberellin acts by removing the repressor from the transcription complex* (**Figure 37.7**). It does this by binding to a receptor protein, which in turn binds to the repressor. Binding of the gibberellin–receptor complex stimulates poly-ubiquitination of the repressor, targeting it for breakdown in the proteasome (see Figure 16.24). The gibberellin receptor contains a region called an **F-box** that facilitates protein–protein interactions necessary for protein breakdown. While animal genomes have few F-box containing proteins, plant genomes have hundreds, an indication that this type of gene regulation is common in plants. As you will see, this regulatory mechanism underlies the effects of another important plant hormone: auxin.

37.2 RECAP

Gibberellins are plant hormones that affect stem growth, fruit size, seed germination, and many other aspects of plant development; the effects vary from species to species.

- How were gibberellins shown to be plant hormones? **See p. 777**

- How do gibberellins contribute to the germination of barley seeds? **See Figure 37.6**

- Explain how gibberellins work at the molecular level. **See pp. 778–779 and Figure 37.7**

Most other hormones, like the gibberellins, have multiple effects within the plant, and they often interact with one another to regulate developmental processes. In controlling stem elongation, for example, gibberellins interact with another hormone, auxin, to which we now turn.

37.3 What Does Auxin Do?

Auxin was discovered as a result of investigations into **phototropism**, the growth of plants toward light (as in shoots) or away from it (as in roots). Ever the curious biologist, Charles Darwin wanted to know how plants bend toward a source of light. In 1880, he and his son Francis published the results of their experiments on what part of a plant senses light.

---
yourBioPortal.com
GO TO Animated Tutorial 37.1 • Tropisms

The Darwins worked with canary grass (*Phalaris canariensis*) seedlings grown in the dark. A young grass seedling has a **coleoptile**—a cylindrical sheath a few cells thick that protects the delicate shoot as it pushes through the soil (see Figure 37.1A). When the seedling breaks through the soil surface, the coleoptile soon stops growing, and the shoot emerges unharmed. The coleoptiles of grasses are phototropic—they grow toward the light.

To find the light-receptive region of the coleoptile, the Darwins "blindfolded" the coleoptiles of dark-grown canary grass seedlings in various places, and then illuminated them from one side (**Figure 37.8**). The coleoptile grew toward the light whenever its tip was exposed. If the top millimeter or more of the coleoptile was covered, however, the coleoptile showed no phototropic response. Thus, the Darwins were able to conclude that the tip contains the photoreceptor that responds to light. The actual bending toward the light, however, takes place in a growing region a few millimeters below the tip. Therefore, the Darwins reasoned, *some type of signal must travel from the tip of the coleoptile to the growing region.* Later, others demonstrated that this signal is a chemical substance by showing that it can move through certain permeable materials, such as gelatin, but not through impermeable materials, such as a metal sheet.

INVESTIGATING LIFE

37.8 The Darwins' Phototropism Experiment

Charles Darwin and his son Francis wanted to know how plants bend toward the light. They grew canary grass seedlings (coleoptiles) in the dark. To discover what part of the coleoptile responds to light, they covered up ("blindfolded") different regions of each coleoptile and then exposed the seedlings to light from one side. The Darwins discovered that the tip of the seedling senses the light and that growth occurs below the tip. Their observations led them to hypothesize the existence of a growth-promoting signal produced by the coleoptile tip.

HYPOTHESIS Only part of the coleoptile senses the light that triggers phototropism.

METHOD

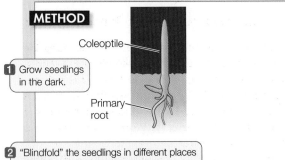

1 Grow seedlings in the dark.

Coleoptile

Primary root

2 "Blindfold" the seedlings in different places and expose to light on one side.

Light

Blindfold

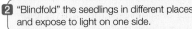

RESULTS

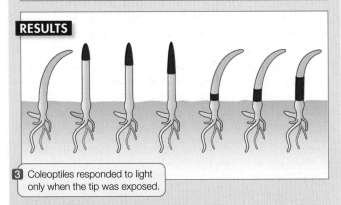

3 Coleoptiles responded to light only when the tip was exposed.

CONCLUSION The part of the coleoptile that senses light is in the tip, and it sends a signal from the tip to the growing region.

INVESTIGATING LIFE

37.9 Went's Experiment

Previous experiments had indicated that the tip of a coleoptile produces a growth-inducing substance. Went verified this conclusion by placing agar blocks containing the substance contained within coleoptile tips on one side of a decapitated coleoptile. In the absence of light, the coleoptile bent away from the side with the substance. The substance was later identified as auxin.

HYPOTHESIS A growth hormone can be isolated from a coleoptile tip.

METHOD **RESULTS**

Experiment

1 Remove the coleoptile tip and place on agar.

2 Place the agar on one edge of another decapitated coleoptile.

3 The coleoptile curves away from the agar as it grows.

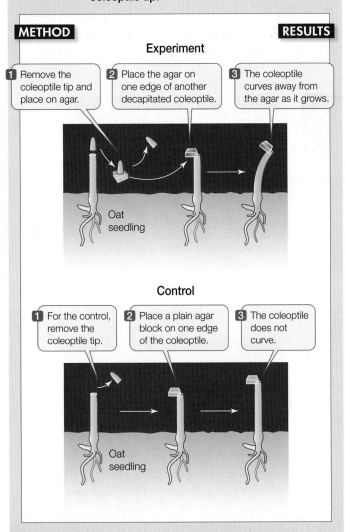

Oat seedling

Control

1 For the control, remove the coleoptile tip.

2 Place a plain agar block on one edge of the coleoptile.

3 The coleoptile does not curve.

Oat seedling

CONCLUSION A growth hormone diffused from the tip into the agar, and from the agar into another plant. It had an effect on the growth of the plant similar to that of a coleoptile tip.

In the 1920s, the Dutch botanist Frits Went followed up on the Darwins' experiment. He removed coleoptile tips and placed their cut surfaces on a block of agar. Then he placed pieces of that agar on decapitated coleoptiles—positioned to cover only one side (**Figure 37.9**). As they grew, the coleoptiles curved away from the side with the agar, showing that the agar contained a substance that stimulated elongation of cells on that side of the coleoptile. This substance had diffused into the agar block from the isolated coleoptile tips. Eventually, the hormone indole-3-acetic acid (see Table 37.1) was isolated from similar agar blocks and was nicknamed *auxin* (from the Latin "to increase").

Auxin transport is polar and requires carrier proteins

The experiments we have just described showed that in coleoptiles auxin movement is strictly *polar*—that is, it is unidirectional along a line from apex to base. Auxin transport is polar in other organs as well. For example, in a leaf petiole, which connects the leaf blade to the stem, auxin moves from the leaf blade end toward the stem. In roots, auxin moves unidirectionally toward the root tip. How does this directional transport occur?

Polar transport depends on four biochemical conditions that should be familiar from earlier chapters (**Figure 37.10**):

- *Diffusion across a plasma membrane.* Polar molecules diffuse across the plasma membrane less readily than nonpolar molecules.
- *Membrane protein asymmetry.* Active transport carriers for auxin are located only at the basal (bottom) end of the plasma membrane.

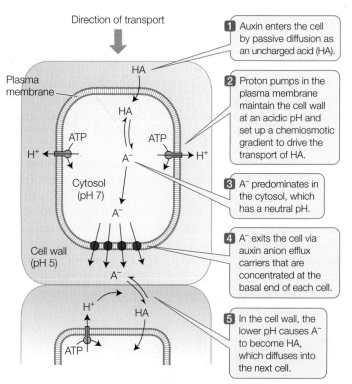

37.10 Polar Transport of Auxin Proton pumps set up a chemiosmotic gradient directing A⁻ toward the basally placed auxin active transport carriers, leading to a net movement of auxin in a basal direction.

- *Proton pumping/chemiosmosis.* An ATP-driven proton pump removes H⁺ from the cell, thereby increasing the intracellular pH and decreasing the pH in the cell wall. Proton pumping also sets up an electrochemical gradient, with potential energy to drive the transport of auxin by the carriers noted above.
- *The ionization of a weak acid.* Indole-3-acetic acid (which, recall, is the chemical name for auxin; see Table 37.1) is a weak acid:

$$A^- + H^+ \rightleftharpoons HA$$

When the pH is low, the increased H⁺ concentration drives this reaction to the right and HA (non-ionized auxin) is the predominant form. When the pH is higher, there is more A⁻ (ionized auxin).

Auxin transport mediates responses to light and gravity

While polar auxin transport distributes the hormone along the longitudinal axis of the plant, *lateral* (side-to-side) redistribution of auxin is responsible for plant movements. This redistribution is carried out by auxin carrier proteins that move from the base of the cell to one side; because of this, auxin exits the cell only on that side of the cell, rather than at the base, and moves sideways within the tissue.

When light strikes a grass coleoptile on one side, auxin at the tip moves laterally toward the shaded side. The asymmetry thus established is maintained as polar transport moves auxin down the coleoptile, so that in the growing region below, the auxin concentration is highest on the shaded side. Cell elongation is thus speeded up on that side, causing the coleoptile to bend toward the light (**phototropism; Figure 37.11A**). If you have noticed a houseplant bending toward a window, you have observed phototropism.

Light is not the only signal that can cause redistribution of auxin. Auxin moves to the lower side of a shoot that has been tipped sideways, causing more rapid growth in the lower side and, hence, an upward bending of the shoot. Such growth in a direction determined by gravity is called **gravitropism (Figure 37.11B)**. The upward gravitropic response of shoots is defined as *negative gravitropism;* that of roots, which bend downward, is *positive gravitropism.*

How does a plant cell sense light and gravity and respond with an asymmetric distribution of auxin? Different mechanisms have been proposed:

- *The phototropic response.* As you will see later in the chapter, plants have a membrane receptor called phototropin that absorbs blue light. This receptor was discovered in a genetic screen in *Arabidopsis* for mutant plants that failed to bend toward light. When the receptor is activated, a signal transduction pathway results in redistribution of auxin transport carriers so that the hormone is transported to the cells on the shaded side. This results in bending toward light (see Figure 37.11).
- *The gravitropic response.* Some types of plant cells contain starch that is stored in large plastids called amyloplasts. These plastids tend to settle on the downward side of a cell

The figure (37.10) contains the following labels and callouts:

Direction of transport

Plasma membrane

HA

HA

ATP   ATP

H⁺   H⁺

A⁻

Cytosol (pH 7)

A⁻

Cell wall (pH 5)

A⁻

H⁺   HA

ATP

1 Auxin enters the cell by passive diffusion as an uncharged acid (HA).

2 Proton pumps in the plasma membrane maintain the cell wall at an acidic pH and set up a chemiosmotic gradient to drive the transport of HA.

3 A⁻ predominates in the cytosol, which has a neutral pH.

4 A⁻ exits the cell via auxin anion efflux carriers that are concentrated at the basal end of each cell.

5 In the cell wall, the lower pH causes A⁻ to become HA, which diffuses into the next cell.

(A) Phototropism

1. Auxin moves to the shaded side within the tip.

2. The redistributed auxin moves down the coleoptile.

3. A higher auxin concentration causes more rapid growth on the shaded side. The tip curves toward the light.

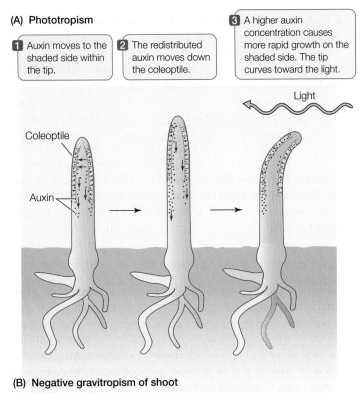

Light

Coleoptile

Auxin

(B) Negative gravitropism of shoot

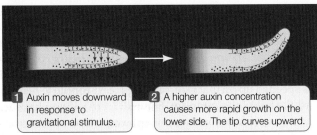

1. Auxin moves downward in response to gravitational stimulus.

2. A higher auxin concentration causes more rapid growth on the lower side. The tip curves upward.

37.11 Plants Respond to Light and Gravity (A) Phototropism and (B) gravitropism occur in shoot apices in response to a redistribution of auxin.

in response to gravity. How gravity-induced plastid movement is sensed is not well understood, but it may be through disturbance of endoplasmic reticulum membranes on the downward side of the cell. This in turn triggers auxin transport to the bottom side of the root or shoot, which causes bending in the appropriate direction.

Auxin affects plant growth in several ways

Like the gibberellins, auxin has many roles in plant development. It affects the vegetative and reproductive growth of plants in a number of ways.

ROOT INITIATION Cuttings from the shoots of some plants can produce roots and develop into entire new plants. For this to occur, certain undifferentiated cells in the interior of the shoot, originally destined to function only in food storage, must set off on a new mission: they must change their cell fate and become organized into the apical meristem of a new root. These changes are similar to those that take place in the pericycle of a root when

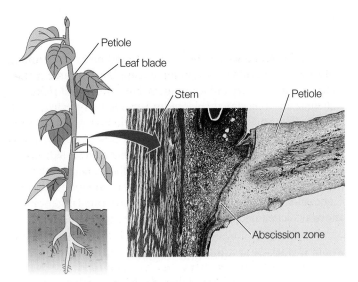

Petiole

Leaf blade

Stem

Petiole

Abscission zone

37.12 Changes Occur when a Leaf Is About to Fall The breakdown of cells in the abscission zone of the petiole causes the leaf to fall.

a lateral root forms (see Figure 34.13A). Shoot cuttings of many species can be made to develop roots by dipping the cut surfaces into an auxin solution. These observations suggest that in an intact plant the plant's own auxin plays a role in the initiation of lateral roots. Commercial preparations that enhance the rooting of plant cuttings typically contain synthetic auxins.

LEAF ABSCISSION In contrast to its stimulatory effect on root initiation, auxin inhibits the detachment of old leaves from stems. This detachment process, called **abscission**, is the cause of autumn leaf fall. Most leaves consist of a blade and a petiole that attaches the blade to the stem. Abscission results from the breakdown of a specific part of the petiole, the *abscission zone* (**Figure 37.12**). If the blade of a leaf is cut off, the petiole falls from the plant more rapidly than if the leaf had remained intact. If the cut surface is treated with an auxin solution, however, the petiole remains attached to the plant, often longer than an intact leaf would have. The timing of leaf abscission in nature appears to be determined in part by a decrease in the movement through the petiole of auxin produced in the blade.

APICAL DOMINANCE Auxin helps maintain **apical dominance**, a phenomenon in which apical buds inhibit the growth of axillary buds (see Figure 34.1), resulting in the growth of a single main stem with minimal branching. A diffusion gradient of auxin from the apical tip of the shoot down the stem results in lower branches receiving less auxin and therefore branching more. The effect of the auxin gradient is apparent in conifers: next time you see a decorated tree during the winter holidays, think of auxin and apical dominance.

High

Auxin

Low

Apical dominance can be demonstrated by an experiment with a young seedling. If the plant remains intact, the stem elongates and the axillary buds remain inactive. Removal of the apical bud—the major site of auxin production—results in growth of the axillary buds. If the cut surface of the stem is treated with auxin, however, the axillary buds do not grow. The apical buds of branches also exert apical dominance: the axillary buds on the branch are inactive unless the apex of the branch is removed. That is why gardeners prune shrubs to encourage branching.

In the experiments on leaves and stems just discussed, removal of a particular part of the plant elicits a response—abscission or loss of apical dominance—and that response is prevented by treatment with auxin. These results are consistent with other data showing that the excised part of the leaf or stem is an auxin source and that auxin in the intact plant delays the abscission of leaves and helps maintain apical dominance.

FRUIT DEVELOPMENT Fruit development normally depends on prior fertilization of the ovule (egg), but in many species treatment of an unfertilized ovary with auxin or gibberellins causes **parthenocarpy**—fruit formation without fertilization. Parthenocarpic fruits form spontaneously in some cultivated varieties of plants, including seedless grapes, bananas, and some cucumbers.

CELL EXPANSION The expansion of plant cells is what causes plant growth. Because the plant cell wall normally prevents expansion of the protoplast (see Section 35.1), the cell wall plays a key role in controlling the rate and direction of plant cell growth. Auxin acts on cell walls to regulate this process.

The expansion of a plant cell is driven primarily by the uptake of water, which enters the cytoplasm of the cell and accumulates in its central vacuole (see Section 35.1). Growth of the vacuole accounts for most of the increase in volume of a growing cell, and the vacuole often makes up more than 90 percent of the volume of a mature cell. As the vacuole expands, it presses the cytoplasm against the cell wall, and the wall resists this force (the basis of turgor pressure). The cell wall is an extensively cross-linked network of polysaccharides and proteins, dominated by cellulose fibrils (see Figure 34.5). If the cell is to expand, some adjustments must be made in the wall structure to allow the wall to "give" under turgor pressure. Think of a balloon (the cell surrounded by a membrane) inside a box (the cell wall). How does the cell wall "box" loosen to allow expansion?

The **acid growth hypothesis** offers a possible explanation for auxin-induced cell expansion (**Figure 37.13**). The hypothesis holds that protons (H^+) are pumped from the cytoplasm into the cell wall, lowering the pH of the wall and activating enzymes called *expansins* that catalyze changes in the cell wall structure such that the polysaccharides adhere to each other less strongly. This loosens the cell wall, making it easier to stretch as the cell expands. Auxin is believed to have two roles in this process: to increase the synthesis of the proton pumps, and to guide their insertion into the plasma membrane. Auxin may also increase the activity of proton pump proteins already in the plasma membrane. Several lines of evidence support the acid growth hypothesis. For example, adding acid to the cell wall to lower the pH stimulates cell expansion even in the absence of auxin. Conversely, when a buffer is used to prevent the wall from becoming more acidic, auxin-induced cell expansion is blocked. The model works more or less well depending on species; in some plants auxin stimulates secretion of new cell wall components quickly enough to account for even rapid changes in growth rate.

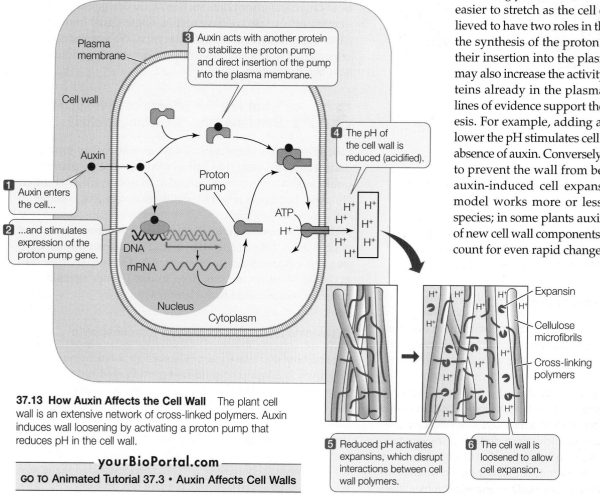

37.13 How Auxin Affects the Cell Wall The plant cell wall is an extensive network of cross-linked polymers. Auxin induces wall loosening by activating a proton pump that reduces pH in the cell wall.

Plasma membrane

Cell wall

Auxin

1 Auxin enters the cell...

2 ...and stimulates expression of the proton pump gene.

DNA

mRNA

Nucleus

Cytoplasm

3 Auxin acts with another protein to stabilize the proton pump and direct insertion of the pump into the plasma membrane.

Proton pump

ATP

H^+

4 The pH of the cell wall is reduced (acidified).

H^+ H^+
H^+ H^+
H^+ H^+
H^+
H^+

Expansin

Cellulose microfibrils

Cross-linking polymers

5 Reduced pH activates expansins, which disrupt interactions between cell wall polymers.

6 The cell wall is loosened to allow cell expansion.

yourBioPortal.com

GO TO Animated Tutorial 37.3 • Auxin Affects Cell Walls

(A) Repression: Auxin absent

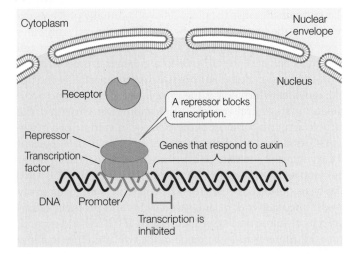

(B) Activation: Auxin present

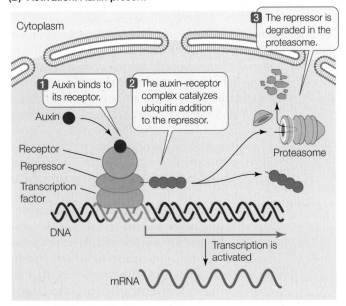

37.14 Signal Transduction Pathway for Auxin As with gibberellins (see Figure 37.7), the auxin response involves the release from inhibition of transcription.

At the molecular level, auxin and gibberellins act similarly

Given that auxin induces so many different physiological responses in plants, biologists expected that the hormone might act through several different signal transduction pathways. However, genetic screens have revealed a single, relatively simple mechanism that accounts for many of auxin's actions. This mechanism is similar to that involved in the action of gibberellins (see Figure 37.7).

Like gibberellin, auxin acts by countering the inhibition of genes involved in the cellular response to the hormone. In the absence of auxin, a repressor blocks transcription of these genes. Auxin binds to a receptor, which then binds to the repressor. This stimulates addition of ubiquitin to the repressor and causes it to be degraded by the proteasome (**Figure 37.14**), thereby allow-

ing transcription. The molecular response takes longer, and is longer-lasting than the rapid acid growth response in the cell wall.

37.3 RECAP

Auxin regulates stem elongation (cell expansion) and mediates phototropism and gravitropism; it also plays roles in apical dominance, leaf abscission, and root initiation. The acid growth hypothesis explains auxin-induced cell wall loosening. Similar molecular mechanisms explain the effects of auxin and gibberellin on gene expression.

- What is the evidence for polar transport of auxin and how does it occur? See pp. 779–780 and Figures 37.9 and 37.10

- Explain why, even though auxin moves *away* from the lighted side of a coleoptile tip, the coleoptile bends *toward* the light. See p. 781 and Figure 37.11

- How does auxin cause cell wall loosening? See p. 783 and Figure 37.13

- What are the similarities between the signal transduction pathways for auxin and gibberellin? See p. 784 and Figures 37.7 and 37.14

How can a single hormone, such as auxin or a gibberellin, have so many effects? As we have seen, a single signal transduction pathway may affect more than one gene. We learn about other important plant hormones in the next section, and they, too, have multiple effects.

37.4 What Are the Effects of Cytokinins, Ethylene, and Brassinosteroids?

Like animal cells, plant cells differentiate after they form from undifferentiated stem cells (called meristem cells in plants). But unlike animal cells, which generally do not divide after differentiation, plant cells retain the ability to divide. For example, in leaf abscission (see Figure 37.12) differentiated parenchyma cells in the petiole resume division, forming a specialized, weak layer of cells. Also, cells of the phloem and cortex can resume division and form secondary meristems. What stimulates these cells to divide? An answer came from studies of cells isolated from the plant and cultured in the laboratory.

Cytokinins are active from seed to senescence

Like bacteria and yeast, plant cells such as parenchyma cells can be grown in a liquid or solidified growth medium containing sugars and salts. The cells will divide continuously until they run out of nutrients. In the early days of plant cell culturing, scientists experimented with many supplements to determine the optimal chemical environment for growth. The best supplement was coconut milk, the fluid that surrounds the developing embryo in coconut fruit. Investigators suspected that a molecule in the fluid must stimulate plant cell division.

A clue to the identity of the molecule came when Folke Skoog at the University of Wisconsin tested various pure substances that might substitute for coconut milk. DNA was among the substances tested, and it did not work; however, heating DNA at high pressure in an autoclave produced a mixture that strongly promoted plant cell division. A derivative of adenine called *kinetin* was identified as the active ingredient. Because it stimulated cell division (cytokinesis), it was called a **cytokinin**.

Kinetin does not exist in cells, but it gave scientists a hint as to what type of molecule might be the active ingredient in coconut milk. In 1963, an adenine derivative called **zeatin** was extracted from corn endosperm, the "coconut milk of corn" (see Table 37.1). Since then, over 150 different cytokinins have been isolated, and most are derivatives of adenine.

Cytokinins have a number of different effects, in many cases interacting with auxin:

- Adding an appropriate combination of auxin and cytokinins to a growth medium induces rapid proliferation of cultured plant cells.
- Cytokinins can cause certain light-requiring seeds to germinate even when kept in constant darkness.
- In cell cultures, a high cytokinin-to-auxin ratio promotes the formation of shoots; a low ratio promotes the formation of roots.
- Cytokinins usually inhibit the elongation of stems, but they cause lateral swelling of stems and roots (the fleshy roots of radishes are an extreme example).
- Cytokinins stimulate axillary buds to grow into branches; the auxin-to-cytokinin ratio controls the extent of branching (bushiness) of a plant.
- Cytokinins delay the senescence of leaves. If leaf blades are detached from a plant and floated on water or a nutrient solution, they quickly turn yellow and show other signs of senescence. If instead they are floated on a solution containing a cytokinin, they remain green and senesce much more slowly. Roots contain abundant cytokinins, and cytokinin transport to the leaves delays senescence.

Cytokinin signaling appears to act through a pathway that includes proteins with amino acid sequences similar to proteins in *two-component systems* in bacteria (see Figure 7.3). Indeed, this system was one of the first of its kind discovered in eukaryotes. The two components in such a system are:

- A *receptor* that can act as a protein kinase, phosphorylating itself as well as a target protein
- A *target protein*, generally a transcription factor, that can act as an *effector*

Genetic screens in *Arabidopsis* for abnormalities in the response to cytokinin have identified the receptor (AHK; *Arabidopsis histidine kinase*) and target effector (ARR; *Arabidopsis response regulator*), the latter acting as a transcription factor when phosphorylated. The cytokinin signal transduction pathway also includes a third protein (AHP; *Arabidopsis histidine phosphotransfer protein*), which transfers phosphates from the receptor to the effector (**Figure 37.15**). The plant genome has over 20 genes that are expressed in response to this signaling pathway.

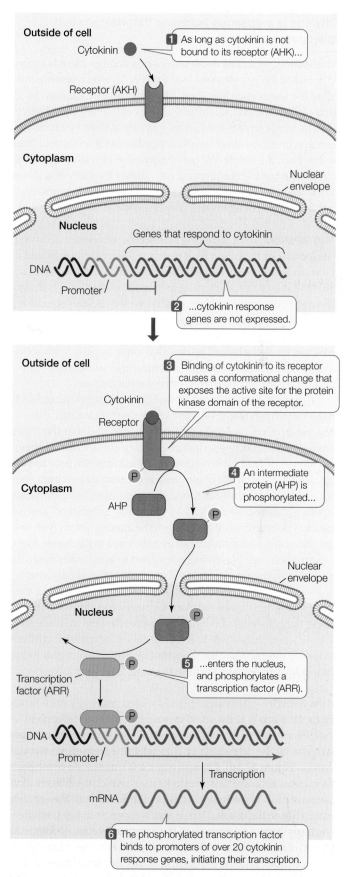

37.15 The Cytokinin Response Pathway Plant cells respond to cytokinins using a signal transduction pathway related to bacterial two-component systems.

Ethylene is a gaseous hormone that hastens leaf senescence and fruit ripening

Whereas the cytokinins delay senescence, another plant hormone promotes it: the gas **ethylene** (see Table 37.1), which is sometimes called the senescence hormone. Ethylene can be produced by all parts of the plant, and, like all plant hormones, it has several effects.

Back when streets were lit by gas rather than by electricity, leaves on trees near street lamps dropped earlier than those on trees farther from the lamps. We now know why: ethylene, a combustion product of the illuminating gas, caused the early abscission. While auxin delays leaf abscission, ethylene strongly promotes it; thus the balance of auxin and ethylene controls abscission.

FRUIT RIPENING By promoting senescence, ethylene also speeds the ripening of fruit. As a fruit ripens, it loses chlorophyll and its cell walls break down; ethylene promotes both of these processes. Ethylene also causes an increase in its own production. Thus, once ripening begins, more and more ethylene forms, and because it is a gas, it diffuses readily throughout the fruit and even to neighboring fruits on the same or other plants. The old saying "one rotten apple spoils the barrel" is true. That rotten apple is a rich source of ethylene, which speeds the ripening and subsequent rotting of the other fruit in a barrel or other confined space.

Farmers in ancient times poked holes in developing figs to make them ripen faster. We now know that wounding causes an increase in ethylene production by the fruit and that the raised ethylene level promotes ripening. Today commercial shippers and storers of fruit hasten ripening by adding ethylene to storage chambers. This use of ethylene is the single most important use of a natural plant hormone in agriculture and commerce. Ripening can also be delayed by the use of "scrubbers" and adsorbents that remove ethylene from the atmosphere in fruit storage chambers. This strategy can even be used in the home. Many supermarkets sell plastic bags designed to keep fruits fresh; the bags are impregnated with a substance that binds ethylene.

As flowers senesce, their petals may abscise, decreasing their value in the cut-flower industry. Growers and florists often immerse the cut stems of ethylene-sensitive flowers in dilute solutions of silver thiosulfate before sale. Silver salts inhibit ethylene action by interacting directly with the ethylene receptor—thus they delay senescence, keeping flowers "fresh" for longer.

STEM GROWTH Although it is associated primarily with senescence, ethylene is active at other stages of plant development, as well. The stems of many eudicot seedlings form an **apical hook** that protects the delicate shoot apex while the stem grows through the soil (**Figure 37.16**). The apical hook is maintained through an asymmetrical production of ethylene gas, which inhibits the elongation of cells on the inner surface of the hook. Once the seedling breaks through the soil surface and is exposed to light, ethylene synthesis stops, and the cells of the inner surface are no longer inhibited. These cells now elongate, and the hook unfolds, raising the shoot apex and the expanding leaves into the sun.

Ethylene also inhibits stem elongation in general, promotes lateral swelling of stems (as do the cytokinins), and decreases the sensitivity of stems to gravitropic stimulation. Together,

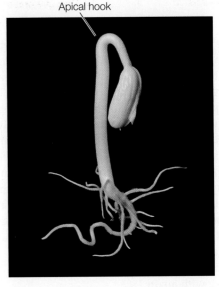

Apical hook

37.16 The Apical Hook of a Eudicot Asymmetrical production of ethylene is responsible for the apical hook of this bean seedling. The ethylene concentration was highest on the right side, so more rapid growth on the left caused and maintained the hook.

these three phenomena constitute the *triple response*, a well-characterized stunted growth habit observed when plants are treated with ethylene.

THE ETHYLENE SIGNAL TRANSDUCTION PATHWAY The mechanism of ethylene action has been worked out by analyzing *Arabidopsis* mutants that have ethylene-related defects. Some of these mutants do not respond to applied ethylene, and others act as if they have been exposed to ethylene even though they have not. Researchers studied the mutant genes and compared their protein products to other known proteins; thus they worked out some of the details of the signal transduction pathway through which ethylene acts (**Figure 37.17**).

The pathway includes two membrane proteins in the endoplasmic reticulum. The first is an ethylene receptor (labeled A in the figure) and the second is a channel protein (C). In the absence of ethylene, a protein kinase (B) keeps C inactive by phosphorylation. When receptor A binds ethylene it inactivates B. Without B to inactivate it, C activates a transcription factor (D), which then moves into the nucleus, where it turns on the genes that produce ethylene's effects in the cell. In other words, ethylene turns off the "off" signal.

Brassinosteroids are plant steroid hormones

In animals, steroid hormones such as cortisol and estrogen are formed from cholesterol (see Figure 3.22). Animal steroids are widespread and have been well studied for many decades. In contrast, plant steroid hormones are a relatively recent discovery. In the 1970s, biologists isolated a steroid (see Table 37.1) from the pollen of rape, a member of the Brassicaceae (mustard family). When applied to various plant tissues, this **brassinosteroid** stimulated cell elongation, pollen tube elongation, and vascular tissue differentiation, but it inhibited root elongation. Since then, dozens of chemically related, growth-affecting brassinosteroids have been found in plants.

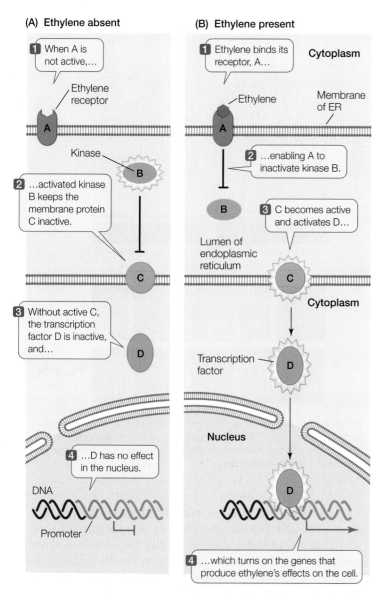

(A) Ethylene absent

1 When A is not active,...

Ethylene receptor

A

Kinase

B

2 ...activated kinase B keeps the membrane protein C inactive.

C

3 Without active C, the transcription factor D is inactive, and...

D

4 ...D has no effect in the nucleus.

DNA

Promoter

(B) Ethylene present

1 Ethylene binds its receptor, A...

Cytoplasm

Ethylene

Membrane of ER

A

2 ...enabling A to inactivate kinase B.

B

3 C becomes active and activates D...

Lumen of endoplasmic reticulum

C

Cytoplasm

Transcription factor

D

Nucleus

D

4 ...which turns on the genes that produce ethylene's effects on the cell.

37.17 The Signal Transduction Pathway for Ethylene This diagram shows the roles of four proteins (A, B, C, and D) in the signal transduction pathway through which ethylene exerts its many effects.

Mutant plants that either do not make brassinosteroids or have defects in brassinosteroid reception and signal transduction are usually dwarf, infertile, and slow to develop. These effects can be reversed by adding small amounts of brassinosteroids, indicating that brassinosteroids are true hormones. These hormones have diverse effects, which vary among plants. Brassinosteroids can:

- enhance cell elongation and cell division in shoots
- promote xylem differentiation
- promote growth of pollen tubes during reproduction
- promote seed germination
- promote apical dominance and leaf senescence

The signaling pathway for these plant steroids differs sharply from those for steroid hormones in animals. In animals, steroids diffuse through the plasma membrane and bind to receptors in the cytoplasm. In contrast, the receptor for brassinosteroids is an integral protein in the plasma membrane

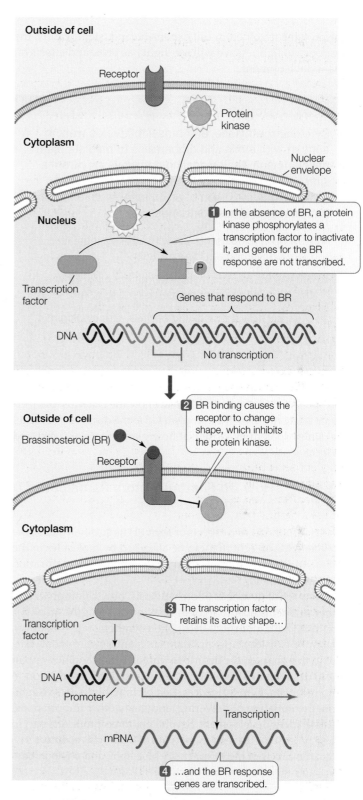

Outside of cell

Receptor

Protein kinase

Cytoplasm

Nuclear envelope

Nucleus

1 In the absence of BR, a protein kinase phosphorylates a transcription factor to inactivate it, and genes for the BR response are not transcribed.

Transcription factor

P

Genes that respond to BR

DNA

No transcription

Outside of cell

Brassinosteroid (BR)

2 BR binding causes the receptor to change shape, which inhibits the protein kinase.

Receptor

Cytoplasm

Transcription factor

3 The transcription factor retains its active shape...

DNA

Promoter

Transcription

mRNA

4 ...and the BR response genes are transcribed.

37.18 The Brassinosteroid Signal Transduction Pathway Begins at the Plasma Membrane Unlike the receptors for animal steroid hormones, the brassinosteroid (BR) receptor is a membrane protein. The signal transduction pathway concludes by activating certain genes.

(**Figure 37.18**). Binding of a brassinosteroid by the receptor inactivates a protein kinase that would otherwise inactivate a transcription factor. The genes activated by this pathway code for proteins involved in cell expansion and, significantly, the response to light. As we will see, light has profound effects on plant development.

RECAP 37.4

Cytokinins, ethylene, and brassinosteroids work in concert with auxin and gibberellins to mediate plant development. Their signaling pathways vary from a simple two-component receptor–effector system (cytokinin) to inhibition of an inhibitor of an effector (ethylene and brassinosteroids).

- How do cytokinins interact with auxin to regulate a plant's development? **See p. 785**

- What is the role of ethylene in fruit ripening? How is this knowledge used commercially? **See p. 786**

- Describe how the signaling pathways for cytokinins and brassinosteroids differ. **See p. 787 and Figures 37.15 and 37.18**

A plant's response to light—the energy source for photosynthesis—is crucial to its survival. We saw how the Darwins' pioneering investigations of phototropism led to the discovery of auxin. Let's now look more closely at how plants sense and respond to light.

37.5 How Do Photoreceptors Participate in Plant Growth Regulation?

Plants respond to two aspects of light: (1) its *quality*—that is, the wavelengths of light that can be absorbed by molecules in the plant; and (2) its *quantity*—that is, the intensity and duration of light exposure.

Chapter 10 describes photosynthesis: how chlorophyll and other pigments absorb light at certain wavelengths (quality), and how light intensity affects photosynthetic rate (quantity). Here, we consider how light affects plant development. Earlier in this chapter, we described phototropism and how auxin mediates a plant stem's bending toward light. In addition to phototropism, light influences seed germination, shoot elongation, the initiation of flowering, and many other important aspects of plant development. Several photoreceptors take part in these processes. Three or more types of **blue-light receptors** mediate the effects of higher-intensity blue light, and **phytochrome** mediates the effects of red light.

Phototropins, cryptochromes, and zeaxanthin are blue-light receptors

Charles and Francis Darwin showed that the apical tip of a growing coleoptile receives light as a signal and then redistributes auxin to stimulate cell elongation below the tip on the shaded side. You may recall from Chapter 10 that an *action spectrum* involves exposing plants to different wavelengths of light to determine what wavelengths are most effective in driving

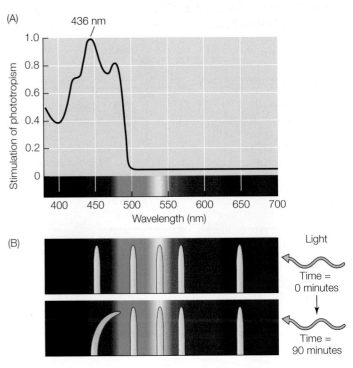

37.19 Action Spectrum for Phototropism (A) The action spectrum for bending of a coleoptile toward light is similar to the absorption spectrum for the receptor, phototropin. (B) After 90 minutes, only the coleoptiles exposed to blue light bend.

a given process (e.g., photosynthesis). For photosynthesis, such studies showed that the most effective wavelengths are those absorbed by chlorophylls (see Figure 10.6). When an action spectrum was obtained for phototropism of coleoptiles, blue light (peak 436 nm) was found to be the most effective at inducing the coleoptile to curve (**Figure 37.19**). What is the blue-light-absorbing receptor/pigment? Biologists have used a genetic approach to answer this question, once again employing the model plant *Arabidopsis*.

Researchers recovered blue-light-insensitive *Arabidopsis* mutants from a genetic screen and identified the gene for a blue-light receptor protein located in the plasma membrane called **phototropin**. Phototropin protein has a flavin mononucleotide associated with it that absorbs blue light, leading to a change in the shape of the protein. This change exposes an active site for a protein kinase, which in turn initiates a signal transduction cascade that ultimately results in stimulation of cell elongation by auxin.

Phototropin is also involved in chloroplast movements in relation to light, and participates with another type of blue-light receptor, the plastid pigment **zeaxanthin**, in the light-induced opening of stomata (see Figure 35.9).

Yet another class of blue-light receptors is the **cryptochromes**, which absorb blue and ultraviolet light. These yellow pigments are located primarily in the plant cell nucleus and affect seedling development and flowering. The exact mechanism of cryptochrome action is not yet known. Strong blue light inhibits cell elongation through the action of cryptochromes, although the most rapid responses are mediated by phototropins.

Phytochromes mediate the effects of red and far-red light

A number of physiological and developmental events in plants are controlled by light, a process called **photomorphogenesis**. For example:

- A bean seedling germinating below ground has an elongated stem, a pale yellow, folded leaf, and a hook that protects the leaf (see Figures 37.2 and 37.16)—it is **etiolated**. As the seedling reaches the surface of the soil, it undergoes several light-induced changes: the apical hook straightens, the rudimentary leaf unfolds, and chlorophyll is made so that photosynthesis can begin. Even very dim light will induce these changes.

- Lettuce seeds spread on the soil will germinate only in response to light. Even just a flash of dim light will suffice.

- Adult cocklebur plants flower when they are exposed to long nights. If there is a brief light flash in the middle of the night, they do not flower.

Action spectra of the above processes show that they are induced by red light (650–680 nm). This indicates that plants must have a photoreceptor pigment that absorbs red light and initiates photomorphogenesis.

What is especially remarkable about these red light responses is that *they are reversible by far-red light* (710–740 nm). For example, if lettuce seeds are exposed to brief, alternating periods of red and far-red light in close succession, they respond only to the final exposure. If it is red, they germinate; if it is far-red, they remain dormant (**Figure 37.20**). This reversibility of the effects of red and far-red light regulates many other aspects of plant development, including flowering and seedling growth.

The basis for the effects of red and far-red light resides in a bluish photoreceptor pigment protein in the cytosol of plants called **phytochrome**. Phytochrome exists in two interconvertible "isoforms" or states. The molecule undergoes a conformational change upon absorbing light at particular wavelengths. The default or "ground" state, which absorbs principally red light, is called P_r. When P_r absorbs a photon of red light it is converted into P_{fr}. The P_{fr} form preferentially absorbs far-red light; when it does so, it is converted back to P_r.

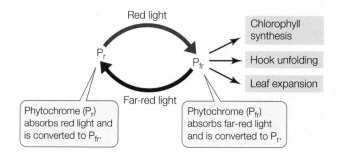

Red light

P_r → P_{fr}

Far-red light

Chlorophyll synthesis

Hook unfolding

Leaf expansion

Phytochrome (P_r) absorbs red light and is converted to P_{fr}.

Phytochrome (P_{fr}) absorbs far-red light and is converted to P_r.

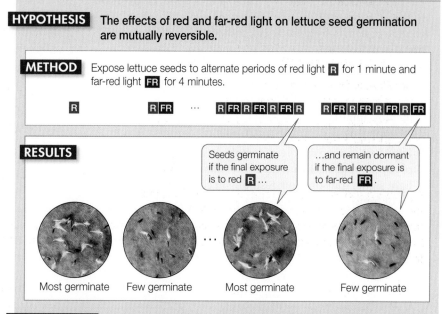

INVESTIGATING LIFE

37.20 Sensitivity of Seeds to Red and Far-Red Light

Lettuce seeds will germinate if exposed to a brief period of light. An action spectrum indicated that red light was most effective in promoting germination, but far-red light would reverse the stimulation if presented right after the red light flash. Harry Borthwick and his colleagues asked what would be the effect of repeated alternating flashes of red and far-red light. In each case, the final exposure determined the germination response. This observation led to the conclusion that a single, photoreversible molecule was involved. That molecule turned out to be phytochrome.

HYPOTHESIS The effects of red and far-red light on lettuce seed germination are mutually reversible.

METHOD Expose lettuce seeds to alternate periods of red light **R** for 1 minute and far-red light **FR** for 4 minutes.

R   **R FR** … **R FR R FR R FR R**   **R FR R FR R FR R FR**

RESULTS

Seeds germinate if the final exposure is to red **R** …

…and remain dormant if the final exposure is to far-red **FR**.

Most germinate   Few germinate   …   Most germinate   Few germinate

CONCLUSION Red light and far-red light reverse each other's effects.

Go to **yourBioPortal.com** for original citations, discussions, and relevant links for all INVESTIGATING LIFE figures.

P_{fr}, not P_r, is the active form of phytochrome—the form that triggers important biological processes in various plants. As we have seen, these processes include seed germination, shoot development after etiolation, and flowering.

For a plant in nature, the ratio of red to far-red light determines whether a phytochrome-mediated response will occur. For example, during daylight, the ratio is about 1.2:1; because there is more red than far-red light, the P_{fr} form predominates. But for a plant growing in the shade of other plants, the ratio is as low as 0.13:1, and phytochrome is mostly in the P_r form. The low ratio of red to far-red light in the shade results from absorption of red light by chlorophyll in the leaves overhead, so less of the red light gets through to the plants below. Shade-intolerant species respond by stimulating cell elongation in the stem and thus growing taller to escape the shade. Shade cast by other plants also prevents germination of seeds that require red light to germinate (see Figure 37.20). The reflective properties of the soil can also affect the red to far-red ratio—and thus plant behavior. For example, cotton seedlings grow more slowly on soils (such as clay) that reflect more red than far-red light.

Phytochrome stimulates gene transcription

How does phytochrome, or more specifically, P_{fr}, work? Phytochrome is a cytoplasmic protein composed of two subunits (**Figure 37.21**). Each subunit has a protein chain and a nonprotein pigment from the plastid called a *chromophore*. In *Arabidopsis*, there is a gene family that encodes five slightly different phytochromes, each functioning in different photomorphogenic responses.

Gene transcription is stimulated when P_r is converted to the P_{fr} isoform. When P_r absorbs red light, the chromophore changes shape, which leads to a change in the conformation of the protein itself from the P_r form to the P_{fr} form. Conversion to the P_{fr} form exposes two important regions of the phytochrome protein (see Figure 37.21), both of which affect transcriptional activity:

- Exposure of a *nuclear localization sequence* (see Figure 14.20) results in movement of P_{fr} from the cytosol to the nucleus. Once in the nucleus, P_{fr} binds to transcription factors and thereby stimulates expression of genes involved in photomorphogenesis.

- Exposure of a *protein kinase* domain causes P_{fr} protein to phosphorylate itself and other proteins involved in red-light signal transduction, resulting in changes in the activity of transcription factors.

The effect of activating these transcription factors is quite large: In *Arabidopsis*, phytochrome affects an amazing 2,500 genes (10 percent of the entire genome!) by either increasing or decreasing their expression. Some of these genes are related to other hormones. For example, when P_{fr} is formed in seed germination, genes for gibberellin synthesis are activated and genes for gibberellin breakdown are repressed. As a result, gibberellins accumulate and seed reserves are mobilized.

Circadian rhythms are entrained by light reception

The timing and duration of biological activities in living organisms are governed in all eukaryotes and some prokaryotes by what is commonly called a "biological clock"—an oscillator within cells that alternates back and forth between two states at roughly 12-hour intervals. The major outward manifestations of this clock are known as **circadian rhythms** (Latin *circa*, "about," and *dies*, "day"). Think of your own life: in all probability you sleep at night, and you are awake during the day. In plants, circadian rhythms influence, for example, the opening (during the day) and closing (at night) of stomata in *Arabidopsis*, and the raising toward the sun (during the day) and lowering (at night) of leaves in bean plants. From these two examples, it is obvious that circadian rhythms are ecologically useful adaptations, in that they relate the plant's physiology to its environment.

Two qualities characterize circadian rhythms, as well as other regular biological cycles: the **period** is the length of one cycle, and the **amplitude** is the magnitude of the change over the course of a cycle. The circadian rhythms of plants have several noteworthy characteristics:

- The period of a circadian rhythm is remarkably insensitive to temperature, although lowering the temperature may drastically reduce the amplitude.

- Circadian rhythms are highly persistent; they may continue for days, even in the absence of environmental cues, such as light–dark periods.

- Circadian rhythms can be *entrained*, within limits, by light–dark cycles that do not exactly correspond to 24 hours. That is, the period of a rhythm can be made to coincide (within limits) with that of the light–dark cycle to which the organism is exposed.

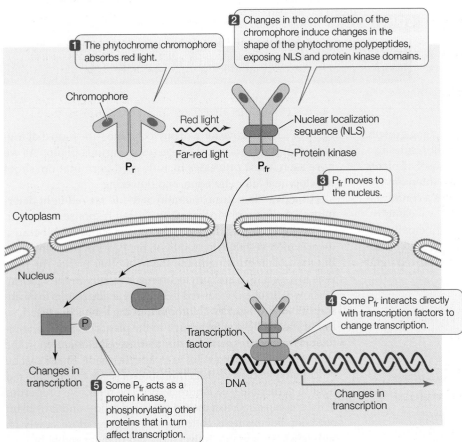

1 The phytochrome chromophore absorbs red light.

2 Changes in the conformation of the chromophore induce changes in the shape of the phytochrome polypeptides, exposing NLS and protein kinase domains.

Chromophore

Red light

Far-red light

Nuclear localization sequence (NLS)

Protein kinase

P_r

P_{fr}

3 P_{fr} moves to the nucleus.

Cytoplasm

Nucleus

P

Transcription factor

DNA

Changes in transcription

4 Some P_{fr} interacts directly with transcription factors to change transcription.

5 Some P_{fr} acts as a protein kinase, phosphorylating other proteins that in turn affect transcription.

Changes in transcription

37.21 Phytochrome Stimulates Gene Transcription Phytochrome is composed of two polypeptide chains, each with a chromophore pigment. This pair of polypeptides undergoes a conformational change upon absorbing light. When phytochrome absorbs red light, it converts to the P_{fr} form, which activates transcription of phytochrome-responsive genes.

Consider what happens when a person abruptly moves across many time zones: what was the night becomes the day, and gradually the person's sleep–wakefulness circadian rhythm entrains to the new environmental cues. Similar entrainment occurs in plants adapting to day length as the seasons progress during the year. The action spectrum for plant entrainment indicates that phytochrome (and to a lesser extent, blue-light receptors) is very likely involved. At sundown phytochrome is mostly in the active P_{fr} form. But as the night progresses, P_{fr} gradually gets converted back to the inactive P_r form. By dawn phytochrome is mostly in the P_r state, but as daylight begins, it rapidly converts to P_{fr}. The switch to the P_{fr} state resets the plant's biological clock. However long the night, the clock is still reset at dawn every day. Thus while the total period measured by the clock is consistent, the clock adjusts to changes in day length over the course of the year.

37.5 RECAP

Light controls a number of physiological and developmental events in plants, a process called photomorphogenesis. Pigment photoreceptors such as phototropin, cryptochromes, and phytochrome mediate the effects of light on plant growth and development. Phytochrome exists in two interconvertible states; conversion from one state to the other is controlled by the ratio of red to far-red light. Circadian rhythms are influenced by light reception.

- Give the evidence for blue-light receptors in plants. **See p. 788 and Figure 37.19**

- Why does red light affect seed germination differently from far-red light? **See p. 789 and Figure 37.20**

- What are circadian rhythms? How are they related to photoreception? **See p. 790**

Photoreceptors also play a regulatory role in flowering. In addition to light, another environmental cue—temperature—regulates flowering. We will examine these topics and others in the next chapter, which focuses on reproduction in flowering plants.

CHAPTER SUMMARY

37.1 How Does Plant Development Proceed?

- As sessile organisms, plants maximize their ability to grow by using meristems, forming new organs and growing throughout life.

- The environment, photoreceptors, hormones, and the plant's genome all regulate plant development.

- Seed **dormancy**, which has adaptive advantages, is maintained by a variety of mechanisms. In nature, dormancy is broken by, for example, abrasion, fire, leaching, and low temperatures. When dormancy ends and the seed **imbibes** water, it **germinates** and develops into a **seedling**. Review Figure 37.1, **WEB ACTIVITIES 37.1 and 37.2**

- Plants have several hormones, each of which regulates multiple aspects of development. Interactions among these hormones are often complex. Review Table 37.1

- **Hormones** and **photoreceptors** act through signal transduction pathways to regulate seedling development. Before the germinating embryo can begin photosynthesis, it relies on energy reserves in the **cotyledons** or the **endosperm**.

- **Genetic screens** using the model organism *Arabidopsis thaliana* have contributed greatly to our understanding of signaling in plants. Review Figure 37.3

37.2 What Do Gibberellins Do?

- The embryos of cereal seeds secrete **gibberellins**, which cause the **aleurone layer** to synthesize and secrete digestive enzymes that break down macromolecules stored in the endosperm. Review Figure 37.6, **WEB ACTIVITY 37.3**

- Dozens of gibberellins exist. These hormones regulate the growth of stems and some fruits.

- Gibberellins act through the breakdown of transcriptional repressors. **Review Figure 37.7**

37.3 What Does Auxin Do?

SEE ANIMATED TUTORIAL 37.1

- In **coleoptiles**, auxin is made in cells at the tip and moves down to the growing region. Review Figures 37.8 and 37.9, **ANIMATED TUTORIAL 37.2**

- Auxin transport is polar. Auxin active transport carriers—membrane proteins confined to the basal ends of cells—cause auxin to move from the tip to the base of the shoot. Review Figure 37.10

- Lateral movement of auxin, mediated by auxin transport carriers, is responsible for **phototropism** and **gravitropism**. Review Figure 37.11

- Auxin plays roles in root formation, leaf **abscission**, **apical dominance**, and **parthenocarpic fruit development**. Certain synthetic auxins are used as selective herbicides.

- The **acid growth hypothesis** explains how auxin promotes cell expansion by disrupting interactions between cell wall microfibrils. Review Figure 37.13, **ANIMATED TUTORIAL 37.3**

- The molecular mechanism underlying auxin action is similar to that of gibberellin; as long as auxin is not bound to its receptor, transcription is repressed. When the auxin–receptor complex binds to a transcriptional repressor, the repressor is degraded and transcription is initiated. Review Figure 37.14

37.4 What Are the Effects of Cytokinins, Ethylene, and Brassinosteroids?

- **Cytokinins** are adenine derivatives. They promote plant cell division, promote seed germination in some species, inhibit stem elongation, promote lateral swelling of stems and roots, stimulate the growth of axillary buds, promote the expansion of leaf tissue, and delay leaf senescence.

- **Cytokinins** act on plant cells by a signal transduction pathway that is similar to bacterial two-component systems. **Review Figure 37.15**

- A balance between auxin and **ethylene** controls leaf abscission. Ethylene promotes senescence and fruit ripening. It causes the formation of a protective **apical hook** in eudicot seedlings. In stems, it inhibits elongation, promotes lateral swelling, and causes a loss of gravitropic sensitivity.

- Ethylene acts on cells by a protein kinase pathway located in the endoplasmic reticulum. **Review Figure 37.17**

- Dozens of different **brassinosteroids** affect cell elongation, pollen tube elongation, vascular tissue differentiation, and root elongation. Some effects of light are mediated by changes in the action and levels of brassinosteroids. These steroids act at a plasma membrane receptor. **Review Figure 37.18**

37.5 How Do Photoreceptors Participate in Plant Growth Regulation?

- **Phototropins** are blue-light photoreceptors for phototropism and chloroplast movements. **Zeaxanthin** acts in conjunction with the phototropins to mediate the light-induced opening of stomata. **Cryptochromes** are blue-light photoreceptors that control seedling development, stem elongation, and floral initiation.

- **Phytochromes** exist in the cytosol in two interconvertible forms, P_r and P_{fr}. The relative amounts of these two forms are a function of the ratio of red to far-red light. Phytochromes affect seedling growth, flowering, and etiolation. **Review Figure 37.20**

- The phytochrome signal transduction pathway affects transcription in two different ways; the P_{fr} form interacts directly with some transcription factors, and influences transcription indirectly through interactions with protein kinases. **Review Figure 37.21**

- **Circadian rhythms** are activities that occur on a near-24-hour cycle. Light can entrain these activities through photoreceptors such as phytochrome.

SELF-QUIZ

1. Which of the following is *not* an advantage of seed dormancy?
   a. It makes the seed more likely to be digested by birds that disperse it.
   b. It counters the effects of year-to-year variations in the environment.
   c. It increases the likelihood that a seed will germinate in the right place.
   d. It favors dispersal of the seed.
   e. It may result in germination at a favorable time of year.

2. Which of the following does *not* occur in seed germination?
   a. Imbibition of water
   b. Metabolic changes
   c. Growth of the radicle
   d. Mobilization of nutrient reserves
   e. Extensive mitotic divisions

3. To mobilize its nutrient reserves, a germinating barley seed
   a. becomes dormant.
   b. undergoes senescence.
   c. secretes gibberellins into its endosperm.
   d. converts glycerol and fatty acids into lipids.
   e. takes up proteins from the endosperm.

4. The gibberellins
   a. are responsible for phototropism and gravitropism.
   b. are gases at room temperature.
   c. are produced only by fungi.
   d. cause flowering in plants.
   e. inhibit the synthesis of digestive enzymes by barley seeds.

5. In coleoptile tissue, auxin
   a. is transported from base to tip.
   b. is transported from tip to base.
   c. can be transported toward either the tip or the base, depending on the orientation of the coleoptile with respect to gravity.
   d. is transported by simple diffusion, with no preferred direction.
   e. is not transported, because auxin is used where it is made.

6. Which process is *not* directly affected by auxin?
   a. Apical dominance
   b. Leaf abscission
   c. Synthesis of digestive enzymes by barley seeds
   d. Root initiation
   e. Cell elongation

7. Signal transduction for both auxin and gibberellins involves
   a. binding of the hormone to a nuclear receptor.
   b. degradation of a repressor of gene transcription.
   c. production of a small molecule second messenger.
   d. light absorption followed by chemical changes.
   e. breakdown of the hormone.

8. Which statement about cytokinins is *not* true?
   a. They promote cell division in tissue cultures.
   b. They delay the senescence of leaves.
   c. They usually promote the elongation of stems.
   d. They act by a receptor with protein kinase activity.
   e. They were discovered as a breakdown product of DNA.

9. Ethylene
   a. causes the triple response in seedlings growing underground.
   b. is liquid at room temperature.
   c. delays the ripening of fruits.
   d. generally promotes stem elongation.
   e. inhibits the swelling of stems, in opposition to cytokinin's effects.

10. Phytochrome
    a. is the only photoreceptor pigment in plants.
    b. exists in two forms interconvertible by light.
    c. is a pigment that is colored red or far-red.
    d. is a green-light receptor.
    e. is the photoreceptor for phototropism in coleoptiles.

FOR DISCUSSION

1. Describe the circumstances under which it would be advantageous for a species to have the dormancy of its seeds broken by fire.

2. Cocklebur fruits contain two seeds each that are kept dormant by two different mechanisms. Why might having two mechanisms of dormancy be advantageous to cockleburs?

3. Supermarkets sell plastic bags that are impregnated with activated charcoal, which binds gases. The bags are designed to keep fruit fresh. How do they work?

4. Corn stunt virus causes a great reduction in the growth rate of infected corn plants. Diseased plants take on a dwarfed form. Since their appearance is reminiscent of the genetically dwarfed corn, you suspect that the virus may inhibit the synthesis of gibberellins by corn plants. Describe two experiments you might conduct to test this hypothesis, only one of which should require chemical measurement.

ADDITIONAL INVESTIGATION

The semi-dwarf wheat and rice plants that led to the Green Revolution described in the chapter opening have mutations in the signal transduction pathway for gibberellins. You wish to use genetic engineering to make corn plants that are semi-dwarf.

a. How would you do a genetic screen to identify the genes in corn involved in gibberellin signaling?

b. Assuming that the signal transduction pathway is similar to that in *Arabidopsis*, what gene would you select for inactivation?

c. Besides short stature, what other effects would you expect for the signal transduction mutant strain? How would you use other hormones to overcome them?

WORKING WITH DATA (GO TO yourBioPortal.com)

The Darwins' Phototropism Experiment   In this exercise based on Figure 37.8, you will read excerpts from a book by Charles Darwin, *The Power of Movement in Plants*, in which he describes the experiments he and his son Francis undertook that ultimately led to the isolation by others of the plant hormone auxin. You will see how they planned their experiments and controls, and analyze the results.

The language of flowers

In the recent film *Kate and Leopold*, a Victorian English nobleman named Leopold is transplanted to modern-day New York, where he meets and falls in love with Kate. At one point, Leopold sees a bundle of flowers at a florist' shop and is amazed that this bouquet would be given to a woman. It's all wrong, he explains: the lavender implies distrust; the orange lily stands for extreme hatred. Better to send amaryllis, which symbolizes great beauty.

During the Victorian era in England (1837–1891) floral symbolism reached its peak of popularity. Social convention discouraged open displays of emotion, so flowers were often used to convey messages people dared not speak aloud. This botanical language was so elaborate that dictionaries were written to describe the specific "meanings" of flowers and their colors.

A student at Cambridge University might "tell" a woman that she was beautiful with a calla lily. He might indicate he would be patient by presenting her with a daisy. If the woman found her suitor attractive, she could tell him so with a camellia; a geranium, on the other hand, would say, "Let's just be friends." Colors had meaning, too. A red rose symbolized love, while yellow was associated with jealousy and white with innocence.

By the early twentieth century the rules of social communication were sufficiently relaxed that intricate floral communication was no longer necessary. Nevertheless, certain flowers continue to have symbolic meaning. Poppies are worn in the British Commonwealth to memorialize soldiers who died in battle. Lilies are often used at funerals to symbolize life and, for Christians, resurrection. The Hindu god Vishnu is often shown with a lotus flower, symbolizing that he is the pure source of all creation.

Floral symbolism flourishes even in the United States. Consider the poinsettia, *Euphorbia pulcherrima*, a bright red shrub native to Central America that was used by the Aztecs as a source of red dye. The plant was brought to the U.S. by the first U.S. ambassador to Mexico, John Roberts Poinsett, an amateur botanist. Some years later a much shorter strain of the plant was developed by a Californian plant breeder named Paul Ecke. By 1950, his son, Paul Ecke, Jr., began promoting this now portable plant as a holiday decoration, blanketing television specials with offers of free plants during the period between Thanksgiving and Christmas. The campaign was successful: over 100 million poinsettia plants are now sold in the U.S. during the winter holidays every year, making it the best-selling potted plant.

Floral Message A girl holds a single flower, perhaps wondering what message it conveys.

Flowers Have Diverse Forms and Meanings
The language of flowers still had some popularity in the early twentieth century, as demonstrated by these Edwardian postcards.

You may be surprised to learn that the brightly colored poinsettia "flowers" are not flowers at all. The red (or sometimes pale yellow) parts of the plant that we most notice and appreciate are actually leaves. The poinsettia has a single tiny yellow female flower, without petals, surrounded by male flowers.

The main task of flowers is not to convey messages to humans. Flowers are reproductive equipment: they produce gametophytes, female and male, which in turn produce the gametes that give rise to the next sporophyte generation. Wildflowers (those not "improved" by plant breeders) may have pleasing shapes and colors, but these are in aid not of poetry but of pollination, which is crucial to angiosperm reproduction.

IN THIS CHAPTER we contrast sexual and asexual reproduction in plants, focusing on the details of sexual reproduction. We consider angiosperm gametophytes, pollination, double fertilization, embryo development, and the roles of fruits in seed dispersal. We examine the transition from the vegetative state to the flowering state, a key event in angiosperm development. We conclude by considering the role of asexual reproduction in nature and in agriculture.

38.1 How Do Angiosperms Reproduce Sexually?

Flowers—the hallmark of angiosperms—contain sex organs; thus it is no surprise that almost all angiosperms reproduce sexually. But many reproduce asexually as well; some even reproduce asexually most of the time. What are the advantages and disadvantages of these two kinds of reproduction?

The relative benefits of sexual versus asexual production are a matter of whether genetic recombination will be advantageous. As we have seen, sexual reproduction produces new combinations of genes and diverse phenotypes (see Section 11.4). Asexual reproduction, in contrast, produces a clone of genetically identical individuals.

Many plants can reproduce either sexually or asexually. For example, strawberry plants can reproduce perfectly well by flowers and seeds (sexual reproduction), but they also reproduce asexually by a stem called a *runner* that spreads over the surface of the soil, sprouting new plants at intervals. For the strawberry plant it might be advantageous to reproduce sexually when possible; this generates genetic diversity, and the seeds that are produced facilitate dispersal to far-flung sites. However, too much diversity can be a drawback for farmers, and they generally propagate this crop asexually to deliver predictably plump and tasty strawberries to the market.

We will return to asexual reproduction later in this chapter. Our concern for now is sexual reproduction.

The flower is an angiosperm's structure for sexual reproduction

Sexual reproduction involves mitosis and meiosis, and the alternation of haploid and diploid generations (see Chapter 11):

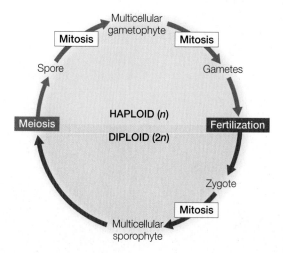

In angiosperms, the plant that we see in nature is a sporophyte and male and/or female gametophytes are contained in the flowers. A complete flower consists of four concentric groups of organs arising from modified leaves: the *carpels*, *stamens*, *petals*, and *sepals*.

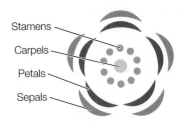

The parts of the flower are usually borne on a stem tip, and derive from a meristem. The differentiation of the meristem into the various organs of the flower is controlled by specific transcription factors (see Figure 19.14). As we discussed in the introductory essay for this chapter, flower parts are very diverse in form.

The carpels and stamens are, respectively, the female and male sex organs. Flowers usually have both stamens and carpels; such flowers are termed *perfect* (**Figure 38.1A**). *Imperfect* flowers, on the other hand, are those with only male or only female sex organs. Male flowers have stamens but not carpels,

and female flowers have carpels but not stamens. Some plants, such as corn, bear both male and female flowers on an individual plant; such species are called **monoecious** ("one house") (**Figure 38.1B**). In **dioecious** species, on the other hand, individual plants bear either male-only or female-only flowers; an example is bladder campion (**Figure 38.1C**).

Flowering plants have microscopic gametophytes

Figure 38.2 offers a detailed look at the gametophytes central to angiosperm reproduction. The haploid gametophytes—the gamete-producing structures—develop from haploid spores in the flower:

- Female gametophytes (megagametophytes), which are called **embryo sacs**, develop in megasporangia.
- Male gametophytes (microgametophytes), which are called **pollen grains**, develop in microsporangia.

FEMALE GAMETOPHYTE Locate the ovule in the flower shown in Figure 38.2. Within the ovule, a megasporocyte—a cell within the megasporangium—divides meiotically to produce four haploid megaspores. In most flowering plants, all but one of these megaspores then undergo apoptosis. The surviving megaspore usually goes through three mitotic divisions without cytokinesis, producing eight haploid nuclei, all initially contained within a single cell—three nuclei at one end, three at the other, and two in the middle. Subsequent cell wall formation leads to an elliptical, seven-celled megagametophyte with a total of eight nuclei:

- At one end of the elliptical megagametophyte are three tiny cells: the **egg** and two cells called *synergids*. The egg is the female gamete, and the synergids participate in fertilization

(A) Perfect: lily

Stamens

Carpels

38.1 Perfect and Imperfect Flowers (A) A lily is an example of a perfect flower, meaning one that has both male and female sex organs. (B) Imperfect flowers are either male or female. Corn is a monoecious species: both types of imperfect flowers are borne on the same plant. (C) Bladder campion is a dioecious species; some bladder campion plants bear male imperfect flowers while others bear female imperfect flowers.

(B) Imperfect monoecious: corn

Male flower with stamens

Female flower with carpels

(C) Imperfect dioecious: bladder campion

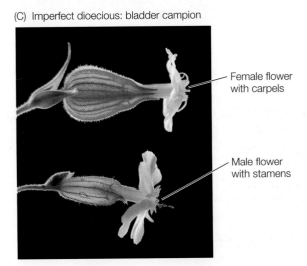

Female flower with carpels

Male flower with stamens

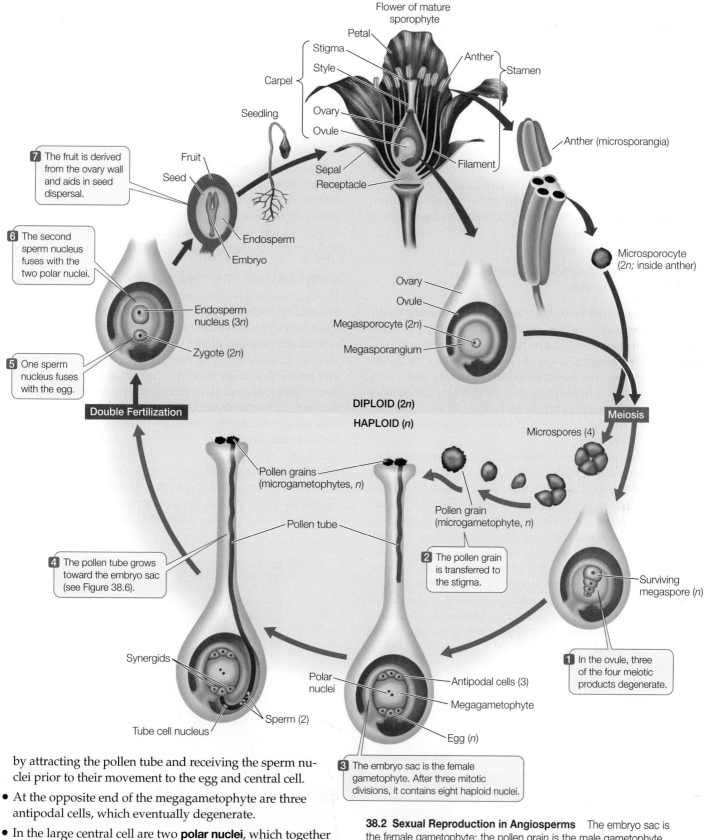

Flower of mature
sporophyte

Petal

Stigma

Style

Carpel

Anther

Stamen

Seedling

Ovary

Ovule

Anther (microsporangia)

Fruit

Seed

Sepal

Receptacle

Filament

7 The fruit is derived from the ovary wall and aids in seed dispersal.

Endosperm

Embryo

6 The second sperm nucleus fuses with the two polar nuclei.

Endosperm nucleus (3n)

Zygote (2n)

Microsporocyte (2n; inside anther)

Ovary

Ovule

Megasporocyte (2n)

Megasporangium

5 One sperm nucleus fuses with the egg.

DIPLOID (2n)

HAPLOID (n)

Meiosis

Double Fertilization

Microspores (4)

Pollen grains (microgametophytes, n)

Pollen grain (microgametophyte, n)

Pollen tube

4 The pollen tube grows toward the embryo sac (see Figure 38.6).

2 The pollen grain is transferred to the stigma.

Surviving megaspore (n)

Synergids

Polar nuclei

Antipodal cells (3)

Megagametophyte

1 In the ovule, three of the four meiotic products degenerate.

Tube cell nucleus

Sperm (2)

Egg (n)

3 The embryo sac is the female gametophyte. After three mitotic divisions, it contains eight haploid nuclei.

by attracting the pollen tube and receiving the sperm nuclei prior to their movement to the egg and central cell.

- At the opposite end of the megagametophyte are three antipodal cells, which eventually degenerate.

- In the large central cell are two **polar nuclei**, which together combine with a sperm nucleus.

38.2 Sexual Reproduction in Angiosperms The embryo sac is the female gametophyte; the pollen grain is the male gametophyte. The male and female nuclei meet and fuse within the embryo sac. Angiosperms have double fertilization, in which a zygote and an endosperm nucleus form from separate fusion events—the zygote from one sperm and the egg, and the endosperm from the other sperm and two polar nuclei.

The embryo sac (megagametophyte) is the entire seven-cell, eight-nucleus structure.

MALE GAMETOPHYTE The pollen grain (microgametophyte) consists of fewer cells and nuclei than the embryo sac. The development of a pollen grain begins when a microsporocyte within the anther divides meiotically. Each resulting haploid microspore develops a spore wall, within which it normally undergoes one mitotic division before the anthers open and release these two-celled pollen grains. The two cells are the tube cell and the generative cell. Further development of the pollen grain, which we will describe shortly, is delayed until the pollen arrives at a stigma (the receptive part of the carpel). In angiosperms, the transfer of pollen from the anther to the stigma is referred to as **pollination**.

Pollination in the absence of water is an evolutionary adaptation

As Chapter 28 describes, the union of gametes in aquatic plants is accomplished in the water. Fertilization of mosses and ferns also requires at least a film of water for movement of gametes. While there are mechanisms to ensure fertilization if and when the two gametes meet, fertilization is clearly a low-probability event. The evolution of pollen made it possible for male gametes to reach the female gametophyte without an aqueous conduit. With this selective advantage, pollen-bearing plants were able to colonize the land.

In the first land plants, wind was the primary vehicle by which pollen reached its destination, and many plant species are wind-pollinated today. Wind-pollinated flowers have sticky or featherlike stigmas, and they produce pollen grains in great numbers. Pollen transport by wind is, however, a relatively chancy means of achieving pollination, explaining why about 75 percent of all angiosperms rely upon animals—including insects, birds, and bats—for pollen transport. Pollen transport by animals greatly increases the probability that pollen will get to the female gametophyte. Suitably pigmented, shaped, and scented flowers attract the pollinating animal, resulting in a pollen transfer from flower to flower within the same plant species (**Figure 38.3**).

Flower color is one of several adaptations that attract pollinators. Bees, for example, are attracted to blue and yellow flowers (bees cannot sense red but are attracted to patterns exhibited by pigments visible in ultraviolet light; see Figure 56.10). Many birds, on the other hand, are attracted to red flowers (bird-pollinated plants also are often shaped to fit their

pollinator's beak.) In both cases, the animals may derive nutrition from the flowers in the form of carbohydrate-rich nectar and/or pollen—a mutually beneficial situation.

Flowering plants prevent inbreeding

You may recall from discussions of Mendel's work (see Section 12.1) that some plants can reproduce sexually by both cross-pollination and self-pollination. Self-pollination increases the chances of successful pollination, but leads to homozygosity, which reduces genetic diversity. Because diversity is the raw material of evolution by natural selection, homozygosity can be selectively disadvantageous. Most plants have evolved mechanisms that prevent self-fertilization. The two primary means to prevent self-fertilization are (1) physical separation of male and female gametophytes, and (2) genetic self-incompatibility.

SEPARATION OF MALE AND FEMALE GAMETOPHYTES Self-fertilization is prevented in dioecious species, which bear only male or female flowers on a particular plant. Pollination in dioecious species is accomplished only when one plant pollinates another. In monoecious plants, which bear both male and female flowers on the same plant, the physical separation of the male and female flowers is often sufficient to prevent self-fertilization. Some monoecious species prevent self-fertilization by staggering the development of male and female flowers so they do not bloom at the same time, making these species functionally dioecious.

GENETIC SELF-INCOMPATIBILITY A pollen grain that lands on the stigma of the same plant will fertilize the female gamete (review Figure 38.2) *only if the plant is self-compatible*, meaning capable of self-pollination. To prevent self-fertilization, many plants are

(A)

(B)

38.3 Flowers and Pollinators (A) Flies are attracted to some flowers (in this case, the tropical plant *Stapelia gigantea*) by chemicals emitted from the flower. (B) Other flowers, such as these *Cavendishia* sp. flowers, have red pigments and a shape that attracts certain birds.

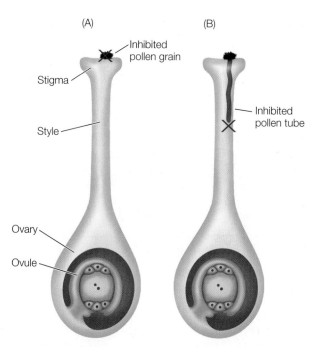

38.4 Self-Incompatibility In a self-incompatible plant, pollen is rejected if it expresses an *S* allele that matches one of the *S* alleles of the stigma and style. Self pollen may (A) fail to germinate or (B) its pollen tube may die before reaching an ovule. In either case, the egg cannot be fertilized by a sperm from the same plant.

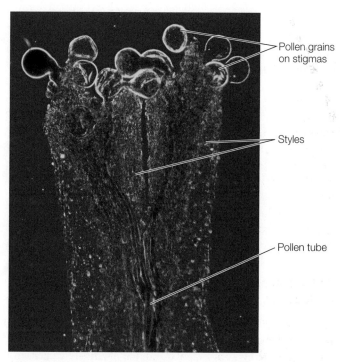

38.5 Pollen Tubes Begin to Grow Staining pollen with a fluorescent dye allows them to be seen through a fluorescence microscope. These pollen grains have landed on the stigmas of a crocus.

self-incompatible, which depends upon the ability of a plant to determine whether pollen is genetically similar or genetically different from "self." Rejection of "same-as-self" pollen prevents self-fertilization. How does it occur?

Self-incompatibility in plants is controlled by a cluster of tightly linked genes called the *S* locus (for self-incompatibility). The *S* locus encodes proteins in the pollen and style that interact during the recognition process. A self-incompatible species typically has many alleles of the *S* locus, and when the pollen carries an allele that matches one of the alleles of the recipient pistil, the pollen is rejected. Depending on the type of self-incompatibility system, the rejected pollen either fails to germinate or is prevented from growing through the style (**Figure 38.4**); either way, self-fertilization is prevented.

A pollen tube delivers sperm cells to the embryo sac

When a functional pollen grain lands on the stigma of a compatible pistil, it germinates. A key event is water uptake by pollen from the stigma: pollen loses most of its water as it matures. Germination involves the development of a **pollen tube** (**Figure 38.5**). The pollen tube either traverses the spongy tissue of the style or, if the style is hollow, grows on the inner surface of the style until it reaches an ovule. The pollen tube typically grows at the rate of 1.5–3 mm/hr, taking just an hour or two to reach its destination, the female gametophyte.

The growth of the pollen tube is guided in part by a chemical signal in the form of a small protein produced by the synergids within the ovule. If one synergid is destroyed, the ovule still attracts pollen tubes, but destruction of both synergids renders the ovule unable to attract pollen tubes, and fertilization does not occur. The attractant appears to be species-specific: in some cases, isolated female gametophytes attract only pollen tubes of the same species.

Angiosperms perform double fertilization

In most angiosperm species, the mature pollen grain consists of two cells, the tube cell and the generative cell. The larger tube cell encloses the much smaller generative cell. Guided by the tube cell nucleus, the pollen tube eventually grows through the style tissue and reaches the embryo sac. The generative cell, meanwhile, has undergone one mitotic division and cytokinesis to produce two haploid **sperm cells** (**Figure 38.6, steps 1 and 2**).

Two fertilization events now occur. One of the two synergids degenerates when the pollen tube arrives and the two sperm cells are released into its remains. (**Figure 38.6, step 3**). Each sperm cell then fuses with a different cell of the embryo sac (**Figure 38.6, steps 4 and 5**). One sperm cell fuses with the egg cell, producing the diploid zygote. The nucleus of the other fuses with the two polar nuclei in the central cell, forming a **triploid (3n) nucleus**. While the zygote nucleus begins mitotic division to form the new sporophyte embryo, the triploid nucleus undergoes rapid mitosis to form a specialized nutritive tissue, the **endosperm**. The endosperm will later be digested by the developing embryo as a source of nutrients, energy, and carbon-based anabolic building blocks (since it often begins its development

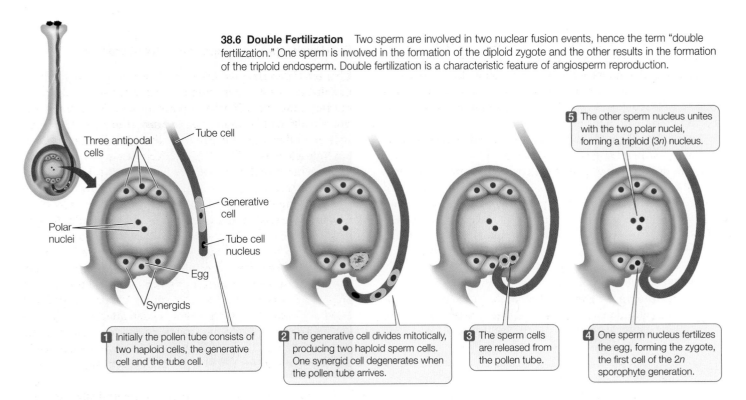

38.6 Double Fertilization Two sperm are involved in two nuclear fusion events, hence the term "double fertilization." One sperm is involved in the formation of the diploid zygote and the other results in the formation of the triploid endosperm. Double fertilization is a characteristic feature of angiosperm reproduction.

Three antipodal cells

Tube cell

Generative cell

Polar nuclei

Tube cell nucleus

Egg

Synergids

1 Initially the pollen tube consists of two haploid cells, the generative cell and the tube cell.

2 The generative cell divides mitotically, producing two haploid sperm cells. One synergid cell degenerates when the pollen tube arrives.

3 The sperm cells are released from the pollen tube.

4 One sperm nucleus fertilizes the egg, forming the zygote, the first cell of the 2n sporophyte generation.

5 The other sperm nucleus unites with the two polar nuclei, forming a triploid (3n) nucleus.

underground and thus cannot perform photosynthesis right away). The remaining cells of the male and female gametophytes, the antipodal cells, and the remaining synergid eventually degenerate, as does the pollen tube nucleus.

Double fertilization is so named because it involves two nuclear fusion events:

• One sperm nucleus fuses with the egg cell nucleus.

• The other sperm nucleus fuses with the two polar nuclei.

The fusion of a sperm cell nucleus with the two polar nuclei to form endosperm is one of the defining characteristics of angiosperms.

Embryos develop within seeds

Fertilization initiates the highly coordinated growth and development of the embryo, endosperm, integuments, and carpel. The *integuments*—tissue layers immediately surrounding the megasporangium—develop into the seed coat, and the carpel ultimately becomes the wall of the fruit that encloses the seed.

The first step in the formation of the embryo is a mitotic division of the zygote that gives rise to two daughter cells. These two cells face different fates. An asymmetrical (uneven) distribution of cytoplasm within the zygote causes one daughter cell to produce the embryo proper and the other daughter cell to produce a supporting structure, the **suspensor** (**Figure 38.7**). The suspensor pushes the embryo against or into the endosperm, thereby facilitating the transfer of nutrients from the endosperm into the embryo.

yourBioPortal.com

GO TO Web Activity 38.1 • Early Development of a Eudicot

38.7 Early Development of a Eudicot The embryo develops through intermediate stages, including a characteristic heart-shaped stage, to reach the torpedo stage.

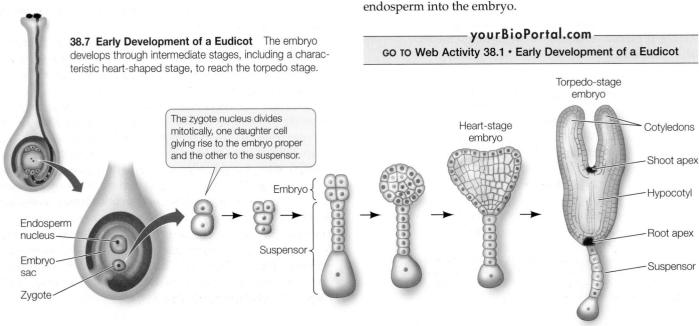

The zygote nucleus divides mitotically, one daughter cell giving rise to the embryo proper and the other to the suspensor.

Endosperm nucleus

Embryo sac

Embryo

Zygote

Suspensor

Heart-stage embryo

Torpedo-stage embryo

Cotyledons

Shoot apex

Hypocotyl

Root apex

Suspensor

The asymmetrical division of the zygote establishes polarity as well as the longitudinal axis of the new plant. A long, thin suspensor and a more spherical or globular embryo are distinguishable after just four mitotic divisions. The suspensor soon ceases to elongate, and the primary meristems and first organs begin to form within the embryo.

In eudicots, the initially globular embryo develops into the characteristic heart stage as the cotyledons ("seed leaves") start to grow. Further elongation of the cotyledons and of the main axis of the embryo gives rise to the torpedo stage, during which some of the internal tissues begin to differentiate (see Figure 34.7). Between the cotyledons is the shoot apex; at the other end of the axis is the root apex. Each of the apical regions contains a cluster of meristematic cells that continue to divide to give rise to new organs throughout the life of the plant.

During seed development, large amounts of nutrients are moved in from other parts of the parent plant, and the endosperm accumulates starch, lipids, and proteins. In many species, the cotyledons absorb the nutrient reserves from the surrounding endosperm and grow very large in relation to the rest of the embryo (**Figure 38.8A**). In others, the cotyledons remain thin (**Figure 38.8B**) and draw on the reserves in the endosperm as needed when the seed germinates.

In the late stages of embryonic development, the seed loses water—sometimes as much as 95 percent of its original water content. This helps the seed remain viable during the time between the seed's dispersal from the parent plant and its eventual germination.

What keeps seeds viable when they have lost water? It appears that as water leaves, sugars and certain protective proteins become more concentrated inside the seeds, creating a very viscous fluid similar to glass. The membranes and proteins of the cells inside the seed retain their integrity in this viscous state. Once the embryo has become desiccated, it is incapable of further development; it remains dormant until internal and external conditions are right for germination (as we saw in Section 37.1).

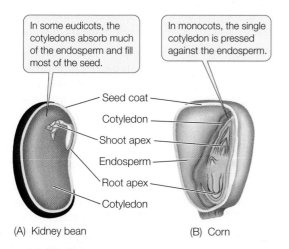

In some eudicots, the cotyledons absorb much of the endosperm and fill most of the seed.

In monocots, the single cotyledon is pressed against the endosperm.

Seed coat
Cotyledon
Shoot apex
Endosperm
Root apex
Cotyledon

(A) Kidney bean   (B) Corn

38.8 Variety in Angiosperm Seeds In some seeds, such as kidney beans (A), the nutrient reserves of the endosperm are absorbed by the cotyledons. In others, such as corn (B), the reserves in the endosperm will be drawn upon after germination.

Seed development is under hormonal control

Chapter 37 describes the role of the hormone gibberellin in the mobilization of stored macromolecules in the seed endosperm during germination. The development of seeds is under the control of a different hormone, **abscisic acid** (**ABA**). Most plant tissues make this hormone, and like other plant hormones it has multiple effects (see Table 37.1). (Unfortunately, its name is misleading, because it does not directly control leaf abscission.) During early seed development the ABA level is low, and it rises as the seed matures. This increase stimulates the endosperm to synthesize seed storage proteins. It also stimulates the synthesis of proteins that prevent cell death as the seeds dry.

ABA also keeps the developing seed from germinating on the plant before it dries. Premature germination, termed **vivipary**, is undesirable in seed crops (such as wheat) because the grain is damaged if it has started to sprout. Viviparous seedlings are also unlikely to survive if they remain attached to the parent plant and are unable to establish themselves in the soil. Mutants of corn that are insensitive to ABA have viviparous seeds, indicating the importance of ABA in preventing precocious germination.

The general effect of ABA in preventing germination extends to seed dormancy. Seeds stay dormant if their ABA level is high and germinate when the level goes down, as usually occurs as dormancy is broken.

Fruits assist in seed dispersal

In angiosperms the ovary wall—together with its seeds—develops into a fruit after fertilization has occurred. Fruits have two main functions:

- They protect the seed from damage by animals and infection by microbial diseases
- They aid in seed dispersal

A **fruit** may consist of only the mature ovary and seeds, or it may include other parts of the flower. Some species produce fleshy, edible fruits such as peaches and tomatoes, while the fruits of other species are dry or inedible.

Fruits are clearly important for carrying seeds, with their embryos, away from the parent plant. Why has this characteristic been selected for during evolution? As products of sexual reproduction, seeds are genetically diverse, and dispersal spreads this diversity around. But if a plant has successfully grown to reproduce, its environment would presumably be favorable for the next generation, too. Some offspring do indeed stay near the parent, as is the case in many tree species, where the seeds essentially fall to the ground. However, this strategy has several disadvantages. If the species is a perennial, offspring that germinate near their parent will be competing with their parent for resources, which may be too limited to support a dense population. Furthermore, even though the local conditions were good enough for the parent to produce at least some seed, there is no guarantee that conditions will still be good the next year, or that they won't be even better elsewhere. Thus, in many cases, seed dispersal is vital to a species' survival.

38.9 Dispersing Fruit   (A) A milkweed seed pod. Silky filaments catch the wind currents and carry the brown seeds with them. (B) Animals who rub up against the "hook-and-loop" surface of burdock fruit walk away with it attached to their fur, thus making the animals unwitting agents of dispersal. This feature of the fruit is said to have inspired the invention of Velcro.

(A) *Asclepias syriaca*

(B) *Arctium* sp.

Some fruits help disperse seeds over substantial distances, increasing the probability that at least a few of the many seeds produced by a plant will find suitable conditions for germination and growth to sexual maturity. Various plants, including milkweed and dandelion, produce a fruit with a "parachute" that may be blown some distance from the parent plant by the wind (**Figure 38.9A**). Still other fruits move by hitching rides with animals—either on them, as with burrs stuck to an animal's fur (or to your hiking socks) (**Figure 38.8B**), or inside them, as with berries eaten by birds. Water disperses some fruits; coconuts have been known to travel thousands of miles between islands. Seeds swallowed whole along with fruits such as berries travel through the animal's digestive tract and are deposited some distance from the parent plant. In some species, seeds must pass through an animal in order to break dormancy.

38.1 RECAP

Flowers contain the organs for sexual reproduction in angiosperms. Plants that use pollen for reproduction have several selective advantages, among them the ability to accomplish fertilization without water, which allowed plants to colonize land. After fertilization, the flower develops into seed(s) and fruit. The selective advantages of seeds and fruits include long-term viability and multiple modes of dispersal.

- What are the relationships between an ovule and an ovary, and between a fruit and a seed? See p. 796 and Figure 38.2

- How do plants prevent self-pollination? See pp. 798–799 and Figure 38.4

- Describe the roles of the two sperm nuclei in double fertilization. See p. 799 and Figure 38.6

- How is plant development controlled by the hormone abscisic acid? See p. 801

We have now traced the sexual life cycle of angiosperms from the flower, to the fruit, to the dispersal of seeds. Seed germination and the vegetative development of the seedling are pre-

sented in Chapter 37. The next section covers the rest of the angiosperm life cycle—the transition from the vegetative to the flowering state—and how this transition is regulated.

38.2 What Determines the Transition from the Vegetative to the Flowering State?

The act of flowering is one of the major events in a plant's life. It represents a reallocation of energy and materials away from making more plant parts (vegetative growth) to making flowers and gametes (reproductive growth). Once a plant is old enough, it can respond to internal or external signals to initiate reproduction. This can happen right at maturity as part of a predetermined developmental program (as in a dandelion plant in the summer) or in response to environmental cues such as light or temperature (as with most ornamental flowers).

Plants fall into three categories depending upon when they mature and initiate flowering, and what happens after they flower:

- **Annuals** complete their lives in one year. This class includes many crops important to the human diet, such as corn, wheat, rice, and soybean. When the environment is suitable, they grow rapidly, with little or no secondary growth. After flowering, they use most of their materials and energy to develop seeds and fruits, and the rest of the plant withers away.

- **Biennials** take two years to complete their lives. They are much less common than annuals and include carrots, cabbage, onions, and Queen Anne's lace. Typically, biennials produce just vegetative growth during the first year and store carbohydrates in underground roots (carrot) and stems (onion). In the second year, they use most of the stored carbohydrates to produce flowers and seeds rather than vegetative growth, and the plant dies after seeds form.

- **Perennials** live three or more—sometimes many more—years. Maple trees, whose leaves symbolize Canada, can live up to 400 years. Perennials include many trees and shrubs, as well as wildflowers. Typically these plants flower every year, but stay alive and keep growing for another season; the reproductive cycle repeats each year. However, some perennials (e.g., century plant) grow vegetatively for many years, flower once, and die.

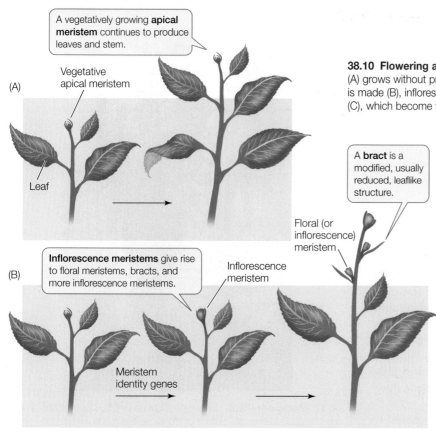

A vegetatively growing **apical meristem** continues to produce leaves and stem.

(A)

Vegetative apical meristem

Leaf

38.10 Flowering and the Apical Meristem A vegetative apical meristem (A) grows without producing flowers. Once the transition to the flowering state is made (B), inflorescence meristems give rise to bracts and to floral meristems (C), which become the flowers.

A **bract** is a modified, usually reduced, leaflike structure.

Floral (or inflorescence) meristem

(B)

Inflorescence meristems give rise to floral meristems, bracts, and more inflorescence meristems.

Inflorescence meristem

Meristem identity genes

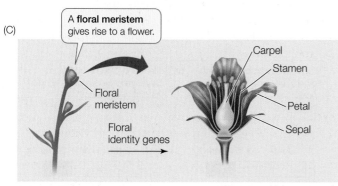

(C)

A **floral meristem** gives rise to a flower.

Carpel

Stamen

Floral meristem

Petal

Floral identity genes

Sepal

No matter what type of life cycle they have, angiosperms all make the transition to flowering. This transition entails significant developmental changes, to which we now turn.

Apical meristems can become inflorescence meristems

The first visible sign of a transition to the flowering state may be a change in one or more apical meristems in the shoot system. As described in Chapter 34, meristems have a pool of undetermined cells. During vegetative growth, an apical meristem continually produces leaves, axillary buds, and stem tissues (**Figure 38.10A**) in a kind of unrestricted growth called *indeterminate growth* (see Section 34.4).

Flowers may appear singly or in an orderly cluster that constitutes an **inflorescence**. If a vegetative apical meristem becomes an **inflorescence meristem**, it ceases production of leaves and axillary buds and produces other structures: smaller leafy structures called bracts, as well as new meristems in the angles between the bracts and the stem (**Figure 38.10B**). These new meristems may also be inflorescence meristems, or they may

be **floral meristems**, each of which gives rise to a flower.

Each floral meristem typically produces four consecutive whorls or spirals of organs—the sepals, petals, stamens, and carpels discussed earlier in the chapter—separated by very short internodes, keeping the flower compact (**Figure 38.10C**). In contrast to vegetative apical meristems and some inflorescence meristems, floral meristems are responsible for *determinate growth*—growth of limited duration, like that of leaves.

A cascade of gene expression leads to flowering

How do apical meristems become floral meristems or inflorescence meristems, and how do inflorescence meristems give rise to floral meristems? How does a floral meristem give rise, in short order, to four different floral organs (sepals, petals, stamens, and carpels)? How does each flower come to have the correct number of each of the floral organs? Numerous genes are expressed and interact to produce these results. We'll refer here to some of the genes whose actions have been most thoroughly studied in *Arabidopsis* and snapdragons (*Antirrhinum*) (see Figure 38.10):

- Expression of a group of **meristem identity genes** initiates a cascade of further gene expression that leads to flower formation. The expression of the genes *LEAFY* and *APETALA1* is both necessary and sufficient for flowering. How do we know this? There are two types of evidence, genetic and molecular. For example, a mutated allele of the gene *APETALA1* leads to continued vegetative growth, even if all other conditions are suitable for flowering. On the other hand, if the wild-type *APETALA1* gene is coupled to an active promoter and introduced into an apical meristem, the plant will flower regardless of the environment. This is powerful evidence that *APETALA1* plays a role in switching meristem cells from a vegetative to a reproductive fate.

- Meristem identity gene products trigger the expression of **floral organ identity genes**, which work in concert to specify the successive whorls of the flower (see Figure 19.14). Floral identity genes are homeotic genes whose products are transcription factors that determine whether cells in the floral meristem will be sepals, petals, stamens, or carpels. An example is the gene *AGAMOUS*, which causes florally determined cells to form stamens and carpels in the "ABC" system described in Section 19.5.

How is this cascade of events initiated? Depending on the species, plants respond to either internal or external cues. Among external clues, the best studied are photoperiod (day length) and temperature. We begin with photoperiod, as it has a fascinating history and clear experimental support.

Photoperiodic cues can initiate flowering

In 1920, W. W. Garner and H. A. Allard of the U.S. Department of Agriculture studied the behavior of a newly discovered mutant tobacco plant. The mutant, named Maryland Mammoth, had large leaves and exceptional height (**Figure 38.11**). Normally tobacco is an annual that flowers in the summer and then stops growing. In contrast, Maryland Mammoth plants remained vegetative and continued to grow.

Garner and Allard now tried to figure out why the mutant plants did not flower in the summer. It wasn't that they *could not flower*: the scientists found that the plants would flower in December in the greenhouse under natural light. To determine what induces flowering in December, they tested several likely environmental variables, such as temperature. The key variable proved to be day length. By moving plants between light and dark rooms at different times to vary the day length artificially, the scientists were able to establish a direct link between flowering and day length.

Maryland Mammoth plants did not flower if exposed to more than 14 hours of light per day, but flowering commenced once the daylight period became shorter than 14 hours, as in December. Thus the **critical day length** for Maryland Mammoth tobacco is 14 hours (**Figure 38.12**). Control of an organism's responses by the length of day or night is called **photoperiodism**.

Plants vary in their responses to photoperiodic cues

Plants that flower in response to photoperiodic stimuli fall into two main classes:

- **Short-day plants** (**SDPs**) flower only when the day is shorter than a critical maximum. They include poinsettias and chrysanthemums, as well as Maryland Mammoth tobacco. Thus, for example, we see chrysanthemums in nurseries in the fall, and poinsettias in winter, as noted in the opening of this chapter.

- **Long-day plants** (**LDPs**) flower only when the day is longer than a critical minimum. Spinach and clover are examples of LDPs. For example, spinach tends to flower and become bitter in the summer, and is therefore normally planted in early spring.

While there are variations on these two patterns, photoperiodic control of flowering serves an important role: it synchronizes the flowering of plants of the same species in a local population, and this promotes cross-pollination and successful reproduction.

The length of the night is the key photoperiodic cue determining flowering

The terms "short-day plant" and "long-day plant" became entrenched before scientists determined that *photoperiodically sen-*

38.11 Mammoth Plant   Wild-type tobacco (left) is much smaller than the Maryland Mammoth mutant of the same age (right), which does not respond to an environmental cue to stop growing and flower.

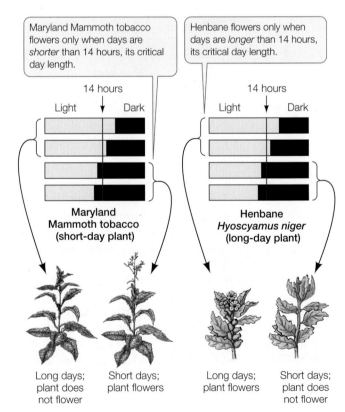

Maryland Mammoth tobacco flowers only when days are *shorter* than 14 hours, its critical day length.

Henbane flowers only when days are *longer* than 14 hours, its critical day length.

14 hours

Light ▾ Dark

14 hours

Light ▾ Dark

Maryland Mammoth tobacco (short-day plant)

Henbane *Hyoscyamus niger* (long-day plant)

Long days; plant does not flower

Short days; plant flowers

Long days; plant flowers

Short days; plant does not flower

38.12 Day Length and Flowering   By artificially varying the day length in a 24-hour period, Garner and Allard showed that the flowering of Maryland Mammoth tobacco is initiated when the days become shorter than a critical length. Maryland Mammoth tobacco is thus called a short-day plant. Henbane, a long-day plant, shows an inverse pattern of flowering.

INVESTIGATING LIFE

38.13 Night Length and Flowering

Short-day plants (SDP) flower only when the day is shorter than a critical maximum. But what environmental cue initiates SDP flowering: day length or night length? To find out, Karl Hamner and James Bonner carried out greenhouse experiments using cocklebur, a SDP.

HYPOTHESIS Short-day plants measure day length.

METHOD Divide plants into two groups. Expose groups to different light conditions: one group to a constant daylight period and varied periods of darkness, the other to varied periods of daylight and fixed periods of darkness.

RESULTS

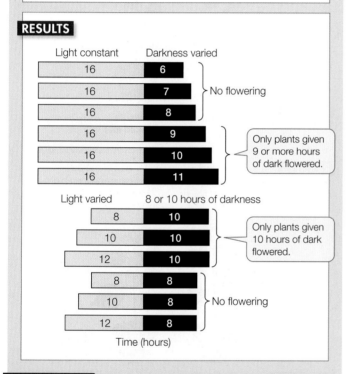

Time (hours)

CONCLUSION The data do not support the hypothesis. Short-day plants measure the length of the night and thus could more accurately be called long-night plants.

FURTHER INVESTIGATION: How would you perform these experiments using long-day plants and what would be the results?

Go to **yourBioPortal.com** for original citations, discussions, and relevant links for all INVESTIGATING LIFE figures.

sitive plants actually measure the length of the night (darkness), rather than the length of the day. This was demonstrated by Karl Hamner of the University of California at Los Angeles and James Bonner of the California Institute of Technology (**Figure 38.13**).

Working with cocklebur, an SDP, Hamner and Bonner ran a series of experiments using two sets of conditions:

• One group of plants was exposed to a constant light period—either shorter or longer than the critical day length—and the dark period was varied.

• A second group of plants was exposed to a constant dark period—and the light period was varied.

Plants flowered under all treatments in which the dark period exceeded 9 hours, regardless of the length of the light period. Hamner and Bonner thus concluded that the length of the *night* is critical to flowering. For cocklebur, the **critical night length** is about 9 hours. It is thus more accurate to call cocklebur a "long-night plant" than a short-day plant.

In cocklebur, a single long night is sufficient to trigger full flowering some days later, even if the intervening nights are short. Most plants are less sensitive than cocklebur and require from two to several nights of appropriate length to induce flowering. For some plants a single shorter night in a series of long ones inhibits flowering, even if the short night comes only one day before flowering would have commenced.

Through other experiments Hamner and Bonner gained some insight into how plants measure night length. They grew SDPs and LDPs under a variety of light/dark conditions. In some experiments, the dark period was interrupted by a brief exposure to light; in others, the light period was interrupted briefly by darkness. Interruptions of the light period by darkness had no effect on the flowering of either short-day or long-day plants. Even a brief interruption of the dark period by light, however, completely nullified the effect of a long night. An SDP flowered *only if the long nights were uninterrupted*. The investigators hypothesized that something must accumulate during that long night that could be broken down by a flash of light in the middle of the night.

To find out what that "something" might be, Hamner and Bonner tested flashes of interrupting light at various wavelengths. You may recall from Section 37.5 that several photoreceptors play roles in regulating plant growth, and that these are sensitive to different wavelengths. In the interrupted-night experiments, the most effective wavelengths of light were in the red range (**Figure 38.14**), and the effect of a red-light interruption of the night could be fully reversed by a subsequent exposure to far-red light, indicating that a phytochrome is the photoreceptor. Where does this occur and what happens downstream from the reception? Once again, elegant experiments provided the answer.

The flowering stimulus originates in a leaf

Early experiments indicated that reception of the photoperiodic stimulus occurs within the leaf. For example, in spinach, an LDP, flowering would occur if the leaves were exposed to long-day periods of light while the bud meristem was masked to simulate short days. Flowering could *not* occur when its leaves were masked to simulate short days while the bud was exposed to long-day periods of light.

INVESTIGATING LIFE

38.14 Interrupting the Night

Knowing that plants measure night duration, the question became whether the dark hours to which a plant is exposed must be continuous. Using SDPs and LDPs as test subjects, Hamner and Bonner interrupted the night with light of different wavelengths.

HYPOTHESIS Red light participates in the photoperiodic timing mechanism.

METHOD Grow plants under short-day conditions, but interrupt the night with light of different wavelengths.

Short-day plants	Light/dark combinations	Long-day plants
Flowering		No flowering

RESULTS

No flowering	R	Flowering
Flowering	FR	No flowering
Flowering	R FR	No flowering
No flowering	R FR R	Flowering
Flowering	R FR R FR	No flowering

CONCLUSION When plants are exposed to red (R) and far-red (FR) light in alternation, the final treatment determines the effect. Phytochrome is the photoreceptor.

FURTHER INVESTIGATION: How would you show that interrupting the day with a brief period of darkness had no effect on flowering?

Go to **yourBioPortal.com** for original citations, discussions, and relevant links for all INVESTIGATING LIFE figures.

─── **yourBioPortal.com** ───
GO TO Animated Tutorial 38.2 • The Effect of Interrupted Days and Nights

INVESTIGATING LIFE

38.15 The Flowering Signal Moves from Leaf to Bud

The receptor for photoperiod, phytochrome, is in the leaf but flowering occurs in the bud meristem. To investigate whether there is a diffusible substance that travels from leaf to bud, James Knott exposed only the leaf to the photoperiodic stimulus.

HYPOTHESIS The leaves measure the photoperiod.

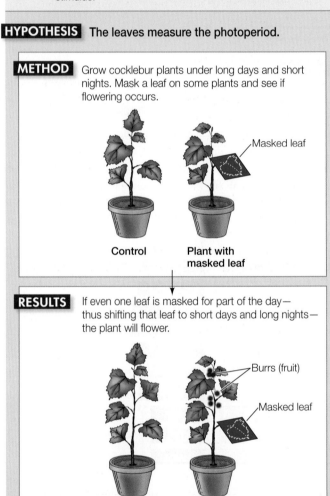

METHOD Grow cocklebur plants under long days and short nights. Mask a leaf on some plants and see if flowering occurs.

Masked leaf

Control — Plant with masked leaf

RESULTS If even one leaf is masked for part of the day—thus shifting that leaf to short days and long nights—the plant will flower.

Burrs (fruit)

Masked leaf

CONCLUSION The leaves measure the photoperiod. Therefore, some signal must move from the induced leaf to the flowering parts of the plant.

FURTHER INVESTIGATION: How would you show experimentally that the flowering signal is the same in different species of plants?

Go to **yourBioPortal.com** for original citations, discussions, and relevant links for all INVESTIGATING LIFE figures.

These "masking" experiments were extended to SDP plants as well (**Figure 38.15**). Because the receptor of the stimulus (in the leaf) is physically separated from the tissue on which the stimulus acts (the bud meristem), the inference can be drawn that a systemic signal travels from the leaf through the plant's tissues to the bud meristem. Other evidence that a diffusible chemical travels from the leaf to the bud meristem signal includes the following:

• If a photoperiodically induced leaf is immediately removed from a plant after the inductive dark period, the plant does not flower. If, however, the induced leaf remains attached to the plant for several hours, the plant will flower. This re-

sult suggests that something is synthesized in the leaf in response to the inductive dark period, and then moves out of the leaf to induce flowering.

• If two or more cocklebur plants are grafted together and if one plant is exposed to inductive long nights and its graft partners are exposed to noninductive short nights, all the plants flower.

- In several species, if an induced leaf from one species is grafted onto another, noninduced plant of a different species, the recipient plant flowers.

Although the transmissible signal was long ago given a name, **florigen** ("flower inducing"), the nature of the signal has only recently been explained.

Florigen is a small protein

The characterization of florigen was made possible by genetic and molecular studies of the model organism *Arabidopsis*, an LDP. Three genes are involved (**Figure 38.16**):

- **FT** (**FLOWERING LOCUS T**) *codes for florigen.* A small protein (20 kDa molecular weight), FT can travel through plasmodesmata. FT is synthesized in phloem companion cells of the leaf and diffuses into the adjacent sieve elements, where it enters the phloem flow to the apical meristem. If *FT* is coupled to an active promoter and expressed at high levels in the shoot meristem, flowering is induced even in the absence of an appropriate photoperiodic stimulus.

- **CO** (**CONSTANS**) *codes for a transcription factor that activates the synthesis of FT.* Like *FT*, *CO* is expressed in leaf companion cells. If *CO* is experimentally overexpressed in the leaf, flowering is induced. Overexpression of *CO* in the apical meristem does not, however, induce flowering, indicating

that its essential activity is in the leaf. CO protein is expressed all the time but is unstable; an appropriate photoperiodic stimulus stabilizes CO so that there is enough to turn on FT synthesis.

- **FD** (**FLOWERING LOCUS D**) *codes for a protein that binds to FT protein when it arrives in the apical meristem.* The FD protein is a transcription factor that when complexed with FT protein, activates promoters for meristem identity genes, such as *APETALA1* (see Figure 38.10). The expression of FD primes meristem cells to change from a vegetative fate to a reproductive fate once florigen arrives.

Before florigen was isolated, grafting experiments indicated that many different plant species could be induced to flower by the same chemical signal. A photoperiod-induced leaf from one species can induce flowering when grafted onto an uninduced plant of another species. Results of molecular experiments confirm that the *FT* gene is involved in photoperiod signaling in many species:

- Transgenic plants (e.g., tobacco and tomato) that express the *Arabidopsis FT* gene at high levels flower regardless of day length.

- Transgenic *Arabidopsis* plants that express high levels of *FT* homologs from other plants (e.g., rice and tomato) flower regardless of day length.

While the molecular basis of the action of florigen has been elucidated, commercial applications of this knowledge have been harder to realize. It was hoped that florigen might be a very small molecule, like auxin or gibberellin that could be sprayed on economically important plants to induce flowering

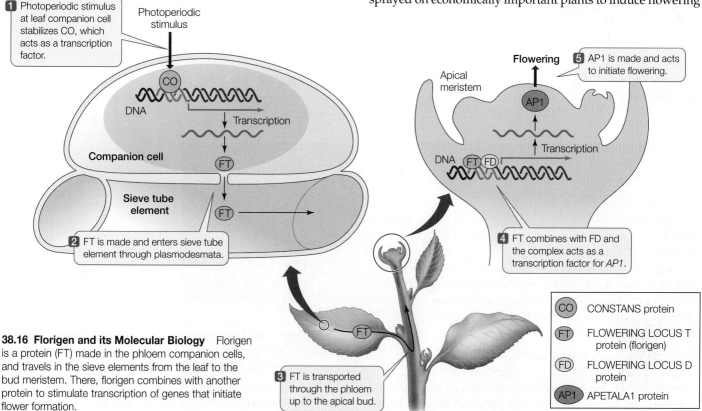

38.16 Florigen and its Molecular Biology Florigen is a protein (FT) made in the phloem companion cells, and travels in the sieve elements from the leaf to the bud meristem. There, florigen combines with another protein to stimulate transcription of genes that initiate flower formation.

at will. The fact that florigen is a protein that cannot readily enter cells from the outside environment makes the development of commercial florigen treatments unlikely.

We have considered the photoperiodic regulation of flowering, from photoreceptors in the leaf to florigen that travels from the induced leaf to the sites of flower formation. In some plant species, however, flowering is induced by other stimuli.

Flowering can be induced by temperature or gibberellin

TEMPERATURE In some plant species, notably certain cereal grains, the environmental signal for flowering is cold temperature, a phenomenon called **vernalization** (Latin *vernus*, "spring"). In both wheat and rye, we distinguish two categories of flowering behavior. Spring wheat, for example, is a typical annual plant: it is sown in the spring and flowers in the same year. Winter wheat is sown in the fall, grows to a seedling, overwinters (often covered by snow), and flowers the following summer. If winter wheat is not exposed to cold in its first year, it will not flower normally the next year.

How vernalization leads to flowering has been elucidated from model organisms such as *Arabidopsis*. In strains of *Arabidopsis* that require vernalization to flower (**Figure 38.17**), a gene called *FLC* (FLOWERING LOCUS C) encodes a transcription factor that blocks the FT–FD florigen pathway (see Figure 38.16) by inhibiting expression of FT and FD. Cold temperature inhibits the synthesis of FLC protein, allowing FT and FD proteins to be expressed and flowering to proceed. Similar proteins control some steps in vernalization in cereals.

GIBBERELLIN *Arabidopsis* plants do not flower if they are genetically deficient in the hormone gibberellin, or if they are treated with an inhibitor of gibberellin synthesis. These observations implicate gibberellins in flowering. Direct application of gibberellins to buds in *Arabidopsis* results in activation of the meristem identity gene *LEAFY*, which in turn promotes the transition to flowering.

Some plants do not require an environmental cue to flower

A number of plant species and strains do not require a photoperiod, vernalization, or gibberellin to flower, but instead flower on cue from an "internal clock." For example, flowering in some strains of tobacco will be initiated in the terminal bud when the stem has grown four phytomers in length (recall that stems are composed of repeating units called phytomers; see Figure 34.1). If such a bud and a single adjacent phytomer is removed and planted, the cutting will flower because the bud has already received the cue for flowering. But the rest of the shoot below the bud that has been removed will not flower because it is only three phytomers long. After it grows an additional phytomer, it flowers. These results suggest that there is something about the *position* of the bud (atop four phytomers of stem) that determines its transition to flowering.

The bud might "know" its position by the concentration of some substance that forms a positional gradient along the length of the plant. Such a gradient could be formed if the root makes a diffusible inhibitor of flowering whose concentration diminishes with plant height. When the plant reaches a certain height, the concentration of the inhibitor would become sufficiently low at the tip of the shoot to allow flowering. What this inhibitor might be is unclear, but there is evidence that it acts by decreasing the amount of FLC, allowing the FT–FD pathway to proceed (just as cold acts on FLC in vernalization). A positional gradient that acts on FLC would be consistent with other mechanisms affecting flowering, which all converge on *LEAFY* and *APETALA1*:

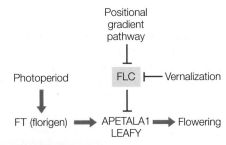

38.17 Vernalization figure diagram

Winter-annual *Arabidopsis* without vernalization

Winter-annual *Arabidopsis* with vernalization

38.17 Vernalization A genetic strain of *Arabidopsis* (winter-annual *Arabidopsis*) requires vernalization for flowering. Without it, the plant is large and vegetative (left), but with the cold period it is smaller and flowers (right).

38.2 RECAP

Flowering of some angiosperms is controlled by night length, a phenomenon called photoperiodism. Gibberellins can induce flowering in some species, as can exposure to low temperatures (vernalization). Some species flower when their stems have grown by a certain amount, independent of environmental cues. All pathways to flowering converge on the meristem identity genes.

- What are the differences between apical meristems, inflorescence meristems, and floral meristems? What genes control the transitions between them? See p. 803 and Figure 38.10

- Explain why "short-day plant" is a misleading term. See p. 805 and Figure 38.13

- What is the evidence for florigen? What is its molecular mechanism of action? See p. 807 and Figures 38.15 and 38.16

We have seen how environmental factors interact with genes to control flowering in angiosperms. The function of flowers is sexual reproduction, which maintains beneficial genetic variation in a population. Many angiosperms, however, also benefit from being able to reproduce asexually.

38.3 How Do Angiosperms Reproduce Asexually?

Although sexual reproduction takes up most of the space in this chapter, asexual reproduction accounts for many of the individual plants present on Earth. This fact suggests that in some circumstances asexual reproduction must be advantageous.

We have noted that genetic recombination is one of the advantages of sexual reproduction. Self-fertilization is a form of sexual reproduction, but offers fewer opportunities for genetic recombination than does cross-fertilization. A diploid, self-fertilizing plant that is heterozygous for a certain locus can produce both kinds of homozygotes for that locus plus the heterozygote among its progeny, but it cannot produce any progeny carrying alleles that it does not itself possess. Nevertheless many self-fertilizing plant species produce viable and vigorous offspring.

Asexual reproduction eliminates genetic recombination altogether. A plant that reproduces asexually produces progeny genetically identical to the parent (clones). What, then, is the advantage of asexual reproduction? If a plant is well adapted to its environment, asexual reproduction allows it to pass on to all its progeny a superior combination of alleles, which might otherwise be separated by sexual recombination.

Many forms of asexual reproduction exist

Stems, leaves, and roots are considered *vegetative organs* and are distinguished from flowers, the reproductive parts of the plant. Asexual reproduction is often accomplished through the modification of a vegetative organ, which is why the term **vegetative reproduction** is sometimes used to describe asexual reproduction in plants. Another type of asexual reproduction, *apomixis*, involves flowers but no fertilization.

VEGETATIVE REPRODUCTION Often the stem is the organ that is modified for vegetative reproduction. As noted earlier, strawberries produce horizontal stems, called *stolons* or runners, which grow along the soil surface, form roots at intervals, and establish potentially independent plants. Asexual reproduction by *tip layers* is accomplished when the tips of upright branches sag to the ground and develop roots, as in blackberry and forsythia.

Some plants, such as potatoes, form enlarged fleshy tips of underground stems, called *tubers*, that can produce new plants (from the "eyes"). *Rhizomes* are horizontal underground stems that can give rise to new shoots. Bamboo is a striking example of a plant that reproduces vegetatively by means of rhizomes. A single bamboo plant can give rise to a stand—even a forest—of plants constituting a single, physically connected entity.

Whereas stolons and rhizomes are horizontal stems, bulbs and corms are short, vertical, underground stems. Lilies and onions form bulbs (**Figure 38.18A**), short stems with many fleshy, highly modified

(A) *Allium* sp.

(B)

Storage leaves grow in layers from the stem of this onion.

The short stem is visible at the bottom of the bulb.

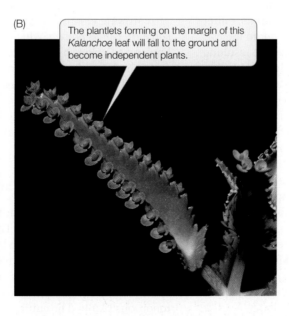

The plantlets forming on the margin of this *Kalanchoe* leaf will fall to the ground and become independent plants.

38.18 Vegetative Organs Modified for Reproduction (A) Bulbs are short stems with large leaves that store nutrients and can give rise to new plants. (B) In *Kalanchoe*, new plantlets can form on leaves.

leaves that store nutrients. These storage leaves make up most of the bulb. Bulbs are thus large underground buds. They can give rise to new plants by dividing or by producing new bulbs from axillary buds. Crocuses, gladioli, and many other plants produce corms, underground stems that function very much as bulbs do. Corms are disclike and consist primarily of stem tissue; they lack the fleshy modified leaves that are characteristic of bulbs.

Stems are not the only vegetative organs modified for asexual reproduction. Leaves may also be the source of new plantlets, as in some succulent plants of the genus *Kalanchoe* (**Figure 38.18B**). Many kinds of angiosperms, ranging from grasses to trees such as aspens and poplars, form interconnected, genetically homogeneous populations by means of suckers—shoots produced by roots. What appears to be a whole stand of aspen trees, for example, may be a clone derived from a single tree by suckers. This is why the leaves of a whole stand of aspens typically turn yellow at the same time.

Plants that reproduce vegetatively often grow in physically unstable environments such as eroding hillsides. Plants with stolons or rhizomes, such as beach grasses, rushes, and sand verbena, are common pioneers on coastal sand dunes. Rapid vegetative reproduction enables these plants, once introduced, not only to multiply but also to survive burial by the shifting sand; in addition, the dunes are stabilized by the extensive network of rhizomes or stolons that develops. Vegetative reproduction is also common in some deserts, where the environment is often not suitable for seed germination and the establishment of seedlings.

APOMIXIS   Some plants produce flowers but use them to reproduce asexually rather than sexually. Dandelions, blackberries, some citrus, and some other plants reproduce by the asexual production of seeds, called **apomixis**. As described earlier, in alternation of generations meiosis typically reduces the number of chromosomes in gametes by half, and fertilization restores the sporophytic (diploid) number of chromosomes in the zygote. In a female gametophyte undergoing apomixis, either meiosis begins and the chromosomes do not undergo meiosis II, or meiosis does not occur at all. In either case, the resulting gamete is diploid. Cells within the ovule simply develop into the embryo and the ovary wall develops into a fruit. The result of apomixis is a fruit with seeds that are genetically identical to the parent plant.

Apomixis would be considered an oddity of the plant reproductive world were it not for its potential use in propagating crop plants. You may recall from Chapter 12 that many crop plants (such as corn) are grown as hybrids because the progeny of a cross between two inbred lines are often superior to either of their parents. The explanation for this phenomenon, called *hybrid vigor*, is not completely understood. One hypothesis attributes the superiority of hybrids to the suppression of undesirable recessive alleles from one parent by dominant alleles from the other. Another hypothesis states that certain advantageous combinations of alleles can be obtained by crossing two inbred strains.

Unfortunately, once a farmer has obtained a hybrid with desirable characteristics, (s)he cannot use those plants for further crosses with themselves (selfing) to get more seeds for the next generation. You can imagine the genetic chaos when a hybrid, which is heterozygous at many of its loci (e.g., *AaBbCcDdEe*, etc.), is crossed with itself: there will be many new combinations of alleles (e.g., *AabbCCDdee*, etc.), resulting in highly variable progeny. The only way to reliably reproduce the hybrid is to maintain populations of the original parents to cross again each year.

However, if a hybrid carried a gene for apomixis, it could reproduce asexually, and its offspring would be genetically identical to itself. So the search is on for a gene for apomixis that could be introduced into desirable crops and allow them to be propagated indefinitely. (A recently published detective novel, *Day of the Dandelion,* explores this idea.)

Researchers recently found a strain of *Arabidopsis* that exhibits apomixis as a result of a mutation in a single gene called *dyad*. In normal plants, *dyad* is essential for chromosome organization, specifically synapsis, during meiosis I (see Figure 11.17). In the apomictic strain, meiosis I resembles mitosis, and the chromosomes replicate again before what would be meiosis II. The result is diploid cells that are genetically identical to the parent instead of the genetically recombined haploid gametes that normally result from meiosis. Scientists are trying to isolate and transfer such apomictic genes into corn and other cereal crops with the hope that plant breeders can use apomixis to propagate plants with desirable hybrid traits (such as high yields, and disease- and insect-resistance) without compromising their hybrid vigor.

Vegetative reproduction has a disadvantage

Vegetative reproduction is highly efficient in an environment that is stable over the long term. A change in the environment, however, can leave an asexually reproducing species at a disadvantage.

A striking example is provided by the demise of the English elm, *Ulmus procera*, which was apparently introduced into England as a clone by the ancient Romans. This tree reproduces asexually by suckers and is incapable of sexual reproduction. In 1967, Dutch elm disease first struck the English elms. After two millennia of clonal growth, the population lacked genetic diversity, and no individuals carried genes that would protect them against the disease. Today the English elm is all but gone from England.

Vegetative reproduction is important in agriculture

Farmers and gardeners take advantage of some natural forms of vegetative reproduction. They have also developed new types of asexual reproduction by manipulating plants. One of the oldest methods of vegetative reproduction used in agriculture consists of simply making cuttings of stems, inserting them in soil, and waiting for them to form roots and thus become autonomous plants. The cuttings are usually encouraged to root

38.19 Grafting Grafting—attaching a piece of a plant to the root or root-bearing stem of another plant—is a common horticultural technique. The "host" root or stem is the stock; the upper grafted piece is the scion. In the photo, a Bing cherry scion is being grafted onto a hardier stock.

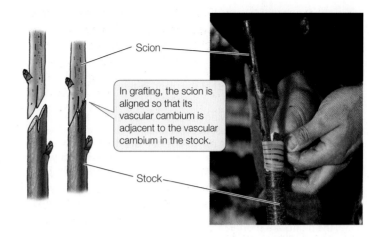

Scion

In grafting, the scion is aligned so that its vascular cambium is adjacent to the vascular cambium in the stock.

Stock

by treatment with a plant hormone, auxin, as described in Section 37.3.

Horticulturists reproduce many woody plants by **grafting**—attaching a bud or a piece of stem from one plant to the root or root-bearing stem of another plant. The part of the resulting plant that comes from the root-bearing "host" is called the **stock**; the part grafted on is the **scion** (**Figure 38.19**).

For a graft to succeed, the vascular cambium of the scion must associate with that of the stock. By cell division, both cambiums form masses of wound tissue. If the two masses meet and connect, the resulting continuous cambium can produce xylem and phloem, allowing transport of water and minerals to the scion and of photosynthate to the stock. Grafts are most often successful when the stock and scion belong to the same or closely related species. Much fruit grown for market in the United States is produced on grafted trees. Another example is wine grapes. The roots of most grape strains are susceptible to soil pests, and so grape varieties are grafted onto root stocks that have pest resistance.

Scientists in universities and commercial laboratories have been developing new ways to produce useful plants through tissue culture. Because many plant cells are totipotent, cultures of undifferentiated tissue can give rise to entire plants, as can small pieces of tissue cut directly from a parent plant. Tissue cultures sometimes are used commercially to produce new plants. This is common in the forestry industry, where uniformity of trees is desirable.

Culturing tiny bits of apical meristem can produce plants free of viruses. Because apical meristems lack developed vascular tissues, viruses tend not to enter them. Treatment with hormones causes a single apical meristem to give rise to 20 or more shoots; thus, a single plant can give rise to millions of genetically identical plants within a year by repeated meristem

culturing. Using this approach, strawberry and potato producers are able to start each year's crop from virus-free plants.

38.3 RECAP

Angiosperms may reproduce asexually by means of modified stems, roots, or leaves, or by apomixis. Asexual reproduction is advantageous when a plant has a superior genotype well adapted to its environment, but decreases the genetic diversity of plant populations.

- How does apomixis differ from sexual reproduction? **See p. 810**

- Explain how vegetative reproduction of plants is advantageous to humans. **See pp. 810–811**

We have seen how angiosperms reproduce sexually and asexually. A disadvantage of asexual reproduction is that its genetic inflexibility may leave a population unable to cope with new challenges. In the next chapter we focus on the mechanisms that have evolved in plants to cope with biological and physical challenges in their environment.

CHAPTER SUMMARY

38.1 How Do Angiosperms Reproduce Sexually?

- Sexual reproduction promotes genetic diversity in a population. The flower is an angiosperm's structure for sexual reproduction.

- Flowering plants have microscopic gametophytes. The megagametophyte is the **embryo sac**, which typically contains eight nuclei in a total of seven cells. The microgametophyte is the **pollen grain**, which usually contains two cells. **Review Figure 38.2, ANIMATED TUTORIAL 38.1**

- Following **pollination**, the pollen grain delivers **sperm cells** to the embryo sac by means of a **pollen tube**.

- Most angiosperms exhibit **double fertilization**: one sperm nucleus fertilizes the **egg**, forming a zygote, and the other

sperm nucleus unites with the two **polar nuclei** to form a triploid **endosperm**. **Review Figure 38.6**

- Plants have both physical and genetic methods of preventing inbreeding. Physical separation of the gametophytes and genetic **self-incompatibility** prevent self-pollination.

- The zygote develops into an embryo (with an attached **suspensor**), which remains quiescent in the seed until conditions are right for germination. **Review Figure 38.7, WEB ACTIVITY 38.1**

- Ovules develop into seeds, and the ovary wall and the enclosed seeds develop into a **fruit**.

- The hormone **abscisic acid** promotes seed development and dormancy.

38.2 What Determines the Transition from the Vegetative to the Flowering State?

- In **annuals** and **biennials**, flowering and seed formation usually leads to death of the rest of the plant. **Perennials** live a long time and typically reproduce repeatedly.

- For a vegetatively growing plant to flower, an apical meristem in the shoot system must become an **inflorescence meristem**, which in turn must give rise to one or more **floral meristems**. These events are under the influence of **meristem identity genes** and **floral organ identity genes**. Review Figure 38.10

- Some plants flower in response to **photoperiod**. **Short-day plants** flower when the nights are longer than a **critical night length** specific to each species; **long-day plants** flower when the nights are shorter than a critical night length. Review Figure 38.13

- The mechanism of photoperiodic control involves phytochromes and a biological clock. Review Figure 38.14, **ANIMATED TUTORIAL 38.2**

- A flowering signal, called **florigen**, is formed in a photoperiodically induced leaf and is translocated to the sites where flowers will form. Review Figures 38.15 and 38.16

- In some angiosperm species, exposure to low temperatures—**vernalization**—is required for flowering; in others internal signals (one of which is gibberellin in some plants) induce flowering. All of these stimuli converge on the meristem identity genes.

38.3 How Do Angiosperms Reproduce Asexually?

- Asexual reproduction allows rapid multiplication of organisms that are well suited to their environment.

- **Vegetative reproduction** involves the modification of a vegetative organ—usually the stem—for reproduction.

- Some plant species produce seeds asexually by **apomixis**.

- Horticulturists often **graft** different plants together to take advantage of favorable properties of both **stock** and **scion**. Review Figure 38.19

SELF-QUIZ

1. Sexual reproduction in angiosperms
   a. is by way of apomixis.
   b. requires the presence of petals.
   c. can be accomplished by grafting.
   d. gives rise to genetically diverse offspring.
   e. cannot result from self-pollination.

2. The typical angiosperm female gametophyte
   a. is called a microspore.
   b. has eight nuclei.
   c. has eight cells.
   d. is called a pollen grain.
   e. is carried to the male gametophyte by wind or animals.

3. Pollination in angiosperms
   a. always requires wind.
   b. never occurs within a single flower.
   c. always requires help by animal pollinators.
   d. is also called fertilization.
   e. makes most angiosperms independent of external water for reproduction.

4. Which statement about double fertilization is *not* true?
   a. It is found in most angiosperms.
   b. It takes place in the microsporangium.
   c. One of its products is a triploid nucleus.
   d. One sperm nucleus fuses with the egg nucleus.
   e. One sperm nucleus fuses with two polar nuclei.

5. The suspensor
   a. gives rise to the embryo.
   b. is heart-shaped in eudicots.
   c. separates the two cotyledons of eudicots.
   d. ceases to elongate early in embryonic development.
   e. is larger than the embryo.

6. Which statement about photoperiodism is *not* true?
   a. It is related to the biological clock.
   b. A phytochrome plays a role in the timing process.
   c. It is based on measurement of the length of the night.
   d. Some plants do not flower in response to photoperiod.
   e. It is limited to plants.

7. Florigen is
   a. produced in the leaves and transported to the apical bud.
   b. produced in the roots and transported to the shoots.
   c. produced in the coleoptile tip and transported to the base.
   d. the same as gibberellin.
   e. activated by prolonged (more than a month) high temperature.

8. Which statement about vernalization is *not* true?
   a. It decreases the abundance of an inhibitor of flowering.
   b. Vernalization involves exposure to cold temperatures.
   c. It only occurs in crop plants such as cereals.
   d. In the vernalized state, the synthesis of FLC protein is inhibited.
   e. If winter wheat is not exposed to cold, it will not flower.

9. Which of the following does *not* participate in asexual reproduction?
   a. Stolon
   b. Rhizome
   c. Zygote
   d. Tuber
   e. Corm

10. Apomixis involves
    a. sexual reproduction.
    b. complete meiosis.
    c. fertilization.
    d. a diploid embryo.
    e. no production of a seed.

FOR DISCUSSION

1. Which method of reproduction might a farmer prefer for a crop plant that reproduces both sexually and asexually? Why?

2. Thompson Seedless grapes are produced by vines that are triploid. Think about the consequences of this chromosomal condition for meiosis in the flowers. Why are these grapes seedless? Describe the role played by the flower in fruit formation when no seeds are being formed. How do you suppose Thompson Seedless grapes are propagated?

3. Poinsettias are popular ornamental plants that typically bloom just before Christmas. Their flowering is photoperiodically controlled. Are they long-day or short-day plants? Explain.

4. You plan to induce the flowering of a crop of long-day plants in the field by using artificial light. Is it necessary to keep the lights on continuously from sundown until the point at which the critical day length is reached? Why or why not?

ADDITIONAL INVESTIGATION

The isolation of *dyad*, the *Arabidopsis* gene that controls apomixis, offers possibilities for crop plant breeding. How would you investigate the possibility of using the mutant allele of this gene to produce hybrid corn plants that can be propagated and retain their hybrid nature?

Sharing plants' defensive strategies

The tropical rainforest of the eastern slopes of the Andes teems with plant life. This region of the Amazon Basin is host to about 40 species of *Cinchona*, a genus of trees that grow to a height of about 20 meters. In this moist environment, *Cinchona* trees grow rapidly and thrive, despite the many natural enemies that threaten their survival and growth.

Unlike animals, which can sometimes escape their enemies, plants must confront their enemies in place. Over evolutionary time, plants with the ability to fight off attackers have survived and passed on that ability to their offspring. In many instances, that ability comes in the form of defensive chemicals. In the case of *Cinchona*, one of those chemicals is quinine, a bitter molecule that is toxic to insects. It may be familiar to you as an ingredient in tonic water.

The Quechua, a group of people native to the Andes forests, have a long history of putting local plants to medicinal use. In the tropics, malaria has long been, and still is, a common and lethal disease. Even today, about 400 million cases arise worldwide, and over 1.5 million people die from malaria each year. Centuries ago, the Quechua found that a tea made from the bark of *Cinchona* trees was highly effective in treating malaria. Legend has it that the tree got its name from the Peruvian countess of Cinchon, who was cured of malaria in 1638 when her physician got some bark from the Quechua. Use of the bark extract quickly spread around the world.

In 1820, the active ingredient of *Cinchona* bark, quinine, was isolated and became the mainstay of malaria treatment. Malaria is caused by an apicomplexan parasite that infects red blood cells. Quinine kills the parasite by interfering with its ability to break down hemoglobin. Unfortunately, mutations render some parasites resistant to quinine and its chemical derivatives. Over time, treatment of billions of people with quinine drugs has selected for parasites with genes for quinine resistance. An urgent need for alternative treatments has led scientists to another plant and its defensive chemical.

Artemisia annua, or sweet wormwood, grows in forests all over the world. It synthesizes a molecule called artemisinin that helps defend the plant against insects. For over 3000 years, people in Asia have made a curative tea from sweet wormwood. In 1972 Japanese chemists isolated artemisinin and found that it works by reacting with iron in red blood cells to form free radicals, which damage lipids and DNA in the infecting par-

The Source of Quinine *Cinchona* trees from the Amazon rainforest synthesize a defensive chemical, quinine, that has been used to treat people with malaria.

A Substitute for Quinine Sweet wormwood (*Artemisia annua*) grows in forests throughout the world. It synthesizes a defensive chemical, called artemisinin, that is now being used to treat people with malaria whose infection is resistant to quinine drugs.

asite. Since this mechanism of action differs from that of quinine, it was thought that artemisinin might be effective in treating quinine-resistant malaria—and it is. Indeed, during the Vietnam War, drinking *Artemisia* tea helped Vietnamese soldiers cope with the quinine-resistant malaria that struck American soldiers. Since 2000, artemisinin has become a mainstay of malaria treatment. Millions of people take it every day.

Until recently, hundreds of such plant chemicals were contemplated only in the context of plant biochemistry. Today we view them as adaptations arising from a plant's interactions with its environment.

IN THIS CHAPTER we describe the biological (biotic) and nonbiological (abiotic) environmental challenges faced by plants and how plants deal with them. We begin by examining interactions between plants and plant pathogens, such as fungi, and then consider plant interactions with herbivores. The chapter concludes by considering plant adaptations to abiotic factors such as water, temperature, salinity, and heavy metals.

39.1 How Do Plants Deal with Pathogens?

Botanists know of dozens of diseases, with many different genetic strains, that can kill a wheat plant, each of them caused by a different pathogen. Plant pathogens—which include bacteria, fungi, protists, and viruses—are part of nature, and for that reason alone they merit our study in biology. Because we humans depend on plants for our food, however, the stakes in our effort to understand plant pathology are especially high. That is why, just as medical schools have departments of pathology, many universities in agricultural regions have departments of plant pathology.

Successful infection by a pathogen can have significant effects on a plant, reducing photosynthesis and causing massive cell and tissue death. Like the responses of the human immune system (see Chapter 42), the responses by which plants fight off disease are varied and fascinating. Plants and pathogens have evolved together in a continuing "arms race": pathogens have evolved mechanisms with which to attack plants, and plants have evolved mechanisms for defending themselves against those attacks. Each set of mechanisms uses information from the other. For example, the pathogen's enzymes may break down the plant's cell walls, and the breakdown products may signal to the plant that it is under attack. In turn, the plant's defenses alert the pathogen that it, too, is under attack.

An arms race of global importance is under way to combat wheat rust, a fungus that can devastate wheat crops (**Figure 39.1**). In 1999, a new genetic strain of the wheat rust fungus *Puccinia graminis* was identified in a wheat field in Uganda. Although many strains of wheat have natural resistance to other strains of rust, the new fungal strain, called Ug99, overcomes resistance in almost all of them. It has spread to the Middle East, and there is a good probability that it has already reached Asia. Over 90 percent of the wheat strains in its path are susceptible. Scientists are racing to discover wheat genes that confer resistance to Ug99 and to implement genetic crosses to get this resistance into the wheat strains under widespread cultivation. Failure in this arms race could have disastrous consequences for the global food supply.

What determines the outcome of a battle between a plant and a pathogen? The key to success for the plant is to respond to information from the pathogen quickly and massively. Plants use both mechanical and chemical defenses in this effort. These defenses can either be **constitutive**, always present in the plant, or **induced**, produced in reaction to damage or stress.

Mechanical defenses include physical barriers

A plant's first line of defense is its outer surfaces, which can prevent the entry of pathogens. As Section 34.3 describes, the or-

39.1 Wheat Rust A field of wheat infected with the wheat rust fungus (right) has a much reduced growth rate and has produced much less grain than an uninfected field of wheat that is resistant to the fungus (left). Inset: Wheat rust causes cell death in leaves.

gans of a growing plant that are exposed to the outside environment are covered with cutin, suberin, and waxes. These substances not only prevent water loss by evaporation, but can also prevent fungal spores and bacteria from entering the underlying tissues. Some fungi get around this defense, however, by secreting enzymes that hydrolyze components of these substances, breaking them down to gain entry.

Much more important to the plant are the induced resistance mechanisms initiated when a pathogen lands on a plant. As we discuss these mechanisms, refer to the overview in **Figure 39.2**.

Plants can seal off infected parts to limit damage

While animals generally repair tissues that have been damaged by pathogens, plants do not. Instead, plants seal off and sacrifice damaged tissues so that the rest of the plant does not become infected. Plants have the option of discarding damaged tissues because most plants, unlike most animals, can replace damaged parts by growing new ones.

Before we look at the details of the defensive process, note that a key response by plant cells to invasion by pathogens is the rapid deposition of additional polysaccharides, as well as a cell wall protein called *extensin*, on the inside of the cell wall. These macromolecules not only reinforce the mechanical barrier formed by the cell wall, but also block the plasmodesmata,

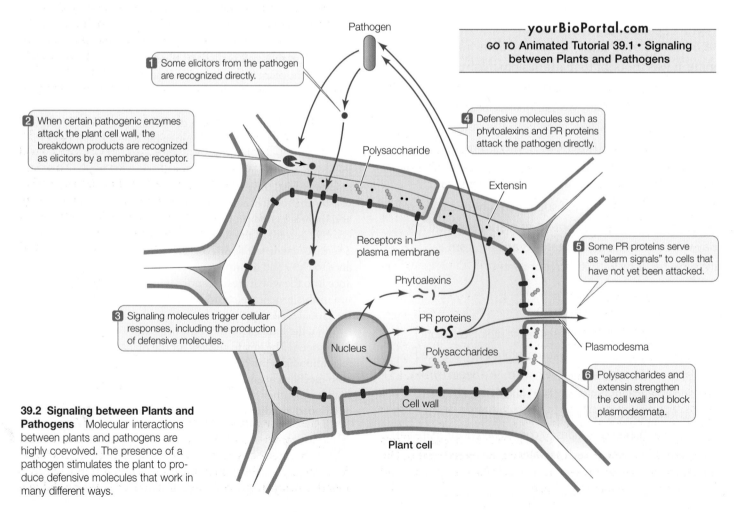

yourBioPortal.com
GO TO Animated Tutorial 39.1 • Signaling between Plants and Pathogens

1 Some elicitors from the pathogen are recognized directly.

2 When certain pathogenic enzymes attack the plant cell wall, the breakdown products are recognized as elicitors by a membrane receptor.

3 Signaling molecules trigger cellular responses, including the production of defensive molecules.

4 Defensive molecules such as phytoalexins and PR proteins attack the pathogen directly.

5 Some PR proteins serve as "alarm signals" to cells that have not yet been attacked.

6 Polysaccharides and extensin strengthen the cell wall and block plasmodesmata.

Pathogen

Polysaccharide

Extensin

Receptors in plasma membrane

Phytoalexins

PR proteins

Polysaccharides

Nucleus

Cell wall

Plasmodesma

Plant cell

39.2 Signaling between Plants and Pathogens Molecular interactions between plants and pathogens are highly coevolved. The presence of a pathogen stimulates the plant to produce defensive molecules that work in many different ways.

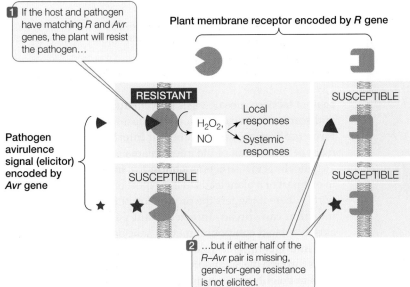

1 If the host and pathogen have matching *R* and *Avr* genes, the plant will resist the pathogen...

Plant membrane receptor encoded by *R* gene

Pathogen avirulence signal (elicitor) encoded by *Avr* gene

RESISTANT

H_2O_2, NO

Local responses

Systemic responses

SUSCEPTIBLE

SUSCEPTIBLE

SUSCEPTIBLE

2 ...but if either half of the *R–Avr* pair is missing, gene-for-gene resistance is not elicited.

39.3 Gene-for-Gene Resistance If a gene in a pathogen that codes for an elicitor "matches" a gene in a plant that codes for a receptor, the receptor binds the elicitor, and a defensive response results.

The signal transduction pathway started by receptor–elicitor binding is mediated by the production of nitric oxide and the toxic peroxide H_2O_2. Together, these substances initiate local defenses and, later, systemic defenses (defenses in parts of the plant distant from the attack site).

Receptor–elicitor binding evokes the hypersensitive response

When infected by certain fungi and bacteria, plants try to contain the infection locally by what is called the **hypersensitive response**. This three-pronged response involves the production of defensive compounds as well as physical isolation of the infection site (see Figure 39.2, labels 3 and 4).

PHYTOALEXINS **Phytoalexins** are antibiotics produced by infected plants that are toxic to many fungi and bacteria. Most are small molecules, and each is made by only a few plant species. They are produced by infected cells, and by their immediate neighbors, within hours of the onset of infection. Because their antimicrobial activity is nonspecific, phytoalexins can destroy many species of fungi and bacteria in addition to the one that originally triggered their production. Phytoalexins also kill the plant cells that produced them, thus sealing off the infection.

Phytoalexins are an example of an induced plant defense: they are not normally present in plants, but are synthesized rapidly when a bacterial or fungal infection occurs. Physical injuries and viral infections can also induce the production of phytoalexins.

Neither the mechanism of phytoalexin induction by pathogens, nor the specific effects of phytoalexins on invading organisms, is clear. Camalexin, a phytoalexin made by the model organism *Arabidopsis thaliana*, is being used to investigate these phenomena. It is synthesized by the plant from the amino acid tryptophan:

Tryptophan   Camalexin

PATHOGENESIS-RELATED PROTEINS Plants produce several types of **pathogenesis-related proteins**, or **PR proteins**. Some are enzymes that break down the cell walls of pathogens. Chitinase, for example, is a PR protein that breaks down chitin, which is found in many fungal cell walls. In some cases, the breakdown

---

limiting the ability of viral pathogens to move from cell to cell. The polysaccharides also serve as a base on which lignin may be laid down. Lignin enhances the mechanical barrier, and the toxicity of lignin precursor chemicals makes the cell inhospitable to some pathogens. These lignin building blocks are only one example of the toxic substances that plants use as chemical defenses.

Plant responses to pathogens may be genetically determined

Plant pathogens cause the host plant to activate various chemical defense responses. Several distinctive molecules called *elicitors* have been identified that trigger these plant defenses. These molecules vary in character, from peptides made by bacteria to cell wall fragments from fungi. Elicitors can also be derived from fragments of plant cell wall components broken down by pathogens (see Figure 39.2, labels 1 and 2). Pathogen genes that code for elicitors are called **avirulence genes (*Avr*)**; there are hundreds of such genes. When an elicitor enters a plant cell, it may encounter a receptor protein in the cytoplasm. If the receptor binds to the elicitor, a signal transduction pathway is set in motion that leads to the plant's defensive response. If the plant has no receptor to bind to the elicitor, the plant does not defend itself.

Over 50 years ago, plant pathologist Harold Flor at North Dakota State University studied the susceptibility of various genetic strains of cereal grain plants to various strains of rust fungus. He proposed that susceptibility is determined by a genetic relationship between the pathogen and the plant (**Figure 39.3**). Today we know that plants have **resistance genes (*R*)**, and we know that these resistance genes code for receptors. We also know that pathogens have *Avr* genes that code for elicitors. If the *R* and *Avr* genes match, there is resistance: the molecules they encode will bind to each other and set off a response. Flor's idea, known as **gene-for-gene resistance**, has been borne out by molecular studies, and it has had a great influence on crop plant breeding for resistance to pathogens.

39.4 Sealing Off the Pathogen and the Damage These necrotic lesions on the leaves of a broad bean plant are a response to "chocolate spot" fungus, *Botrytis fabae*.

products of the pathogen's cell walls serve as elicitors that trigger further defensive responses. Other PR proteins may serve as alarm signals to plant cells that have not yet been attacked (Figure 39.2, label 5). In general, PR proteins appear not to be rapid-response weapons; rather, they act more slowly, perhaps after other mechanisms have blunted the pathogen's attack.

PHYSICAL ISOLATION   A third component of the hypersensitive response seals off the damage and the pathogen from the rest of the plant. Cells around the site of infection undergo apoptosis, preventing the spread of the pathogen by depriving it of nutrients. Some of these cells produce phytoalexins and other chemicals before they die. The dead tissue, called a *necrotic lesion*, contains and isolates what is left of the infection (**Figure 39.4**). The rest of the plant remains free of the infecting pathogen.

Systemic acquired resistance is a form of long-term "immunity"

The hypersensitive response is not the only defensive response initiated by receptor–elicitor binding. **Systemic acquired resistance** is a general increase in the resistance of the entire plant to a wide range of pathogens. It is not limited to the pathogen that originally triggered it, or to the site of the original infection, and its effect may last as long as an entire growing season.

This defensive response is initiated by salicylic acid, a defensive chemical produced during the local hypersensitive response. Since ancient times, people in Asia, Europe, and the Americas have used willow (*Salix*) leaves and bark to relieve pain and fever. The active ingredient in willow is salicylic acid, the substance from which aspirin is derived. It now appears that all plants contain at least some salicylic acid.

Systemic acquired resistance is accompanied by the synthesis of PR proteins. Treatment of plants with salicylic acid or aspirin leads to the production of PR proteins and to resistance to pathogens. It provides substantial protection, for example,

against tobacco mosaic virus (a well-studied plant pathogen) and some other viruses. In some cases salicylic acid inhibits virus replication, and in others it interferes with the movement of viruses out of the infected area.

Salicylic acid also acts as a hormone. In some cases, infection in one part of a plant leads to the export of salicylic acid to other parts, where it triggers the production of PR proteins before the infection can spread. Infected plant parts also produce the closely related compound methyl salicylate (also known as oil of wintergreen). This volatile substance travels to other plant parts through the air and may trigger the production of PR proteins in neighboring plants that have not yet been infected.

Plants develop specific immunity to RNA viruses

Before we leave the topic of plant defenses against pathogens, let's consider a recently discovered defense mechanism directed against a specific pathogen type: RNA viruses (viruses that have RNA instead of DNA as their hereditary material). The plant uses its own enzymes to convert some of the single-stranded RNA of the invading virus into *double-stranded RNA* (dsRNA) and to chop that dsRNA into small pieces called *small interfering RNAs* (siRNAs). Some of the viral RNA is transcribed, forming mRNAs that advance the infection. However, the siRNAs interact with another cellular component to degrade those mRNAs, blocking viral replication. This phenomenon is an example of *RNA interference* (RNAi) (see Section 18.4).

The immunity conferred by RNAi spreads quickly throughout the entire plant through plasmodesmata. However, the establishment of immunity depends on the extent of the original infection and the speed of the plant's response. Plant viruses are continuing their side of the arms race: most have evolved mechanisms to confound RNA interference.

39.1 RECAP

In the hypersensitive response to infection by pathogens, plants produce two types of chemical defenses and seal off infected areas. Systemic acquired resistance, providing a longer-lasting, more general immunity, may follow.

- Name two types of defensive compounds produced by plant cells when they are infected by bacteria or fungi. **See p. 817 and Figure 39.2**

- How do *R* and *Avr* genes determine which pathogens a plant can resist? **See p. 817 and Figure 39.3**

- How do infected plant cells signal infection to other parts of the plant, or other plants? **See p. 818**

Not all biological threats to plants come from pathogens. Another threat comes from the many animals, from inchworms to elephants, that eat plants.

39.2 How Do Plants Deal with Herbivores?

Herbivores—animals that eat plants—depend on plants for energy and nutrients. Their foraging activities cause physical damage to plants, and they often spread disease among plants as well. While the majority of herbivores are insects (**Figure 39.5**), every major class of vertebrates includes at least a few herbivores (see also Section 56.2, which discusses herbivory in the ecological context of species interactions). Plants cannot evade their consumers by running away, but they have many other ways of protecting themselves against herbivory.

While in most cases, the physical damage caused by herbivores is severe, sometimes limited herbivory may be harmless or even beneficial. Before turning to the ways that plants resist herbivory, let's examine those few cases in which herbivory enhances plant growth.

Herbivory increases the growth of some plants

How detrimental is herbivory to plants? How well have plants adapted to their place in the food web? Like plants and pathogens, plants and herbivores have evolved together, each acting as an agent of natural selection on the other. This coevolution has led to arms races in some cases, but it has favored increased photosynthetic production in some plant species subjected to herbivory.

Removal of some leaves from a plant usually increases the rate of photosynthesis in the remaining leaves for several reasons:

- Nitrogen obtained from the soil by the roots no longer needs to be divided among as many leaves.
- The export of sugars and other photosynthetic products from the leaves may be enhanced because the demand for those products in the roots is undiminished, while the sources for those products—leaves—have been decreased.
- The removal of older or dead leaves makes more light available to the younger, more active leaves or leaf parts.

Grasses are especially tolerant of herbivory because, unlike most other plants, which grow from shoot apical meristems, grasses grow from the base of the shoot and leaf, so their growth is not cut short by grazing.

In western North America, mule deer and elk graze many plants, including a wildflower called scarlet gilia (*Ipomopsis aggregata*). Although grazing removes about 95 percent of the aboveground plant, the scarlet gilia quickly regrows not one, but up to four replacement stems (**Figure 39.6**). Grazed individuals produce three times as many fruits by the end of the growing season as do ungrazed plants.

Mechanical defenses against herbivores are widespread

All parts of the plant body offer some resistance against herbivores. In addition, plants have a number of constitutive anatom-

(A) *Locusta migratoria*

(B) *Manduca sexta*

39.5 Insect Herbivores The great majority of herbivores are insects. (A) Some herbivores, such as this locust, are generalists that will attack nearly any plant. (B) Others are specialists, like this tobacco hornworm, which feeds only on tobacco plants.

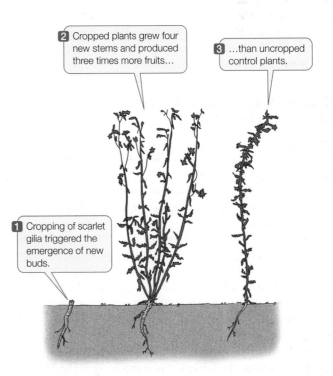

2 Cropped plants grew four new stems and produced three times more fruits…

3 …than uncropped control plants.

1 Cropping of scarlet gilia triggered the emergence of new buds.

39.6 Overcompensation for Being Eaten Experiments confirm that herbivory increases the growth of some plants.

TABLE 39.1
Secondary Metabolites Used in Defense

CLASS	TYPE	ROLE	EXAMPLE
Nitrogen-containing	Alkaloids	Neurotoxin	Nicotine in tobacco
	Glycosides	Inhibit electron transport	Dhurrin in sorghum
	Nonprotein amino acids	Disrupt protein structure	Canavanine in jack bean
Ephedrine (an alkaloid)			
Nitrogen–sulfur-containing	Glucosinolates	Inhibit respiration	Methylglucosinolate in cabbage
Methylglucosinolide			
Phenolics	Coumarins	Block cell division	Umbelliferone in carrots
	Flavonoids	Phytoalexins	Capsidol in peppers
	Tannins	Inhibit enzymes	Gallotannin in oak trees
Umbelliferone			
Terpenes	Monoterpenes	Neurotoxins	Pyrethrin in chrysanthemums
	Diterpenes	Disrupt reproduction and muscle function	Gossypol in cotton
	Triterpenes	Inhibit ion transport	Digitalis in foxglove
	Sterols	Block animal hormones	Spinasterol in spinach
Pyrethrin	Polyterpenes	Deter feeding	Latex in *Euphorbia*

ical features, such as trichomes, thorns, spines, or hairs, that are specialized for defense. An example of an induced mechanical defense is the production of latex. When they are injured by an herbivore attack, some plants, such as *Euphorbia* species, produce a thick, white aqueous suspension of cellular debris, oils, and resins called *latex*. Insects trapped by this sticky substance starve to death.

Plants produce chemical defenses against herbivores

Many plants attract, resist, and inhibit other organisms with special chemicals known as secondary metabolites. You learned about two of these chemicals, quinine and artemisinin, in the opening of this chapter.

Primary metabolites are substances—such as proteins, nucleic acids, carbohydrates, lipids, and their building blocks—that are produced and used by all living organisms, including plants. They and their metabolic products are used in basic cellular processes such as photosynthesis, respiration, and nutrient uptake. **Secondary metabolites** are substances that are not used for basic cellular processes. Each is found in only certain plants or plant groups.

The more than 10,000 known secondary metabolites range in molecular mass from about 70 to more than 390,000 daltons, but most have a low molecular mass (**Table 39.1**). Some are produced by only a single plant species, while others are characteristic of entire genera or even families. The effects of defensive secondary metabolites on animals are diverse. Some act on the nervous systems of herbivorous insects, mollusks, or mammals. Others mimic the natural hormones of insects, causing some larvae to fail to develop into adults. Still others damage the digestive tracts of herbivores. Some secondary metabolites are toxic to fungal pests. As we saw at the opening of this chapter, humans make use of many secondary metabolites as pharmaceuticals and pesticides.

The secondary metabolite nicotine was one of the first insecticides to be used by farmers and gardeners. This molecule kills insects by acting as an inhibitor of nervous system function. Yet commercial varieties of tobacco and related plants that produce nicotine are still attacked, with moderate damage, by pests such as the tobacco hornworm (see Figure 39.5B). Given that observation, does nicotine really deter herbivores? Biologists answered this question conclusively with a study that used tobacco plants in which an enzyme involved in nicotine biosynthesis had

INVESTIGATING LIFE

39.7 Nicotine Is a Defense against Herbivores

The secondary metabolite nicotine, made by tobacco plants, is an insecticide, yet most commercial varieties of tobacco are susceptible to insect attack. Ian Baldwin demonstrated that a tobacco strain with a reduced nicotine concentration was more susceptible to insect damage.

HYPOTHESIS Nicotine helps protect tobacco plants against insects.

METHOD

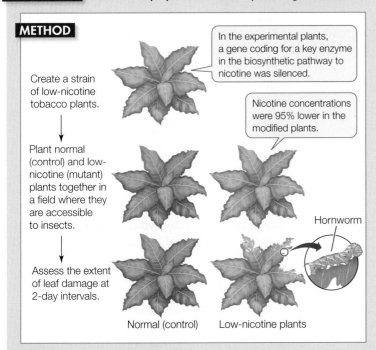

Create a strain of low-nicotine tobacco plants.

In the experimental plants, a gene coding for a key enzyme in the biosynthetic pathway to nicotine was silenced.

Nicotine concentrations were 95% lower in the modified plants.

Plant normal (control) and low-nicotine (mutant) plants together in a field where they are accessible to insects.

Hornworm

Assess the extent of leaf damage at 2-day intervals.

Normal (control)   Low-nicotine plants

RESULTS The low-nicotine plants suffered more than twice as much leaf damage as did the wild-type controls.

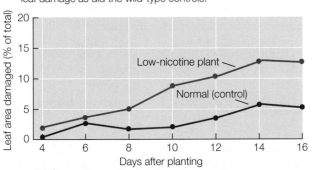

Low-nicotine plant

Normal (control)

Leaf area damaged (% of total)

Days after planting

CONCLUSION Nicotine provides tobacco plants with at least some protection against insects.

FURTHER INVESTIGATION: Treatment of tobacco plants with jasmonate (a hormone) elicits the production of nicotine and other compounds. How would you modify this experiment to determine whether nicotine is the only insecticidal compound produced by tobacco?

Go to **yourBioPortal.com** for original citations, discussions, and relevant links for all INVESTIGATING LIFE figures.

---

been silenced, lowering the nicotine concentration in the plants by more than 95 percent. These low-nicotine plants suffered much more damage from insect herbivory than normal plants did (**Figure 39.7**).

Some secondary metabolites play multiple roles

Canavanine is a secondary metabolite that has two important roles in the plants that produce it. The first is as a nitrogen-storing compound in seeds. The second role is defensive, and is based on its chemical structure. Canavanine is an amino acid that is not found in proteins, but is very similar to the amino acid arginine, which is found in almost all proteins:

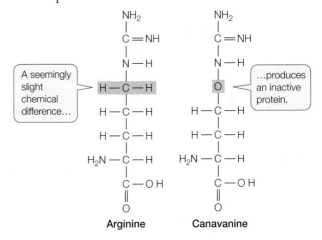

A seemingly slight chemical difference...

...produces an inactive protein.

Arginine   Canavanine

When an insect larva consumes canavanine-containing plant tissue, the canavanine is incorporated into the insect's proteins in some of the places where the mRNA codes for arginine, because the enzyme that charges the tRNA specific for arginine fails to discriminate accurately between the two amino acids (see Section 14.5). The structure of canavanine, however, is different enough from that of arginine that some of the resulting proteins end up with a modified tertiary structure, and hence reduced biological activity. These defects in protein structure and function lead to developmental abnormalities that kill the insect.

In plants that produce them, canavanine and other secondary metabolites are constitutive defenses—that is, they are present regardless of whether the plant is under attack. Other chemical defenses come into play only when an herbivore strikes.

Plants respond to herbivory with induced defenses

The first step in a plant's response to herbivory is to somehow sense the event. Two mechanisms for plant perception of herbivore damage have been described: *membrane signaling* and *chemical signaling*.

MEMBRANE SIGNALING The plasma membrane is the part of the plant cell that is in contact with the environment. Within the first minute after an herbivore strikes, changes in the electrical potential of the plasma membrane occur in the damaged area. As you will see later when you study the animal nervous system (see Section 45.2), such changes can be rapidly transmitted as a signal along the plasma membrane. In the case of plants responding to herbivory, the continuity of the symplast (see Figure 35.6) ensures that the signal travels over much of the plant within 10 minutes.

CHEMICAL SIGNALING   When an insect chews on a plant, substances in the insect's saliva combine with fatty acids derived from the consumed plant tissue. The resulting compounds act as elicitors (see Section 39.1) to trigger both local and systemic responses to the herbivore. In corn, the herbivore-produced elicitor has been named volicitin for its ability to induce production of volatile signals that can travel to other plant parts—and to neighboring corn plants—and simulate their defense responses.

SIGNAL TRANSDUCTION PATHWAY   The perception of damage from herbivory initiates a signal transduction pathway in the plant that involves the plant hormone *jasmonic acid* (*jasmonate*) (**Figure 39.8**):

Jasmonic acid

When the plant senses an herbivore-produced elicitor, it makes jasmonate, which triggers many plant defenses, including the synthesis of a protease inhibitor. The inhibitor, once in an insect's gut, interferes with the digestion of proteins and thus stunts the insect's growth. Jasmonates also "call for help" by triggering the formation of volatile compounds that attract insects that prey on the herbivores attacking it.

Why don't plants poison themselves?

Why don't the chemicals that are so toxic to herbivores and pathogens kill the plants that produce them? Plants that produce toxic defensive chemicals generally use one of the following measures to protect themselves:

- The toxic substance is isolated in a special compartment within the cell.
- The toxic substance is produced only after the plant's cells have already been damaged.
- The plant uses modified enzymes or modified receptors that do not react with the toxic substance.

Isolation of the toxic substance is the most common means of avoiding exposure. Plants store their toxins in vacuoles if the toxins are water-soluble. If they are hydrophobic, the toxins may be dissolved in latex and stored in specialized tubes called

39.8 A Signal Transduction Pathway for Induced Defenses   The chain of events initiated by herbivory that leads to the production of a defensive chemical can consist of many steps. These steps may include the synthesis of one or two hormones, binding of receptors, gene activation, and, finally, synthesis of defensive compounds.

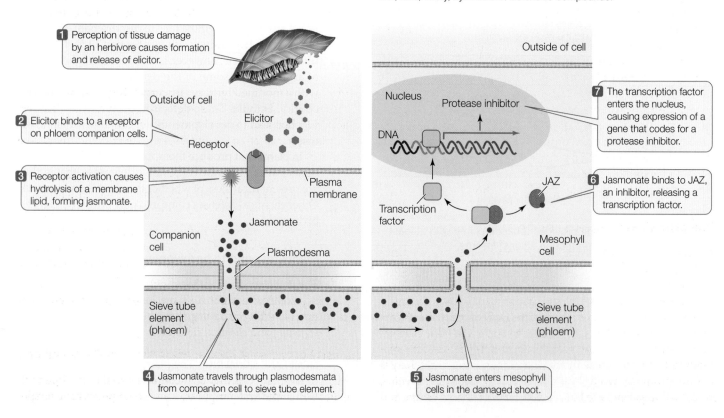

39.9 Disarming a Plant's Defenses This beetle is inactivating a milkweed's defense system by cutting its laticifer supply lines.

interrupts the latex supply to a downstream portion of the leaf. The beetles then move to the relatively latex-free portion and eat their fill.

Does this behavior of the beetles negate the adaptive value of latex protection? Not entirely. Great numbers of potential insect pests are still effectively deterred by the latex. And evolution proceeds. Over time, milkweed plants producing higher concentrations of toxins may be selected by virtue of their ability to kill even the beetles that cut their laticifers.

39.2 RECAP

Many plants use secondary metabolites as defenses against herbivory. Other defenses are induced by herbivory through a signal transduction pathway involving the hormone jasmonate.

- Describe one example of a secondary metabolite and how it affects herbivores. **See pp. 820–821**

- What role does jasmonate play in plant defense? **See p. 822 and Figure 39.8**

- What are three ways in which a plant avoids being poisoned by its own defensive chemicals? **See pp. 822–823**

laticifers, or dissolved in waxes on the epidermal surface. Such compartmentalized storage keeps the toxins away from the mitochondria, chloroplasts, and other parts of the plant's metabolic machinery.

Some plants store the precursors of toxic substances in one type of tissue, such as the epidermis, and store the enzymes that convert those precursors into the active toxin in another type, such as the mesophyll. When an herbivore chews part of the plant, cells are ruptured, the enzymes come into contact with the precursors, and the toxin is produced. The only part of the plant that is damaged by the toxin is that which was already damaged by the herbivore. Plants such as sorghum and some legumes, which respond to herbivory by producing cyanide—an inhibitor of cellular respiration—are among those that use this type of protective measure.

The third protective measure is used by the canavanine-producing plants described earlier. In these plants, unlike most other plants, the enzyme that charges the arginine tRNA discriminates correctly between arginine and canavanine, so canavanine is not incorporated into their proteins. Some herbivores, however, can evade canavanine poisoning in a similar manner, demonstrating that no plant defense is perfect. Like plants and pathogens, plants and herbivores evolve together in a continuing arms race, which the plants don't always win.

The plant doesn't always win the arms race

Milkweeds such as *Asclepias syriaca* store their defensive chemicals in laticifers. When damaged, a milkweed releases copious amounts of toxic latex from its laticifers, which run alongside the veins in its leaves. Field studies have shown that most insects that feed on neighboring plants of other species do not attack laticiferous plants, but there are exceptions. One population of beetles that feeds on *A. syriaca* exhibits a remarkable prefeeding behavior: these beetles cut a few veins in the leaves before settling down to dine (**Figure 39.9**). Cutting the veins, with their adjacent laticifers, causes massive latex leakage and

A plant's survival depends not only on successful defenses against pathogens and herbivores, but also on coping with a sometimes hostile physical environment. In the next section we consider how plants deal with climate-imposed stresses.

39.3 How Do Plants Deal with Climatic Extremes?

Plants are threatened by many aspects of the physical environment, such as drought, waterlogged soils, and extreme temperatures. How do plants survive these environmental challenges? Plants cope with environmental stresses through adaptation or acclimation.

- *Adaptation* is genetically encoded resistance to stress. A plant may have structures or biochemical properties that aid in its survival in the face of environmental challenges.

- *Acclimation* is increased tolerance for environmental extremes because of prior exposure to them. An individual plant previously exposed to extreme cold, for example, may be more likely to survive the subsequent winter.

Let's begin by describing some adaptations of plants to extremes of water availability and temperature.

Desert plants have special adaptations to dry conditions

Many plants, especially those living in deserts, must cope with extremely limited water supplies. Some desert plants have no spe-

39.10 Desert Annuals Avoid Drought The seeds of many desert annuals lie dormant for long periods awaiting conditions appropriate for germination. When they do receive enough moisture to germinate, they grow and reproduce rapidly before the short wet season ends. During the long dry spells, only dormant seeds remain alive.

39.12 Succulence The *Aloe* plant stores water in its fleshy leaves.

cial structural adaptations for water conservation. Instead, these desert annuals, called *drought avoiders*, simply evade periods of drought. Drought avoiders carry out their entire life cycle—from seed to seed—during a brief period in which rainfall has made the surrounding desert soil sufficiently moist (**Figure 39.10**). Deciduous plants, particularly in Africa and South America, shed their leaves in response to drought as a way to conserve water.

LEAF STRUCTURES Most desert plants grow in their dry environment year-round. Plants adapted to dry environments are called **xerophytes** (from the Greek *xeros*, "dry"). Three structural adaptations are found in the leaves of many xerophytes:

- Specialized leaf anatomy that reduces water loss

- A thick cuticle and a profusion of trichomes over the leaf epidermis, which retard water loss

- Diffraction and reflection of sunlight by trichomes, which decrease the intensity of the light impinging on the leaves, thus decreasing the risk of damage to the photosynthetic apparatus by excess light

In some species the stomata are strategically located in sunken cavities below the leaf surface (known as **stomatal crypts**), where they are sheltered from the drying effects of air currents (**Figure 39.11**). Hairs surrounding the stomata slow air currents as well.

Cacti and similar plants have spines rather than typical leaves, and photosynthesis is confined to the fleshy stems. The spines may help plants cope with desert condidtions by reflecting incident radiation, or by dissipating heat.

WATER-STORING STRUCTURES Succulence—the possession of fleshy, water-storing leaves or stems—is an adaptation to dry environments (**Figure 39.12**). Other adaptations of succulents include a reduced number of stomata and CAM photosynthesis, which separates the light-requiring and CO_2-assimilating reactions of photosynthesis to conserve water (see Section 10.4).

Stomata

A section through a leaf's surface shows stomata sunken in crypts protected by hairs.

Protective hairs

Lower surface of leaf

39.11 Stomatal Crypts Stomata in the leaves of some xerophytes are located in sunken cavities called stomatal crypts. The hairs covering these crypts trap moist air.

39.13 Mining Water with Deep Taproots In Death Valley, California, the root of this mesquite tree must reach far beneath the dunes for its water supply.

ROOT SYSTEMS THAT MAXIMIZE WATER UPTAKE Roots may also be adapted to dry environments. Cacti have shallow but extensive fibrous root systems that effectively intercept water at the soil surface following even light rains. Mesquite trees (*Prosopis*; **Figure 39.13**) obtain water through taproots that grow to great depths, reaching water supplies far underground, as well as from condensation on their leaves. The Atacama Desert in northern Chile often goes several years without measurable rainfall, but the landscape there has many surprisingly large mesquite trees.

One of the most successful desert plants of the southwestern United States, creosote bush (*Larrea tridentata*), displays a range of xerophytic features. It has a deep taproot, a shallow and ex-

tensive root system that can absorb water quickly after rare rain events, and small wax-covered leaves. The plant owes its name to its ability to produce noxious resins that smell like the wood preservative creosote. These natural resins not only help to seal in water, but also render the leaves virtually indigestible to browsing mammals—another adaptation to the stresses of desert life.

CHANGES IN WATER POTENTIAL Xerophytes and other plants that must cope with inadequate water supplies may accumulate high concentrations of the amino acid proline or of secondary metabolites in their vacuoles. These accumulations lower the water potential in the plant's cells below that in the soil, which results in the uptake of water by the cells via osmosis (see Section 35.1). Plants living in salty environments share this and several other adaptations with xerophytes, as we will see shortly.

In water-saturated soils, oxygen is scarce

For some plants, the environmental challenge is the opposite of that faced by xerophytes: too much water. Some plants live in environments so wet that the diffusion of oxygen to their roots is severely limited. Since most plant roots require oxygen to support respiration and ATP production, most plants cannot tolerate saturated soil conditions for long.

Some species, however, are adapted to life in a water-saturated habitat. Their roots grow slowly and hence do not penetrate deeply. Because the oxygen concentration in saturated soil is too low to support aerobic respiration, their roots carry on alcoholic fermentation (an anaerobic process; see Section 9.4), which provides ATP for the activities of the root system. This adaptation explains why their growth is slow: fermentation is much less efficient in producing ATP than aerobic respiration.

The root systems of some plants adapted to swampy environments, such as cypresses and some mangroves, have **pneumatophores**, which are extensions that grow out of the water and up into the air (**Figure 39.14**). Pneumatophores have lenticels that allow oxygen to diffuse through them, aerating the submerged parts of the root system.

Pneumatophores are root extensions that grow out of the water, under which the rest of the roots are submerged.

39.14 Coming Up for Air The roots of these mangroves obtain oxygen through pneumatophores.

Large air spaces are found in the leaf parenchyma and in the petioles of many submerged or partly submerged aquatic plants. Tissue containing such air spaces is called **aerenchyma** (**Figure 39.15**). Aerenchyma stores oxygen produced by photosynthesis and permits its ready diffusion to parts of the plant where it is needed for cellular respiration. Aerenchyma also imparts buoyancy. Furthermore, because it contains far fewer cells than most other plant tissues, metabolism in aerenchyma proceeds at a lower rate, so the need for oxygen is much reduced.

Many plants, rather than facing continual water deficits or excesses, live in fluctuating environments with unpredictable rainfall. We now turn to the mechanisms plants use to respond to those challenges.

Plants can acclimate to drought stress

When the weather is abnormally dry, the water content of the soil is reduced, and less water is available to plants. Water deficits in plant cells have two major effects: a reduction in membrane integrity as the polar–nonpolar forces that orient the lipid bilayer proteins are reduced, and changes in the three-dimensional structures of proteins. Plant growth is reduced when the structure of plant cells is compromised in these ways. Indeed, inadequate water supply is the single most important factor that limits production of our most important food crops.

Plants can, however, acclimate to drought stress to maintain their structure and function. How do they do it? When plants sense a water deficit in their roots, a signal transduction pathway is set in motion that initiates several measures to conserve water and maintain cellular integrity. This signal transduction pathway begins with the production of a hormone, abscisic acid, in the roots. This hormone travels from the roots to the shoot, where it causes stomatal closure and initiates gene transcrip-

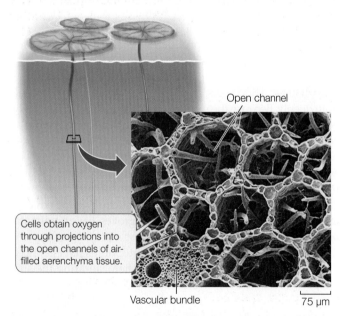

39.15 **Aerenchyma Lets Oxygen Reach Submerged Tissues** This scanning electron micrograph of a cross section of a petiole of the yellow water lily shows the structure of the air-filled channels that make up aerenchyma tissue.

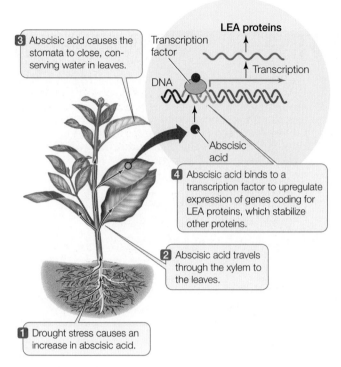

3 Abscisic acid causes the stomata to close, conserving water in leaves.

4 Abscisic acid binds to a transcription factor to upregulate expression of genes coding for LEA proteins, which stabilize other proteins.

2 Abscisic acid travels through the xylem to the leaves.

1 Drought stress causes an increase in abscisic acid.

39.16 **A Signal Transduction Pathway in Response to Drought Stress** Acclimation to drought stress begins in the root with the production of the hormone abscisic acid.

tion that leads to other physiological events that conserve water and cellular integrity (**Figure 39.16**).

Plant genes whose expression is altered by drought stress have been identified, largely through research using DNA microarrays (see Figure 18.9), proteomics (see Figure 17.17), and other molecular approaches. One group of proteins whose production is upregulated during drought stress is the *late embryogenesis abundant* (LEA; pronounced "lee-yuh") proteins. These hydrophobic proteins accumulate in maturing seeds as they dry out. The LEA proteins bind to membranes and other cellular proteins to stabilize them, preventing their aggregation during desiccation. The importance of LEA proteins in coping with drought stress was demonstrated by Ray Wu and his colleagues at Cornell University, who showed that transgenic rice plants expressing a high level of a LEA protein in their leaves and roots grew better than normal plants under drought conditions (**Figure 39.17**). Genes that code for LEA proteins are widely distributed among plants, bacteria, and invertebrates; in all these organisms, high-level expression results in drought tolerance. As described at the opening of Chapter 35, which discusses the discovery of *HARDY*, another gene that promotes water use efficiency in plants, genetic research holds the promise of solutions to many agricultural problems, including those that may be presented by global climate change.

Plants have ways of coping with temperature extremes

Temperatures that are too high or too low can stress plants and even kill them. Plants differ in their sensitivity to heat and cold, but all plants have their limits. Any temperature extreme can damage cellular membranes.

INVESTIGATING LIFE

39.17 A Molecular Response to Drought Stress

Understanding the responses of plants to drought conditions is vital for agriculture. Ray Wu and colleagues transformed rice cells with a gene that codes for a LEA protein that is expressed in seeds as they mature and dry out. The investigators then measured the response of the transgenic rice plants to 28 days of drought.

HYPOTHESIS LEA proteins protect plants from the effects of drought stress.

METHOD

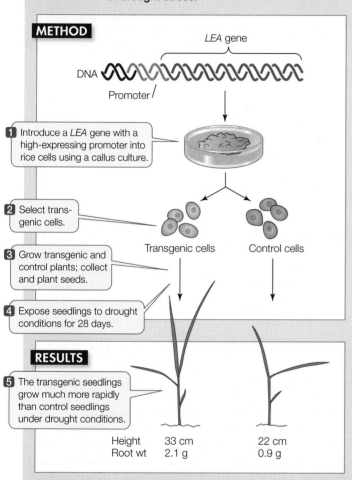

1. Introduce a *LEA* gene with a high-expressing promoter into rice cells using a callus culture.

2. Select transgenic cells.

3. Grow transgenic and control plants; collect and plant seeds.

4. Expose seedlings to drought conditions for 28 days.

RESULTS

5. The transgenic seedlings grow much more rapidly than control seedlings under drought conditions.

	Transgenic	Control
Height	33 cm	22 cm
Root wt	2.1 g	0.9 g

CONCLUSION Plants with higher concentrations of a LEA protein grow better under drought conditions.

Go to **yourBioPortal**.com for original citations, discussions, and relevant links for all INVESTIGATING LIFE figures.

- High temperatures destabilize membranes and denature many proteins, especially some of the enzymes of photosynthesis.

- Low temperatures cause membranes to lose their fluidity and alter their permeabilities to solutes.

- Freezing temperatures may cause ice crystals to form, damaging membranes.

Therefore, it is not surprising that many plants living in hot environments have adaptations similar to those of xerophytes. These adaptations include hairs and spines that dissipate heat, leaf forms that intercept less direct sunlight, and CAM photosynthesis, which allows plants to perform some metabolic processes in the cool of night (see Section 10.4).

The plant response to heat stress is similar to the response to drought stress in that new proteins are made, often under the direction of an abscisic acid–mediated signal transduction pathway. Within minutes of experimental exposure to raised temperatures (typically a 5°C–10°C increase), plants synthesize several kinds of **heat shock proteins**. Among these proteins are chaperonins (see Figure 3.12), which help other proteins maintain their structures and avoid denaturation. Threshold temperatures for the production of heat shock proteins vary, but 39°C is sufficient to induce them in most plants.

Low temperatures above the freezing point can cause *chilling injury* in many plants, including crops such as rice, corn, and cotton, as well as tropical plants such as bananas. Many plant species can acclimate to cooler temperatures through a process called **cold-hardening**, which requires repeated exposure to cool, but not injurious, temperatures over many days. A key change during the hardening process is an increase in the proportion of unsaturated fatty acids in membranes, which allows them to retain their fluidity and function normally at cooler temperatures. Plants have a greater ability to modify the degree of saturation of their membrane lipids than animals do. In addition, low temperatures induce the formation of proteins similar to heat shock proteins, which protect against chilling injury.

If ice crystals form within plant cells, they can kill the cells by puncturing organelles and plasma membranes. Furthermore, the growth of ice crystals outside the cells can draw water from the cells and dehydrate them. Freeze-tolerant plants have a variety of adaptations to cope with these problems, including the production of *antifreeze proteins* that slow the growth of ice crystals.

39.3 RECAP

Plants that live in continually dry or water-saturated environments have structural adaptations to cope with those conditions. Mechanisms that protect plants from drought stress are initiated by a signal transduction pathway involving abscisic acid. Heat shock proteins help plants acclimate to high and low temperatures.

- Describe two structural adaptations for growth in water saturated soils. **See pp. 825–826**

- What is the role of abscisic acid in acclimation to drought stress? **See p. 826 and Figure 39.16**

- What environmental conditions induce the formation of heat shock proteins, and what functions do those proteins serve? **See p. 827**

Just as climatic extremes can limit plant growth, the presence of certain substances, such as salt and heavy metals, can make an environment inhospitable to plant growth.

39.4 How Do Plants Deal with Salt and Heavy Metals?

A number of toxic solutes are found in soils, but worldwide, no toxic substance restricts angiosperm growth more than salt (sodium chloride). Saline—salty—habitats support, at best, limited types of vegetation. Saline habitats are found in diverse locales, from hot, dry, deserts to moist, cool coastal marshes. Along the seashore, saline environments are created by ocean spray. The ocean itself is a saline environment, as are estuaries, where fresh and salt water meet and mingle. Salinization of agricultural land is an increasing global problem. Even where crops are irrigated with fresh water, sodium ions from the water accumulate in the soil to ever greater concentrations as the water evaporates.

Saline environments pose an osmotic challenge for plants. Because of its high salt concentration, a saline environment has an unusually negative water potential. To obtain water from such an environment, a plant must have an even more negative water potential; otherwise, water will diffuse out of its cells, and the plant will wilt and die. Plants in saline environments are also challenged by the potential toxicity of sodium, which inhibits enzymes and protein synthesis.

Most halophytes accumulate salt

Halophytes—plants adapted to saline habitats—are found in a wide variety of flowering plant groups. Most halophytes share one adaptation: they take up sodium and, usually, chloride ions and transport those ions to their leaves. The accumulated ions are stored in the central vacuoles of leaf cells, away from more sensitive parts of the cells. Nonhalophytes accumulate relatively little sodium, even when placed in a saline environment; of the sodium that is absorbed by their roots, very little is transported to the shoot. The increased salt concentration in the tissues of halophytes lowers their water potential and allows them to take up water from their saline environment.

Some halophytes have other adaptations to life in saline environments. Some, for example, have **salt glands** in their leaves. These glands excrete salt, which collects on the leaf surface until it is removed by rain or wind (**Figure 39.18**). This adaptation, which reduces the danger of poisoning by accumulated salt, is found in some desert plants, such as *Frankenia palmeri*, and in some mangroves growing in seawater.

Salt glands can play multiple roles, as in the desert shrub *Atriplex halimus*. This shrub has glands that secrete salt into small bladders on the leaves. By lowering the water potential of the leaves, this salt not only helps them obtain water from the roots, but also reduces their transpirational loss of water to the atmosphere.

The adaptations we have just discussed are specific to halophytes. Several other adaptations are shared by halophytes and

39.18 Excreting Salt This saltwater mangrove plant has special salt glands that excrete salt, which appears here as crystals on the leaves.

xerophytes, including thick cuticles, succulence, and CAM photosynthesis.

Some plants can tolerate heavy metals

Salt is not the only toxic solute found in soils. High concentrations of some heavy metal ions, such as chromium, mercury, lead, and cadmium, are toxic to most plants; many of these ions are more toxic than sodium at equivalent concentrations.

Some geographic sites are naturally rich in heavy metals as a result of normal geological processes. In other places, acid rain leads to the release of toxic aluminum ions in the soil. Human activities, notably the mining of metallic ores, leave localized areas—known as tailings—with high concentrations of heavy metals and low concentrations of nutrients. Such sites are hostile to most plants, and seeds falling on them generally do not produce adult plants.

Most mine tailings rich in heavy metals, however, are not completely barren. They may support healthy plant populations that differ genetically from populations of the same species on the surrounding normal soils. How do these plants survive?

Initially, botanists believed that some plants were able to tolerate heavy metals by excluding them: that by not taking up the metal ions, the plants avoided being poisoned. Further investigations have shown, however, that tolerant plants growing on mine tailings do take up heavy metals, accumulating concentrations that would kill most plants. Over 200 plant species have been identified as **hyperaccumulators** that store large quantities of metals such as arsenic (As), cadmium (Cd), nickel (Ni), aluminum (Al), and zinc (Zn).

Perhaps the best-studied hyperaccumulator is alpine pennycress (*Thlaspi caerulescens*). Before the advent of chemical analysis, miners used to use the presence of this plant as an indicator of mineral-rich deposits. A *Thlaspi* plant may accumulate as much as 30,000 ppm Zn (most plants contain 100 ppm) and 1,500 ppm Cd (most plants contain 1 ppm). Studies of *Thlaspi*

39.19 Phytoremediation Plants that accumulate heavy metals can be used to clean up contaminated soils. Here, poplars are being used to remove contaminants from an air base.

grown in the contaminated soil, where they act as natural "vacuum cleaners" by taking up the contaminants (**Figure 39.19**). The plants are then harvested and disposed of to remove the contaminants. Perhaps the most dramatic use of phytoremediation occurred after an accident at the nuclear power plant at Chernobyl, Ukraine (then part of the Soviet Union), in 1986, when sunflower plants were used to remove uranium from the nearby soil. Phytoremediation is now widely used in cleaning up land after strip mining.

After finding plants that accumulate valuable metals such as Ni, cobalt (Co), and silver (Ag), some scientists have proposed using those plants for *phytomining*. As in phytoremediation, the plants would be used to take up metals from the soil, but the metals would be extracted from the plants after they are harvested.

and other hyperaccumulators have revealed the presence of several common mechanisms:

- Increased ion transport into the roots
- Increased rates of translocation of ions to the leaves
- Accumulation of ions in vacuoles in the shoot
- Resistance to the ions' toxicity

Knowledge of these hyperaccumulation mechanisms and the genes underlying them has led to the emergence of **phytoremediation**, a form of bioremediation (see Section 18.6) that uses plants to clean up environmental pollution. Some phytoremediation projects use natural hyperaccumulators, while others use genes from hyperaccumulators to create transgenic plants that grow more rapidly and are better adapted to a particular polluted environment. In either case, the plants are

39.4 RECAP

Halophytes have a number of adaptations to saline habitats, most of which involve mechanisms that lower their water potential. Some plants can tolerate heavy-metal-rich soils that are toxic to most other plants.

- What are some of the roles of salt glands in halophyte leaves? See p. 828
- How are plants used for phytoremediation? See p. 829

CHAPTER SUMMARY

39.1 How Do Plants Deal with Pathogens?

- Plants and pathogens have evolved together in a continuing "arms race": pathogens have evolved mechanisms for attacking plants, and plants have evolved mechanisms for defending themselves against those attacks. **Review Figure 39.2, ANIMATED TUTORIAL 39.1**
- Plants strengthen their cell walls and block plasmodesmata when attacked, limiting the ability of viral pathogens to move from cell to cell.
- **Gene-for-gene resistance** depends on a match between a plant's **resistance (R) genes** and a pathogen's **avirulence (Avr) genes. Review Figure 39.3**
- In the **hypersensitive response** to infection by bacteria or fungi, cells produce two kinds of defensive molecules: **phytoalexins** and **PR proteins**. Some cells around the infected area die, sealing off the pathogens and the damage they have caused.

- The hypersensitive response is often followed by **systemic acquired resistance**, in which salicylic acid activates further synthesis of defensive compounds.
- Plants use RNA interference to develop specific immunity to invading RNA viruses.

39.2 How Do Plants Deal with Herbivores?

- Herbivory increases the productivity of some plants. **Review Figure 39.6**
- Some plants produce **secondary metabolites** as defenses against herbivores. **Review Table 39.1, Figure 39.7**
- Hormones, including **jasmonates**, participate in signal transduction pathways leading to the production of defensive compounds. **Review Figure 39.8**
- Plants protect themselves against their own toxic defensive chemicals by isolating them in specialized compartments, by producing them only after the plant has already been damaged, or by using modified enzymes or receptors that are not affected by the toxic substance.

39.3 How Do Plants Deal with Climatic Extremes?

- Plants cope with environmental stresses by **adaptation** (genetically encoded resistance) or **acclimation** (increased tolerance).

- **Xerophytes** are plants that are adapted to dry environments.

- Some xerophytic adaptations are structural, including thickened cuticles, specialized trichomes, **stomatal crypts**, **succulence**, and long taproots.

- Some plants accumulate solutes, making their water potential lower so they can tolerate drought.

- Adaptations to water-saturated habitats include **pneumatophores**, extensions of roots allow oxygen uptake from the air, and **aerenchyma**, in which oxygen can be stored and diffuse throughout the plant.

- A signal transduction pathway involving abscisic acid initiates a plant's response to drought stress. **Review Figures 39.16 and 39.17**

- Membranes and proteins can be damaged by extremely high or low temperatures. Plants respond to extreme temperatures by producing **heat shock proteins**.

- Some plants undergo **cold-hardening**, an acclimation process that includes changes in membrane lipids and production of heat shock proteins.

- Some plants resist freezing by producing antifreeze proteins.

39.4 How Do Plants Deal with Salt and Heavy Metals?

- Most **halophytes** accumulate salt. Some have **salt glands** that excrete salt to the leaf surface.

- Some plants living in soils that are rich in heavy metals are **hyperaccumulators** that take up large amounts of those metals into their tissues.

- **Phytoremediation** is the use of hyperaccumulating plants or their genes to clean up environmental pollution.

SEE WEB ACTIVITY 39.1 for a concept review of this chapter.

SELF-QUIZ

1. Which of the following is *not* a common defense against bacteria and fungi?
   a. Lignin formation
   b. Phytoalexins
   c. A waxy covering
   d. The hypersensitive response
   e. Mycorrhizae

2. Plants sometimes protect themselves from their own toxic secondary metabolites by
   a. producing special enzymes that destroy the toxin.
   b. storing precursors of the toxic substances in one compartment and the enzymes that convert those precursors to toxic products in another compartment.
   c. storing the toxic substances in mitochondria or chloroplasts.
   d. distributing the toxic substances to all cells of the plant.
   e. performing crassulacean acid metabolism.

3. Herbivory
   a. is an attack by plants on animals.
   b. always reduces plant growth.
   c. usually increases the rate of photosynthesis in the remaining leaves.
   d. reduces the rate of transport of photosynthetic products from the remaining leaves.
   e. is always lethal to the grazed plant.

4. Which statement about secondary metabolites is *not* true?
   a. They may be used in defense against fungi.
   b. Some are poisonous to herbivores.
   c. Some are amino acids that are normally part of proteins.
   d. Water soluble molecules are stored in vacuoles.
   e. Some mimic the hormones of animals.

5. Which statement about latex is *not* true?
   a. It is sometimes contained in laticifers.
   b. It is typically white.
   c. It is often toxic to insects.
   d. It is a rubbery solid.
   e. Milkweeds produce it.

6. Which of the following is *not* an adaptation to dry environments?
   a. Increased solute concentration in the vacuoles
   b. Hairy leaves
   c. A heavier cuticle over the leaf epidermis
   d. Sunken stomata
   e. A root system that grows each rainy season and dies back when it is dry

7. Some plants adapted to swampy environments meet the oxygen needs of their roots by means of a specialized tissue called
   a. parenchyma.
   b. aerenchyma.
   c. collenchyma.
   d. sclerenchyma.
   e. chlorenchyma.

8. Halophytes
   a. may accumulate abscisic acid in their vacuoles.
   b. may have water potentials that are lower than those of other plants.
   c. only accumulate sodium.
   d. have low root-to-shoot ratios.
   e. rarely accumulate sodium.

9. Which of the following is *not* involved in the response to drought stress?
   a. Abscisic acid
   b. Closing of aquaporins
   c. LEA gene expression
   d. Closing of stomata
   e. Jasmonate

10. Plants that tolerate heavy metals commonly
    a. differ genetically from other members of their species.
    b. do not take up the heavy metals.
    c. are tolerant to all heavy metals.
    d. are slow to colonize an area rich in heavy metals.
    e. weigh more than plants that are sensitive to heavy metals.

FOR DISCUSSION

1. How might plant adaptations affect the evolution of herbivores? How might the adaptations of herbivores affect plant evolution?

2. The stomata of the common oleander (*Nerium oleander*) are located in sunken crypts in its leaves. Whether or not you know what an oleander is, you should be able to describe an important feature of its natural habitat. What is that feature?

3. In the coming decades, climate change may have significant effects on the growth and productivity of plants, in particular the crop plants on which we depend for our food. Discuss the physiological effects, and possible genetic responses in terms of plant breeding, of the following:

   a. In Pakistan, reduced rainfall causes a reduction in wheat yields.

   b. In the Mekong Delta of Vietnam, rising sea level inundates rice fields, causing a drastic reduction in yields.

   c. Increased temperature and humidity in western Canada causes an increase in wheat rust.

ADDITIONAL INVESTIGATION

The tobacco hornworm (*Manduca sexta*) is adapted to feeding on nicotine-producing plants. Using the genetically modified tobacco plants described in Figure 39.7, how might you test the hypothesis that dietary nicotine protects the tobacco hornworm against its parasite *Cotesia congregata*?

WORKING WITH DATA (GO TO yourBioPortal.com)

Nicotine Is a Defense against Herbivores To test the hypothesis that nicotine is a plant defense chemical, Baldwin and colleagues generated transgenic plants that expressed a low amount of nicotine (Figure 39.7). In this exercise, you will analyze data from the original research paper, which included the effects of the hormone jasmonic acid.

40 Physiology, Homeostasis, and Temperature Regulation

Cool it!

"A new world record!" These words convey the thrill of world-class athletic competition. But as records are broken by mere centimeters or by fractions of a second, are we reaching absolute limits to human performance? We can assess many physiological limits to extreme performance—maximum breathing rate, for example, or the maximum rate at which the heart can supply blood to the muscles. A less obvious physiological limit is temperature.

The 2008 New York City Marathon took place on a cold, clear, windy day in November. For the third time, the first-place woman in this 41-km race was world record holder Paula Radcliffe. Radcliffe had also been expected to win the women's marathon back in the 2004 Olympics. But that race took place on a hot, humid day in Athens. Overcome by heat stress, Radcliffe collapsed 6 km from the finish line. The critical difference in the two races was probably temperature.

Thermal stress can have more serious consequences than losing a race. Every year some athletes die of heat stroke, which can occur when internal body temperature exceeds 41°C. This elevated internal temperature results in the failure of major organs and, in more than 20 percent of cases, death. Soldiers in the desert are at extreme risk of heat stroke, as are firefighters. Agricultural, industrial, and construction workers are also subject to the adverse affects of heat. Biologists at Stanford University developed a technology to cool individuals in such situations, and in the process discovered a way to enhance athletic performance.

Working muscles produce heat, which is carried by the blood to skin surfaces, where it is lost to the environment. Not all skin surfaces are equally good at dissipating heat, however. Because fur impedes heat loss, mammals evolved efficient bare-skin heat-loss portals such as the nose, tongue, footpads, and parts of the face. These areas have specialized blood vessels that can act like radiators to disperse heat or close down to conserve heat. Humans are not furred, but we retain these ancestral blood vessels in our hands, feet, and face (which is why we blush). The Stanford team designed a device to amplify heat extraction from these areas.

The heat extractor is a chamber that encases the hand and is sealed at the wrist. The hand is in contact with a

Limits to Performance Paula Radcliffe, photographed here during her winning performance at the 2008 New York City Marathon, collapsed from heat stress during the 2004 Olympic marathon. When the body's internal temperature is subjected to extreme heat, its homeostatic mechanisms may fail.

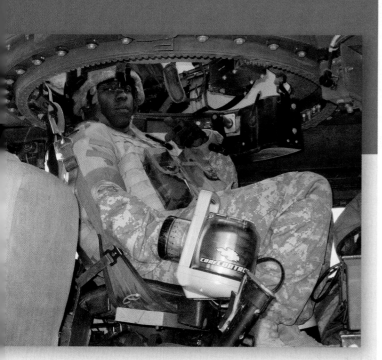

A Cooling Glove The heat extractor increases heat loss and allows the body to perform at a higher level in severe conditions.

cooled surface. A mild vacuum in the chamber pulls more blood into the hand, enabling the cool surface to extract more heat. With this device, an active individual's body temperature rises more slowly, and cools more rapidly during rest. An unexpected benefit is that cooling reduces fatigue and greatly increases exercise capacity. In one study, college freshmen improved their push-up performance at a rate of 5 push-ups a day without cooling, but 9 push-ups a day with cooling. Some men and women in the study achieved more than 800 push-ups in a workout session.

Human beings survive in environments that are extremely hot or extremely cold because we have the physiological and behavioral means of regulating our internal body temperature—an example of *homeostasis*, the maintenance of a "steady state" in our internal environment.

IN THIS CHAPTER we will explore the internal environment that serves the needs of all of the body's cells. We will survey the cell and tissue types that make up physiological systems and discuss how these systems maintain the internal environment within certain physiological limits, a condition called homeostasis. Homeostasis will be described using one important example, the regulation of body temperature.

40.1 How Do Multicellular Animals Supply the Needs of Their Cells?

All animal cells need nutrients and oxygen from the environment and must eliminate carbon dioxide and other waste products of metabolism to the environment. The cells of very small or very thin aquatic animals meet these needs by direct exchanges with the external environment. In such animals, no cell is far from direct contact with the water it lives in; the water contains nutrients, absorbs waste, and provides a relatively unchanging physical environment. Most cells of larger animals do not have direct contact with the external environment, and their needs must be served by an environment that is wholly internal to the animal.

An internal environment makes complex multicellular animals possible

The cells of multicellular animals exist within an **internal environment** of extracellular fluid (ECF). A human, for example, is about 60 percent water. Two-thirds of that water is contained within cells, and one-third makes up the ECF that is our internal environment. About 20 percent of that extracellular fluid, or 3 liters, is the blood plasma that circulates in our blood vessels. The rest—about 11 liters—is the **interstitial fluid** that bathes every cell of the body (**Figure 40.1**). Individual cells get their nutrients from this interstitial fluid and dump their waste products into it. As long as conditions in this internal environment are held within certain limits, the cells are protected from changes or harsh conditions in the external environment. A stable internal environment makes it possible for an animal to occupy habitats that would kill its cells if they were exposed to it directly. How is the internal environment kept constant?

As multicellular organisms evolved, cells became specialized for maintaining specific aspects of the internal environment. In turn, the internal environment enabled these specializations, since each cell did not have to provide for all of its own needs. Some cells evolved to be the interface between the internal and the external environments and to provide the necessary transport functions to get nutrients in and move wastes out. Other cells became specialized to provide internal functions such as circulation of the extracellular fluids, energy storage, movement, and information processing. The evolution of physiological systems to maintain the internal environment made it possible for multicellular animals to become larger, thicker, and more complex, and to occupy many different habitats.

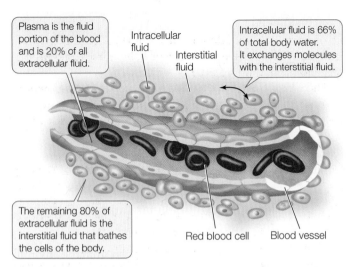

Plasma is the fluid portion of the blood and is 20% of all extracellular fluid.

Intracellular fluid

Interstitial fluid

Intracellular fluid is 66% of total body water. It exchanges molecules with the interstitial fluid.

The remaining 80% of extracellular fluid is the interstitial fluid that bathes the cells of the body.

Red blood cell   Blood vessel

40.1 The Internal Environment The "internal environment" is the extracellular fluid, or ECF. ECF, which accounts for about one-third of total body water, is made of the blood plasma and the interstitial fluid. The physiological composition of the ECF must remain stable within narrow limits, and maintaining that stability is the job of the body's organ systems.

The composition of the internal environment is constantly being challenged by the external environment and by the metabolic activity of the cells of the body. The maintenance of stable conditions (within a narrow range) in the internal environment is called **homeostasis**. If a physiological system fails to function properly, homeostasis is compromised and cells are damaged and can die. To avoid the loss of homeostasis, physiological systems must be controlled and regulated in response to changes in both the external and internal environments.

Physiological systems maintain homeostasis

The activities of all physiological systems are controlled—speeded up or slowed down—by actions of the nervous and endocrine systems. But to *regulate* the internal environment, information is required. As an analogy, think of driving a car (**Figure 40.2**). To regulate the speed of your car, you have to know both how fast you are going and how fast you want to go. The desired speed is a **set point**, or reference point; the reading on your speedometer is **feedback information**. Any difference between the set point and the feedback information is an **error signal**. Error signals suggest corrective actions, such as stepping on the accelerator or brake.

Some components of physiological systems are called **effectors** because they *effect* changes in the internal environment. Effectors are **controlled systems** because their activities are controlled by commands from regulatory systems. **Regulatory systems**, in contrast, obtain, process, and integrate information, then issue commands to controlled systems.

Important components of any regulatory system are the **sensors** that provide the feedback information to be compared with the internal set point. How is information from the sensors used?

Negative feedback is the most common use of sensory information in regulatory systems. Negative feedback information is used to counteract the influence that created an error signal. Whatever force is pushing the system away from its set point must be "negated." In our car analogy, the recognition that you are going too fast is negative feedback that causes you to release the accelerator and press the brake.

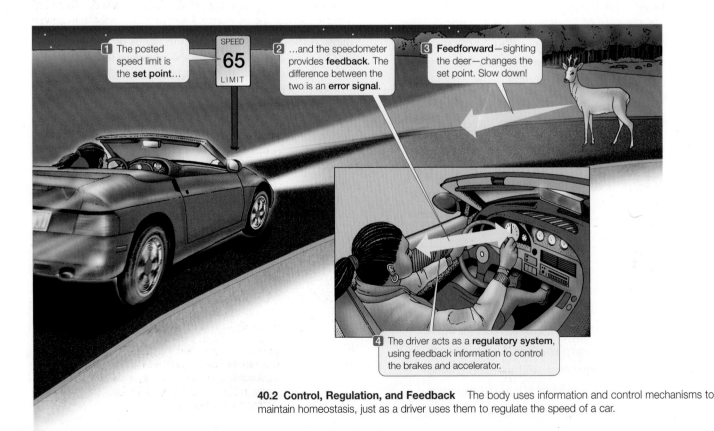

1 The posted speed limit is the **set point**…

SPEED **65** LIMIT

2 …and the speedometer provides **feedback**. The difference between the two is an **error signal**.

3 **Feedforward**—sighting the deer—changes the set point. Slow down!

4 The driver acts as a **regulatory system**, using feedback information to control the brakes and accelerator.

40.2 Control, Regulation, and Feedback The body uses information and control mechanisms to maintain homeostasis, just as a driver uses them to regulate the speed of a car.

Although not as common as negative feedback, **positive feedback** is also seen in physiological systems. Rather than returning a system to a set point, positive feedback amplifies a response (i.e., it *increases* the deviation from the set point). Examples of regulatory systems that use positive feedback are the responses that empty body cavities, such as urination and defecation. Another example is sexual behavior, in which a little stimulation causes more behavior, which causes more stimulation, and so on. Positive feedback responses tend to reach a limit and terminate rapidly. The birth process is a good example. Contractions of the uterus stretch the birth canal, which stimulates more and stronger contractions until the baby is delivered, at which time contractions cease.

Feedforward information is another feature of regulatory systems. The function of feedforward information is to change the set point. Seeing a deer in the road when you are driving is an example of feedforward information; this information takes precedence over the posted speed limit, and you slow down. Before the start of a race, hearing the command "on your mark" is feedforward information that raises your heart rate before you begin to run. Feedforward information predicts a change in the internal environment before that change occurs.

These principles of control and regulation help organize our thinking about physiological systems. Once we understand how a system works, we can then ask how it is regulated. The example we will explore in this chapter is the regulation of body temperature. But first we need to become acquainted with the important structural features that all physiological systems have in common.

Cells, tissues, organs, and systems are specialized to serve homeostatic needs

Each physiological system is composed of discrete organs, such as the liver, heart, lungs, and kidneys. These organs are made up of assemblages of cells known as **tissues**. Although there are many specialized cell types, there are only four kinds of tissues: *epithelial, muscle, connective,* and *nervous*.

The word "tissue" is often used in a general way to refer to a piece of an organ, such as "lung tissue" or "kidney tissue." As we will see, an organ always consists of more than one of the four kinds of tissues.

EPITHELIAL TISSUES **Epithelial tissues** are sheets of densely packed, tightly connected *epithelial cells* (**Figure 40.3**). Epithelial cells create boundaries between the inside and the outside of the body and between body compartments; they line the blood vessels and make up various ducts and tubules.

Filtration and transport are important functions of epithelial cells; they both act as barriers and provide transport across those barriers. They control what molecules and ions can move between the blood and the interstitial fluid. They can selectively transport ions and molecules from one side of an epithelial membrane to the other. Examples are the absorption of nutrient molecules from your gut and the secretion of acid into your stomach. Some epithelial cells, like those in the lungs or at the skin's surface, are extremely thin (*squamous*) to facilitate movement of substances across them.

The skin is epithelial tissue that receives much wear and tear. Accordingly, epithelial cells in the deepest layer of the skin have a high rate of cell division, producing new cells that move progressively to the skin surface, die, and are shed. A cross section of the skin reveals the layering of cells, from the newly formed ones on the innermost germinal layer to the dead ones on the surface. Because of this appearance, the skin is called a stratified epithelium (see Figure 40.3A). In contrast, gut epithelium consists of a single layer of tall, closely packed cells called a simple columnar epithelium. The epithelial cells of your gut are replaced about every 5 days; those in your skin are renewed every 1 to 2 months.

(A) Squamous cells

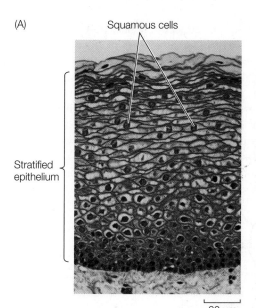

Stratified epithelium

30 μm

(B) Columnar epithelium

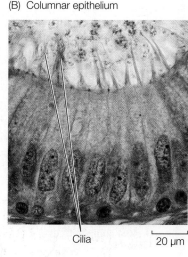

Cilia

20 μm

(C) Cuboidal epithelial cells

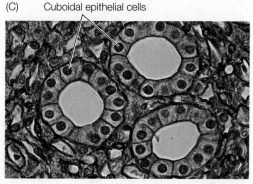

25 μm

40.3 Epithelial Tissue (A) Epithelial cells make up the outer layers of skin. They are stratified, from extremely thin (squamous) older cells at the surface to rapidly dividing new layers that will rise to the surface as older cells are shed. (B) Ciliated columnar epithelium from the male reproductive duct (the vas deferens). (C) A single layer of cuboidal epithelial cells forms a tubule in the kidney. These cells have many molecular transport functions.

Epithelial cells have many other roles. Some secrete hormones, milk, mucus, digestive enzymes, or sweat. Others have cilia that move substances over surfaces or through tubes (see Figure 40.3B). Epithelial cells can also provide information to the nervous system. Smell and taste receptors, for example, are epithelial cells that detect specific chemicals.

MUSCLE TISSUES **Muscle tissues** consist of elongated cells that contract to generate forces and cause movement. Muscle tissues are the most abundant tissues in the body, and they use most of the energy produced in the body. All muscle cells contain long protein polymers called actin and myosin which interact to cause muscle cells to contract and exert force. There are three types of muscle tissues (**Figure 40.4**).

(A)

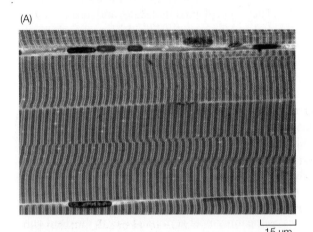

15 μm

(B)

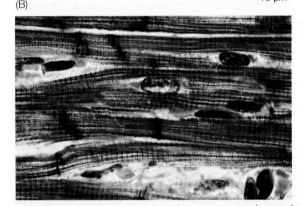

15 μm

(C)

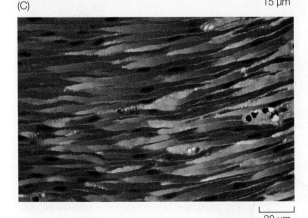

30 μm

- **Skeletal muscles** (so named because they mostly attach to bones) are responsible for locomotion and other body movements such as facial expressions, shivering, and breathing.
- **Cardiac muscle** makes up the heart and is responsible for the beating of the heart and the pumping of blood. Individual cardiac muscle cells are branched, and the interweaving of these branches gives heart muscle structural strength.
- **Smooth muscle** is responsible for involuntary generation of forces in many hollow internal organs such as the gut, bladder, and blood vessels.

Skeletal muscles are under both voluntary and involuntary control, as will be described in detail in Section 48.1. Cardiac and smooth muscles are under involuntary control; they are controlled by physiological regulatory systems.

CONNECTIVE TISSUES In contrast to densely packed epithelial and muscle tissues, **connective tissues** are generally dispersed populations of cells embedded in an *extracellular matrix* that they secrete (**Figure 40.5**). The composition and properties of the matrix differ among types of connective tissues.

Protein fibers are an important component of the extracellular matrix of connective tissue cells. The dominant protein in the extracellular matrix is *collagen* (see Figure 5.25), which makes up about 25 percent of total body protein. Collagen fibers are strong and resistant to stretch, giving strength to the skin and to the connections between bones and between bones and muscles. The fibers provide a netlike framework for organs, giving them shape and structural strength.

Elastin is another type of protein fiber in the extracellular matrix of connective tissues. It is so named because it can be stretched to several times its resting length and then recoil. Fibers composed of elastin are most abundant in tissues that are regularly stretched, such as the walls of the lungs and the large arteries.

Cartilage and bone are connective tissues that provide rigid structural support. In *cartilage*, a network of collagen fibers is embedded in a flexible matrix consisting of a protein–carbohydrate complex, along with a specific type of cell called a *chondrocyte*. Cartilage, which lines the joints of vertebrates, is resistant to compressive forces. Since it is flexible, it provides structural support for flexible structures such as external ears and noses. The extracellular matrix in *bone* also contains many collagen fibers, but it is hardened by the deposition of the mineral calcium phosphate. We discuss cartilage and bone in greater detail in Section 48.3.

40.4 Muscle Cells Contain Protein Filaments The filaments of two specific proteins—actin and myosin—interact to cause contraction and generate force in muscle tissue. (A) The regular arrangement of actin and myosin filaments results in the striated (striped) appearance of skeletal muscle. (B) The individual cells of cardiac muscle are branched and form a strong structural meshwork. (C) The actin and myosin filaments of smooth muscle are not regularly arranged and thus it does not have a striated appearance.

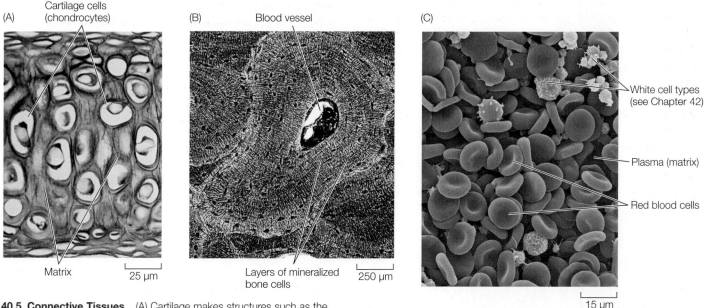

(A) Cartilage cells (chondrocytes)

Matrix

25 μm

(B) Blood vessel

Layers of mineralized bone cells

250 μm

(C)

White cell types (see Chapter 42)

Plasma (matrix)

Red blood cells

15 μm

40.5 Connective Tissues (A) Cartilage makes structures such as the ear stiff but flexible. Cartilage cells, or chondrocytes, secrete an extracellular matrix rich in collagen and elastin fibers. In this micrograph, the elastin fibers are stained dark blue. (B) Bone is the mineral-rich connective tissue of the vertebrate skeleton. (C) Blood is unique among the connective tissues, consisting of blood cells floating in an extracellular matrix of plasma.

Adipose cells form loose connective tissue that stores lipids. Adipose tissue, or "fat," is a major source of stored energy. It also cushions organs, and layers of adipose tissue under the skin can provide a barrier to heat loss.

Blood is a connective tissue consisting of cells dispersed in an extensive liquid extracellular matrix, the blood *plasma*. We present many of the proteins and cellular elements of blood in Section 42.1, and we will discuss blood again in Section 50.4.

NERVOUS TISSUES The two basic cell types in **nervous tissues** are *neurons* and *glial cells* (**Figure 40.6**). Neurons come in many shapes and sizes, and all neurons encode information as electrical signals. These signals can travel over long extensions called *axons* to communicate with other neurons, muscle cells, or secretory cells through the release of chemicals called *neurotransmitters*. Neurons control the activities of most organ systems. Glial cells do not generate or conduct electrical signals, but they provide a variety of supporting functions for neurons. There are more glial cells than neurons in the nervous system.

Chapters 45, 46, and 47 detail the many fascinating properties of nervous tissues.

Organs consist of multiple tissues

Organs include more than one kind of tissue, and most organs include all four (**Figure 40.7**). The wall of the gut is a good example. Its inner surface is lined with a sheet of columnar epithelial cells. Different epithelial cells secrete mucus, enzymes, or stomach acid. Beneath the epithelial lining is connective tissue. Within this connective tissue are blood vessels, neurons, and glands (clusters of secretory epithelial

(A)

Cell body of neuron

Axon

20 μm

(B) Astrocytes

Capillaries

60 μm

40.6 Nervous Tissue Includes Neurons and Glial Cells (A) This human neuron consists of a cell body, a number of processes that receive input from other neurons, and one long axon that sends information to other cells. (B) A section through human brain tissue shows astrocytes, a type of glial cell. Glial cells provide support and protection for neurons, including creating a barrier that protects the brain from many chemicals circulating in the blood.

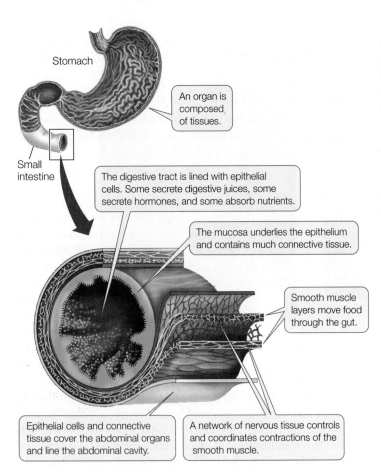

Stomach

An organ is composed of tissues.

Small intestine

The digestive tract is lined with epithelial cells. Some secrete digestive juices, some secrete hormones, and some absorb nutrients.

The mucosa underlies the epithelium and contains much connective tissue.

Smooth muscle layers move food through the gut.

Epithelial cells and connective tissue cover the abdominal organs and line the abdominal cavity.

A network of nervous tissue controls and coordinates contractions of the smooth muscle.

40.7 Tissues Form Organs Most organs contain more than one of the four tissue types. The organs of the human digestive system, such as the stomach and small intestine, are made up of all four.

cells). Concentric layers of smooth muscle tissue enable the gut to contract to mix food with digestive juices. A network of neurons between the muscle layers controls these movements.

An individual organ is usually part of an **organ system**—a group of organs that work together to carry out certain functions. The stomach and small intestine, for example, are part of the digestive system. The digestive system is the subject of Chapter 51.

40.1 RECAP

The internal environment provides for the needs of all the cells that make up a complex animal. Organs and organ systems control the composition of the internal environment. The activities of organs and organ systems are regulated to maintain homeostasis of the internal environment.

- Explain the difference between negative and positive feedback control mechanisms. **See pp. 834–835 and Figure 40.2**

- Describe a key function of each of the four kinds of tissue found in animals. **See pp. 835–837 and Figures 40.3–40.7**

Subsequent chapters will describe each of the organ systems mentioned above in much greater detail. The remainder of this chapter focuses on the mechanisms of homeostasis, using one important variable of the internal environment—its temperature—as our example.

40.2 How Does Temperature Affect Living Systems?

Temperatures vary enormously over the face of Earth, from the boiling hot springs of Yellowstone National Park to the interior of Antarctica, where the temperature can fall below –80°C. Cells, however, can function over only a narrow range of temperatures. If cells cool below 0°C, ice crystals form and damage cell structures. Some animals have adaptations, such as antifreeze molecules in their blood, that help them resist freezing; others can survive freezing. Generally, however, cells must remain above 0°C to stay alive.

The upper temperature limit for survival in most cells is about 45°C (although some specialized algae can grow in hot springs at 70°C, and some archaea live at near 100°C). In general, proteins begin to denature and lose their function as temperatures rise above 40°C. Therefore, most cellular functions are limited to the range between 0°C and 40°C, which approximates the thermal limits for life. Each particular species, however, usually has much narrower limits. To stay within those limits in spite of environmental conditions, animals have evolved thermoregulatory adaptations that give them certain thermal tolerances that determine their distribution ranges. When environments change rapidly, as may be happening with global climate warming, animals may find themselves in situations that exceed their thermal tolerances.

Q_{10} is a measure of temperature sensitivity

Even between 0°C and 40°C, changes in tissue temperature create problems for animals. Most physiological processes, like the biochemical reactions that constitute them, are temperature-sensitive, going faster at higher temperatures (see Figure 8.21). The temperature sensitivity of a reaction or process is described in terms of Q_{10}, a factor calculated by dividing the rate of a process or reaction at a certain temperature, R_T, by that rate at a temperature 10°C lower, R_{T-10}:

$$Q_{10} = \frac{R_T}{R_{T-10}}$$

Q_{10} can be measured for a simple enzymatic reaction or for a complex physiological process, such as rate of oxygen consumption. If a reaction or process is not temperature-sensitive, it has a Q_{10} of 1. Most biological Q_{10} values are between 2 and 3. A Q_{10} of 2 means that the reaction rate doubles as temperature increases by 10°C, and a Q_{10} of 3 indicates a tripling of the rate over a 10° temperature range (**Figure 40.8**).

Changes in body temperature can disrupt an animal's physiology because not all of the biochemical reactions that constitute the metabolism of an animal have the same Q_{10}. These biochemical reactions are linked together in complex networks. The products of one reaction are the reactants for other reactions.

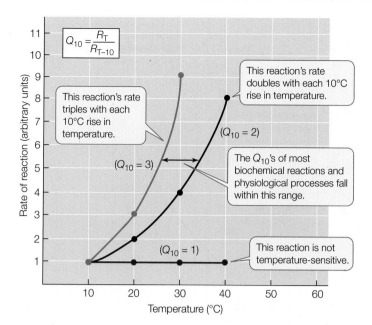

$$Q_{10} = \frac{R_T}{R_{T-10}}$$

This reaction's rate triples with each 10°C rise in temperature.

This reaction's rate doubles with each 10°C rise in temperature.

$(Q_{10} = 2)$

$(Q_{10} = 3)$

The Q_{10}'s of most biochemical reactions and physiological processes fall within this range.

$(Q_{10} = 1)$

This reaction is not temperature-sensitive.

40.8 Q_{10} and Reaction Rate The larger the Q_{10} of a reaction or process, the faster its rate rises in response to an increase in temperature.

Because different reactions have different Q_{10}'s, changes in tissue temperature will shift the rates of some reactions more than others, disrupting the overall network. To maintain homeostasis, organisms must be able to compensate for or prevent changes in body temperature.

Animals acclimatize to seasonal temperatures

The body temperature of some animals (especially aquatic animals) is coupled to environmental temperature. The body temperature of a fish in a pond, for example, will be the same as the water temperature, which might range from 4°C in winter to 24°C in summer. If we bring that fish into the laboratory in the summer and measure its metabolism at different temperatures, we will demonstrate a Q_{10} relationship and can predict what the fish's metabolic rate will be in its pond in the winter. However, if we bring that fish back into the laboratory in the winter and measure its metabolic rate at winter pond temperature, we will find the rate to be much higher than we predicted. The fish's biochemistry and physiology have *acclimatized* to the seasonal change in water temperature so that the fish can remain active at winter temperatures. For example, it may express isozymes with different temperature optima. The ability to acclimatize means that metabolic functions are less sensitive to long-term changes in temperature than to short-term changes.

40.2 RECAP

Cells can survive only within a narrow range of temperatures, but even changes within that range can be disruptive because different physiological processes have different temperature sensitivities.

● Plot a $Q_{10} = 2.5$ curve for a physiological process. **See p. 838 and Figure 40.8**

● Explain how a change in body temperature can disrupt physiological processes. **See pp. 838–839**

We have seen how animals are affected by the temperature of their environment. Now let's take a look at the adaptations that allow animals to control and regulate their body temperatures.

40.3 How Do Animals Alter Their Heat Exchange with the Environment?

Many of us learned to think of animals as being either "cold-blooded" or "warm-blooded," which implies a comparison with our own body temperature and sets mammals and birds apart from other animals. This simple classification breaks down when we realize that mammals that hibernate become cold, and that many reptiles and insects can be quite warm when they are active. Physiologists sometimes classify animals according to whether they have a constant body temperature (homeotherms) or a variable body temperature (poikilotherms). But a deep-sea fish has a constant body temperature. Should it be classified with mammals?

A thermal classification system that avoids such irrational results is one based on the source of heat that predominantly determines the temperature of the animal. **Ectotherms** are animals whose body temperatures are determined primarily by external sources of heat. **Endotherms** regulate their body temperatures by producing heat metabolically or by using active mechanisms of heat loss.

Mammals and birds are endotherms most of the time; other animals are ectotherms most of the time. Like the homeotherm/poikilotherm classification, the endotherm/ectotherm scheme is not perfect. Therefore we have a third category; a **heterotherm** is an animal that behaves sometimes as an endotherm and other times as an ectotherm. For example, a mammal that hibernates is a perfect endotherm over the summer, but during the winter it has bouts of hibernation during which its internal heat production falls and it behaves much like an ectotherm. At times some ectotherms can produce internal heat and act like endotherms.

Endotherms produce heat metabolically

Section 8.1 described how transfers of energy in biological systems are always inefficient. With every transfer of energy—from food molecules to ATP, from ATP to biological work—some of the energy is lost as heat. This is true for both ectotherms and endotherms, so why do endotherms produce more heat? The answer is that the cells of endotherms are less efficient at using energy than are the cells of ectotherms.

In a resting endotherm, most of the energy expended goes into pumping ions across membranes. K+ is the dominant positive ion inside cells, and Na+ is the dominant positive ion outside cells. To the extent that cell membranes permit, these ions diffuse down their concentration gradients. To maintain their proper concentrations inside and outside cells, the ions must be transported back "uphill," which requires expending energy.

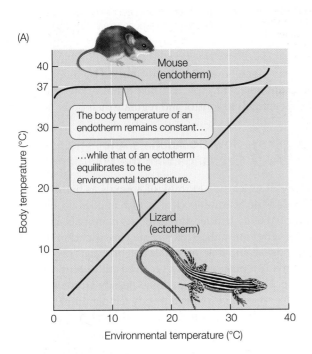

(A)

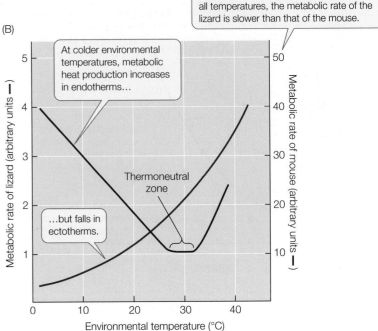

(B)

Notice the difference in the two scales. At all temperatures, the metabolic rate of the lizard is slower than that of the mouse.

At colder environmental temperatures, metabolic heat production increases in endotherms...

Thermoneutral zone

...but falls in ectotherms.

40.9 Ectotherms and Endotherms React Differently to Environmental Temperatures (A) At the same environmental temperature, an ectotherm and an endotherm of approximately the same body size (here, a lizard and a mouse) have different body temperatures. (B) The metabolic rates of the lizard and mouse react in opposite manners to cooler temperatures. (The mouse's metabolic rate rises again at higher temperatures because, after a certain point, it takes metabolic energy to dissipate heat by sweating or panting.)

While this is true for both ectotherms and endotherms, the cells of endotherms tend to be more "leaky" to ions than are those of ectotherms. Thus endotherms must expend more energy (and thus release more heat) than do ectotherms to maintain ion concentration gradients. This is akin to running on a treadmill: the faster the treadmill goes (analogous to leaking ions), the faster you have to run (analogous to pumping ions) to remain in the same position.

We can speculate that a mutation allowing seemingly faulty or leaky ion channels may have led to the evolution of endothermy. Such a mutation in a small ectotherm may have promoted sufficient heat production to allow this ectotherm to remain active for a longer time after the sun went down. Thus, for the first endotherms an entirely new nocturnal world of ecological opportunities opened, one in which there was less competition from similar-sized ectotherms. Major differences between endotherms and ectotherms are their resting metabolic rates—the sum total of all energy expenditures in their bodies when at rest—and their responses to changes in environmental temperature.

Ectotherms and endotherms respond differently to changes in temperature

Let's compare how two similar-sized animals, a lizard (an ectotherm) and a mouse (an endotherm) respond to changes in temperature. We put each animal in a closed chamber and measure its body temperature and metabolic rate as we change the temperature of the chamber from 37°C to 0°C.

The body temperature of the lizard equilibrates with that of the chamber, whereas the body temperature of the mouse remains stable (**Figure 40.9A**). The metabolic rate of the lizard (already much lower than that of the mouse) decreases as the temperature is lowered (**Figure 40.9B**). In contrast, the mouse's metabolic rate increases as the chamber temperature falls below 25°C. The increase in the mouse's metabolism produces enough heat to prevent its body temperature from falling. In other words, the mouse can regulate its body temperature by increasing its metabolic rate; the lizard cannot.

This experiment might lead us to conclude that the ectotherm cannot regulate its body temperature, but observations of the lizard in nature do not support this conclusion. In nature, unlike in the laboratory, the lizard's body temperature is sometimes considerably different than the environmental temperature. The desert habitat where the lizard lives can fluctuate by 40°C in a few hours. During its daily activities, however, the lizard maintains a fairly stable body temperature by using behavior to alter its heat exchange with the environment (**Figure 40.10A**). Its behavioral strategies include spending time in a burrow, basking in the sun, seeking shade, climbing vegetation, and changing its orientation with respect to the sun. While the lizard can regulate its body temperature quite well, it does so by behavioral mechanisms rather than by altering its internal metabolic heat production.

Behavioral thermoregulation is not the exclusive domain of ectotherms (**Figure 40.10B**). Endotherms usually select the most comfortable thermal environment possible. They may change posture, orient to the sun, move between sun and shade, and move between still air and moving air, the same as the ectotherm in our field experiment. Examples of more complex

(A)

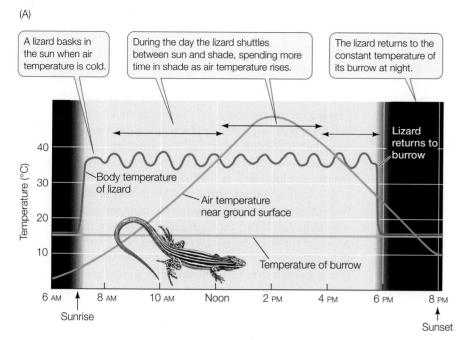

A lizard basks in the sun when air temperature is cold.

During the day the lizard shuttles between sun and shade, spending more time in shade as air temperature rises.

The lizard returns to the constant temperature of its burrow at night.

Lizard returns to burrow

Body temperature of lizard

Air temperature near ground surface

Temperature of burrow

Sunrise

Sunset

(B) *Loxodonta africana*

(C) *Maccaca fuscata*

40.10 Using Behavior to Regulate Body Temperature (A) The body temperature of a lizard (an ectotherm) depends on environmental heat, but the lizard can regulate its temperature by moving from place to place within its environment. (B) When air temperatures on the African savanna soar, an elephant (an endotherm) may thermoregulate by showering itself with water. (C) Japanese macaques are social primates and will huddle together for warmth.

thermoregulatory behaviors include nest construction and social behaviors such as huddling. Humans put on or remove clothing and burn fossil fuels to generate the energy to heat or cool buildings.

Energy budgets reflect adaptations for regulating body temperature

Both ectotherms and endotherms can influence their body temperatures by altering four avenues of heat exchange between their bodies and the environment (**Figure 40.11**):

- **Radiation**: Heat transfers from warmer objects to cooler ones via the exchange of infrared radiation (what you feel when you stand in front of a fire).
- **Convection**: Heat transfers to a surrounding medium such as air or water as that medium flows over a surface (the wind-chill factor).
- **Conduction**: Heat transfers directly when objects of two different temperatures come into contact (think of putting an icepack on a sprained ankle).
- **Evaporation**: Heat transfers away from a surface when water evaporates on that surface (the effect of sweating).

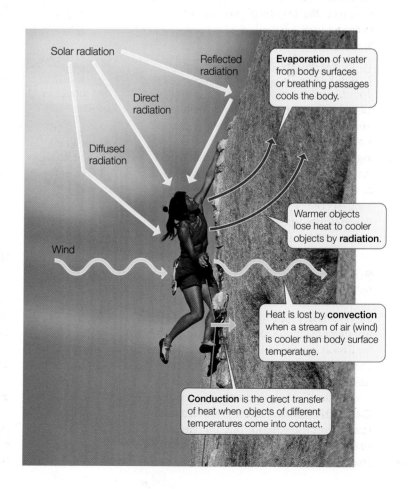

Solar radiation

Reflected radiation

Direct radiation

Diffused radiation

Wind

Evaporation of water from body surfaces or breathing passages cools the body.

Warmer objects lose heat to cooler objects by **radiation**.

Heat is lost by **convection** when a stream of air (wind) is cooler than body surface temperature.

Conduction is the direct transfer of heat when objects of different temperatures come into contact.

40.11 Animals Exchange Heat with the Environment An animal's body temperature is determined by the balance between internal heat production and four avenues of heat exchange with the environment: radiation, convection, conduction, and evaporation.

The total balance of heat production and heat exchange can be expressed as an **energy budget**, based on the simple fact that if the body temperature of an animal is to remain constant, the heat entering the animal must equal the heat leaving it. The heat coming in usually comes from metabolism and solar radiation (R_{abs}, for radiation absorbed). Heat leaves the body via the four mechanisms listed above: radiation emitted (R_{out}), convection, conduction, and evaporation. The energy budget takes the following form:

$$\overbrace{\text{metabolism} + R_{abs}}^{\text{heat}_{in}} = \overbrace{R_{out} + \text{convection} + \text{conduction} + \text{evaporation}}^{\text{heat}_{out}}$$

Anyone who has experienced a very hot environment knows that heat can also *enter* the body through convection (e.g., the hot desert wind) and conduction (e.g., a hot car seat). In that case, the values of those factors become negative in the energy budget equation.

The energy budget is a useful concept because any adaptation that influences the ability of an animal to deal with its thermal environment must affect one or more components of the budget. So the energy budget gives us the ability to quantify and compare the thermal adaptations of animals. One interesting observation is that all of the components on the right side of the energy budget equation—that is, the heat-loss side—depend on the surface temperature of the animal. One way surface temperature can be controlled is by altering the flow of blood to the skin.

Both ectotherms and endotherms control blood flow to the skin

Heat exchange between the internal environment and the skin occurs largely through blood flow. As described at the beginning of this chapter, when body temperature rises because of exercise, blood flow to the skin increases, and the skin surface becomes warm. The heat that the blood brings from the body core to the skin is lost to the environment through the four avenues listed above, which helps bring the body temperature back to normal. In contrast, when body temperature is too low or the environment is too cold, the blood vessels supplying the skin constrict, reducing heat loss to the environment.

The control of blood flow to the skin can be an important adaptation for an ectotherm such as the marine iguana (a reptile) of the Galápagos archipelago (**Figure 40.12**). The Galápagos are volcanic islands that lie on the equator but are bathed by cold ocean currents. The iguanas bask on hot black lava rocks on the shore, then enter the cold ocean water to feed on seaweed. When the iguanas are feeding, they cool to the temperature of the sea. This cooling lowers their metabolism, making them slower, more vulnerable to predators, and incapable of efficient digestion. They therefore alternate between feeding in the cold sea and basking on the hot rocks. It is advantageous for iguanas to retain body heat as long as possible while swimming and to warm up as fast as possible when basking. They can accomplish these changes in heat transfer rates by changing their heart rate and the rate of blood flow to their skin.

What about furred mammals? Fur acts as *insulation* to keep body heat in, making it possible for mammals to live in very cold climates. When they are active, however, mammals still must get rid of excess heat, and it does little good to transport that heat to the skin under the fur. Thus, as mentioned at the beginning of this chapter, mammals have special blood vessels for transporting heat to their hairless skin surfaces. Heat loss from these areas of skin is tightly controlled by the opening and closing of these special blood vessels. When you are cold, the blood flow to your hands and feet decreases and they can feel very cold, but when you exercise, these same surfaces can get very hot quickly.

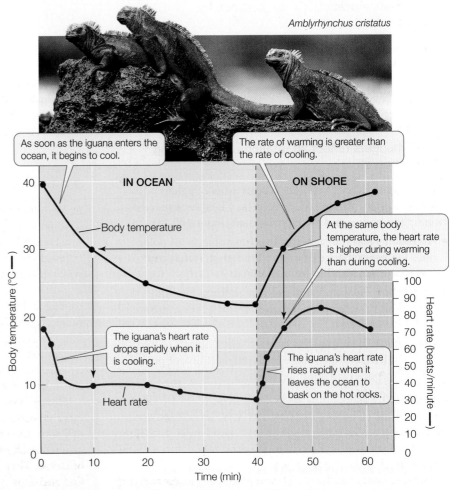

Amblyrhynchus cristatus

As soon as the iguana enters the ocean, it begins to cool.

The rate of warming is greater than the rate of cooling.

IN OCEAN — ON SHORE

Body temperature

At the same body temperature, the heart rate is higher during warming than during cooling.

The iguana's heart rate drops rapidly when it is cooling.

The iguana's heart rate rises rapidly when it leaves the ocean to bask on the hot rocks.

Heart rate

40.12 Some Ectotherms Regulate Blood Flow to the Skin Galápagos marine iguanas control blood flow to the skin to alter their heating and cooling rates.

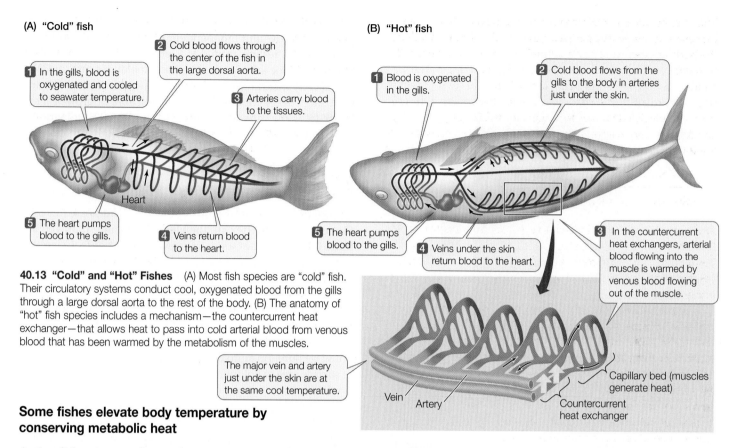

(A) "Cold" fish

1 In the gills, blood is oxygenated and cooled to seawater temperature.

2 Cold blood flows through the center of the fish in the large dorsal aorta.

3 Arteries carry blood to the tissues.

5 The heart pumps blood to the gills.

4 Veins return blood to the heart.

Heart

(B) "Hot" fish

1 Blood is oxygenated in the gills.

2 Cold blood flows from the gills to the body in arteries just under the skin.

5 The heart pumps blood to the gills.

4 Veins under the skin return blood to the heart.

3 In the countercurrent heat exchangers, arterial blood flowing into the muscle is warmed by venous blood flowing out of the muscle.

The major vein and artery just under the skin are at the same cool temperature.

Vein   Artery

Countercurrent heat exchanger

Capillary bed (muscles generate heat)

40.13 "Cold" and "Hot" Fishes (A) Most fish species are "cold" fish. Their circulatory systems conduct cool, oxygenated blood from the gills through a large dorsal aorta to the rest of the body. (B) The anatomy of "hot" fish species includes a mechanism—the countercurrent heat exchanger—that allows heat to pass into cold arterial blood from venous blood that has been warmed by the metabolism of the muscles.

Some fishes elevate body temperature by conserving metabolic heat

Active fishes can produce substantial amounts of metabolic heat, but they have difficulty retaining any of that heat. Blood pumped from the heart goes directly to the gills, where it comes very close to the surrounding water to exchange respiratory gases. So any heat that the blood picks up from metabolically active muscles is lost to the surrounding water as it flows through the gills. It is thus surprising that some large, rapidly swimming fishes, such as bluefin tuna and great white sharks, can maintain temperature differences as great as 10° to 15°C between their bodies and the surrounding water. The heat comes from their powerful swimming muscles, and the ability of these "hot" fishes to conserve that heat is based on the remarkable arrangements of their blood vessels.

In the usual ("cold") fish circulatory system, oxygenated blood from the gills collects in a large dorsal vessel, the aorta, which travels through the center of the fish, distributing blood to all organs and muscles (**Figure 40.13A**). "Hot" fishes have a smaller central dorsal aorta, and most of their oxygenated blood is transported in large vessels just under the skin (**Figure 40.13B**). The cold blood from the gills is thus kept close to the surface of the fish. Smaller vessels transporting this cold blood into the muscle mass run parallel to vessels transporting warm blood from the muscle mass back toward the heart. Since the vessels carrying the cold blood into the muscle are in close contact with the vessels carrying warm blood away, heat flows from the warm to the cold blood by conduction and is therefore retained in the muscle mass.

Because heat is exchanged between blood vessels carrying blood in opposite directions, this adaptation is called a **countercurrent heat exchanger**. It keeps the heat within the muscle

mass, enabling these fishes to have an internal body temperature considerably higher than the water temperature. Why is it advantageous for the fish to be warm? Each 10°C rise in muscle temperature increases the fish's sustainable power output almost threefold, giving it a faster foraging capability!

Some ectotherms regulate heat production

Some ectotherms raise their body temperature by producing heat. For example, the powerful flight muscles of many insects must reach 35° to 40°C before the insects can fly, and they must maintain these high temperatures during flight. Such insects produce the required heat by contracting their flight muscles in a manner analogous to shivering in mammals. The heat-producing ability of insects can be quite remarkable. Probably the most impressive case is a species of scarab beetle that lives mostly underground in mountains north of Los Angeles, California. To mate, these beetles come aboveground, and males fly in search of females. They undertake this mating ritual at night, in winter, and only during snowstorms.

Honey bees regulate temperature as a group. They live in large colonies consisting mostly of female worker bees that maintain the hive and rear the larval offspring of the single queen bee. During winter, worker bees cluster around the brood of larvae. They adjust their individual metabolic heat production and density of clustering so that the brood temperature remains remarkably constant, at about 34°C, even as the outside air temperature drops below freezing (**Figure 40.14**).

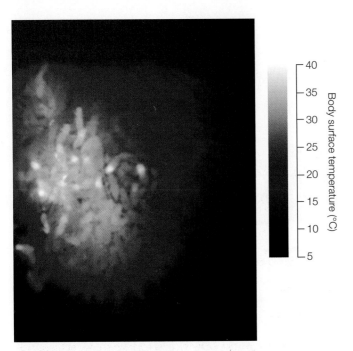

40.14 Bees Keep Warm in Winter Honey bee colonies survive winter cold because workers generate metabolic heat. In this infrared photograph of the center of an overwintering hive, individual bees are discernible by the heat their bodies produce as they cluster around their queen.

40.3 RECAP

Animals that metabolically produce their own heat are called endotherms. Those that depend on environmental sources of heat are called ectotherms. Heat exchange between an animal and its environment occurs via radiation, convection, conduction, and evaporation.

- In terms of the energy budget relationship, why is the control of blood flow to the skin so important for thermoregulation? **See p. 842**

- Explain how countercurrent heat exchange makes it possible for some fishes to have a body temperature higher than that of the surrounding water. **See p. 843 and Figure 40.13**

Endotherms must keep their body temperatures within a critical physiological range. Let's look more closely at the evolutionary adaptations that enable endothermic mammals to maintain this optimal temperature range.

40.4 How Do Mammals Regulate Their Body Temperatures?

As we saw in Figure 40.9, endotherms can respond to changes in environmental temperature by changing their metabolic rate. Physiologists determine metabolic rate by measuring the rate at which an animal consumes O_2 and produces CO_2. Within a narrow range of environmental temperatures, called the **ther-**

moneutral zone (see Figure 40.9B), the metabolic rate of endotherms is low and independent of temperature. The metabolic rate of a resting animal at a temperature within the thermoneutral zone is known as the **basal metabolic rate**, or **BMR**. It is usually measured in animals that are quiet but awake and not using energy for digestion, reproduction, or growth. Thus the BMR is the rate at which a resting animal is consuming just enough energy to carry out its minimal body functions.

Basal metabolic rates are correlated with body size and environmental temperature

As you might expect, the BMR of an elephant is greater than that of a mouse. After all, the elephant is more than 100,000 times larger than the mouse. However, the BMR of the elephant is only about 7,000 times greater than that of the mouse. That means that a gram of mouse tissue uses energy at a rate 15 times greater than a gram of elephant tissue (**Figure 40.15**). Across all of the endotherms, BMR per gram of tissue increases as animals get smaller.

Why should this disproportionate difference exist? We don't know for sure. As animals get bigger, they have a smaller ratio of surface area to volume (see Figure 5.2). Since heat production is related to the volume, or mass, of the animal, but its capacity to dissipate heat is related to its surface area, it was once reasoned that larger animals evolved lower metabolic rates to avoid overheating. This explanation is insufficient because the relationship between body mass and metabolic rate holds for even very small organisms and for ectotherms, in which overheating is not a problem. Other hypotheses have also been advanced. For example, a larger animal has a greater proportion of support tissue (skin, bone), which is not as metabolically active. The real answer is probably a mixture of different causative factors.

In an endotherm, the metabolic rate versus environmental temperature curve represents the integrated response of all of the animal's thermoregulatory adaptations (**Figure 40.16**). The

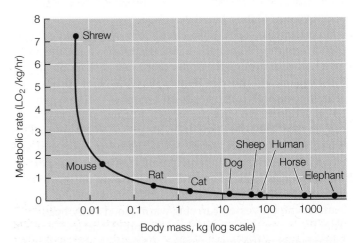

40.15 The Mouse-to-Elephant Curve On a weight-specific basis, the metabolic rate of small endotherms is much greater than that of larger endotherms. This graph plots O_2 consumption per kilogram of body mass (a measure of metabolic rate) against a logarithmic plot of body mass.

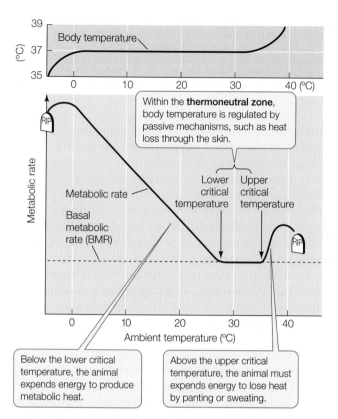

Within the **thermoneutral zone**, body temperature is regulated by passive mechanisms, such as heat loss through the skin.

Below the lower critical temperature, the animal expends energy to produce metabolic heat.

Above the upper critical temperature, the animal must expends energy to lose heat by panting or sweating.

40.16 Environmental Temperature and Mammalian Metabolic Rates Outside the thermoneutral zone, maintaining a constant body temperature requires expending energy. Outside extreme limits (0°C and 40°C in this instance), the animal cannot maintain its body temperature and dies.

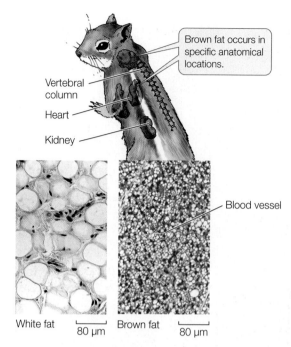

40.17 Brown Fat In many mammals, specialized brown fat tissue produces heat. When looking through a microscope at similar magnifications, we see that white fat cells (left) contain large droplets of lipid but have few organelles and limited blood supply, whereas brown fat cells (right) are packed with mitochondria and richly supplied with blood.

thermoneutral zone is bounded by a *lower* and an *upper critical temperature*. Within its thermoneutral zone, an endotherm's thermoregulatory adaptations do not require much energy and could be considered passive; such adaptations include changing posture, fluffing fur, and controlling blood flow to the skin. Outside its thermoneutral zone, however, an endotherm's thermoregulatory responses are active and require considerable metabolic energy, as shown on the left side of Figure 40.16.

—— **yourBioPortal**.com ——
GO TO Web Activity 40.1 • Thermoregulation in an Endotherm

Endotherms respond to cold by producing heat and adapt to cold by reducing heat loss

When environmental temperatures fall below the lower critical temperature, endotherms must produce heat to compensate for the heat they lose to the environment. Mammals can accomplish this in two ways: shivering and nonshivering heat production. Birds use only shivering heat production.

Shivering uses the contractile machinery of skeletal muscles to consume ATP without causing large movements. Shivering muscles pull against each other so that little movement other than a tremor results. The energy from the conversion of ATP to ADP in this process is released as heat. "Shivering heat production" is perhaps too narrow a term, however; increased mus-

cle tone and increased body movements also contribute to increased heat production in cold environments.

Most *nonshivering heat production* occurs in a specialized adipose tissue called **brown fat** (**Figure 40.17**). This tissue looks brown because of its abundant mitochondria and rich blood supply. In brown fat cells, a protein called *thermogenin* uncouples proton movement from ATP production, allowing protons to leak across the inner mitochondrial membrane rather than having to pass through the ATP synthase and generate ATP (review the discussion of the chemiosmotic mechanism and Figure 9.9). As a result, metabolic fuels are consumed without producing ATP, but heat is still released. Brown fat is abundant in newborns of many mammalian species (including humans), in some adult mammals that are small and acclimatized to cold, and in mammals that hibernate.

In spite of their ability to produce heat, endotherms in cold climates have evolved adaptations to reduce their heat loss and therefore remain within their thermoneutral zones as much as possible. Heat is lost from the body surface, and cold-climate species have anatomical adaptations that give them smaller surface-to-volume ratios than their warm-climate relatives. These adaptations include rounder body shapes and shorter appendages (**Figure 40.18**).

The most significant means of decreasing heat loss is to increase thermal insulation. Animals adapted to cold climates have much thicker layers of fur, feathers, or fat than do their warm-climate relatives. Fur and feathers are good insulators because they trap a layer of still, warm air close to the skin surface. If that air is displaced by water, insulation is drastically reduced. In many species, oil secretions spread through fur or

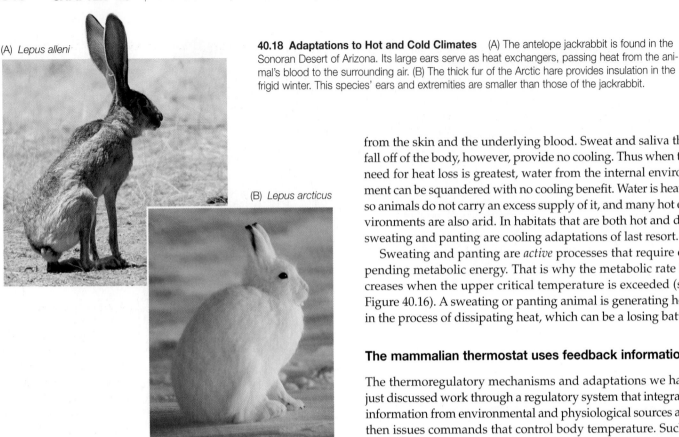

(A) *Lepus alleni*

(B) *Lepus arcticus*

40.18 Adaptations to Hot and Cold Climates (A) The antelope jackrabbit is found in the Sonoran Desert of Arizona. Its large ears serve as heat exchangers, passing heat from the animal's blood to the surrounding air. (B) The thick fur of the Arctic hare provides insulation in the frigid winter. This species' ears and extremities are smaller than those of the jackrabbit.

feathers by grooming are critical for resisting wetting and maintaining a high level of insulation.

Decreasing blood flow to the skin is an important thermoregulatory adaptation in the cold. Constriction of blood vessels in the skin, and especially in the appendages, greatly improves an animal's ability to conserve heat. Countercurrent heat exchange like we saw in the "hot" fishes is also an important adaptation in the appendages of endotherms. Blood flowing out to the paw of a wolf, the hoof of a caribou, or the foot of a bird parallels the flow of the blood returning. Heat is transferred from the outgoing to the returning blood, thus retaining heat in the animal's core.

Evaporation of water can dissipate heat, but at a cost

As environmental temperature rises within an endotherm's thermoneutral zone, the animal dissipates more of its metabolic heat by increasing blood flow to the skin. When the temperature exceeds the upper critical temperature, however, overheating becomes a problem. For an exercising animal, overheating can occur even at low environmental temperatures. Large mammals, especially those in hot habitats such as elephants, rhinoceroses, and water buffaloes, have little or no insulating fur and seek out water to wallow in when the air temperature is high (see Figure 40.10B). Having water in contact with the skin greatly increases heat loss because the heat-absorbing capacity of water is much greater than that of air.

Evaporation from external or internal body surfaces through sweating or panting can also cool an endotherm. A gram of water absorbs about 580 calories of heat when it evaporates. If this evaporation occurs on the skin, most of that heat is absorbed

from the skin and the underlying blood. Sweat and saliva that fall off of the body, however, provide no cooling. Thus when the need for heat loss is greatest, water from the internal environment can be squandered with no cooling benefit. Water is heavy, so animals do not carry an excess supply of it, and many hot environments are also arid. In habitats that are both hot and dry, sweating and panting are cooling adaptations of last resort.

Sweating and panting are *active* processes that require expending metabolic energy. That is why the metabolic rate increases when the upper critical temperature is exceeded (see Figure 40.16). A sweating or panting animal is generating heat in the process of dissipating heat, which can be a losing battle.

The mammalian thermostat uses feedback information

The thermoregulatory mechanisms and adaptations we have just discussed work through a regulatory system that integrates information from environmental and physiological sources and then issues commands that control body temperature. Such a regulatory system is based on feedback information, and can be thought of as a thermostat like the one in your home.

The major thermoregulatory integrative center of mammals is at the bottom of the brain in a structure called the **hypothalamus**. If you slide your tongue back as far as possible along the roof of your mouth, it will be just a few centimeters below your hypothalamus. The hypothalamus is a key part of many regulatory systems, including thermoregulation in all vertebrates.

In many vertebrates and all mammals, the temperature of the hypothalamus itself is the major negative feedback signal, and damage to the hypothalamus can disrupt thermoregulation. The hypothalamus generates a set point like a setting on a home thermostat. When the temperature of the hypothalamus exceeds or drops below that set point, thermoregulatory responses (the controlled system) are activated to reverse the direction of temperature change.

In mammals, experiments show that directly cooling the hypothalamus increases metabolic heat production and stimulates constriction of the blood vessels that supply the skin, thus causing body temperature to rise. Conversely, mild warming of the hypothalamus stimulates dilation of the blood vessels, while stronger hypothalamic heating stimulates sweating or panting. Consequently, heating the hypothalamus causes the overall body temperature to fall (**Figure 40.19**).

───── **yourBioPortal.com** ─────
GO TO Animated Tutorial 40.1 • The Hypothalamus

The mammalian thermoregulatory system has adjustable set points and integrates sources of information in addition to hypothalamic temperature. For example, temperature sensors in the skin register environmental temperature; change in skin tem-

INVESTIGATING LIFE

40.19 The Hypothalamus Regulates Body Temperature

In this laboratory experiment, a mammal's hypothalamus was subjected directly to temperature manipulation. The body's responses to the experimenters' manipulations were as expected if the hypothalamus is indeed the mammalian "thermostat."

HYPOTHESIS Heating or cooling the mammalian hypothalamus results in corresponding and predictable changes in body temperature.

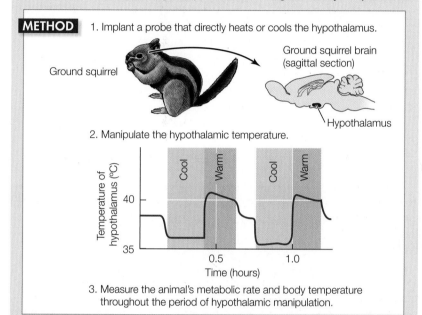

METHOD

1. Implant a probe that directly heats or cools the hypothalamus.

Ground squirrel

Ground squirrel brain (sagittal section)

Hypothalamus

2. Manipulate the hypothalamic temperature.

3. Measure the animal's metabolic rate and body temperature throughout the period of hypothalamic manipulation.

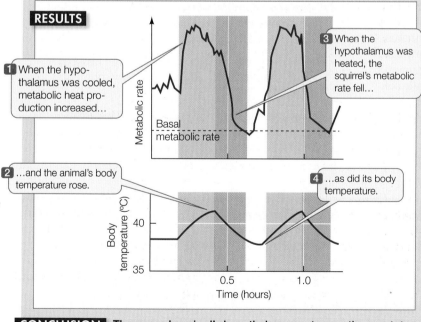

RESULTS

1 When the hypothalamus was cooled, metabolic heat production increased...

2 ...and the animal's body temperature rose.

3 When the hypothalamus was heated, the squirrel's metabolic rate fell...

4 ...as did its body temperature.

Basal metabolic rate

CONCLUSION The ground squirrel's hypothalamus acts as a thermostat. When cooled it activates metabolic heat production; when warmed, it suppresses metabolic heat production and favors heat loss.

Go to **yourBioPortal.com** for original citations, discussions, and relevant links for all INVESTIGATING LIFE figures.

perature is feedforward information that shifts the hypothalamic set point for thermoregulatory responses. The set point for metabolic heat production is higher when skin is cold and lower when skin is warm.

Hypothalamic set points are higher during wakefulness than during sleep, and they are higher during the active part of the daily cycle than during the inactive part, even if the animal is awake at both times. Even when an endotherm is kept under constant environmental conditions, its body temperature displays a daily cycle of changes in set point. This kind of cycle is controlled by an internal *circadian rhythm*; we discuss these endogenous bodily rhythms in Chapter 54.

Fever helps the body fight infections

Fever is an adaptive response that helps the body fight pathogens. A fever is a rise in body temperature in response to molecules called **pyrogens**. *Exogenous pyrogens* come from foreign substances such as bacteria or viruses that invade the body. *Endogenous pyrogens* are produced by cells of the immune system in response to infection.

The presence of a pyrogen in the body causes a rise in the hypothalamic set point for the metabolic heat production response. As a result, you shiver, put on a sweater, or crawl under a blanket, and your body temperature rises until it matches the new set point. At the higher body temperature you no longer feel cold, and you may not feel hot, but someone touching your forehead will say that you are "burning up." Taking aspirin lowers your set point to normal. Now you feel hot, take off clothes, and even sweat until your elevated body temperature returns to normal. Although modest fevers help the body fight infections, extreme fevers can be dangerous and must be controlled, usually with fever-reducing drugs.

Turning down the thermostat

Hypothermia is a below-normal body temperature. It can result from starvation (lack of metabolic fuel), exposure to extreme cold, serious illness, or anesthesia. In each of these cases, the drop in body temperature is unregulated. However, many birds and mammals undergo regulated drops in body temperature to survive periods of cold and food scarcity, an adaptation known as *regulated hypothermia*.

Hummingbirds, for example, are very small endotherms with a high metabolic rate. Just getting through a single day without food could exhaust their metabolic reserves. Hummingbirds and other small endotherms can extend the pe-

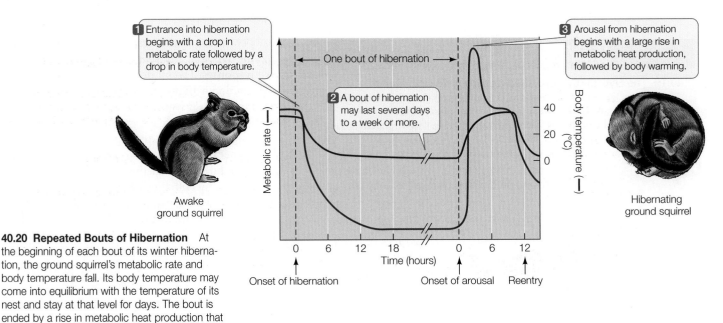

40.20 Repeated Bouts of Hibernation   At the beginning of each bout of its winter hibernation, the ground squirrel's metabolic rate and body temperature fall. Its body temperature may come into equilibrium with the temperature of its nest and stay at that level for days. The bout is ended by a rise in metabolic heat production that returns body temperature to a normal level.

riod over which they can survive without food by dropping their body temperature during the portion of day or night when they are normally inactive. This adaptive hypothermia is called **daily torpor**. Body temperature can drop 10° to 20°C during daily torpor, lowering metabolic rate and saving energy.

Regulated hypothermia that lasts for days or even weeks, with body temperature falling close to the ambient temperature, is called **hibernation (Figure 40.20)**. Many species of mammals, including bats, bears, and ground squirrels, hibernate, but only one species of bird (the poorwill) has been shown to hibernate. The metabolic rate needed to sustain a hibernating animal may be only one-fiftieth its basal metabolic rate, and many animals maintain body temperatures close to the freezing point. Arousal from hibernation occurs when the hypothalamic set point returns to the normal level for a mammal.

The ability of hibernators to reduce their thermoregulatory set point so dramatically probably evolved as an extension of the set point decrease that accompanies sleep even in nonhibernating species of mammals and birds.

40.4 RECAP

Within the thermoneutral zone, an endotherm controls its body temperature by altering insulation and blood flow to the skin. When the temperature drops below the thermoneutral zone, the animal increases metabolic heat production. Above this zone, it dissipates heat by panting or sweating.

- Describe how endotherms produce heat. How does heat production change with body size? **See p. 844 and Figure 40.15**

- Why is dependence on evaporative water loss a dangerous strategy for dealing with hot environments? **See p. 846**

- What is the nature of negative feedback information and feedforward information used by the mammalian thermostat? **See pp. 847–848 and Figure 40.19**

CHAPTER SUMMARY

40.1 How Do Multicellular Animals Supply the Needs of Their Cells?

- Multicellular animals provide for the needs of all their cells by maintaining a stable **internal environment**, which consists of the two types of extracellular fluid: the interstitial fluid, and the plasma of blood. **Review Figure 40.1**

- The regulation of physiological systems is mostly through **negative feedback**. **Feedforward information** functions to change **set points**. **Review Figure 40.2**

- The four types of **tissues** are assemblages of cells. **Epithelial tissues** provide barriers and have secretory and transport func-

tions. The three types of **muscle tissue** (skeletal, cardiac, and smooth muscle) are able to contract and are the source both voluntary and involuntary movement. **Connective tissues**, including cartilage, bone, adipose tissue, and blood, are supportive tissues made up of cells embedded in an extracellular matrix. **Nervous tissues** process and communicate information; they contain two cell types, neurons and glial cells.

Organs are made up of tissues, and most organs contain all four kinds of tissue. Organs are grouped into **organ systems**. **Review Figure 40.7**

40.2 How Does Temperature Affect Living Systems?

- Life is sustained within a narrow range of environmental temperatures. Q_{10} is a measure of the sensitivity of a life process to temperature. A Q_{10} of 2 means that the reaction rate doubles as temperature increases by 10°C. **Review Figure 40.8**
- Animals can acclimatize to seasonal changes in temperature through biochemical and physiological adaptations.

40.3 How Do Animals Alter Their Heat Exchange with the Environment?

- **Endotherms** can produce considerable metabolic heat to compensate for heat loss to the environment. **Ectotherms** generally do not. **Review Figure 40.9**
- **Energy budgets** describe all pathways for heat exchange between an organism and its environment. The four avenues of heat exchange are **radiation, convection, conduction,** and **evaporation. Review Figure 40.11**
- Skin temperature is an important variable, and it can be influenced by blood flow. Circulatory system adaptations such as **countercurrent heat exchange** can conserve metabolic heat. Review Figures 40.12 and 40.13

40.4 How Do Mammals Regulate Their Body Temperatures?

- Within the **thermoneutral zone**, mammals have a **basal metabolic rate** (BMR) that scales with body size. **Review Figures 40.15 and 40.16, WEB ACTIVITY 40.1**
- In mammals, control of thermoregulatory effectors relies on commands from a regulatory center in the **hypothalamus**. This thermostat uses its own temperature as a major negative feedback signal, and skin temperature as a feedforward signal. **Review Figure 40.19, ANIMATED TUTORIAL 40.1**

SELF-QUIZ

1. Which of the following characterizes the protein elastin?
   a. It functions predominantly in muscle tissue to resist excess stretching.
   b. It is found predominantly in epithelial tissue.
   c. It is found in the extracellular matrix of connective tissue.
   d. It is the most abundant protein in the body.
   e. It is responsible for the elasticity of the long extensions of neurons.

2. If the Q_{10} of the metabolic rate of an animal is 2, then
   a. the animal is better acclimatized to a cold environment than if its Q_{10} is 3.
   b. the animal is an ectotherm.
   c. the animal consumes half as much oxygen per hour at 20°C as it does at 30°C.
   d. the animal's metabolic rate is not at basal levels.
   e. the animal produces twice as much heat at 20°C as it does at 30°C.

3. Which statement about brown fat is true?
   a. It produces heat without producing ATP.
   b. It insulates animals acclimatized to cold.
   c. It is a major source of heat production for birds.
   d. It is found only in hibernators.
   e. It provides fuel for muscle cells.

4. Which of the following is the most important and most general characteristic of endotherms adapted to cold climates compared with those adapted to warm climates?
   a. Higher basal metabolic rates
   b. Higher Q_{10} values
   c. Brown fat
   d. Greater insulation
   e. Ability to hibernate

5. Which of the following would cause a decrease in the hypothalamic temperature set point for metabolic heat production?
   a. Entering a cold environment
   b. Taking an aspirin when you have a fever
   c. Arousing from hibernation
   d. Getting an infection that causes a fever
   e. Cooling the hypothalamus

6. Mammalian hibernation
   a. occurs when animals run out of metabolic fuel.
   b. is a regulated decrease in body temperature.
   c. is less common than hibernation in birds.
   d. can occur at any time of year.
   e. lasts for several months, during which body temperature remains close to the environmental temperature.

7. Which of the following is an important difference between an ectotherm and an endotherm of similar body size?
   a. An ectotherm has higher Q_{10} values.
   b. Only an ectotherm uses behavioral thermoregulation.
   c. Only an endotherm can constrict and dilate the blood vessels to the skin to alter heat flow.
   d. Only an endotherm can have a fever.
   e. At a body temperature of 37°C, an ectotherm has a lower metabolic rate than an endotherm.

8. How would you describe the role of skin temperature in the human thermoregulatory system?
   a. It provides feedforward information.
   b. It acts as a set point for metabolic heat production.
   c. It provides positive feedback information.
   d. It provides an error signal.
   e. It provides negative feedback information.

9. What is the biggest difference between a "cold" fish such as a trout and a "hot" fish such as a tuna?
   a. The temperature of the blood leaving the heart
   b. The temperature of the blood entering the gills
   c. The arrangement of blood vessels in the gills
   d. The temperature of the brain
   e. The volume of blood flowing in arteries just under the skin

10. Which of the following statements about the thermoneutral zone is true?
    a. Metabolic heat production is variable.
    b. Skin blood flow is variable.
    c. The environmental temperature equals body temperature.
    d. The lower boundary (lower critical temperature) is lower for small than for large endotherms.
    e. It is the range of hypothalamic temperatures that do not alter metabolic heat production.

FOR DISCUSSION

1. What is the advantage of feedforward information for homeostasis? Can you suggest what some sources of feedforward information could be for regulation of breathing, blood pressure, secretion of digestive juices, and elimination of wastes?

2. In some epithelial tissues there are "tight junctions" between the individual cells that prevent anything from passing between them (see Figure 6.7); in other cases the junctions between epithelial cells are quite loose. What are the possible advantages in different organs of loose versus tight junctions between epithelial cells? Give some examples in which these differences would be important.

3. Newton's law of cooling describes how a physical object comes into thermal equilibrium with its environment. The law is expressed

$$HL = K(T_o - T_a)$$

HL is the rate of heat loss, K is the thermal conductance constant (how easily an object loses heat), T_o is the temper-

ature of the object, and T_a is the ambient temperature. Compare this expression with the metabolic rate/temperature curve for endotherms. In Newton's law of cooling, K is a constant reflecting the properties of the object. What would K represent for an endotherm? Using a version of Newton's law that replaces T_o with T_b (body temperature), explain why the metabolic rate curve projects to zero at an ambient temperature that equals body temperature.

4. The range of temperatures compatible with life is about 0°C to 40°C. Endotherms have regulated body temperatures much closer to the upper limit of this range than to the lower. What are the advantages of living so close to the upper limit?

5. Discuss what it means when we say that the metabolic rate of mammals scales to the ¾ power of body mass. In contrast, heart size of mammals scales according to the first power of body mass. What does this difference imply for the functions of the hearts of mammals of different sizes?

ADDITIONAL INVESTIGATION

1. The text described the drop in body temperature of a hibernator as regulated hypothermia—a turning down of the thermostat. Yet we also saw that if we put an ectotherm in a cold environment, its body temperature will fall. What experiment could you do to prove that the mammalian hibernator did not just simply turn off or inactivate its thermoregulatory system in order to behave like an ectotherm in the cold?

2. The observations on the Galápagos marine iguana showed that its body temperature rose faster in air than it fell in water. The inference was that the iguana was influencing its gain or loss of heat by altering the blood flow to its skin. However, the thermal properties of air and water are different, and in the case described in the text, the animal was breathing when in air but not when diving in the water. What experiment could you do to strengthen the argument that the iguana was actively altering the flow of heat across its skin?

WORKING WITH DATA (GO TO yourBioPortal.com)

A Hibernator's Thermostat In this exercise based on the experiments outlined in Figure 40.19, you will plot data gathered from hibernating and non-hibernating ground squirrels to graph the relationship between hypothalamic temperature and metabolic rate. You will then analyze these data and draw conclusions about the role of the hypothalamus in mammalian hibernation.

Juiced

The use of performance-enhancing drugs—particularly *anabolic steroids*—has become a scandal in athletics. Olympic champions have lost medals, professional athletes have been suspended, coaches have lost their jobs, and exceptional performances have been expunged from the record books. The recent history of baseball in the United States has been termed "the steroid era" because of the huge impact the extensive use of performance-enhancement drugs has had on the game and the controversy this has created. Gains in performance and new records raise the question of whether an aspiring athlete can succeed without using performance-enhancing drugs. The U.S. Congress has passed laws against non-medical use of steroids, and Major League Baseball has instituted penalties for players who break the law. To date, over 100 Major League players have admitted to or been implicated in the use of steroids, and many have been suspended for up to 80 games.

You have probably heard of one anabolic steroid: testosterone, the male sex hormone. Shortly before puberty, the male reproductive system increases its production of this important chemical signal. Testosterone enters cells, where it binds to receptors and alters gene expression. Cells that have these receptors are those involved in the development of male secondary sexual characteristics, such as a deep voice, facial and body hair, and increased muscle and bone mass. Anabolic steroids are used therapeutically to treat conditions such as delayed puberty, erectile dysfunction, and the loss of muscle mass that occurs with certain diseases.

When a muscle is exercised, an interaction between the exercise and the steroids results in growth of that muscle. Body builders who abuse anabolic steroids typically use them in doses 10 to100 times greater than normal levels or therapeutic doses. The resulting extreme growth of skeletal muscle mass occurs in women as well as men. Both sexes have receptors for testosterone, but women normally have much lower concentrations of testosterone in their blood than men do. When female body builders use these hormones, they develop male muscle patterns. They also develop deep voices and body and facial hair, and because these steroids generate negative feedback

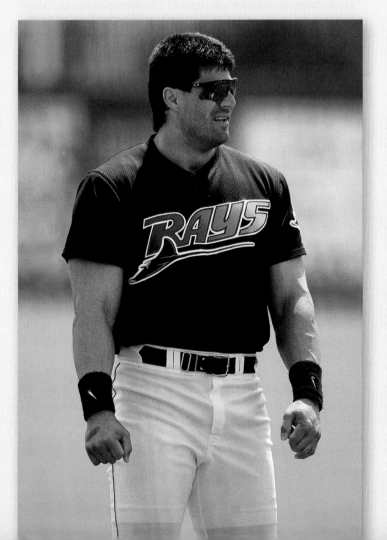

How Baseball Got Big Jose Canseco was the American League Rookie of the Year in 1986. In 1988, he became the first major league player to hit 40 home runs and steal 40 bases in a season and was named the American League's Most Valuable Player. In 2005 he wrote a book called *Juiced: Wild Times, Rampant 'Roids, Smash Hits, and How Baseball Got Big* in which he admitted to using anabolic steroids and implicated many other players.

Anabolic Steroids Build Big Muscles Anabolic steroids greatly enhance the development of skeletal muscle in response to exercise. Steroids have this effect on women as well as men.

information that controls female reproductive physiology, their breast tissue diminishes, they stop menstruating, and they become infertile. Similar negative feedback in males causes infertility. Behavioral changes—"roid rage"—are also common. The most serious side effects, for both men and women, are greatly increased risks of cancer and of heart, liver, and kidney diseases.

Despite the risks, athletes seeking an advantage have frequently turned to anabolic steroids. Athletic governing organizations administer blood and urine tests to detect their use; in their turn, illicit drug makers constantly seek to design new forms of anabolic steroids that produce the desired physical results but are not detectable.

IN THIS CHAPTER we will examine how hormones control and regulate anatomical, developmental, physiological, and behavioral changes in animals. First we examine hormonal control of invertebrate life cycles. Next we discuss the general characteristics of hormones and their receptors. Then we describe the functions, control, and mechanisms of action of mammalian hormones.

41.1 What Are Hormones and How Do They Work?

In multicellular animals, physiological control and regulation require information and cell-to-cell communication. Most intercellular communication is by means of chemical signals that bind to receptors, as described in Chapter 7. **Hormones** are chemical signals that are released by certain types of cells and that influence the activities of other cells at a distance. In this and subsequent chapters you will learn about hormones and other examples of chemical signals, including growth factors, morphogens, cytokines, and neurotransmitters. These general names come from the context in which the chemical signal operates—endocrine system, growth and development, immune system, or nervous system—but the principles of their function are the same: one cell releases a chemical signal that travels to and binds to a receptor, causing a cellular response.

The information that animals use to develop, grow, and function comes from four major sources: the genome, the endocrine system, the immune system, and the nervous system. In each of these systems, information is encoded in the specificity of chemical signals and their receptors. In earlier chapters we learned a lot about genetic information. In this and following chapters we discuss the endocrine, immune, and nervous systems. Lest you think that all signaling is chemical, however, keep in mind that there are also receptors in the nervous system that encode physical sources of information, such as temperature, pressure, and light. And the nervous system uses electrical signals called action potentials to get information from place to place in the body. Regardless of the system, the processing of information depends on which cells have receptors for the signals and how those cells respond and interact with other cells.

Some analogies might help distinguish how the immune, endocrine, and nervous informational systems operate. The immune system (the topic of Chapter 42) operates like an army of private security guards. The various cellular agents make their rounds of the body, and when they detect a security breach, they sound their alarms—cytokines—which activate the body's defenses. The nervous system (see Chapters 45–47) operates like a telephone system with a central integration and command center that sends signals along specific wires to specific receivers. The endocrine system is more like a radio or TV network that broadcasts signals that can be picked up by anyone who has an appropriate receiver that is turned on and tuned in.

In this chapter we focus on the **endocrine system**, which includes a variety of cells that produce and release hormonal chemical signals into the extracellular fluid.

Chemical signals can act locally or at a distance

Endocrine cells secrete chemical signals; **target cells** have receptors for those signals. Chemicals secreted into the extracellular fluid diffuse locally and may diffuse into the blood. Endocrine signals that enter the blood are called hormones, and they can activate target cells far from their site of release (**Figure 41.1A**). Testosterone is an example of a **hormone**.

Some endocrine signals are released in such tiny quantities, or are so rapidly inactivated by enzymes, or are taken up so efficiently by local cells that they never diffuse into the blood in sufficient amounts to act on distant cells. Because these signals affect only target cells near their release site, they are called **paracrines** (*para*, "near"; **Figure 41.1B**). An example of a paracrine is histamine, one of the mediators of inflammation. The most local action an endocrine signal can have is when it binds to receptors on or in the same cell that secreted it. When a chemical signal influences the cell that secreted it, it has **autocrine** function (**Figure 41.1C**). Hormones and paracrines can have autocrine functions as a means of providing negative feedback to control their rates of secretion.

Some endocrine cells exist as single cells within a tissue. Hormones of the digestive tract, for example, are secreted by isolated endocrine cells in the walls of the stomach and small intestine. Many hormones, however, are secreted by aggregations of endocrine cells forming secretory organs called **endocrine glands**. The name "endocrine" reflects the fact that these glands secrete their products directly into the extracellular fluid, which they pass into the blood. In contrast, **exocrine glands,** such as sweat glands or salivary glands, have ducts that carry their products to the surface of the skin or into a body cavity such as the gut. A single endocrine gland may secrete multiple hormones.

To complete our overview of intercellular chemical communication, we must mention neurotransmitters, which we discuss in detail in Chapters 45–47, and pheromones, which we discuss in Chapter 54. Neurons, the cells of the nervous system, conduct information over long distances as electrical signals, but where a neuron communicates that information to another cell, be it another neuron, a muscle cell, or a secretory cell, it does so by releasing chemical signals called **neurotransmitters**. Most neurotransmitters act very locally and frequently act on the neuron that released them. Some neurotransmitters, however, diffuse into the blood and are therefore referred to as **neurohormones**. **Pheromones** are chemical signals that an animal releases into the environment to communicate information to other individuals of the same species.

Hormonal communication has a long evolutionary history

Intercellular chemical signaling was critical for the evolution of multicellularity. A protist, the slime mold *Dictyostelium*, uses a chemical signal (cAMP) to coordinate the aggregation of individual cells to form a multicellular fruiting structure (see Figure 27.32). The most primitive of the multicellular animals—the sponges—do not have nervous systems, but they do have intercellular chemical communication. And as discussed in Chapter 37, plant growth is regulated by a variety of hormones.

Studying the evolution of hormonal signaling reveals an interesting generalization: the signal molecules themselves are highly conserved. We find the same chemical compounds over broad groups of organisms, but their functions differ. As organisms have evolved to occupy different environments and have different lifestyles, hormone–receptor systems have evolved to serve different functions—for example, the hormone prolactin (**Figure 41.2**). Another important example is the hormonal control of molting and metamorphosis—critical events in the lives of arthropods, the most diverse animal group on Earth (see Chapter 32). The hormones involved represent an ancient system of hormonal communication that is genetically related to the anabolic steroid system discussed at the opening of this chapter.

HORMONAL CONTROL OF MOLTING IN ARTHROPODS
Insects, like all arthropods, have a rigid exoskeleton. Therefore, their growth is episodic, punctu-

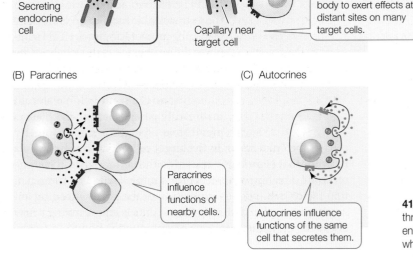

(A) Hormones

Capillary near secreting cell

Target cells

Only cells that posess receptor molecules that recognize and bind a specific hormone will respond to that hormone.

Larger blood vessels

Receptor molecules

Secreting endocrine cell

Hormones travel in the blood throughout the body to exert effects at distant sites on many target cells.

Capillary near target cell

(B) Paracrines

Paracrines influence functions of nearby cells.

(C) Autocrines

Autocrines influence functions of the same cell that secretes them.

41.1 Chemical Signaling Systems Hormones (A) are distributed throughout the body by the blood. Paracrines and autocrines do not enter the bloodstream; paracrines (B) simply diffuse to nearby cells, while autocrines (C) influence the cells that release them.

Fish

Required for osmoregulation in freshwater species. In saltwater species that return to fresh water to spawn (e.g., salmon), prolactin production in adults may play a role in generating the drive to return to natal streams.

Amphibians

Alters the osmoregulatory properties of the skin for animals that enter fresh water. In some species, creates a "water drive" that returns adults to breeding locations. Stimulates oviduct development and production of egg jelly in females. In some species, controls development of sexual characteristics.

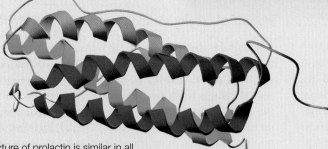

The structure of prolactin is similar in all vertebrate groups.

Birds

In some species, stimulates nesting activity, incubation behavior, and parental care in both sexes. Stimulates the epithelial cells of the upper GI tract to proliferate and slough off to form "crop milk" to nourish the young.

Mammals

In females, stimulates growth of the mammary glands and milk production. In humans, it is responsible for the sensation of sexual gratification as well as the male refractory period following sexual intercourse.

41.2 Prolactin's Structure Is Conserved, but Its Functions Have Evolved The hormone prolactin is found in all vertebrate groups and has a long evolutionary history. Its probable function in early vertebrates was in regulating the body's salt and water balance (osmoregulation). It maintains this function in some species, and has evolved in others to control a number of physiological processes, most of which are associated with reproduction.

nius looks like a miniature adult but lacks certain adult features. The juvenile bug molts five times before developing into a mature adult; a blood meal triggers each episode of molting and growth.

Rhodnius is an amazingly hardy experimental animal—it survives for quite a long time even after its head is cut off. Wigglesworth's studies revealed that, if decapitated within an hour after a blood meal, *Rhodnius* can survive for up to a year, but it never molts. If decapitated a week after its blood meal, however, it does molt. Wigglesworth hypothesized that the time lag meant that the substance that triggers molting diffuses slowly from the head. He tested this hypothesis with the experiment described in **Figure 41.3**.

We now know that two hormones working in sequence regulate molting in arthropods: *prothoracicotropic hormone* (PTTH) and *ecdysone*. Cells in the brain produce PTTH, which is why it has also been called "brain hormone." PTTH is transported to and stored in paired structures called the *corpora cardiaca* attached to the brain. After appropriate stimulation (which for *Rhodnius* is a blood meal), PTTH is released and diffuses through the extracellular fluid to an endocrine gland, the prothoracic gland. PTTH stimulates the prothoracic gland to release the hormone ecdysone. Ecdysone diffuses to target tissues and stimulates molting.

Ecdysone is a lipid-soluble steroid that readily passes through the plasma membrane of its target cells (mostly cells of the epidermis). In the target cells, ecdysone binds to a receptor that is probably related to the vertebrate testosterone receptor. The hormone–receptor complex acts as a transcription factor that induces expression of the genes encoding enzymes involved in digesting the old cuticle and secreting a new one. The related testosterone receptor, when bound to testos-

ated with *molts* (shedding of the exoskeleton). Each growth stage between two molts is called an *instar*.

The British physiologist Sir Vincent Wigglesworth was a pioneer in the study of the hormonal control of growth and development in insects. Wigglesworth conducted a series of experiments on the bloodsucking bug *Rhodnius prolixus*. Upon hatching, *Rhod-*

INVESTIGATING LIFE

41.3 A Diffusible Substance Triggers Molting

The bloodsucking bug *Rhodnius prolixus* develops from hatchling to adult in a series of five molts (instars) that are triggered by ingesting blood. Sir Vincent Wigglesworth's experiments demonstrated that a blood meal stimulates production of some molt-inducing substance in the insect's head.

HYPOTHESIS The substance that controls molting in *R. prolixus* is produced in the head segment and diffuses slowly through the body.

OBSERVATION

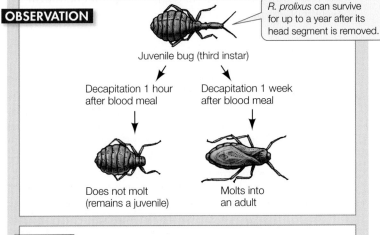

R. prolixus can survive for up to a year after its head segment is removed.

Juvenile bug (third instar)

Decapitation 1 hour after blood meal

Decapitation 1 week after blood meal

Does not molt (remains a juvenile)

Molts into an adult

METHOD

1. Decapitate third-instar juveniles at different times after blood meal.

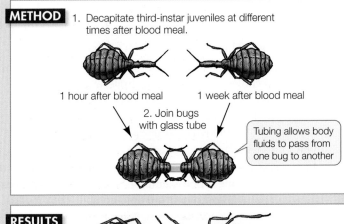

1 hour after blood meal   1 week after blood meal

2. Join bugs with glass tube

Tubing allows body fluids to pass from one bug to another

RESULTS

Both bugs molt into adults

CONCLUSION A blood meal stimulates production of some substance within the insect's head that then diffuses slowly through the body, triggering a molt.

Go to **yourBioPortal.com** for original citations, discussions, and relevant links for all INVESTIGATING LIFE figures.

terone, also plays important roles in development and growth, illustrating the evolutionary conservatism of both the chemical signal and its general domain of function.

The control of molting by PTTH and ecdysone is a general arthropod hormonal control mechanism, and it exemplifies how the endocrine system works with the nervous system to integrate diverse information and induce long-term effects. The nervous system of an arthropod receives various types of information about the animal's environment (such as day length, temperature, social cues, and nutrition) that help determine the optimal timing for the stages of growth and development. When conditions are right, the brain (part of the nervous system) signals the prothoracic gland (part of the endocrine system), which produces the hormone (ecdysone) that orchestrates the physiological processes involved in development and molting. Later in this chapter we will see similar links between the nervous system and endocrine glands in vertebrates.

HORMONAL CONTROL OF MATURATION IN ARTHROPODS

Wigglesworth's experiments with *Rhodnius* yielded another curious result: regardless of the instar used, decapitated bugs that molted always molted directly into an adult. Additional experiments by Wigglesworth demonstrated that another hormone determines whether a bug molts into another juvenile instar or into an adult.

Because the head of *Rhodnius* is long, it is possible to remove just the front part of it, which contains the brain, while leaving the rear part intact. When fourth-instar bugs that had had a blood meal a week earlier were partly decapitated in this way, they molted into fifth-star juveniles, not into adults.

This experiment was followed by more experiments using glass tubes to connect individual bugs. When an unfed, completely decapitated fifth-instar bug was connected to a fed, partly decapitated fourth-instar bug (with only the front part of its head removed), both bugs molted into juvenile forms. A substance from the rear part of the head of the fourth-instar bug prevented both bugs from molting into adults.

The substance responsible for preventing maturation is **juvenile hormone**, which is released continuously from the *corpora allata* (structures that are attached to the corpora cardiaca, which release PTTH). As long as juvenile hormone is present, *Rhodnius* molts into another juvenile instar. Normally *Rhodnius* stops producing juvenile hormone during the fifth instar, then molts into an adult.

The control of development by juvenile hormone is more complex in insects, such as butterflies, that undergo complete metamorphosis. These animals undergo dramatic developmental changes in their life cycles. The fertilized egg hatches into a *larva*, which feeds and molts several times, becoming bigger each time. After a fixed number of molts, it enters an inactive stage called *pupation*. The pupa undergoes major body reorganization and finally emerges as an adult.

An excellent example of complete metamorphosis is provided by the silkworm moth *Hyalophora cecropia*

Endocrine cells in the brain produce PTTH, which is transported to the corpus cardiacum, where it is released.

Brain

Corpus cardiacum

Corpus allatum

Prothoracic gland

PTTH stimulates the prothoracic gland to secrete **ecdysone** (red).

The corpus allatum produces **juvenile hormone** (blue) in declining amounts.

Each release of ecdysone stimulates a molt.

First-instar larva

Molt

Second-instar larva

Molt

Third-instar larva

Molt

Fourth-instar larva

Molt

Fifth-instar larva

Pupation

Cocoon

Pupa

Metamorphosis

When juvenile hormone reaches a low level, the larva spins a cocoon and molts into a pupa.

Adult

The pupa does not produce juvenile hormone, so it metamorphoses into an adult.

41.4 Hormonal Control of Metamorphosis Three hormones control molting and metamorphosis in the silkworm moth *Hyalophora cecropia*.

─── **yourBioPortal.com** ───

GO TO Animated Tutorial 41.1 • Complete Metamorphosis

identified chemically. That is not surprising when you consider the tiny amounts of certain hormones that exist in an organism. In one of the earliest studies of ecdysone, biochemists produced only 250 milligrams of pure ecdysone (about one-fourth the weight of an apple seed) from 4 tons of silkworms!

Hormones can be divided into three chemical groups

Now that we have seen examples of the roles hormones can play in long-term physiological and developmental processes, we can step back and ask some general questions about them. What kinds of hormones exist? What is their chemical nature, and what is their mode of action? There is enormous diversity in the chemical structure of hormones, but most of them can be classified into three groups:

- The majority of hormones are *peptides* or *proteins*. These hormones (insulin is an example) are water-soluble and thus are easily transported in the blood without carrier molecules. Peptide and protein hormones are packaged in vesicles within the cells that make them; they are released by exocytosis. Their receptors are on cell surfaces.

- *Steroid hormones* (such as testosterone) are synthesized from the steroid cholesterol, are lipid-soluble, and easily pass through cell membranes. Steroid hormones diffuse out of the cells that make them and are usually bound to carrier molecules in the blood. Their receptors are mostly intracellular

- *Amine hormones* are mostly synthesized from the amino acid tyrosine (thyroxine is one example). Some amine hormones are water-soluble and others are lipid-soluble; their modes of release differ accordingly.

Hormone receptors can be membrane-bound or intracellular

Water-soluble hormones cannot pass readily through plasma membranes, so their receptors are located on the surfaces of target cells. These receptors are large transmembrane glycoprotein complexes with three domains: a *binding domain* that projects outside the plasma membrane, a *transmembrane domain* that anchors the receptor in the membrane, and a *cytoplasmic domain* that extends into the cytoplasm of the cell. When a hormone binds to the binding domain, the cytoplasmic domain initiates the target cell's response through a second messenger-activated cascade, eventually activating protein kinases or protein phosphatases (see Figures 7.7 and 7.8). In most cases these protein kinases and phosphatases activate

(**Figure 41.4**). As long as juvenile hormone is present in high concentrations, larvae molt into larger larvae. When the level of juvenile hormone falls, larvae spin cocoons and molt into pupae. Because no juvenile hormone is produced in pupae, they molt into adults.

The existence and function of insect hormones were experimentally demonstrated many years before the hormones were

41.5 The Fight-or-Flight Response When a person is suddenly faced with a threatening situation, the brain sends a signal to the adrenal glands, which almost instantaneously release the hormone epinephrine. Epinephrine circulates around the body and induces the various components of the fight-or-flight response in different tissues.

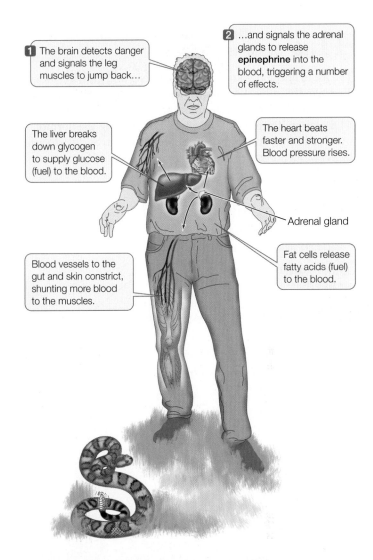

1 The brain detects danger and signals the leg muscles to jump back…

2 …and signals the adrenal glands to release **epinephrine** into the blood, triggering a number of effects.

The liver breaks down glycogen to supply glucose (fuel) to the blood.

The heart beats faster and stronger. Blood pressure rises.

Adrenal gland

Blood vessels to the gut and skin constrict, shunting more blood to the muscles.

Fat cells release fatty acids (fuel) to the blood.

or inactivate enzymes in the cytoplasm, which leads to the cell's response, but the signaling cascade initiated by the receptor can also generate signals that enter the nucleus and alter gene expression (see Figure 7.12).

Lipid-soluble hormones can diffuse through plasma membranes, and therefore their receptors are usually inside cells, in either the cytoplasm or the nucleus (although some membrane-bound receptors for lipid-soluble hormones have recently been described). In most cases, the complex formed by the lipid-soluble hormone and its receptor acts by altering gene expression in the cell's nucleus (see Figure 7.9).

Hormone action depends on the nature of the target cell and its receptors

Wherever a hormone encounters a cell with a receptor to which it can bind, it binds to that receptor and triggers a response. The nature of the response depends on the responding cell and its receptors. Thus the same hormone can cause different responses in different types of cells.

Consider the amine hormone *epinephrine*. Suppose you are walking in the forest and almost trip over a rattlesnake. You jump back. Your heart starts to thump, and an entire set of protective actions are set in motion. The jump and the initial heart thumping are driven by your nervous system, which reacts very quickly. Simultaneously with these muscular responses, however, your nervous system stimulates endocrine cells in the adrenal glands just above your kidneys to secrete epinephrine. Within seconds, epinephrine is diffusing into your blood and circulating around your body to activate the many components of the **fight-or-flight response (Figure 41.5)**.

Epinephrine binds to receptors in your heart, causing a faster and stronger heartbeat. Your heart is now pumping more blood. Epinephrine also binds to receptors in certain blood vessels. By causing constriction of blood vessels supplying your skin, kidneys, and digestive tract (digesting lunch can wait!), the hormone diverts more blood to the muscles needed for your escape from danger.

Epinephrine also binds to cells in the liver and to receptors on fat cells. In the liver, epinephrine stimulates the breakdown of glycogen into glucose for a quick energy supply (see Figure 7.20). In fatty tissue, it stimulates the breakdown of fats to yield fatty acids—another source of energy. These are just some of the many actions that are triggered by one hormone; they all contribute to increasing your chances of surviving a dangerous situation. In each case the cellular response depends on the cell's receptors and associated intracellular signaling cascade.

41.1 RECAP

Hormones are chemical signals released by endocrine cells into the extracellular fluid, where they diffuse into the blood and travel to distant target cells. The receptors for water-soluble hormones are on the surfaces of target cells; receptors for most lipid-soluble hormones are inside the target cells.

- What is the role of juvenile hormone in metamorphosis? **See pp. 855–856 and Figures 41.3 and 41.4**

- Describe the different methods by which water-soluble and lipid-soluble hormones reach their receptors. **See p. 856**

- Do you understand why a single hormone can have diverse effects in the body?

Since the nervous system and the endocrine system are the two major information systems of the body, it is not surprising that their activities are coordinated. Let's look next at how this coordination works.

41.2 How Do the Nervous and Endocrine Systems Interact?

The list of hormones known to exist is long and growing longer. To make the subject manageable, we will focus primarily on the endocrine system of humans and other mammals (**Figure 41.6**). We will begin our survey by considering the hormones involved in the integration of nervous system and endocrine system functions.

The pituitary connects the nervous and endocrine systems

The **pituitary gland** sits in a depression at the bottom of the skull, just over the back of the roof of the mouth (**Figure 41.7A**). It is attached by a stalk to a part of the brain called the **hypothalamus,** which is involved in many physiological regulatory systems. Through its close connection with the hypothalamus, the pituitary serves as the interface between the nervous system and the endocrine system and is involved in the hormonal control of many physiological processes.

Pineal gland
Melatonin: regulates biological rhythms

Thyroid gland (see Figures 41.10 and 41.11)
Thyroxine (T_3 and T_4): increases cell metabolism; essential for growth and neural development
Calcitonin: stimulates incorporation of calcium into bone

Parathyroid glands (on posterior surface of thyroid; see Figure 41.10)
Parathyroid hormone (PTH): stimulates release of calcium from bone and absorption of calcium by gut and kidney

Adrenal gland (see Figure 41.12)
Cortex
Cortisol: mediates metabolic responses to stress
Aldosterone: involved in salt and water balance
Sex steroids

Medulla
Epinephrine (adrenaline) and *norepinephrine* (noradrenaline): stimulate immediate fight or flight reactions

Gonads (see Chapter 43)
Testes (male)
Testosterone: development and maintenance of male sexual characteristics

Ovaries (female)
Estrogens: development and maintenance of female sexual characteristics
Progesterone: supports pregnancy

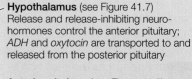

Hypothalamus (see Figure 41.7)
Release and release-inhibiting neurohormones control the anterior pituitary; *ADH* and *oxytocin* are transported to and released from the posterior pituitary

Anterior pituitary (see Figure 41.8)
Thyrotropin (TSH): activates the thyroid gland
Follicle-stimulating hormone (FSH): in females, stimulates maturation of ovarian follicles; in males, stimulates spermatogenesis
Luteinizing hormone (LH): in females, triggers ovulation and ovarian production of estrogens and progesterone; in males, stimulates production of testosterone
Corticotropin (ACTH): stimulates adrenal cortex to secrete cortisol
Growth hormone (GH): stimulates protein synthesis and growth
Prolactin: stimulates milk production
Melanocyte-stimulating hormone (MSH)
Endorphins and *enkephalins*: pain control

Posterior pituitary (see Figure 41.7)
Receives and releases two hypothalamic hormones:
Oxytocin: stimulates contraction of uterus, flow of milk, interindividual bonding
Antidiuretic hormone (ADH; also known as vasopressin): promotes water conservation by kidneys

Thymus (diminishes in adults)
Thymosin: activates immune system T cells

Pancreas (islets of Langerhans)
Insulin: stimulates cells to take up and use glucose
Glucagon: stimulates liver to release glucose
Somatostatin: slows release of insulin and glucagon and digestive tract functions

Other organs include cells that produce and secrete hormones

Organ	Hormone
Adipose tissue	Leptin
Heart	Atrial natriuretic peptide
Kidney	Erythropoietin
Stomach	Gastrin
Intestine	Secretin, cholecystokinin
Skin	Vitamin D (cholecalciferol)
Liver	Somatomedins, insulin-like growth factors

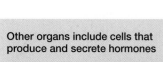

41.6 The Endocrine System of Humans Cells that produce and secrete hormones may be organized into discrete endocrine glands, or they may be embedded in the tissues of other organs, such as the digestive tract or kidneys. The hypothalamus is part of the brain, but it includes cells that secrete neurohormones into the extracellular fluid.

yourBioPortal.com
GO TO Web Activity 41.1 • The Human Endocrine Glands

The pituitary has two parts with different developmental origins. The *anterior pituitary* originates as an outpocketing of the roof of the embryonic mouth cavity, whereas the *posterior pituitary* originates as an outpocketing of the floor of the developing brain. Thus the anterior pituitary originates from gut epithelial tissue and the posterior pituitary from neural tissue. Both parts interact with the nervous system but in different ways. The anterior pituitary contains endocrine cells controlled by neurohormones secreted by the hypothalamus. The posterior pituitary contains axons from hypothalamic neurons.

THE POSTERIOR PITUITARY Long axons extend into the **posterior pituitary** from neurons in the hypothalamus. The ends (*terminals*) of those axons release two hormones produced by the neurons, antidiuretic hormone and oxytocin (**Figure 41.7B**). These neurohormones are packaged in vesicles that are transported down the axons. The vesicles are stored until an action potential stimulates their release.

The main action of **antidiuretic hormone** (**ADH**) in mammals and birds is to increase the amount of water conserved by the kidneys. When ADH secretion is high, the kidneys produce only a small volume of highly concentrated urine. When ADH secretion is low, the kidneys produce a large volume of dilute urine. The posterior pituitary increases its release of ADH when blood pressure falls or the blood becomes too salty. ADH is also known as *vasopressin* because at high concentrations it causes the constriction of peripheral blood vessels as a means of elevating blood pressure.

When a woman is about to give birth, her posterior pituitary releases **oxytocin**, which stimulates the uterine contractions that deliver the baby (see Figure 44.16). Oxytocin also brings about the flow of milk from the mother's breasts. The baby's suckling stimulates neurons in the mother's brain that cause the secretion of oxytocin. Even the sight and sound of a baby can cause a nursing mother to secrete oxytocin and release breast milk. This is a good example of how the nervous system integrates information and contributes to the control of hormonally mediated processes.

Hormones, in turn, can influence the nervous system. Oxytocin, for example, promotes bonding (see the story that opens Chapter 7). If oxytocin release is experimentally blocked, mammalian mothers, from rats to sheep, will reject their newborn offspring, but if a virgin rat is given a dose of oxytocin, she will adopt strange pups as if they were her own. Oxytocin promotes pair bonding and trust in a variety of animals. In humans, its secretion rises with intimate sexual contact. Not surprisingly, oxytocin has been nicknamed the "cuddle hormone."

THE ANTERIOR PITUITARY The **anterior pituitary** releases four peptide and protein hormones that act as **tropic hormones**, meaning they control the activities of other endocrine glands

yourBioPortal.com

GO TO Animated Tutorial 41.2 • The Hypothalamic–
Pituitary–Endocrine Axis

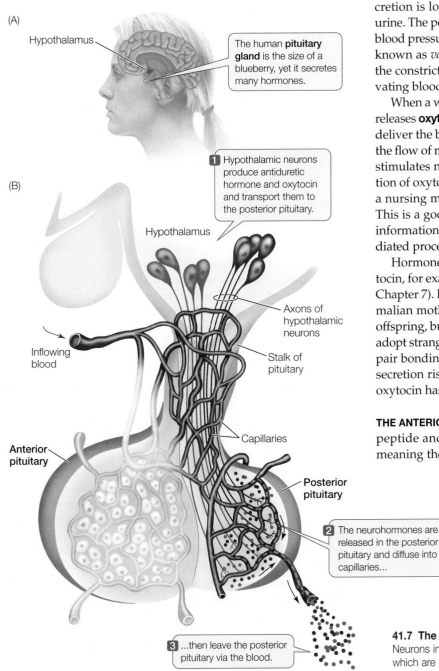

(A)

Hypothalamus

The human **pituitary gland** is the size of a blueberry, yet it secretes many hormones.

(B)

1 Hypothalamic neurons produce antidiuretic hormone and oxytocin and transport them to the posterior pituitary.

Hypothalamus

Axons of hypothalamic neurons

Inflowing blood

Stalk of pituitary

Anterior pituitary

Capillaries

Posterior pituitary

2 The neurohormones are released in the posterior pituitary and diffuse into capillaries...

3 ...then leave the posterior pituitary via the blood.

41.7 The Posterior Pituitary Releases Neurohormones
Neurons in the hypothalamus produce two peptide neurohormones, which are stored and released by the posterior pituitary.

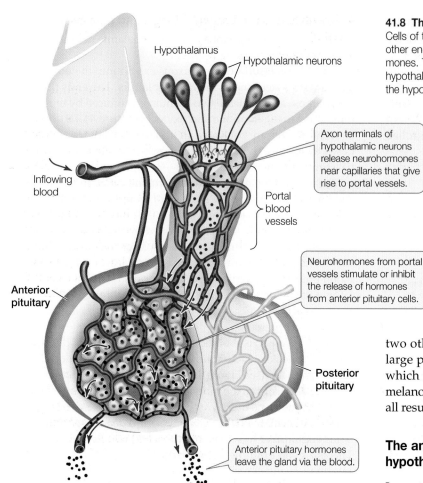

41.8 The Anterior Pituitary Is Controlled by the Hypothalamus
Cells of the anterior pituitary produce four tropic hormones that control other endocrine glands, as well as several other peptide and protein hormones. These cells are controlled by neurohormones produced in the hypothalamus and delivered through portal blood vessels that run between the hypothalamus and the anterior pituitary through the pituitary stalk.

Hypothalamus
Hypothalamic neurons

Axon terminals of hypothalamic neurons release neurohormones near capillaries that give rise to portal vessels.

Inflowing blood

Portal blood vessels

Neurohormones from portal vessels stimulate or inhibit the release of hormones from anterior pituitary cells.

Anterior pituitary

Posterior pituitary

Anterior pituitary hormones leave the gland via the blood.

(Figure 41.8). These four hormones are *thyrotropin, luteinizing hormone, follicle-stimulating hormone,* and *corticotropin.* Each tropic hormone is produced by a different type of pituitary cell. We will say more about the tropic hormones when we describe their target glands (thyroid, testes, ovaries, and adrenal cortex) later in this chapter and in Chapter 43.

Other peptide and protein hormones produced by the anterior pituitary are *growth hormone, prolactin* (see Figure 41.2) *melanocyte-stimulating hormone, enkephalins,* and *endorphins.*

Growth hormone (GH) acts on a wide variety of tissues to promote growth. One of its important effects is to stimulate cells to take up amino acids. Growth hormone also promotes growth by stimulating the liver to produce chemical signals called *somatomedins* or *insulin-like growth factors* (*IGFs*), which stimulate the growth of bone and cartilage. Thus growth hormone can be considered a tropic hormone because it stimulates endocrine cells in the liver.

Overproduction of growth hormone in children causes *gigantism*, in which affected individuals may grow to nearly 8 feet tall. Underproduction causes *pituitary dwarfism*, in which individuals fail to reach normal adult height. Beginning in the late 1950s, children with serious growth hormone deficiencies were treated with growth hormone extracted from pituitaries of human cadavers. The treatment was successful in stimulating substantial growth, but a year's supply of the hormone for one individual required up to 50 cadaver pituitaries. In the 1980s,

scientists using recombinant DNA technology isolated the gene for human growth hormone and introduced it into bacteria that could be grown in large quantities, making it possible to purify enough of the hormone to make it widely available.

Endorphins and **enkephalins** are the body's natural painkillers. In the brain, these molecules act as neurotransmitters in pathways that control pain. Their production in the anterior pituitary is normally quite small and probably has no significant effect. They are a by-product of the production of two other anterior pituitary hormones. One gene encodes a large parent molecule called *pro-opiomelanocortin,* or POMC, which is cleaved to produce several peptides. Corticotropin, melanocyte-stimulating hormone, endorphins, and enkephalins all result from the cleavage of POMC.

The anterior pituitary is controlled by hypothalamic neurohormones

In contrast to the posterior pituitary, the anterior pituitary makes and secretes its own hormones, but its secretion of hormones is under the control of neurohormones from the hypothalamus. The hypothalamus senses and receives information about conditions in the body and in the external environment, and it communicates that information to the anterior pituitary by releasing neurohormones. If the connection between the hypothalamus and the pituitary is experimentally cut, the release of pituitary hormones no longer changes when conditions in the internal or external environment change. In experiments in which pituitary cells were maintained in culture, adding extracts of hypothalamic tissue stimulated some of those cells to release their hormones into the culture medium. Therefore, scientists hypothesized that secretions from hypothalamic cells control the activities of anterior pituitary cells.

Hypothalamic neurons do not extend into the anterior pituitary as they do into the posterior pituitary. Remember that the posterior pituitary develops from neural tissue whereas the anterior pituitary develops from gut tissue. A special set of **portal blood vessels** bridges the gap between the hypothalamus and the anterior pituitary (see Figure 41.8). It was thus proposed that secretions from neurons in the hypothalamus enter the blood and are conducted down the portal vessels to the anterior pituitary, where they stimulate the release of anterior pituitary hormones.

In the 1960s, two large teams of scientists, led by Roger Guillemin and Andrew Schally, initiated the search for these hypothalamic secretions. Because the amounts of such neurohormones in any individual mammal would be tiny, massive numbers of hypothalami from pigs and sheep were collected from slaughterhouses and shipped to laboratories in refrigerated trucks. One extraction effort began with the hypothalami from 270,000 sheep and yielded only 1 milligram of purified **thyrotropin-releasing hormone (TRH)**. TRH was the first hypothalamic *releasing hormone* (that is, release-stimulating hormone) to be isolated and characterized. It turned out to be a simple tripeptide consisting of glutamine, histidine, and proline. It causes certain anterior pituitary cells to release the tropic hormone thyrotropin, which in turn stimulates the activity of the thyroid gland.

Soon after discovering TRH, Guillemin's and Schally's teams identified **gonadotropin-releasing hormone (GnRH)**, which stimulates certain anterior pituitary cells to release the tropic hormones that control the activity of the gonads (the ovaries and the testes). For these discoveries, Guillemin and Schally received the 1977 Nobel Prize in Medicine. Many other hypothalamic neurohormones, including both releasing hormones and release-inhibiting hormones, are now known. The major hypothalamic neurohormones that control anterior pituitary function are:

- Thyrotropin-releasing hormone
- Gonadotropin-releasing hormone
- Prolactin-releasing and release-inhibiting hormones
- Growth hormone–releasing hormone

- Growth hormone release-inhibiting hormone (somatostatin)
- Corticotropin-releasing hormone

Negative feedback loops regulate hormone secretion

As well as being controlled by hypothalamic releasing and release-inhibiting hormones, the endocrine cells of the anterior pituitary are also under direct and indirect negative feedback control by the hormones of the target glands they stimulate (**Figure 41.9**). For example, the hormone cortisol, produced by the adrenal gland in response to corticotropin secreted by the anterior pituitary, reaches the pituitary in the circulating blood and inhibits further release of that tropic hormone. Cortisol also acts as a negative feedback signal to the hypothalamus, inhibiting the release of corticotropin-releasing hormone. In some cases, a tropic hormone also exerts negative feedback control on the hypothalamic cells that produce the corresponding releasing hormone.

41.2 RECAP

The pituitary is the interface between the nervous system and the endocrine system. The posterior pituitary releases two neurohormones. The anterior pituitary, under the control of other neurohormones from the hypothalamus, releases hormones that control other endocrine glands.

- Describe the anatomical and functional relationships between the brain and the two parts of the pituitary.
- What are the tropic hormones of the anterior pituitary, and how do they influence the endocrine system? **See pp. 859–860 and Figure 41.8**

Now that we know some of the mechanisms by which endocrine systems are controlled, we will take a more detailed look at the functions of the major endocrine glands of the mammalian body.

41.3 What Are the Major Mammalian Endocrine Glands and Hormones?

Hormones help regulate functions in all mammalian physiological systems. In this section we will examine a few major examples of hormonal action in physiological processes. We will see many more in the chapters that follow.

The thyroid gland secretes thyroxine

The **thyroid gland** wraps around the front of the windpipe (*trachea*) and expands into a lobe on either side (see Figure 41.6). There are two cell types in the thyroid gland, each of which produces a specific hormone. Thyroxine is produced, stored, and released by epithelial cells that make up round, colloid-contain-

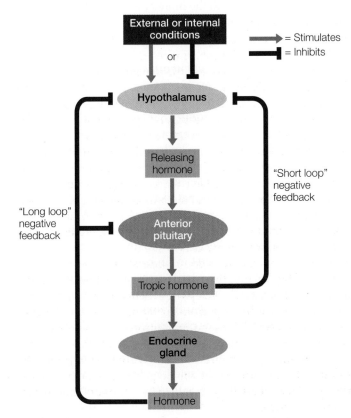

41.9 Multiple Feedback Loops Control Hormone Secretion
Multiple negative feedback loops regulate the chain of command from hypothalamus to anterior pituitary to endocrine glands.

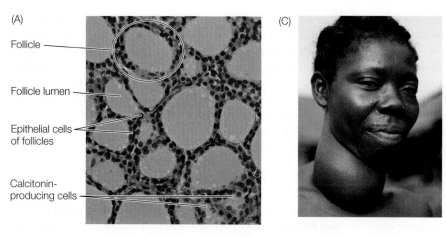

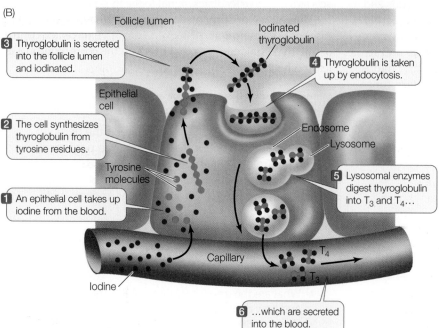

41.10 The Thyroid Gland Consists of Many Follicles (A) Cross section through a thyroid gland, showing numerous follicles bounded by epithelial cells. Calcitonin-secreting cells are located in the spaces between the follicles. (B) The epithelial cells of the follicle synthesize thyroglobulin and secrete it into the lumen of the follicle, where it is iodinated and stored until it is processed by the epithelial cells to generate T_3 and T_4. (C) Iodine deficiency can result in hypothyroid goiter. In this condition, a lack of functional thyroxine results in oversynthesis of thyroglobulin and subsequent enlarged follicles.

and if they are iodinated at only three sites, they are triiodothyronine, or T_3:

The thyroid usually releases about four times as much T_4 as T_3; however, T_3 is the more active hormone in the cells of the body. Circulating T_4 can be converted to T_3 by an enzyme within target cells. Therefore, each target cell can set its own sensitivity to thyroid hormones by controlling the conversion of T_4 to the more active T_3. When you read about thyroxine, keep in mind that the actions discussed are primarily those of T_3.

THYROXINE REGULATES CELL METABOLISM
Thyroxine in mammals plays many roles in regulating cell metabolism. Thyroxine is lipid-soluble, so it enters cells readily and binds to receptors in the nucleus. The receptor is found in most cells of the body, and when combined with thyroxine, it acts as a transcription factor that stimulates the transcription of numerous genes whose products are enzymes involved in energy metabolic pathways, transport proteins, and structural proteins. As a result, thyroxine elevates the metabolic rates of most cells and tissues. Exposure to cold for several days leads to an increased release of thyroxine, an increased conversion of T_4 to T_3, and therefore an increased basal metabolic rate (see Section 40.4). Thyroxine is especially crucial during development and growth, as it promotes amino acid uptake and protein synthesis. Insufficient thyroxine in a human fetus or growing child greatly retards physical and mental development, resulting in a condition known as *cretinism*.

The tropic hormone **thyrotropin**, or **thyroid-stimulating hormone (TSH)**, produced by the anterior pituitary, activates the thyroxine-producing follicle cells in the thyroid. Thyrotropin-releasing hormone (TRH), produced in the hypothalamus and transported to the anterior pituitary through the portal blood vessels, activates the TSH-producing pituitary cells. The hypothalamus uses environmental information, such as temperature or

ing structures called *follicles* (**Figure 41.10A and B**). Calcitonin is produced in cells in spaces between the follicles.

Thyroxine begins as the glycoprotein **thyroglobulin**, which the follicle cells synthesize as long chains consisting largely of tyrosine residues. As these thyroglobulin molecules are exported into the follicle for storage, the tyrosine residues are iodinated with one or two atoms of iodine. When the thyroid gland is stimulated to release thyroxine, the follicle cells take up thyroglobulin from the follicle by endocytosis. These bits of thyroglobulin are then cleaved to form smaller molecules consisting of only two tyrosine residues, and these molecules leave the follicle cells. If these molecules are iodinated at the maximum of four sites on the tyrosine residues, the hormone is tetraiodothyronine, or T_4;

day length, to determine whether to increase or decrease its secretion of TRH. This sequence of steps is regulated by a negative feedback loop like the one described earlier for cortisol (see Figure 41.9). Circulating thyroxine inhibits the response of pituitary cells to TRH, so less TSH is released when thyroxine levels are high, and more TSH is released when thyroxine levels are low. Circulating thyroxine also exerts negative feedback on the production and release of TRH by the hypothalamus.

A *goiter* is an enlarged thyroid gland (**Figure 41.10C**) that can be associated with either **hyperthyroidism** (excess production of thyroxine) or **hypothyroidism** (thyroxine deficiency). The negative feedback loop whereby thyroxine controls TSH release helps explain how two very different conditions can result in the same symptom.

The most common cause of *hyperthyroid* goiter is an autoimmune disease involving an antibody to the TSH receptor. This antibody binds to and activates the TSH receptors on the follicle cells, causing uncontrolled production and release of thyroxine. Blood levels of TSH are quite low because of the negative feedback from high levels of thyroxine, but the thyroid remains maximally stimulated and grows bigger. People with hyperthyroidism have high metabolic rates, are jumpy and nervous, usually feel hot, and may develop a buildup of fat behind the eyeballs, which causes their eyes to bulge.

Hypothyroid goiter results when there is not enough circulating thyroxine to turn off TSH production. The most common cause of this condition is a deficiency of dietary iodine, without which the follicle cells cannot make thyroxine. Without sufficient thyroxine, TSH levels remain high, and the thyroid continues to produce large amounts of thyroglobulin. Because sufficient iodine is not available, however, the thyroglobulin is poorly iodinated. When it is broken down by the follicle cells, it does not yield functional thyroxine (T_3 or T_4). TSH levels remain high and stimulate more and more synthesis of thyroglobulin, and the thyroid gets bigger. The symptoms of hypothyroidism are low metabolism, intolerance of cold, and general physical and mental sluggishness.

Goiter affects about 5 percent of the world's population. The addition of iodine to table salt has greatly reduced the incidence of hypothyroid goiter in industrialized nations, but the condition is still common in the other parts of the world.

Three hormones regulate blood calcium concentrations

The regulation of calcium concentration in the blood is a crucial and difficult task. It is crucial because shifts in blood calcium concentration above or below a narrow range can cause serious problems. When blood calcium falls below this range, the nervous system becomes overly excited, resulting in muscle spasms and even seizures. When blood calcium rises above this range, the nervous system becomes depressed and muscles—including the heart—weaken. Regulation of blood calcium is difficult

because only about 0.1 percent of the calcium in the body is located in the extracellular fluid. About 1 percent is in cells, and almost 99 percent is in the bones. Therefore, the body must maintain a tiny pool of calcium in the blood at a precise concentration, and that tiny pool can be influenced greatly by relatively small shifts in the much larger pools of calcium in the cells and bones.

The body has multiple mechanisms for changing blood calcium levels, including:

- Deposition or absorption of bone
- Excretion or retention of calcium by the kidneys
- Absorption of calcium from the digestive tract

These mechanisms are controlled by the hormones calcitonin, parathyroid hormone, and calcitriol (synthesized from vitamin D).

CALCITONIN REDUCES BLOOD CALCIUM Calcitonin released by the thyroid lowers the concentration of calcium in the blood, mainly by regulating bone turnover (**Figure 41.11**). Bone is continuously remodeled through a dynamic process that involves

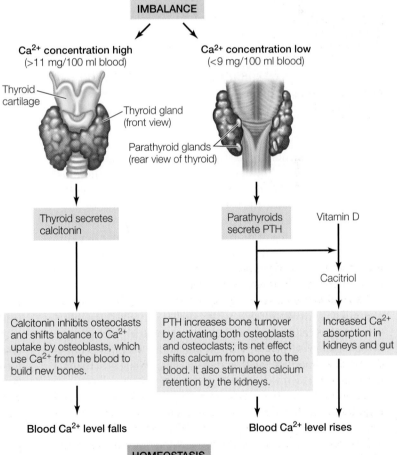

IMBALANCE

Ca^{2+} concentration high (>11 mg/100 ml blood)

Ca^{2+} concentration low (<9 mg/100 ml blood)

Thyroid cartilage

Thyroid gland (front view)

Parathyroid glands (rear view of thyroid)

Thyroid secretes calcitonin

Parathyroids secrete PTH

Vitamin D

Cacitriol

Calcitonin inhibits osteoclasts and shifts balance to Ca^{2+} uptake by osteoblasts, which use Ca^{2+} from the blood to build new bones.

PTH increases bone turnover by activating both osteoblasts and osteoclasts; its net effect shifts calcium from bone to the blood. It also stimulates calcium retention by the kidneys.

Increased Ca^{2+} absorption in kidneys and gut

Blood Ca^{2+} level falls

Blood Ca^{2+} level rises

HOMEOSTASIS

Ca^{2+} concentration between 9 and 11 mg/100 ml blood

41.11 Hormonal Regulation of Calcium Calcitonin, parathyroid hormone (PTH), and vitamin D regulate calcium levels in the blood.

yourBioPortal.com

GO TO Animated Tutorial 41.3 • Hormonal Regulation of Calcium

both resorption of old bone and synthesis of new bone, as we discuss in Section 48.3. Cells called *osteoclasts* break down bone and release calcium into the blood, and cells called *osteoblasts* take up calcium from the blood and deposit it in new bone. Calcitonin decreases the activity of osteoclasts and thereby favors removal of calcium from the blood and its deposition in bone by osteoblasts. The turnover of bone in adult humans is not very high, so calcitonin does not play a major role in calcium homeostasis in adults. It is probably more important in young individuals whose bones are actively growing.

PARATHYROID HORMONE INCREASES BLOOD CALCIUM The **parathyroid glands** are four tiny structures embedded in the posterior surface of the thyroid gland (see Figure 41.11). Their single hormone product, **parathyroid hormone** (also called **PTH** or *parathormone*), is the most important hormone in the regulation of blood calcium levels. Circulating calcium activates receptors in the plasma membrane of the parathyroid cells. When these receptors are active, they inhibit the synthesis and release of PTH. A fall in blood calcium removes this inhibition and triggers the synthesis and release of PTH.

PTH stimulates bone turnover by actions on both osteoclasts and osteoblasts. The end result of these actions of PTH is a net increase of calcium in the blood.

Another mechanism by which PTH raises blood concentrations of calcium is by stimulating the kidneys to reabsorb it rather than excrete it in the urine. In addition, PTH activates the synthesis of calcitriol from vitamin D, which in turn causes the digestive tract to increase its absorption of calcium from food.

VITAMIN D IS A HORMONE A *vitamin* is a substance that the body requires in small quantities but cannot synthesize, and must therefore obtain from the diet. By this definition, **vitamin D** is not a vitamin, because the body can and does synthesize it.

It had long been known that fragile bones were common among people living at high latitudes, where winter days are short and the winter diet often lacked meat, fish, dairy products, and fresh vegetables. Since the condition could be reversed by taking cod-liver oil (which, as it turned out, contains large amounts of vitamin D), scientists assumed that a dietary vitamin was involved. We now know that vitamin D is synthesized in skin cells, where cholesterol is converted into vitamin D (also called *calciferol*) by ultraviolet light. Vitamin D circulates in the blood and acts on distant cells; thus it is actually a hormone.

PTH lowers blood phosphate levels

Bone minerals are made up primarily of calcium and phosphate. Thus, when PTH stimulates the release of calcium from bone, it also causes the release of phosphate. Excessive blood concentrations of both elements can be dangerous. The normal concentrations of calcium and phosphate in the blood are just below the levels at which they precipitate out of solution as calcium phosphate salts. Thus even a small rise may cause such precipitation, leading to maladies such as kidney stones and calcium deposits in the arteries (hardening of the arteries). To reduce this risk, PTH acts on the kidneys to increase the elimination of phosphate via the urine.

Insulin and glucagon regulate blood glucose concentrations

Before the 1920s, *diabetes mellitus* was a fatal disease, characterized by weakness, lethargy, and a dramatic loss of body mass. The disease was known to be connected somehow with the **pancreas**—a large gland located just below the stomach (see Figure 41.6)—and with abnormal glucose metabolism, but the links were not clear.

Today we know that diabetes mellitus is caused by a lack of the protein hormone **insulin** (in type I or juvenile-onset diabetes) or by a lack of insulin responsiveness in target tissues (in type II or adult-onset diabetes). Glucose enters cells by diffusion, not active transport, but cell membranes are not very permeable to glucose. Therefore there are proteins called glucose transporters in cell membranes that facilitate the movement of glucose into cells. The glucose transporters that are most common in muscle and adipose tissue are controlled by insulin. When insulin binds to its receptor on the cell membrane, it causes these glucose transporters to move from cytoplasmic vesicles to the cell membrane, making the cell more permeable to glucose. When insulin is not present, these transporters are returned to the cytoplasmic pool through endocytosis.

In the absence of insulin or insulin responsiveness, glucose entry into cells is impaired, and so much glucose accumulates in the blood that it starts to spill over into the urine; urine production also goes up. A high concentration of glucose in the blood increases urine output by two mechanisms. First, it causes water to move from cells into the blood by osmosis, and this increase in blood volume results in increased urine production. Second, the increased glucose in the tubules of the kidneys pulls more water into the urine by osmosis. Thus diabetic individuals can become dehydrated, but more importantly they suffer from a lack of metabolic fuel. Because glucose uptake by muscle and adipose tissue is impaired in the absence of insulin, muscle cells must depend on fat and protein for fuel and adipose tissue cannot replenish its stores of triglycerides. As a result, the body of the untreated diabetic can waste away.

For centuries, the prospects for diabetics were bleak. A change came almost overnight in 1921, when the physician Frederick Banting and a medical student, Charles Best, at the University of Toronto, discovered that they could reduce the symptoms of diabetes by injecting an extract prepared from pancreatic tissue. The active component of this extract was found to be a small protein hormone—insulin—consisting of just 51 amino acids. In the United States today, insulin replacement therapy, using manufactured insulin, allows 1.5 million people with type I diabetes to lead almost normal lives.

ISLETS OF LANGERHANS Insulin is produced in clusters of endocrine cells in the pancreas. These clusters are called **islets of Langerhans** after the German medical student who discovered them. They contain three types of cells, each of which produces a specific hormone:

- Beta (β) cells produce and secrete insulin.
- Alpha (α) cells produce and secrete glucagon, a hormone that has effects mostly opposite from those of insulin.
- Delta (δ) cells produce a hormone called somatostatin.

The rest of the pancreas is made up of exocrine tissue, which produces enzymes and other secretions that travel through ducts to the gut, where they participate in digestion.

After a meal, the concentration of glucose in the blood rises as glucose is absorbed from the food in the gut. This increase stimulates the β cells of the islets to release insulin, which causes target cells throughout the body to use the circulating glucose as fuel and to convert it into storage products, such as glycogen and fat. When the gut contains no more food, the glucose concentration in the blood falls and the islets stop releasing insulin. As a result, most cells shift to using glycogen and fat, rather than glucose, as fuel. If the concentration of glucose in the blood falls substantially below normal, the islet α cells release **glucagon**, which stimulates the liver to break down stored glycogen and release glucose into the blood. These actions are discussed in greater detail in Section 51.4.

SOMATOSTATIN **Somatostatin** is released from the δ cells of the pancreas in response to rapid increases of glucose and amino acids in the blood. This hormone has paracrine functions within the islets: it inhibits the release of both insulin and glucagon. Its actions outside the pancreas slow the digestive activities of the gut, extending the period during which nutrients are absorbed. Somatostatin is also produced in very small amounts by cells in the hypothalamus. This somatostatin is transported in the portal blood vessels to the anterior pituitary, where it acts as a neurohormone to inhibit the release of growth hormone and thyrotropin.

The adrenal gland is two glands in one

An **adrenal gland** sits above each kidney, just below the middle of your back. Functionally and anatomically, each adrenal gland consists of a gland within a gland (**Figure 41.12**). The core, or **adrenal medulla**, produces two hormones: **epinephrine** (also known as *adrenaline*) and, to a lesser degree, **norepinephrine** (or *noradrenaline*). The medulla develops from nervous tissue and is under the control of the nervous system. Surrounding the medulla is the **adrenal cortex**, which produces steroid hormones. The cortex is under hormonal control, largely by corticotropin produced by the anterior pituitary.

THE ADRENAL MEDULLA The adrenal medulla produces epinephrine and norepinephrine in response to stressful situations, arousing the body to action. As we saw earlier in this chapter, epinephrine increases heart rate and blood pressure and diverts blood flow to active muscles and away from the gut and skin. Norepinephrine has similar functions.

Epinephrine and norepinephrine are both water-soluble, and both bind to the same set of receptors on the surfaces of target cells. These *adrenergic receptors* are of two general types, α-adrenergic and β-adrenergic; each type stimulates different actions

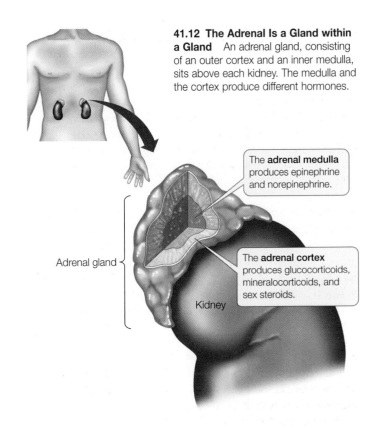

41.12 The Adrenal Is a Gland within a Gland An adrenal gland, consisting of an outer cortex and an inner medulla, sits above each kidney. The medulla and the cortex produce different hormones.

The **adrenal medulla** produces epinephrine and norepinephrine.

The **adrenal cortex** produces glucocorticoids, mineralocorticoids, and sex steroids.

Adrenal gland

Kidney

within cells (**Figure 41.13**). The α-adrenergic receptors are the most common on target cells of the sympathetic nervous system, and they respond more strongly to norepinephrine than to epinephrine. The β-adrenergic receptors respond about equally to both epinephrine and norepinephrine. One class of the β-adrenergic receptors are found on cells not innervated by the sympathetic fibers, so they are positioned to respond to circulating epinephrine. Therefore, drugs called *beta blockers* (so named because they selectively block β-adrenergic receptors) can reduce the fight-or-flight response to epinephrine without disrupting the physiological regulatory functions of norepinephrine. Beta blockers are commonly prescribed to reduce the symptoms of anxiety, such as dry mouth and elevated heart rate (palpitations).

THE ADRENAL CORTEX The cells of the adrenal cortex use cholesterol to produce three classes of steroid hormones, collectively called **corticosteroids**:

- The *glucocorticoids* influence blood glucose concentrations as well as other aspects of fat, protein, and carbohydrate metabolism.
- The *mineralocorticoids* influence the salt and water balance of the extracellular fluid.
- The *sex steroids* play roles in sexual development, sexual behavior, and anabolism.

In adults, the adrenal cortex secretes sex steroids in only negligible amounts. The major producers of sex steroids are the gonads, as we will see in the following section.

Aldosterone, the main mineralocorticoid (**Figure 41.14A**), stimulates the kidneys to conserve sodium and excrete potas-

(A) Epinephrine

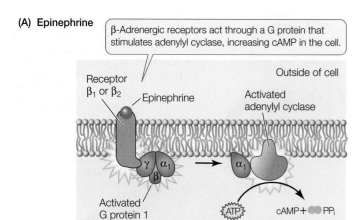

β-Adrenergic receptors act through a G protein that stimulates adenylyl cyclase, increasing cAMP in the cell.

(B) Norepinephrine

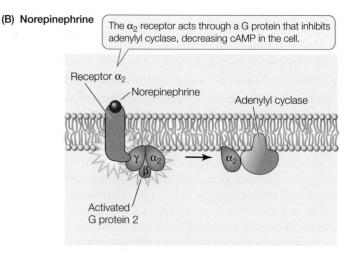

The α$_2$ receptor acts through a G protein that inhibits adenylyl cyclase, decreasing cAMP in the cell.

41.13 Hormones Can Activate a Variety of Signal Transduction Pathways Epinephrine and norepinephrine bind to G protein-linked adrenergic receptors that act through different signal transduction pathways. Epinephrine acts equally on both α- and β-adrenergic receptors; norepinephrine acts mostly on α-adrenergic receptors.

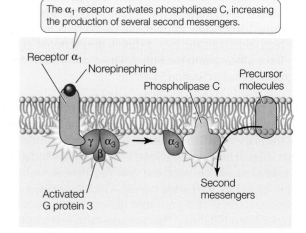

The α$_1$ receptor activates phospholipase C, increasing the production of several second messengers.

sium, as we discuss in Chapter 51. If the adrenal glands are removed from an animal, sodium must be added to its diet, or its sodium will be depleted and it will die.

The main glucocorticoid in humans is **cortisol (Figure 41.14B)**, which is critical for mediating the body's metabolic responses to stress. Within minutes of a stressful stimulus (one provoking fear or anger, for example), your blood cortisol level rises. Cells not critical for your fight-or-flight responses are stimulated by cortisol to decrease their use of blood glucose and shift instead to using fats and proteins for energy. This is no time to feel sick, have allergic reactions, or heal wounds, so cortisol also blocks immune system reactions. That is why cortisol and drugs that mimic its action are useful for reducing inflammation and allergic responses.

Cortisol release is controlled from the anterior pituitary by **corticotropin** (also called *adrenocorticotropic hormone*, or **ACTH**), whose release is controlled in turn by **corticotropin-releasing hormone** from the hypothalamus. Because the cortisol response to stress has this chain of steps, each involving secretion, diffusion, circulation, and cell activation, it is much slower than the epinephrine response to stress. Furthermore, in many cases cortisol acts as a transcription factor to change gene expression, and those changes take time.

Turning off the responses to stress activated by cortisol is as important as turning them on. A study of stress in rats showed that old rats could turn on these stress responses as effectively as young rats, but they had lost the ability to turn them off as rapidly. As a result, they suffered from the well-known consequences of stress seen in humans: digestive system problems, cardiovascular

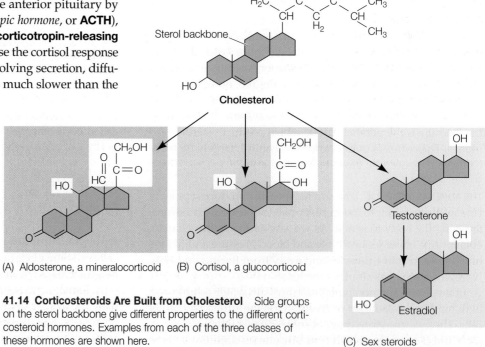

41.14 Corticosteroids Are Built from Cholesterol Side groups on the sterol backbone give different properties to the different corticosteroid hormones. Examples from each of the three classes of these hormones are shown here.

(A) Aldosterone, a mineralocorticoid   (B) Cortisol, a glucocorticoid   (C) Sex steroids

problems, strokes, impaired immune system function, and increased susceptibility to cancers and other diseases. Further research showed that these stress responses are turned off by the negative feedback action of cortisol on the hypothalamus, which causes a decrease in the release of corticotropin-releasing hormone (see Figure 41.9). Repeated activation of this negative feedback mechanism, either through repeated stress or prolonged medical use of cortisol, leads to a gradual loss of cortisol-sensitive cells in the hypothalamus, and therefore to a decreased ability to terminate the stress response.

Sex steroids are produced by the gonads

The **gonads**—the testes of the male and the ovaries of the female—produce hormones as well as sperm and ova. The male steroids are collectively called **androgens**, and the dominant one is *testosterone*. The female steroids are **estrogens** and **progesterone**. The dominant estrogen is *estradiol*, which is synthesized from testosterone. Males and females both synthesize testosterone, but females have an enzyme (aromatase) that converts testosterone to estradiol (**Figure 41.14C**).

The sex steroids have important developmental effects: they determine whether a mammalian embryo develops into a phenotypic female or male. After birth, the sex steroids control the maturation of the reproductive organs and the development and maintenance of secondary sexual characteristics, such as breasts and facial hair. The roles they play in adult sexual behavior and reproduction are described in Chapter 43.

EARLY DEVELOPMENT   Sex steroids begin to exert their effects in the human embryo in the seventh week of development. Until that time, the embryo has the potential to develop into either sex. In mammals and birds, the instructions for sex determination reside in the genes. In mammals, individuals that receive two X chromosomes normally become females, and individuals that receive an X and a Y chromosome normally become males (see Section 12.4).

These genetic instructions are carried out through the production and action of the sex steroids. In humans, the presence of a Y chromosome normally causes the undifferentiated embryonic gonads to begin producing androgens in the seventh week of development. In response to those androgens, the reproductive system develops into that of a male. If androgens are not produced at that time, female reproductive structures develop (**Figure 41.15**). In other words, androgens are required to trigger male development in humans, and the default condition is female.

INITIATION OF PUBERTY   Sex steroids have dramatic effects at **puberty**—the time of sexual maturation in humans. Sex steroids are produced at low levels by the juvenile gonads, but their production increases rapidly at the beginning of puberty—around the age of 12 to 13 years. Why does this sudden increase occur?

In the juvenile, as in the adult, the production of sex steroids by the gonads is controlled by the tropic hormones **luteinizing hormone (LH)** and **follicle-stimulating hormone (FSH)**, which together are called the **gonadotropins**. The production of these

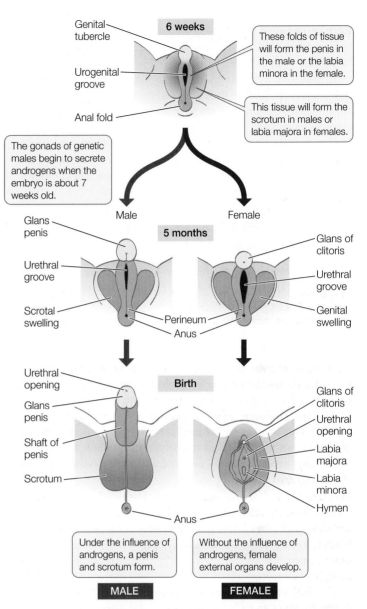

41.15 Sex Steroids Direct the Development of Human Sex Organs The sex organs of early human embryos are undifferentiated. Androgens promote the development of male sex organs. In the absence of androgens, female sex organs form.

tropic hormones by the anterior pituitary is under the control of the hypothalamic neurohormone gonadotropin-releasing hormone (GnRH). Before puberty, the gonads can respond to gonadotropins and the anterior pituitary can respond to GnRH, but the hypothalamus produces only very low levels of GnRH. Puberty is initiated by a reduction in the sensitivity of hypothalamic GnRH-producing cells to negative feedback from sex steroids and from gonadotropins. As a result, GnRH release increases, stimulating increased production of gonadotropins and hence increased production of sex steroids.

In females, increasing levels of LH and FSH at puberty stimulate the ovaries to increase their production of the female sex hormones. The increased circulating levels of these hormones initiate the development of the traits of a sexually mature woman: enlarged breasts, vagina, and uterus; broadened hips; increased subcutaneous fat; pubic hair; and initiation of the menstrual cycle.

In males, an increasing level of LH stimulates groups of cells in the testes to increase their synthesis of testosterone, which in turn initiates the physiological, anatomical, and psychological changes associated with adolescence. The voice deepens, hair begins to grow on the face and body, and the testes and penis grow larger. As we saw at the beginning of this chapter, testosterone also help bones and skeletal muscles grow. FSH in males stimulates production of sperm.

Melatonin is involved in biological rhythms and photoperiodicity

The **pineal gland** is situated between the two hemispheres of the brain and is connected to the brain by a stalk. It produces the amine hormone **melatonin** from the amino acid tryptophan. The pineal gland releases melatonin in the dark and therefore marks the length of the night. Exposure to light inhibits the release of melatonin.

In vertebrates, melatonin is involved in biological rhythms, including **photoperiodicity**—the phenomenon whereby seasonal changes in day length cause physiological changes. Many species, for example, come into reproductive condition when the days begin to lengthen (**Figure 41.16**). Humans are not strongly photoperiodic, but melatonin in humans may play a role in synchronizing daily biological rhythms to the daily cycle of light and dark.

(A)

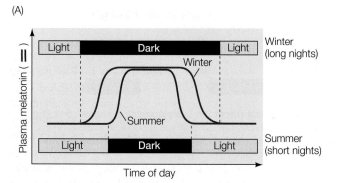

(B)

41.16 Melatonin Regulates Seasonal Changes (A) Melatonin release occurs in the dark and is inhibited by exposure to light. The duration of daily melatonin release thus changes as day length (photoperiod) changes, inducing dramatic seasonal physiological changes in some animals. (B) In winter, these Siberian hamsters are white and do not reproduce. In summer, they are mottled brown and breed.

Many chemicals may act as hormones

We have discussed the major endocrine glands and their hormones in this chapter, but many more hormones exist. As we discuss the organ systems of the body in the chapters that follow, we will frequently describe hormones that their tissues produce as well as hormones that control their functions.

41.3 RECAP

In mammals, the major endocrine glands include the hypothalamus, pituitary gland, thyroid gland, parathyroid glands, pancreas, adrenal glands, gonads, and pineal gland. Each of these glands secretes, and responds to, hormones that play crucial roles in controlling physiology and development.

- Describe how thyroxine is produced and how its production and release are controlled. **See pp. 861–862 and Figure 41.10**

- How is the concentration of calcium in the blood regulated? **See pp. 863–864 and Figure 41.11**

- What are the hormonal bases for the two forms of diabetes? **See p. 864**

- What changes in the feedback control of sex steroids result in puberty? **See pp. 867–868**

Studying endocrine systems is not easy. Many hormones are released in very small quantities, and some disappear from the extracellular fluid rapidly. A hormone's receptors may be found on diverse cells around the body, and those cells can respond in different ways to the same hormone. How have we overcome these difficulties to learn how hormones work?

41.4 How Do We Study Mechanisms of Hormone Action?

We can break the study of hormone actions into different sets of problems. First, we must be able to detect, identify, and measure hormones. Second, we must be able to identify and characterize the receptors for hormones. Third, we must understand the signal transduction pathways activated by hormones in different tissues.

Hormones can be detected and measured with immunoassays

As we have seen, testosterone has many dramatic and diverse effects, yet its concentration in the blood of adult human males is only about 30 to 100 *billionths of a gram* per milliliter. Measuring hypothalamic neurohormones requires calibrations in the range of *trillionths* of a gram per milliliter.

The ability to detect and measure minute quantities of hormone was an important breakthrough. Rosalyn Yalow developed a method called *radioimmunoassay* because it used radioactive la-

TOOLS FOR INVESTIGATING LIFE

41.17 An Immunoassay Allows Measurement of Small Concentrations

An immunoassay uses labeled and unlabeled samples of a purified hormone (or other substance to be measured) and an antibody to that hormone to develop a standard curve against which a sample of unknown concentration can be measured.

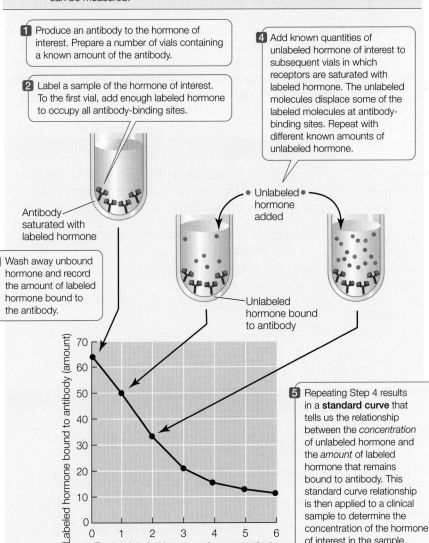

1 Produce an antibody to the hormone of interest. Prepare a number of vials containing a known amount of the antibody.

2 Label a sample of the hormone of interest. To the first vial, add enough labeled hormone to occupy all antibody-binding sites.

Antibody saturated with labeled hormone

3 Wash away unbound hormone and record the amount of labeled hormone bound to the antibody.

4 Add known quantities of unlabeled hormone of interest to subsequent vials in which receptors are saturated with labeled hormone. The unlabeled molecules displace some of the labeled molecules at antibody-binding sites. Repeat with different known amounts of unlabeled hormone.

Unlabeled hormone added

Unlabeled hormone bound to antibody

5 Repeating Step 4 results in a **standard curve** that tells us the relationship between the *concentration* of unlabeled hormone and the *amount* of labeled hormone that remains bound to antibody. This standard curve relationship is then applied to a clinical sample to determine the concentration of the hormone of interest in the sample.

(Graph: y-axis "Labeled hormone bound to antibody (amount)" from 0 to 70; x-axis "Total unlabeled hormone (concentration)" from 0 to 6)

Hormones are not simple on–off switches, and an important characteristic of a hormone is the time course over which it acts. This time course can be measured by the hormone's *half-life* in the blood (defined as the length of time it takes for one-half of the hormone molecules to disappear). Soon after endocrine cells are stimulated to secrete their hormone, the hormone reaches its maximum concentration in the blood. By taking a series of blood samples and using immunoassays, researchers can determine how long it takes for the circulating hormone to drop to half of that maximum concentration. The fight-or-flight response to epinephrine, as we have seen, is relatively quick in its onset and termination; the half-life of epinephrine in the blood is only 1 to 3 minutes. The effects of other hormones, such as cortisol and thyroxine, are expressed over much longer periods, and their half-lives are on the order of days or weeks.

Immunoassays have also facilitated the measurement of dose–response relationships. To evaluate a drug or a natural hormone for therapeutic use, it is critical to know the sensitivity of the body to that drug or hormone. Being able to measure the concentrations of drugs or hormones in the blood makes it possible to construct dose–response curves that help physicians adjust dosages appropriately (**Figure 41.18**).

A hormone can act through many receptors

Different receptors may be involved in mediating the actions of a single hormone. Because there are slight differences among the receptors for a particular hormone, it is possible to create drugs that are very selective in blocking or stimulating specific responses. A number of receptors have been identified, isolated, and purified through biochemical separation techniques. For example, a hormone can be bound to a substrate such as

bels (she used radioactive isotopes of iodine) to track interactions between an antigen (the hormone or other substance of interest to be measured) and an antibody made to that antigen. Today we are more likely to use nonradioactive labels, so the technique is called simply **immunoassay** (**Figure 41.17**). Being able to measure hormones in the blood made it possible to study many important hormonal mechanisms.

41.18 Dose-Response Curves Quantify the Body's Response to a Hormone Between the threshold and maximum doses, a dose-response curve frequently has an S shape. Anything that changes the responsiveness of a system—such as an increase or decrease in the number of receptors in target cells—affects the position of the curve.

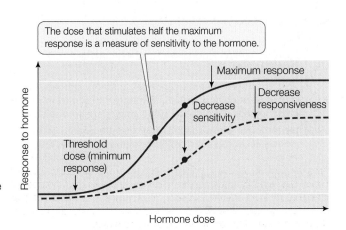

The dose that stimulates half the maximum response is a measure of sensitivity to the hormone.

Maximum response

Decrease responsiveness

Decrease sensitivity

Threshold dose (minimum response)

Response to hormone

Hormone dose

resin beads packed into a glass column. When an extract of cells suspected of containing receptors to that hormone is added to the column, the receptors bind to the hormone molecules on the beads. The hormone–receptor complexes can then be washed off the beads and the receptors isolated. This technique is called **affinity chromatography**.

As more receptors are isolated and characterized, researchers discover that they frequently exist in families with common structural features. These common features result from common nucleotide sequences in the receptors' genes. Genomic analyses have led to the discovery of many new receptors. Investigators "scan" the genome for sequences that bear homologies to known receptor gene sequences; when they get a "hit," they have found a candidate gene for a new receptor. They can then identify the molecule to which that receptor binds (its ligand), describe where the receptor occurs within the body, and characterize the receptor's physiological effects.

Knowing the molecular identity of a receptor and being able to measure its concentration makes it possible to study that receptor's regulation. We saw above that the release of hormones can be under negative feedback control. Similarly, the abundance of receptors for a hormone can be under feedback control. In some cases, continuous high concentrations of a hormone can decrease the number of its receptors, a process known as **downregulation**. **Upregulation** of receptors can occur when hormone secretion is suppressed. The regulation of receptor abundance is an important mechanism controlling the sensitivity of the body to hormonal signaling.

An example of downregulation occurs in type II diabetes mellitus, which is characterized by elevated insulin concentrations in the blood and a loss of insulin receptors. Although genetic factors are probably involved, a possible immediate cause of the disease is an overstimulation of pancreatic release of insulin by excessive carbohydrate intake, which leads to downregulation of the insulin receptors. An example of upregulation may be seen in people who have been on a regular dose of beta blockers (see page 865). As the activity of the β-adrenergic receptors is blocked over time, more receptors are produced. If the person goes off the medication suddenly, the effects of the receptors are amplified, resulting in heightened anxiety. Changes in dosage in the long-term use of such medications thus are usually gradual and carefully supervised.

41.4 RECAP

Studying the mechanisms of hormone action requires the ability to measure hormone concentrations and to identify and characterize hormone receptors.

- Describe how an immunoassay is performed. See p. 869 and Figure 41.17
- How are receptors for a particular hormone identified? See pp. 869–870

CHAPTER SUMMARY

41.1 What Are Hormones and How Do They Work?

- **Endocrine cells** secrete chemical signals that induce responses in other cells that have receptors for those molecules. In some cases endocrine cells are aggregated into **endocrine glands**.
- **Hormones** are endocrine signals that are secreted from a cell, circulate in the blood, and bind to **target cells** distant from the secreting cell. Review Figure 41.1
- The chemical structures of hormones are highly conserved, but through evolution, hormones have acquired different functions in different animal groups. Review Figure 41.2
- Two hormones, PTTH and ecdysone, control molting in arthropods. A third hormone, **juvenile hormone**, prevents maturation. When an insect stops producing juvenile hormone, it molts into an adult. Review Figures 41.3 and 41.4, **ANIMATED TUTORIAL 41.1**
- Most hormones are either peptides, proteins, steroids, or amines. Peptide and protein hormones and some amines are water-soluble; steroids and some amines are lipid-soluble.
- Receptors for water-soluble hormones are located on the cell surface. Receptors for most lipid-soluble hormones are inside the cell.
- Hormones can cause different responses in different target cells. Review Figure 41.5

41.2 How Do the Nervous and Endocrine Systems Interact?

- In humans, the major endocrine glands are distributed around the body. Review Figure 41.6, **WEB ACTIVITY 41.1**
- The **pituitary gland** is the interface between the nervous and endocrine systems. The **anterior pituitary** develops from embryonic mouth tissue; the **posterior pituitary** develops from the developing brain.
- The posterior pituitary secretes two **neurohormones**: antidiuretic hormone (ADH) and oxytocin. The anterior pituitary secretes **tropic hormones** (thyrotropin, corticotropin, luteinizing hormone, and follicle-stimulating hormone) as well as **growth hormone**, **prolactin**, melanocyte-stimulating hormone, **endorphins**, and **enkephalins**.
- The anterior pituitary is controlled by neurohormones produced by cells in the **hypothalamus** and transported through **portal blood vessels** to the anterior pituitary. Review Figures 41.7 and 41.8, **ANIMATED TUTORIAL 41.2**
- Hormone release is controlled by negative feedback loops. Review Figure 41.9

41.3 What Are the Major Mammalian Endocrine Glands and Hormones?

- The **thyroid gland** is controlled by **thyrotropin** and secretes **thyroxine**, which controls cell metabolism. **Review Figure 41.10**

- The level of calcium in the blood is regulated by three hormones. **Calcitonin** from the thyroid lowers blood calcium by promoting bone deposition. **Parathyroid hormone** (**PTH**) raises blood calcium by promoting bone turnover and decreasing calcium excretion. **Vitamin D** promotes calcium absorption from the digestive tract. **Review Figure 41.11, ANIMATED TUTORIAL 41.3**

- The **pancreas** secretes three hormones. **Insulin** stimulates glucose uptake by cells and lowers blood glucose, **glucagon** raises blood glucose, and **somatostatin** slows the rate of nutrient processing.

- The **adrenal gland** has two portions, one within the other. The inner portion, the **adrenal medulla**, releases **epinephrine** and **norepinephrine** in response to stress. The outer portion, the **adrenal cortex**, produces three classes of **corticosteroids**: glucocorticoids, mineralocorticoids, and small amounts of sex steroids. **Review Figures 41.12 and 41.13**

- **Aldosterone** is a mineralocorticoid that stimulates the kidneys to conserve sodium and excrete potassium. **Cortisol** is a glucocorticoid that is released in response to stressful stimuli but acts more slowly than the hormones of the adrenal medulla. **Review Figure 41.14**

- Sex hormones (**androgens** in males, **estrogens** and **progesterone** in females) are produced by the **gonads** in response to tropic hormones. Sex hormones control sexual development, secondary sexual characteristics, and reproductive functions. **Review Figures 41.14 and 41.15**

- The **pineal gland** releases **melatonin**, which is involved in controlling biological rhythms. **Review Figure 41.16**

41.4 How Do We Study Mechanisms of Hormone Action?

- **Immunoassay** techniques are used to measure concentrations of hormones and other substances. **Review Figure 41.17**

- The sensitivity of a cell to hormones can be altered by **downregulation** or **upregulation** of the receptors in that cell.

SEE WEB ACTIVITY 41.2 for a concept review of this chapter.

SELF-QUIZ

1. Before puberty
   a. the pituitary secretes luteinizing hormone and follicle-stimulating hormone, but the gonads are unresponsive.
   b. the hypothalamus does not secrete much gonadotropin-releasing hormone.
   c. males can stimulate massive muscle development through a vigorous training program.
   d. testosterone plays no role in development of the male sex organs.
   e. genetic females will develop male genitals unless estrogen is present.

2. Both epinephrine and cortisol are secreted in response to stress. Which of the following statements is also true for *both* of these hormones?
   a. They act to increase blood glucose availability.
   b. Their receptors are on the surfaces of target cells.
   c. They are secreted by the adrenal cortex.
   d. Their secretion is stimulated by corticotropin.
   e. They are secreted into the blood within seconds of the onset of stress.

3. Growth hormone
   a. can cause adults to grow taller.
   b. stimulates protein synthesis.
   c. is released by the hypothalamus.
   d. can be obtained only from cadavers.
   e. is a steroid.

4. PTH
   a. stimulates osteoblasts to lay down new bone.
   b. reduces blood calcium levels.
   c. stimulates calcitonin release.
   d. is produced by the thyroid gland.
   e. is released when blood calcium levels fall.

5. Steroid hormones
   a. are produced only by the adrenal cortex.
   b. have only cell surface receptors.
   c. are water-soluble.

   d. act by altering the activity of proteins in the target cell.
   e. act by altering gene expression in the target cell.

6. The hormone ecdysone
   a. is released from the posterior pituitary.
   b. stimulates molting in insects.
   c. maintains an insect in larval stages unless PTTH is present.
   d. stimulates the secretion of juvenile hormone from the prothoracic glands.
   e. keeps the insect exoskeleton flexible to permit growth.

7. The posterior pituitary
   a. synthesizes oxytocin.
   b. is under the control of hypothalamic releasing neurohormones.
   c. secretes tropic hormones.
   d. secretes neurohormones.
   e. is under feedback control by thyroxine.

8. Which of the following contributes to the development of goiter?
   a. Inadequate iodine in the diet
   b. Autoimmune antibodies that stimulate the TSH receptor
   c. Lack of feedback from circulating T_3 and T_4
   d. Overproduction of thyroglobulin
   e. All of the above

9. Which of the following is a likely cause of diabetes?
   a. Overproduction of insulin by β cells of the pancreas
   b. Loss of α cells of the pancreas
   c. Loss of insulin receptors
   d. Overproduction of glucagon
   e. Loss of receptors for somatostatin

10. Which statement is true of all hormones?
    a. They are secreted by glands.
    b. They have receptors on cell surfaces.
    c. They may stimulate different responses in different cells.
    d. There is a gene that codes for each hormone.
    e. When the same hormone occurs in different species, it has the same action.

FOR DISCUSSION

1. Explain how both hyperthyroidism and hypothyroidism can cause goiter. Refer to the roles of the hypothalamus and the pituitary in your answer.

2. Neurons, the cells of the nervous system, do not require insulin. Why might this be so, and why is it important?

3. Various side effects of anabolic steroid abuse were mentioned at the opening of this chapter. Some of these effects are due to the direct action of the steroid, but others are due to the negative feedback action of the steroid. Discuss an example of each and explain possible mechanisms.

4. Compare the characteristics you would expect of a hormone signaling system that controls a short-term process, such as digestion, with the characteristics you would expect of a hormone signaling system that controls a long-term process, such as development.

ADDITIONAL INVESTIGATION

Each spring, male deer grow antlers that they use in male–male competition for mates. Each fall they shed those antlers. Although females of most deer species do not grow antlers, the caribou are an exception. Female caribou grow antlers in the spring but do not shed them in the fall. (Over the winter, they use their antlers to defend patches of food from males.) In addition, newborn caribou begin to grow antlers after birth. Assuming that antler growth is controlled by hormones, how would you investigate the control of antler growth in caribou? What hormonal assays might you want to use? What hypotheses would you test with respect to the relationship between time of year and antler growth?

Immunology: Animal Defense Systems

The plague of Athens

As a rule, however, there was no ostensible cause; but people in good health were all of a sudden attacked by violent heats in the head, and redness and inflammation in the eyes, the inward parts, such as the throat or tongue, becoming bloody and emitting an unnatural and fetid breath.

The Athenian historian Thucydides wrote these words in 430 BCE. He was describing a rapidly spreading infectious disease, or plague, that was sweeping the city of Athens in ancient Greece, and would ultimately kill about one-third of its inhabitants.

One year earlier, in 431 BCE, the Spartans had surrounded Athens. Sparta was a rival state with a highly disciplined army, while Athens was near a port and had a powerful navy. The Athenian leader Pericles decided to take advantage of his naval supremacy. He instructed his soldiers to abandon the countryside to the Spartans, relying on the navy to keep the enemies out of the city and to keep the Athenians supplied with food. People living in the countryside poured into Athens.

As Athens became more crowded, declining sanitation and close living conditions resulted in the city becoming an incubator for infectious diseases. In his famous *History of the Peloponnesian War*, Thucydides speculated that the plague came to Athens from Africa. His precise clinical description of the disease has provoked great debate among medical historians regarding its nature. In 1994, a burial ground was found near Athens that apparently contained the remains of several dozen people who died in the plague. PCR (polymerase chain reaction) analyses of dental pulp from the skeletons revealed the presence of DNA sequences from the bacterium that causes typhoid fever, and this has been proposed as the cause of the plague. Other scientists disagree, using arguments based on a comparison of Thucydides' account with the clinical description of typhoid fever today.

Whatever the cause, another passage from Thucydides (who was one of the few lucky survivors of the plague) bears noting:

Yet it was with those who had recovered from the disease that the sick and the dying found most compassion. These knew what it was from experience and now had no fear for themselves; for the same man was never attacked twice, never at least fatally.

Two questions arise from this passage: First, how could a person survive the disease? And second, how could the survivors be resistant to further infection?

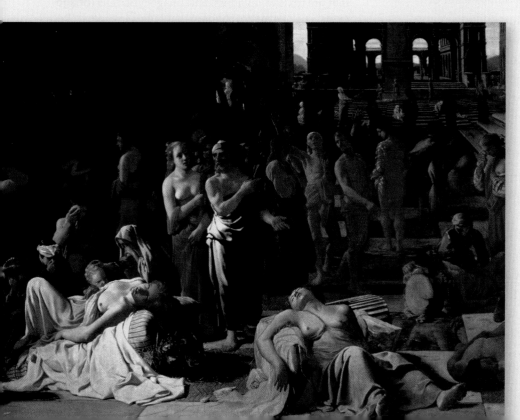

The Peloponnesian War The blockade of Sparta and its Peloponnesian allies on Athens made the latter crowded and vulnerable to the spread of an infectious disease.

Thucydides The Athenian general wrote a detailed history of the Peloponnesian war and the spread of an infectious disease that killed 1 in 3 Athenians. He survived the disease, presumably because of his immune system.

Almost 2,500 years after Thucydides recorded his observations, we have some answers to these questions. These answers come from studies of the mammalian immune system, which recognizes and fights off pathogens. When Thucydides caught the disease, specialized white blood cells called T cells recognized proteins of the infecting agent on the surfaces of his body's cells. This initiated a series of events that resulted in the killing of infected cells and destruction of the disease agent by antibodies, specific proteins made by a different set of white blood cells called B cells. Some descendants of the T and B cells persisted in his body as "memory cells," which defended him when he was exposed to the disease later. The operation of the mammalian immune system and how its components defend the body are the focus of this chapter.

IN THIS CHAPTER we first describe how the innate, nonspecific immune system prevents most pathogens from entering the body and deals with those that do. Then we look at how the specific, adaptive immune system targets invaders such as the typhoid bacterium, which may have killed the ancient Athenians. We see how the reshuffling of genetic material helps animals fight off the staggering diversity of potential invaders, and finally, we see what happens when this complex and crucial system malfunctions.

42.1 What Are the Major Defense Systems of Animals?

Animals have a number of ways of defending themselves against **pathogens**—harmful organisms and viruses that can cause disease. These defense systems are based on the distinction between *self*—the animal's own molecules—and *nonself*, or foreign, molecules. The defensive response involves three phases:

- *Recognition phase.* The organism must be able to discriminate between self and nonself.

- *Activation phase.* The recognition event leads to a mobilization of cells and molecules to fight the invader.

- *Effector phase.* The mobilized cells and molecules destroy the invader.

There are two general types of defense mechanisms:

- **Nonspecific defenses,** or *innate defenses,* provide the first line of defense against pathogens. They typically act very rapidly, and include barriers such as the skin, molecules that are toxic to invaders, and phagocytic cells (**phagocytes,** such as **macrophages**) that ingest invaders. (Recall from Section 6.5 that phagocytosis is a form of endocytosis, in which a cell engulfs a large particle or cell.) This system recognizes broad classes of organisms or molecules and gives a quick response, within minutes or hours. Most animals have nonspecific defenses.

- **Specific defenses** are *adaptive* mechanisms aimed at specific pathogens. For example, a specific defense system can make an antibody protein that will recognize, bind to, and aid in the destruction of a certain virus if that virus ever enters the bloodstream. These systems recognize specific configurations of atoms in a molecule and are typically slow to develop and long-lasting. Specific defense mechanisms are found in vertebrate animals.

Mammals have both kinds of defense mechanism and are the focus of this chapter. In mammals and other vertebrates, the nonspecific and specific mechanisms operate together as a *coordinated* defense system. The nonspecific defenses are the body's first line of defense because the specific defenses often require days or even weeks to become effective.

Blood and lymph tissues play important roles in defense

The components of the mammalian defense system are dispersed throughout the body and interact with almost all of its other tissues and organs. The *lymphoid tissues*, which include the thymus, bone marrow, spleen, and lymph nodes, are essential parts of the defense system (**Figure 42.1**). The blood and lymph are complex systems with nondefensive functions that are discussed in Chapter 50. They each have central roles in defense as well.

The blood and lymph both consist of liquids in which cells are suspended:

- **Blood plasma** is a yellowish solution containing ions, small molecular solutes, and soluble proteins. Suspended in the plasma are red blood cells, white blood cells, and platelets (cell fragments essential to blood clotting). While red blood cells are normally confined to the closed circulatory system (the heart, arteries, capillaries, and veins), white blood cells and platelets are also found in the lymph.

- **Lymph** is a fluid derived from the blood and other tissues that accumulates in intercellular spaces throughout the body. From these spaces, the lymph moves slowly into the vessels of the lymphatic system. Tiny lymph capillaries conduct this fluid to larger ducts that eventually join to-gether, forming one large vessel, the thoracic duct, which joins a major vein (the left subclavian vein) near the heart. By this system of vessels, the lymph is eventually returned to the blood and the circulatory system.

At many sites along the lymph vessels are small, roundish structures called **lymph nodes**, which contain a type of white blood cell called a **lymphocyte**. As lymph passes through a lymph node, the lymphocytes encounter foreign cells and molecules that have entered the body, and if they are recognized, an immune response is initiated.

White blood cells play many defensive roles

One milliliter of human blood typically contains about *5 billion* red blood cells and *7 million* of the larger white blood cells. All of these cells originate from multipotent stem cells (constantly dividing undifferentiated cells that can form several different cell types) in the bone marrow. Examine **Figure 42.2** and you will see that there are two major families of **white blood cells** (also called *leukocytes*): *phagocytes* and *lymphocytes*. Lymphocytes, which include **B cells** and **T cells**, are smaller than phagocytes and are not phagocytic. Each family contains different types of cells with specialized functions. Natural killer cells and some kinds of phagocytes are also referred to collectively as *granulocytes* because they contain numerous granules (vesicles containing defensive enzymes). Defensive proteins and signals play fundamental roles in the interactions and functioning of these cells.

Immune system proteins bind pathogens or signal other cells

Similar to the actions of hormones, which we discuss in Chapter 41, the cells that defend mammalian bodies work together, interacting with one another and with the cells of invading pathogens. These cell–cell interactions are accomplished by a variety of key proteins, including receptors, other cell surface proteins, and signaling molecules. Four of the major players are listed here, and will be discussed in more detail later in the chapter.

Thoracic
duct

Lymph ducts
conduct lymph.

T cells mature
in the **thymus**.

In the **lymph nodes**, lymph is
filtered and white blood cells
inspect it for pathogens.

The spleen
filters circulating
blood.

B cells mature in
the **bone marrow**.

42.1 The Human Lymphatic System A network of ducts and vessels collects lymph from body tissues and carries it toward the heart, where it mixes with blood to be pumped back to the tissues. Other lymphoid tissues, including the thymus, spleen, and bone marrow, are also essential to the body's defense system.

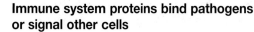

yourBioPortal.com

GO TO Web Activity 42.1 • The Human Defense System

42.2 Blood Cells Multipotent stem cells in the bone marrow can differentiate into red blood cells, platelets, and the various types of white blood cells.

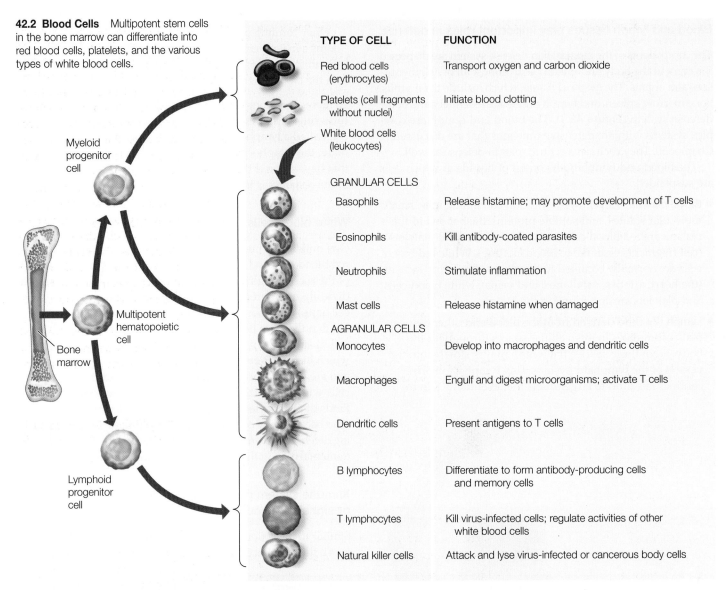

TYPE OF CELL	FUNCTION
Red blood cells (erythrocytes)	Transport oxygen and carbon dioxide
Platelets (cell fragments without nuclei)	Initiate blood clotting
White blood cells (leukocytes)	
GRANULAR CELLS	
Basophils	Release histamine; may promote development of T cells
Eosinophils	Kill antibody-coated parasites
Neutrophils	Stimulate inflammation
Mast cells	Release histamine when damaged
AGRANULAR CELLS	
Monocytes	Develop into macrophages and dendritic cells
Macrophages	Engulf and digest microorganisms; activate T cells
Dendritic cells	Present antigens to T cells
B lymphocytes	Differentiate to form antibody-producing cells and memory cells
T lymphocytes	Kill virus-infected cells; regulate activities of other white blood cells
Natural killer cells	Attack and lyse virus-infected or cancerous body cells

Labels in figure: Myeloid progenitor cell; Bone marrow; Multipotent hematopoietic cell; Lymphoid progenitor cell

─── **yourBioPortal.com** ───
GO TO Animated Tutorial 42.1 • Cells of the Immune System

- **Antibodies** are proteins that bind specifically to certain substances identified by the immune system as nonself or *altered self* (a major histocompatibility protein that has been altered by a pathogen). This binding can inactivate viruses and toxins, and it can act as tag on nonself cells, making them easier for the immune system cells to attack. Antibodies are produced by B cells.

- **Major histocompatibility complex (MHC)** proteins are found in two classes. MHC I proteins are found on the surfaces of most cells in the mammalian body. MHC II proteins are found on most immune system cells. MHC proteins are important self-identifying labels and play a major role in coordinating interactions between lymphocytes and macrophages.

- **T cell receptors** are integral membrane proteins on the surfaces of T cells. They recognize and bind to nonself substances presented with self MHC molecules on the surfaces of other cells.

- **Cytokines** are soluble signaling proteins released by many cell types. They bind to cell surface receptors and alter the behavior of their target cells. Various cytokines activate or inactivate B cells, macrophages, and T cells.

42.1 RECAP

Animals have nonspecific and specific defenses against pathogens. Both kinds of mechanism are based on the ability to differentiate self from nonself. Nonspecific defenses target a broad range of molecules and organisms, while specific defenses target specific configurations of atoms on molecules.

- List the differences between specific and nonspecific defenses. See p. 874

- What are the two classes of white blood cells, and how do they function in vertebrate defense systems? See p. 875 and Figure 42.2

The outcome of a disease—the life or death of the host—often depends on the success of both rapid, nonspecific responses and long-lasting, specific responses to invading pathogens. We turn now to the nonspecific defenses that protect vertebrates from disease.

42.2 What Are the Characteristics of the Nonspecific Defenses?

Nonspecific defenses are general protection mechanisms that attempt to stop pathogens from invading the body. As noted above, they provide first line of defense, in both time and location. Essentially, they are "ready to go," in contrast to specific immunity, which takes time to develop after a pathogen or toxin has been recognized as nonself. In mammals, nonspecific mechanisms include physical barriers as well as cellular and chemical defenses (**Table 42.1**).

Barriers and local agents defend the body against invaders

Consider a pathogenic bacterium that causes disease in internal organs and that lands on human skin. The challenges faced by the bacterium just to reach its target are formidable. First, there is the physical barrier of the skin. Bacteria rarely penetrate intact skin; by the same token, broken skin increases the risk of infection. The saltiness and dryness of skin may not be hospitable to the growth of the invader. On the other hand, there are some bacteria and fungi that normally live and sometimes reproduce in great numbers on our body surfaces without causing disease. They are referred to as **normal flora**. These natural occupants of our bodies form a nonspecific defense because they compete with pathogens for space and nutrients.

Consider a pathogen that lands on the surface of the nose. The mucous membranes found at the inner surfaces of the nose (as well as the digestive, respiratory, and urogenital systems) contain the enzyme **lysozyme**, which attacks the cell walls of many bacteria, causing them to *lyse* (burst open). Mucus in the nose traps airborne microorganisms, and most of those that get past this filter end up trapped in mucus deeper in the respiratory tract. Mucus and trapped pathogens are removed by the beating of cilia in the respiratory passageway, which continuously move a sheet of mucus and its trapped debris up toward the nose and mouth where it can be expelled or swallowed. Sneezing is another way to remove microorganisms from the respiratory tract.

TABLE 42.1
Human Nonspecific Defenses

DEFENSIVE MECHANISM	FUNCTION
Surface barriers	
Skin	Prevents entry of pathogens and foreign substances
Acid secretions	Inhibit bacterial growth on skin
Mucus	Prevents entry of pathogens; produces defensins that kill pathogens
Mucous secretions	Trap bacteria and other pathogens in digestive and respiratory tracts
Nasal hairs	Filter bacteria in nasal passages
Cilia	Move mucus and trapped materials away from respiratory passages
Gastric juice	Concentrated HCl and proteases destroy pathogens in stomach
Acid in vagina	Limits growth of fungi and bacteria in female reproductive tract
Tears, saliva	Lubricate and cleanse; contain lysozyme, which destroys bacteria
Nonspecific cellular, chemical, and coordinated defenses	
Normal flora	Compete with pathogens; may produce substances toxic to pathogens
Fever	Body-wide response inhibits microbial multiplication and speeds body repair processes
Coughing, sneezing	Expels pathogens from upper respiratory passages
Inflammatory response (involves leakage of blood plasma and phagocytes from vessels)	Limits spread of pathogens to neighboring tissues; concentrates defenses; digests pathogens and dead tissue cells; releases chemical mediators that attract phagocytes and lymphocytes to site
Phagocytes (macrophages and neutrophils)	Engulf and destroy pathogens that enter body
Natural killer cells	Attack and lyse virus-infected or cancerous body cells
Antimicrobial proteins	
Interferons	Released by virus-infected cells to protect healthy tissue from viral infection; mobilize specific defenses
Complement proteins	Lyse microorganisms, enhance phagocytosis, and assist in inflammatory and antibody responses

Finally, the mucous membranes produce **defensins**, peptides that are 18–45 amino acids long and contain hydrophobic domains. They are toxic to a wide range of pathogens, including bacteria, microbial eukaryotes, and enveloped viruses. Defensins insert themselves into the plasma membranes of these organisms and make the membranes permeable, thus killing the invaders. Defensins are also produced in phagocytes, where they kill pathogens trapped by phagocytosis. Plants also produce defensins in response to pathogen exposure (see Chapter 39).

Consider a bacterium that lands in the mouth. If it survives the lysozyme in saliva and enters the digestive tract (stomach, small intestine, and large intestine) it is met by other defenses. The gastric juice in the stomach is a deadly environment for many bacteria because of the hydrochloric acid and proteases that are secreted into it. Bacteria cannot normally penetrate the lining of the small intestine, and some pathogens are killed by bile salts secreted into this part of the digestive tract. Any bacteria that survive these lines of defense (which include normal flora) and enter the large intestine are removed quickly with the feces.

All of these barriers and local agents are nonspecific defenses because they act on all invading pathogens in the same way. Most are properties of the tissue, called *epithelial tissue*, that covers surfaces on and within organs. More complex nonspecific defenses await any pathogens that manage to elude this first set of defenses.

Other nonspecific defenses include specialized proteins and cellular processes

Pathogens that penetrate the body's outer and inner surfaces encounter more complex nonspecific defenses. These include the activation of defensive cells and the secretion of various defensive proteins, such as *complement* and *interferon* proteins.

COMPLEMENT PROTEINS   Vertebrate blood contains more than 20 different proteins that make up the antimicrobial **complement system**. This system can be activated by various mechanisms, including both nonspecific and specific defense responses. The proteins act in a characteristic sequence, or cascade, with each protein activating the next:

- First, they attach to specific components on the surface of a microbe or to an antibody that has already bound to the microbe's surface. In either case, binding helps phagocytes recognize and destroy the microbe.
- Then, they activate the inflammation response and attract phagocytes to the site of infection.
- Finally, they lyse invading cells (such as bacteria).

INTERFERONS   When a cell is infected by a pathogen, it produces small amounts of signaling proteins called **interferons** that increase the resistance of neighboring cells to infection. Interferons are a class of cytokines and have been found in many vertebrates. Various molecules, including double stranded (viral) RNA, induce the production of interferons. Thus interferons are particularly important as a first line of nonspecific defense against viruses. Interferons bind to receptors on the plasma membranes of uninfected cells, stimulating a signaling pathway that inhibits viral reproduction if the cells are subsequently infected. In addition, interferons stimulate the cells to hydrolyze bacterial or viral proteins to peptides, an initial step in specific immunity (see Section 42.3).

PHAGOCYTES   Some phagocytes travel freely in the circulatory and lymphatic systems; others can move out of blood vessels and adhere to certain tissues. Pathogenic cells, viruses, or fragments of these invaders can be recognized by phagocytes, which then ingest them by phagocytosis. Defensins, nitric oxide, and reactive oxygen intermediates inside these phagocytes then kill the pathogens.

NATURAL KILLER CELLS   One class of lymphocytes, known as **natural killer cells**, can distinguish virus-infected cells and some tumor cells from their normal counterparts and initiate the apoptosis of these target cells. In addition to this nonspecific action, natural killer cells interact with the specific defense mechanisms by lysing antibody-labeled target cells.

Inflammation is a coordinated response to infection or injury

When tissue is damaged because of infection or injury, the body responds with **inflammation**. This response can happen almost anywhere in the body, internally as well as on the surface. Inflammation is an important phenomenon: it isolates the damaged area to stop the spread of the damage; it recruits cells and molecules to the damaged location to kill the invader; and it promotes healing. The first responders to tissue damage are **mast cells**, which adhere to the skin and the linings of organs, and release numerous chemical signals including:

- **tumor necrosis factor**, a cytokine protein that kills target cells and activates immune cells.
- **prostaglandins**, fatty acid derivatives that play roles in various responses including the widening of blood vessels. Prostaglandins interact with nerve endings and are partly responsible for the pain caused by inflammation.
- **histamine**, an amino acid derivative that leads to itchy, watery eyes, and rashes seen with some types of allergic reactions.

The redness and heat of inflammation result from the dilation and leakiness of blood vessels in the infected or injured area (**Figure 42.3**). Phagocytes enter the inflamed area where they engulf the invaders and dead tissue cells. Phagocytes are responsible for most of the healing associated with inflammation. They produce several cytokines, which (among other functions) signal the brain to produce a fever. This rise in body temperature accelerates lymphocyte production and phagocytosis, thereby speeding the immune response. In some cases, pathogens are temperature-sensitive, and their growth is inhibited. The pain of inflammation results from increased pressure due to swelling, the action of leaked enzymes on nerve endings, and the action of prostaglandins, which increase the sensitivity of the nerve endings to pain.

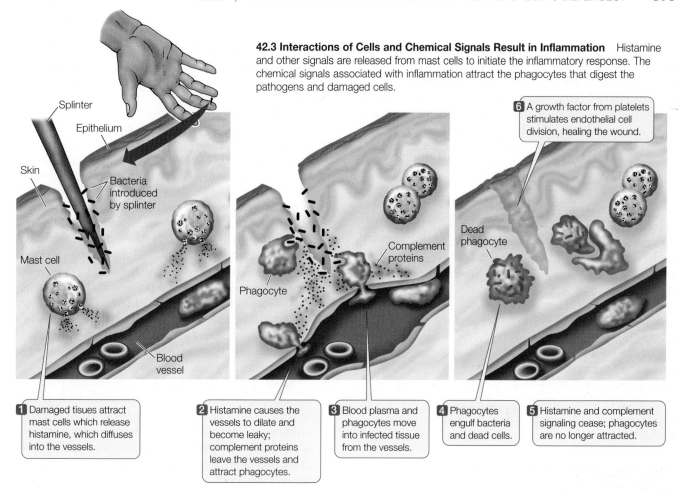

42.3 Interactions of Cells and Chemical Signals Result in Inflammation Histamine and other signals are released from mast cells to initiate the inflammatory response. The chemical signals associated with inflammation attract the phagocytes that digest the pathogens and damaged cells.

Splinter

Epithelium

Skin

Bacteria introduced by splinter

Mast cell

Phagocyte

Blood vessel

Complement proteins

6 A growth factor from platelets stimulates endothelial cell division, healing the wound.

Dead phagocyte

1 Damaged tisues attract mast cells which release histamine, which diffuses into the vessels.

2 Histamine causes the vessels to dilate and become leaky; complement proteins leave the vessels and attract phagocytes.

3 Blood plasma and phagocytes move into infected tissue from the vessels.

4 Phagocytes engulf bacteria and dead cells.

5 Histamine and complement signaling cease; phagocytes are no longer attracted.

---

yourBioPortal.com

GO TO **Web Activity 42.2 • Inflammation Response**

Following inflammation, pus may accumulate. Pus is a mixture of leaked fluid and dead cells: bacteria, neutrophils (the most abundant white blood cells—see Figure 42.2), and damaged body cells. Pus is a normal result of inflammation, and is gradually consumed and further digested by phagocytes called macrophages.

Inflammation can cause medical problems

While inflammation is generally a good thing, sometimes the inflammatory response is inappropriately strong, resulting in some allergies, cases of autoimmunity, and sepsis. In these cases, the response causes more damage than was originally there. We will discuss allergy and autoimmune diseases in Section 42.7. In some cases of severe bacterial infection, the inflammation response does not remain local. Instead, it extends throughout the bloodstream in a condition called *sepsis*. As in a local infection or injury, blood vessels dilate, but they do so throughout the body. The lowering of blood pressure that results is a medical emergency and can be lethal.

The symptoms of swelling, pain, and fever caused by excessive inflammation can be bothersome to the point of incapacitation. Diseases such as rheumatoid arthritis and chronic obstructive pulmonary disease, and accidents such as athletic injuries result in tissue damage and an inflammatory response. In order to manage excessive inflammation, drugs have been developed that act on the various cytokines and signal pathways to reduce inflammation and its symptoms. For example, the bark of the willow tree, *Salix*, has been known for millennia to contain a potent antidote to fever and swelling. In the nineteenth century the active ingredient, salicylic acid, was isolated and a better version, acetylsalicylic acid, was made. This drug—also called aspirin—works by inhibiting an enzyme on the pathway to the synthesis of prostaglandins. Other anti-inflammatory drugs act on the prostaglandin pathway, on the actions of tumor necrosis factor, and on the actions of histamine.

Cell signaling pathways stimulate the body's defenses

An invading pathogen can be regarded as a signal. In response to that signal, the body produces molecules (complement proteins, interferons, and other cytokines) that regulate phagocytosis and other defense processes. Not surprisingly, the link between signal and response is a signal transduction pathway, similar to the ones we considered in Section 7.3. A key group of receptors is called the **toll-like receptors**. The toll protein was first identified in insects, where it is involved in development and in sensing infection. Comparative genomics has revealed at least ten similar receptors in humans.

Toll-like receptors are part of a protein kinase cascade that ultimately results in the transcription of at least 40 genes in-

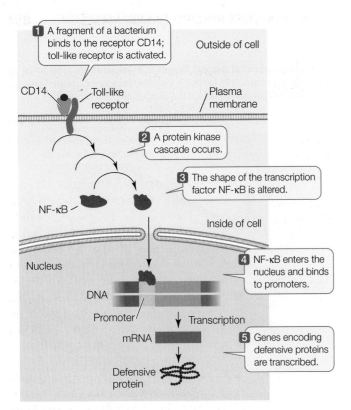

1 A fragment of a bacterium binds to the receptor CD14; toll-like receptor is activated.

Outside of cell

CD14

Toll-like receptor

Plasma membrane

2 A protein kinase cascade occurs.

3 The shape of the transcription factor NF-κB is altered.

NF-κB

Inside of cell

Nucleus

4 NF-κB enters the nucleus and binds to promoters.

DNA

Promoter

Transcription

mRNA

5 Genes encoding defensive proteins are transcribed.

Defensive protein

42.4 Cell Signaling and Defense Binding of a pathogenic molecule to the toll-like receptor initiates a signal transduction pathway which results in the transcription of genes whose products are involved in specific and nonspecific defenses. CD14 is expressed on the surface of white blood cells, including macrophages and monocytes.

volved in both nonspecific and specific defenses (**Figure 42.4**). The CD14 protein (for *c*luster of *d*ifferentiation) is a membrane protein that binds to the toll-like receptor. Among the molecules that stimulate this pathway are bacterial and fungal cell wall fragments. Binding of these fragment molecules to the receptor sets in motion a cascade of molecular changes that results in a change in the three-dimensional structure of the transcription factor NF-κB. (NF-κB stands for *n*uclear *f*actor *kappa* light chain enhancer of activated *B* cells. It is a key transcription factor in the activation of both the nonspecific and specific immune systems.) The shape change allows NF-κB to enter the nucleus, bind to the promoters of genes, and activate the transcription of genes encoding defensive proteins.

42.2 RECAP

Nonspecific defenses are the first line of defense against pathogens. These include physical barriers such as the skin; defensive proteins; and coordinated responses such as inflammation.

- How do complement proteins and interferons defend the body against microbes? See p. 578

- Describe the inflammation response. See pp. 878–879 and Figure 42.3

Often the innate immune system, with its nonspecific defenses, is adequate to prevent or fight off a pathogenic infection. But in

many cases, this system works together with specific immunity, which adapts to a particular pathogen. We now turn to the development and functioning of specific immunity.

42.3 How Does Specific Immunity Develop?

Before the twentieth century, scientists had long suspected that blood was somehow involved in immunity against pathogens. Over a century ago, Emil von Behring and Shibasaburo Kitasato at the University of Marburg in Germany performed a key experiment that pointed to blood as an important factor in specific immunity (**Figure 42.5**). They showed that guinea pigs injected with a sublethal dose of diphtheria toxin developed in their blood serum (the noncellular fluid that remains after blood is clotted) a factor that protected other guinea pigs from a lethal dose of the same toxin. In other words, the recipients had developed **immunity**. Moreover, the immunity was specific: the immune factor made by the guinea pigs protected only against the specific toxin, from one strain of diphtheria-causing bacteria, that they had been injected with.

Based on the animal model, Behring realized that serum protection might work in human diseases as well. It did, and he won the Nobel Prize for his efforts in protecting children against diphtheria. Later, the agent of this immunity was identified as an antibody protein, and the process of developing immunity from antibodies received from another individual was called *passive immunity*.

In this section we outline the main features of the specific immune system, much of which does indeed occur in blood serum. We will consider the two major types of specific responses: the *humoral immune response*, which produces antibodies; and the *cellular immune response*, which destroys infected cells.

— **yourBioPortal.com** —
GO TO Animated Tutorial 42.2 • Pregnancy Test

Adaptive immunity has four key features

Four important features of the adaptive immune system are: specificity; the ability to distinguish self from nonself; the ability to respond to an enormous diversity of nonself molecules; and immunological memory.

SPECIFICITY Lymphocytes (B and T cells) are crucial components of specific immunity. T cell receptors and the antibodies produced by B cells recognize and bind to specific nonself or altered self substances (**antigens**), and this interaction initiates a specific immune response. The specific sites on antigens that the immune system recognizes are called **antigenic determinants** or *epitopes*:

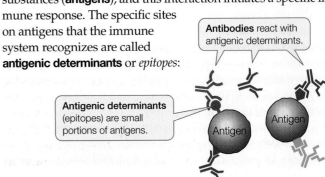

Antibodies react with antigenic determinants.

Antigenic determinants (epitopes) are small portions of antigens.

Antigen

Antigen

INVESTIGATING LIFE

42.5 The Discovery of Specific Immunity

Until the twentieth century, most people did not survive an attack of the bacterium that causes diphtheria, but a few did. Emil von Behring and Shibasaburo Kitasato performed a key experiment using an animal model, and demonstrated that the factor(s) responsible for immunity against diphtheria were in blood serum.

HYPOTHESIS Serum from guinea pigs injected with a sublethal dose of diphtheria toxin protects other guinea pigs that are exposed to a lethal dose of the toxin.

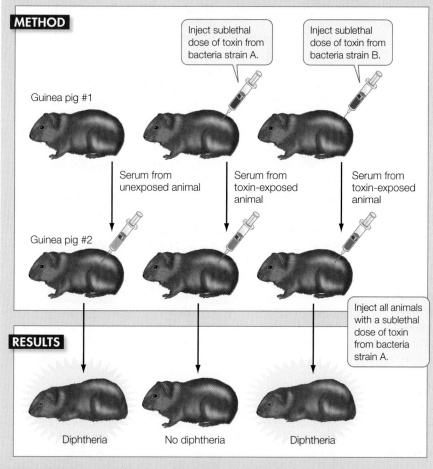

METHOD

Inject sublethal dose of toxin from bacteria strain A.

Inject sublethal dose of toxin from bacteria strain B.

Guinea pig #1

Serum from unexposed animal

Serum from toxin-exposed animal

Serum from toxin-exposed animal

Guinea pig #2

Inject all animals with a sublethal dose of toxin from bacteria strain A.

RESULTS

Diphtheria   No diphtheria   Diphtheria

CONCLUSION Serum of toxin-exposed guinea pigs is protective against later exposure to a lethal dose of the toxin from the same genetic strain of bacteria, but not a different strain.

Go to **yourBioPortal.com** for original citations, discussions, and relevant links for all INVESTIGATING LIFE figures.

An antigenic determinant is a specific portion of a large molecule, such as a certain sequence of amino acids that may be present in a protein. Antigens are usually proteins or polysaccharides, and there can be multiple antigens on a single invading bacterium. A single antigenic molecule can have multiple, different antigenic determinants. The host animal responds to the presence of an antigen with highly specific defenses involving T cell receptors and antibodies. These receptors and soluble proteins bind to the antigenic determinants. Each T cell and each antibody is specific for a single antigenic determinant.

For the remainder of the chapter, we will refer to antigenic determinants simply as "antigens."

DISTINGUISHING SELF FROM NONSELF The human body contains tens of thousands of different proteins, each with a specific three-dimensional structure capable of generating immune responses. Thus every cell in the body bears a tremendous number of antigens. A crucial requirement of an individual's immune system is that it recognize the body's own antigens and not attack them. This is accomplished by clonal deletion, which you will encounter in a few pages.

DIVERSITY Challenges to the immune system are numerous. Pathogens take many forms: viruses, bacteria, protists, fungi, and multicellular parasites. Furthermore, each pathogenic species usually exists as many subtly different genetic strains, and each strain possesses multiple surface features. Estimates vary, but a reasonable guess is that humans can respond specifically to 10 million different antigens. Upon recognizing an antigen, the immune system responds by activating lymphocytes of the appropriate specificity. This capacity is accomplished by a special genetic recombination mechanism described in Section 42.6.

IMMUNOLOGICAL MEMORY After responding to a particular type of pathogen once, the immune system "remembers" that pathogen and can usually respond more rapidly and powerfully to the same threat in the future. This **immunological memory** usually saves us from repeats of childhood diseases such as chicken pox. Vaccination against specific diseases works because the immune system "remembers" the antigens that were introduced into the body.

All four of these features of specific immune defense characterize both the humoral immune response and the cellular immune response.

Two types of specific immune responses interact: an overview

The specific immune system mounts two types of responses against invaders: the humoral immune response and the cellular immune response. B cells that make antibodies are the workhorses of the humoral immune response, and cytotoxic

42.6 The Specific Immune System Humoral immunity involves the production of antibodies by B cells. Cellular immunity involves the activation of cytotoxic T cells that bind to cells expressing the antigen. For further details, see Figure 42.13.

(killer) T cells are the workhorses of the cellular immune response. These two responses operate simultaneously and cooperatively, sharing many mechanisms. A key event early in these two processes is the exposure of the antigen's three-dimensional structure to the immune system. This occurs when an antigenic molecule, or a fragment of the molecule, is inserted into the plasma membrane of a cell and the unique epitope structure protrudes from the membrane, where it is exposed to nearby T or B cells. Many different types of cell can "present" the antigen to the immune system in this way, and they are collectively referred to as **antigen-presenting cells**. **Figure 42.6** provides an overview of antigen presentation and the roles of T and B cells in the immune response. The key player integrating the two responses is the *T-helper* (T_H) *cell*. By binding to the antigen on a presenting cell, the T_H cell stimulates events in both responses.

HUMORAL IMMUNE RESPONSE In the **humoral immune response** (from the Latin *humor*, "fluid"), antibodies react with antigens on pathogens in blood, lymph, and tissue fluids. An animal can produce a staggering diversity of antibodies capable of binding to almost any conceivable antigen the animal encounters. Antibodies are secreted by B cells and travel free in the blood and lymph. A particular B cell also possesses receptors on its surface with the same specificity as the antibodies it produces.

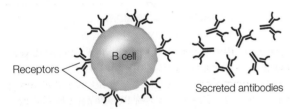

The first time a specific antigen invades the body, it may be presented and then detected by binding to a B cell receptor. This binding activates the B cell, which makes and secretes multiple copies of the antibody.

CELLULAR IMMUNE RESPONSE The **cellular immune response** is directed against antigens that have become established within a cell of the host animal. It detects and destroys virus-infected or mutated cells, such as cancer cells expressing unique proteins caused by mutations.

T cells within the lymph nodes, the bloodstream, and the intercellular spaces carry out the cellular immune response. These T cells have integral membrane proteins—T cell receptors—that recognize and bind to antigens. T cell receptors are rather similar to antibodies in structure and function, each including specific molecular configurations that bind to specific antigens.

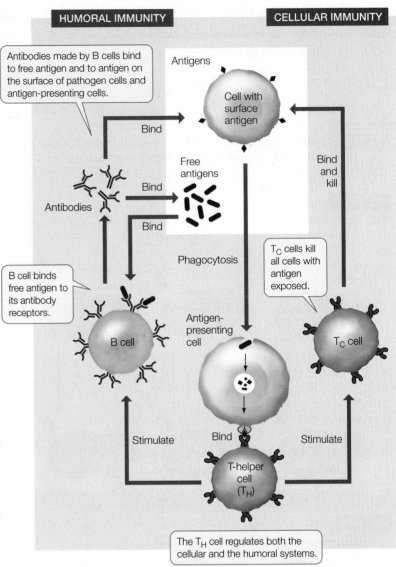

Once a T cell is bound to an antigen, it initiates an immune response that typically results in the total destruction of the antigen-containing cell.

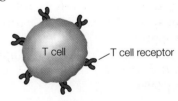

Genetic changes and clonal selection generate the specific immune response

Before the reactions just described for the humoral and cellular immune responses can take place, the body needs to generate a vast diversity of lymphocytes that have the ability to bind different antigens. How does this tremendous diversity arise? How do lymphocytes specific for certain antigens proliferate? The answers lie in the processes of DNA rearrangement and **clonal selection**. These processes generate the enormous numbers of different receptors with antibody-like specificities in the

42.7 Clonal Selection in B Cells The binding of an antigen to a specific receptor on the surface of a B cell stimulates that cell to divide, producing a clone of genetically identical cells to fight that invader.

B and T lymphocytes, even before the body encounters antigens for these receptors.

- *Diversity is generated primarily by DNA changes*—chromosomal rearrangements and other mutations—that occur just after the B and T cells are formed in the bone marrow. Each B cell is able to produce only one kind of antibody; thus there are millions of different B cells. Similarly, there are millions of different T cells with specific T cell receptors. We will describe the mechanisms for the formation of these antibodies and receptors later in the chapter. For now, keep in mind that the adaptive immune system is "predeveloped"—*all of the machinery available to respond to an immense diversity of antigens is already there, even before the antigens are ever encountered.*

- *Antigen binding "selects" a particular B or T cell for proliferation.* When an antigen fits the surface receptor on a B cell and binds to it, that B cell is activated. It divides to form a clone of cells (a genetically identical group derived from a single cell), all of which produce and secrete antibodies with the same specificity as the receptor (**Figure 42.7**). Binding, activation, and proliferation also apply to T lymphocytes. When a foreign or abnormal cell antigen binds to a T cell receptor, that binding activates the proliferation of T cells with that particular receptor. Binding and activation select a particular lymphocyte, while proliferation generates the clone, hence the term "clonal selection."

Immunity and immunological memory result from clonal selection

The first time a vertebrate animal is exposed to a particular antigen there is a time lag (usually several days) before the B cell-produced antibody molecules and T cells specific to that antigen slowly increase. But for years afterward—sometimes for life—the immune system "remembers" that particular antigen. How does this happen?

The answer lies in the fact that activated lymphocytes divide and differentiate to produce *two types* of daughter cells: effector cells and memory cells.

- **Effector cells** carry out the attack on the antigen. Effector B cells, called **plasma cells**, secrete antibodies. Effector T cells release cytokines and other molecules that initiate reactions that destroy nonself or altered cells. Effector cells live only a few days.

- **Memory cells** are long-lived cells that retain the ability to start dividing on short notice to produce more effector and more memory cells. Memory B and T cells may survive in the body for decades, rarely dividing.

These two types of lymphocytes can respond to an antigen in two different ways:

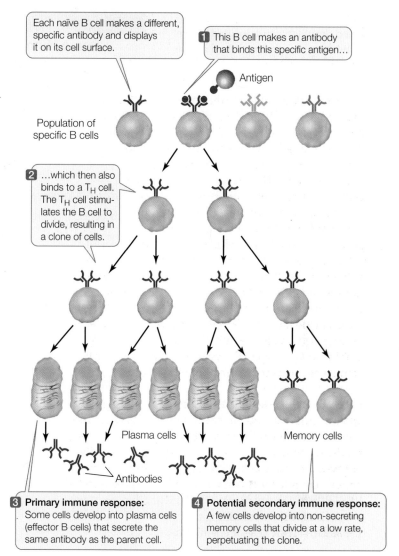

- When the body first encounters a particular antigen, a **primary immune response** is activated, in which the "naïve" (previously unexposed to antigen) lymphocytes that recognize that antigen proliferate to produce clones of effector and memory cells.

- After a primary immune response to a particular antigen, subsequent encounters with the same antigen will trigger a much more rapid and powerful **secondary immune response**. The memory cells that bind with that antigen proliferate, launching a huge army of plasma cells and effector T cells.

Vaccines are an application of immunological memory

You will recall Behring's experiment on using the serum of diphtheria-exposed animals to protect other animals from the disease (see Figure 42.5). As we noted, he later used this method on people to cause passive immunity. Now we can contrast this with *active immunity*, in which the immune system develops a specific response, including memory cells that lead to long-term protection. The animals that survived diphtheria and donated serum in Berhing's experiment had active immunity.

Thanks to active immunity and immunological memory, exposure to many diseases, such as chicken pox, provides a natural immunity to those diseases. Furthermore, it is possible to provide artificial immunity against many life-threatening diseases by **vaccination**: the introduction of antigen into the body in a form that does not cause disease.

Vaccination initiates a primary immune response, generating memory cells without making the person ill. Later, if a pathogen carrying the same antigen attacks, specific memory cells already exist. They recognize the antigen and quickly overwhelm the invaders with a massive production of lymphocytes and antibodies.

Because the antigens used for immunization or vaccination are produced by pathogenic organisms, they must be altered so that they cannot cause disease but are still able to provoke an immune response. There are three principal ways to do this:

- *Inactivation* involves killing the pathogen with heat or chemicals.

- *Attenuation* involves reducing the virulence of a virus by repeatedly infecting cells with it in the laboratory; this results in mutations in the virus that render it nontoxic but still recognized as nonself.

- *Recombinant DNA technology* can be used to produce peptide fragments that bind to and activate lymphocytes but do not have the harmful part of a protein toxin.

For most of the 70 or so bacteria, viruses, fungi, and parasites known to cause serious human diseases, vaccines are already available or will be in the next few years (**Table 42.2**). Vaccination has completely or almost completely wiped out some deadly diseases, including smallpox, diphtheria, and polio, in industrialized countries.

Animals distinguish self from nonself and tolerate their own antigens

Normally, the body is *tolerant* of its own molecules—the same molecules that would generate an immune response in another individual. This occurs primarily during the early differentiation of T and B cells, when they encounter self antigens. Any immature B or T cell that shows the potential to mount an immune response

TABLE 42.2
Some Human Pathogens for which Vaccines are Available

INFECTIOUS AGENT	DISEASE	VACCINATED POPULATION
Bacteria		
Bacillus anthracis	Anthrax	Those at risk in biological warfare
Bordetella pertussis	Whooping cough	Children and adults
Clostridium tetani	Tetanus	Children and adults
Corynebacterium diphtheriae	Diphtheria	Children
Haemophilus influenzae	Meningitis	Children
Mycobacterium tuberculosis	Tuberculosis	All people
Salmonella typhi	Typhoid fever	People in areas exposed to agent
Streptococcus pneumoniae	Pneumonia	Elderly people
Vibrio cholerae	Cholera	People in areas exposed to agent
Viruses		
Adenovirus	Respiratory disease	Military personnel
Hepatitis A	Liver disease	People in areas exposed to agent
Hepatitis B	Liver disease, cancer	All people
Influenza virus	Flu	All people
Measles virus	Measles	Children and adolescents
Mumps virus	Mumps	Children and adolescents
Poliovirus	Polio	Children
Rabies virus	Rabies	People exposed to agent
Rubella virus	German measles	Children
Vaccinia virus	Smallpox	Laboratory workers, military personnel
Varicella-zoster virus	Chicken pox	Children

against self antigens undergoes programmed cell death (apoptosis) within a short time. This process is called **clonal deletion**.

42.3 RECAP

The specific immune system reacts against nonself or altered self molecules called antigens. The system generates amazing diversity. Immunological memory arises from clonal selection.

- How does an antigen initiate a specific immune response? **See pp. 881–882**

- Describe clonal selection. How does it contribute to immunological memory? **See pp. 882–883 and Figure 42.7**

- How do vaccines make use of immunological memory? **See pp. 883–884**

Now that we understand the general features of the specific immune system, let's focus in more detail on the B lymphocytes and the humoral immune response.

42.4 What Is the Humoral Immune Response?

Every day, billions of B cells survive the test of clonal deletion and are released from the bone marrow into the circulation. B cells are the basis for the humoral immune response.

Some B cells develop into plasma cells

A B cell begins by making a receptor protein on its cell surface. As we have seen, if a B cell is activated by antigen binding to this receptor, it gives rise to clones of plasma cells and memory cells. The plasma (effector B) cells secrete antibodies into the bloodstream (see Figure 42.7).

Usually, for a naïve B cell (one that has not yet been exposed to antigen) to develop into an antibody-secreting plasma cell, a T-helper cell (T_H) with the same specificity must also bind to the antigen (see Figure 42.6). The division and differentiation of the B cell is stimulated by the receipt of chemical signals from the T_H cell. Thus, as we will see in Section 42.5, the B cell also functions as an antigen-presenting cell.

As plasma cells develop, the number of ribosomes and the amount of endoplasmic reticulum in their cytoplasms increase greatly (**Figure 42.8**). These increases allow the cells to synthesize and secrete large amounts of antibody proteins—up to 2,000

per second! *All the plasma cells arising from a given B cell produce antibodies that are specific for the antigen that originally bound to the parent B cell.* Thus antibody specificity is maintained as B cells proliferate.

Different antibodies share a common structure

Antibodies belong to a class of proteins called **immunoglobulins**. There are several types of immunoglobulins, but all contain a tetramer consisting of four polypeptide chains (**Figure 42.9**). In each immunoglobulin molecule, two of these polypeptides are identical *light chains*, and two are identical *heavy chains*.

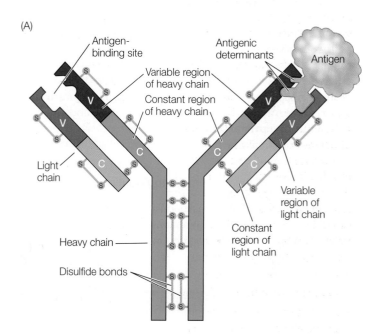

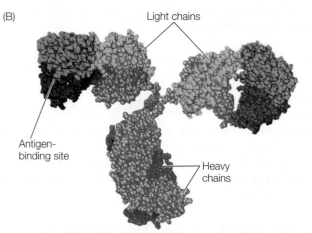

42.9 The Structure of An Immunoglobulin Four polypeptide chains (two light, two heavy) make up an immunoglobulin molecule. Here we show both diagrammatic (A) and space-filling (B) representations of immunoglobulin.

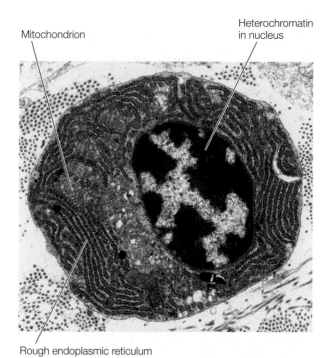

42.8 A Plasma Cell A prominent nucleus with a large amount of heterochromatin (dark brown) and a cytoplasm crowded with rough endoplasmic reticulum (purple) are features of a cell that is actively synthesizing and exporting proteins. In this case, a B cell is making a specific antibody. Whole blocks of genes not needed for this specialized function are kept inactive in the heterochromatin.

Disulfide bonds hold the chains together. Each polypeptide chain has a constant region and a variable region:

- The amino acid sequences of the **constant regions** are similar among the immunoglobulins. They determine the destination and function—the class—of each immunoglobulin.

- The amino acid sequences of the **variable regions** are different for each specific immunoglobulin. Their three-dimensional antigen-binding sites are determined by their secondary structures, and are responsible for antibody specificity.

The two antigen-binding sites on each immunoglobulin molecule are identical, making the antibody *bivalent* (*bi*, "two"; *valent*, "binding"). This ability to bind two antigen molecules at once, along with the existence of multiple epitopes on each antigen, permits antibodies to form large complexes with the antigens. For example, one antibody might bind two molecules of an antigen. Another antibody might bind the same antigen at a different epitope. It may bind one of the antigen molecules that is already bound to the first antibody, along with a third antigen molecule. A third antibody may bind one of the antigen molecules that is already part of the complex, and a fourth antigen molecule. This binding of multiple antigens and multiple antibodies can result in large complexes that are easy targets for ingestion and breakdown by phagocytes.

There are five classes of immunoglobulins

While the variable regions are responsible for the specificity of an immunoglobulin, the constant regions of the heavy chain determine the class of the immunoglobulin—for example, whether it will be an integral membrane receptor or a soluble antibody that is secreted into the bloodstream. The five immunoglobulin classes are described in **Table 42.3**. The most abundant class is IgG; these soluble antibody proteins make up about 80 percent of the total immunoglobulin content of the bloodstream. They are made in greatest quantity during a secondary immune response. IgG molecules defend the body in several ways. For example, after some IgG molecules bind to antigens, they become attached by their heavy chains to macrophages. This attachment permits the macrophages to destroy the antigens by phagocytosis.

Monoclonal antibodies have many uses

The specificity of antibodies suggested to scientists that they might be useful for detecting specific substances in the laboratory. Suppose that a physician wishes to measure the level of the hormone estrogen in a woman's blood. Because hormones are present in such minute amounts, their concentrations cannot usually be measured effectively by conventional chemical methods. A more sensitive method is needed. One way would be to add an antibody specific for estrogen to a sample of the patient's blood and observe how much antigen–antibody complex formed.

However, the immune response to a complex antigen is polyclonal—that is, most antigens carry many different antigenic determinants and will produce a complex mixture of antibodies, each made by a different clone of B cells. Furthermore, as emphasized in our study of biochemistry, many biological molecules share regions of similar structure—all human steroid hormones, for example, have a similar multi-ring structure (see Fig-

		TABLE 42.3		
		Antibody Classes		
CLASS	**GENERAL STRUCTURE**		**LOCATION**	**FUNCTION**
IgG	Monomer		Free in blood plasma; about 80 percent of circulating antibodies	Most abundant antibody in primary and secondary immune responses; crosses placenta and provides passive immunization to fetus
IgM	Pentamer		Surface of B cell; free in blood plasma	Antigen receptor on B cell membrane; first class of antibodies released by B cells during primary response
IgD	Monomer		Surface of B cell	Cell surface receptor of mature B cell; important in B cell activation
IgA	Dimer		Saliva, tears, milk, and other body secretions	Protects mucosal surfaces; prevents attachment of pathogens to epithelial cells
IgE	Monomer		Secreted by plasma cells in skin and tissues lining gastrointestinal and respiratory tracts	Binds to mast cells and basophils to sensitize them to subsequent binding of antigen, which triggers release of histamine that contributes to inflammation and some allergic responses

ure 3.22). A polyclonal group of antibodies targeted to estrogen might be uninformative because some of the antibodies would bind to any steroid hormone present in the blood sample. More useful would be a clone of B cells that produce large amounts of an antibody that binds to only one specific epitope—a **monoclonal antibody**. How could such a clone be produced?

A clone of cells that produce a single antibody can be made artificially by fusing a single B cell (which has a finite lifetime and makes a lot of antibody) with a tumor cell (which has an infinite lifetime and can be grown in culture). The resulting hybrid cell, called a **hybridoma**, makes a specific monoclonal antibody and proliferates in culture (**Figure 42.10**). In order to produce useful monoclonal antibodies by this approach, the hybridoma clones must be screened to isolate those producing antibodies that are specific for the target molecule. Recently, new methods have been developed for producing monoclonal antibodies using recombinant DNA technology. These approaches do not require the injection of live animals with antigen, the isolation of their B cells, or the production and screening of hybridomas.

Monoclonal antibodies have many applications:

- *Immunoassays* use monoclonal antibodies to detect tiny amounts of molecules in tissues and fluids. For example, this technique is used in pregnancy tests to detect human chorionic gonadotropin, the hormone made by the developing embryo.

- *Immunotherapy* uses monoclonal antibodies targeted against antigens on the surfaces of cancer cells. The coupling of a radioactive ligand or toxin to the antibody makes it into a medical "smart bomb." In a related approach, binding of the antibody itself is enough to trigger a cellular immune response that destroys the cancer. This is the case with trastuzumab (Herceptin®), a monoclonal antibody that binds to a growth factor receptor on some breast cancer cells.

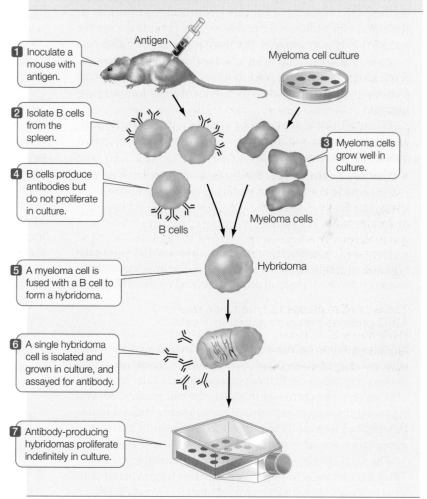

TOOLS FOR INVESTIGATING LIFE

42.10 Creating Hybridomas for the Production of Monoclonal Antibodies
Cancerous myeloma cells and normal B cells can be fused so that the proliferative properties of the myeloma cells are combined with the ability to produce specific antibodies.

1 Inoculate a mouse with antigen.

Antigen

Myeloma cell culture

2 Isolate B cells from the spleen.

3 Myeloma cells grow well in culture.

4 B cells produce antibodies but do not proliferate in culture.

B cells

Myeloma cells

5 A myeloma cell is fused with a B cell to form a hybridoma.

Hybridoma

6 A single hybridoma cell is isolated and grown in culture, and assayed for antibody.

7 Antibody-producing hybridomas proliferate indefinitely in culture.

42.4 RECAP

The humoral immune response is based on the synthesis by B cells of specific antibodies directed against antigens. The specificity of an antibody derives from the amino acid sequence of its variable regions. Monoclonal antibodies are specific to one epitope on an antigen and can be produced artificially for use in diagnostics and therapy.

- How does a B cell respond to an antigen? **See p. 885**

- How is the structure related to the function of an antibody molecule? **See pp. 885–886 and Figure 42.9**

- What are monoclonal antibodies? How are they used? **See pp. 886–887 and Figure 42.10**

By making antibodies, B cells are the major players in the humoral immune response. We now turn to the cellular immune response, where T cells are active at all stages.

42.5 What Is the Cellular Immune Response?

Two types of effector T cells (T-helper cells and cytotoxic T cells) are involved in the cellular immune response, along with proteins of the major histocompatibility complex, the MHC proteins, which underlie the immune system's tolerance for the body's own cells.

yourBioPortal.com
GO TO Animated Tutorial 42.4 • Cellular Immune Response

T cell receptors bind to antigens on cell surfaces

Like B cells, T cells possess specific membrane receptors. The T cell receptor is not an immunoglobulin, however, but a glycoprotein with a molecular weight of about half that of an IgG. It is made up of two polypeptide chains, each encoded by a separate gene (**Figure 42.11**). Thus the two chains have distinct regions with constant and variable amino acid sequences. As in the immunoglobulins, the variable regions provide the site for specific binding to antigens. But there is one major difference: whereas antibodies bind to an intact antigen such as a protein, T cell receptors bind to a piece of an antigen, such as a peptide from the protein displayed on the surface of an antigen-presenting cell.

When a T cell is activated by contact with a specific antigen, it proliferates and forms a clone. Its descendants differentiate into two types of effector T cells:

- **Cytotoxic T cells**, or T_C cells, recognize virus-infected or mutated cells and kill them by inducing lysis (see page 890).

- **T-helper cells** (T_H cells, also called helper T cells) assist both the cellular and the humoral immune responses.

How do T lymphocytes recognize surface antigens on cells without attacking the body's own cells? The answers to these questions involve a group of proteins encoded by the *MHC* genes.

MHC proteins present antigen to T cells

Since recognition between biological molecules involves three-dimensional structure (think of enzymes, receptors, and antibodies), it makes sense that recognition at the cellular level must also involve structure—in this case, the cell surface. Several types of mammalian cell surface proteins are involved in this process, but we will focus here on the products of a cluster of genes called the *major histocompatibility complex*, or MHC.

The MHC proteins are plasma membrane glycoproteins. Their major role is to present antigens or fragments of antigens to a T cell receptor in such a way that it can distinguish be-

TABLE 42.4				
The Interaction between T Cells and Antigen-Presenting Cells				
PRESENTING CELL TYPE	ANTIGEN PRESENTED	MHC CLASS	T CELL TYPE	T CELL SURFACE PROTEIN
Any cell	Intracellular protein fragment	Class I	Cytotoxic T cell (T_C)	CD8
Macrophages and B cells	Fragments from extracellular proteins	Class II	Helper T cell (T_H)	CD4

tween self and nonself antigens. In humans, there are three genetic loci for MHC class I proteins and three for MHC class II proteins. Each of these six loci has as many as 100 different alleles. With so many possible allele combinations, it is not surprising that different people are very likely to have different MHC genotypes. So the unique three-dimensional structure of the MHC protein is important in self-tolerance: the T cell receptor binds to the MHC and the antigen together.

There are two classes of MHC proteins. Both function to present antigens to the different T lymphocytes:

- **Class I MHC** proteins are present on the surface of every nucleated cell in the animal body. When cellular proteins are degraded into small peptide fragments by a proteasome (see Figure 16.24), an MHC I protein may bind to a fragment and carry it to the plasma membrane. There, the MHC I protein is oriented in such a way that the bound fragment is exposed to the outside of the cell, where both the MHC and the fragment can interact with T_C cells. The T_C cells have a surface protein called CD8 that recognizes and binds to MHC I.

- **Class II MHC** proteins are found mostly on the surfaces of B cells, macrophages, and and other antigen-presenting cells (see Figure 42.2). When one of these cells ingests a nonself antigen such as a virus, the antigen is broken down in a phagosome. An MHC II molecule may bind to one of the fragments and carry it to the cell surface, where it is presented to a T_H cell (**Figure 42.12**). T_H cells have a surface protein called CD4 that recognizes and binds to MHC II.

The information on MHC proteins, the cellular origins of antigens, and T lymphocytes is summarized in **Table 42.4**. To accomplish their role in antigen presentation, both MHC I and MHC II proteins have an antigen-binding site, which can hold a peptide of about 10–20 amino acids. The T cell receptor recognizes not just the antigenic fragment, but the MHC I or MHC II molecule to which the fragment is bound.

MHC proteins play a vital role in the selection of T cells during their development:

1. *Binding to self MHC molecules.* The ultimate goal is for the T cell receptor to bind not to the antigen alone but to the antigen–MHC complex. Here, there is positive selection for T

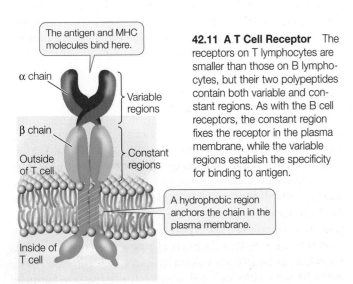

The antigen and MHC molecules bind here.

α chain

Variable regions

β chain

Constant regions

Outside of T cell

A hydrophobic region anchors the chain in the plasma membrane.

Inside of T cell

42.11 A T Cell Receptor The receptors on T lymphocytes are smaller than those on B lymphocytes, but their two polypeptides contain both variable and constant regions. As with the B cell receptors, the constant region fixes the receptor in the plasma membrane, while the variable regions establish the specificity for binding to antigen.

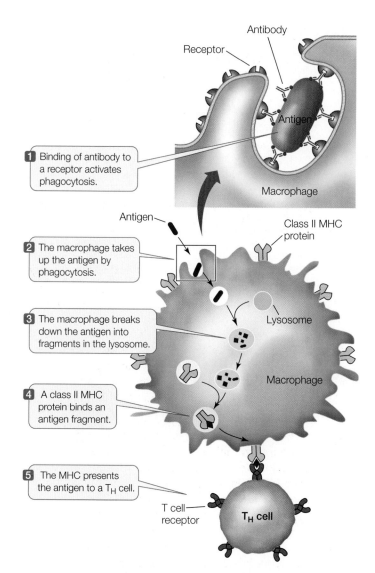

1 Binding of antibody to a receptor activates phagocytosis.

Antibody

Receptor

Antigen

Macrophage

Antigen

2 The macrophage takes up the antigen by phagocytosis.

Class II MHC protein

3 The macrophage breaks down the antigen into fragments in the lysosome.

Lysosome

4 A class II MHC protein binds an antigen fragment.

Macrophage

5 The MHC presents the antigen to a T_H cell.

T cell receptor

T_H cell

42.12 Macrophages Are Antigen-Presenting Cells A fragment of an antigen is displayed by MHC II on the surface of a macrophage. T cell receptors on a specific T-helper cell can then bind to and interact further with the antigen–MHC II complex.

cells that bind. Any T cells that do not recognize self MHC (do not bind to the antigen-presenting cell) are eliminated soon after they develop; the rest of the T cells go on to the next selection step.

2. *Binding to self peptides bound to self MHC.* Here, there is negative selection for T cells that bind, thereby eliminating the further production of T cells that react to self antigens. Both of these selection events occur in the thymus gland (see Figure 41.6).

T-helper cells and MHC II proteins contribute to the humoral immune response

When a T_H cell survives the selection processes and binds to an antigen-presenting macrophage, it releases cytokines, which ac-

tivate the T_H cell to proliferate, producing a clone of T_H cells with the same specificity. The steps to this point constitute the activation phase of the humoral immune response, and they occur in the lymphoid tissues. Next comes the effector phase, in which the T_H cells activate naïve B cells with the same specificity to produce antibodies.

B cells are also antigen-presenting cells. B cells take up antigens bound to their surface immunoglobulin receptors by endocytosis, break them down, and display antigenic fragments on class II MHC proteins. When a T_H cell binds to the displayed antigen–MHC II complex, it releases cytokines that cause the B cell to produce a clone of plasma cells (**Figure 42.13A**). Finally, the plasma cells secrete antibodies, completing the effector phase of the humoral immune response.

Cytotoxic T cells and MHC I proteins contribute to the cellular immune response

Class I MHC proteins play a role in the cellular immune response that is similar to the role played by class II MHC proteins in the humoral immune response. In a virus-infected or mutated cell, foreign or abnormal proteins or peptide fragments combine with MHC I molecules. The resulting complex is displayed on the cell surface and presented to T_C cells. When a T_C cell recognizes and binds to this antigen–MHC I complex, it is activated to proliferate (**Figure 42.13B**).

In the effector phase of the cellular immune response, T_C cells recognize and bind to cells bearing the same antigen–MHC I complex. These bound T_C cells produce a substance called *perforin*, which lyses the bound target cell. In addition, the T_C cells can bind to a specific receptor (called Fas) on the target cell that initiates apoptosis in that cell. These two mechanisms, cell lysis and programmed cell death, work in concert to eliminate the antigen-containing host cell. Because T_C cells recognize self MHC proteins complexed with nonself antigens, they help rid the body of its own virus-infected cells.

In addition to the binding of an antigen–MHC complex to its receptors, a T cell must receive a second signal for activation. This *co-stimulatory* signal occurs after the initial specific binding. It involves the interaction between another receptor on the T cell surface and a protein called B7 on certain antigen-presenting cells. This second binding event leads to T cell activation, including cytokine production and proliferation.

Regulatory T cells suppress the humoral and cellular immune responses

As we describe in Chapter 40, *homeostasis*—the maintenance of internal constancy—is a hallmark of animal biology. A third class of T cells called **regulatory T cells (Tregs)** ensures that the immune response does not spiral out of control. Like T_H and T_C cells, Tregs are made in the thymus gland, express the T cell receptor, and become activated if they bind to antigen–MHC complexes. But Tregs are different in one important way: the antigens that Tregs recognize are *self antigens*. The activation of Tregs causes them to secrete the cytokine *interleukin 10*, which blocks T cell activation and leads to apopto-

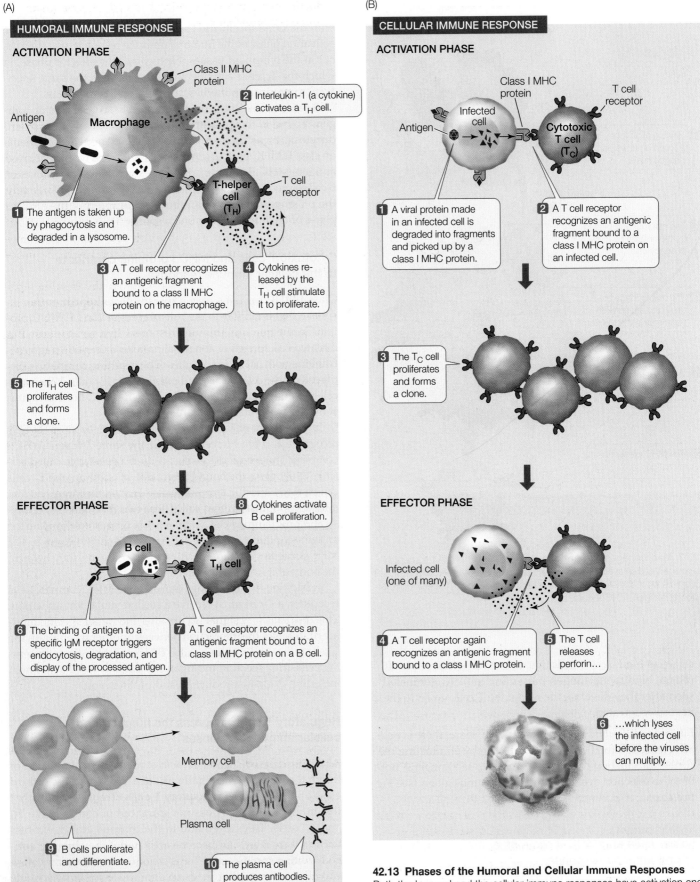

42.13 Phases of the Humoral and Cellular Immune Responses
Both the humoral and the cellular immune responses have activation and effector phases, all of which involve T cells.

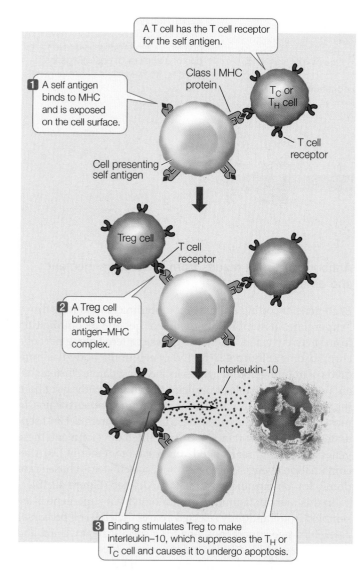

A T cell has the T cell receptor for the self antigen.

Class I MHC protein

T_C or T_H cell

T cell receptor

1 A self antigen binds to MHC and is exposed on the cell surface.

Cell presenting self antigen

Treg cell

T cell receptor

2 A Treg cell binds to the antigen–MHC complex.

Interleukin-10

3 Binding stimulates Treg to make interleukin–10, which suppresses the T_H or T_C cell and causes it to undergo apoptosis.

42.14 Tregs and Tolerance A special class of T cells called regulatory T cells (Tregs) inhibit the activation of the immune system in response to self antigens.

sis of the T_C and T_H cells that are bound to the same antigen-presenting cell (**Figure 42.14**).

The important role of Tregs is to maintain homeostasis by mediating tolerance to "self" antigens. Thus they constitute one of the mechanisms for distinguishing self from nonself. How do we know this? There are two lines of evidence for the role of Tregs. As in many other biological studies, the cause-and-effect relationships were worked out using experimental manipulations and genetics:

- If Tregs are experimentally destroyed in the thymus of a mouse, the mouse grows up with an out-of-control immune system, mounting strong immune responses to self antigens (*autoimmunity*—see Section 42.7).

- In humans, a rare X-linked inherited disease occurs when a gene critical to Treg function is mutated. An infant with this disease, called IPEX (*immune dysregulation, polyen-*

docrinopathy and *e*nteropathy, X-linked), mounts an immune response that attacks the pancreas, thyroid, and intestine. Most affected individuals die within the first few years of life.

MHC proteins are important in tissue transplants

In humans, one consequence of the major histocompatibility complex became important with the development of organ transplant surgery. Because the proteins produced by the MHC are specific to each individual, they act as nonself antigens if transplanted into another individual. An organ or a piece of tissue transplanted from one person to another is recognized as nonself by the host body and soon provokes an immune response; the tissue is then killed, or "rejected," by the host's cellular immune system. But if the transplant is performed immediately after birth, or if it comes from a genetically identical person (an identical twin), the material is recognized as self and is not rejected.

The rejection problem can be overcome by treating a patient with a drug, such as cyclosporin, that suppresses the immune system. Cyclosporin blocks the activation of a transcription factor that is essential for T cell development. However, this approach compromises the ability of transplant recipients to defend themselves against pathogens. This problem must be managed by the use of antibiotics and other drugs to combat infections that develop.

42.5 RECAP

The cellular immune response acts against antigens expressed on the surfaces of virus-infected or mutated body cells. Specific receptors on T cells bind to antigen–MHC complexes displayed on the cell surface. During development, T cells are selected that recognize self MHC proteins. Tregs suppress immune responses to self antigens.

- What are the roles of a T cell receptor in cellular immunity? See p. 888

- What parts do MHC proteins play in the cellular immune response? See pp. 888–889

- What occurs during the cellular immune response to a virus-infected cell? See p. 889 and Figure 42.13

We have alluded to genes that encode various components of the immune system and to the tremendous diversity in these components. We will now consider the genetic mechanisms that make this diversity possible.

42.6 How Do Animals Make So Many Different Antibodies?

Each mature B cell makes one—and only one—specific antibody with a specific amino acid sequence targeted to a single antigen. As we have seen, there are millions of possible epitopes to which a human is exposed or can be exposed. With millions of possible amino acid sequences in immunoglobu-

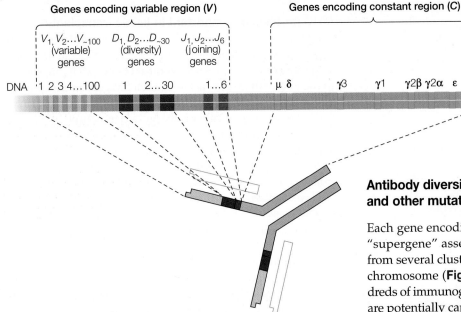

The **variable region** for the heavy chain of a specific antibody is encoded by one *V* gene, one *D* gene, and one *J* gene. Each of these genes is taken from a pool of like genes.

The **constant region** is selected from another pool of genes. The number of possible combinations to make an immunoglobulin heavy chain from these pools of genes is (100 *V*)(30 *D*)(6 *J*)(8 *C*) = 144,000.

42.15 Heavy-Chain Genes Mouse immunoglobulin heavy chains have four domains, each of which is coded for by one of a number of possible genes selected from a cluster of similar genes.

lins, the molecular genetic explanation would be that there are millions of genes, each one coding for one antibody molecule. A simple calculation using approximate numbers shows that this is impossible:

$$\frac{\text{one immunoglobulin}}{\text{heavy chain}} = \frac{500 \text{ amino acids coded}}{\text{for by } 1500 \text{ bp DNA}}$$

$$\frac{\text{one immunoglobulin}}{\text{light chain}} = \frac{200 \text{ amino acids coded}}{\text{for by } 600 \text{ bp DNA}}$$

therefore,

$$\frac{\text{one antibody (2 each identical light}}{\text{and heavy chains)}} = 2100 \text{ bp DNA}$$

$$10 \text{ million different antibodies} = 21 \text{ billion bp DNA}$$

This is 7 times the size of the entire human genome! There must be another way to generate antibody diversity.

It turns out that instead of a single gene encoding each immunoglobulin, the genome of the differentiating B cell has a limited number of alleles for several *regions* or *domains* of the protein, and that *combinations of these alleles* generate diversity. In this section we will describe the unusual process of shuffling this genetic deck to generate the enormous immunological diversity that characterizes each individual mammal. Then we will see how similar events produce five classes of immunoglobulins with different cellular locations or functions in the body.

━━━━━━ **yourBioPortal.com** ━━━━━━

GO TO Animated Tutorial 42.5 • A B Cell Builds an Antibody

Antibody diversity results from DNA rearrangement and other mutations

Each gene encoding an immunoglobulin chain is in reality a "supergene" assembled by means of genetic recombination from several clusters of smaller genes scattered along part of a chromosome (**Figure 42.15**). Every cell in the body has hundreds of immunoglobulin genes located in separate clusters that are potentially capable of participating in the synthesis of both the variable and constant regions of immunoglobulin chains. In most body cells and tissues, these genes remain intact and separated from one another. But during B cell development, these genes are cut out, rearranged, and joined together in DNA recombination events. One gene from each cluster is chosen randomly for joining, and the others are deleted (**Figure 42.16**).

In this manner, a unique immunoglobulin supergene is assembled from randomly selected "parts." Each B cell precursor assembles two supergenes, one for a specific heavy chain and the other, assembled independently, for a specific light chain. This remarkable example of irreversible cell differentiation generates an enormous diversity of immunoglobulins from the same starting genome. It is a major exception to the generalization that all somatic cells derived from the fertilized egg have identical DNA.

In both humans and mice, the gene clusters encoding immunoglobulin heavy chains are on one pair of chromosomes and those for light chains are on two others. Two families of genes encode the variable region of the light chain, and three families encode the variable region of the heavy chain.

Figure 42.15 illustrates the gene families that encode the constant and variable regions of the heavy chain in mice. There are multiple genes that encode each of three parts of the variable region: 100 *V*, 30 *D*, and 6 *J* genes. Each B cell randomly selects one gene from each of these clusters to make the final coding sequence (*VDJ*) of the heavy-chain variable region. So the number of different heavy chains that can be made through this random recombination process is quite large:

$$100 \ V \times 30 \ D \times 6 \ J = 18{,}000 \text{ possible combinations}$$

Now consider that the light chains are similarly constructed, with a similar amount of diversity made possible by random recombination. If we assume that the degree of potential light-

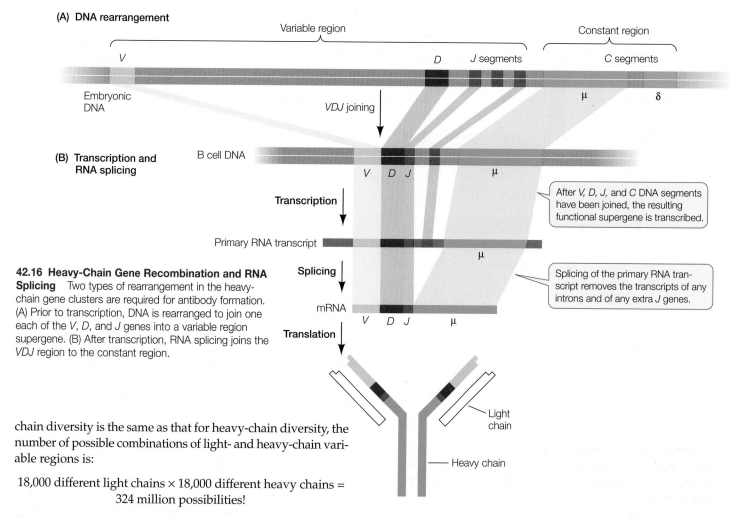

42.16 Heavy-Chain Gene Recombination and RNA Splicing Two types of rearrangement in the heavy-chain gene clusters are required for antibody formation. (A) Prior to transcription, DNA is rearranged to join one each of the *V*, *D*, and *J* genes into a variable region supergene. (B) After transcription, RNA splicing joins the *VDJ* region to the constant region.

chain diversity is the same as that for heavy-chain diversity, the number of possible combinations of light- and heavy-chain variable regions is:

18,000 different light chains × 18,000 different heavy chains = 324 million possibilities!

There are other mechanisms that generate even more diversity:

- When the DNA sequences that encode the *V*, *D*, and *J* regions are rearranged so that they are next to one another, the recombination event is not precise, and errors occur at the junctions. This *imprecise recombination* can create frameshift mutations, generating new codons at the junctions, with resulting amino acid changes.

- After the DNA sequences are cut out and before they are joined, the enzyme *terminal transferase* often adds some nucleotides to the free ends of the DNA pieces. These additional bases create insertion mutations.

- There is a relatively high *spontaneous mutation rate* in immunoglobulin genes. Once again, this process creates many new alleles and adds to antibody diversity.

When we include these possibilities with the millions of combinations that can be made by random DNA rearrangements, it is not surprising that the immune system can mount a response to almost any natural or artificial substance.

Once the pretranscriptional processing is completed, each supergene is transcribed and then translated to produce an immunoglobulin light chain or heavy chain. These combine to form an active immunoglobulin protein.

This genetic system is capable of still other kinds of changes. The B cell or plasma cell can switch the immunoglobulin class it produces while retaining its antigen specificity.

The constant region is involved in immunoglobulin class switching

Table 42.3 describes the different classes of immunoglobulins and their functions. Generally, a B cell makes only one class at a time. But **class switching** can occur, in which a B cell changes the immunoglobulin class it synthesizes. For example, a B cell making IgM can switch to making IgG.

Early in its life, a B cell produces IgM molecules, which are the receptors responsible for its recognition of a specific antigen. At this time, the constant region of the heavy chain is encoded by the first constant region gene, the μ gene (see Figure 42.16). If the B cell later becomes a plasma cell during a humoral immune response, another deletion occurs in the cell's DNA, positioning the variable region genes (consisting of the same *V*, *D*, and *J* genes) next to a constant region gene farther away on the original DNA molecule (**Figure 42.17**). Such a DNA deletion results in the production of a new immunoglobulin with a different constant region of the heavy chain, and therefore a different function (see Table 42.3). However, this immunoglobulin has the same variable regions, and therefore the same antigen specificity, as the IgM produced by the parent B cell. The new immunoglobulin protein falls into one of the four classes (IgA, IgD, IgE, or IgG), depending on which of the constant region genes is placed adjacent to the variable region genes.

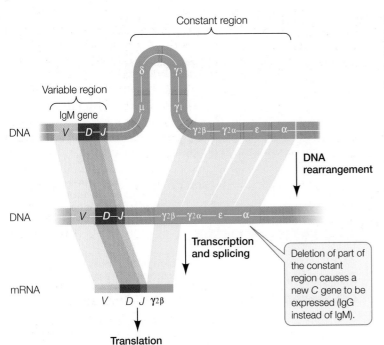

42.17 Class Switching: Exchanging C Regions The supergene produced by joining *V*, *D*, *J*, and *C* genes (see Figure 42.16) may later be modified, causing a different *C* region to be transcribed. This modification, known as class switching, is accomplished by deletion of part of the constant region gene cluster. Shown here is class switching from IgM to IgG.

What triggers class switching, and what determines the class to which a given B cell will switch? T-helper (T_H) cells direct the course of an immune response and determine the nature of the attack on the antigen. These T cells induce class switching by sending cytokine signals. The cytokines bind to receptors on the target B cells, generating signal transduction cascades that result in recombination and altered expression of the immunoglobulin genes.

42.6 RECAP

The immune system can make millions of immunoglobulins with different specificities by rearranging the genes that encode the variable regions of the heavy and light chains. Additional mechanisms create mutations that provide for further diversity in the immunoglobulins. The class of the immunoglobulin molecule can be changed by recombination of the genes coding for the constant region of the heavy chain.

- How can millions of antibodies with different specificities be generated from a relatively small number of genes? **See pp. 892–893 and Figures 42.15 and 42.16**

- What is the role of the constant region of the immunoglobulin in class switching? **See pp. 893–894 and Figure 42.17**

Given the numerous and complex cellular interactions that activate the immune system and generate antibody diversity, you may have perceived many points at which the immune system could fail. We now turn to several situations in which one or more components of this complex system malfunction.

42.7 What Happens When the Immune System Malfunctions?

Sometimes the immune system fails us in one way or another. It may overreact, as in an allergic reaction; it may attack self antigens, as in an autoimmune disease; or it may function weakly or not at all, as in an immune deficiency disease.

Allergic reactions result from hypersensitivity

An **allergic reaction** arises when the human immune system overreacts to (is hypersensitive to) a dose of antigen. Although the antigen itself may present no danger to the host, the inappropriate immune response may produce inflammation and other symptoms, which can cause serious illness or even death. Allergic reactions are the most familiar examples of this phenomenon. Two types of allergic reactions involve immediate hypersensitivity and delayed hypersensitivity.

IMMEDIATE HYPERSENSITIVITY **Immediate hypersensitivity** arises when an allergic individual is exposed to an antigen (in this case referred to as an allergen) from the environment, such as a food, pollen, or the venom of an insect. In response to the allergen, the individual makes large amounts of IgE. When this happens, mast cells in tissues and basophils in the blood bind the constant end of the IgE. If that individual is exposed to the same allergen again, binding of the allergen to the IgE causes the mast cells and basophils to rapidly release a large amount of histamine (**Figure 42.18**). This results in symptoms such as dilation of blood vessels, inflammation, and difficulty breathing. If not treated with antihistamines, a severe allergic reaction can lead to death. It is not known why some people produce excessive amounts of IgE in response to allergens. There is some evidence for genetic factors predisposing people to allergic responses.

Allergy to pollen can be treated using a process called *desensitization*. The process involves injecting small amounts of the allergen (typically just an extract of the offending plant tissue) into the skin—enough to stimulate IgG production but not enough to stimulate IgE production. The next time the person is exposed to the allergen, IgG binds to it, tying it up before IgE can bind it and exert its harmful effects.

Desensitization does not work for food allergens because the IgE response to those substances is so strong that even a small amount of antigen provokes it. The best approach for those with food allergies—there are an estimated 3 million in the U.S.—is to avoid foods containing the allergens. This can be difficult, but food labels listing all the ingredients are helpful. Molecular biologists are beginning to identify the antigens that act as allergens, with the hope of developing vaccines or genetically modified foods that lack the allergenic epitopes.

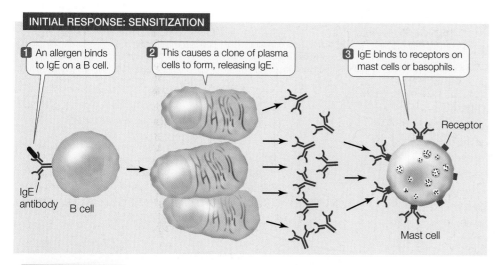

INITIAL RESPONSE: SENSITIZATION

1 An allergen binds to IgE on a B cell.

2 This causes a clone of plasma cells to form, releasing IgE.

3 IgE binds to receptors on mast cells or basophils.

IgE antibody

B cell

Receptor

Mast cell

42.18 An Allergic Reaction An allergen is an antigen that stimulates B cells to make large amounts of IgE antibodies, which bind to mast cells and basophils. When the body encounters the allergen again, these cells produce large amounts of histamine, which have harmful physiological effects.

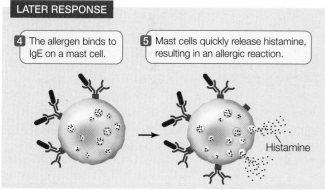

LATER RESPONSE

4 The allergen binds to IgE on a mast cell.

5 Mast cells quickly release histamine, resulting in an allergic reaction.

Histamine

DELAYED HYPERSENSITIVITY **Delayed hypersensitivity** is an allergic reaction that does not begin until hours after exposure to an antigen. In this case, the antigen is taken up by antigen-presenting cells and a T cell response is initiated. An example is the rash that develops after exposure to poison ivy.

Autoimmune diseases are caused by reactions against self antigens

Errors in the selection of T cells in the thymus can result in T cells that bind to antigen–MHC complexes that carry self antigens. Although the precise origin of **autoimmunity** is not known, there are several hypotheses:

- *Failure of clonal deletion.* A clone of lymphocytes making antibodies against self antigens that should have been destroyed by Tregs and apoptosis is not destroyed.

- *Molecular mimicry.* T cells that recognize a nonself antigen, such as a virus, also recognize something on the self antigen that has a similar structure.

Autoimmunity does not always result in disease, but a number of autoimmune diseases are common:

- People with *systemic lupus erythematosis* (SLE) have antibodies to many cellular components, including DNA and nuclear proteins. These antinuclear antibodies can cause serious damage when they bind to normal tissue antigens and

form large circulating antigen–antibody complexes, which become stuck in tissues and provoke inflammation.

- People with *rheumatoid arthritis* have difficulty in shutting down a T cell response. An inhibitory protein called CTLA4 blocks T cells from reacting to self antigens. People with rheumatoid arthritis may have low CTLA4 activity, which results in inflammation of the joints due to the infiltration of excess white blood cells.

- *Hashimoto's thyroiditis* is the most common autoimmune disease in women over 50. Immune cells attack thyroid tissue, resulting in fatigue, depression, weight gain, and other symptoms.

- *Insulin-dependent diabetes mellitus*, or type I diabetes, occurs most often in children. It is caused by an immune reaction against several proteins in the cells of the pancreas that manufacture the protein hormone insulin. This reaction kills the insulin-producing cells, so people with type I diabetes must take insulin daily in order to survive.

AIDS is an immune deficiency disorder

There are a number of inherited and acquired *immune deficiency disorders*. In some individuals, T or B cells never form; in others, B cells lose the ability to give rise to plasma cells. In either case, the affected individual is unable to mount an immune response and thus lacks a major line of defense against pathogens. The T_H cell is perhaps the most central component of the immune system because of its essential roles in both the humoral and cellular immune responses (see Figure 42.6). This cell is the target of **human immunodeficiency virus** (**HIV**), the retrovirus that results in **acquired immune deficiency syndrome** (**AIDS**).

HIV can be transmitted from person to person in body fluids containing the virus (such as blood, semen, vaginal fluid, or breast milk). The recipient tissue is either blood (by transfusion) or a mucous membrane lining an organ (the mucus contains a high concentration of lymphocytes). HIV initially infects macrophages, T_H cells, and antigen-presenting dendritic cells in the blood and tissues. At first there is an immune response to the viral infection, and T_H cells are activated. But because HIV

42.19 The Course of an HIV Infection An HIV infection may be carried, unsuspected, for many years before the onset of symptoms.

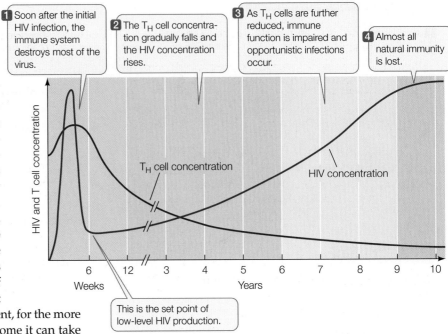

1 Soon after the initial HIV infection, the immune system destroys most of the virus.

2 The T_H cell concentration gradually falls and the HIV concentration rises.

3 As T_H cells are further reduced, immune function is impaired and opportunistic infections occur.

4 Almost all natural immunity is lost.

T_H cell concentration

HIV concentration

HIV and T cell concentration

This is the set point of low-level HIV production.

infects the T_H cells, they are killed both by HIV itself and by T_C cells that lyse infected T_H cells. Consequently, T_H cell numbers decline after the first month or so of infection. Meanwhile, the extensive production of HIV by infected cells activates the humoral immune system. Antibodies bind to HIV and the complexes are removed by phagocytes. The HIV level in blood goes down. There is still a low level of infection, however, because of the depletion of T_H cells (**Figure 42.19**). This process reaches a low, steady-state level called the "set point." This point varies among individuals and is a strong predictor of the rate of progression of the disease. For most people it takes 8–10 years, even without treatment, for the more severe manifestations of AIDS to develop. In some it can take as little as a year; in others, 20 years.

During this dormant period, people carrying HIV generally feel fine, and their T_H cell levels are adequate for them to mount immune responses. Eventually, however, the virus destroys the T_H cells, and their numbers fall to the point where the infected person is susceptible to infections that the T_H cells would normally eliminate. These infections result in conditions such as Kaposi's sarcoma, a skin tumor caused by a herpesvirus; pneumonia caused by the fungus *Pneumocystis jirovecii*; and lymphoma tumors caused by the Epstein–Barr virus. These conditions result from *opportunistic infections* because the pathogens take advantage of the crippled immune system of the host. They lead to death within a year or two.

The molecular biology of HIV and its life cycle have been intensively studied (see Figure 16.6). Drug treatments are focused on inhibiting the viral proteins, such as *reverse transcriptase* that makes cDNA from the viral RNA, and *viral protease* that cuts the large precursor viral protein into its final active proteins. Combinations of such drugs result in long-term survival. Unfortunately, like many medical treatments, HIV drugs are not available to all who need them—particularly in poor regions of the world where AIDS is prevalent. As a result, there are about 2 million deaths per year worldwide from AIDS.

42.7 RECAP

Failures of the immune system include allergic reactions (caused by hypersensitivity to antigens), autoimmune diseases (caused by reactions against self antigens), and immune deficiency disorders.

- How does immediate hypersensitivity develop? See p. 894 and Figure 42.18

- What is an autoimmune disease? Give an example. See p. 895

- Describe the course of events in the human immune system during HIV infection. See pp. 895–896 and Figure 42.19

CHAPTER SUMMARY

42.1 What Are the Major Defense Systems of Animals?

- Animal defenses against pathogens are based on the body's ability to distinguish between self and nonself.
- **Nonspecific (innate) defenses** are inherited mechanisms that protect the body from many kinds of pathogens. They typically act rapidly.
- **Specific defenses** are adaptive mechanisms that respond to specific pathogens. They develop more slowly than nonspecific defenses but are long-lasting.

- Many defenses are implemented by cells and proteins carried in the blood plasma and lymph. **Review Figure 42.1, WEB ACTIVITY 42.1**
- White blood cells fall into two broad groups. **Phagocytes** include **macrophages** that engulf pathogens by phagocytosis. **Lymphocytes**, which include **B cells** and **T cells**, participate in specific responses. **Review Figure 42.2, ANIMATED TUTORIAL 42.1**

42.2 What Are the Characteristics of the Nonspecific Defenses?

- An animal's nonspecific defenses include physical barriers such as the skin, and competing resident microorganisms known as **normal flora**. Review Table 42.1

- The **complement system** consists of more than 20 different antimicrobial proteins that act to alter membrane permeability and kill targeted cells.

- Circulating defensive cells, such as phagocytes and **natural killer cells**, eliminate invaders.

- The **inflammation response** activates several types of cells and proteins that act against invading pathogens. **Mast cells** release **histamines**, which cause blood vessels to dilate and become "leaky." Review Figure 42.3, WEB ACTIVITY 42.2

- A cell signaling pathway involving the **toll-like receptor** stimulates the body's defenses. Review Figure 42.4

42.3 How Does Specific Immunity Develop?

SEE ANIMATED TUTORIAL 42.2

- The specific immune response recognizes specific **antigens**, responds to an enormous diversity of **antigenic determinants**, distinguishes self from nonself, and remembers the antigens it has encountered.

- Each antibody and each T cell is specific for a single antigenic determinant. **T cell receptors** bind to antigens on the surfaces of virus-infected cells and abnormal cells.

- The **humoral immune response** is directed against pathogens in the blood, lymph, and tissue fluids. The **cellular immune response** is directed against an antigen established within a host cell. Both responses are mediated by antigenic fragments being presented on a cell surface along with the proteins of the **major histocompatibility complex**.

- **Clonal selection** accounts for the specificity and diversity of the immune response and for **immunological memory**. Review Figure 42.7

- An activated B or T lymphocyte produces **effector cells** that attack the antigen, and **memory cells** that are long-lived and rarely divide. Effector B cells are called **plasma cells** and secrete specific **antibodies**.

- **Vaccination** is inoculation with modified pathogens or antigens that provoke an immune response but are not pathogenic.

42.4 What Is the Humoral Immune Response?

SEE ANIMATED TUTORIAL 42.3

- B cells are the basis of the humoral immune response. Naïve B cells are activated by binding of antigen and by stimulation by T_H cells with the same specificity, and then form plasma cells. These cells synthesize and secrete specific antibodies.

- The basic unit of an **immunoglobulin** is a tetramer of four polypeptides: two identical light chains and two identical heavy chains, each consisting of a **constant region** and a **variable region**. Review Figure 42.9, WEB ACTIVITY 42.3

- The variable regions determine the specificity of an immunoglobulin, and the constant regions of the heavy chain determine its **class**. There are five classes of immunoglobulins with different body locations and functions. Review Table 42.3

- A **monoclonal antibody** can be made by fusing a B cell with a tumor cell to form a **hybridoma**. Review Figure 42.10

42.5 What Is the Cellular Immune Response?

SEE ANIMATED TUTORIAL 42.4

- T cells are the effectors of the cellular immune response. T cell receptors are somewhat similar in structure to the immunoglobulins, having variable and constant regions. Review Figure 42.11

- There are three types of T cells. **Cytotoxic T cells** recognize and kill virus-infected cells or mutated cells. **T-helper cells** direct both the cellular and humoral immune responses. **Regulatory T cells** inhibit the other T cells from mounting an immune response to self antigens.

- The genes of the major histocompatibility complex (MHC) encode membrane proteins that bind antigenic fragments and present them to T cells. Review Figures 42.12 and 42.13

- Organ transplants are rejected when the host's immune system recognizes MHC proteins on transplanted tissue as nonself and initiates an immune defense attacking the foreign tissue.

42.6 How Do Animals Make So Many Different Antibodies?

SEE ANIMATED TUTORIAL 42.5

- B cell genomes undergo random recombinations of genes coding for regions of the immunoglobulin polypeptide chains so that each cell can produce a specific antibody protein. The immunoglobulin chains derive from "supergenes" that are constructed from different combinations of V, D, J, and C genes. This DNA rearrangement and rejoining yields millions of different immunoglobulin chains. Review Figures 42.15 and 42.16

- Once a B cell becomes a plasma cell, it may undergo **class switching**, in which a deletion of one or more constant region genes results in the production of an immunoglobulin with a different constant region and a different function. Review Figure 42.17

42.7 What Happens When the Immune System Malfunctions?

- An **allergic reaction** is an inappropriate immune response caused by immediate or delayed **hypersensitivity** to certain antigens. Review Figure 42.18

- Autoimmune diseases result when the immune system produces B and T cells that attack self antigens.

- Immune deficiency disorders result from failure of one or another part of the immune system. **AIDS** is an acquired immune deficiency disorder arising from depletion of the T_H cells as a result of infection with HIV. Review Figure 42.19

SELF-QUIZ

1. Phagocytes kill harmful bacteria by
   a. endocytosis.
   b. producing antibodies.
   c. complement proteins.
   d. T cell stimulation.
   e. inflammation.

2. Which statement about immunoglobulins is true?
   a. They help antibodies do their job.
   b. They recognize and bind antigenic determinants.
   c. They encode some of the most important genes in an animal.
   d. They are the chief participants in nonspecific defense mechanisms.
   e. They are a specialized class of white blood cells.

3. Which statement about an antigenic determinant is *not* true?
   a. It is a specific chemical grouping.
   b. It may be part of many different molecules.
   c. It is the part of an antigen to which an antibody binds.
   d. It may be part of a cell.
   e. A single protein has only one on its surface.

4. T cell receptors
   a. are the primary receptors for the humoral immune system.
   b. are carbohydrates.
   c. cannot function unless the animal has previously encountered the antigen.
   d. are produced by plasma cells.
   e. are important in combating viral infections.

5. According to the clonal selection theory,
   a. an antibody changes its shape to match the antigen it meets.
   b. an individual animal contains only one type of B cell.
   c. an individual animal contains many types of B cells, each producing one kind of antibody.
   d. each B cell produces many types of antibodies.
   e. many clones of antiself lymphocytes appear in the bloodstream.

6. Immunological tolerance
   a. depends on exposure to an antigen.
   b. develops late in life and is usually life-threatening.
   c. disappears at birth.
   d. results from the activities of the complement system.
   e. results from DNA splicing.

7. The extraordinary diversity of antibodies results in part from
   a. the action of monoclonal antibodies.
   b. the splicing of protein molecules.
   c. the action of cytotoxic T cells.
   d. the rearrangement of genes.
   e. their remarkable nonspecificity.

8. Which of the following play(s) no role in the B cell response?
   a. T-helper cells
   b. Growth factors
   c. Macrophages
   d. Reverse transcriptase
   e. Products of class II MHC genes

9. The major histocompatibility complex
   a. codes for specific proteins found on the surfaces of cells.
   b. plays no role in T cell immunity.
   c. plays no role in antibody responses.
   d. plays no role in skin graft rejection.
   e. is encoded by a single locus with multiple alleles.

10. Which of the following is important in the immune deficiency caused by HIV?
   a. Pneumonia infection by bacteria
   b. Activated T_H cell infection by HIV
   c. Circulating free HIV in blood plasma
   d. Antibodies made against HIV-infected cells
   e. Increase in T cells

FOR DISCUSSION

1. Describe the part of an antibody molecule that interacts with an antigen. How is it similar to the active site of an enzyme? How does it differ from the active site of an enzyme?

2. Contrast immunoglobulins and T cell receptors with respect to their structures and functions.

3. Discuss the diversity of antibody specificities in an individual in relation to the diversity of enzymes. Does every cell in an animal contain genetic information for all the organism's enzymes? Does every cell contain genetic information for all the organism's immunoglobulins?

4. The gene family determining MHC on the cell surface in humans is on a single chromosome. A father's MHC type is A1, A3, B5, B7, D9, D11. A mother's genotype is A2, A4, B6, B7, D11, D12. Their child is A1, A4, B6, B7, D11, D12. What are the parents' haplotypes—that is, which alleles are linked on each of the two chromosomes of each parent? Assuming there is no recombination among the genes determining the MHC type, can these same two parents have a child who is A1, A2, B7, B8, D10, D11?

ADDITIONAL INVESTIGATION

Development of an effective HIV vaccine requires that the person being vaccinated develop both cellular and humoral immunity against HIV. Explain what this means in terms of studies you would do on people given a potential new vaccine.

Animal Reproduction

Explosive sex

Producer of valuable honey and pollinator of many crucial plants, the common honey bee, *Apis mellifera*, has been the subject of study and fascination for humans throughout recorded history. The unique sex life of these social insects is among their most intriguing aspects.

More than 99 percent of female honey bees do not reproduce. They exist to help just one female—the queen—reproduce. At any given time, there is only one queen in a hive. She lays all of the eggs, which is a full-time job. She also determines whether or not an egg will be fertilized. Fertilized eggs develop into females; unfertilized eggs develop into males. Wait, you say, it's the *male* that fertilizes the egg! Yes, the male's sperm fertilizes the egg, but in this case the female—the queen bee—stores sperm after mating and controls its delivery to her eggs.

Eventually, every hive must have a new queen. Sometimes the old queen dies. Other times, in a phenomenon known as *swarming*, the queen and a retinue of workers leave their current hive to start a new one. Whether the queen dies or leaves, a new queen must be produced. The remaining worker bees enlarge a few cells in the honeycomb that contain fertilized eggs laid by the old queen. The larvae that hatch from those eggs are fed special food that stimulates their growth and development into prospective queens.

The first queen to pupate and emerge from her royal chamber kills any other aspiring queens. She then leaves the hive for her mating flight. Males from all around get the message that a virgin queen is available and congregate around her. While in flight, she mates with 15 to 20 males, and each coupling is an event. After a male manages to insert his penis into her vagina, he literally explodes, leaving inside the female not only his sperm but also his sex organs (the latter will drop out later). This process is repeated with other males, which also die as a result. The queen returns to the hive with a lifetime supply of sperm and sets about laying eggs.

The queen lives for about 2 years and lays as many as 3,000 eggs each day. If she releases sperm, the eggs will be fertilized and develop into sterile females that will devote their lives to feeding her, maintaining the hive, foraging for nectar and pollen, and raising their sisters. Unfertilized eggs develop into males, which hang around the hive doing nothing useful until they take off to search for a receptive queen—an unlikely event. It is thus in the

A Unique Reproductive Strategy Among honey bees and other hymenopteran insects, the only reproductive female is the queen, seen here in the center as she deposits eggs in the comb. The female workers that attend her are sterile.

Swarming Means a New Queen Bee When a honey bee colony swarms, the queen leaves the hive and takes a retinue of workers with her. A new queen will emerge to take over the old hive, at which time a few males will get to perform their one brief function—fertilizing a virgin queen—before they die.

queen's best interest to limit the number of males she produces.

Natural selection has resulted in some amazing adaptations, none more so than those involved in reproduction. Sexual or asexual, bizarre or otherwise, the anatomy, physiology, biochemistry, and behavior surrounding the urge to propagate one's species are fascinating fields of study.

IN THIS CHAPTER we will examine the diverse ways in which animals produce offspring. First we examine asexual reproduction, in which only a single parent is involved. We then turn to sexual reproduction, in which an egg and a sperm unite to create a new diploid individual. Finally, we focus on the anatomy, function, and endocrine control of the human reproductive system, as well as technologies used to limit or enhance human fertility.

43.1 How Do Animals Reproduce without Sex?

Sexual reproduction is a nearly universal trait in animals, although many species can also reproduce asexually and some reproduce only asexually. Offspring produced asexually are genetically identical to one another and to their parents. Asexual reproduction is efficient because no time or energy is wasted on mating and every member of the population can convert resources into offspring. However, asexual reproduction does not generate genetic diversity as sexual reproduction does, and this diversity is the raw material that enables natural selection to shape adaptations in response to environmental change. When such changes occur, the lack of genetic diversity can be disadvantageous to a species.

A variety of animals, mostly invertebrates, reproduce asexually. They tend to be species that are attached to their substrate and cannot search for mates or that live in sparse populations and rarely encounter potential mates. Asexually reproducing species are likely to be found in relatively constant environments where genetic diversity is less important for species success. In fact, asexual reproduction is a good way to preserve a genotype that is successful in a particular environment, as long as that environment does not change. Three common modes of asexual reproduction are *budding, regeneration,* and *parthenogenesis.*

Budding and regeneration produce new individuals by mitosis

Many simple multicellular animals produce offspring by **budding**. New individuals form as outgrowths or buds from the bodies of older animals. A bud grows by mitotic cell division, and the cells differentiate before the bud breaks away from the parent (**Figure 43.1A**). The bud is genetically identical to the parent, and it may grow as large as the parent before it becomes independent.

Regeneration is usually thought of as the replacement of damaged tissues or lost limbs, but in some cases pieces of an organism can regenerate complete individuals. Echinoderms, for example, have remarkable abilities to regenerate. If sea stars (starfishes) are cut into pieces, each piece that includes an arm and a portion of the central disc can grow into a new animal (**Figure 43.1B**). In the early 1900s, oyster fishermen in Narragansett Bay tried to eliminate the sea stars that were preying on their oysters. Whenever they encountered a sea star, they chopped it up with knives and threw it back into the water. As a result, the sea star population increased explosively.

Regeneration can occur when an animal is broken by an outside force such as wave action in the intertidal zone. In some

(A) *Hydra* sp.

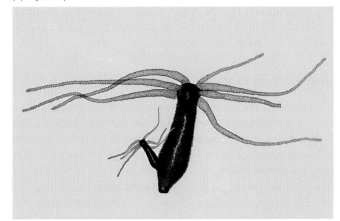

(B) *Fromia* sp.

43.1 Two Forms of Asexual Reproduction (A) Budding: A new individual forms as an outgrowth from an adult hydra. (B) Regeneration: A single severed arm and a piece of the central disc of a mature sea star can regenerate into an entire animal.

(A)

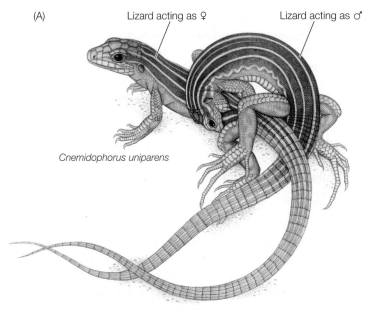

Lizard acting as ♀

Lizard acting as ♂

Cnemidophorus uniparens

(B)

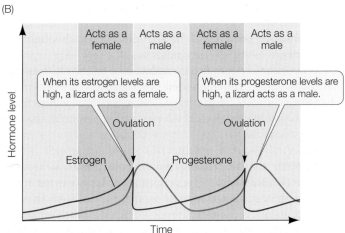

Acts as a female | Acts as a male | Acts as a female | Acts as a male

When its estrogen levels are high, a lizard acts as a female.

When its progesterone levels are high, a lizard acts as a male.

Ovulation

Ovulation

Hormone level

Estrogen

Progesterone

Time

cases, breakage occurs in the absence of external forces. Some species of segmented marine worms develop segments with rudimentary heads bearing sensory organs. The segments then break apart, and each one forms a new worm.

Parthenogenesis is the development of unfertilized eggs

Not all eggs must be fertilized to develop. A common mode of asexual reproduction in arthropods is the development of offspring from unfertilized eggs. This phenomenon, called **parthenogenesis**, also occurs in some species of fishes, amphibians, and reptiles. Most species that reproduce parthenogenetically also engage in sexual reproduction or at least sexual behavior at other times.

In some species, parthenogenesis is part of the mechanism that determines sex. As we saw at the beginning of this chapter, in honey bees (as well as in most ants and wasps), males develop from unfertilized eggs and are haploid. Females develop from fertilized eggs and are diploid.

Parthenogenetic reproduction in some species requires sexual behavior even though sperm are not delivered to the female reproductive tract and eggs are not fertilized. One case that has been investigated extensively by David Crews and his students at the University of Texas is parthenogenetic reproduction in a species of whiptail lizard. This species has no males, but females can act as males, engaging in all aspects of courtship display and mating, although no sperm are produced or transferred (**Figure 43.2**). Whether a specific female acts as a female or as a male depends on cyclical hormonal states. When estrogen levels are high, she acts as a female. When her progesterone level peaks, she acts as a male. The stimulation resulting from the sexual activity triggers the release of eggs from the ovaries.

43.2 Sexual Behavior May Be Required for Asexual Reproduction (A) Parthenogenetic whiptail lizards are all females, but they take turns acting the male role in reproductive behavior. The stimulation from sexual behavior is necessary for ovulation to occur. (B) The stage of the ovarian cycle determines the role an individual whiptail plays.

Asexual reproduction is an efficient way to use resources. However, the fact that sexual reproduction produces genetic diversity must be a tremendous advantage, because most species reproduce sexually.

43.2 How Do Animals Reproduce Sexually?

Given the efficiency of asexual reproduction in perpetuating an organism's genome, the prevalence of sexual reproduction is somewhat surprising. Even the evolution of meiosis—an extremely complicated process in comparison to mitosis—has been the subject of much speculation and debate among evolutionary biologists. And of course, mating behaviors involve costs and risks. Costs include time and energy for finding, attracting, and competing for a mate as well as the opportunity costs of detracting from other activities such as feeding and caring for existing offspring. Risks include increased exposure to predation and the potential for physical damage. Despite these disadvantages, most eukaryotic organisms reproduce sexually. Thus it would seem that the production of genetic diversity is an evolutionary advantage that overwhelms "the cost of sex" (see Section 21.4).

Sexual reproduction requires the joining of two haploid sex cells to form a diploid individual. These haploid cells, or *gametes*, are produced through **gametogenesis**, a process that involves meiotic cell divisions. Two events in meiosis contribute to genetic diversity: *crossing over* between homologous chromosomes and the *independent assortment* of chromosomes (see Sections 11.5 and 12.1). Sexual reproduction itself also contributes to genetic diversity. The genetic variation among the gametes of a single individual and the genetic variation between any two parents produce an enormous potential for genetic variation between any two offspring of a sexually reproducing pair of individuals.

Sexual reproduction in animals consists of three fundamental steps:

- *Gametogenesis*: making gametes
- *Spawning* or *mating*: bringing gametes together
- *Fertilization*: fusing gametes

The process of gametogenesis is very similar across sexually reproducing animal species. Processes of fertilization are also rather similar in widely different species. Therefore, while our discussion of gametogenesis will focus generally on mammals, and our discussion of fertilization will feature sea urchins, the facts would not be dramatically different were we to consider many other animal groups. Adaptations for spawning and mating, in contrast, show incredible anatomical, physiological, and behavioral diversity across species.

Gametogenesis produces eggs and sperm

Gametogenesis occurs in the **gonads**, which are **testes** (singular *testis*) in males and **ovaries** (singular *ovary*) in females. The tiny gametes of males, the **sperm**, move by beating their flagella. The larger gametes of females, called eggs or **ova** (singular *ovum*), are nonmotile.

Gametes are produced from **germ cells**, which have their origin in the earliest cell divisions of the embryo and remain distinct from all the other cells of the body (the *somatic cells*). Germ cells are sequestered in the body of the embryo until its gonads begin to form. The germ cells then migrate to the developing gonads, where they take up residence and proliferate by mitosis, producing **spermatogonia** (singular *spermatogonium*) in males and **oogonia** (singular *oogonium*) in females. Spermatogonia and oogonia are diploid and multiply by mitosis.

In the next step in gametogenesis, meiotic cell division reduces the chromosomes to the haploid number (see Section 11.5). The spermatogonia and oogonia that enter meiosis are **primary spermatocytes** and **primary oocytes**. Although the steps of meiosis are similar in males and females, gametogenesis differs between the sexes.

SPERMATOGENESIS The initial proliferation of male germ cells into spermatogonia proceeds by mitosis in the embryo. As illustrated in **Figure 43.3A**, primary spermatocytes then undergo the first meiotic division to form **secondary spermatocytes**. The second meiotic division produces four haploid **spermatids** for each primary spermatocyte that enters meiosis. In mammals, the progeny of primary spermatocytes remain connected by *cytoplasmic bridges* after each division.

One reason that mammalian spermatocytes remain in cytoplasmic contact throughout their development is the asymmetry of sex chromosomes in males. Half the secondary spermatocytes receive an X chromosome, the other half a Y chromosome. The Y chromosome contains fewer genes than the X chromosome, and some of the products of genes found only on the X chromosome are essential for spermatocyte development. By remaining in cytoplasmic contact, all four spermatocytes can share the gene products of the X chromosomes, although only half of them have an X chromosome.

A spermatid bears little resemblance to a mature sperm. Through further differentiation (*spermiogenesis*), the spermatid becomes compact, streamlined, and grows a flagellum to become motile. We will look at the production of human sperm in the next section.

OOGENESIS Oogonia, like spermatogonia, proliferate through mitosis (**Figure 43.3B**). The resulting primary oocytes immediately enter prophase of the first meiotic division. In many species, including humans, the oocyte experiences developmental arrest at this point and may remain in that state for days,

(A) SPERMATOGENESIS

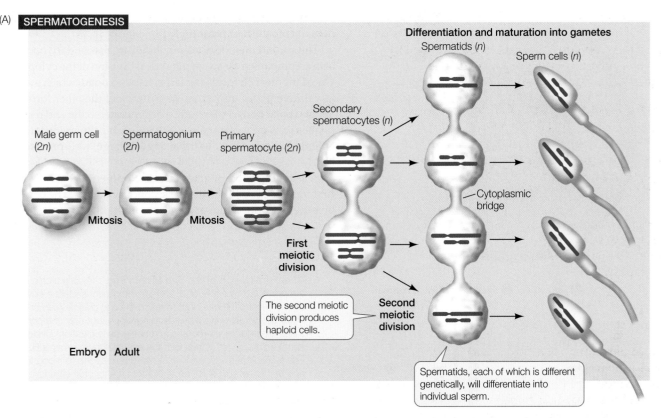

Differentiation and maturation into gametes

Spermatids (n)

Sperm cells (n)

Secondary spermatocytes (n)

Male germ cell (2n)

Spermatogonium (2n)

Primary spermatocyte (2n)

Mitosis   Mitosis

First meiotic division

Cytoplasmic bridge

The second meiotic division produces haploid cells.

Second meiotic division

Embryo   Adult

Spermatids, each of which is different genetically, will differentiate into individual sperm.

(B) OOGENESIS

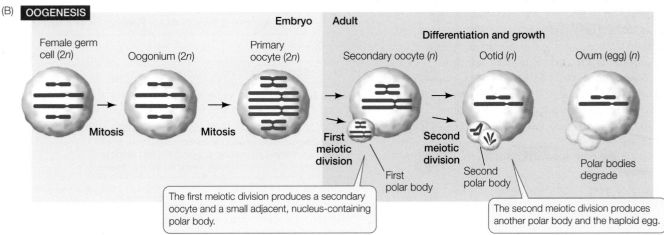

Embryo   Adult

Differentiation and growth

Female germ cell (2n)

Oogonium (2n)

Primary oocyte (2n)

Secondary oocyte (n)

Ootid (n)

Ovum (egg) (n)

Mitosis   Mitosis

First meiotic division

First polar body

Second meiotic division

Second polar body

Polar bodies degrade

The first meiotic division produces a secondary oocyte and a small adjacent, nucleus-containing polar body.

The second meiotic division produces another polar body and the haploid egg.

43.3 Gametogenesis Male and female germ cells proliferate by mitosis and produce diploid spermatogonia and oogonia that mature into primary spermatocytes and oocytes before entering meiosis. (A) Spermatogonia continue to divide by mitosis in adults, producing a steady supply of spermatocytes that divide meiotically to produce haploid spermatids, which differentiate into sperm. In many species, the progeny of spermatocytes remain in contact through cytoplasmic bridges until the sperm mature. (B) In mammals, oogonia cease division in the embryo, and primary oocytes remain arrested in prophase I of meiosis until they are ovulated and fertilized. Each oocyte will produce one haploid ootid which matures into an ovum.

months, or years. In the human female, this period of arrest is at least 10 years (i.e., until puberty), and some primary oocytes remain in prophase I for up to 50 years (i.e., until menopause). In contrast, spermatogenesis continues, uninterrupted, to completion once the primary spermatocyte has differentiated.

During this prolonged prophase I, or shortly before it ends, the primary oocyte grows larger through increased production of ribosomes, RNA, cytoplasmic organelles, and energy stores. At this point, the primary oocyte acquires all the energy, raw materials, and RNA that the ovum will need to survive its first cell divisions after fertilization. In fact, the nutrients in the egg must maintain the embryo until it is either nourished by the maternal circulatory system or can feed on its own.

When a primary oocyte resumes meiosis, its nucleus completes the first meiotic division near the surface of the cell. The daughter cells of this division receive grossly unequal shares of cytoplasm. This asymmetry represents another major difference from spermatogenesis, in which cytoplasm is apportioned equally. The daughter cell that receives almost all the cytoplasm

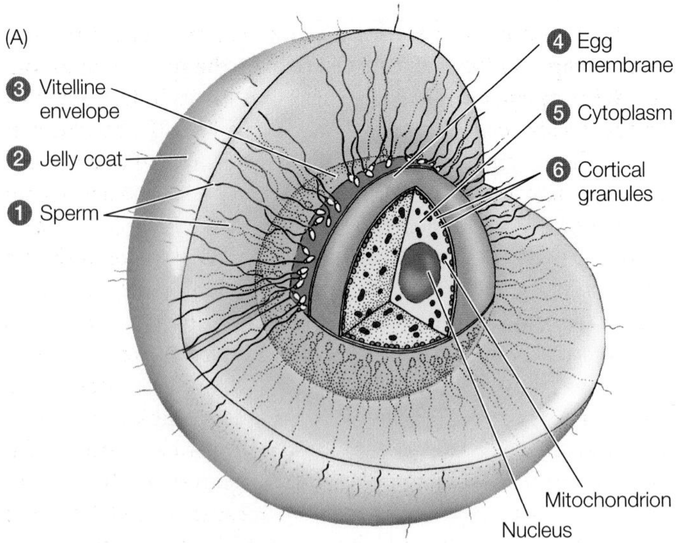

(A)

③ Vitelline envelope

② Jelly coat

① Sperm

④ Egg membrane

⑤ Cytoplasm

⑥ Cortical granules

Mitochondrion

Nucleus

becomes the **secondary oocyte**, and the one that receives almost none forms the **first polar body** (see Figure 43.3B).

The second meiotic division of the large secondary oocyte is also accompanied by an asymmetrical division of the cytoplasm. One daughter cell forms the large, haploid **ootid**, which eventually differentiates into a mature ovum, and the other forms the **second polar body**. Polar bodies degenerate, so the end result of oogenesis is only one mature egg for each primary oocyte that entered meiosis. However, that egg is a large, well-provisioned cell.

A second period of arrested development occurs after the first meiotic division forms the secondary oocyte. The egg may be expelled from the ovary in this condition. In many species,

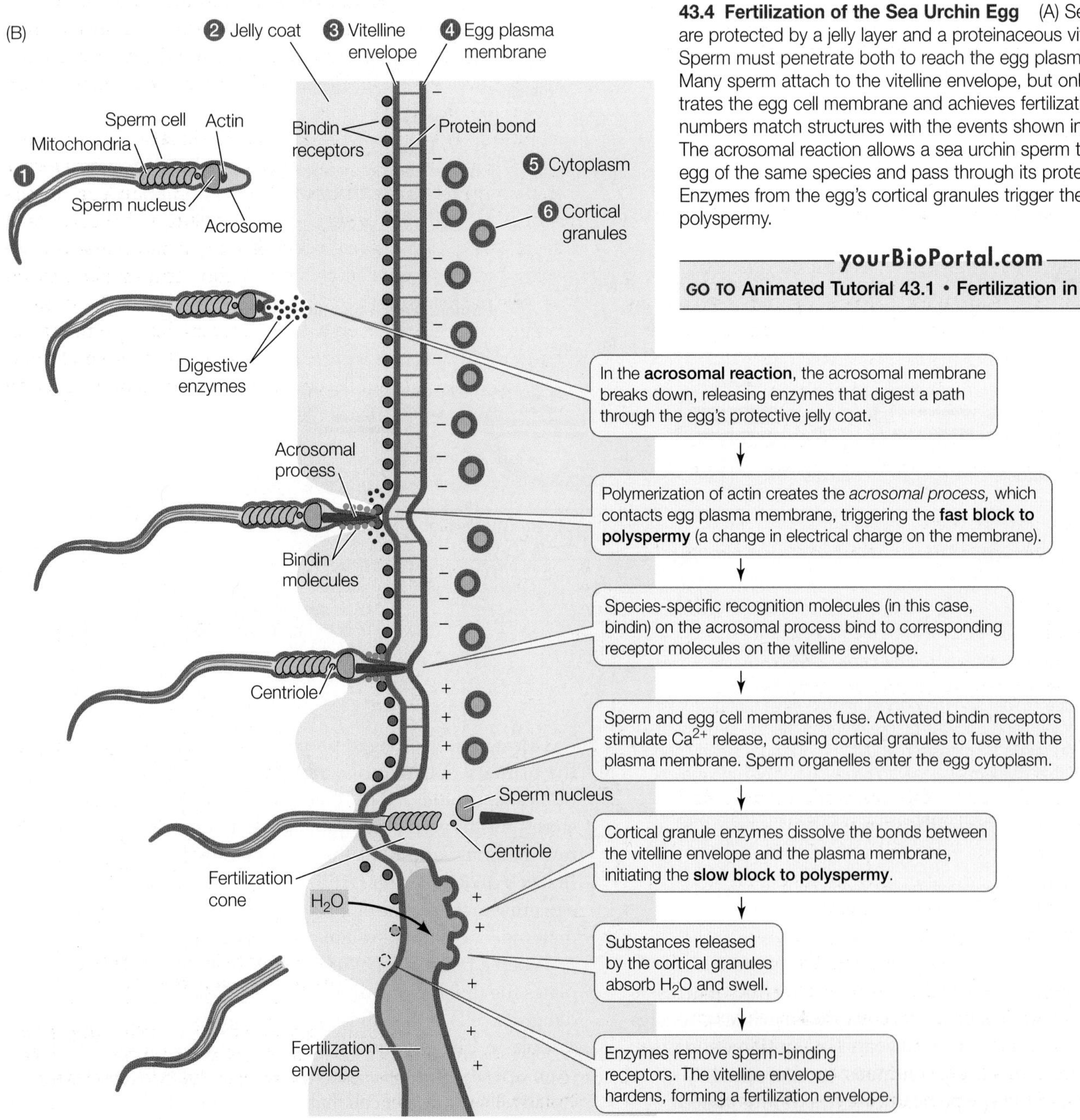

(B)

② Jelly coat   ③ Vitelline envelope   ④ Egg plasma membrane

Sperm cell   Actin

Mitochondria

Sperm nucleus

Acrosome

Bindin receptors

Protein bond

⑤ Cytoplasm

⑥ Cortical granules

①

Digestive enzymes

Acrosomal process

Bindin molecules

Centriole

Sperm nucleus

Centriole

Fertilization cone

H_2O

Fertilization envelope

43.4 Fertilization of the Sea Urchin Egg   (A) Sea urchin eggs are protected by a jelly layer and a proteinaceous vitelline envelope. Sperm must penetrate both to reach the egg plasma membrane. Many sperm attach to the vitelline envelope, but only one penetrates the egg cell membrane and achieves fertilization. Circled numbers match structures with the events shown in panel B. (B) The acrosomal reaction allows a sea urchin sperm to recognize an egg of the same species and pass through its protective layers. Enzymes from the egg's cortical granules trigger the slow block to polyspermy.

──────── **yourBioPortal.com** ────────

GO TO Animated Tutorial 43.1 • Fertilization in a Sea Urchin

In the **acrosomal reaction**, the acrosomal membrane breaks down, releasing enzymes that digest a path through the egg's protective jelly coat.

↓

Polymerization of actin creates the *acrosomal process,* which contacts egg plasma membrane, triggering the **fast block to polyspermy** (a change in electrical charge on the membrane).

↓

Species-specific recognition molecules (in this case, bindin) on the acrosomal process bind to corresponding receptor molecules on the vitelline envelope.

↓

Sperm and egg cell membranes fuse. Activated bindin receptors stimulate Ca^{2+} release, causing cortical granules to fuse with the plasma membrane. Sperm organelles enter the egg cytoplasm.

↓

Cortical granule enzymes dissolve the bonds between the vitelline envelope and the plasma membrane, initiating the **slow block to polyspermy**.

↓

Substances released by the cortical granules absorb H_2O and swell.

↓

Enzymes remove sperm-binding receptors. The vitelline envelope hardens, forming a fertilization envelope.

including humans, the second meiotic division is not completed until the egg is fertilized by a sperm.

Fertilization is the union of sperm and egg

The union of the haploid sperm and the haploid egg in **fertilization** creates a single diploid cell, called a **zygote**, which will develop into an embryo. Fertilization does more, however, than just restore the full genetic complement of the animal. The processes associated with fertilization help the egg and sperm get together, prevent the union of the sperm and egg of different species, and guarantee that only one sperm will enter and activate the egg metabolically. Fertilization involves a complex series of events:

- The sperm and the egg recognize each other.
- The sperm is *activated*, enabling it to gain access to the plasma membrane of the egg.
- The plasma membrane of the egg fuses with the plasma membrane of a single sperm.
- The egg blocks entry of additional sperm.
- The egg is metabolically activated and stimulated to start development.
- The egg and sperm nuclei fuse to create the diploid nucleus of the zygote.

SPECIFICITY IN SPERM–EGG INTERACTIONS Specific recognition molecules mediate interactions between sperm and eggs. These molecules ensure that the activities of sperm are directed toward eggs and not other cells, and they help prevent eggs from being fertilized by sperm from the wrong species. The latter function is particularly important in aquatic species that release eggs and sperm into the surrounding water, as the eggs of such animals may readily be exposed to sperm of other species. The sea urchin is a good example of such a species, and its mechanisms of fertilization have been well studied.

The eggs of sea urchins release chemical attractants that increase the motility of sperm and cause them to swim toward the egg. These chemical attractants are species-specific. For example, eggs of one species of sea urchin release a specific peptide consisting of 14 amino acids. As this peptide diffuses from the egg, it binds to receptors on the sperm of the same species. The sperm respond by increasing their mitochondrial respiration and motility. Before exposure to the peptide, the sperm swim in tight little circles, but after binding to the peptide, they swim energetically up the concentration gradient of the peptide until they reach the egg that is releasing it.

When sperm reach an egg, they must get through two protective layers before they can fuse with the egg plasma membrane. The eggs of sea urchins are covered with a **jelly coat**, which surrounds a proteinaceous **vitelline envelope** (**Figure 43.4A**). The success of a sperm's assault on these protective layers depends on a membrane-enclosed structure at the front of the sperm head called an **acrosome**.

The acrosome, which contains enzymes and other proteins, forms a cap over the sperm nucleus. When the sperm makes contact with an egg of its own species, substances in the jelly coat trigger an *acrosomal reaction*, which begins with the breakdown of the plasma membrane covering the sperm head and the underlying acrosomal membrane (**Figure 43.4B**). The acrosomal enzymes are released and digest a hole through the jelly coat.

As a result of the polymerization of actin triggered by the acrosomal reaction, a structure called the *acrosomal process* extends out of the head of the sperm. The acrosomal process is coated with species-specific recognition molecules called *bindin*, and there are bindin receptors on the vitelline envelope of the egg. The interaction of these two molecules enables the sperm to contact the egg plasma membrane. That contact results in fusion of the sperm and egg plasma membranes and the formation of a *fertilization cone* that engulfs the sperm head, bringing it into the egg cytoplasm. The sperm mitochondrion, which largely constitutes the mid-piece of the sperm, is also drawn into the egg cytoplasm, but it degrades and disappears so that the mitochondria and mitochondrial genes of the new urchin are derived only from the egg.

In animals that practice internal fertilization, mating behaviors help guarantee species specificity, but egg–sperm recognition mechanisms still exist. The mammalian egg is surrounded by a thick layer called the **cumulus**, which consists of a loose assemblage of maternal cells in a gelatinous matrix (**Figure 43.5**). Beneath the cumulus is a glycoprotein envelope called the **zona pellucida**, which is functionally similar to the vitelline envelope of sea urchin eggs. When mammalian sperm are deposited in the female reproductive tract, they are metabolically activated and made capable of an acrosomal reaction

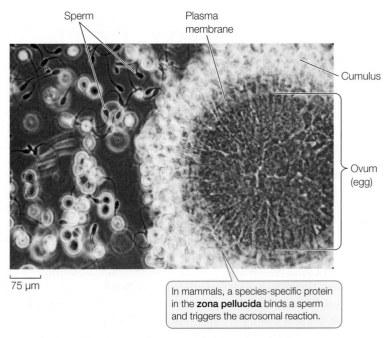

Sperm   Plasma membrane

Cumulus

Ovum (egg)

75 µm

In mammals, a species-specific protein in the **zona pellucida** binds a sperm and triggers the acrosomal reaction.

43.5 Barriers to Sperm This human egg, like other mammalian eggs, is surrounded by the cumulus and zona pellucida, both of which a sperm must penetrate to fertilize the egg. Only one sperm will penetrate the zona pellucida and fuse with the plasma membrane.

should they encounter an egg. An activated sperm can penetrate the cumulus and interact with the zona pellucida.

Unlike the jelly coat of sea urchin eggs, the cumulus of mammalian eggs does not trigger the acrosomal reaction. When sperm make contact with the zona pellucida, a species-specific glycoprotein binds to recognition molecules on the head of the sperm. This binding triggers the acrosomal reaction, releasing acrosomal enzymes that digest a path through the zona pellucida. When the sperm head reaches the egg plasma membrane, other proteins facilitate its adhesion to and fusion with the egg plasma membrane.

The importance of the zona pellucida and its sperm-binding molecules as a species-specific recognition mechanism was revealed in experiments on mammalian eggs and sperm in culture dishes. When the zona was stripped from human eggs and the eggs were exposed to hamster sperm, fertilization took place, resulting in a hamster–human hybrid zygote. The hybrid zygote did not survive its first cell division, but the experiment demonstrated that a recognition mechanism in mammalian species resides in the zona pellucida.

BLOCKS TO POLYSPERMY The fusion of the sperm and egg plasma membranes and the entry of the sperm into the egg initiate a programmed sequence of events. The first responses to sperm entry are **blocks to polyspermy**—that is, mechanisms that prevent more than one sperm from entering the egg. If more than one sperm enters the egg, the embryo is unlikely to survive.

Blocks to polyspermy have been studied extensively in sea urchin eggs, which can be fertilized in a dish of seawater. Within seconds after the sperm membrane contacts the egg membrane, an influx of sodium ions changes the electric charge difference across the egg's plasma membrane. This *fast block to polyspermy* prevents the fusion of any other sperm with the egg plasma membrane, but it is transient. The change in membrane electrical charge lasts only about a minute, but that is enough time to allow a slower block to sperm entry to develop.

The *slow block to polyspermy* involves converting the vitelline envelope to a physical barrier that sperm cannot penetrate. Before fertilization, the vitelline envelope is bonded to the egg plasma membrane. Just under the plasma membrane are vesicles called *cortical granules* (see Figure 43.4), which contain enzymes and other proteins.

The sea urchin egg, like all animal cells, contains calcium ions sequestered in its endoplasmic reticulum. Sperm entry into the egg stimulates the release of calcium ions from the endoplasmic reticulum and into the egg cytoplasm. This increase in cytosolic calcium causes the egg's cortical granules to fuse with the plasma membrane and release their contents. Cortical granule enzymes break the bonds between the vitelline envelope and the plasma membrane, and other proteins released from the cortical granules attract water into the space between them. As a result, the vitelline envelope rises to form a *fertilization envelope*. Cortical granule enzymes also degrade sperm-binding molecules on the surface of the fertilization envelope and cause it to harden, thus preventing additional sperm from contacting the egg's plasma membrane.

In mammals, sperm entry does not cause a rapid change in membrane potential, but it does trigger a release of calcium from the endoplasmic reticulum. As in the sea urchin, the increased calcium causes the cortical granules to fuse with the egg plasma membrane. A fertilization envelope does not form around the mammalian egg, but the cortical granule enzymes destroy the sperm-binding molecules in the zona pellucida. The rise in cytosolic calcium also signals the egg to complete meiosis. The stage is set for the first cell division.

Getting eggs and sperm together

As we have just seen, sexual reproduction requires the production of haploid gametes (gametogenesis) and the joining together of those gametes to form a diploid zygote (fertilization). Spawning and mating behaviors get eggs and sperm close enough together that fertilization can occur. Fertilization can occur externally or internally.

EXTERNAL FERTILIZATION In an aquatic environment, animals can bring their gametes together by simply releasing them into the water. This practice, called spawning, results in **external fertilization**. Many aquatic animals are not very mobile, but they produce huge numbers of gametes that can travel far from the point of release. A female oyster, for example, will release millions of eggs when she spawns, and the number of sperm produced by a male oyster is astronomical.

Numbers alone, however, do not guarantee that gametes will meet. The reproductive activities of the males and females of a population must be synchronized, since released gametes have a limited life span. Seasonal breeders may use day length, changes in temperature, or changes in weather to time the production and release of their gametes. Mutual stimulation is also important. Release of gametes into the water by one individual can stimulate others to spawn.

Behavior can play an important role in bringing gametes together even when fertilization is external. Many species travel great distances to congregate with potential mates and release their gametes at the same time in a suitable environment. Salmon are an extreme example. They hatch and develop in freshwater streams and then migrate to the ocean where they remain for years. When they are grown, they travel hundreds of miles to spawn back in the stream where they hatched. Males and females expend great amounts of energy to swim up the streams to the spawning grounds, where they pair up, prepare a depression in the streambed gravel, and together release their sperm and eggs. As the gametes drift down into the gravel, fertilization occurs.

INTERNAL FERTILIZATION Terrestrial animals cannot simply release their gametes into the environment. Sperm can move only through liquid, and delicate gametes released into air would dry out and die. Terrestrial animals avoid these problems by **internal fertilization**, the release of sperm into the female reproductive tract. Some aquatic animals also practice internal fertilization, but it is ubiquitous in terrestrial animals.

Animals have evolved an astonishing diversity of behavioral and anatomical adaptations for internal fertilization. As we saw above, gametogenesis occurs in the gonads, which are the *primary sex organs*. All additional anatomical components of an animal's reproductive system are called *accessory sex organs*. An obvious accessory sex organ in males of many species is the **penis**, which enables the male to deposit sperm in the female's **vagina**, the entry to her reproductive tract. Accessory sex organs include a variety of glands, tubules, ducts, and other structures.

Copulation is the physical joining of male and female accessory sex organs. Transfer of sperm in internal fertilization can also be indirect. Males of many invertebrate species (for example, mites and scorpions) and a few vertebrates (salamanders) deposit *spermatophores*—packets of sperm—in the environment. The packets protect the sperm from desiccation. When a female mite encounters a spermatophore from a potential mate, she straddles it and opens a pair of plates in her abdomen so that the tip of the spermatophore enters her reproductive tract and allows the sperm to enter.

Male squids and spiders play a more active role in spermatophore transfer. The male spider secretes a drop containing sperm onto a bit of web, then uses a special structure on his foreleg to pick up the sperm-containing web and insert it through the female's genital opening. Male squids use one specialized tentacle to pick up a spermatophore and insert it into the female's genital opening.

Most male insects copulate and transfer sperm to the female's vagina through a penis. The **genitalia**—external sex organs—of insects often have species-specific shapes that ensure that the male and female genitalia match in a lock-and-key fashion. This mechanism ensures a tight, secure fit between the mating pair during the prolonged period of sperm transfer. In some insect species in which females mate with more than one male, the males have elaborate structures on their penises that can scoop sperm deposited by other males out of a female's reproductive tract, replacing it with their own.

An individual animal can function as both male and female

In most species, gametes are produced by individuals that are either male or female. Species that have separate male and female members are called **dioecious** species (from the Greek for "two houses"). In some species, however, a single individual may produce both sperm and eggs. Such species are called **monoecious** ("one house") or **hermaphroditic**, species.

Almost all invertebrate groups contain some hermaphroditic species. An earthworm is an example of a *simultaneous hermaphrodite*, meaning an individual is both male and female at the same time. When two earthworms mate, they exchange sperm, and as a result, the eggs of each are fertilized (see Figure 32.13C). Some vertebrates are *sequential hermaphrodites*, meaning that an individual may function as a male or a female at different times in its life. An example is the anemone fish, or clown fish (**Figure 43.6**), a species that lives in small groups within large

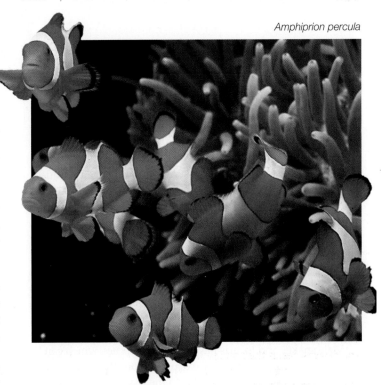

Amphiprion percula

43.6 When Size Determines Sex   Anemone fish (also known as clown fish) live in groups of about a dozen centered on a single sea anemone. All anemone fish are born male, and the largest one in the group becomes a functional female. Thus any one fish may function as a male and as a female at different times in its life.

sea anemones. All anemone fish are born male. The largest one in a group becomes a functional female. If that fish is removed from the group, the next largest male becomes a female. Also, the second-largest anemone fish in the group is the only male in breeding condition.

What is the evolutionary advantage of hermaphroditism? Some simultaneous hermaphrodites, such as parasitic tapeworms, have a low probability of meeting a potential mate—it may be the only tapeworm in the host. Although tapeworms can fertilize their own eggs, most simultaneous hermaphrodites must mate with another individual; but because every member of the population is both male and female, the probability of encountering a possible mate is double what it would be in strictly monoecious species. In some sequential hermaphrodites, all siblings are either male or female at the same time, thus reducing the incidence of inbreeding.

The evolution of vertebrate reproductive systems parallels the move to land

The earliest vertebrates evolved in aquatic environments. The closest living relatives of those earliest vertebrates are modern-day fishes. They remain exclusively aquatic animals, and most practice external fertilization. The most primitive of the fishes, the lampreys and hagfishes, simply release their gametes into the environment. In most fishes, however, mating behaviors bring females and males into close proximity at the time of gamete release. In sharks and rays, fins have evolved into claspers that hold the male and female together and enable sperm to be transferred directly into the female reproductive tract.

Amphibians were the first vertebrates to live in terrestrial environments. They dealt with the challenge of a dry environment by returning to water to reproduce, as most amphibians still do today.

Reptiles were the first vertebrate group to solve the problem of reproduction in the terrestrial environment. Their solution, the **amniote egg**, is shared with the birds (see Section 33.4). A chicken egg is a good example of an amniote egg. It contains a supply of food (yolk) and water for the developing embryo. A hard shell protects the embryo and impedes water loss while allowing the diffusion of oxygen into the egg and carbon dioxide out of the egg (**Figure 43.7A**). The eggshell creates an obvious problem for fertilization. Sperm cannot penetrate the shell, so they must reach the egg before the shell forms. Hence internal fertilization and the evolution of accessory sex organs were necessary for the evolution of the amniote egg.

Male snakes and lizards have paired *hemipenes*, which can be filled with blood and thereby extruded from the male's body. Only one hemipenis is inserted into the female's reproductive tract at a time. It is usually rough or spiny at the end to achieve a secure hold while sperm are transferred down a groove on its surface. Retractor muscles pull the hemipenis back into the male's body when mating is completed. Some evolutionarily ancient bird species have erectile penises that channel sperm along a groove into the female's reproductive tract. Birds with more recent evolutionary origins, however, do not have erectile penises; instead, the male and female simply bring their genital openings close together to transfer sperm. Usually this involves the male standing on the female's back (**Figure 43.7B**).

All mammals practice internal fertilization, and with the exception of the prototherian mammals, the developing embryo is retained for some time in the female reproductive tract. Pro-

totherian mammals (the monotremes; see Figure 33.24) lay eggs. The other mammals (the *therians*) vary enormously as to the developmental stage of their offspring at the time of birth.

Animals with internal fertilization are distinguished by where the embryo develops

Two patterns of care and nurture of the embryo have evolved in animals: *oviparity* (egg laying) and *viviparity* (live bearing).

Oviparous animals lay eggs in the environment, and their embryos develop outside the mother's body. Oviparous terrestrial animals such as insects, reptiles, and birds protect their eggs from desiccation with waterproof membranes or shells. Oviparity is possible because eggs are stocked with abundant nutrients to supply the needs of the embryo. Some oviparous animals engage in various forms of parental behavior to protect their eggs, but until the eggs hatch, the embryos depend entirely on the nutrients stored in the egg.

Viviparous animals retain the embryo within the mother's body during its early developmental stages. Although examples of viviparity exist in all vertebrate groups except the crocodiles, turtles, and birds (even some sharks retain fertilized eggs in their bodies and give birth to free-living offspring), there is a big difference between viviparity in mammals and in other species.

All mammals except the prototherians are viviparous and have a specialized portion of the female reproductive tract, the **uterus** or *womb*, that holds the embryo and interacts with it to produce a **placenta**, which enables the exchange of nutrients and wastes between the blood of the mother and that of the embryo. Very few non-mammalian species have evolved such a connection between the embryo and the mother.

In most non-mammalian viviparous animals, such as garter snakes and the well-known aquarium fish the guppy, fertilized

(A) *Chelonia mydas*

(B) *Merops apiaster*

43.7 The Shelled Egg The shelled egg was a major evolutionary step that allows reptiles and birds to reproduce in the terrestrial environment. (A) A female green sea turtle has deposited her eggs in the sand. (B) The shelled egg requires that sperm meet egg before the shell forms. Terrestrial animals thus must practice internal fertilization, as these European bee-eaters are doing.

eggs are retained in the mother's body until they hatch. These embryos still receive nutrition from stores in the egg, so this reproductive adaptation is called **ovoviviparity**.

43.2 RECAP

Sexual reproduction involves gametogenesis, mating, and fertilization. Fertilization can be external or internal and involves mechanisms for ensuring that only one sperm from the right species enters the egg.

- Describe the steps by which a sea urchin sperm penetrates the egg. **See Figure 43.4**
- Explain how polyspermy is prevented and why it is crucial to do so. **See p. 906 and Figure 43.4**
- What reproductive adaptations made life on land possible? **See p. 908**

Now that we have covered some of the general aspects of gametogenesis and fertilization and have briefly discussed the great diversity of mating systems, we will next consider the human reproductive systems in detail.

43.3 How Do the Human Male and Female Reproductive Systems Work?

In this section we describe the structures and functions of the male and female reproductive systems in mammals, using human beings as our prime example. We also discuss the hormonal regulation of both male and female systems. Our discussion includes the primary sex organs (testes in males and ovaries in females) that produce gametes and serve endocrine functions. It also includes the accessory sex organs: the ducts through which the gametes pass, the various glands that empty into those ducts, and the external genitalia. We also discuss *secondary sexual characteristics*, which are not directly involved in reproduction but are responsible for the major differences in external appearance of men and women and are important in mating.

Male sex organs produce and deliver semen

Semen is the product of the male reproductive system. Besides sperm, semen contains a complex mixture of fluids and molecules that support the sperm and facilitate fertilization. Sperm make up less than 5 percent of the volume of the semen.

The male reproductive organs are diagrammed in **Figure 43.8**. Sperm are produced in the testes, the paired male gonads. The testes of most mammals are located outside the body cavity in a pouch of skin called the **scrotum**.

Why should the testes be located outside the body cavity? The optimal temperature for spermatogenesis in most mammals is slightly lower than the normal body temperature. The scrotum

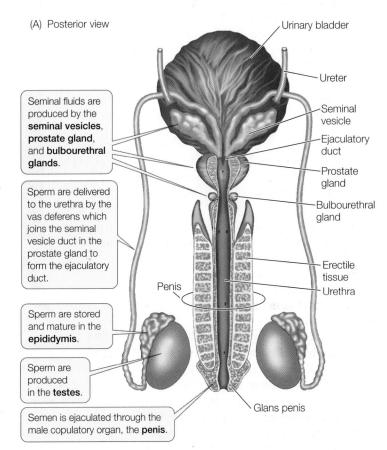

(A) Posterior view

- Urinary bladder
- Ureter
- Seminal vesicle
- Ejaculatory duct
- Prostate gland
- Bulbourethral gland
- Erectile tissue
- Urethra
- Penis
- Glans penis

Seminal fluids are produced by the **seminal vesicles**, **prostate gland**, and **bulbourethral glands**.

Sperm are delivered to the urethra by the vas deferens which joins the seminal vesicle duct in the prostate gland to form the ejaculatory duct.

Sperm are stored and mature in the **epididymis**.

Sperm are produced in the **testes**.

Semen is ejaculated through the male copulatory organ, the **penis**.

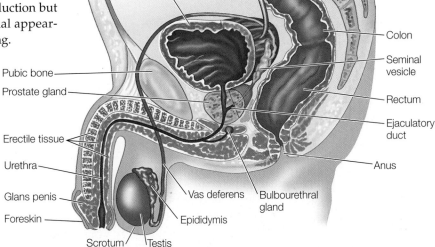

(B) Side view

- Ureter (from kidney)
- Urinary bladder
- Colon
- Seminal vesicle
- Rectum
- Ejaculatory duct
- Anus
- Pubic bone
- Prostate gland
- Erectile tissue
- Urethra
- Glans penis
- Foreskin
- Scrotum
- Testis
- Vas deferens
- Epididymis
- Bulbourethral gland

43.8 Reproductive Tract of the Human Male The male reproductive organs are shown (A) from the rear and (B) from the side.

yourBioPortal.com

GO TO **Web Activity 43.1 • The Human Male Reproductive Tract**

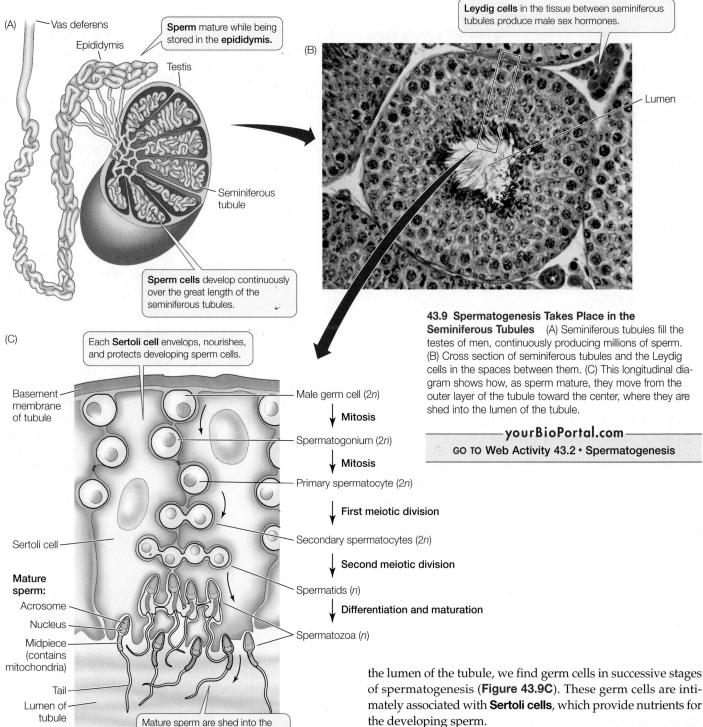

(A)

Vas deferens

Epididymis

Testis

Sperm mature while being stored in the **epididymis.**

Seminiferous tubule

Sperm cells develop continuously over the great length of the seminiferous tubules.

(B)

Leydig cells in the tissue between seminiferous tubules produce male sex hormones.

Lumen

43.9 Spermatogenesis Takes Place in the Seminiferous Tubules (A) Seminiferous tubules fill the testes of men, continuously producing millions of sperm. (B) Cross section of seminiferous tubules and the Leydig cells in the spaces between them. (C) This longitudinal diagram shows how, as sperm mature, they move from the outer layer of the tubule toward the center, where they are shed into the lumen of the tubule.

yourBioPortal.com

GO TO Web Activity 43.2 • Spermatogenesis

(C)

Each **Sertoli cell** envelops, nourishes, and protects developing sperm cells.

Basement membrane of tubule

Sertoli cell

Mature sperm:

Acrosome

Nucleus

Midpiece (contains mitochondria)

Tail

Lumen of tubule

Mature sperm are shed into the lumen of the seminiferous tubule.

Male germ cell (2n)

↓ **Mitosis**

Spermatogonium (2n)

↓ **Mitosis**

Primary spermatocyte (2n)

↓ **First meiotic division**

Secondary spermatocytes (2n)

↓ **Second meiotic division**

Spermatids (n)

↓ **Differentiation and maturation**

Spermatozoa (n)

keeps the testes at this optimal temperature. Muscles in the scrotum contract in a cold environment, bringing the testes closer to the warmth of the body; in a hot environment they relax, cooling the testes by suspending them farther from the body.

Spermatogenesis takes place within the **seminiferous tubules**, which are tightly coiled in each testis (**Figure 43.9A**). Between the seminiferous tubules are clusters of *Leydig cells*, or *interstitial cells*, which produce testosterone (**Figure 43.9B**). Spermatogonia reside in the outer regions of the tubule, just under the basement membrane. Moving inward from these outer layers toward

the lumen of the tubule, we find germ cells in successive stages of spermatogenesis (**Figure 43.9C**). These germ cells are intimately associated with **Sertoli cells**, which provide nutrients for the developing sperm.

When the second meiotic division is complete, each primary spermatocyte has given rise to four spermatids (see Figure 43.3A), which develop into spermatozoa as they continue to migrate toward the lumen of the seminiferous tubule. The nucleus becomes compact, and the surrounding cytoplasm is lost. A flagellum, or sperm tail, develops. The mitochondria, which will provide energy for tail motility, become condensed into a midpiece between the head and the tail. An acrosome forms over the nucleus in the head of the sperm. Immature sperm are shed into the lumen of the seminiferous tubule.

From the seminiferous tubules, sperm move into the **epididymis** (see Figure 43.8), where they mature and become motile. The epididymis connects to the **urethra** via the **vas deferens** (plural *vasa deferentia*) and the **ejaculatory duct**. The urethra originates in the bladder, runs through the penis, and opens to the outside of the body at the tip of the penis. It serves as the common final duct for the urinary and reproductive systems. The components of the semen other than sperm come from several accessory glands. About 60 percent of the volume of semen is secreted by the paired **seminal vesicles**, which empty into the vas deferens just before it joins the urethra. Seminal fluid is thick because it contains mucus and fibrinogen, a protein also found in the blood, where it can polymerize to form blood clots. Seminal fluid also contains the monosaccharide fructose, an energy source for the sperm.

The **prostate gland** surrounds the urethra and contributes about 30 percent of the volume of the semen. Prostate fluid is alkaline, so it neutralizes the acidity in the male and female reproductive tracts and makes those environments more hospitable to sperm. The prostate also secretes a clotting enzyme that causes the fibrinogen from the seminal vesicles to convert the semen into a coagulum (gelatinous mass), facilitating its propulsion into and retention in the upper regions of the female reproductive tract. Another enzyme in the prostate fluid, profibrinolysin, is inactive when secreted but is activated shortly after it enters the female reproductive tract. Active fibrinolysin dissolves the clotted semen and liberates the sperm.

The **bulbourethral glands** produce a small volume of an alkaline, mucoid secretion that helps neutralize acidity in the urethra and lubricate it to facilitate the passage of semen at the climax of sexual intercourse. Secretions of the bulbourethral glands precede the climax of the sex act and can carry with them residual sperm from prior sexual activity. Therefore, it is possible for pregnancy to occur even if the penis is withdrawn from the vagina just before climax (a rather ineffective birth control practice known as *coitus interruptus*).

The penis and the scrotum are the male genitalia. The shaft of the penis is covered with normal skin, but the highly sensitive tip, the **glans penis**, is covered with thinner, more sensitive skin that is especially responsive to sexual stimulation. A fold of skin called the *foreskin* covers the glans of the human penis. The procedure known as *circumcision* removes a portion of the foreskin.

Sexual stimulation triggers responses in the nervous system that result in penile **erection**. Nerve endings release a gaseous neurotransmitter, nitric oxide (NO), onto blood vessels leading into the penis. NO stimulates production of the second messenger cGMP (see Figure 7.17), which causes these vessels to dilate. The increased blood flow that results fills and swells shafts of spongy, erectile tissue located along the length of the penis. The enlargement of these blood-filled cavities compresses the vessels that normally carry blood out of the penis. As a result, the erectile tissue becomes more and more engorged with blood. The penis becomes hard and erect, facilitating its insertion into the vagina. Many species of mammals, but not humans, have a bone in the penis, but these species still depend on erectile tissue for copulation.

At the climax of copulation, about 2 to 6 milliliters of semen are propelled through the vasa deferentia and the urethra in two steps, emission and ejaculation. During **emission**, rhythmic contractions of smooth muscles in the vasa deferentia and accessory glands move the semen into the urethra at the base of the penis.

Ejaculation is caused by contractions of other muscles at the base of the penis surrounding the urethra. These contractions force the coagulum of semen through the urethra and out of the penis. The muscle contractions of ejaculation are accompanied by feelings of intense pleasure known as **orgasm**. They are also accompanied by transient increases in heart rate, blood pressure, breathing, and skeletal muscle contractions throughout the body.

After ejaculation, NO release decreases and enzymes break down cGMP, causing the blood vessels flowing into the penis to constrict. The blood pressure in the erectile tissue decreases, relieving the compression of the blood vessels leaving the penis, and the erection declines.

Erectile dysfunction (ED), or *impotence*, is the inability to achieve or sustain an erection. ED may have different causes, including cardiovascular disease. Drugs used to treat ED act by inhibiting the breakdown of cGMP, thus enhancing the effect of NO released in the penis, which improves the ability to achieve and maintain an erection.

Male sexual function is controlled by hormones

Spermatogenesis and maintenance of male secondary sexual characteristics such as facial hair and a deep voice depend on testosterone produced by the Leydig cells of the testes. As described in Section 41.3, increased production of testosterone at puberty results from an increased release of gonadotropin-releasing hormone (GnRH) by the hypothalamus, which stimulates anterior pituitary cells to increase their secretion of luteinizing hormone (LH) and follicle-stimulating hormone (FSH) (**Figure 43.10**). Higher levels of LH stimulate the Leydig cells

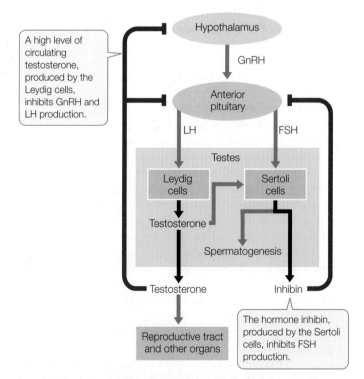

43.10 Male Reproductive Hormones The male reproductive system is under hormonal control by the hypothalamus and the anterior pituitary. Red lines indicate inhibition; green lines indicate stimulation.

to increase their production and release of testosterone. Testosterone exerts negative feedback on the anterior pituitary and the hypothalamus. At the time of puberty, the sensitivity of the hypothalamus to negative feedback from testosterone declines, and the level of circulating testosterone increases.

Increased testosterone in pubertal boys causes the development of pubic and facial hair, a deeper voice, enlarged genitals, and an increased growth rate. Testosterone also promotes increased muscle mass and maturation of the testes. Continued production of testosterone after puberty is essential for the maintenance of secondary sexual characteristics and the production of sperm.

Spermatogenesis is controlled by the influence of FSH and testosterone on the Sertoli cells in the seminiferous tubules. The Sertoli cells also produce a hormone called *inhibin*, which exerts negative feedback on the anterior pituitary cells that produce and secrete FSH.

Female sex organs produce eggs, receive sperm, and nurture the embryo

When a mammalian egg matures, it is released from the ovary directly into the body cavity. But the egg does not go far. The ovaries are close to the *fimbria,* the undulating, fringed openings of the **oviducts** (also known as the *Fallopian tubes*). The fimbria draw the released egg into an oviduct (**Figure 43.11**), where fertilization takes place. Whether or not the egg is fertilized, cilia lining the oviduct propel the egg slowly toward the uterus, a muscular, thick-walled cavity shaped in humans like an upside-down pear. The uterus is where the embryo develops if the egg is fertilized. At the bottom, the uterus narrows into a region called the **cervix**, which leads into the vagina.

In humans, two sets of skin folds surround the opening of the vagina and the opening of the urethra, through which urine passes. The inner, more delicate folds are the *labia minora;* the outer, thicker folds are the *labia majora.* At the anterior tip of the labia minora is the *clitoris,* a small bulb of erectile tissue that has the same developmental origins as the penis. The clitoris is highly sensitive and plays an important role in sexual response. The labia minora and the clitoris become engorged with blood in response to sexual stimulation.

The external opening of an infant's vagina is usually, but not always, partly covered by a thin membrane, the *hymen*. Eventually the hymen can be torn by vigorous physical activity or by first sexual intercourse; it can sometimes make first intercourse difficult or painful for the woman.

To fertilize an egg, sperm deposited in the vagina swim and are propelled by contractions of the female reproductive tract through the cervical opening, the uterus, and most of the oviduct. The egg (actually a secondary oocyte) is fertilized in the upper region of the oviduct. Fertilization stimulates the completion of the second meiotic division, after which the haploid

(A) Front view

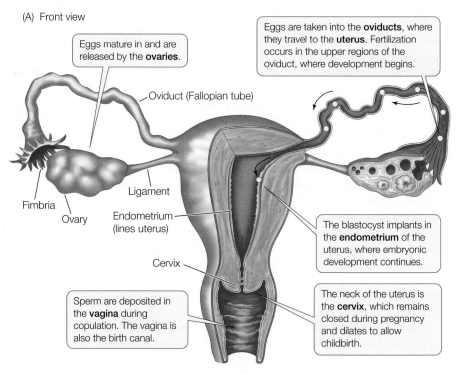

Eggs mature in and are released by the **ovaries**.

Eggs are taken into the **oviducts**, where they travel to the **uterus**. Fertilization occurs in the upper regions of the oviduct, where development begins.

Oviduct (Fallopian tube)

Ligament

Fimbria

Ovary

Endometrium (lines uterus)

Cervix

The blastocyst implants in the **endometrium** of the uterus, where embryonic development continues.

Sperm are deposited in the **vagina** during copulation. The vagina is also the birth canal.

The neck of the uterus is the **cervix**, which remains closed during pregnancy and dilates to allow childbirth.

(B) Side view

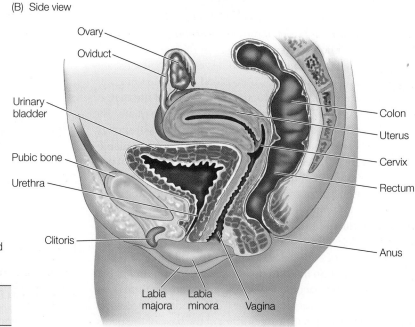

Ovary

Oviduct

Urinary bladder

Pubic bone

Urethra

Clitoris

Labia majora

Labia minora

Vagina

Colon

Uterus

Cervix

Rectum

Anus

43.11 Reproductive Tract of the Human Female The female reproductive organs are shown (A) from the front and (B) from the side.

yourBioPortal.com

GO TO **Web Activity 43.3 • The Human Female Reproductive Tract**

nuclei of the sperm and the egg can fuse to produce a diploid zygote nucleus.

Still in the oviduct, the zygote undergoes its first few cell divisions to become a **blastocyst**. The blastocyst moves down the oviduct to the uterus, where it attaches itself to the epithelial lining of the uterus, or **endometrium**. Once attached, the blastocyst burrows into the endometrium—a process called *implantation*—and interacts with it to form the placenta, as we will see in Chapter 44. The placenta nurtures the embryo and produces hormones that help sustain pregnancy.

As an egg matures in the ovary, the endometrium thickens. If no blastocyst arrives in the uterus, the endometrium regresses or is sloughed off. Thus the female reproductive cycle actually consists of two linked cycles: an *ovarian cycle* that produces eggs and hormones, and a *uterine*, or *menstrual*, *cycle* that prepares the endometrium for the arrival of a blastocyst.

The ovarian cycle produces a mature egg

An **ovarian cycle** is about 28 days long in the human female, but it varies considerably among individuals. During the first half of each cycle, at least one primary oocyte matures into a secondary oocyte (egg) and is expelled from the ovary (*ovulation*). During the second half of the cycle, cells in the ovary that were associated with the maturing oocyte develop endocrine functions and then regress if the egg is not fertilized. The progression of these events is shown diagrammatically in **Figure 43.12A**.

A newborn baby girl has about a million primary oocytes in each ovary. By the time she reaches puberty, she has only about 200,000; the rest have degenerated. During a woman's fertile years, her ovaries go through about 450 ovarian cycles. During each cycle, several oocytes begin to mature, but usually only one matures completely and is released; the others degenerate. At around the age of 50, a woman reaches **menopause**—the end of fertility—and may have few if any oocytes left in each ovary.

Each primary oocyte in the ovary is surrounded by a layer of ovarian cells. An oocyte and its surrounding cells constitute the functional unit of the ovary, the **follicle (Figure 43.12B)**. Between puberty and menopause, 6 to 12 follicles begin to mature each month. In each follicle, the oocyte enlarges and the surrounding follicular cells proliferate. After about a week, one follicle is larger than the rest, and it continues to grow, while the others cease to develop. In the enlarged follicle, the follicular cells nurture the growing egg, supplying it with nutrients, growth factors, and hormonal stimulation.

In humans, after 2 weeks of follicular growth, **ovulation** occurs: the follicle ruptures and the egg is released. Following ovulation, the follicle cells that remain in the ovary continue to proliferate and form a mass of endocrine tissue about the size of a marble. This structure is the *corpus luteum* (plural *corpora lutea*). It functions as an endocrine gland, producing estrogen and progesterone for about 2 weeks. It then degenerates, unless a blastocyst implants in the endometrium.

The uterine cycle prepares an environment for the fertilized egg

The **uterine cycle** parallels the ovarian cycle and consists of a buildup and then a breakdown of the endometrium (**Figure 43.13**). About 5 days into the ovarian cycle, the endometrium starts to grow in preparation for receiving a blastocyst. The uterus

43.12 The Ovarian Cycle (A) The ovarian cycle progresses from the development of a follicle to ovulation and finally to growth and degeneration of the corpus luteum. (B) This micrograph shows a mature mammalian follicle; the oocyte is in the center.

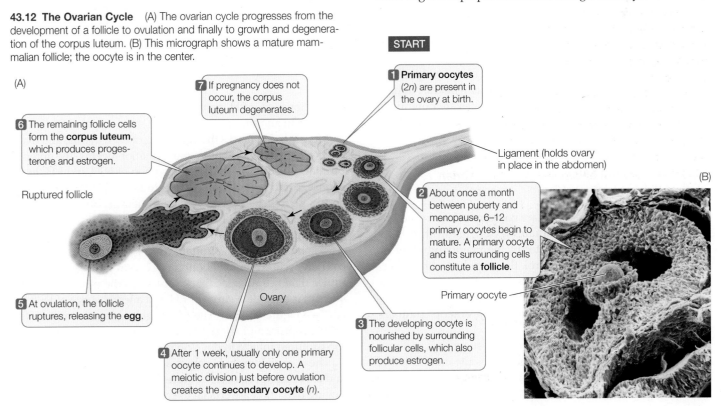

(A)

START

1 **Primary oocytes** (*2n*) are present in the ovary at birth.

2 About once a month between puberty and menopause, 6–12 primary oocytes begin to mature. A primary oocyte and its surrounding cells constitute a **follicle**.

3 The developing oocyte is nourished by surrounding follicular cells, which also produce estrogen.

4 After 1 week, usually only one primary oocyte continues to develop. A meiotic division just before ovulation creates the **secondary oocyte** (*n*).

5 At ovulation, the follicle ruptures, releasing the **egg**.

6 The remaining follicle cells form the **corpus luteum**, which produces progesterone and estrogen.

7 If pregnancy does not occur, the corpus luteum degenerates.

Ruptured follicle

Ovary

Ligament (holds ovary in place in the abdomen)

Primary oocyte

(B)

43.13 The Ovarian and Uterine Cycles
During a woman's ovarian and uterine cycles, coordinated changes occur in (A) gonadotropin release by the anterior pituitary, (B) the ovary, (C) the release of female sex steroids, and (D) the uterus. The cycles begin with the onset of menstruation; ovulation is at midcycle (yellow bar).

─── **yourBioPortal.com** ───
GO TO Animated Tutorial 43.2 •
The Ovarian and Uterine Cycles

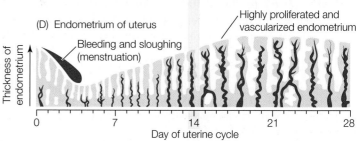

FSH and LH are under control of GnRH from the hypothalamus and the ovarian hormones estrogen and progesterone (part C).

FSH stimulates the development of follicles; the LH surge causes ovulation and then the development of the corpus luteum.

Estrogen and progesterone stimulate the development of the endometrium in preparation for pregnancy.

(A) Gonadotropins (from anterior pituitary)

Estrogen inhibits LH and FSH release | Estrogen stimulates LH and FSH release | Estrogen inhibits LH and FSH release

LH surge triggers ovulation.
Luteinizing hormone (LH)
Follicle-stimulating hormone (FSH)

(B) Events in ovary (ovarian cycle)

Oocyte maturation — Developing follicle — Ovulation (day 14) — Corpus luteum — Developing oocyte

(C) Ovarian hormones and the uterine cycle

Estrogen
Progesterone

(D) Endometrium of uterus

Bleeding and sloughing (menstruation)
Highly proliferated and vascularized endometrium

Thickness of endometrium

0   7   14   21   28
Day of uterine cycle

---

attains its maximal state of preparedness about 5 days after ovulation and remains in that state for another 9 days. If a blastocyst has not arrived by that time, the endometrium begins to break down, and the sloughed-off tissue, including blood, flows from the body through the vagina—the process of **menstruation** (from *menses*, the Latin word for "months").

The uterine cycles of most mammals other than humans do not include menstruation; instead, the uterine lining typically is resorbed. In these species, the most obvious correlate of the ovarian cycle is a state of sexual receptivity called **estrus** around the time of ovulation. You may be aware of the bloody discharge that occurs in dogs at the time of estrus. This discharge is not the same as menstruation—in fact it is exactly the opposite. Bleeding in dogs occurs during the *proliferation* of the uterine lining, which occurs just before ovulation. When the female mammal comes into estrus, or "heat," she actively solicits male attention and may be aggressive to other females. Humans are unusual among mammals in that females are potentially sexually receptive throughout their ovarian cycles and at all seasons of the year.

Hormones control and coordinate the ovarian and uterine cycles

The ovarian and uterine cycles are coordinated and timed by the same hormones that initiate sexual maturation. Gonadotropins (FSH and LH) secreted by the anterior pituitary are the central elements of this control. Before puberty (that is, before about 11 years of age), the secretion of FSH and LH is low and the ovaries are inactive. At puberty, the hypothalamus increases its release of GnRH, stimulating the anterior pituitary to secrete FSH and LH.

In response to FSH and LH, ovarian tissue grows and produces estrogen. The rise in estrogen causes the maturation of the accessory sex organs and the development of female secondary sexual characteristics. Between puberty and menopause,

interactions of GnRH, gonadotropins, and sex steroids control the ovarian and uterine cycles.

Menstruation marks the beginning of each uterine and ovarian cycle. A few days before menstruation begins, the anterior pituitary begins to increase its secretion of FSH and LH. In response, some 10 to 20 follicles begin to mature in the ovaries, and these follicles steadily increase their production of estrogen. After about a week, all but one of the follicles wither away.

Estrogen exerts negative feedback control on gonadotropin release by the anterior pituitary during the first 12 days of the ovarian cycle. Then, on about day 12, estrogen exerts positive rather than negative feedback control on the pituitary (**Figure 43.14**). As a result, a surge of LH and a lesser surge of FSH occur (see Figure 43.13A). The LH surge triggers the mature follicle to rupture and release its egg, and it stimulates the cells of the ruptured follicle to develop into a corpus luteum.

The corpus luteum becomes an endocrine gland. Estrogen and especially progesterone secreted by the corpus luteum following ovulation are crucial to continued growth and mainte-

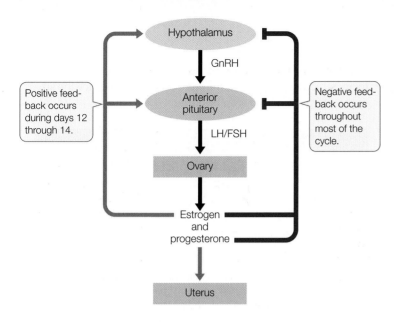

43.14 Hormones Control the Ovarian and Uterine Cycles The ovarian and uterine cycles are under a complex series of positive and negative feedback controls involving several hormones.

nance of the endometrium. In addition, these sex steroids exert negative feedback control on the pituitary, inhibiting gonadotropin release and thus preventing new follicles from beginning to mature.

If the egg is not fertilized, the corpus luteum degenerates on about day 26 of the cycle. Without production of progesterone by the corpus luteum, the endometrium sloughs off and menstruation occurs. The decrease in circulating steroids also releases the hypothalamus and pituitary from negative feedback control, so GnRH, FSH, and LH all begin to increase. The increase in these hormones induces the next round of follicle development, and the ovarian cycle begins again.

In pregnancy, hormones from the extraembryonic membranes take over

If the egg is fertilized and a blastocyst arrives in the uterus and implants in the endometrium, a new hormone comes into play. A layer of cells covering the blastocyst begins to secrete **human chorionic gonadotropin**, or **hCG** (**Figure 43.15A**). This gonadotropin, a molecule similar to LH, stimulates the corpus luteum to continue to produce estrogen and progesterone to support the growth and maintenance of the endometrium and thereby prevent menstruation. Because it is present only in the blood of pregnant women, the presence of hCG is the basis for pregnancy testing. Pregnancy tests use an antibody to detect hCG in urine; they take only minutes, and can be done at home.

43.15 Pregnancy and Childbirth (A) When a fertilized ovum implants in the uterus, cells surrounding it produce human chorionic gonadotropin, which acts like LH and keeps the corpus luteum functioning as an endocrine gland. The ovarian and uterine cycles are put on hold for the duration of pregnancy. (B) Both mechanical and hormonal signals are involved in stimulating the uterine contractions of labor and delivery. (C) A new person is delivered into the world headfirst.

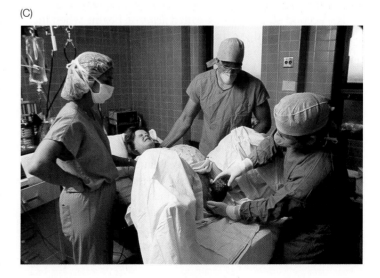

Blastocyst and endometrial tissues form the placenta, which produces estrogen and progesterone (eventually replacing the corpus luteum as the most important source of these sex steroids). Continued high levels of estrogen and progesterone prevent the pituitary from secreting gonadotropins; thus, the ovarian cycle ceases for the duration of pregnancy. (This mechanism underlies the action of birth control pills, which contain synthetic hormones resembling estrogen and progesterone that exert negative feedback control on the hypothalamus and pituitary.)

Childbirth is triggered by hormonal and mechanical stimuli

Throughout pregnancy, the muscles of the uterine wall periodically undergo slow, weak, rhythmic contractions called *Braxton Hicks contractions*. These contractions become stronger during the third trimester of pregnancy and are sometimes called *false labor contractions*. True labor contractions usually mark the beginning of childbirth. Both hormonal and mechanical stimuli contribute to the onset of labor.

Progesterone inhibits and estrogen stimulates contractions of uterine muscle. Toward the end of the third trimester, the estrogen–progesterone ratio shifts in favor of estrogen. The onset of labor is marked by increased secretion of the hormone oxytocin by the posterior pituitaries of both mother and fetus. Oxytocin is a powerful stimulant of uterine muscle contraction. Manufactured oxytocin is used to induce labor when that is necessary.

Mechanical stimuli come from the stretching of the uterus by the fully grown fetus and the pressure of the fetal head on the cervix. These mechanical stimuli increase the release of oxytocin by the mother's posterior pituitary, which in turn increases the activity of uterine muscle, which causes even more pressure on the cervix. This positive feedback loop converts the weak, slow, rhythmic Braxton Hicks contractions into stronger labor contractions (**Figure 43.15B**).

In the early stage of labor, hormonal changes and pressure created by the contractions cause the cervix to dilate (expand) until it is large enough to allow the baby to pass through. Gradually the contractions become more frequent and more intense. This stage of labor lasts an average of 12 to 15 hours in a first pregnancy; it is usually 8 hours or less in subsequent ones.

The second stage of labor, called *delivery*, begins when the cervix is fully dilated to a diameter of about 10 centimeters (**Figure 43.15C**). The baby's head moves into the vagina; passage of the fetus through the vagina is assisted by the mother's bearing down ("pushing") with her abdominal and other muscles. Once the head and shoulders of the baby clear the cervix, the rest of its body eases out rapidly, but it is still connected to the placenta by the umbilical cord. Once the baby clears the birth canal, it starts breathing and is independent of its mother's circulation. The umbilical cord may then be clamped and cut. The segment still attached to the baby dries up and sloughs off in a few days, leaving behind its distinctive signature, the belly button—more properly called the *umbilicus*. The detachment and expulsion of the placenta and fetal membranes take from a few minutes to an hour, and may be accompanied by uterine contractions. If the baby suckles at the breast immediately following birth, its suckling stimulates additional secretion of oxytocin, which augments uterine contractions that reduce the size of the uterus and help stop bleeding.

43.3 RECAP

The reproductive systems of men and women produce gametes and hormones, and these functions are controlled by hypothalamic and anterior pituitary hormones. In women, the hormonal control of reproductive functions produces linked ovarian and uterine cycles.

- Describe the path the human sperm and ovum take in moving from their respective gonads to the point at which fertilization occurs. **See Figures 43.8 and 43.11**

- In males, increased production of GnRH at puberty stimulates the release of what two hormones of the anterior pituitary? What effect do these hormones have? **See p. 911**

- Explain the events in the ovarian cycle that result in release of a single ovum each month. What events prepare the uterus to receive the egg? **See Figures 43.12 and 43.13**

Understanding the physiology of human reproduction has led to numerous methods and technologies for controlling it, either to prevent unwanted pregnancies or to overcome infertility.

43.4 How Can Fertility Be Controlled?

Sexual issues and sexual behavior are dominant aspects of our society, and reproductive technologies have had huge impacts on our sexual and reproductive lives.

Human sexual responses have four phases

The responses of both women and men to sexual stimulation consist of four phases: excitement, plateau, orgasm, and resolution. As sexual *excitement* begins in a woman, her heart rate and blood pressure rise, muscular tension increases, breasts swell, and nipples become erect. Her external genitals, including the sensitive clitoris, swell as they become filled with blood, and the walls of the vagina secrete lubricating fluid that facilitates copulation. In the *plateau* phase, her blood pressure and heart rate rise further and her breathing becomes rapid. The sensitivity once focused in the clitoris spreads over the external genitals, and the clitoris itself becomes even more sensitive. *Orgasm* may last as long as a few minutes, and unlike men, some women can experience several orgasms in rapid succession. During the *resolution* phase, blood drains from the genitals, and body physiology returns to close to normal.

In the man, the excitement phase is marked by an increase in blood pressure, heart rate, and muscle tension (just like in the woman) and by penile erection. In the plateau phase, breathing becomes rapid, the diameter of the glans increases, and a few drops of a lubricating fluid from the bulbourethral gland may ooze from the penis. Continued stimulation of the nerve endings in the glans and in the skin along the shaft of the penis eventually trigger orgasm. Massive spasms of the muscles in the genital area and contractions in the accessory reproductive organs result in ejaculation. Within a few minutes after ejaculation, the penis shrinks to its former size and body physiology returns to resting conditions.

The male sexual response also includes a *refractory period* immediately after orgasm. During this period, which may last from minutes to hours, a man cannot achieve a full erection or another orgasm, regardless of the intensity of sexual stimulation. This refractory period is believed to be controlled by the anterior pituitary hormone prolactin (see Figure 41.2), which is released during orgasm. We say "believed" because the study shown in **Figure 43.16** is not conclusive; it compares the data from a single subject who displayed an unusual physiological characteristic with averaged data from a group of typical individuals. In Chapter 1 we learned the importance of statistical tests to establish whether an observed result could be due to chance alone, but we can't perform the necessary tests with a sample size of only one. This problem frequently crops up in medical science when there is an effort to find an explanation for a highly unusual trait or disease. In such cases, data from even a single individual is of interest and can lead to additional studies and experiments.

Humans use a variety of methods to control fertility

According to a recent study, almost half of the more than 6 million pregnancies that occur in the United States each year are unintended. For women of college age, a single act of unprotected intercourse in the 2 days prior to ovulation carries a chance of conception as high as 50 percent.

The only failure-proof methods of preventing pregnancy are complete abstinence from sexual activity or the surgical removal of the gonads. Since those options are not acceptable to most people, they turn to other methods to prevent pregnancy. Many of these methods prevent fertilization or implantation (*conception*) and are therefore referred to as **contraception**. Most methods are used by the woman, some are used by the man; they also vary enormously in their effectiveness. **Table 43.1** lists some of the most commonly used contraceptive methods and their relative failure rates.

Once a fertilized egg is successfully implanted in the uterus, any termination of the pregnancy is called an **abortion**. A *spontaneous abortion* is the medical term for what most people call a

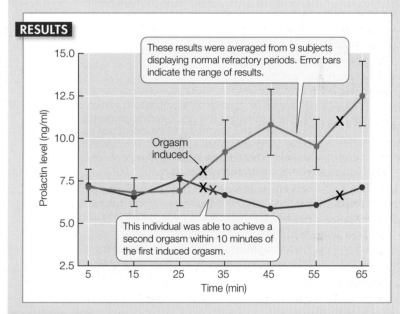

INVESTIGATING LIFE

43.16 Prolactin and the Male Refractory Period

Men experience a refractory period immediately after achieving orgasm, during which most men cannot achieve a full erection. A study done in Germany on a single individual who was able to achieve multiple orgasms indicated that the hormone prolactin may control the onset of refraction.

HYPOTHESIS Release of prolactin during copulation induces a refractory period in males.

METHOD
1. Catheterize subjects and withdraw blood samples (9 normal subjects; 1 multiorgasmic subject).
2. Induce orgasms 30 minutes apart.
3. Withdraw blood samples following orgasms.
4. Analyze all blood samples for levels of prolactin.

RESULTS

These results were averaged from 9 subjects displaying normal refractory periods. Error bars indicate the range of results.

Orgasm induced

This individual was able to achieve a second orgasm within 10 minutes of the first induced orgasm.

CONCLUSION In normal men, blood levels of prolactin rise following ejaculation. The multiorgasmic male showed no increase in prolactin levels, which may explain the lack of a refractory period in this individual.

Go to **yourBioPortal.com** for original citations, discussions, and relevant links for all INVESTIGATING LIFE figures.

miscarriage. Spontaneous abortions are common early in pregnancy and are usually the result of either a chromosomal abnormality in the fetus or to a breakdown in the process of implantation. Many spontaneous abortions occur before the woman even realizes she is pregnant.

Abortions that result from medical intervention may be performed either for therapeutic purposes or for fertility control. A therapeutic abortion may be necessary to protect the health of the mother, or it may be performed because prenatal testing reveals that the fetus has a severe defect. Of the approximately 3 million unintended pregnancies in the United States each year, almost half are ended by abortion.

TABLE 43.1
Methods of Contraception

METHOD	MODE OF ACTION	FAILURE RATE[a]	COMMENTS
Unprotected	No form of birth control	85	High risk of pregnancy, especially for women 15–30.
NONTECHNOLOGICAL METHODS			
Rhythm method	The couple abstains from intercourse between days 10 and 20 of the ovarian cycle (peak fertility).	15–35	High failure rate due to miscalculation and/or variation of individual cycles.
Coitus interruptus	The man withdraws his penis prior to ejaculation, so no sperm is deposited in the vagina.	20–40	Requires self-control, especially by the male partner. Very high failure rate.
BARRIER METHODS[b]			
Condom	A sheath of impermeable material (often latex) is fitted over the erect penis. Semen is trapped in the condom, so no sperm is deposited in the vagina.	15	If fitted correctly, an intact condom can prevent pregnancy and provide protection against sexually transmitted diseases (STDs), including HIV (AIDS).
Spermicidal jellies	Applied inside the vagina, these chemical compounds kill or immobilize sperm.	25	Used alone, spermicidal compounds have a fairly high failure rate.
Diaphragms, cervical caps	Inserted by the woman prior to intercourse, these items work by blocking the cervix so that sperm cannot pass into the uterus.	10–15	Approximately the same failure rate as condom use by males, but do not protect against STDs. Can be used in conjunction with spermicidal jelly for extra protection.
HORMONE-BASED CONTRACEPTIVES			
Oral hormones ("the pill")	A daily pill for females containing a combination of synthetic estrogens and progesterone (progestin). These hormones mimic pregnancy to the extent that the ovarian cycle and ovulation are suspended. The uterine cycle is allowed to continue by including a week of non-hormone administration every 21–28 days.	0–3	Requires medical consultation and prescription. Taken correctly, oral contraceptives are extremely effective. In the U.S., more than 12 million women use them each year; they are sometimes prescribed to treat menstrual disorders.
Non-orally administered hormones	Making use of same hormonal actions as the pill, these methods include long-acting injections, patches that release hormones transdermally (through the skin), and a hormone-containing vaginal ring.	<1	Same as oral contraceptives. A slightly lower failure rate because the woman does not have to remember to take a daily pill.
Progestin-only pill (Plan B®)	An oral contraceptive meant to be taken within 72 hours *after* unprotected sex. A high dose of progestin in two pills prevents ovulation in the same manner birth control pills do.	5–40[c]	Not an "abortion pill," this drug will not terminate an existing pregnancy. Currently available to women over 17 without a prescription.
IMPLANTATION BLOCKERS			
Intrauterine device (IUD)	A medical professional inserts a small plastic or metal device into the uterus. The resulting inflammation reaction (see Chapter 42) releases prostaglandins, which prevent implantation of the fertilized egg.	0.5–5	A highly effective contraceptive, it is the most widely used birth control device in China (and hence the world). With medical monitoring, can remain in place for several years.
Mifepristone (RU-486)	Also known as the "morning after" or "abortion pill," this drug blocks progesterone receptors necessary to maintain the endometrium during implantation and pregnancy.	0.5–6	Prevents implantation when taken up to several days after unprotected intercourse. Can end a pregnancy up to the time of the first missed menstrual period. In the U.S., available from specialized providers.
STERILIZATION			
Vasectomy	The vasa deferentia (see Figure 43.8A) are cut and tied off so that sperm can no longer pass into the urethra. Sperm continue to be produced but are reabsorbed by the man's body. Male hormone levels and sexual responses are not affected.	0–0.15	A simple surgical procedure performed under local anesthetic in a doctor's office. Although it can theoretically be reversed, vasectomy should be considered permanent.
Tubal ligation	The oviducts (see Figure 43.11A) are tied off so that eggs cannot reach the uterus and sperm cannot reach the egg. As with vasectomy, hormone levels and sexual responses are not affected.	0–0.05	This surgical procedure is somewhat more complex than vasectomy. It is often performed in conjunction with childbirth when a woman has decided that her family is complete.

[a] "Failure rate" refers to the number of pregnancies per 100 women per year.

[b] All of these barrier methods are routinely available without medical prescription.

[c] Failure rate varies widely depending on when taken.

In a medical abortion, the cervix is dilated and some of the endometrium, along with the implanted fetus, is removed from the uterus. When performed in the first trimester of a pregnancy, a medical abortion carries less risk of death to the mother than a full-term pregnancy. The risk rises after the first 12 weeks of pregnancy but remains less than that of a full-term pregnancy through the second trimester.

Reproductive technologies help solve problems of infertility

About 15 percent of the couples in the United States are infertile; they can't have children. There are many reasons for infertility, and they are equally distributed between men and women. A number of technologies have been developed to overcome barriers to both conceiving and bearing a child.

The simplest treatment available is **artificial insemination**, in which the physician positions sperm in the female's reproductive tract. This technique is useful if the male's sperm count is low, if his sperm lack motility, or if problems in the female's reproductive tract prevent the normal movement of sperm up to and through the oviducts. Artificial insemination is used widely in the production of domesticated animals such as cattle.

More recent advances, called **assisted reproductive technologies**, or **ARTs**, involve procedures that remove unfertilized eggs from the ovary, combine them with sperm outside the body, and then place fertilized eggs or egg–sperm mixtures in the appropriate location in the female's reproductive tract for development to take place. The first successful ART was *in vitro fertilization* (IVF). In IVF, the female is treated with hormones that stimulate many follicles in her ovaries to mature. Eggs are collected from these follicles, and sperm are collected from the male. Eggs and sperm are combined in a culture medium outside the body, where fertilization takes place. The resulting embryos can be injected into the mother's uterus in the blastocyst stage or kept frozen for implantation later. The first "test-tube baby" resulting from IVF was born in England in 1978. Since then, more than 3 million babies have been produced by this ART.

A major cause of failure of IVF is failure of sperm to gain access to the egg plasma membrane (see Figure 43.4). To solve this problem, methods have been developed to inject a sperm cell directly into the cytoplasm of an egg. In *intracytoplasmic sperm injection (ICSI)*, an egg is held in place by suction applied to a polished glass pipette. A slender, sharp pipette is then used to penetrate the egg and inject a sperm (**Figure 43.17**). This ART was used successfully for the first time in 1992 by researchers in Belgium; now thousands of these procedures are performed in U.S. clinics each year, with a success rate of about 25 percent.

IVF, coupled with techniques of genetic analysis, can eliminate the risk that adults who are carriers of genetic diseases will produce affected children. It is now possible to take a cell from a human embryo at the 4- or 8-cell stage (see Figure 44.4) without damaging its developmental potential. The sampled cell can

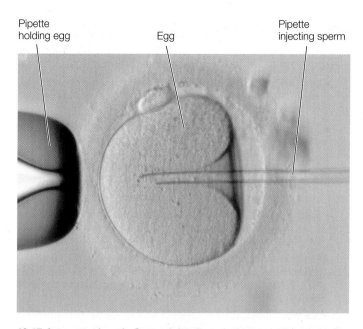

43.17 Intracytoplasmic Sperm Injection In this procedure, sperm are injected directly into a mature egg cell. The fertilized egg is then placed in the female reproductive tract, where it can implant and develop into a fetus.

be subjected to molecular analysis to determine whether it carries the harmful gene. This procedure, called preimplantation genetic diagnosis (PGD), makes it possible to determine whether an embryo produced by IVF carries the genetic defect of concern.

43.4 RECAP

Controlling fertility is an important aspect of modern human life. Decreasing the probability of pregnancy is achieved through methods that prevent sperm and egg from meeting and from preventing implantation. Pregnancies can be facilitated through medical technology.

- What are the four phases of the human sexual response? See p. 916

- Which method of contraception is the only one to offer protection against sexually transmitted diseases (STDs)? See Table 43.1

The fertilized egg of a sexually reproducing organism is a single cell containing all the genetic information needed to create a new organism. Chapters 19 and 20 introduced some of the molecular aspects of the process of development in multicellular animals. Chapter 44 describes the physical events of animal development.

CHAPTER SUMMARY

43.1 How Do Animals Reproduce Without Sex?

- Asexual reproduction produces offspring that are genetically identical to their parent and to one another; it produces no genetic diversity.
- Means of asexual reproduction include **budding**, **regeneration**, and **parthenogenesis**. Review Figures 43.1 and 43.2

43.2 How Do Animals Reproduce Sexually?

- Sexual reproduction consists of three basic steps: gametogenesis; spawning and mating; and fertilization.
- **Gametogenesis** and fertilization are similar in all animals, but spawning and mating includes a great variety of anatomical, physiological, and behavioral adaptations.
- In sexually reproducing species, genetic diversity is created by crossing over and independent assortment of chromosomes during gametogenesis. Fertilization also contributes to genetic diversity.
- Gametogenesis occurs in **testes** and **ovaries**. In spermatogenesis (the production of **sperm**) and oogenesis (the production of eggs), the **germ cells** proliferate mitotically, undergo meiosis, and mature into gametes.
- Each **primary spermatocyte** can produce four haploid sperm through the two divisions of meiosis. **Review Figure 43.3A**
- **Primary oocytes** immediately enter prophase of the first meiotic division, and in many species, including humans, their development is arrested at this point. Each **oogonium** produces only one egg. **Review Figure 43.3B**
- Fertilization involves sperm activation, species-specific binding of sperm to egg, the acrosomal reaction, digestion of a path through the protective coverings of the egg, and fusion of sperm and egg plasma membranes. Fusion of these two membranes triggers **blocks to polyspermy**, which prevent additional sperm from entering the egg, and in mammals, signal the egg to complete meiosis and begin development. **Review Figure 43.4, ANIMATED TUTORIAL 43.1**
- **External fertilization** is common in aquatic species. **Internal fertilization** is necessary in terrestrial species and usually involves **copulation**.
- The shelled egg of amniotes is an important evolutionary adaptation that allows reptiles and birds to reproduce in the terrestrial environment.
- **Hermaphroditic** or **monoecious** species have both male and female reproductive systems in the same individual, either sequentially or simultaneously. **Dioecious** species have separate male and female individuals.
- Animals can be classified as **oviparous** or **viviparous**, depending on whether the early stages of development occur outside or inside the mother's body.

43.3 How Do the Human Male and Female Reproductive Systems Work?

- Men produce **semen** and deliver it into the woman's reproductive tract. Semen consists of sperm suspended in seminal fluid, which nourishes the sperm and facilitates fertilization.
- Sperm are produced in the **seminiferous tubules** of the testes, mature in the **epididymis**, and are delivered to the **urethra** through the **vasa deferentia**. Other components of semen are produced in the **seminal vesicles**, **prostate gland**, and **bulbourethral gland**. Review Figures 43.8 and 43.9, **WEB ACTIVITY 43.1**
- All components of the semen join in the urethra at the base of the **penis** and are ejaculated through the erect penis by muscle contractions at the culmination of copulation.
- Spermatogenesis depends on testosterone secreted by the Leydig cells of the testes, which are under the control of hormones produced in the anterior pituitary and the hypothalamus. The production of these hormones is controlled by negative feedback from testosterone and another hormone, inhibin, produced by the **Sertoli cells** of the testes. **Review Figure 43.10, WEB ACTIVITY 43.2**
- Eggs mature in the woman's ovaries and are released into the **oviducts**. Sperm deposited in the **vagina** during copulation move up through the **cervix** and **uterus** into the oviducts. Fertilization occurs in the upper regions of the oviducts. Review Figure 43.11, **WEB ACTIVITY 43.3**
- The maturation and release of eggs constitute an **ovarian cycle**. In women, this cycle takes about 28 days. The **uterine cycle** prepares the uterus for receipt of a blastocyst. If no blastocyst is implanted, the lining of the uterus sloughs off in the process of **menstruation**. Review Figure 43.13, **ANIMATED TUTORIAL 43.2**
- Both the ovarian and the uterine cycles are under the control of hypothalamic and pituitary hormones, which in turn are under the feedback control of estrogen and progesterone. **Review Figure 43.14**
- Childbirth is initiated by hormonal and mechanical stimuli that increase the contraction of uterine muscle. **Review Figure 43.15**

43.4 How Can Fertility Be Controlled?

- Human sexual responses consist of four phases: excitement, plateau, orgasm, and resolution. In addition, men have a refractory period during which renewed excitement is not possible. **Review Figure 43.16**
- Methods of **contraception** include abstention from copulation and the use of technologies that decrease the probability of fertilization. **Review Table 43.1**
- **Assisted reproductive technologies** (ARTs) have been developed to increase fertility.

SELF-QUIZ

1. A species in which the individual possesses both male and female reproductive systems is termed
   a. dioecious.
   b. parthenogenetic.
   c. hermaphroditic.
   d. monoecious.
   e. ovoviviparous.

2. The major advantage of internal fertilization is that
   a. it ensures paternity.
   b. it permits the fertilization of many gametes.
   c. it reduces the incidence of destructive competitive interactions among the members of a group.
   d. it increases the number of sperm having access to each egg.
   e. it gives the developing organism a greater degree of protection during the early phases of development.

3. Which statement about human oocytes is *true*?
   a. By birth, the human female infant has produced a lifetime supply of oocytes.
   b. At the onset of puberty, ovarian follicles produce new oocytes in response to hormonal stimulation.
   c. At the onset of menopause, a woman stops producing oocytes.
   d. Oocytes are produced by a woman throughout adolescence.
   e. Oocytes produced by a woman are stored in the oviducts.

4. Spermatogenesis and oogenesis differ in that
   a. spermatogenesis produces gametes with greater stores of raw materials than those produced by oogenesis.
   b. spermatocytes remain in prophase of the first meiotic division longer than oocytes.
   c. oogenesis produces four equally functional haploid cells per meiotic event and spermatogenesis does not.
   d. spermatogenesis produces many gametes with meager energy reserves, whereas oogenesis produces relatively few, well-provisioned gametes.
   e. spermatogenesis begins before birth in humans, whereas oogenesis does not start until the onset of puberty.

5. Semen contains all of the following except
   a. fructose.
   b. mucus.
   c. clotting enzymes.
   d. substances to lower the pH of the uterine environment.
   e. an active clot-dissolving enzyme.

6. During oogenesis in mammals, the second meiotic division occurs
   a. in the formation of the primary oocyte.
   b. in the formation of the secondary oocyte.
   c. before ovulation.
   d. after fertilization.
   e. after implantation.

7. One of the major differences between the sexual responses of men and women is
   a. the increase in blood pressure in men.
   b. the increase in heart rate in women.
   c. the presence of a refractory period in women.
   d. the presence of a refractory period in men after orgasm.
   e. the increase in muscle tension in men.

8. Which of the following statements about the ovarian and uterine cycles is *false*?
   a. Falling estrogen and progesterone levels induce menstruation.
   b. A sudden rise in LH induces ovulation.
   c. Estrogen levels rise in the first half of the ovarian cycle and progesterone levels rise in the second half.
   d. If fertilization occurs, the corpus luteum secretes hCG.
   e. Estrogen is produced by follicle cells

9. Contractions of muscles in the uterine wall and milk let-down are stimulated by
   a. progesterone.
   b. estrogen.
   c. prolactin.
   d. oxytocin.
   e. human chorionic gonadotropin.

10. Which of the following methods of contraception is *most likely* to fail?
    a. Rhythm method
    b. Birth control pills
    c. Diaphragm
    d. Vasectomy
    e. Condom

FOR DISCUSSION

1. In the deep ocean, there are species of fish in which the male is much smaller than the female and actually lives attached to her body. In terms of the selective pressures that operate on sexual and asexual reproduction and in terms of the deep-sea environment, what factors do you think resulted in the evolution of this extreme sexual dimorphism?

2. What are two main differences between the immediate products of the first and second meiotic divisions in spermatogenesis and oogenesis? Why do these differences exist?

3. At the beginning of each ovarian cycle in humans, 6 to 12 follicles begin to develop in response to rising levels of FSH, but after a week, only one follicle continues to develop, and the others wither away. Given that follicles produce estrogen, that estrogen stimulates follicle cells to produce FSH receptors, and that estrogen exerts negative feedback on FSH production in the pituitary, can you explain how a single follicle is "selected" to grow?

4. Compare the actions of LH and FSH in the ovaries and testes.

5. Ovarian and uterine events during the month following ovulation differ depending on whether fertilization occurs. Describe the differences and explain their hormonal controls.

ADDITIONAL INVESTIGATION

No male contraceptive methods exist other than vasectomy or the condom. How would you go about developing a pharmacological method to block sperm production that could lead to a male pill without affecting the maintenance of male secondary sexual characteristics or male sexual behavior?

Animal Development

Go with the flow

Place your hand over your heart. Now place it over your liver. Next over your appendix. Surely you put your hand first on the left side of your chest, then on your right side just under your ribs, and finally on the right side of your lower abdomen. But in Chapter 31 you learned that vertebrates (including you) are bilaterally symmetrical (left arm/right arm, left kidney/right kidney, and so on). Clearly, however, our bilateral symmetry is not absolute.

Some of our internal organs—including the heart, liver, appendix, stomach, and the lobes of the lungs—are oriented differently with respect to the left and right sides of the body. How does a developing embryo know which side is left and which is right?

Clues to answering this question came from the fact that in about 1 out of every 7,000 people, the arrangement of the internal organs is reversed, a condition known as *situs inversus*, Latin for "location inverted." This developmental difference arises when the very early embryo goes from being a single layer of cells to multiple layers of cells.

As you will learn in this chapter, to get from a single layer of cells to the next stage with two layers of cells, a pore or slit forms as cells in one area of the embryo migrate inward from the surface. Other cells from the surface migrate toward and through this opening to take up positions underneath. The place where the inward movement of cells starts is called the node. Cells of the node have motile cilia that sweep extracellular fluid through the opening.

Cells bordering the node also have one nonmotile cilium each—a primary cilium. When the primary cilia are bent by the flow of extracellular fluid, they initiate signaling cascades that determine the pattern of internal organ development. Since the fluid driven by the motile cilia tends to flow from right to left, the signaling cascades are not expressed symmetrically, and this initiates the left–right organization of organ development.

Among individuals carrying a mutation that eliminates motility of the nodal cilia, half have the normal orientation of the internal organs and half have *situs inversus*. Most people with this condition lead normal, healthy lives. They may not even know about it

Go with the Flow The internal organs of humans are not all symmetrical, and some individuals are born with the mirror-image pattern of what is seen in most people— a condition called *situs inversus*. The left–right asymmetry of the internal organs is initiated by asymmetrical stimulation of primary cilia at a very early stage in development.

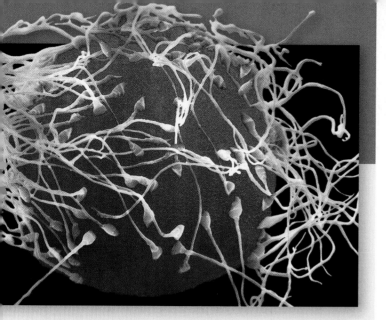

When Sperm Meet Egg Development begins with the fertilization of an egg by a sperm. Once one sperm fuses with the egg, all other sperm are blocked. An animal egg typically is much larger than the sperm; the egg cytoplasm is loaded with informational molecules and nutrients that will direct development and nourish the growing embryo.

unless a routine physical exam reveals that the organs are not where they should be.

As frequently happens in biology, we just pushed the question back one more step. Why do the nodal cilia beat in such a way that the extracellular fluid flows from right to left? The root cause of this asymmetry likely originates in some of the early cell divisions of the fertilized egg. The fertilized egg goes through an initial series of cell divisions that subdivide the egg cytoplasm into a mass of undifferentiated cells. Although this mass of cells shows no hints of the eventual body plan, an uneven distribution of molecules from the cytoplasm of the fertilized egg can provide information that directs the fates of cells and sets up the body plan.

IN THIS CHAPTER we will see how a single cell becomes a multicellular animal through orderly cell movements that create multiple layers and set up cell–cell interactions. The regional and temporal differences in gene expression that control cell differentiation, described in Chapters 19 and 20, lead to the emergence of the body plan of the animal. We will discuss these early developmental steps in four organisms that have been studied extensively: sea urchins, frogs, chickens, and humans.

44.1 How Does Fertilization Activate Development?

Fertilization is the joining of sperm and egg. You might therefore think of it as the event that begins development. Keep two things in mind, however. First, in animals that reproduce asexually, development proceeds without fertilization. And second, in animals where fertilization does occur, it is preceded by critical events in the maturing egg that will influence subsequent development. Thus, in studying fertilization we are really asking how it activates or restarts multicellular development in sexually reproducing animals.

Fertilization does far more than just restore a full diploid complement of maternal and paternal genes. The fusion of sperm and egg plasma membranes accomplishes several things:

- It sets up blocks to the entry of additional sperm.
- It stimulates ion fluxes across the egg membrane.
- It changes the egg's pH.
- It increases egg metabolism and stimulates protein synthesis.
- It initiates the rapid series of cell divisions that produce a multicellular embryo.

Section 43.2 described the mechanisms of fertilization. Here we take a closer look at the cellular and molecular interactions of sperm and egg that initiate the first steps of development.

The sperm and the egg make different contributions to the zygote

In most species, eggs are much larger than sperm. Egg cytoplasm is well stocked with organelles, nutrients, and a variety of molecules, including transcription factors and mRNAs. The sperm is little more than a DNA delivery vehicle. Nearly everything the embryo needs during its early stages of development comes from the mother. In addition to providing its haploid nucleus, the sperm makes another important contribution to the zygote in most species—a centriole.

The centriole becomes the centrosome of the zygote, which organizes the mitotic spindles for subsequent cell divisions (see Figure 11.10). The centriole is also the origin of the primary cilia of cells, which are important in cell signaling, as we saw in the opening story about *situs inversus*.

Cytoplasmic factors in the egg play important roles in setting up the signaling cascades that orchestrate the major events of development: *determination, differentiation, and morphogenesis*.

Rearrangements of egg cytoplasm set the stage for determination

The unique attributes of amphibian eggs make them ideal models for illustrating how rearrangements of egg cytoplasm set the stage for determination. The molecules in the cytoplasm of the amphibian egg are not homogeneously distributed. The entry of the sperm into the egg stimulates rearrangements of the egg cytoplasm that introduce additional organization to the egg cytoplasm. This rearrangement establishes the polarity of the zygote, and when cell divisions begin, the informational molecules that will guide development are not divided equally among daughter cells.

Rearrangement of egg cytoplasm following fertilization is easily observed in some frog species because of pigments in the cytoplasm. The nutrients in an unfertilized frog egg are dense yolk granules that are concentrated by gravity in the lower half of the egg, called the **vegetal hemisphere**. The haploid nucleus of the egg is located at the opposite end, in the **animal hemisphere**. The outermost (*cortical*) cytoplasm of the animal hemisphere is heavily pigmented, and the underlying cytoplasm has more diffuse pigmentation. The vegetal hemisphere is not pigmented. Because of these differences, it is easy to observe how the cytoplasm is rearranged when a frog egg is fertilized.

The frog egg is radially symmetrical. You can turn it on its vegetal–animal pole axis, and all sides are the same. Sperm-binding sites are localized on the surface of the animal hemisphere, so that is where the sperm enters the egg. When a sperm enters the egg, bilateral symmetry is imposed by creating an anterior–posterior axis. Cortical cytoplasm rotates toward the site of sperm entry. This rotation brings different regions of cytoplasm into contact with each other on opposite sides of the egg, producing a band of diffusely pigmented cytoplasm on the side opposite the site of sperm entry. This band, called the **gray crescent**, marks the location of important developmental events in some species of amphibians (**Figure 44.1**).

The one non-nuclear organelle that the sperm contributes to the egg—the centriole—initiates the cytoplasmic reorganization revealed by the appearance of the gray crescent. The centriole organizes the microtubules in the vegetal hemisphere cytoplasm

into a parallel array that guides the movement of the cortical cytoplasm. These microtubules also appear to be directly responsible for movement of specific organelles and proteins, because these organelles and proteins move from the vegetal

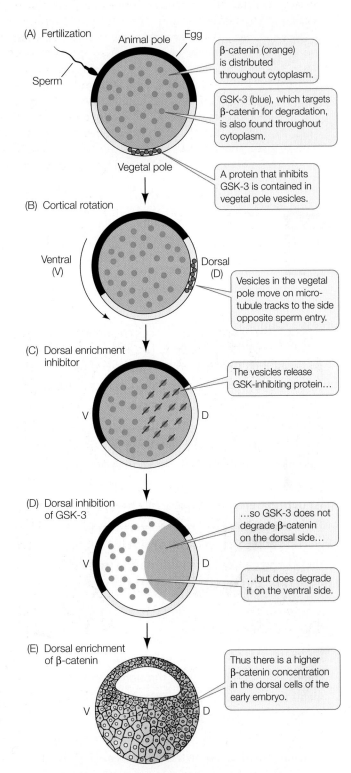

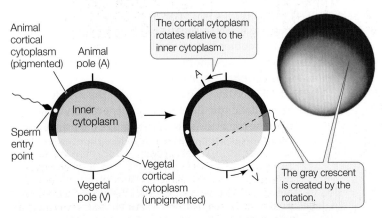

44.1 The Gray Crescent Rearrangement of the cytoplasm of frog eggs after fertilization creates the gray crescent.

44.2 Cytoplasmic Factors Set Up Signaling Cascades Cytoplasmic movement changes the distributions of critical developmental signals. In the frog zygote, the interaction of the protein kinase GSK-3, its inhibitor, and the protein β-catenin are crucial in specifying the dorsal–ventral axis of the embryo.

hemisphere to the gray crescent region even faster than the cortical cytoplasm rotates.

The movement of cytoplasm, proteins, and organelles changes the distribution of critical developmental signals. A key transcription factor in early development is β-catenin, which is produced from maternal mRNA (mRNA produced and stored in the egg while it was maturing in the ovary). Beta-catenin is found throughout the egg cytoplasm. Also present throughout the egg cytoplasm is a protein kinase called glycogen synthase kinase-3 (GSK-3), which phosphorylates and thereby targets β-catenin for degradation. An inhibitor of GSK-3 is segregated in the vegetal cortex of the egg. After sperm entry, this inhibitor is moved along microtubules to the gray crescent, where it prevents the degradation of β-catenin. As a result, the concentration of β-catenin is higher on the dorsal than on the ventral side of the developing embryo (**Figure 44.2**).

Beta-catenin plays a major role in the cell–cell signaling cascade that begins the process of cell determination and the formation of the embryo. But before cell–cell signaling can occur, multiple cells must be in place. Let's turn to the early series of cell divisions that transform the zygote into a multicellular embryo.

Cleavage repackages the cytoplasm

Transformation of the diploid zygote into a mass of cells occurs through a rapid series of cell divisions called **cleavage**. Because the cytoplasm of the zygote is not homogeneous, these first cell divisions result in the differential distribution of nutrients and cytoplasmic determinants in the early embryo.

In most animals, cleavage proceeds with rapid DNA replication and mitosis but with no cell growth and little gene expression. The embryo becomes a solid ball of smaller and smaller cells. Eventually, this ball forms a central fluid-filled cavity called a **blastocoel**, at which point the embryo is called a **blastula**. Its individual cells are called **blastomeres**. The pattern of cleavage in different species influences the form of their blastulas.

- **Complete cleavage** occurs in most eggs that have little **yolk** (stored nutrients). In this pattern, early cleavage furrows divide the egg completely and the blastomeres are of similar size. The frog egg undergoes complete cleavage, but because its vegetal pole contains more yolk, the division of the cytoplasm is unequal and the blastomeres in the animal hemisphere are smaller than those in the vegetal hemisphere (**Figure 44.3A**).

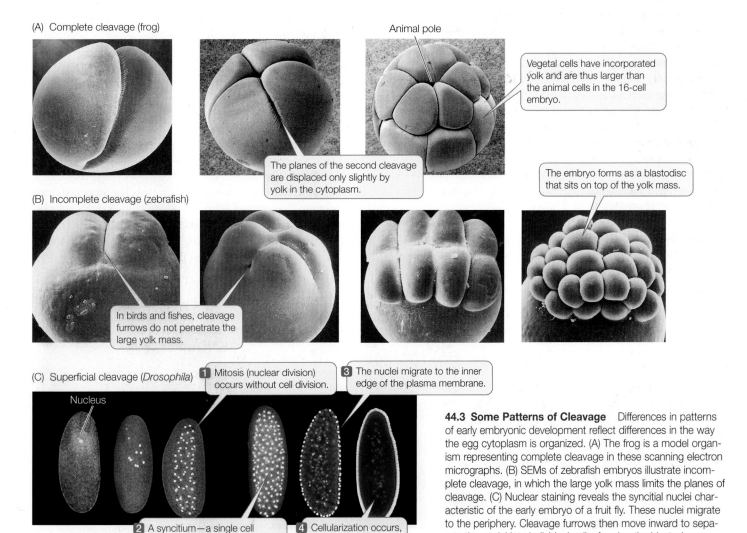

(A) Complete cleavage (frog)

Animal pole

> Vegetal cells have incorporated yolk and are thus larger than the animal cells in the 16-cell embryo.

> The planes of the second cleavage are displaced only slightly by yolk in the cytoplasm.

> The embryo forms as a blastodisc that sits on top of the yolk mass.

(B) Incomplete cleavage (zebrafish)

> In birds and fishes, cleavage furrows do not penetrate the large yolk mass.

(C) Superficial cleavage (*Drosophila*)

1 Mitosis (nuclear division) occurs without cell division.

3 The nuclei migrate to the inner edge of the plasma membrane.

Nucleus

2 A syncitium—a single cell with many nuclei—is produced.

4 Cellularization occurs, creating a blastoderm

44.3 Some Patterns of Cleavage Differences in patterns of early embryonic development reflect differences in the way the egg cytoplasm is organized. (A) The frog is a model organism representing complete cleavage in these scanning electron micrographs. (B) SEMs of zebrafish embryos illustrate incomplete cleavage, in which the large yolk mass limits the planes of cleavage. (C) Nuclear staining reveals the syncitial nuclei characteristic of the early embryo of a fruit fly. These nuclei migrate to the periphery. Cleavage furrows then move inward to separate the nuclei into individual cells, forming the blastoderm.

- **Incomplete cleavage** occurs in many species in which the egg contains a lot of yolk and the cleavage furrows do not penetrate it all. **Discoidal cleavage** is a type of incomplete cleavage common in fishes, reptiles, and birds, the eggs of which contain a dense yolk mass. The embryo forms as a disc of cells, called a **blastodisc**, that sits on top of the yolk mass (**Figure 44.3B**).

- **Superficial cleavage** is a variation of incomplete cleavage that occurs in insects such as the fruit fly (*Drosophila*). Early in development, cycles of mitosis occur without cell division, producing a *syncytium*—a single cell with many nuclei (**Figure 44.3C**). The nuclei eventually migrate to the periphery of the egg, after which the plasma membrane of the egg grows inward, partitioning the nuclei into individual cells surrounding a core of yolk.

The positions of the mitotic spindles during cleavage are not random but are defined by cytoplasmic factors produced from the maternal genome and stored in the egg (see Section 19.4). The orientation of the mitotic spindles can determine the planes of cleavage and the arrangement of the blastomeres.

In complete cleavage, if the mitotic spindles of successive cell divisions form parallel or perpendicular to the animal–vegetal axis of the zygote, a pattern of **radial cleavage** occurs as seen in the frog: the first two cell divisions are parallel to the animal–vegetal axis and the third is perpendicular to it (see Figure 44.3A). **Spiral cleavage** results when the mitotic spindles are at oblique angles to the animal–vegetal axis. In spiral cleavage, each new cell layer is shifted to the left or right, depending on the orientation of the mitotic spindles. Most mollusks have spiral cleavage, reflected in some species by a coiling shell pattern (as seen in snails).

Early cell divisions in mammals are unique

Several features of early cell divisions in placental mammals (eutherians) are so different from those seen in other animal groups that some biologists think it is inappropriate to call it cleavage. But whether you call it cleavage or not, it is still the sequence of early cell divisions that produces a body of undifferentiated cells that will become the embryo. This process in mammals is very slow. Cell divisions are 12 to 24 hours apart, compared with tens of minutes to a few hours in non-mammalian species. Also, the cell divisions of mammalian blastomeres are not in synchrony with each other. Because the blastomeres do not undergo mitosis at the same time, the number of cells in the embryo does not increase in the regular (2, 4, 8, 16, 32, etc.) progression typical of other species.

The pattern of mammalian cleavage is *rotational*: the first cell division is parallel to the animal–vegetal axis, yielding two blastomeres. In the second cell division, those two blastomeres divide at right angles to one other: one divides parallel to the animal–vegetal axis, while the other divides perpendicular to it (**Figure 44.4A**).

Another unique feature of the slow, rotational mammalian cleavage is that gene products expressed during cleavage play roles in cleavage. In animals such as sea urchins and frogs, gene transcription does not occur in the blastomeres, and cleavage is therefore directed exclusively by molecules that were present in the egg before fertilization.

As in other animals that have complete cleavage, the early cell divisions in a mammalian zygote produce a loosely associated ball of cells. After the 8-cell stage, however, the behavior of the mammalian blastomeres changes. They change shape to maximize their surface contact with one another, form tight junctions, and become a compact mass of cells (**Figure 44.4B**).

At the transition from the 16- to the 32-cell stage (the fourth division), the cells separate into two groups. The **inner cell mass** will become the embryo, while the surrounding outer cells become an encompassing sac called the **trophoblast**. Trophoblast cells secrete fluid, creating a cavity—the *blastocoel*—with the in-

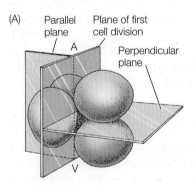

(A)   Parallel plane   Plane of first cell division

A

Perpendicular plane

V

44.4 Becoming a Blastocyst (A) Mammals have rotational cleavage, in which the plane of the first cleavage is parallel to the animal–vegetal (A–V) axis, but the second cell division involves two planes (beige) at right angles to each other. (B) Scanning electron micrographs show that asynchronous cell division results in an asymmetrical blastocyst at about the 32-cell stage. (C) Seen in cross section under a light microscope, the mammalian blastocyst consists of an inner cell mass adjacent to a fluid-filled blastocoel and surrounded by trophoblast cells.

(B)

8-cell stage

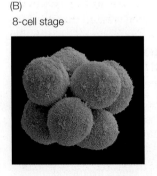

16-cell stage

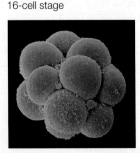

32-cell stage

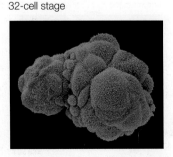

(C)   Blastocyst (cross section)   Trophoblast (outer cells)

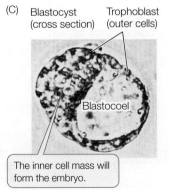

Blastocoel

The inner cell mass will form the embryo.

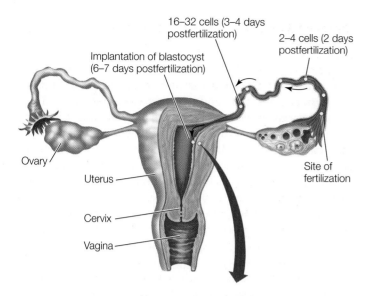

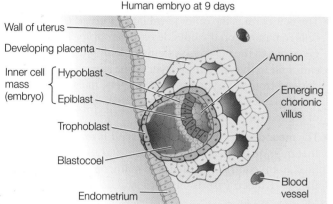

44.5 A Human Blastocyst at Implantation Adhesion molecules and proteolytic enzymes secreted by trophoblast cells allow the blastocyst to burrow into the endometrium. Once the blastocyst is implanted in the wall of the uterus, the trophoblast cells send out numerous projections—the chorionic villi—which increase the embryo's area of contact with the mother's bloodstream.

bryo, a connection develops between the circulatory systems of the embryo and the mother. As we will see later in this chapter, the structures that provide this connection are the placenta and the umbilical cord. Thus, the mammalian blastocyst must produce both the embryo (from the inner cell mass) and its support structures (from the trophoblast).

Fertilization in mammals occurs in the upper reaches of the mother's oviduct, and cleavage occurs as the zygote travels down the oviduct to the uterus. When the blastocyst arrives in the uterus, the trophoblast adheres to the lining of the uterus (the *endometrium*), beginning the process of **implantation**. In humans, implantation begins about 6 days after fertilization and is aided by adhesion molecules and enzymes secreted by the trophoblast (**Figure 44.5**).

As the blastocyst moves down the oviduct to the uterus, it must not embed itself in the oviduct (Fallopian tube) wall, or the result will be an *ectopic*, or *tubal*, *pregnancy*—a very dangerous condition. Early implantation is prevented by the zona pellucida, which surrounded the egg and remains around the cleaving ball of cells (see Section 43.2). At about the time the blastocyst reaches the uterus, it hatches from the zona pellucida, and implantation can occur.

Specific blastomeres generate specific tissues and organs

Cleavage results in a repackaging of the egg cytoplasm into a large number of small cells surrounding the fluid-filled blastocoel. Except in mammals, little cell differentiation and little if any gene expression occur during cleavage. Nevertheless, cells in different regions of the blastula possess different complements of the nutrients and cytoplasmic determinants that were present in the egg.

The blastocoel prevents cells from different regions of the blastula from coming into contact and interacting, but that will soon change. During the next stage of development, the cells of the blastula will move around and come into new associations with one another, communicate instructions to one another, and begin to differentiate. In many animals, these movements of the blastomeres are so regular and well orchestrated that it is possible to label a specific blastomere with a dye and identify the tissues and organs that form from its progeny. Such labeling experiments produce **fate maps** of the blastula (**Figure 44.6**).

Blastomeres become **determined**—committed to specific fates—at different times in different species. In some species,

ner cell mass at one end (**Figure 44.4C**). At this stage, the mammalian embryo is called a **blastocyst**, distinguishing it from the blastulas of other animal groups.

Why is mammalian cleavage so different? A key factor is that mammalian eggs contain no yolk and must derive all nutrients from the mother. Mammals are *viviparous*: the embryo develops within the uterus of the mother. To support the developing em-

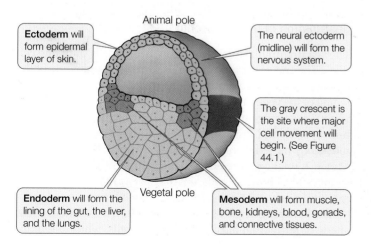

44.6 Fate Map of a Frog Blastula Colors indicate the portions of the blastula that will form the three germ layers and subsequently the frog's tissues and organs.

such as roundworms, the fates of blastomeres are restricted as early as the two-cell stage. If one of these blastomeres is experimentally removed, a particular portion of the embryo will not form. This type of development has been called **mosaic development** because each blastomere seems to contribute a specific set of "tiles" to the final "mosaic" that is the adult animal.

In contrast to mosaic development, the loss of some cells during cleavage in **regulative development** does not affect the developing embryo, because the remaining cells compensate for the loss. Regulative development is typical of many vertebrate species, including humans. The pluripotent cells of the mammalian blastocyst (the inner cell mass) are known as *embryonic stem cells* and are the subject of much research, particularly because of their therapeutic potential (see Section 19.2).

If some blastomeres can change their fate to compensate for the loss of other cells during cleavage and blastula formation, can those cells form an entire embryo? To a certain extent, yes. During cleavage or early blastula formation in mammals, for example, if the blastomeres are physically separated into two groups, both groups can produce complete embryos. Since the two embryos come from the same zygote, they will be *monozygotic twins*—genetically identical.

Non-identical twins occur when two separate eggs are fertilized by two separate sperm. Thus, while identical twins are always of the same sex, non-identical twins have a 50 percent chance of being the same sex. In about 1 out of 50,000 human pregnancies, genetic or environmental factors cause the inner cell mass to split partially. The result is twins that are *conjoined* at some point on their bodies, usually sharing some of their organs and limbs.

44.1 RECAP

The egg is stocked with nutrients and informational molecules that power and direct the early stages of development. Fertilization activates the egg and stimulates rearrangement of the cytoplasm, setting up the body axes and positional information that initiate signaling cascades, which control determination and differentiation.

- Explain how β-catenin becomes concentrated in only certain blastomeres. **See p. 925 and Figure 44.2**

- In general terms, describe the difference between complete and incomplete cleavage. **See pp. 925–926 and Figure 44.3**

- What does a fate map tell us? How are fate maps constructed? **See pp. 927–928 and Figure 44.6**

Of the next stage of development—gastrulation—the developmental biologist Louis Wolpert once said, "It is not birth, marriage, or death, but gastrulation which is the most important time in your life." During gastrulation, cell movements create new cell-to-cell contacts, which in turn sets up signaling cascades. Signaling cascades initiate the differentiation of cells and tissues and set the stage for the emergence of the body plan.

44.2 How Does Gastrulation Generate Multiple Tissue Layers?

The blastula is typically a fluid-filled ball of cells. How does this simple ball of cells become an embryo made up of multiple tissue layers with head and tail ends and dorsal and ventral sides? **Gastrulation** is the process whereby the blastula is transformed by massive movements of cells into an embryo with multiple tissue layers and distinct body axes. The resulting spatial relationships between tissues make possible the inductive interactions between cells that trigger differentiation and organ formation (see Figure 19.10).

During gastrulation, three **germ layers** (also called *cell layers* or *tissue layers*) form (see Figure 44.6):

- The **endoderm** is the innermost germ layer, created as some blastomeres move to the inside of the embryo. The endoderm gives rise to the lining of the digestive tract, respiratory tract, pancreas, and liver.

- The **ectoderm** is the outer germ layer, formed from those cells remaining on the outside of the embryo. The ectoderm gives rise to the nervous system, including the eyes and ears; and to the epidermal layer of the skin and structures derived from skin, such as hair, feathers, nails or claws, sweat glands, oil glands, and even teeth and other tissues of the mouth.

- The **mesoderm** is the middle layer and is made up of cells that migrate between the endoderm and the ectoderm. The mesoderm contributes tissues to many organs, including the heart, blood vessels, muscles, and bones.

Some of the most interesting and important challenges in animal development have dealt with two related questions: what directs the cell movements of gastrulation, and what is responsible for the resulting patterns of cell differentiation and organ formation? Scientists have made significant progress in answering both these questions at the molecular level. In the following discussion, we will begin with sea urchin gastrulation because it is the simplest to conceptualize in spatial terms. We will then describe the more complex pattern of gastrulation in frogs, which in turn will help elucidate the still more complex patterns in reptiles, birds, and mammals.

Invagination of the vegetal pole characterizes gastrulation in the sea urchin

The sea urchin blastula is a hollow ball of cells only one cell thick. The end of the blastula stage is marked by slowing of the rate of mitosis; the beginning of gastrulation is marked by a flattening of the vegetal hemisphere (**Figure 44.7**). Some cells at the vegetal pole bulge into the blastocoel, break away from neighboring cells, and migrate into the cavity. These cells become **mesenchyme**—cells of the middle germ layer, the mesoderm. Mesenchymal cells are not organized in tightly packed sheets or tubes like epithelial cells are; they act as independent units, migrating into and among the other tissue layers.

The flattening at the vegetal pole results from changes in the shape of individual blastomeres. These cells, which are origi-

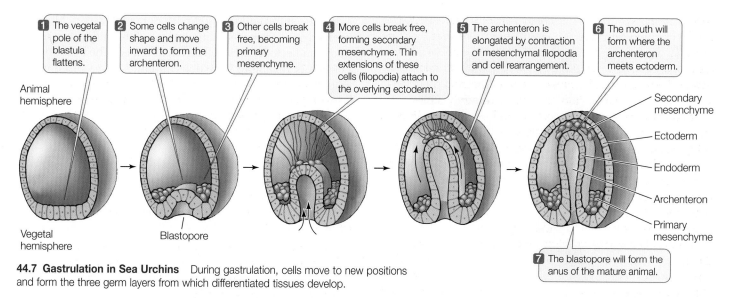

1 The vegetal pole of the blastula flattens.

2 Some cells change shape and move inward to form the archenteron.

3 Other cells break free, becoming primary mesenchyme.

4 More cells break free, forming secondary mesenchyme. Thin extensions of these cells (filopodia) attach to the overlying ectoderm.

5 The archenteron is elongated by contraction of mesenchymal filopodia and cell rearrangement.

6 The mouth will form where the archenteron meets ectoderm.

Animal hemisphere

Vegetal hemisphere

Blastopore

Secondary mesenchyme

Ectoderm

Endoderm

Archenteron

Primary mesenchyme

7 The blastopore will form the anus of the mature animal.

44.7 Gastrulation in Sea Urchins During gastrulation, cells move to new positions and form the three germ layers from which differentiated tissues develop.

---

yourBioPortal.com

GO TO Animated Tutorial 44.1 • Gastrulation

---

nally rather cuboidal, become wedge-shaped, with smaller outer edges and larger inner edges. As a result, the vegetal pole bulges inward, or *invaginates*, as if someone were poking a finger into a hollow ball (see Figure 44.7). The invaginating cells become endoderm and form the primitive gut, called the **archenteron**. At the tip of the archenteron, more cells enter the blastocoel to form more mesoderm.

Changes in cell shapes cause the initial invagination of the archenteron, but eventually it is pulled by the mesenchyme cells. These cells, attached to the tip of the archenteron, send out extensions called filopodia that adhere to the overlying ectoderm. When the filopodia contract, they pull the archenteron toward the ectoderm at the opposite end of the embryo from where the invagination began. The mouth of the animal forms where the archenteron makes contact with this overlying ectoderm. The opening created by the invagination of the vegetal pole is called the **blastopore**; it will become the anus of the animal.

What mechanisms control the various cell movements of sea urchin gastrulation? The immediate answer is that specific properties of particular blastomeres change. For example, some vegetal cells change shape and bulge into the blastocoel, and these cells become mesenchyme. Once they lose contact with their neighboring cells on the surface of the blastula, they send out filopodia that then move along an extracellular matrix of proteins laid down by the cells lining the blastocoel.

A deeper understanding of gastrulation requires that we discover the molecular mechanisms whereby different blastomeres develop different properties. Cleavage systematically divides up the cytoplasm of the egg. The sea urchin blastula at the 64-cell stage is radially symmetrical, but it has *polarity*, as described in Section 19.4. It consists of tiers of cells. As in the frog blastula, the top is the animal pole and the bottom the vegetal pole.

If different tiers of blastula cells are separated, they show different developmental potentials; only cells from the vegetal pole are capable of initiating the development of a complete larva

(see Figure 19.8). It has been proposed that these differences are due to uneven distribution of various transcriptional regulatory proteins in the egg cytoplasm. As cleavage progresses, these proteins end up in different groups of cells. Therefore, specific sets of genes are activated in different cells, determining their different developmental capacities.

Let's turn now to gastrulation in the frog, in which several key signaling molecules have been identified.

Gastrulation in the frog begins at the gray crescent

Amphibian blastulas have considerable yolk and are more than one cell thick; therefore, gastrulation is more complex in amphibians than in sea urchins. Variation is considerable among different species of amphibians, but in this brief account we will use results from studies done on different species to produce a generalized picture of amphibian development.

Amphibian gastrulation begins when certain cells in the gray crescent region change their shapes and cell adhesion properties. These cells bulge inward toward the blastocoel while they remain attached to the outer surface of the blastula by slender necks. Because of their shape, these cells are called *bottle cells*.

Bottle cells mark the spot where the **dorsal lip** of the blastopore will form (**Figure 44.8**). As the bottle cells move inward, the dorsal lip is created, and a sheet of cells moves over it into the blastocoel. This process is called **involution**. One group of involuting cells is the prospective endoderm; these cells form the primitive gut, or archenteron. Another group will move between the endoderm and the outermost cells to form the mesoderm. These rearrangements are due to changes in cell properties called **convergent extension**. The cells elongate in the direction of movement, but they also intercalate (move in between each other). If they just elongated, the migrating group of cells would become much narrower; by intercalating, they maintain the width of the migrating cell group.

As gastrulation proceeds, cells from the animal hemisphere flatten and move toward the site of involution in a process called **epiboly**. The blastopore lip widens and eventually forms a com-

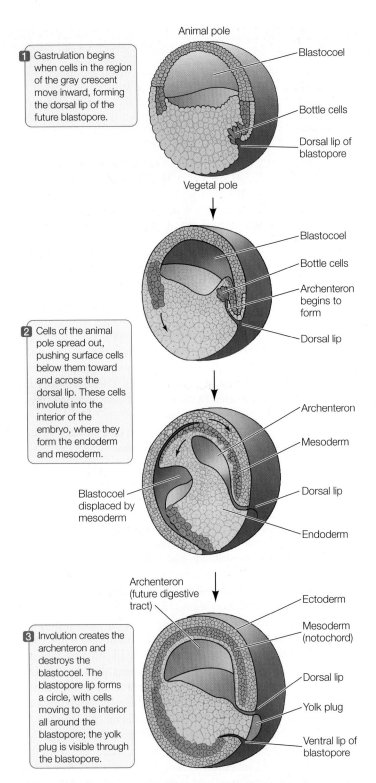

1 Gastrulation begins when cells in the region of the gray crescent move inward, forming the dorsal lip of the future blastopore.

Animal pole

Blastocoel

Bottle cells

Dorsal lip of blastopore

Vegetal pole

Blastocoel

Bottle cells

Archenteron begins to form

Dorsal lip

2 Cells of the animal pole spread out, pushing surface cells below them toward and across the dorsal lip. These cells involute into the interior of the embryo, where they form the endoderm and mesoderm.

Archenteron

Mesoderm

Dorsal lip

Blastocoel displaced by mesoderm

Endoderm

Archenteron (future digestive tract)

Ectoderm

Mesoderm (notochord)

3 Involution creates the archenteron and destroys the blastocoel. The blastopore lip forms a circle, with cells moving to the interior all around the blastopore; the yolk plug is visible through the blastopore.

Dorsal lip

Yolk plug

Ventral lip of blastopore

44.8 Gastrulation in the Frog Embryo The colors in this diagram are matched to those in Figure 44.6, the frog fate map.

plete circle surrounding a "plug" of yolk-rich cells. As cells continue to move inward through the blastopore, the archenteron grows, gradually displacing the blastocoel.

As gastrulation comes to an end, the amphibian embryo consists of three germ layers: ectoderm on the outside, endoderm on the inside, and mesoderm in between. The embryo also has

a dorsal–ventral and anterior–posterior organization. Most importantly, the fates of specific regions of the endoderm, mesoderm, and ectoderm have been determined. The beautiful experiments revealing how determination takes place in the amphibian embryo are an old but exciting story.

The dorsal lip of the blastopore organizes embryo formation

In the early 1900s, the German biologist Hans Spemann was studying the development of salamander eggs. He was interested in finding out whether the nuclei of blastomeres remain capable of directing the development of complete embryos. With great patience and dexterity, he formed loops from single hairs taken from a baby (in fact, his daughter) and tied them around fertilized eggs along the plane of the first cell division, effectively dividing the eggs in half, with the nucleus restricted to one side. That side went through cell divisions and developed into a salamander; the other half simply degenerated. Up until the 16-cell stage, if one nucleus escaped to the other side of the constriction, twin salamanders could develop. Thus, each of the nuclei of the blastula (at least up to the 16-cell stage) was capable of directing and supporting development of the whole organism.

But, as often happens in science, Spemann's bisection experiments revealed a new phenomenon. Sometimes the half of the blastula receiving an escaped nucleus did not develop. When his loops bisected the gray crescent, both halves of the zygote developed into a complete embryo. When he tied the loops so the gray crescent was on only one side of the constriction, however, only that half of the zygote developed into a complete embryo (**Figure 44.9**). The half lacking gray crescent material underwent cell division, but even if it contained a nucleus, it became a clump of undifferentiated cells that Spemann called a "belly piece." Spemann hypothesized that cytoplasmic factors unequally distributed in the fertilized egg were necessary for gastrulation and the development of a normal salamander.

To further test the hypothesis that cells receiving different complements of cytoplasmic factors had different developmental fates, Spemann transplanted pieces of early gastrulas to various locations on other gastrulas. Guided by fate maps (see Figure 44.6), he was able to take a piece of ectoderm he knew would develop into skin and transplant it to a region that normally becomes part of the nervous system, and vice versa.

When he performed these transplants in early gastrulas—when the blastopore was just beginning to form—the transplanted pieces always developed into tissues that were appropriate for the location where they were placed. Transplanted cells destined to become epidermis in their original location developed into nervous system tissue, and transplanted cells destined to become nervous system tissue in their original location developed into host epidermis. Thus, Spemann learned that the fates of the transplanted cells had not been determined before the transplantation (see Figure 19.2).

In late gastrulas, however, the same experiment yielded opposite results. Transplanted cells destined to become epidermis in their original location produced patches of skin cells in the host

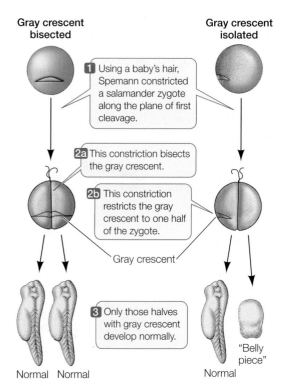

Gray crescent bisected

Gray crescent isolated

1 Using a baby's hair, Spemann constricted a salamander zygote along the plane of first cleavage.

2a This constriction bisects the gray crescent.

2b This constriction restricts the gray crescent to one half of the zygote.

Gray crescent

3 Only those halves with gray crescent develop normally.

Normal   Normal

"Belly piece"

Normal

44.9 Gastrulation and the Gray Crescent Spemann's research revealed that gastrulation and subsequent normal development in salamanders depends on cytoplasmic determinants localized in the gray crescent.

nervous system, and the transplanted cells from regions that would develop into nervous system tissue produced neural tissue in the skin of the recipient. At some point during gastrulation, the fates of the embryonic cells had become determined.

Spemann's next experiment, done with his student Hilde Mangold, produced momentous results: they transplanted the dorsal lip of the blastopore (**Figure 44.10**). When this small piece of tissue was transplanted into the presumptive belly area of another gastrula, it stimulated a second site of gastrulation—and a second complete embryo formed belly-to-belly with the original embryo. Because the dorsal lip of the blastopore was apparently capable of inducing the host tissue to form an entire embryo, Spemann and Mangold dubbed the dorsal lip tissue the **primary embryonic organizer**, or simply the **organizer**. For more than 80 years, the organizer has been an active area of research.

─── **yourBioPortal.com** ───

GO TO Animated Tutorial 44.2 • Tissue Transplants Reveal the Process of Determination

Transcription factors underlie the organizer's actions

With the advent of modern molecular methods, the primary embryonic organizer has been studied intensively to discover the molecular mechanisms involved

INVESTIGATING LIFE

44.10 The Dorsal Lip Induces Embryonic Organization

In a classic experiment, Hans Spemann and Hilde Mangold transplanted the dorsal blastopore lip mesoderm of an early gastrula stage salamander embryo. The results showed that the cells of this embryonic region, which they dubbed "the organizer," could direct the formation of an entire embryo.

HYPOTHESIS Cytoplasmic factors in the early dorsal blastopore lip organize cell differentiation in amphibian embryos.

METHOD

1. Excise a patch of mesoderm tissue from above the dorsal blastopore lip of an early gastrula stage salamander embryo (the donor).
2. Transplant the donor tissue onto a recipient embryo at the same stage. The donor tissue is transplanted onto a region of ectoderm that should become epidermis (skin).

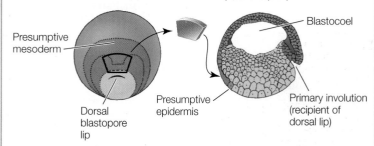

Presumptive mesoderm

Dorsal blastopore lip

Presumptive epidermis

Blastocoel

Primary involution (recipient of dorsal lip)

RESULTS

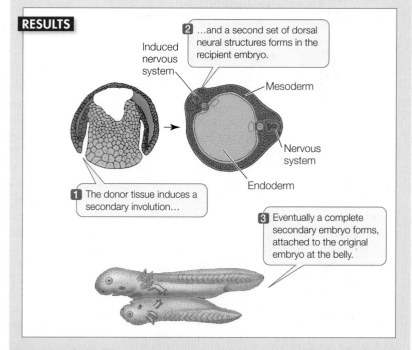

2 ...and a second set of dorsal neural structures forms in the recipient embryo.

Induced nervous system

Mesoderm

Nervous system

Endoderm

1 The donor tissue induces a secondary involution...

3 Eventually a complete secondary embryo forms, attached to the original embryo at the belly.

CONCLUSION The cells of the dorsal blastopore lip can induce other cells to change their developmental fates.

Go to **yourBioPortal.com** for original citations, discussions, and relevant links for all INVESTIGATING LIFE figures.

in its action. The distribution of the transcription factor β-catenin in the late blastula corresponds to the location of the organizer in the early gastrula, so β-catenin is a candidate for the initiator of organizer activity. To prove that a protein is an inductive signal, it has to be shown that it is both *necessary* and *sufficient* for the proposed effect. In other words, the effect should not occur if the candidate protein is not present (necessity), and the candidate protein should be capable of inducing the effect where it would otherwise not occur (sufficiency).

The criteria of necessity and sufficiency have been satisfied for β-catenin. If β-catenin mRNA transcripts are depleted by injections of antisense RNA into the egg (see Section 18.4), gastrulation does not occur. If β-catenin is experimentally overexpressed in another region of the blastula, it can induce a second axis of embryo formation, as the transplanted dorsal lip did in the Spemann–Mangold experiments. Thus, β-catenin appears to be both necessary and sufficient for the formation of the primary embryonic organizer—but it is only one component of a complex signaling process.

How the presence of β-catenin creates the organizer, and how the organizer then induces the beginnings of the body plan, involves a complex series of interactions between transcription factors and growth factors that control gene expression. What follows is only a portion of this complex and still emerging story. What you should take from this description is not the names of the genes and gene products involved. Rather, we hope you will gain a basic appreciation for how signaling molecules interact to produce different combinations of signals that convey positional and temporal information. This information guides cells into different paths of determination and differentiation.

Studies of early gastrulas revealed that primary embryonic organizer activity is generated by the interaction of β-catenin with signals coming from the vegetal cells. Together, they activate the expression of the transcription factor **Goosecoid**. Expression of the *goosecoid* gene depends on two signaling pathways.

The first of these pathways involves a *goosecoid*-promoting transcription factor called Siamois. The *siamois* gene is normally repressed by a ubiquitous transcription factor called Tcf-3, but in cells in which β-catenin is present, an interaction between Tcf-3 and β-catenin induces *siamois* expression (**Figure 44.11**). But Siamois protein alone is not sufficient for *goosecoid* expression.

The second pathway involves mRNAs from the original egg cytoplasm for a family of proteins called transforming growth factor-β (TGF-β). TGF-β interacts with the Siamois protein to control *goosecoid* transcription. Thus, you can see that it is a complex combination of factors that determines which cells become the primary organizer.

The organizer changes its activity as it migrates from the dorsal lip

Organizer cells begin the process of formation of the dorsal lip of the blastopore. Specifically, these cells are at the center of the dorsal lip and involute, moving forward on the midline (i.e., the middle of the anterior–posterior axis). The first organizer cells to enter the embryo migrate anteriorly to become the head endoderm and head mesoderm. Here, they induce neighboring

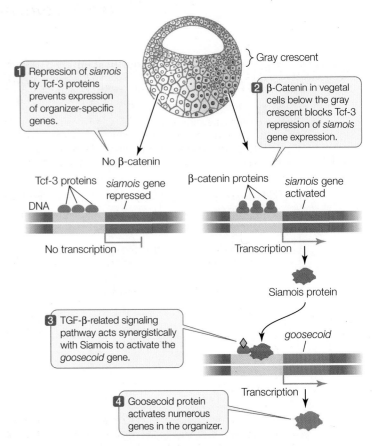

44.11 Molecular Mechanisms of the Organizer In amphibians, the organizing potential of the gray crescent depends on the activity of the *goosecoid* gene, which in turn is activated by signaling pathways set up in the vegetal cells below the gray crescent.

cells to participate in making structures of the head. Later organizer cells that involute into the embryo will induce structures of the trunk, and the last of the organizer cells to move inward from the dorsal lip will induce structures of the tail. How does the nature of the organizer cells change to enable them to induce head, trunk, or tail structures?

Inductive tissue interactions can suppress as well as activate. As we learned above, the early organizer cells express the transcription factor Goosecoid, which activates genes encoding soluble signals. As these cells move forward in the blastocoel, they come into contact with new populations of cells that produce a number of different growth factors. For head structures to form, certain of these growth factors have to be suppressed. The anteriormost organizer cells, under the influence of Goosecoid, produce and release antagonists to those growth factors.

The induction of trunk structures requires suppression of a different set of growth factors. In organizer cells that involute later than the head organizers, Goosecoid is no longer the dominant transcription factor, and these cells express different growth factor antagonists. The induction of tail structures requires still different activities of the organizer cells that involute last. Thus, the organizer cells express appropriate sets of growth factor antagonists at the right times to achieve different patterns of differentiation on the anterior–posterior axis.

The initiation of the development of the nervous system also involves a suppressive tissue interaction. For a long time it was thought that the involuting organizer cells actively induced the

INVESTIGATING LIFE

44.12 Differentiation Can Be Due to Inhibition of Transcription Factors

When organizer cells involute to underlie dorsal ectoderm along the embryo midline, that overlying ectoderm becomes neural tissue rather than skin (epidermis). But do the organizer cells cause dorsal ectoderm to become neural tissue, or do they *prevent* this ectoderm from becoming skin?

HYPOTHESIS The default state of amphibian dorsal ectoderm is neural; it is induced by underlying mesoderm to become epidermis.

METHOD

1. Excise the animal caps of late-stage frog blastulas and disperse the cells in culture medium so there is no cell-to-cell contact. From the culture, extract molecules of BMP4 (secreted by mesoderm cells) and molecules of an inhibitor of BMP4.

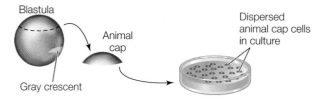

2. Prepare four separate cultures of embryonic ectodermal cells. Incubate with no additions (control); with BMP4 from step 1; with BMP4 inhibitor from step 1; and with both molecules.

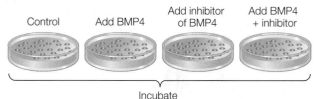

3. After incubation, extract mRNAs from the ectodermal cells and analyze for the presence of mRNAs for marker proteins NCAM (neural cell adhesion molecule, a neural protein) and/or keratin (an epidermal protein).

RESULTS The control ectoderm (no inductive factors added) expresses the neural marker. In the presence of mesodermal BMP4, ectoderm expresses the epidermal marker. If BMP4 is inhibited, ectoderm expresses the neural marker.

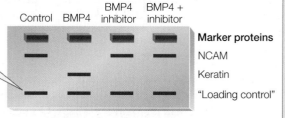

This control message is from a gene expressed in all cells and verifies that each sample contains similar amounts of mRNA.

CONCLUSION The default state of amphibian dorsal ectoderm is neural. BMP4 protein from mesoderm can induce ecotoderm cells to differentiate into epidermis. Thus the organizer cells must secrete an inhibitor of BMP4.

Go to **yourBioPortal.com** for original citations, discussions, and relevant links for all INVESTIGATING LIFE figures.

overlying ectoderm to form neural tissue rather than becoming epidermis. We now know, however, that epidermis is not the default state of the dorsal ectoderm. Rather, the underlying mesoderm secretes factors called BMP proteins that induce the ectoderm to become epidermis. The role of the involuting organizer cells is to block that induction, allowing the overlying ectodermal cells to follow what is really their default pathway—differentiation into neural tissue (**Figure 44.12**).

Reptilian and avian gastrulation is an adaptation to yolky eggs

The eggs of reptiles and birds contain a mass of yolk, and the blastulas of these groups develop as a disc of cells on top of the yolk (see Figure 44.3B). We will use the chicken egg to show how gastrulation proceeds in a flat disc of cells rather than in a ball of cells.

Cleavage in the chick results in a flat, circular layer of cells called a blastodisc (**Figure 44.13**). Between the blastodisc and the yolk mass is a fluid-filled space. Some cells from the blastodisc break free and move into this space. These cells come together to form a continuous layer called the **hypoblast**, which will later contribute to *extraembryonic membranes* that will support and nourish the developing embryo. The overlying cells make up the **epiblast**, from which the embryo proper will form. Thus, the avian blastula is a flattened structure consisting of an upper epiblast and a lower hypoblast, which are joined at the margins of the blastodisc. The blastocoel is the fluid-filled space between the epiblast and hypoblast.

Gastrulation begins with a thickening in the posterior region of the epiblast, caused by the movement of cells toward the midline and then forward along the midline (see Figure 44.13). The result is a midline ridge called the *primitive streak*. A depression called the *primitive groove* forms along the length of the primitive streak. The primitive groove functions as the blastopore, and cells migrate through it into the blastocoel to become endoderm and mesoderm.

In the chick embryo, no archenteron forms, but the endoderm and mesoderm migrate forward to form the gut and other structures. At the anterior end of the primitive groove is a thickening called **Hensen's node**, which in birds, reptiles, and mammals is the equivalent of the dorsal lip of the amphibian blastopore. Many signaling molecules that have been identified in the frog organizer are also expressed in Hensen's node. Some cells that pass over

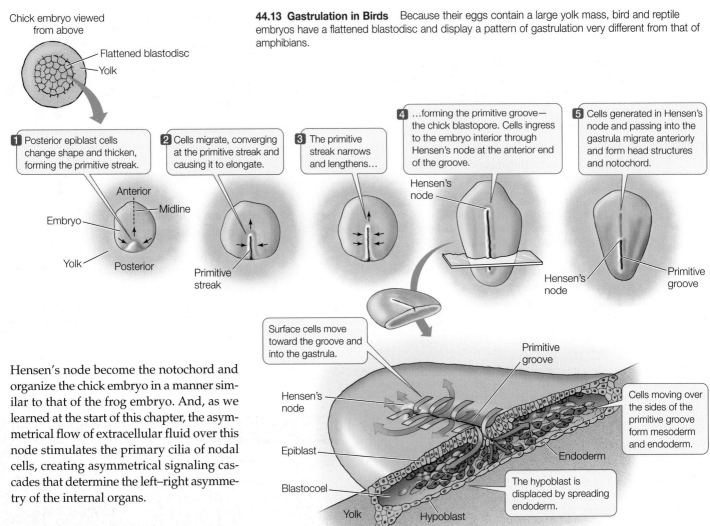

Chick embryo viewed from above

Flattened blastodisc

Yolk

44.13 Gastrulation in Birds Because their eggs contain a large yolk mass, bird and reptile embryos have a flattened blastodisc and display a pattern of gastrulation very different from that of amphibians.

1 Posterior epiblast cells change shape and thicken, forming the primitive streak.

2 Cells migrate, converging at the primitive streak and causing it to elongate.

3 The primitive streak narrows and lengthens…

4 …forming the primitive groove— the chick blastopore. Cells ingress to the embryo interior through Hensen's node at the anterior end of the groove.

5 Cells generated in Hensen's node and passing into the gastrula migrate anteriorly and form head structures and notochord.

Anterior

Midline

Embryo

Yolk

Posterior

Primitive streak

Hensen's node

Hensen's node

Primitive groove

Surface cells move toward the groove and into the gastrula.

Primitive groove

Hensen's node

Cells moving over the sides of the primitive groove form mesoderm and endoderm.

Epiblast

Endoderm

Blastocoel

The hypoblast is displaced by spreading endoderm.

Yolk

Hypoblast

Cross section through chick embryo

Hensen's node become the notochord and organize the chick embryo in a manner similar to that of the frog embryo. And, as we learned at the start of this chapter, the asymmetrical flow of extracellular fluid over this node stimulates the primary cilia of nodal cells, creating asymmetrical signaling cascades that determine the left–right asymmetry of the internal organs.

Placental mammals retain the avian–reptilian gastrulation pattern but lack yolk

Mammalian embryos (with the exception of monotremes) derive their nourishment from the maternal circulation, and therefore mammalian eggs do not have large amounts of yolk to constrain their patterns of cleavage and early development. Nevertheless, mammals and birds evolved from reptilian ancestors, so it is not surprising that they share certain patterns of early development. Earlier we described the development of the mammalian inner cell mass (the equivalent of the avian epiblast) and the outer trophoblast.

As in avian development, in placental mammals the inner cell mass splits into an upper layer called the epiblast and a lower layer called the hypoblast. The embryo forms from the epiblast, while the hypoblast contributes to the extraembryonic membranes that will encase the developing embryo and help form the placenta (see Figure 44.5). The epiblast also contributes to the extraembryonic membranes; specifically, it splits off an upper layer of cells that will form the amnion. The amnion will grow to surround the developing embryo as a membranous sac filled with amniotic fluid. Gastrulation occurs in the mammalian epiblast just as it does in the avian epiblast. A primitive groove forms, and epiblast cells migrate through the groove to become layers of endoderm and mesoderm.

44.2 RECAP

The cell movements of gastrulation convert the blastula into an embryo with three tissue layers. New contacts between cells set up inductive signaling interactions that determine cell fates. Dorsal lip tissue is the source of organizer cells that induce development of preliminary head, trunk, and tail structures.

- Describe and compare the cell movements that occur during gastrulation in a sea urchin, a frog, and a bird. **See Figures 44.7, 44.8, and 44.13**

- Explain the molecular basis for the inductive capabilities of the organizer. **See pp. 931–932 and Figures 44.11 and 44.12**

We have described how the fertilized egg develops into an embryo with three germ layers and how cellular signals trigger different patterns of differentiation. In the next section we will describe how organs and organ systems develop.

44.3 How Do Organs and Organ Systems Develop?

Gastrulation produces an embryo with three germ layers that are positioned to influence one another through inductive tissue interactions. During the next phase of development, called **organogenesis**, many organs and organ systems develop simultaneously and in coordination with one another. An early process of organogenesis in chordates that is directly related to gastrulation is neurulation. **Neurulation** is the initiation of the nervous system. We will examine neurulation in the amphibian embryo, but it occurs in a similar fashion in reptiles, birds, and mammals.

The stage is set by the dorsal lip of the blastopore

As we learned in the previous section, one group of cells that passes over the dorsal lip of the blastopore moves anteriorly and becomes the endodermal lining of the digestive tract. The other group of cells that involutes over the dorsal lip becomes *chordamesoderm*, so named because it forms a rod of mesoderm—the **notochord**—that extends down the center of the embryo. These cells also have important organizer functions (see Figure 44.8). The notochord gives structural support to the developing embryo and is eventually replaced by the vertebral column. The organizing capacity of the chordamesoderm enables the overlying ectoderm to become neural ectoderm (see Figure 44.12). It does this by expressing signaling molecules (one appropriately called Noggin and another one called Chordin) that initiate differentiation of the different divisions of the nervous system.

Neurulation involves the formation of an internal neural tube from an external sheet of cells. The first signs of neurulation are flattening and thickening of the ectoderm overlying the notochord; this thickened area forms the *neural plate* (**Figure 44.14A**). The edges of the neural plate that run in an anterior–posterior direction continue to thicken to form ridges or folds. Between these neural folds, a groove forms and deepens as the folds roll over it to converge on the midline. The folds fuse, forming a cylinder, the **neural tube**, and a continuous overlying layer of epidermal ectoderm (**Figure 44.14B–D**).

Cells from the most lateral portions of the neural plate do not become part of the neural tube, but disassociate from it and come to lie between the neural tube and the overlying epidermis. These **neural crest cells** migrate outward to lead the development of the connections between the central nervous system (brain and spinal cord) and the rest of the body.

The neural tube develops bulges at the anterior end, which become the major divisions of the brain; the rest of the tube becomes the spinal cord. In humans, failure of the neural folds to fuse in this posterior region results in a birth defect known as *spina bifida*. If the folds fail to fuse at the anterior end, an infant can develop without a forebrain—a condition called *anencephaly*. Although several genetic factors can cause these defects, other factors are environmental, including maternal diet. The incidence of neural tube defects in the United States in the early 1900s was as high as 1 in 300 live births; today it is less than 1

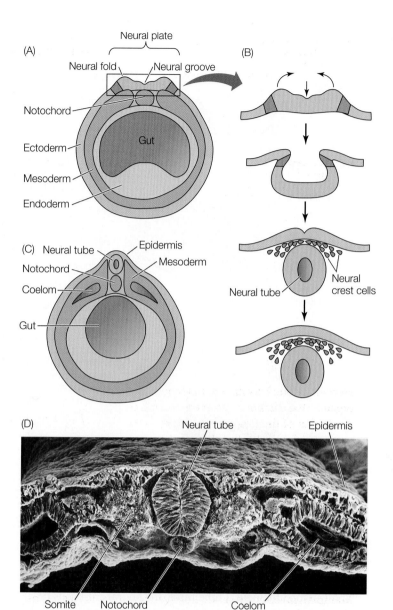

44.14 Neurulation in a Vertebrate (A) At the start of neurulation, the ectoderm of the neural plate (green) is flat. (B) The neural plate invaginates and folds, forming a tube. (C,D) The completely formed neural tube seen in (C) diagrammatic form and (D) in a scanning electron micrograph of a chick embryo.

in 1,000. A major factor in this improvement has been the inclusion of folic acid (a B vitamin, also known as folate) in the mother's diet. It is essential for pregnant women to ingest sufficient folic acid.

Body segmentation develops during neurulation

The vertebrate body plan, like that of arthropods, consists of repeating segments that are modified during development. These segments are most evident as the repeating patterns of vertebrae, ribs, nerves, and muscles along the anterior–posterior axis.

As the neural tube forms, mesodermal tissues gather along the sides of the notochord to form separate, segmented blocks

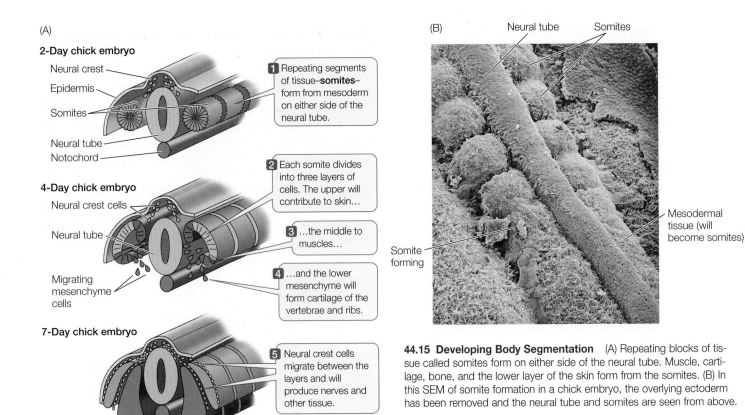

(A)

2-Day chick embryo

Neural crest
Epidermis
Somites
Neural tube
Notochord

1 Repeating segments of tissue—**somites**—form from mesoderm on either side of the neural tube.

4-Day chick embryo

Neural crest cells
Neural tube
Migrating mesenchyme cells

2 Each somite divides into three layers of cells. The upper will contribute to skin…

3 …the middle to muscles…

4 …and the lower mesenchyme will form cartilage of the vertebrae and ribs.

7-Day chick embryo

5 Neural crest cells migrate between the layers and will produce nerves and other tissue.

(B)

Neural tube   Somites

Mesodermal tissue (will become somites)

Somite forming

44.15 Developing Body Segmentation (A) Repeating blocks of tissue called somites form on either side of the neural tube. Muscle, cartilage, bone, and the lower layer of the skin form from the somites. (B) In this SEM of somite formation in a chick embryo, the overlying ectoderm has been removed and the neural tube and somites are seen from above.

of cells called **somites** (**Figure 44.15**). The somites produce cells that will become the vertebrae, ribs, muscles of the trunk and limbs, and the lower layer of the skin.

Nerves that connect the brain and spinal cord with tissues and organs throughout the body are also arranged segmentally. The somites help guide the organization of these peripheral nerves, but the nerves are not of mesodermal origin. As we saw above, when the neural tube fuses, the neural crest cells break loose and migrate inward between the epidermis and the somites and through the somites. These neural crest cells have diverse fates, including the development of peripheral nerves.

As development progresses, the different segments of the body change. Regions of the spinal cord differ, regions of the vertebral column differ in that some vertebrae grow ribs of various sizes and others do not, forelegs arise in the anterior part of the embryo, and hind legs arise in the posterior region.

Hox genes control development along the anterior–posterior axis

How is mesoderm in the anterior part of a mouse embryo programmed to produce forelegs rather than hind legs? In Section 19.5, we saw how homeotic genes control body segmentation in *Drosophila*. We also learned that all homeotic genes contain a DNA sequence called the *homeobox*. Some of the genes directing gastrulation in the frog are homeobox genes—for example, *goosecoid* and *siamois*. In vertebrates, the homeotic genes that control differentiation along the anterior–posterior body axis are called **Hox genes**.

In mammals, four Hox gene complexes reside on different chromosomes in clusters of about 10 genes each. Remarkably, the temporal and spatial expression of these genes follows the same pattern as their linear order on their chromosome. That is, the Hox genes closest to the 3′ end of each gene complex are expressed first and in the anterior of the embryo. The Hox genes at the 5′ end of the gene complex are expressed later and in a more posterior part of the embryo. As a result, different segments of the embryo receive different combinations of Hox gene products, which serve as transcription factors (**Figure 44.16**; see also Figure 20.2).

Whereas Hox genes give cells information about their position on the anterior–posterior body axis, other genes provide information about their dorsal–ventral position. Tissues in each segment of the body differentiate according to their dorsal–ventral location. The notochord provides many of these signals. One example of a dorsal–ventral difference is seen in the spinal cord; sensory nerve connections develop in the dorsal region, and motor nerve connections in the ventral region. The protein Sonic hedgehog (named for a video-game character), which is expressed in the mammalian notochord, induces cells in the overlying neural tube (i.e., the ventralmost cells of the neural tube) to become motor neurons.

After the development of body segmentation, the formation of organs and organ systems progresses rapidly. The development of an organ involves extensive inductive interactions of the kind we saw in the example of the vertebrate eye (see Figure 19.10). These inductive interactions are a current focus of study for developmental biologists.

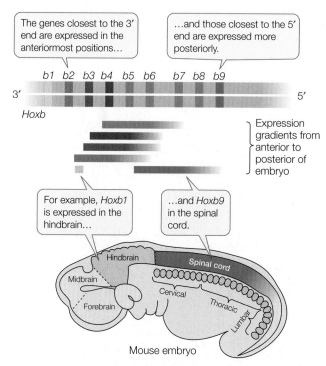

The genes closest to the 3' end are expressed in the anteriormost positions...

...and those closest to the 5' end are expressed more posteriorly.

Expression gradients from anterior to posterior of embryo

For example, *Hoxb1* is expressed in the hindbrain...

...and *Hoxb9* in the spinal cord.

Mouse embryo

44.16 Hox Genes Control Body Segmentation Hox genes are expressed along the anterior–posterior axis of the embryo in the same order as their arrangement between the 3' and 5' ends of the gene complex. As a result of gene duplication during evolution, vertebrates have four copies of the Hox gene complex shown.

44.3 RECAP

Gastrulation sets up tissue interactions that initiate organogenesis. Neurulation is initiated by organizer mesoderm that forms the notochord.

- Describe the formation of the neural tube in vertebrates. **See p. 935 and Figure 44.14**

- How do somites relate to segmentation of the body axis? **See pp. 935–936 and Figure 44.15**

- Using information from this chapter and from Chapters 19 and 20, explain what Hox genes are and how they instruct patterns of differentiation along the body axis. **See Figures 19.19, 20.1, and 44.16**

You may be aware that in mammals the circulatory systems of the fetus and mother are separate and that nourishment reaches the fetus through the placenta and the umbilical cord. In the next section we will examine the developmental events that result in the creation of the placenta.

44.4 How is the Growing Embryo Sustained?

There is more to a developing reptile, bird, or mammal than the embryo itself. As mentioned earlier, the embryos of these vertebrates are surrounded by several **extraembryonic membranes**, which originate from the embryo but are not part of it. Extraembryonic membranes function in nutrition, gas exchange, and waste removal. In mammals, they interact with tissues of the mother to form the placenta.

Extraembryonic membranes form with contributions from all germ layers

The chicken provides a good example of how extraembryonic membranes form from the germ layers created during gastrulation. In the chick, four membranes form—the *yolk sac*, the *allantoic membrane*, the *amnion*, and the *chorion*. The **yolk sac** is the first to form, and it does so by extension of the hypoblast layer along with some adjacent mesoderm. The yolk sac grows to enclose the entire body of yolk in the egg (**Figure 44.17**). It constricts at the top to create a tube that is continuous with the gut of the embryo. However, yolk does not pass through this tube. Yolk is digested by the cells of the yolk sac, and the nu-

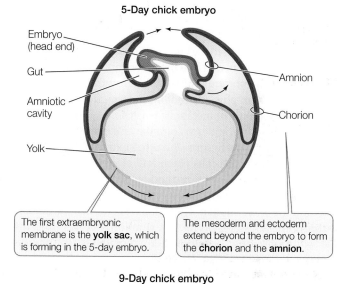

5-Day chick embryo

Embryo (head end)

Gut

Amniotic cavity

Yolk

Amnion

Chorion

The first extraembryonic membrane is the **yolk sac**, which is forming in the 5-day embryo.

The mesoderm and ectoderm extend beyond the embryo to form the **chorion** and the **amnion**.

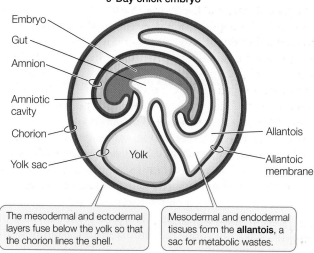

9-Day chick embryo

Embryo

Gut

Amnion

Amniotic cavity

Chorion

Yolk sac

Yolk

Allantois

Allantoic membrane

The mesodermal and ectodermal layers fuse below the yolk so that the chorion lines the shell.

Mesodermal and endodermal tissues form the **allantois**, a sac for metabolic wastes.

44.17 The Extraembryonic Membranes In birds, reptiles, and mammals, the embryo constructs four extraembryonic membranes. The yolk sac encloses the yolk, and the amnion and chorion enclose the embryo. Fluids secreted by the amnion fill the amniotic cavity, providing an aqueous environment for the embryo. The chorion, along with the allantoic membrane, mediates gas exchange between the embryo and its environment. The allantois stores the embryo's waste products.

yourBioPortal.com

GO TO Web Activity 44.1 • Extraembryonic Membranes

44.18 The Mammalian Placenta In humans and most other mammals, nutrients and wastes are exchanged between maternal and fetal blood in the placenta, which forms from the chorion and tissues of the uterine wall. The embryo is attached to the placenta by the umbilical cord. Embryonic blood vessels invade the placental tissue to form fingerlike chorionic villi. Maternal blood flows into the spaces surrounding the villi, and placental blood flows through the villi so nutrients and respiratory gases can be exchanged between the maternal and fetal blood.

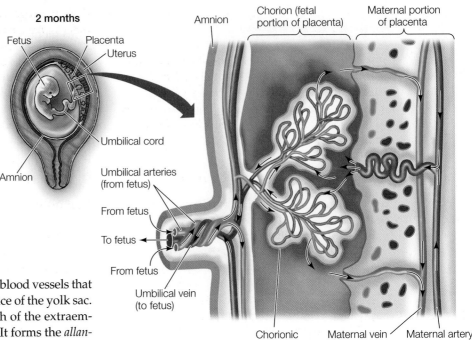

trients are transported to the embryo through blood vessels that form from mesoderm and line the outer surface of the yolk sac. The **allantoic membrane** is also an outgrowth of the extraembryonic endoderm plus adjacent mesoderm. It forms the *allantois*, a sac for storage of metabolic wastes.

Ectoderm and mesoderm combine and extend beyond the limits of the embryo to form the other extraembryonic membranes. Two layers of cells extend all along the inside of the eggshell, both over the embryo and below the yolk sac. Where they meet, they fuse, forming two membranes, the inner **amnion** and the outer **chorion**. The amnion surrounds the embryo, forming the amniotic cavity. The amnion secretes fluid into the cavity, providing a protective environment for the embryo. The outer membrane, the chorion, forms a continuous membrane just under the eggshell (see Figure 44.17). It limits water loss from the egg and also works with the enlarged allantoic membrane to exchange respiratory gases between the embryo and the outside world.

Extraembryonic membranes in mammals form the placenta

In placental mammals, the first extraembryonic membrane to form is the trophoblast, which is already apparent by the fifth cell division (see Figure 44.4). When the blastocyst reaches the uterus and hatches from its encapsulating zona pellucida, the trophoblast cells interact directly with the endometrium. Adhesion molecules expressed on the surfaces of these cells attach them to the uterine wall. By secreting proteolytic enzymes, the trophoblast burrows into the endometrium, beginning the process of implantation (see Figure 44.5). Eventually, the entire trophoblast is within the wall of the uterus. The trophoblast cells then send out numerous projections, or villi, to increase the surface area of contact with maternal blood.

Meanwhile, the hypoblast cells proliferate to form what in the bird would be the yolk sac. But there is no yolk in eggs of placental mammals, so the yolk sac contributes mesodermal tissues that interact with trophoblast tissues to form the chorion. The chorion, along with tissues of the uterine wall, produces the **placenta**, the organ that exchanges nutrients, respiratory

gases, and metabolic wastes between the mother and the embryo (**Figure 44.18**).

At the same time the yolk sac is forming from the hypoblast, the epiblast produces the amnion, which grows to enclose the entire embryo in a fluid-filled amniotic cavity. The rupturing of the amnion and chorion and the loss of the **amniotic fluid** (a process called "water breaking") herald the onset of labor in humans.

An allantois also develops in mammals, but its importance depends on how well nitrogenous wastes can be transferred across the placenta. In humans the allantois is minor; in pigs it is important. In humans and other mammals, allantoic tissues contribute to the formation of the umbilical cord, by which the embryo is attached to the chorionic placenta. It is through the blood vessels of the umbilical cord that nutrients and oxygen from the mother reach the developing fetus, and wastes, including carbon dioxide and urea, are removed (see Figure 44.18).

44.4 RECAP

The extraembryonic membranes of reptiles, birds, and mammals sustain the growing embryo. In reptiles and birds, these membranes surround the embryo within the shelled egg. In mammals the extraembryonic membranes form the placenta, an organ that exchanges nutrients, respiratory gases, and metabolic wastes between the mother and the embryo.

- Describe each of the four extraembryonic membranes and their functions in the developing chick egg. See pp. 937–938 and Figure 44.17

- Explain the role of the trophoblast in the early development of a mammalian embryo. See p. 938

44.5 What Are the Stages of Human Development?

In humans, **gestation**, or pregnancy, lasts about 266 days, or 9 months. In smaller mammals gestation is shorter—for example, 21 days in mice—and in larger mammals it is longer—for example, 330 days in horses and 600 days in elephants. The events of human gestation can be divided into three periods of roughly 3 months each, called *trimesters*.

Organ development begins in the first trimester

Implantation of the human blastocyst begins about 6 days after fertilization. After implantation, gastrulation occurs, tissues differentiate, the placenta forms, and organs begin to develop. The heart begins to beat during week 4, and limbs are formed by week 8 (**Figure 44.19 A,B**). By the end of the first trimester, most organs have started to form. The embryo is about 8 centimeters long and weighs about 40 grams (less than 2 ounces); it would fit neatly in a teaspoon. At about this point in time, the human embryo is medically and legally referred to as a **fetus**. (This distinction is not made for other mammals; developing mice, for example, remain embryos until they are born.)

The first trimester is a time of rapid cell division and tissue differentiation. Signal transduction cascades and the resulting branching sequences of developmental processes are in their early stages. Therefore, the first trimester is the period during which the embryo is most sensitive to damage from radiation, drugs, chemicals, and pathogens that can cause birth defects. An embryo can be damaged before the mother even knows she is pregnant. A classic and tragic case is that of thalidomide, a drug widely prescribed in Europe in the late 1950s to treat nausea. Women who took this drug in the fourth and fifth weeks of pregnancy, when the embryo's limbs are beginning to form, gave birth to children with missing or severely malformed arms and legs.

Organ systems grow and mature during the second and third trimesters

During the second trimester the fetus grows rapidly to a weight of about 600 grams. The limbs of the fetus elongate, and the fingers, toes, and facial features become well formed (**Figure 44.19C**). Eyebrows and fingernails grow and the fetus's nervous system develops rapidly. Fetal movements are first felt by the mother early in the second trimester, and they become progressively stronger and more coordinated.

The fetus grows rapidly during the third trimester (**Figure 44.19D**). As the trimester approaches its end, internal organs mature. The digestive system begins to function, the liver stores glycogen, the kidneys produce urine, and the brain undergoes cycles of sleep and waking. A human infant is born as soon as the last of its critical organs—the lungs—mature.

Although the first-trimester embryo is the most susceptible to adverse effects of drugs, chemicals, and diseases, the potential for serious effects from exposure to environmental factors

(A) 4 weeks

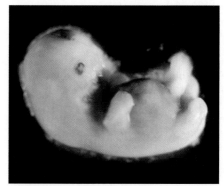

Actual length ~0.4 cm (4 mm)

(B) 8 weeks

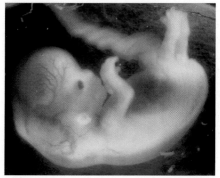

Actual length ~3 cm

(C) 4 months

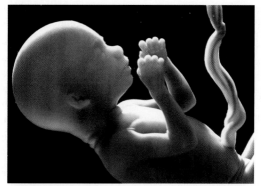

Actual length ~10 cm

(D) 9 months

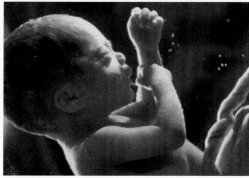

Actual length ~40 cm

44.19 Stages of Human Development (A) At 4 weeks of gestation, most of the embryo's organ systems have been formed and the heart is beating. (B) The body structures of this 8-week-old embryo are forming rapidly, and it is visibly a male. The umbilical cord attaches the embryo to the placenta (upper left). (C) At 4 months, the fetus has fully formed limbs with fingers and toes, and moves freely within the amniotic cavity. (D) This fetus is well along in its ninth month. Soon its lungs will be mature enough to trigger the onset of contractions and birth.

exists throughout pregnancy and continues after birth. Severe protein malnutrition, alcohol, and cigarette smoke are examples of factors that can cause low birth weight, mental retardation, and other developmental complications.

Developmental changes continue throughout life

Development does not end with birth. Obviously, growth continues until adult size is reached, and even when growth stops, organs of the body continue to repair and renew themselves through cycles of cell replacement by the progeny of undifferentiated stem cells. In humans especially, enormous developmental changes occur in the brain in the years between birth and adolescence. Especially in the early years, there is a great deal of plasticity in the organization of the nervous system as the connections between neurons develop.

For example, a child born with misaligned eyes (a condition known as *strabismus*) will use mostly one eye. The connections to the brain from that eye will become strong while connections from the other eye remain weak, and the child will develop with reduced visual acuity and depth perception. If eye alignment is corrected in the first 3 years of life, the connections between the eyes and the brain can improve and the child is likely to develop normal vision. After the age of 3, correcting the connections between the eyes and the brain is less likely to result in improvement, and visual impairments may persist. Thus plasticity in human visual system development declines during early childhood. However, recent data indicate that it is not lost entirely and may be reactivated even in adulthood.

44.5 RECAP

Human gestation lasts 9 months and can be divided into 3 trimesters. At the end of the first trimester, the fetus is very small but most of its organs have begun to form. In the second trimester, limbs elongate and the fetus moves. By the end of the third trimester, most organs have begun to function.

- Why is a first-trimester embryo particular sensitive to environmental risks? See p. 939

CHAPTER SUMMARY

44.1 How Does Fertilization Activate Development?

- The sperm and the egg contribute differentially to the zygote. The sperm contributes a haploid nucleus and, in most species, a centriole. The egg contributes a haploid nucleus, nutrients, ribosomes, mitochondria, mRNAs, and proteins.

- In amphibians, the cytoplasmic contents of the egg are not distributed homogeneously, and they are rearranged after fertilization to set up the major axes of the future embryo. The nutrient molecules are generally found in the **vegetal hemisphere**, whereas the nucleus is found in the **animal hemisphere**. **Review Figures 44.1 and 44.2**

- **Cleavage** is a period of rapid cell division. Except in mammals, little if any gene expression occurs during cleavage. Cleavage can be complete or incomplete, and the pattern of cell divisions depends on the orientation of the mitotic spindles. The result of cleavage is a ball or mass of cells called a **blastula**. **Review Figure 44.3**

- Early cell divisions in mammals are unique in being slow and allowing for gene expression early in the process. These cell divisions produce a **blastocyst** composed of an **inner cell mass** that becomes the embryo and an outer cell mass that becomes the **trophoblast**. At the time of **implantation**, the trophoblast secretes molecules that help the blastocyst implant in the uterine wall. **Review Figures 44.4 and 44.5**

- A **fate map** can be created by labeling specific **blastomeres** and observing what tissues and organs are formed by their progeny. **Review Figure 44.6**

- Some species undergo **mosaic development**, in which the fate of each cell is determined during early divisions. Other species, including vertebrates, undergo **regulative development**, in which remaining cells can compensate for cells lost in early cleavages.

44.2 How Does Gastrulation Generate Multiple Tissue Layers?

- **Gastrulation** involves massive cell movements that produce three **germ layers** and place cells from various regions of the blastula into new associations with one another. **Review Figure 44.7, ANIMATED TUTORIAL 44.1**

- The initial step of sea urchin and amphibian gastrulation is inward movement of certain blastomeres. The site of inward movement becomes the **blastopore**. Cells that move into the blastula become the **endoderm** and **mesoderm**; cells remaining on the outside become the **ectoderm**. Cytoplasmic factors in the vegetal pole cells are essential to initiate development. **Review Figures 44.7 and 44.8**

- The **dorsal lip** of the amphibian blastopore is a critical site for cell determination. It has been called the **primary embryonic organizer** because it induces determination in cells that pass over it during gastrulation. **Review Figures 44.8, 44.9, and 44.10, ANIMATED TUTORIAL 44.2**

- The protein β-catenin activates a signaling cascade that induces the primary embryonic organizer and sets up the anterior–posterior body axis. **Review Figures 44.2, and 44.11**

- Gastrulation in reptiles and birds differs from that in sea urchins and frogs because the large amount of yolk in reptile and bird eggs causes the blastula to form a flattened disc of cells. **Review Figure 44.13**

- Although their eggs have no yolk, placental mammals have a pattern of gastrulation similar to that of reptiles and birds.

44.3 How Do Organs and Organ Systems Develop?

- Gastrulation is followed by **organogenesis**, the process whereby tissues interact to form organs and organ systems.

- In the formation of the vertebrate nervous system, one group of cells that migrates over the blastopore lip is determined to become the **notochord**. The notochord organizes the overlying ectoderm to thicken, form parallel ridges, and fold in on itself to form a **neural tube** below the epidermal ectoderm. The nervous system develops from this neural tube. Review Figure 44.14

- The notochord and **neural crest cells** participate in the segmental organization of mesoderm into structures called **somites** along the body axis. Rudimentary organs and organ systems form during these stages. Review Figure 44.15

- In vertebrates, **Hox genes** determine the pattern of anterior–posterior differentiation along the body axis in mammals. Other genes, such as *sonic hedgehog*, contribute to dorsal–ventral differentiation. Review Figure 44.16

44.4 How is the Growing Embryo Sustained?

- The embryos of reptiles, birds, and mammals are protected and nurtured by four **extraembryonic membranes**. In birds and reptiles, the **yolk sac** surrounds the yolk and provides nutrients to the embryo, the **chorion** lines the eggshell and participates in gas exchange, the **amnion** surrounds the embryo and encloses it in an aqueous environment, and the **allantois** stores metabolic wastes. Review Figure 44.17, **WEB ACTIVITY 44.1**

- In mammals, the chorion and the trophoblast cells interact with the maternal uterus to form a **placenta**, which provides the embryo with nutrients and gas exchange. The amnion encloses the embryo in an aqueous environment. Review Figure 44.18

44.5 What Are the Stages of Human Development?

- Human pregnancy, or **gestation**, can be divided into three trimesters. The embryo forms in the first trimester; during this time, it is most vulnerable to environmental factors that can lead to birth defects. During the second and third trimesters the **fetus** grows, the limbs elongate, and the organ systems mature.

- Development continues throughout childhood and throughout life.

SELF-QUIZ

1. Fertilization involves all of the following *except*
   a. equal contributions of cell organelles from sperm and egg.
   b. joining of sperm and egg haploid nuclei.
   c. induction of rearrangements of the egg cytoplasm.
   d. sperm binding to specific sites on the egg surface.
   e. metabolic activation of the egg.

2. Which of the following does *not* occur during cleavage in frogs?
   a. A high rate of mitosis
   b. Reduction in the size of cells
   c. Expression of genes critical for blastula formation
   d. Orientation of cleavage planes at right angles
   e. Unequal division of cytoplasmic determinants

3. How does cleavage in mammals differ from cleavage in frogs?
   a. Slower rate of cell division
   b. Formation of tight junctions
   c. Expression of the embryo's genome
   d. Early separation of cells that will not contribute to the embryo
   e. All of the above

4. Which statement about gastrulation is *true*?
   a. In frogs, gastrulation begins in the vegetal hemisphere.
   b. In sea urchins, gastrulation produces the notochord.
   c. In birds, cells from the surface of the blastodisc move down through the primitive groove to form the hypoblast.
   d. In mammals, gastrulation occurs in the hypoblast.
   e. In sea urchins, gastrulation produces only two germ layers.

5. Which of the following was a conclusion from the experiments of Spemann and Mangold?
   a. Cytoplasmic determinants of development are homogeneously distributed in the amphibian zygote.
   b. In the late blastula, certain regions of cells are determined to form skin or nervous tissue.
   c. The dorsal lip of the blastopore can be isolated and will form a complete embryo.
   d. The dorsal lip of the blastopore can initiate gastrulation.
   e. The dorsal lip of the blastopore gives rise to the neural tube.

6. Which of the following is true of human development?
   a. Most organs begin to form during the second trimester.
   b. Gastrulation takes place in the oviducts.
   c. Genetic diseases can be detected by sampling cells from the chorion.
   d. Implantation occurs through interactions of the zona pellucida with the uterine lining.
   e. Exposure to drugs and chemicals is most likely to cause birth defects when it occurs in the third trimester.

7. Which of the following characterizes neurulation?
   a. The notochord forms a neural tube.
   b. The neural tube is formed from ectoderm.
   c. A neural tube forms around the notochord.
   d. The neural tube forms somites.
   e. In birds, the neural tube forms from the primitive groove.

8. Which statement about trophoblast cells is *true*?
   a. They are capable of producing monozygotic twins.
   b. They are derived from the hypoblast of the blastocyst.
   c. They are endodermal cells.
   d. They secrete proteolytic enzymes.
   e. They prevent the zona pellucida from attaching to the oviduct.

9. Which of the following is part of the embryonic contribution to placenta formation?
   a. Amnion
   b. Chorion
   c. Ectoderm
   d. Allantois
   e. Zona pellucida

10. When is the developing human most susceptible to the occurrence of birth defects from radiation or chemical insults?
    a. At the time of birth
    b. During the third trimester
    c. During the first trimester
    d. When it is a zygote
    e. During the final stages of organ formation

FOR DISCUSSION

1. If you found a protein that was localized to a small group of cells in the frog blastula, how would you determine whether that protein played a role in development? Address the issues of sufficiency and necessity.

2. During gastrulation in birds, the *sonic hedgehog* gene is expressed only on the left side of Hensen's node. What might be the cause of this expression pattern, and what is its significance?

3. Much of the early work of describing animal development was done on sea urchins, amphibians, and chickens. Most of the recent work on the molecular mechanisms of animal development has been done on nematodes, fruit flies, zebrafish, and mice. Why do you think there has been a shift in the animal models used by developmental biologists?

4. If all the mitochondria and mitochondrial DNA in the embryo come from the egg, what implications does this have for using mitochondrial DNA for molecular evolutionary studies?

5. There is currently much controversy over therapeutic cloning as a way of obtaining embryonic stem cells to treat diseases. Given that human development is regulative—in other words, twinning can occur if an early blastocyst is divided into two cell masses—can you think of a way to guarantee a source of isogenic (i.e., identically matching a person's own body) stem cells for an individual without resorting to therapeutic cloning? Assume isolated cells can be preserved indefinitely in a frozen state.

ADDITIONAL INVESTIGATION

It is hypothesized that the differential development of the different body segments is due to the differential expression of Hox genes along the anterior–posterior body axis. For example, in mammals, ribs develop in anterior body trunk segments but not in posterior segments. Using the mouse as a model, how would you test this hypothesis?

Fear and survival in the brain

Charles Whitman was a normal and responsible child. He became the youngest Eagle Scout in the country. He was a fine son and husband and received commendations as a U.S. Marine. While in the service, however, he began having unexplained fits of anger and other personality disorders. He was discharged from the Marines and entered the University of Texas. Several times he visited campus doctors and complained about having violent thoughts. Then, on August 1, 1966, after killing his wife and mother, Whitman barricaded himself inside the top floor of the clock tower on campus along with several high-powered rifles. From this vantage point, he killed 14 people and wounded 38 others before being shot and killed by Austin police. In his suicide note, he requested that proceeds from his insurance be donated to a mental health foundation. He also requested an autopsy, and that autopsy revealed a tumor pressing on his amygdala.

The *amygdala* (Latin for "almond") is the brain's center for the emotion and memory of fear. When the cells of this structure are activated, your heart beats faster, your breathing becomes rapid and shallow, and your hands get cold and clammy. If you encounter a threatening situation, your amygdala is activated. If you are alone at night and hear a strange noise, your amygdala is activated. If you are faced with attempting something physically dangerous, your amygdyla is activated. Without an amygdala, you would never be scared—and *not* being scared could be hazardous to your health.

A rare case of brain damage left a woman without a functional amygdala. When shown pictures of faces registering different emotions, she could not pick out the ones that were threatening or scary. She could not recall ever having a frightening experience. In tests where she was administered mild electrical shocks, she developed no anticipatory fear; even though she knew that seeing a red card meant she was about to receive a shock, she never reacted to the red card. In real life, she would not have the reflex to pull away from a threat.

People with a damaged amygdala frequently have trouble engaging in normal social relationships. They cannot "read" the nature, mood, or intentions of other people by looking at their faces. The pres-

Fear Factor   The fear response—muscle tension, racing heart, cold sweat—kicks in when you are faced with potential pain, even if that experience ultimately will be beneficial.

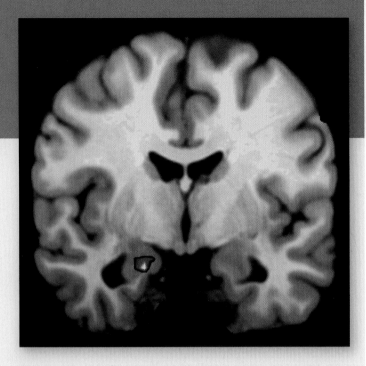

Source of the Fear Response Frightening situations—or even memories of such a situation—activate a brain region called the amygdala, as shown in this functional magnetic resonance image (fMRI) of the brain of a person experiencing fear.

sure on Charles Whitman's amygdala may have been a factor in the emotions that drove him to mass murder; this diagnosis remains a matter for medical speculation.

Our nervous systems enable us to experience the world around us and to react to it. But in between sensing and reacting, there is much interpretation based on memory, learning, emotions, and beliefs—all of which are based on the activities of cells in the nervous system. To understand how our eyes see, how our fingers play the piano, or how our emotions affect our behavior, we have to understand how cells in different parts of our nervous systems work and interact.

IN THIS CHAPTER we will first describe the cells that make up nervous systems—neurons and glia —and explain how neurons process and transmit information by generating and transmitting electrical signals called action potentials. We will examine the membrane properties and events underlying action potentials and the mechanisms of cell-to-cell communication where neurons come together. Finally, we will consider how changes in the properties of neurons might explain learning and memory.

45.1 What Cells Are Unique to the Nervous System?

Nervous systems are composed of two types of cells: *nerve cells,* or **neurons,** and *glial cells,* or **glia** (see Figure 40.6). Neurons are *excitable*: they can generate and transmit electrical signals, which are known as nerve impulses, or **action potentials**. Many neurons have a long extension called an **axon** that enables them to conduct action potentials over long distances. Glia do not conduct action potentials; rather, they support neurons physically, immunologically, and metabolically. A **nerve** (as distinct from a neuron) is a bundle of axons that come from many different neurons. Many axons are wrapped by glia to electrically isolate them and increase their speed of conduction of action potentials.

Nervous systems can process information because their neurons are organized into **neural networks**. These networks include three functional categories of neurons, which can be thought of as being involved with input, output, and integration:

- **Afferent neurons** carry sensory information into the nervous system. That information comes from specialized **sensory neurons** that transduce (convert) various kinds of sensory stimuli (e.g., light, heat, pressure) into action potentials.
- **Efferent neurons** carry commands to physiological and behavioral *effectors* such as muscles and glands.
- **Interneurons** integrate and store information and communicate between afferent and efferent neurons.

Neural networks can be simple, like the one that causes your leg to kick when a physician taps your knee; or they can be exceedingly complex, like the network that enables you to read and remember this chapter.

Neural networks range in complexity

Simple animals such as cnidarians (e.g., sea anemones) can process information with simple neural networks that do little more than provide direct lines of communication from sensory cells to effectors; there is little or no integration or processing of signals (**Figure 45.1A**). The cnidarian's *nerve net* is most developed around the tentacles and the oral opening, where it facilitates detection of food or danger and causes tentacles to extend or retract.

Animals that are more complex and actively move about in search for food and mates need to process and integrate larger amounts of information. Even earthworms fit this description, and their increased need for information processing is met by higher numbers of neurons organized into clusters called **ganglia**. Ganglia serving different functions may be distributed around the body, as in earthworms or squid (**Figure 45.1B,C**).

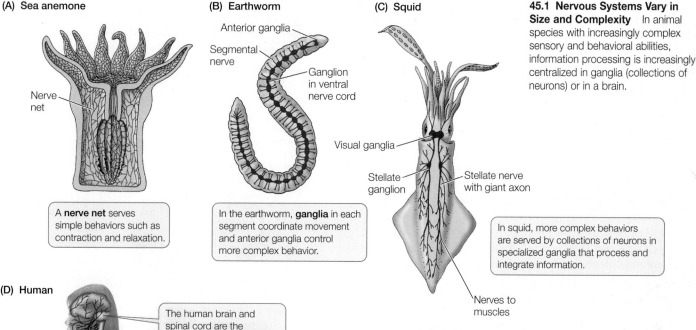

(A) Sea anemone

Nerve net

A **nerve net** serves simple behaviors such as contraction and relaxation.

(B) Earthworm

Anterior ganglia

Segmental nerve

Ganglion in ventral nerve cord

In the earthworm, **ganglia** in each segment coordinate movement and anterior ganglia control more complex behavior.

(C) Squid

Visual ganglia

Stellate ganglion

Stellate nerve with giant axon

Nerves to muscles

In squid, more complex behaviors are served by collections of neurons in specialized ganglia that process and integrate information.

45.1 Nervous Systems Vary in Size and Complexity In animal species with increasingly complex sensory and behavioral abilities, information processing is increasingly centralized in ganglia (collections of neurons) or in a brain.

(D) Human

The human brain and spinal cord are the **central nervous system**…

…which communicates to the cells and organs of the body via the **peripheral nervous system**.

In animals that are bilaterally symmetrical, ganglia frequently come in pairs, one on each side of the body. Also, as animals increase in complexity, some ganglia may become enlarged or fused together at the anterior end, forming a larger, centralized integrative center, or **brain**. The small nervous systems of invertebrates can be remarkably complex. Consider the nervous systems of spiders, which have programmed within them the thousands of precise movements necessary to construct a beautiful web without prior experience or opportunities to learn the specific web architecture of their species.

In vertebrates, most cells of the nervous system are found in the brain and the **spinal cord**, the sites of most information processing, storage, and retrieval (**Figure 45.1D**). Therefore, the brain and spinal cord are called the **central nervous system** (**CNS**). Information is transmitted from sensory cells to the CNS and from the CNS to effectors via neurons that extend or reside outside the brain and the spinal cord; these neurons and their supporting cells are called the **peripheral nervous system** (**PNS**). Vertebrates differ greatly in their behavioral complexity and in their physiological specializations, and their nervous systems reflect this diversity. **Figure 45.2** shows the brains of four vertebrate species of similar body mass drawn to the same scale.

The human nervous system contains an estimated 10^{11} neurons. Information is passed from one neuron to another where they come into close proximity at structures called **synapses**. The neuron sending the information is the **presynaptic neuron**, and the neuron receiving the information is the **postsynaptic neuron**. A given neuron in the brain can have 1,000 or more synapses. Thus the human brain can contain 10^{14} synapses (10^{11} neurons $\times 10^3$ synapses per neuron).

Synapses are not constant but instead can be highly plastic. They can increase or decrease in number and size. They can become more or less sensitive. Therein lies the incredible ability of the human brain to process information, to learn, to do complex tasks, to remember, and to have emotions. The astronomical number of neurons and synapses is divided into thousands of distinct but interacting networks that function in parallel. Before we can understand how even one of these circuits works, however, we must understand the properties of individual neurons.

Neurons are the functional units of nervous systems

Nervous systems of different species vary enormously in structure and function, but neurons behave similarly in animals as different as squid and humans. Their plasma membranes generate action potentials and conduct these signals from one lo-

45.2 Brains Vary in Size and Complexity The brains of four verte-brate species—all of which may have a similar body mass—show immense differences. Note that the brainstem, which is involved in physi-ological regulation and stereotypic behavior, differs less among these species than does the cerebrum, which in higher vertebrates is responsi-ble for complex behavior and learning.

cation on a neuron to the most distant reaches of that cell—a distance that can be more than a meter in a human and many meters in a whale. Moreover, this transmission of action poten-tials can be rapid—up to 100 meters/sec, which is 360 km/hr—making it possible to sense, process, and act on information very quickly.

Most neurons have four regions—a *cell body*, *dendrites*, one or more *axons*, and *axon terminals* (**Figure 45.3A**)—but the vari-ation among different types of neurons is considerable (**Figure 45.3B**). The *cell body* contains the nucleus and most of the cell's organelles. Many projections may sprout from the cell body. Most of these projections are shrublike **dendrites** (from the Greek *dendron*, "tree"), which bring information from other neu-

45.3 Neurons (A) A generalized diagram of a neuron. (B) Neurons from different parts of the mammalian nervous system are specifically adapted to their functions.

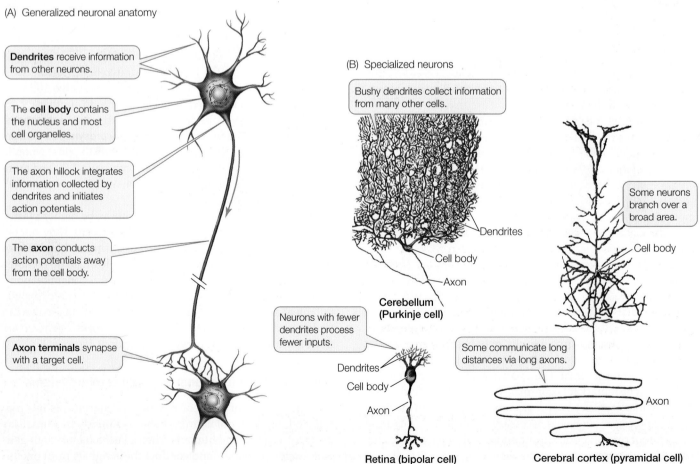

Human

Cerebrum

Olfactory lobe

The brain region that processes olfactory information is relatively more developed in the pig than in the human.

In humans, much of the brain is given over to integration of complex behaviors, learning, and memory.

Cerebellum

Brain stem

Pig

The cerebellum integrates sensory and motor information.

Olfactory lobe

Brain stem structures serve basic physiological functions.

Alligator

Shark

A shark has little complex behavior. Its brain deals primarily with sensory and motor information.

(A) Generalized neuronal anatomy

Dendrites receive information from other neurons.

The **cell body** contains the nucleus and most cell organelles.

The axon hillock integrates information collected by dendrites and initiates action potentials.

The **axon** conducts action potentials away from the cell body.

Axon terminals synapse with a target cell.

(B) Specialized neurons

Bushy dendrites collect information from many other cells.

Dendrites

Cell body

Axon

Cerebellum (Purkinje cell)

Neurons with fewer dendrites process fewer inputs.

Dendrites

Cell body

Axon

Retina (bipolar cell)

Some neurons branch over a broad area.

Cell body

Some communicate long distances via long axons.

Axon

Cerebral cortex (pyramidal cell)

rons or sensory cells to the cell body. The degree of branching of the dendrites differs among different types of neurons.

In most neurons, one projection—the axon—is much longer than the others. Axons usually carry action potentials away from the cell body. The length of the axon also differs among different types of neurons—some axons are remarkably long, such as those that run from the spinal cord to the toes. Axons are the "telephone lines" of the nervous system. Information received by dendrites can cause the axon to generate an action potential that is conducted down the axon toward its target cell. At the target cell, the axon divides into a spray of fine nerve endings. At the tip of each of these tiny nerve endings is a swelling, called an **axon terminal**, that comes very close to the membrane of the target cell to form a synapse.

As we will discuss in Section 45.3, synapses can be either chemical or electrical. Most synapses in vertebrates are chemical synapses. At chemical synapses a space only about 25 nanometers wide (about 1/2000th of a human hair) separates the *presynaptic* and *postsynaptic membranes*. An action potential arriving at an axon terminal causes it to release chemical messenger molecules called **neurotransmitters**. The neurotransmitters diffuse across the space and bind to receptors on the plasma membrane of the postsynaptic or target cell. This binding alters the activity of the postsynaptic neuron. Some neurotransmitter–receptor combinations inhibit activity of the postsynaptic neuron, and other neurotransmitter–receptor combinations excite it. Neurons integrate information by summing excitatory and inhibitory inputs.

Glia are also important components of nervous systems

Glia are another class of nervous system cells. There are many more glia than neurons in the human brain. Like neurons, glia come in several forms and have diverse functions. They do not generate or transmit electrical signals, but they can release neurotransmitters. Some glia physically support and orient the neurons and help them make the right contacts during embryonic development. Others supply neurons with nutrients, maintain the extracellular environment, and insulate axons. Still others consume debris and foreign particles and provide immune functions for the nervous system.

In the CNS, glia called **oligodendrocytes** wrap around the axons of neurons, covering them with concentric layers of insulating plasma membrane. In the PNS, glia called **Schwann cells** perform this function (**Figure 45.4**). **Myelin** is the covering produced by oligodendrocytes and Schwann cells, and it gives many parts of the nervous system a glistening white appearance. Not all axons are myelinated, but those that are can conduct action potentials more rapidly than axons that are not myelinated.

Diseases that affect myelin can be devastating because they impair conduction of action potentials. The most common demyelinating disease is *multiple sclerosis*—meaning literally "multiple scars"—which occurs in about 1 in 700 people in the United States. Individuals with this autoimmune disease produce antibodies to proteins in the myelin in the CNS. The symptoms and damage from the disease depend on where in the CNS the antibody attacks take place. Motor impairment is common. An example of a demyelinating disease that affects the PNS is Guillain-Barre Syndrome. Environmental factors such as pesticide exposure can also damage myelin. There are no known cures for demyelinating diseases.

Glia called **astrocytes** (because they look like stars; see Figure 40.6B) contribute to the **blood–brain barrier**, which protects the brain from toxic chemicals in the blood. Blood vessels throughout the body are very permeable to many chemicals, including toxic ones, which would reach the brain if this barrier did not exist. Astrocytes help form the blood–brain barrier by surrounding the smallest, most permeable blood vessels in the brain. The barrier is not perfect, however. Since it consists of plasma membranes, it is permeable to fat-soluble substances such as anesthetics and alcohol, which explains why these substances have such rapid and marked effects on the nervous system.

The blood–brain barrier usually prevents antibodies in the general circulation from entering the CNS. To provide the CNS with immune defenses, **microglia**, which originate during development from stem cells in the bone marrow, come to reside

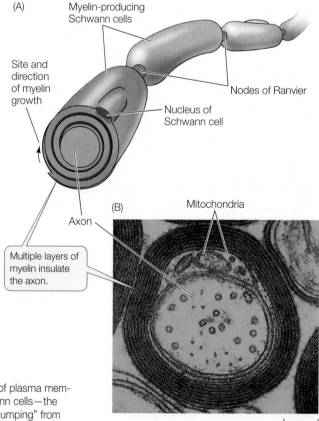

(A) Myelin-producing Schwann cells

Site and direction of myelin growth

Nodes of Ranvier

Nucleus of Schwann cell

Axon

Multiple layers of myelin insulate the axon.

(B) Mitochondria

0.1 µm

45.4 Wrapping Up an Axon (A) Schwann cells produce layers of myelin, a type of plasma membrane that provides electrical insulation to the axon. At the intervals between Schwann cells—the nodes of Ranvier—the axon is exposed. Action potentials travel along the axon by "jumping" from node to node. (B) A myelinated axon, seen in cross section through an electron microscope.

in the CNS and act as macrophages and mediators of inflammatory responses.

The one feature common to all nervous systems is that they process information in the form of action potentials. In the next section we will focus on how action potentials are generated and transmitted by nervous systems.

45.2 How Do Neurons Generate and Transmit Electrical Signals?

All animal cells have more potassium ions (K^+) inside and more sodium ions (Na^+) in the extracellular fluid. Both inside and outside the cell, the positive charges of these ions are balanced by negatively charged ions so that, individually, the cell interior and the extracellular fluid are both electrically neutral. However, *across the cell membrane* there is an electrical charge difference, with the inside being negative to the outside. Why is this so?

The reason is that there are *leak currents* in the cell membrane due to channels that allow only certain ions—mostly K^+—to "leak" passively across. Because there are more potassium ions inside the cell than outside, K^+ diffuses out of the cell down a concentration gradient. But when K^+ leaks out of the cell, it leaves behind an unbalanced negative charge that tends to pull K^+ back into the cell. An equilibrium is reached when the tendency for K^+ to diffuse out is countered by the electrical charge pulling K^+ back in. The result is an electric charge difference—a **membrane potential**—across the plasma membrane, with the inside of the cell negative to the outside.

Membrane potentials exist in all cells. In neurons, the steady state membrane potential is called the **resting potential**, because neurons can also be active, generating rapid, large changes in membrane potential. These sudden large shifts are called *action potentials* (sometimes referred to as *nerve impulses*). An action potential is generated by sudden openings and rapid closings of ion channels. Before describing the properties of ion channels and action potentials in detail, a review of some simple concepts of electricity may be useful.

Simple electrical concepts underlie neural function

Voltage (electric potential difference) is a force that causes electrically charged particles to move between two points. Voltage is to the flow of electrically charged particles as pressure is to the flow of water. If the negative and positive poles of a battery are connected by a wire, an electric current will flow through the wire because there is a voltage difference between the two poles. This flow of electric current can be used to do work, just as a current of water can be used to do work.

In wires, electric current is carried by electrons, but in solutions and across cell membranes, electric current is carried by ions. The major ions that carry electric charges across the plasma membranes of neurons are sodium (Na^+), potassium (K^+), calcium (Ca^{2+}), and chloride (Cl^-). Recall that ions with opposite charges attract one another and that those with like charges repel one another. How do these basic principles of bioelectricity establish the resting potential of the neural plasma membrane? And how is the flow of ions through membrane channels turned on and off to generate action potentials? We address these questions next.

Membrane potentials can be measured with electrodes

We can record electrical events in a cell using electrodes. **Figure 45.5** shows how this technique is applied across an unstimulated axon to measure the resting potential, which is usually between –60 and –70 mV. The minus sign indicates that the inside of the cell is electrically negative compared with the outside.

The resting potential provides a means for neurons to respond to a stimulus. Because of the voltage difference across the membrane, and the different ion concentrations on either side of the membrane, ions would cross the membrane if they could. For example, Na^+ ions are more abundant outside the cell than inside and the inside of the resting cell is negatively charged. Therefore, if the membrane suddenly became permeable to Na^+, those positively charged ions would rush into the cell. Any chemical or physical stimulus that changes the permeability of the plasma membrane to ions will produce a change in the cell's membrane potential. The most extreme change in membrane potential is the action potential, a sudden and rapid reversal in the voltage across a portion of the plasma membrane. For one or two milliseconds, positively charged ions flow into the cell, making the inside of the cell *more positive* than the outside.

Ion transporters and channels generate membrane potentials

The plasma membranes of neurons, like those of all other cells, are lipid bilayers that are impermeable to ions but contain many

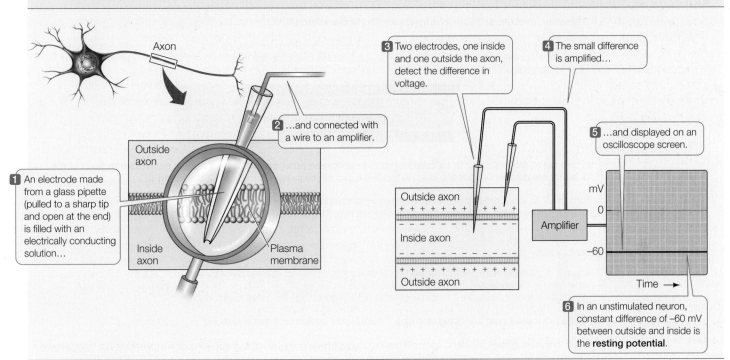

TOOLS FOR INVESTIGATING LIFE

45.5 Measuring the Membrane Potential

An electrode can be made from a glass pipette with a very sharp tip filled with a solution that conducts electric charges. If one electrode is placed inside the plasma membrane of an axon and another is placed just outside the axon, the difference in voltage can be measured.

Axon

1 An electrode made from a glass pipette (pulled to a sharp tip and open at the end) is filled with an electrically conducting solution...

2 ...and connected with a wire to an amplifier.

Outside axon

Inside axon

Plasma membrane

3 Two electrodes, one inside and one outside the axon, detect the difference in voltage.

4 The small difference is amplified...

Outside axon
+ + + + + + + + + + +
− − − − − − − − − −

Inside axon
− − − − − − − − − −
+ + + + + + + + + + +

Outside axon

Amplifier

5 ...and displayed on an oscilloscope screen.

mV
0
−60
Time →

6 In an unstimulated neuron, constant difference of −60 mV between outside and inside is the **resting potential**.

---

protein molecules that serve as ion transporters and channels. Ion transporters and channels are responsible for the distribution of charges across the membrane that create resting and action potentials.

Ion transporters require energy to move ions against their concentration or electrical gradients and are therefore called ion pumps. A major ion transporter in the plasma membranes of neurons (and all other cells) is the **sodium–potassium pump**, so called because it actively expels Na⁺ ions from inside the cell, exchanging them for K⁺ ions from outside the cell (**Figure 45.6A**). The Na⁺–K⁺ pump is an *antiporter* (see Section 6.3) and is also known as *sodium–potassium ATPase*, a term emphasizing that it is an enzyme complex requiring ATP to do its work. The Na⁺–K⁺ pump keeps the concentration of K⁺ inside the cell greater than that of the extracellular fluid, and the concentration of Na⁺ inside the cell less than that of the extracellular fluid. The concentration differences established by this antiporter mean that K⁺ would diffuse out of the cell and Na⁺ would dif-

45.6 Ion Transporters and Channels (A) The sodium–potassium pump is in an antiporter that actively moves K⁺ to the inside of a neuron and Na⁺ to the outside. (B) Ion channels allow specific ions to diffuse down their concentration gradients; K⁺ tends to leave neurons when potassium channels are open, and Na⁺ tends to enter neurons when sodium channels are open. Leak channels like the K⁺ channel shown are always open and create the resting membrane potential. Gated channels like the Na⁺ channels shown are opened by chemical or electrical stimulation.

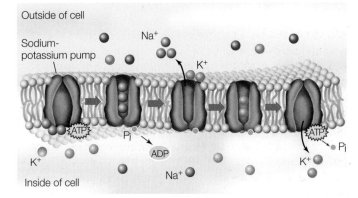

(A) Na⁺–K⁺ pump (ATPase)

Outside of cell

Sodium-potassium pump

Na⁺

K⁺

ATP

Pᵢ

ADP

K⁺

Na⁺

Pᵢ

ATP

K⁺

Inside of cell

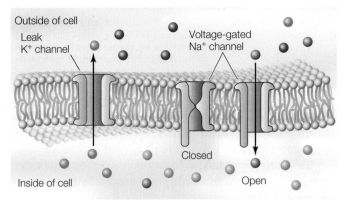

(B) Na⁺–K⁺ channels

Outside of cell

Leak K⁺ channel

Voltage-gated Na⁺ channel

Closed

Open

Inside of cell

TOOLS FOR INVESTIGATING LIFE

45.7 Using the Nernst Equation

The Nernst equation calculates membrane potential when only one type of ion can cross a membrane that separates solutions with different concentrations of that ion.

1. Measure concentrations of ions inside and outside a neuron.

To measure the concentration of ions in a neuron, the neuron (and its axon) must be big. Squid have giant neurons that control their escape response (see Figure 45.1C). It is possible to sample the cytoplasm of these axons, which are about 1 mm in diameter.

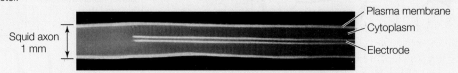

Squid axon 1 mm

Plasma membrane
Cytoplasm
Electrode

2. Use the Nernst equation to calculate what the membrane potential would be if it were permeable to each of the ions that are differently concentrated on the two sides of the membrane: Na^+, K^+, Ca^{2+}, and Cl^-.

The Nernst equation predicts the membrane potential resulting from membrane permeability to a single type of ion that differs in concentration on the two sides of the membrane. The equation is written

$$E_{ion} = 2.3 \frac{RT}{zF} \log \frac{[ion]_o}{[ion]_i}$$

where E is the equilibrium (resting) membrane potential (the voltage across the membrane in mV), R is the universal gas constant, T is the absolute temperature, z is the charge on the ion (+1, +1, +2, or –1), and F is the Faraday constant. The subscripts o and i indicate the ion concentrations outside and inside the cell, respectively.

At this point you could just "plug and play," but do you understand this equation?

A concentration difference of ions across a membrane creates a *chemical* force that pushes the ions across the membrane; however, the resulting unbalanced *electrical charges* will pull the ions back the other way. At *equilibrium*, the work done moving ions in each direction will be the same.

The *chemical* energy pushing the ions will equal $2.3\ RT \log\ [ion]_o/[ion]_i$
The *electrical* energy pulling the ions will equal zEF. So, at equilibrium:

$$zEF = 2.3\ RT \log \frac{[ion]_o}{[ion]_i}$$

Rearranging the equation to solve for E, we get the Nernst equation:

$$E_{ion} = 2.3 \frac{RT}{zF} \log \frac{[ion]_o}{[ion]_i}$$

We can simplify the equation by picking a temperature—let's use "room temperature," or 20°C—and solving for $2.3\ RT/F$. At 20°C, $2.3\ RT/F$ equals 58. Thus:

$$E_{ion} = 58/z \log \frac{[ion]_o}{[ion]_i}$$

3. Measuring ion concentrations in squid giant axon cytoplasm and in seawater, then solving the Nernst equation for each ion, we find:

Ion	Ion concentration (mM)		Predicted membrane
	in squid axon	in seawater	potential (mV)
K^+	400	20	–75
Na^+	50	460	+56
Ca^{2+}	0.5	10	+38
Cl^-	50	560	–60

4. Since the measured membrane potential is –66 mV, it is clear that the resting potential of the axon is due to permeability of the membrane to more than just one type of ion.

fuse in if the ions could cross the lipid bilayer. How do these concentration gradients relate to the electric gradients we discussed above?

Ion channels permit the diffusion of ions across membranes. These channels are water-filled pores formed by proteins that cross the lipid bilayer and are generally *selective*—they allow some types of ions to pass through more easily than others (**Figure 45.6B**). Thus, there are potassium channels, sodium channels, chloride channels, and calcium channels, and there are different kinds of each. Ions can diffuse through these channels in either direction. The direction and magnitude of the net movement of ions through a channel depend on the concentration gradient of that ion type across the plasma membrane, as well as on as the voltage difference across that membrane. These two motive forces acting on an ion are termed its **electrochemical gradient**. Although the electrochemical gradient drives the movement of ions through channels across bilayers, movement through channels can be modified by gates that open and close channels.

Potassium channels are the most common open, or leak, channels in the plasma membranes of resting (non-stimulated) neurons. As a consequence, resting neurons are more permeable to K^+ than to any other ion. Thus, open potassium channels are largely responsible for the resting membrane potential. Because the potassium channels make the plasma membrane permeable to K^+, and because the $Na^+–K^+$ pump keeps the concentration of K^+ inside the cell much higher than that outside the cell, K^+ tends to diffuse down its electrochemical gradient, out of the cell, through the channels. As these positively charged potassium ions diffuse out of the cell, they leave behind unbalanced negative charges, generating an electric potential across the membrane that tends to pull K^+ back into the cell.

The membrane potential at which the net diffusion of K^+ out of the cell ceases (that is, the point at which K^+ diffusion out due to the concentration gradient is balanced by its movement in due to the negative electric potential) is the **potassium equilibrium potential**. The value of the potassium equilibrium potential can be calculated from the concentrations of K^+ on the two sides of the membrane using the **Nernst equation** (**Figure 45.7**). This equation, developed in the late 1800s, illustrates that the nature of ion channels in neural membranes was hypothesized long before their specific structures and properties were described.

In the late 1940s, A. L. Hodgkin and A. F. Huxley at the University of Cambridge set out to study the electrical properties of axonal membranes. With the techniques available at that time, the necessary measurements could be made only if you had a very large axon to work with. Such an axon exists in nature, in the huge neuron that controls the escape response of squid. Hodgkin and Huxley used electrodes to measure the voltage across the plasma membrane of this large axon, as seen in Figure 45.7, and to pass electric current into it to change its resting potential. They also changed the concentrations of Na^+ and K^+ both inside and outside the squid axon and measured the resulting changes in membrane potential. On the basis of their many careful experiments, Hodgkin and Huxley developed virtually all of our basic concepts about the electrical properties of neurons, and received the Nobel prize in 1963.

We now know that, in general, the resting potential is less negative than the Nernst equation predicts because resting neurons are also slightly permeable to other ions, such as Na^+ and Cl^-. Another equation, called the Goldman equation, takes all of the ions that can cross the membrane into account and therefore can calculate the membrane potential accurately.

Ion channels and their properties can now be studied directly

Because Hodgkin and Huxley were working long before there were laboratory techniques that could investigate ion channels, they could only hypothesize their properties. These hypotheses could not be tested until the late 1970s, when B. Sakmann and E. Neher developed a technique called **patch clamping**, for which they won the Nobel prize in 1991. Patch clamping, described in **Figure 45.8**, is widely used by neurobiologists, en-

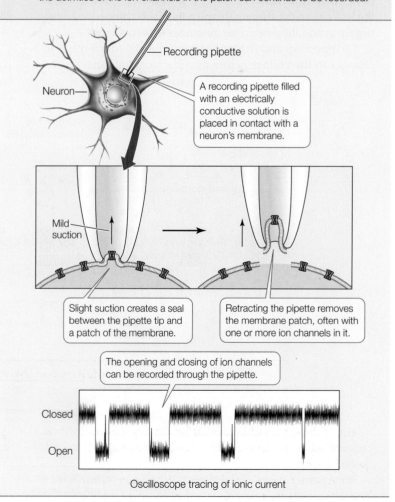

TOOLS FOR INVESTIGATING LIFE

45.8 Patch Clamping

The patch clamp is a glass micropipette filled with an electrically conductive solution that has the same composition as extracellular fluids. When this pipette/electrode is positioned against the membrane of a cell and slight suction is applied, a seal forms. If a single ion channel (or a few ion channels) are within the patch of membrane bounded by the seal, the openings and closings of individual channels can be recorded by the electrode. If the pipette is retracted, it can tear the patched membrane away from the cell, and the activities of the ion channels in the patch can continue to be recorded.

Recording pipette

Neuron

A recording pipette filled with an electrically conductive solution is placed in contact with a neuron's membrane.

Mild suction

Slight suction creates a seal between the pipette tip and a patch of the membrane.

Retracting the pipette removes the membrane patch, often with one or more ion channels in it.

The opening and closing of ion channels can be recorded through the pipette.

Closed

Open

Oscilloscope tracing of ionic current

abling them to record in real time the tiny electrical currents caused by the openings and closings of single ion channels.

Gated ion channels alter membrane potential

The ion channels called leak channels are always open, but other ion channels in the plasma membranes of neurons behave as if they contain "gates"; they are open under some conditions and closed under other conditions. **Voltage-gated channels** open or close in response to a change in the voltage across the plasma membrane. **Chemically gated channels** open or close depending on the presence or absence of a specific molecule that binds to the channel protein, or to a separate receptor that in turn alters the channel protein. **Mechanically gated channels** open or close in response to mechanical force applied to the plasma membrane. Gated channels play important roles in neural function.

Openings and closings of gated channels alter the resting potential. Imagine what happens, for example, if sodium channels in the plasma membrane open. Na^+ diffuses into the neuron down its electrochemical gradient. As a result, the inside of the cell becomes less negative. When the inside of a neuron becomes less negative (or more positive) in comparison to its resting condition, its plasma membrane is **depolarized** (**Figure 45.9**).

An opposite change in the resting potential occurs if gated K^+ channels open. When K^+ efflux from the neuron increases over the normal leak current (the movement of K+ through the leak channels), the membrane potential becomes even more negative, and the plasma membrane is **hyperpolarized**.

The openings and closings of ion channels, which result in changes in the voltage across the plasma membrane, are the basic mechanisms by which neurons respond to stimuli, be they electrical, chemical, or mechanical. How do such local changes in membrane potential get communicated to other parts of the cell?

A local change in membrane potential causes a flow of ions that spreads the change in membrane potential to adjacent regions of the membrane. For example, when Na^+ enters a neuron through open sodium channels at one location, those positively charged ions are attracted to adjacent areas on the inside of the membrane that are more negative, and thus there is a rapid flow of electric current (movement of ions) away from the site of the open Na^+ channels. However, this local flow of electric current decays as it spreads and therefore does not spread very far. Electric currents do not spread far in cells because cell membranes are not completely impermeable to ions. An electric current traveling along a membrane is like water flowing through a leaky hose.

Graded changes in membrane potential can integrate information

Even though the flow of electric current along plasma membranes can only extend over short distances, it can cause graded changes in membrane potentials locally. A **graded membrane**

45.9 Membranes Can Be Depolarized or Hyperpolarized The resting potential is produced by leak K^+ channels. A shift from the resting potential to a less negative membrane potential, as occurs when Na^+ enters the cell through a gated sodium channel, is called depolarization. Hyperpolarization occurs when the membrane potential becomes more negative, as when additional K^+ leaves the cell through gated K^+ channels, which occurs extensively in your brain when you fall asleep.

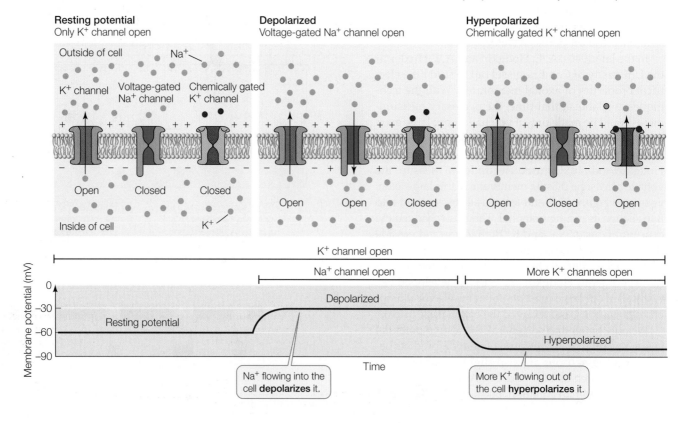

potential is a change from the resting potential. Such changes can be due to chemical or mechanical influences on ion channels. Graded potentials are a means of integrating inputs to a cell because the membrane can respond to those inputs with proportional amounts of depolarization or hyperpolarization.

Graded potentials can transmit signals over very short distances and play an important role at the neuromuscular junction (see Section 45.3). In the next chapter we will learn how they play important roles in sensory systems. However, axons are too long to transmit information as a continuous flow of electric current (as telephone wires do). Therefore axons code information as discrete action potentials that travel along their membranes. Graded potentials, however, play an important role in the generation of action potentials.

Sudden changes in Na⁺ and K⁺ channels generate action potentials

Action potentials are sudden, transient, large changes in membrane potential. In unmyelinated axons they can be conducted at speeds of up to 2 meters per second, but in myelinated axons the conduction velocity can be 100 meters per second. Think of running the 100-meter dash—the world record is slightly under 10 seconds.

If we place the tips of a pair of electrodes on either side of the plasma membrane of a resting axon and measure the voltage difference, the reading might be about –60 mV, as we saw in Figure 45.5. If these electrodes are in place when an action potential travels down the axon, they register a rapid change in membrane potential, from –60 mV to about +50 mV. The membrane potential then rapidly returns to its resting level of –60 mV as the action potential passes (**Figure 45.10**).

The action potential is generated by the actions of voltage-gated Na⁺ and K⁺ channels in the plasma membrane of the axon. At the resting potential, most of these channels are closed (balloon 1 in Figure 45.10). Depolarization of the membrane causes them to open. For example, if a neuron is stimulated sufficiently

45.10 The Course of an Action Potential Action potentials result from rapid changes in voltage-gated Na⁺ and K⁺ channels.

yourBioPortal.com
GO TO Animated Tutorial 45.2 •
The Action Potential

1 Leak K⁺ channels create the resting potential. Gated channels are closed.

2 Activation gates of some Na⁺ channels open, depolarizing the cell to threshold.

3 Additional voltage-gated Na⁺ channel activation gates open, causing a rapid spike of depolarization—an action potential.

4 Na⁺ channel inactivation gates close; gated K⁺ channels open, repolarizing and even hyperpolarizing the cell.

5 All gated channels close. The cell returns to its resting potential. Na⁺ inactivation gates reopen.

to cause the plasma membrane of its cell body to depolarize, that depolarization can spread by local current flow to the **axon hillock**, the region of the cell body at the base of the axon (see Figure 45.3). Voltage-gated Na⁺ channels are concentrated in the axon hillock. When the plasma membrane in this area depolarizes, some of these voltage-gated channels open briefly—for less than a millisecond (balloon 2 in Figure 45.10). When these channels open, Na⁺ rushes into the axon and depolarizes the membrane even more, causing more Na⁺ channels to open—a *positive feedback* effect. When the membrane is depolarized about 5 to 10 mV above the resting potential, a **threshold** is reached; a large number of sodium channels open (balloon 3 in Figure 45.10), and the membrane potential becomes positive—an action potential. The rising phase of the action potential halts abruptly in 1 to 2 milliseconds, and the membrane potential rapidly becomes negative once again.

What causes the axon to return to resting potential? There are two contributing factors: the voltage-gated Na⁺ channels close, and voltage-gated K⁺ channels open (balloon 4 in Figure 45.10). The voltage-gated K⁺ channels open more slowly than the Na⁺ channels and stay open longer, allowing K⁺ to carry excess positive charges out of the axon. As a result, the membrane potential returns to a negative value and usually becomes even more negative than the resting potential until the voltage-gated K⁺ channels close (balloon 5 in Figure 45.10).

Another feature of the voltage-gated Na⁺ channels is that once they open and close, they have a **refractory period** of 1 to 2 milliseconds during which they cannot open again. This property can be explained by the channels having two gates, an **activation gate** and an **inactivation gate** (see Figure 45.10). Under resting conditions, the activation gate is closed and the inactivation gate is open. Depolarization of the membrane to the threshold level causes both gates to change state, but the activation gate responds faster. As a result, the channel is open for a brief time between the opening of the activation gate and the closing of the inactivation gate. Inactivation gates remain closed for 1 to 2 milliseconds before they spontaneously open again, thus explaining why the membrane has a refractory period before it can fire another action potential. By the time the inactivation gate reopens, the activation gate is closed, and the membrane is poised to generate another action poten-

tial. Another contribution to the refractory period is the duration of the opening of the voltage-gated K⁺ channels, as we saw above. The dip in the membrane potential following an action potential is called the *after-hyperpolarization* or *undershoot*.

The difference in the concentration of Na⁺ across the plasma membrane and the negative resting potential constitute the "bat-

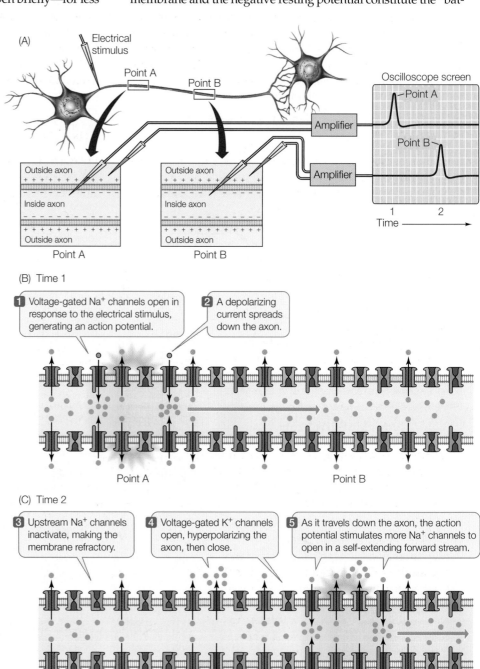

45.11 Action Potentials Travel along Axons   (A) There is no loss of signal as an action potential travels along an axon. (B) When an action potential is stimulated in one region of membrane, electric current flows to adjacent areas of membrane and depolarizes them. (C) The advancing wave of depolarization causes more Na⁺ channels to open, and the action potential is generated anew in the next section of membrane. Meanwhile, in the region where the action potential has just fired, the Na⁺ channels are inactivated and the voltage-gated K⁺ channels are still open, rendering this section of the axon refractory. Hence the action potential cannot "back up," but moves continuously forward along the axon, regenerating itself as it goes.

tery" that drives action potentials. How rapidly does the battery run down? It might seem that a substantial number of ions would have to cross the membrane for the membrane potential to change from –60 mV to +50 mV and back to –60 mV again. In fact, only a vanishingly small percentage of the Na$^+$ concentrated outside the plasma membrane moves through the channels during the passage of an action potential. Thus the effect of a single action potential on the concentration gradients of Na$^+$ and K$^+$ is very small, and it is possible in most cases for the sodium–potassium pump to keep the "battery" charged, even when the neuron is generating many action potentials every second.

Action potentials are conducted along axons without loss of signal

Action potentials can travel over long distances with no loss of signal. If we place two pairs of electrodes at two different locations along an axon, we can record an action potential at those two locations as it travels along the axon (**Figure 45.11A**). The magnitude of the action potential does not change between the two recording sites. This constancy is possible because an action potential is an all-or-none, self-regenerating event.

- An action potential is *all-or-none* because of the interaction between the voltage-gated Na$^+$ channels and the membrane potential. If the membrane is depolarized slightly, some voltage-gated Na$^+$ channels open. Some sodium ions cross the plasma membrane and depolarize it even more, opening more voltage-gated Na$^+$ channels, and so on, until the membrane reaches threshold and generates an action potential. This positive feedback mechanism ensures that action potentials always rise to their maximum value.

- An action potential is *self-regenerating* because it spreads by local current flow to adjacent regions of the plasma membrane. The resulting depolarization brings those neighboring areas of membrane to threshold. So when an action potential occurs at one location on an axon, it stimulates the adjacent region of axon to generate an action potential, and so on down the length of the axon.

We can use an electrode to stimulate an axon, causing it to depolarize and to fire an action potential that is then conducted along the axon. **Figure 45.11B** shows the changes in the ion channels in the membrane that are responsible for conducting the action potential along the axon without a reduction in amplitude. Normally, an action potential is propagated in only one direction—away from the

body. It cannot reverse itself because the voltage-gated Na$^+$ channels in the region of the membrane it came from are still in their refractory period.

Action potentials do not travel along all axons at the same speed. They travel faster in large-diameter axons than in small-diameter axons because the resistance to current flow decreases as an axon's diameter gets bigger. They travel faster in myelinated than in nonmyelinated axons because they can "jump" from one node to another without traversing the intervening space (**Figure 45.12**). Invertebrates mostly depend on increased axon diameter for fast conduction, but vertebrates mostly depend on myelination of axons to increase conduction velocity.

Action potentials can jump along axons

In vertebrate nervous systems, increasing the speed of action potentials by increasing the diameter of axons is not feasible because of the huge number of axons involved. Each of our eyes,

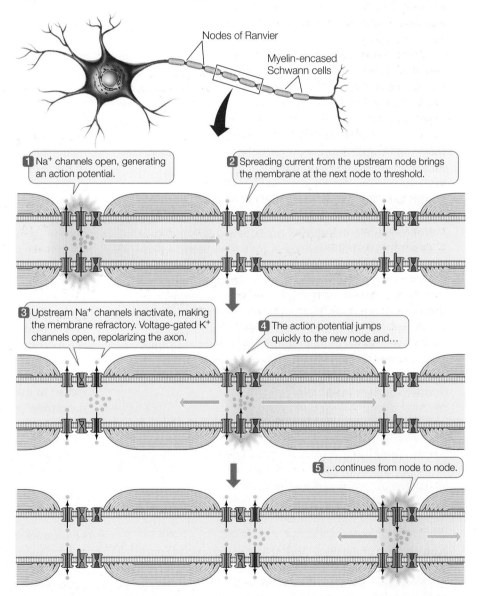

Nodes of Ranvier

Myelin-encased Schwann cells

1 Na$^+$ channels open, generating an action potential.

2 Spreading current from the upstream node brings the membrane at the next node to threshold.

3 Upstream Na$^+$ channels inactivate, making the membrane refractory. Voltage-gated K$^+$ channels open, repolarizing the axon.

4 The action potential jumps quickly to the new node and...

5 ...continues from node to node.

45.12 Saltatory Action Potentials Action potentials appear to jump from node to node in myelinated axons.

for example, has about a million axons connecting it to the brain. These axons conduct action potentials at about the same speed as does the squid giant axon—about 20 meters per second—yet the diameter of each is 200 times smaller than the squid axon's diameter. Vertebrates have evolved a different way of increasing conduction velocity of axons, and that adaptation is *myelination*.

When glia wrap themselves around axons, covering them with concentric layers of myelin (see Figure 45.4), they leave regularly spaced gaps, called **nodes of Ranvier**, where the axon is not covered (see Figure 45.12). The leakage of ions across the regions of the plasma membrane that are wrapped in myelin is reduced, so electric current can spread farther along the inside of a myelinated axon than it can along a nonmyelinated axon. Additionally, voltage-gated ion channels are clustered at the nodes of Ranvier. Thus an axon can fire action potentials only at nodes, and those action potentials cannot be propagated through the adjacent patch of membrane covered with myelin. The positive charges that flow into the axon at the node do, however, flow down the inside of the axon in the form of electric current. When the current reaches the next node, the plasma membrane at that node is depolarized to threshold and fires another action potential. Action potentials therefore appear to jump from node to node along the axon.

The speed of conduction is increased in these myelin-wrapped axons because electric current flows much faster through the cytoplasm than ion channels can open and close. This form of rapid impulse propagation is called **saltatory conduction** (Latin *saltare*, "to jump").

45.2 RECAP

Neurons have membrane potentials due to ionic concentration differences across their membranes and because leak channels make the membrane differentially permeable to ions. Changes in ion channel permeabilities cause graded changes in membrane potentials. Sudden openings and closings of gated ion channels in the membrane produce action potentials. Action potentials are rapid, all-or-none changes in membrane potential that are conducted along axons from the neuron body to the axon terminals.

- How are membrane resting potentials generated and altered? See pp. 949–950 and Figures 45.6 and 45.9

- How are action potentials generated? See pp. 953–955 and Figure 45.10

- How are action potentials transmitted along axons? See pp. 955–956 and Figures 45.11 and 45.12

Now that we understand how action potentials are generated and transmitted along axons, let's address the question of what happens when the action potential gets to the axon terminals. How is it communicated to the next cell, which could be another neuron, a muscle cell, or perhaps a secretory cell?

45.3 How Do Neurons Communicate with Other Cells?

Neurons communicate with each other and with other cells at synapses. The most common type of synapse in the nervous system is the **chemical synapse**, in which neurotransmitters released from a presynaptic cell induce changes in a postsynaptic cell. In **electrical synapses** the action potential spreads directly from presynaptic to postsynaptic cell. We will begin this section with a discussion of the synapses between neurons and muscle cells. We will then consider the diversity in synapses, how they integrate information, and how they are involved in learning and memory.

The neuromuscular junction is a model chemical synapse

Neuromuscular junctions are synapses between neurons and the skeletal muscle cells they innervate. They are excellent models for how chemical synaptic transmission works. Like other neurons, a motor neuron has only one axon, but close to its target cell that axon can branch into numerous axon terminals that form many synapses with muscle cells. At each axon terminal an enlarged knob or buttonlike structure contains vesicles filled with neurotransmitter molecules. The neurotransmitter used by all vertebrate neuromuscular synapses is **acetylcholine (ACh)**. ACh is released by exocytosis when the membrane of a vesicle fuses with the presynaptic membrane of the axon terminal.

Where does the neurotransmitter come from? Some neurotransmitters, such as ACh, are synthesized in the axon terminal and packaged in vesicles. The enzymes required for ACh biosynthesis, however, are produced in the cell body of the motor neuron and are transported along microtubules down the axon to the terminals. In contrast, peptide neurotransmitters are produced in the cell body and packaged into membrane-bound vesicles by the Golgi apparatus. These vesicles are rapidly transported down the axon to the terminals.

The postsynaptic membrane of the neuromuscular junction is a modified part of the muscle cell plasma membrane called a **motor end plate (Figure 45.13)**. It appears as a depression in the muscle cell membrane, and the terminals of the motor neuron sit in the depression. The space between the presynaptic membrane and the postsynaptic membrane is the **synaptic cleft**, which in chemical synapses is about 20 to 40 nanometers wide. ACh released into the cleft by the presynaptic cell diffuses across to the postsynaptic membrane.

The arrival of an action potential causes the release of neurotransmitter

Neurotransmitter is released when an action potential arrives at the axon terminal and causes the opening of voltage-gated Ca^{2+} channels in the presynaptic membrane. Because the Ca^{2+} concentration is greater outside the cell than inside, Ca^{2+} enters the axon terminal near the sites of vesicle exocytosis. The increase in Ca^{2+} inside the axon terminal causes the vesicles con-

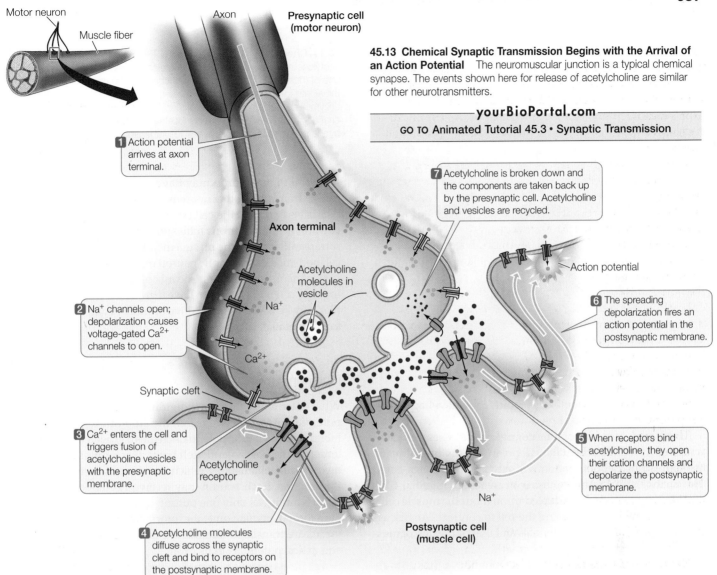

Motor neuron

Muscle fiber

Axon

Presynaptic cell
(motor neuron)

45.13 Chemical Synaptic Transmission Begins with the Arrival of an Action Potential The neuromuscular junction is a typical chemical synapse. The events shown here for release of acetylcholine are similar for other neurotransmitters.

── **yourBioPortal.com** ──
GO TO Animated Tutorial 45.3 • Synaptic Transmission

1 Action potential arrives at axon terminal.

Axon terminal

Acetylcholine molecules in vesicle

Na^+

2 Na^+ channels open; depolarization causes voltage-gated Ca^{2+} channels to open.

Ca^{2+}

Synaptic cleft

3 Ca^{2+} enters the cell and triggers fusion of acetylcholine vesicles with the presynaptic membrane.

Acetylcholine receptor

4 Acetylcholine molecules diffuse across the synaptic cleft and bind to receptors on the postsynaptic membrane.

7 Acetylcholine is broken down and the components are taken back up by the presynaptic cell. Acetylcholine and vesicles are recycled.

Action potential

6 The spreading depolarization fires an action potential in the postsynaptic membrane.

5 When receptors bind acetylcholine, they open their cation channels and depolarize the postsynaptic membrane.

Na^+

Postsynaptic cell
(muscle cell)

taining neurotransmitter to fuse with the presynaptic membrane and empty their contents into the synaptic cleft.

In neuromuscular synapses, vesicle fusion and emptying is all-or-none. The vesicle membrane is incorporated into the presynaptic membrane, which actually gets larger as a result—at least until the extra membrane is recycled through endocytosis. The recycled membrane is processed through endosomes to become new vesicles that are then refilled with neurotransmitter.

Synaptic functions involve many proteins

The description above of the release of neurotransmitter from the presynaptic membrane may seem simple, but it involves hundreds of proteins that are responsible for various aspects of the process: vesicle formation, transport of neurotransmitter into vesicles, anchoring of vesicles to cytoskeletal elements, docking of the vesicles with the presynaptic membrane, fusion of the vesicular and cell membranes, and endocytosis of the vesicle membrane for recycling.

Some of these proteins are the targets of toxins. For example, toxins from bacteria of the genus *Clostridium* are proteases that destroy several of the proteins necessary for the docking of

vesicles to the presynaptic membrane. These toxins cause botulism and tetanus, frequently fatal diseases that involve muscle impairment because of loss of transmitter release. Poisons can also become medicines, however. You have surely heard about the use of botulinum toxin (Botox) for decreasing muscle spasms for cosmetic (removal of wrinkles) or therapeutic purposes.

The postsynaptic membrane responds to neurotransmitter

When ACh is released at a synapse, some of it diffuses across the synaptic cleft and binds to ACh receptors on the postsynaptic membrane (**Figure 45.14**). The postsynaptic membrane of the motor end plate is highly folded. ACh receptors are on the crests of the folds, and voltage-gated cation channels are at the bottoms of the folds and in the surrounding muscle cell membrane (see Figure 45.13). The ACh receptors are channels that allow both Na^+ and K^+ to flow through, but since the electrochemical gradients favor a net influx of Na^+, the response of the motor end plate to ACh is to depolarize. That graded potential reflecting the number of receptors activated spreads to the depths of the folds of the motor end plate membrane and to sur-

> The acetylcholine receptor-mediated channel is normally closed.

> When ACh binds at specific sites on the receptor, the channel opens, allowing Na⁺ to enter the postsynaptic cell.

> Acetylcholinesterase breaks down ACh, causing the channel to close once again.

Outside of cell

Na⁺

ACh

Acetylcholinesterase

ACh receptor

Inside of cell

45.14 Chemically Gated Channels   The motor end plate contains acetylcholine receptors, which are chemically gated ion channels. When one of these receptors binds ACh, its channel pore opens and Na^+ ions move into the postsynaptic cell, depolarizing its plasma membrane. The enzyme acetylcholinesterase breaks down ACh in the synapse, closing the channel; the breakdown products (acetate and choline) are then taken up by the presynaptic membrane and resynthesized into more ACh.

rounding muscle cell membrane, which contain voltage-gated Na^+ channels.

If the axon terminal of a motor neuron releases sufficient amounts of ACh to adequately depolarize a motor end plate, that spreading depolarization will activate the voltage-gated Na^+ channels and cause the firing of an action potential. This action potential is then conducted throughout the muscle cell's system of membranes, causing the cell to contract. We discuss muscle membrane action potentials and the contraction of muscle cells in greater detail in Section 48.1.

How much neurotransmitter is enough? Neither a single ACh molecule nor the contents of an entire vesicle (about 10,000 ACh molecules) will bring the plasma membrane of a muscle cell to threshold. However, a single action potential in an axon terminal releases the contents of about 100 vesicles, which is more than enough to fire an action potential in the muscle cell and cause it to contract.

Synapses between neurons can be excitatory or inhibitory

In vertebrates, the synapses between motor neurons and muscle cells are always **excitatory**; that is, motor end plates always respond to ACh with a graded potential that is less negative than the resting potential. However, synapses between neurons can also be inhibitory. For example, recall that there are more chloride ions (Cl⁻) outside the cell than inside it. If the receptor on the postsynaptic membrane is a Cl⁻ channel opened by a neurotransmitter, the effect of that neurotransmitter will be to cause Cl⁻ ions to enter the postsynaptic cell and hyperpolarize it. Hyperpolarization will take the membrane farther from the threshold potential for the voltage-gated Na^+ channels, and therefore make it less likely that the cell will fire action potentials. A synapse that causes hyperpolarization of the postsynaptic membrane is **inhibitory**.

Recall that a neuron may have many dendrites. Axon terminals from many other neurons may form synapses with those dendrites and with the cell body. The axon terminals of different presynaptic neurons may store and release different neurotransmitters, and the plasma membrane of the dendrites and cell body of a postsynaptic neuron may have receptors for a variety of neurotransmitters. The mix of synaptic activity impinging on a cell will cause it to have a graded membrane potential that may be either more positive or more negative than its resting potential.

The postsynaptic cell sums excitatory and inhibitory input

What determines when an individual neuron will fire an action potential? As we just learned, the sum of excitatory and inhibitory postsynaptic potentials creates a graded membrane potential in the postsynaptic cell body. This summation ability is the major mechanism by which the nervous system integrates information. Each neuron may receive 1,000 or more synaptic inputs, but it has only one output: an action potential in a single axon. At any one time, the information from all of the active inputs are translated into the rate at which that neuron generates action potentials in its axon.

For most neurons, summation takes place in the axon hillock at the base of the axon. The plasma membrane of the axon hillock is not insulated by glia and has many voltage-gated Na^+ channels. Excitatory and inhibitory postsynaptic potentials from synapses anywhere on the dendrites or the cell body may spread to the axon hillock by local current flow. If the resulting graded potential depolarizes the axon hillock to threshold, it fires an action potential. Because postsynaptic potentials decrease in strength as they spread from the site of the synapse, a synapse at the tip of a dendrite has less influence than a synapse on the cell body, near the axon hillock.

Excitatory and inhibitory postsynaptic potentials are summed over space and over time. **Spatial summation** adds up the simultaneous influences of synapses at different sites on the postsynaptic cell (**Figure 45.15A**). **Temporal summation** adds up postsynaptic potentials generated at the same site in a rapid sequence (**Figure 45.15B**).

Synapses can be fast or slow

Most neurotransmitter receptors induce changes in postsynaptic cells by opening or closing ion channels. How they do so is the basis for grouping receptors into two general categories:

- **Ionotropic receptors** are ion channels themselves. Neurotransmitter binding to an ionotropic receptor causes a direct

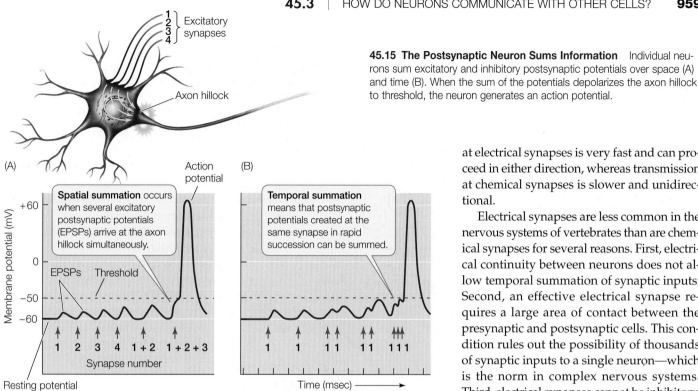

45.15 The Postsynaptic Neuron Sums Information Individual neurons sum excitatory and inhibitory postsynaptic potentials over space (A) and time (B). When the sum of the potentials depolarizes the axon hillock to threshold, the neuron generates an action potential.

change in ion movement across the plasma membrane of the postsynaptic cell. These proteins enable fast, short-lived responses.

The ACh receptor of the motor end plate is an example of an ionotropic receptor. It consists of five subunits, each of which extends through the plasma membrane. When assembled, the subunits create a central pore that allows ions to pass through (see Figure 45.14). Of several different kinds of subunits, only one kind has the ability to bind ACh. Each functional receptor has two of the ACh-binding subunits and three other subunits.

- **Metabotropic receptors** are not ion channels, but they induce signaling cascades in the postsynaptic cell that secondarily lead to changes in ion channels (see Figure 7.10A). Postsynaptic cell responses mediated by metabotropic receptors are generally slower and longer-lived than those induced by ionotropic receptors.

Metabotropic receptors are also transmembrane proteins, but instead of acting as ion channels, they initiate an intracellular signaling process that can result in the opening or closing of an ion channel. An example is shown in Figure 7.19.

Electrical synapses are fast but do not integrate information well

Electrical synapses are different from chemical synapses because they couple neurons electrically. Electrical synapses contain numerous *gap junctions* (see Figure 7.21A). At these synapses, the presynaptic and postsynaptic cell membranes are separated by a space of only 2 to 3 nanometers, and membrane proteins called *connexons* link the two neurons by forming pores that connect the cytoplasm of the two cells. Ions and small molecules can pass directly from cell to cell through these pores. Transmission

at electrical synapses is very fast and can proceed in either direction, whereas transmission at chemical synapses is slower and unidirectional.

Electrical synapses are less common in the nervous systems of vertebrates than are chemical synapses for several reasons. First, electrical continuity between neurons does not allow temporal summation of synaptic inputs. Second, an effective electrical synapse requires a large area of contact between the presynaptic and postsynaptic cells. This condition rules out the possibility of thousands of synaptic inputs to a single neuron—which is the norm in complex nervous systems. Third, electrical synapses cannot be inhibitory. Thus, electrical synapses are useful for rapid communication, but they are less useful for processes of integration and learning.

The action of a neurotransmitter depends on the receptor to which it binds

More than 50 neurotransmitters are now recognized, and more will surely be discovered. ACh, as we have seen, is an important neurotransmitter because it is how the nervous system commands muscles to contract. ACh also plays roles in certain synapses between neurons in the CNS, but it accounts for only a small percentage of the total neurotransmitter content of the CNS.

The workhorse neurotransmitters of the CNS are simple amino acids: glutamate (excitatory) and glycine and γ-aminobutyrate (GABA) (inhibitory). Another important group of neurotransmitters in the CNS is the monoamines, which are derivatives of amino acids. They include dopamine and norepinephrine (derivatives of tyrosine) and serotonin (a derivative of tryptophan). Peptides also function as neurotransmitters; for example, *endorphins* and *enkephalins* are the body's opiates and modulate the sensation of pain. Another peptide, *substance P*, transmits pain sensations. Even a gas, nitric oxide, is used by neurons as an intercellular messenger (see Figure 7.17).

Neurotransmission is complex in part because each neurotransmitter has multiple receptor types. ACh, for example, has two receptor types: *nicotinic receptors*, which are ionotropic, and *muscarinic receptors*, which are metabotropic. Both types of ACh receptors are found in the CNS, where nicotinic receptors tend to be excitatory and muscarinic receptors tend to be inhibitory. ACh actions can differ outside the CNS as well. ACh acting through nicotinic receptors causes the smooth muscle of the gut to increase its motility, but ACh acting through muscarinic receptors causes cardiac muscle to hyperpolarize and therefore to slow down. We could give many more examples of neurotrans-

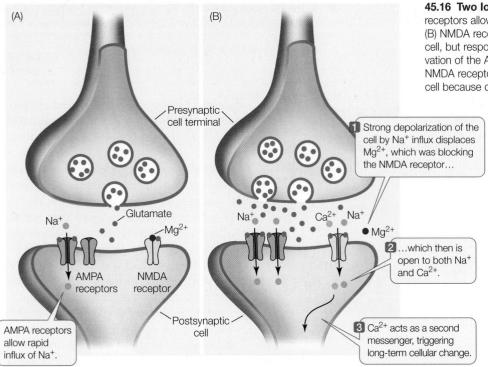

45.16 Two Ionotropic Glutamate Receptors (A) AMPA receptors allow rapid influx of Na⁺ into the postsynaptic cell. (B) NMDA receptors allow both Na⁺ and Ca²⁺ to enter the cell, but respond to synaptic input more slowly. Strong activation of the AMPA receptor leads to unblocking of the NMDA receptor, which results in longer-term effects on the cell because of Ca²⁺ entry.

1 Strong depolarization of the cell by Na⁺ influx displaces Mg^{2+}, which was blocking the NMDA receptor…

2 …which then is open to both Na⁺ and Ca²⁺.

3 Ca²⁺ acts as a second messenger, triggering long-term cellular change.

AMPA receptors allow rapid influx of Na⁺.

mitters that have different effects in different tissues, but the important thing to remember is that the action of a neurotransmitter depends on the receptor to which it binds.

── **yourBioPortal.com.com** ──
GO TO Web Activity 45.1 • Neurotransmitters

Glutamate receptors may be involved in learning and memory

Glutamate is a neurotransmitter that can bind to a variety of receptors, including both metabotropic and ionotropic receptors. The glutamate receptors are divided into several classes because they can be differentially activated by other chemicals that mimic the action of glutamate. One class of ionotropic glutamate receptors is the *NMDA receptors*, which can be activated by the chemical N-methyl-D-aspartate. Another class of ionotropic glutamate receptors is activated by a different chemical, abbreviated as *AMPA* (which stands for α-amino-3-hydroxy-5-methyl-4-isoxazole propionate, demonstrating why biologists are fond of abbreviations).

Glutamate is an excitatory neurotransmitter, so activation of ionotropic glutamate receptors always results in Na⁺ entry into the neuron and depolarization. But the *timing* of the response to activation by these different types of receptors differs significantly. The AMPA receptors allow a rapid influx of Na⁺ into the postsynaptic cell, whereas the NMDA receptors allow a slower and longer-lasting influx of Na⁺. The NMDA receptors require that the cell be somewhat depolarized through the action of other receptors before their pores will open and permit Na⁺ influx. When they do open, these receptors also allow Ca²⁺ as well as Na⁺ to enter the cell. Calcium ions act as second messengers in

the cell and can trigger a variety of cellular changes, such as activation of certain protein kinases.

Figure 45.16 shows how the AMPA and NMDA receptors can work in concert. At resting potential, the NMDA receptor is blocked by a magnesium ion (Mg^{2+}). Strong depolarization of the neuron due to other inputs—such as the activation of AMPA receptors—displaces Mg^{2+} from the NMDA receptors and allows Na⁺ and Ca²⁺ to pass through them when they are activated by glutamate. These special properties of the NMDA receptor are probably involved in learning and memory.

Most of the synaptic events we have studied so far happen very quickly. It is therefore a special challenge to understand how the messages carried by action potentials can result in long-term events such as learning and memory. Our understanding of these processes has been greatly enhanced by investigation of a phenomenon called **long-term potentiation (LTP)**, which has been studied extensively by neurobiologists working with slices of brain kept alive in dishes of culture medium. Using these brain slice preparations, it is possible to stimulate and record from specific brain regions, and even specific neurons.

In the studies leading to the discovery of LTP, experimenters repeatedly stimulated synaptic inputs to a particular neuron and observed the usual action potential response. When the neuron was stimulated many times in rapid succession, however, they found that the properties of the neuron changed. The magnitude of the postsynaptic response was enhanced, or *potentiated*, and this change lasted for days or weeks.

How does potentiation of a synapse occur? The answer, at least for some areas of the brain, now seems clear. With low-frequency stimulation, the glutamate released by presynaptic cells activates only AMPA receptors, and the postsynaptic membrane simply responds with action potentials. With higher frequency stimulation, however, NMDA receptors are also activated, allowing both Na⁺ and Ca²⁺ ions to enter the postsynaptic neuron. The Ca²⁺ ions induce long-term changes in the postsynaptic membrane that make it more sensitive to synaptic input (**Figure 45.17**).

To turn off responses, synapses must be cleared of neurotransmitter

Turning off the action of neurotransmitters is as important as turning it on. If released neurotransmitter molecules simply remained in the synaptic cleft, the postsynaptic membrane would become saturated with neurotransmitter, and receptors would be constantly activated. As a result, the postsynaptic cell would remain hyperpolarized or depolarized and would be unresponsive to short-term changes in the presynaptic cell. The more rapidly neurons can respond to input, the more information they can process in a given time. Thus, neurotransmitter must be cleared from the synaptic cleft shortly after it is released by the axon terminal.

Neurotransmitter action may be terminated in several ways. First, enzymes may destroy the neurotransmitter. ACh, for example, is rapidly destroyed by the enzyme acetylcholinesterase, which is present in the synaptic cleft in close association with the ACh receptors on the postsynaptic membrane (see Figure 45.14). Some of the most deadly nerve gases developed for chemical warfare work by inhibiting acetylcholinesterase. As a result, ACh lingers in the synaptic cleft, causing the victim to die of spastic (contracted) muscle paralysis. Some agricultural insecticides, such as malathion, also inhibit acetylcholinesterase and can poison farm workers if used without safety precautions.

Neurotransmitter also may simply diffuse away from the cleft, or be taken up via active transport by nearby cell membranes. The drug commonly prescribed under the brand name Prozac to treat depression slows the reuptake of the neurotransmitter serotonin, thus enhancing its activity at the synapse.

The diversity of receptors makes drug specificity possible

Many drugs used to treat the nervous system act by modulating specific synaptic interactions. Drugs that mimic or potentiate the effect of a neurotransmitter are called *agonists*; those that block the actions of neurotransmitters are called *antagonists*. For example, morphine is an agonist at the endorphin receptor and therefore blocks pain. Propranolol, known as a β-blocker, is an antagonist of certain adrenergic receptors and therefore decreases panic attacks and anxiety. A major emphasis in neurobiology is to identify neurotransmitter receptor subtypes and design drugs that selectively bind to them to have highly specific effects on nervous system activity.

INVESTIGATING LIFE

45.17 Repeated Stimulation Can Cause Long-Term Potentiation

When a cell receives low-frequency synaptic input, the resulting postsynaptic response remains constant. If that same synaptic pathway is stimulated briefly at a high frequency, however, the subsequent sensitivity of the postsynaptic cell to the original level of synaptic input is potentiated for a longer period.

HYPOTHESIS Repeated stimulation can change the properties of a synapse.

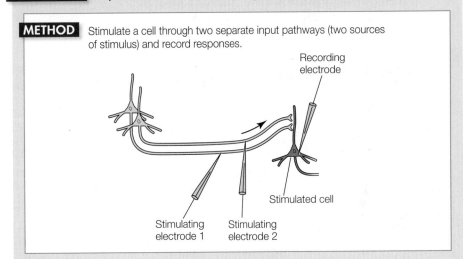

METHOD Stimulate a cell through two separate input pathways (two sources of stimulus) and record responses.

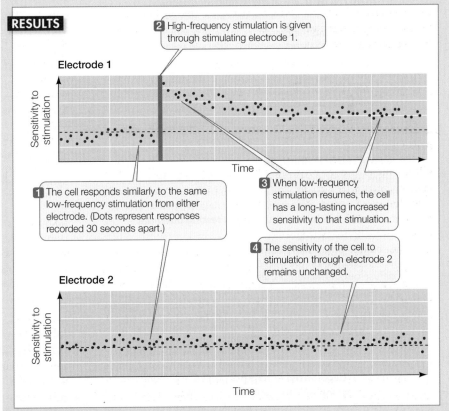

RESULTS

2 High-frequency stimulation is given through stimulating electrode 1.

Electrode 1

1 The cell responds similarly to the same low-frequency stimulation from either electrode. (Dots represent responses recorded 30 seconds apart.)

3 When low-frequency stimulation resumes, the cell has a long-lasting increased sensitivity to that stimulation.

4 The sensitivity of the cell to stimulation through electrode 2 remains unchanged.

Electrode 2

CONCLUSION High-frequency stimulation of synapses can result in a long-lasting change in the sensitivity of these synapses (long-term potentiation, or synaptic memory).

Go to **yourBioPortal.com** for original citations, discussions, and relevant links for all INVESTIGATING LIFE figures.

45.3 RECAP

Chemical synapses involve the release of neurotransmitter molecules stored in vesicles in the presynaptic terminal. Action potentials reaching that terminal cause the fusion of vesicles with the presynaptic membrane, releasing neurotransmitter that can then bind to receptors on the postsynaptic membrane and influence its membrane potential. There is a great diversity of neurotransmitters and their receptors.

- Describe the role of Ca^{2+} channels in synaptic events. See pp. 956–957 and Figure 45.13

- How can some synapses be excitatory and others inhibitory? See p. 958

- How do neurons integrate the input from various synapses? See p. 958 and Figure 45.15

CHAPTER SUMMARY

45.1 What Cells Are Unique to the Nervous System?

- Nervous systems include **neurons** and **glia**. Neurons are organized in circuits with sensory inputs, integration, and outputs to effectors. Glia serve support functions. Review Figures 45.1 and 45.3

- In vertebrates, the **brain** and **spinal cord** form the **central nervous system (CNS)**, which communicates with the rest of the body via the **peripheral nervous system (PNS)**. The CNS increases in complexity from invertebrates to vertebrates and from fishes to mammals. Review Figures 45.1 and 45.2

- Neurons generally receive information via their **dendrites**, of which there can be many, and transmit information via their single **axons**, which end in **axon terminals**. Review Figure 45.3

- Where neurons and their target cells meet, information is transmitted across specialized junctions called **synapses**.

- Glia include **Schwann cells** and **oligodendrocytes,** both of which generate **myelin** sheets on axons. Glia also include **astrocytes**, which support neurons metabolically and contribute to the **blood–brain barrier**. Review Figure 45.4

45.2 How Do Neurons Generate and Transmit Electrical Signals?

SEE ANIMATED TUTORIAL 45.1

- Neurons have an electric charge difference across their plasma membranes, the **membrane potential**. The membrane potential is created by ion transporters and channels. When a neuron is not active, its membrane potential is a **resting potential**. Review Figures 45.5 and 45.6

- The **sodium–potassium pump** concentrates K^+ on the inside of a neuron and Na^+ on the outside. Potassium channels allow K^+ to diffuse out of the neuron, leaving behind unbalanced negative charges. Review Figures 45.6 and 45.7

- **Patch clamping** allows the study of single ion channels. Review Figure 45.8

- The resting potential is perturbed when ion channels open or close, changing the permeability of the plasma membrane to charged ions. Through this mechanism, the plasma membrane can become **depolarized** or **hyperpolarized** and therefore have a **graded membrane potential** response to input. Review Figure 45.9

- An **action potential** is a rapid reversal in charge across a portion of the plasma membrane resulting from the sequential opening and closing of **voltage-gated channels** of Na^+ and K^+. These changes in voltage-gated channels occur when the plasma membrane depolarizes to a **threshold** level. Review Figure 45.10, **ANIMATED TUTORIAL 45.2**

- Action potentials are all-or-none, self-regenerating events. They are conducted down axons because local current flow depolarizes adjacent regions of membrane and brings them to threshold. Review Figure 45.11

- In myelinated axons, action potentials appear to jump between **nodes of Ranvier**, patches of axonal plasma membrane that are not covered by myelin. Review Figure 45.12

45.3 How Do Neurons Communicate with Other Cells?

- Neurons communicate with each other and with other cells by transmitting information over **chemical synapses** (with neurotransmitters) or **electrical synapses**.

- The neuromuscular junction is a well-studied chemical synapse between a motor neuron and a skeletal muscle cell. Its neurotransmitter is **acetylcholine (ACh)**, which causes depolarization of the postsynaptic membrane when it binds to its receptor. Review Figure 45.13, **ANIMATED TUTORIAL 45.3**

- When an action potential reaches an axon terminal, it causes the release of neurotransmitters, which diffuse across the **synaptic cleft** and bind to receptors on the postsynaptic membrane. Review Figures 45.13 and 45.14

- Synapses between neurons can be either **excitatory** or **inhibitory**. A postsynaptic neuron integrates information by summing excitatory and inhibitory postsynaptic potentials in both space and time. Review Figure 45.15

- **Ionotropic receptors** are ion channels or directly influence ion channels. **Metabotropic receptors** influence the postsynaptic cell through various signal transduction pathways and result in the opening or closing of ion channels. The actions of ionotropic synapses are generally faster than those of metabotropic synapses.

- There are many different neurotransmitters and even more types of receptors. The action of a neurotransmitter depends on the receptor to which it binds. **SEE WEB ACTIVITY 45.1**

- With repeated stimulation, a neuron can become more sensitive to its inputs through **long-term potentiation (LTP)**. The properties of the NMDA glutamate receptor appear to explain LTP. Review Figures 45.16 and 45.17

SELF-QUIZ

1. The rising phase of an action potential is due to the
   a. closing of K⁺ channels.
   b. opening of chemically gated Na⁺ channels.
   c. closing of voltage-gated Ca²⁺ channels.
   d. opening of voltage-gated Na⁺ channels.
   e. spread of positive current along the plasma membrane.

2. The resting potential of a neuron is due mostly to
   a. local current spread.
   b. open Na⁺ channels.
   c. synaptic summation.
   d. open K⁺ channels.
   e. open Cl⁻ channels.

3. Which statement about synaptic transmission is *not* true?
   a. The synapses between neurons and skeletal muscle cells use ACh as their neurotransmitter.
   b. A single vesicle of neurotransmitter can cause a muscle cell to contract.
   c. The release of neurotransmitter at the neuromuscular junction causes the motor end plate to depolarize.
   d. In vertebrates, the synapses between motor neurons and muscle fibers are always excitatory.
   e. Inhibitory synapses cause the resting potential of the postsynaptic membrane to become more negative.

4. Which statement accurately describes an action potential?
   a. Its magnitude increases along the axon.
   b. Its magnitude decreases along the axon.
   c. All action potentials in a single neuron are of the same magnitude.
   d. During an action potential, the membrane potential of a neuron remains constant.
   e. An action potential permanently shifts a neuron's membrane potential away from its resting value.

5. A neuron that has just fired an action potential cannot be immediately restimulated to fire a second action potential. The short interval of time during which restimulation is not possible is called
   a. hyperpolarization.
   b. the resting potential.
   c. depolarization.
   d. repolarization.
   e. the refractory period.

6. Graded membrane potentials
   a. can be more negative than resting potential.
   b. can be less negative than resting potentials.
   c. integrate the many synaptic inputs to a cell.
   d. are important means of summing sensory inputs.
   e. are all of the above.

7. The binding of an inhibitory neurotransmitter to the postsynaptic receptors results in
   a. depolarization of the membrane.
   b. generation of an action potential.
   c. hyperpolarization of the membrane.
   d. increased permeability of the membrane to sodium ions.
   e. increased permeability of the membrane to calcium ions.

8. The difference between slow and fast synapses is
   a. the width of the synaptic cleft.
   b. the size of the synapse.
   c. whether or not the neurotransmitter acts directly on ion channels.
   d. the density of receptors on the postsynaptic membrane.
   e. the amount of neurotransmitter that is released.

9. Whether a synapse is excitatory or inhibitory depends on the
   a. type of neurotransmitter.
   b. presynaptic axon terminal.
   c. size of the synapse.
   d. nature of the postsynaptic receptors.
   e. concentration of neurotransmitter in the synaptic space.

10. Which of the following is a likely mechanism for long-term potentiation?
    a. When glutamate binds to postsynaptic AMPA receptors, it activates intracellular changes.
    b. When glutamate binds to NMDA receptors, it allows magnesium ions to enter the cell, which initiate intracellular changes.
    c. When sufficient glutamate is released by the presynaptic neuron, it dislodges Mg²⁺ from the AMPA receptors and allows Ca²⁺ to leave the cell.
    d. When sufficient glutamate is released, both AMPA and NMDA receptors are activated, and NMDA receptors allow Ca²⁺ as well as Na⁺ to enter the cell, thus initiating intracellular changes.
    e. When both glutamate and ACh are released together, they create a long-lasting depolarization of the postsynaptic cell.

FOR DISCUSSION

1. The language of the nervous system consists of one "word," the action potential. How can this single message convey a diversity of information, how can that information be quantitative, and how can it be integrated?

2. If you stimulate an axon in the middle, action potentials are conducted in both directions. Yet when an action potential is generated at the axon hillock, it goes only toward the axon terminals and does not backtrack. Explain why action potentials are bidirectional in the first example and unidirectional in the second.

3. The nature of synapses presents various opportunities for plasticity in the nervous system. Discuss at least four synaptic mechanisms that could be altered to change the response of a neuron to a specific input.

4. Benzodiazepines are drugs that potentiate the effects of GABA on its receptor. What effects would you expect these drugs to have?

ADDITIONAL INVESTIGATION

The patch clamping technique shown in Figure 45.8 produces an "inside-out" patch because the cytoplasmic side of the membrane faces away from the pipette tip and the surface side of the membrane is exposed to the solution in the pipette. Another patch preparation is the "outside-out patch" in which the cytoplasmic side of the membrane is exposed to the solution in the pipette. Describe what kind of experiments you could conduct with the "inside-out" patch and what kind you could conduct with the "outside-out" patch.

Out of range

A rattlesnake can see to strike and kill a running rodent in complete darkness. How can this be, when "seeing" means using the eyes to detect light, and "complete darkness" means the absence of any light? It is possible because these definitions are based on human capabilities. What we call "light" is actually only a small portion (red, orange, yellow, green, blue, indigo, and violet) of the spectrum of electromagnetic radiation. Other animals see wavelengths we cannot. Insects and birds see patterns on flowers that reflect ultraviolet wavelengths invisible to us. Similarly, rattlesnakes can "see" infrared wavelengths that are invisible to us (although at high enough levels of intensity, humans feel these wavelengths as heat).

It is not the snake's eyes that perceive infrared wavelengths. Rattlesnakes and their relatives have *pit organs* containing high densities of infrared-sensitive neurons. The two pits, located between the nostrils and the eyes on each side of the skull, are positioned in such a way that sensory receptor cells in the pits receive directional information. The fields of "view" of the bilateral pits are overlapping and thus convey a three-dimensional perspective. Information from the pit organs goes to the same region of the brain as information from the eyes, so rattlesnakes actually do "see" the world in a range of electromagnetic radiation that is different from the human visual spectrum.

Our definition of silence is as arbitrary as our definition of darkness. "Sound" is actually pressure waves in the environment, and many animals are sensitive to pressure waves with frequencies (*pitches*) we cannot hear. Elephants communicate in sound waves that are below human hearing range; such long waves travel great distances, an advantage to large animals that roam over extensive areas.

Bats emit incredibly loud, brief sound pulses, the frequencies of which are above our range of hearing, and a flying bat hears echoes of these pulses bouncing off objects in the environment. The pulses are so loud and the echoes so weak that it is rather like a construction worker trying to overhear

Sensing Infrared Radiation The "hole" to the right of this diamondback rattlesnake's eye is one of its bilateral pit organs. Pit organs detect infrared radiation from the snake's preferred prey—small rodents—with unerring precision, even in total darkness. The forked tongue also provides positional information, picking up molecular signals that are transmitted to the brain by a specialized organ in the roof of the snake's mouth.

Echolocating around an Obstacle Course A bat's ability to echolocate using sound waves is so precise that, in a totally dark room strung with fine wires, bats can capture small insects while avoiding the wires.

a whispered conversation while using a pneumatic drill. Why don't the loud pulses "drown out" the weak echoes for the bat? Small muscles in the bat's ears contract to dampen their hearing sensitivity while the sounds are being emitted, but relax in time for the bat to hear the echo—a truly remarkable ability, since the pulses are emitted at rates of 20 to 80 per second.

Our senses are our windows on the world. "Reality" is what our eyes see, our ears hear, our noses smell, and what we touch and taste. But human beings sense only a limited range of the information available. Animals with different ranges of sensitivity process different sources of information and may perceive "reality" quite differently.

IN THIS CHAPTER we will examine how sensory receptor cells convert environmental stimuli into the electrochemical signals of nervous systems. We will examine in detail the diversity of cells responsible for our senses and see how they are incorporated into sensory systems that provide the central nervous system with information about the world around and within us. We will also learn about the unusual sensory abilities of other animals.

46.1 How Do Sensory Cells Convert Stimuli into Action Potentials?

Sensory receptor cells, sometimes simply called *sensors* or *receptors*, transduce (convert) physical and chemical stimuli such as light and sound waves, touch, and odorant and taste molecules into neural signals. These signals are then transmitted to the central nervous system for processing and interpretation. We will examine several sensory systems. In each case, we ask the same general question: How do sensory receptor cells transduce energy from a stimulus into a change in membrane potential?

Sensory receptor proteins act on ion channels

As discussed in Section 45.2, cells use energy to create gradients of charged ions across their plasma membranes. In this way, cells are like batteries: batteries store potential energy by separating electrical charges between their poles, and cells store potential energy in the ionic gradients across their plasma membranes. If the two poles of a battery are connected by a wire and a switch, current flows through the wire when the switch is closed. Ion channels are like switches. When they open, charged ions can flow down their electrochemical gradient. When the ion channel is selective for one type of ion, the flow of that charged ion creates a change in the electrical potential across the membrane.

Sensory transduction begins with a **receptor protein** or some other mechanism in a sensory cell that can detect a specific stimulus modality such as heat, light, chemicals, mechanical force (including sound waves), or electrical fields. The receptor protein then directly or indirectly opens or closes ion channels in that sensory cell, leading either to an action potential or to the release of neurotransmitter.

Section 45.3 notes that synaptic receptor proteins are either *ionotropic* or *metabotropic*; the same distinction can be applied to sensory receptor proteins. Ionotropic sensory receptor proteins are either ion channels themselves or directly affect the opening of an ion channel. Examples are receptors that respond to physical force (*mechanoreceptors*) and those that respond to temperature (*thermoreceptors*). *Electrosensors* most likely have no receptor protein at all, but they are grouped with the ionotropic receptors; the plasma membrane of electrosensory cells is sensitive to the voltage across it and releases neurotransmitter in response to slight changes in membrane potential. Metabotropic sensory receptor proteins influence ion channels indirectly, through G proteins and second messengers as described in Chapter 7. Examples are most *chemoreceptors* and *photoreceptors* (**Figure 46.1**).

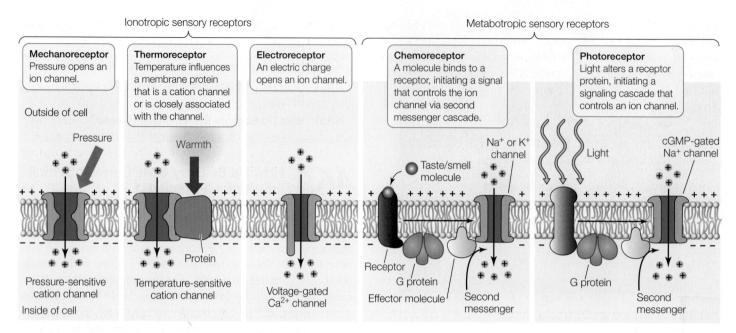

46.1 Sensory Cell Membrane Receptor Proteins Respond to Stimuli The receptor proteins in mechanoreceptors are ion channels. The activated receptor proteins of metabotropic chemoreceptors and photoreceptors initiate signal transduction cascades that eventually open or close ion channels.

Sensory transduction involves changes in membrane potentials

A change in the resting membrane potential of a sensory receptor cell in response to a stimulus is called a **receptor potential**. Receptor potentials are graded membrane potentials that spread over only short distances. To travel long distances in the nervous system, receptor potentials must generate action potentials, which they can do in two ways:

- The receptor potential may generate action potentials within the receptor cell itself.

- The receptor potential may trigger the release of a neurotransmitter that induces a postsynaptic neuron to generate action potentials.

A good example of a sensory receptor cell that can generate action potentials is the stretch receptor of a crayfish (**Figure 46.2**). By placing an electrode in the cell body of a crayfish stretch receptor, we can record the receptor potential that results from stretching the muscle to which the cell's dendrites are attached. These receptor potentials spread to the base of the cell's axon (the axon hillock), which contains voltage-gated Na⁺ channels. Action potentials generated here travel down the axon to the central nervous system (CNS). The rate at which action potentials are fired by the axon depends on the magnitude of the

receptor potential; that, in turn, depends on how much the muscle is stretched.

In a receptor cell that does not fire action potentials, the spreading receptor potential reaches a presynaptic patch of plasma membrane and induces the release of a neurotransmitter. The intensity of the stimulus influences how much neurotransmitter is released. That neurotransmitter binds to receptor proteins on an associated sensory neuron, altering its membrane

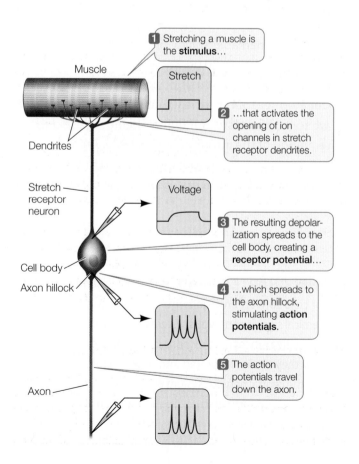

1 Stretching a muscle is the **stimulus**…

2 …that activates the opening of ion channels in stretch receptor dendrites.

3 The resulting depolarization spreads to the cell body, creating a **receptor potential**…

4 …which spreads to the axon hillock, stimulating **action potentials**.

5 The action potentials travel down the axon.

46.2 Stimulating a Sensory Cell Produces a Receptor Potential Signal transduction in the stretch receptor of a crayfish can be investigated by measuring the membrane potential at different places on the stretch receptor neuron while stretching the muscle innervated by that sensory neuron.

potential and causing it to increase or decrease its rate of firing action potentials. In a few cases, this second cell also responds by changing the rate at which it releases neurotransmitter onto another neuron. Eventually, however, the stimulation of the sensory cell is coded as a change in firing of action potentials in a sensory circuit.

Sensation depends on which neurons receive action potentials from sensory cells

All sensory systems process information in the form of action potentials. But the sensations we perceive—such as heat, pressure, light, smell, and sound—differ because the messages from different kinds of sensory cells arrive at different places in the CNS. Action potentials arriving in the visual cortex of the brain are interpreted as light, in the auditory cortex as sound, in the olfactory bulb as smell, and so forth.

A small patch of skin on your arm contains some sensory receptor cells that increase their firing rates when the skin is warmed and others that increase their activity when the skin is cooled. Other types of sensory cells in the same patch of skin respond to touch, movement of hairs, irritants such as mosquito bites, and painful stimuli. These receptor cells transmit their messages through axons that enter the CNS at the spinal cord. The synapses made by those axons in the spinal cord and the subsequent pathways of transmission determine whether the stimulation of the patch of skin on your arm is perceived as warmth, cold, touch, tickle, itch, or pain. So even though the action potentials carried by all of these sensory axons look the same, the connectivity of each axon is specific for a given sensory modality.

How is the intensity of the stimulus encoded if, as Section 45.2 describes, each action potential is an all-or-none event? Intensity of sensation is coded as the frequency of the action potentials.

Some sensory cells transmit information about internal conditions in the body, but we may not be consciously aware of that information. The brain continuously receives information about body temperature, blood carbon dioxide and oxygen concentrations, arterial pressure, muscle tension, and the positions of the limbs—all of which are important for homeostasis. All sensory cells produce information that the nervous system can use, but that information does not always result in conscious sensation.

Some sensory receptor cells are assembled with other types of cells into *sensory organs*, such as eyes, ears, and noses, that enhance the ability of the sensory cells to collect, filter, and amplify stimuli. We therefore refer to **sensory systems**, which include the sensory cells, the associated structures, and the neural networks that process the information.

Many receptors adapt to repeated stimulation

Some sensory cells give gradually diminishing responses to maintained or repeated stimulation. This phenomenon is known as **adaptation**, and it enables an animal to ignore background or unchanging conditions while remaining sensitive to changes or to new information. (Note that this use of the term "adaptation" is different from its application in an evolutionary context.) When you dress, you feel each item of clothing touch your skin, but the sensation of clothes touching your skin is not constantly on your mind throughout the day. You are immediately aware, however, when a seam rips, your shoe comes untied, or someone touches your back.

Animals can discriminate between continuous and changing stimuli partly because some sensory cells adapt; it is also a result of information processing by the CNS. Some sensory cells adapt very little or very slowly; examples are some types of pain receptors and the mechanoreceptors for balance. You do not want to ignore pain that is signaling that something is wrong in your body, and to maintain equilibrium you must continuously know the tensions and forces on all of your joints and muscles.

46.1 RECAP

Sensory receptor cells have receptor proteins that respond to specific stimuli from the external or internal environment by opening or closing ion channels, which results in the generation of action potentials in sensory neurons.

- Explain the difference between ionotropic and metabotropic sensory receptor proteins. **See p. 965 and Figure 46.1**

- How are we able to perceive action potentials—which are all essentially the same—as different sensations? **See p. 967**

Now that we have a general view of how sensory systems code and process information, we will discuss how sensory systems gather and filter stimuli, transduce specific stimuli into action potentials, and transmit action potentials to the central nervous system. We will look in more depth at specific sensory modalities, beginning with chemosensation, the basis of smell and taste.

46.2 How Do Sensory Systems Detect Chemical Stimuli?

A colony of corals responds to a small amount of meat extract in seawater by extending bodies and tentacles and searching for food; a solution of a single amino acid can stimulate this response. Conversely, a small amount of seawater in which corals were crushed will stimulate a defensive retraction of the coral polyps. Humans also react strongly to certain chemical stimuli. When we smell freshly baked bread, we salivate and feel hungry; when we smell rotting meat, we feel nauseated.

All animals receive information about chemical stimuli through **chemoreceptors**, which are receptor proteins that bind to various molecules — their *ligands*—and are responsible for smell and taste. Chemoreceptors are also responsible for monitoring aspects of the internal environment such as the level of carbon dioxide in the blood. Information from chemoreceptors can cause powerful physiological and behavioral responses during activities such as feeding, mating, fighting, and recognizing individuals.

Arthropods are good models for studying chemoreception

A chemical signal used in communication among members of a species is a **pheromone**. Arthropods use pheromones to attract mates. These signals demonstrate the sensitivity of chemosensory systems, and one of the best-studied examples is that of the silkworm moth *Bombyx mori*.

To attract a mate, the female silkworm moth releases a pheromone called *bombykol* from a gland at the tip of her abdomen. The male silkworm moth has receptors for this molecule on his antennae (**Figure 46.3**). Each feathery antenna carries about 10,000 bombykol-sensitive hairs. A single molecule of bombykol may be sufficient to generate action potentials in the antennal nerve that transmits the signal to the CNS. Because of the male's great sensitivity, the sexual message of a female moth is likely to reach any male within a huge downwind area. When approximately 200 hairs per second are activated, the male orients upwind in search of the female. Because the rate of firing in the male's sensory nerves is proportional to the bombykol concentration in the air, he can follow the airborne concentration gradient and home in on the signaling female.

Olfaction is the sense of smell

The sense of smell, **olfaction**, depends on chemoreceptors. In vertebrates, the olfactory sensors are neurons embedded in a layer of epithelial tissue at the top of the nasal cavity. The axons from these neurons extend to the olfactory integration area of the brain (the olfactory bulb), whereas their dendrites end in ol-

factory cilia on the surface of the nasal epithelium. A protective layer of mucus covers the epithelium. Molecules from the environment must diffuse through this mucus to reach the receptor proteins on the olfactory cilia. When you have a cold, the amount of mucus in your nose increases, and the epithelium swells. With this in mind, study **Figure 46.4**, and you will easily understand why respiratory infections can cause you to lose your sense of smell.

Humans have a sensitive olfactory system, but we are unusual among mammals in that we depend more on vision than on olfaction. A typical dog's nasal epithelium is 15 to 20 times larger than a human's and has around 1 billion receptors, compared with around 20 million in humans. For some scents, the threshold sensitivity of the dog is 100 million times lower. A dog's nose reveals a huge amount of information not available to us.

An **odorant** is a molecule that activates an olfactory receptor protein. Odorants bind to receptor proteins on the olfactory cilia of the olfactory neurons. Olfactory receptor proteins are specific for particular odorant molecules. When an odorant molecule binds to its receptor on an olfactory neuron, it activates a G protein. The G protein in turn activates an enzyme that causes an increase of a second messenger (cAMP in vertebrates) in the cytoplasm (see Figure 7.19). The second messenger binds to cation channels in the sensory cell's plasma membrane and opens them, causing an influx of Na^+. The sensory neuron depolarizes to threshold and fires action potentials.

The olfactory world has an enormous number of odors, and accordingly, there are a large number of olfactory receptor proteins. In the 1990s, Linda Buck and Richard Axel discovered in mice a family of about 1,000 genes (about 3 percent of the genome) that code for olfactory receptor proteins. Humans have about one-third that number of functional olfactory receptor genes. Each receptor protein that is expressed is found in a limited number of receptor cells in the olfactory epithelium, and each cell expresses just one receptor type. Using a combination of patch clamping (see Figure 45.8) and molecular techniques, the investigators were able to match specific gene prod-

46.3 Some Scents Travel Great Distances Mating in silkworm moths of the genus *Bombyx* is coordinated by a pheromone called bombykol.

(A)

The female moth releases a pheromone from a gland at the tip of her abdomen. The pheromone can travel thousands of meters downwind.

(B)

A male moth detects this pheromone in the air passing over his antennae, which are covered with chemosensitive hairs.

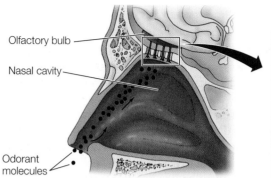

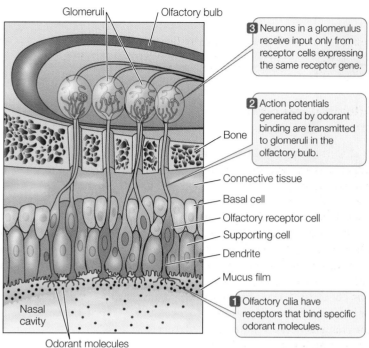

46.4 Olfactory Receptors Communicate Directly with the Brain The receptor cells of the human olfactory system are embedded in epithelial tissues lining the nasal cavity and send their axons to the olfactory bulb of the brain.

Labels for img_1:
Glomeruli
Olfactory bulb
3 Neurons in a glomerulus receive input only from receptor cells expressing the same receptor gene.
2 Action potentials generated by odorant binding are transmitted to glomeruli in the olfactory bulb.
Bone
Connective tissue
Basal cell
Olfactory receptor cell
Supporting cell
Dendrite
Mucus film
1 Olfactory cilia have receptors that bind specific odorant molecules.
Nasal cavity
Odorant molecules

ucts with the odorants they detect. For their discoveries of the molecular nature of the olfactory system, Buck and Axel received the Nobel prize in 2004.

Olfactory sensitivity enables discrimination of many more odorants than there are olfactory receptors. An odorant molecule can be quite complex, and different regions of that molecule may bind to different receptor proteins. The next stage of processing olfactory information is in the olfactory bulb, where axons from neurons expressing the same receptor protein cluster together on olfactory bulb neurons, forming structures called *glomeruli* (see Figure 46.4). Therefore, a complex odorant molecule can activate a unique combination of glomeruli in the olfactory bulb, so an olfactory system with hundreds of different receptor proteins can discriminate an astronomically large number of smells. How does the olfactory receptor cell signal the intensity of a smell? The more odorant molecules that bind to receptors, the greater the frequency of action potentials and the greater the intensity of the perceived smell.

In vertebrates, odorants and other chemoreceptor molecules can stimulate strong responses. Many vertebrate species have separate olfactory epithelia that are dedicated to the detection of pheromones.

The vomeronasal organ contains chemoreceptors

The **vomeronasal organ** (**VNO**) is a small, paired tubular structure embedded in the nasal epithelium of amphibians, reptiles, and many mammals (although probably not humans). In mammals, the VNO is located on the septum dividing the two nostrils (see Figure 53.5).

The vomeronasal organ has a pore that opens into the nasal cavity. When the animal sniffs, the VNO pulsates and draws a sample of nasal fluid over the chemoreceptors embedded in its walls. The information from these chemoreceptors goes to an accessory olfactory bulb in the brain, and information from there goes to brain regions involved in sexual and other instinctive behaviors.

In snakes, the VNO opens into the roof of the mouth cavity. Each time the snake's forked tongue darts in and out, the forks fit into the VNO openings and present the chemoreceptors located there with a sample of molecules from the surrounding air (see the chapter-opening photo). Thus the snake uses its tongue to smell its environment, not to taste it. Why doesn't the snake simply use the flow of air to and from its lungs, as we do, to smell the environment? In reptiles, air flows to and from the lungs slowly (and can even stop entirely for long periods of time), but the tongue can dart in and out many times in a second. It is a quick source of olfactory information.

Studies on mice have led to the hypothesis that the mammalian VNO is a specialized olfactory organ that detects pheromones. Lawrence Katz and his colleagues at Duke University recorded the activity of neurons in the mouse accessory olfactory bulb, which receives input from chemosensors in the VNO. These accessory olfactory neurons were activated when a mouse attached to recording electrodes sniffed another mouse placed in the same cage. However, the neurons fired differentially, depending on the gender and strain of the "intruder" mouse. Other studies on mouse behavior have supported a role for the VNO in gender identification and sexual behaviors that are linked to pheromone perception (see Section 53.2).

Gustation is the sense of taste

The sense of taste, or **gustation**, in humans and other vertebrates depends on clusters of chemoreceptors called **taste buds**. The taste buds of terrestrial vertebrates are confined to the mouth cavity, but some fishes have taste buds in the skin that enhance their ability to sense their environment. Some fishes living in murky water are very sensitive to small amounts of amino acids in the water around them and can find food without the use of vision. The duck-billed platypus, a prototherian mammal (see Figure 33.24), has similar talents as a result of taste buds on the sensitive skin of its bill.

A human tongue has approximately 10,000 taste buds. The taste buds are embedded in the epithelium, and most are found on the papillae (**Figure 46.5**). (Look at your tongue in a mirror—the papillae make it look fuzzy.) Each papilla has many taste

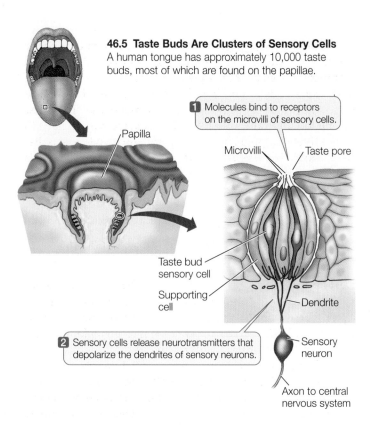

46.5 Taste Buds Are Clusters of Sensory Cells
A human tongue has approximately 10,000 taste buds, most of which are found on the papillae.

Papilla

1 Molecules bind to receptors on the microvilli of sensory cells.

Microvilli

Taste pore

Taste bud sensory cell

Supporting cell

Dendrite

2 Sensory cells release neurotransmitters that depolarize the dendrites of sensory neurons.

Sensory neuron

Axon to central nervous system

buds, mostly on the sides. The outer surface of a taste bud has a pore that exposes the tips of the sensory receptor cells. Microvilli (tiny hairlike projections) increase the surface area of these cells where their tips converge at the pore. These sensory cells generate action potentials and release neurotransmitter at their bases, where they form synapses with sensory neurons that convey the signals into the central nervous system.

The tongue does a lot of hard work, so its epithelium, along with cells of its taste buds, are shed and replaced at a rapid rate. Individual taste bud cells last about 10 days before they are replaced, but the sensory neurons associated with them live on, always forming new synapses as new taste buds form.

You may have heard that humans can perceive only four tastes: sweet, salty, sour, and bitter. However, taste buds can distinguish among a variety of sweet-tasting molecules and a variety of bitter-tasting molecules. Recently, small families of genes for receptor proteins responding to sweet and bitter tastes have been discovered. In addition, we now recognize a fifth taste, *umami*, which is a savory, meaty taste that originates from receptors for amino acids, including monosodium glutamate (MSG). The full complexity of the chemosensitivity that enables us to enjoy the subtle flavors of food comes from the combined activation of gustatory and olfactory receptors, which is why you lose much of your sense of taste when you have a cold.

Gustation begins with receptor proteins in the membranes of the microvilli of the taste bud sensory cells (see Figure 46.5). The nature of these proteins and the mechanisms by which they depolarize the sensory receptor cell differ for the different basic tastes. Saltiness receptors are ionotropic and simply respond to Na^+ diffusing through open Na^+ channels depolarizing the sensory cell. Sourness receptors are also probably ionotropic. Depolarization of these receptors is similarly due to a direct

effect of H^+ ions on Na^+ channels. In contrast, sweetness and bitterness involve families of receptor proteins similar to those involved in olfaction, and they are metabotropic. The bitter taste probably evolved as a protective mechanism enabling animals to detect and avoid toxic plant compounds such as quinine, caffeine, and nicotine. Since plants have evolved many such molecules to repel herbivorous predators, a variety of receptors is essential. Similarly, a large number of molecules in food could indicate nutritional value, so a variety of receptors is of value. The diversity of sweet receptors helps explain why it has been possible to invent many different artificial sweeteners.

Regardless of the mechanism of taste transduction by the sensory cells of the taste buds, all these cells release neurotransmitter onto sensory neurons. These neurons then conduct that information to the CNS, where it is interpreted as specific taste sensations.

46.2 RECAP

All animals receive information about chemical stimuli through chemoreceptors, which have diverse structures and bind to a tremendous variety of stimulus molecules. Chemoreceptors are the basis of the sensations of olfaction and gustation and the reception of pheromones. They also monitor some aspects of an animal's internal environment.

- Why are we able to distinguish so many different smells? Why do some people and some animals experience more or different odors than others? See pp. 968–969 and Figure 46.4

- Describe how different substances in food are transduced into action potentials in taste buds. See pp. 969–970 and Figure 46.5

46.3 How Do Sensory Systems Detect Mechanical Forces?

Let's turn now to **mechanoreceptors**, the sensory cells that respond to mechanical forces. Physical distortion of a mechanoreceptor's plasma membrane causes ion channels to open, altering the membrane potential of the cell to create a graded potential, which in turn leads to the release of neurotransmitter or the generation of action potentials. The rate of action potentials tells the CNS the strength of the stimulus to the mechanoreceptor. A considerable diversity of mechanosensory cells and mechanisms has evolved. Involved in many sensory systems, their functions range from interpreting skin sensations to sensing blood pressure to hearing and maintaining balance.

Many different cells respond to touch and pressure

Human skin (and that of other mammals) is packed with diverse mechanoreceptors that generate varied sensations (**Figure 46.6**). The most important tactile receptors, found in both hairy and nonhairy skin, are *Merkel's discs*, which adapt rather

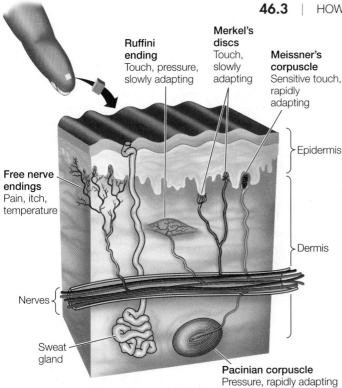

Ruffini ending
Touch, pressure, slowly adapting

Merkel's discs
Touch, slowly adapting

Meissner's corpuscle
Sensitive touch, rapidly adapting

Epidermis

Free nerve endings
Pain, itch, temperature

Dermis

Nerves

Sweat gland

Pacinian corpuscle
Pressure, rapidly adapting

46.6 The Skin Feels Many Sensations Even a very small patch of skin contains a variety of sensory cells, making the skin a multi-modal receptor that can sense temperature, pressure, texture, pain, touch, and itch.

A *two-point spatial discrimination test* reveals that the density of tactile mechanoreceptor cells varies across the body's surface. By touching someone's skin with two toothpicks simultaneously, you can determine how far apart two stimuli have to be before the person can tell whether the sensations are produced by one toothpick or by two. On the back, the stimuli have to be relatively far apart before they are perceived as two discrete stimuli. But when the same test is applied to the person's lips or fingertips, the person can identify as separate two stimuli that are quite close together, meaning receptor density is much greater in these regions.

Mechanoreceptors are found in muscles, tendons, and ligaments

An animal receives information from mechanoreceptors about the position of its limbs and the stresses on its muscles and joints. These mechanoreceptors supply information continuously to the CNS, and this information is essential for postural control and the coordination of movements.

The mechanoreceptors found in skeletal muscle are called *muscle spindles*. These are **stretch receptors**, modified muscle cells that are embedded in connective tissue in muscles and innervated by sensory neurons (**Figure 46.7**). Whenever the muscle is stretched, muscle spindles are also stretched, and the neurons

slowly and provide continuous information about things touching the skin. *Meissner's corpuscles*, found primarily in nonhairy skin, are very sensitive but adapt rapidly, so they provide information about *changes* in things touching the skin. The rapid adaptation of Meissner's corpuscles is why you roll a small object between your fingers (rather than holding it still) to discern its shape and texture: as you roll it, the object continues to stimulate Meissner's corpuscles.

Two other kinds of mechanoreceptors are found deeper in the skin. *Ruffini endings* adapt slowly and are good at providing information about vibrating stimuli of low frequencies. *Pacinian corpuscles*, which adapt rapidly, provide information about vibrating stimuli of higher frequencies. Even deeper in the skin, dendrites of sensory neurons wrap around hair follicles. When the surface hairs are displaced, those neurons are stimulated.

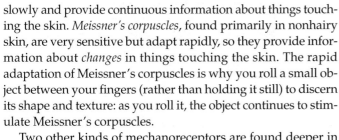

Muscle

Tendon

Load

(A) Muscle spindles

Muscle

Muscle spindle

Sensory neuron

Firing of sensory neuron

Stretch

Time

1 Muscle spindles are stretch receptors. When muscle spindles are stretched…

2 …sensory neurons associated with them transmit action potentials to the CNS. These signals stimulate motor neurons that initiate muscle contraction.

(B) Golgi tendon organs

Muscle

Golgi tendon organ

Sensory neuron

Tendon

Firing of sensory neuron

Load on muscle

Time

1 Golgi tendon organs sense load and measure the force of muscle contraction. When contraction becomes too forceful…

2 …the sensory neurons send action potentials to the CNS that inhibit motor neurons, and the muscle relaxes.

46.7 Stretch Receptors Stretch receptors provide information about the stresses on muscles and joints in an animal's limbs. (A) Signals from muscle spindles to the CNS initiate muscle contraction. (B) Golgi tendon organs in tendons and ligaments inhibit a contraction that becomes too forceful, triggering a reduction in muscle tension and protecting the muscle from tearing.

(A)

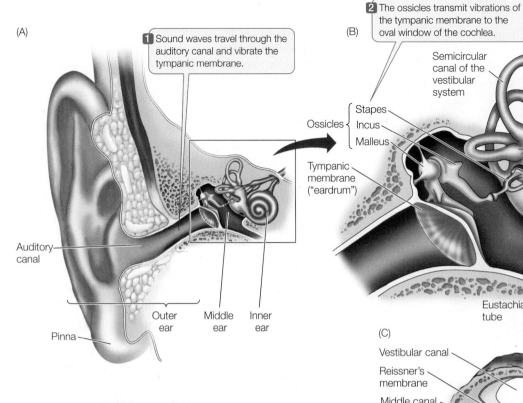

1 Sound waves travel through the auditory canal and vibrate the tympanic membrane.

Auditory canal

Pinna

Outer ear

Middle ear

Inner ear

(B)

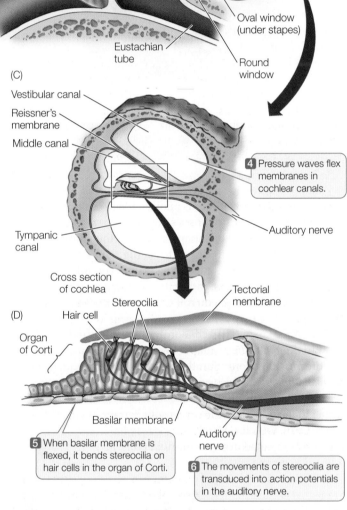

2 The ossicles transmit vibrations of the tympanic membrane to the oval window of the cochlea.

3 Vibrations at oval window create pressure waves in fluid-filled cochlear canals.

Semicircular canal of the vestibular system

Auditory nerve

Ossicles { Stapes Incus Malleus

Tympanic membrane ("eardrum")

Cochlea

Oval window (under stapes)

Eustachian tube

Round window

(C)

Vestibular canal

Reissner's membrane

Middle canal

4 Pressure waves flex membranes in cochlear canals.

Tympanic canal

Auditory nerve

Cross section of cochlea

(D)

Tectorial membrane

Stereocilia

Hair cell

Organ of Corti

Basilar membrane

Auditory nerve

5 When basilar membrane is flexed, it bends stereocilia on hair cells in the organ of Corti.

6 The movements of stereocilia are transduced into action potentials in the auditory nerve.

transmit action potentials to the central nervous system. The CNS uses this information to adjust the strength of the muscle contraction to match the load put on the muscle. Thus a bartender can hold a beer mug in the same position as he fills it from the tap, as seen in Figure 46.7A. In Figure 46.2, we saw how crayfish stretch receptors transduce physical force into action potentials. The actions of muscle spindles are similar.

Another type of mechanoreceptor, the *Golgi tendon organ*, is found in tendons and ligaments and provides information about the force generated by a contracting muscle (see Figure 46.7B). When a contraction becomes too forceful, action potentials from the Golgi tendon organ inhibit the spinal cord motor neurons innervating that muscle, causing it to relax and protecting it from tearing. (You may recall a cell organelle called the Golgi apparatus. What these two cellular structures have in common is their discovery by the Italian anatomist Camillo Golgi, who received a Nobel prize in 1906.)

Auditory systems use hair cells to sense sound waves

The stimuli that animals perceive as sounds are pressure waves. **Auditory systems** use mechanoreceptors to convert pressure waves into receptor potentials. Auditory systems include special structures that gather sound waves, direct them to the sensory organ, and amplify their effect on the mechanoreceptors.

Human hearing provides a good example of an auditory system. The organs of hearing are the ears. The two prominent structures on the sides of our heads are the *pinnae*. The pinna of an ear collects sound waves and directs them into the *auditory canal*, which leads to the actual hearing apparatus in the *middle ear* and the *inner ear* (**Figure 46.8A**). If you have ever watched a cat or dog change the orientation of its ears to focus on a particular sound, then you have witnessed the role of pinnae in hearing.

46.8 Structures of the Human Ear (A) The pinnae direct sound waves down the auditory canal to impinge on the tympanic membrane. The tympanic membrane mechanically transmits these pressure waves into movements of the ossicles in the middle ear. (B) The ossicles transmit their movement into pressure waves in the fluid of the cochlea at the oval window. (C) The cochlea is divided into fluid-filled chambers; pressure waves from the ossicles cause the membranes between the chambers to flex. (D) Flexing of the basilar membrane bends stereocilia on hair cells in the organ of Corti.

yourBioPortal.com

GO TO Web Activity 46.1 • Structures of the Human Ear

The eardrum, or **tympanic membrane**, covers the end of the auditory canal. The tympanic membrane vibrates in response to pressure waves traveling down the auditory canal. The middle ear, an air-filled cavity, lies on the other side of the tympanic membrane.

MIDDLE EAR The middle ear is open to the throat at the back of the mouth through the *eustachian tube*. Because the eustachian tube is also filled with air, pressure equilibrates between the middle ear and the outside world. When you have a cold or allergy, the tube can become blocked by mucus or by tissue swelling, so you have difficulty "clearing your ears," or equilibrating the pressure in the middle ear with the outside air pressure, which you have to do when ascending or descending in a plane or when scuba diving.

The middle ear contains three delicate bones called the **ossicles**, individually named the *malleus* (Latin, "hammer"), *incus* ("anvil"), and *stapes* ("stirrup") (**Figure 46.8B**). The ossicles transmit the vibrations of the tympanic membrane to another flexible membrane called the **oval window**. The ossicles act as a lever—like a hammer pulling out a nail—translating a large movement of the tympanic membrane into a smaller movement of the oval window, but a movement of greater force. Also, because the oval window is much smaller than the tympanic membrane, the pressure the stapes transmits to the oval window is more than 20 times greater than the pressure exerted by the sound wave on the tympanic membrane. Behind the oval window lies the fluid-filled inner ear. Movements of the oval window impart pressure changes to that enclosed fluid. These pressure waves are transduced into action potentials. To see how, let's take a closer look at the inner ear.

INNER EAR The inner ear is a bony structure consisting of two sets of canals. One is the organ of balance, the **vestibular system**, and the other is the organ of hearing, the **cochlea**. The cochlea (Latin and Greek, "snail" or "spiral shell") is a long, tapered, coiled canal. A cross section of the cochlea reveals that it is composed of three parallel canals separated by two membranes: **Reissner's membrane** and the **basilar membrane** (**Figure 46.8C**). Sitting on the basilar membrane is the **organ of Corti**, which transduces pressure waves into action potentials. The organ of Corti contains *hair cells* with *stereocilia*. The tips of the hair cells are embedded in a gelatinous overhanging shelf called the *tectorial membrane* (**Figure 46.8D**).

Stereocilia are fingerlike extensions of the cell plasma membrane that are stiffened by cross-linked actin filaments. They are not motile, but because their tips are attached to the more rigid tectorial membrane, stereocilia bend when the basilar membrane flexes. The response of the hair cell is a graded membrane potential. The hair cells do not fire action potentials, but the changes in their membrane potential alter the rate at which the hair cells release neurotransmitter onto sensory neurons whose axons make up the auditory nerve and transmit action potentials to the brain.

What causes the basilar membrane to flex, and how does this mechanism distinguish sounds of different frequencies? In **Figure 46.9**, the cochlea is shown uncoiled to make it easier to un-

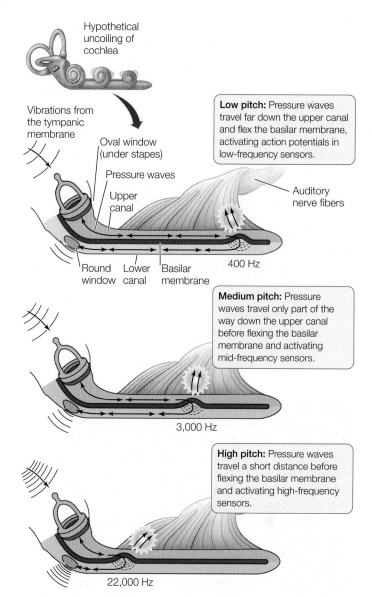

Hypothetical uncoiling of cochlea

Vibrations from the tympanic membrane

Oval window (under stapes)

Pressure waves

Upper canal

Round window   Lower canal   Basilar membrane

Low pitch: Pressure waves travel far down the upper canal and flex the basilar membrane, activating action potentials in low-frequency sensors.

Auditory nerve fibers

400 Hz

Medium pitch: Pressure waves travel only part of the way down the upper canal before flexing the basilar membrane and activating mid-frequency sensors.

3,000 Hz

High pitch: Pressure waves travel a short distance before flexing the basilar membrane and activating high-frequency sensors.

22,000 Hz

46.9 Sensing Pressure Waves in the Inner Ear Pressure waves of different frequencies flex the basilar membrane at different locations. Information about sound frequency is specified by which hair cells are activated. For simplicity, this representation illustrates the cochlea as uncoiled, and leaves out the middle ear.

yourBioPortal.com
GO TO Animated Tutorial 46.1 • Sound Transduction in the Human Ear

derstand its structure and function. The upper and lower canals separated by the basilar membrane are joined at the distal end of the cochlea (the end farthest from the oval window), making one continuous canal that turns back on itself. Just as the oval window is a flexible membrane at the beginning of the upper canal of the cochlea, the **round window** is a flexible membrane at the end of the lower canal.

Air is highly compressible but fluids are not. Therefore, a pressure wave can travel through air without much displacement of the air, whereas a pressure wave in fluid displaces that fluid. When the stapes pushes on the oval window, the fluid in the upper canal of the cochlea is displaced. If this movement of the oval window occurs slowly, the cochlear fluid pressure wave travels down the upper canal, around the bend, and back through the lower canal. At the end of the lower canal, the displacement pressure is dissipated by the outward bulging of the round window.

BASILAR MEMBRANE AND PITCH If the oval window vibrates in and out rapidly, the waves of fluid pressure create traveling waves, or *flexions*, in the basilar membrane. The basilar membrane is not uniform; it is thicker and stiffer at its base and wider and thinner at its apical end. Pressure waves in the cochlear fluid have different frequencies and set up different patterns of traveling waves. High-frequency waves cause maximal flexion at the basal end of the basilar membrane, whereas low-frequency pressure waves result in maximal flexion at its apical end. Thus different pitches of sound flex the basilar membrane at different locations and activate different sets of hair cells. Action potentials stimulated by the mechanoreceptors at different positions along the organ of Corti travel to different regions of the *auditory cortex* along the auditory nerve.

HEARING LOSS There are two general types of hearing loss, or deafness. *Conduction deafness* is caused by the loss of function of the tympanic membrane and/or the ossicles of the middle ear. Repeated infections of the middle ear can cause scarring of the tympanic membrane and stiffening of the connections between the ossicles. The consequence is less efficient conduction of sound waves from the tympanic membrane to the oval window. With increasing age, the ossicles stiffen, resulting in a gradual loss of the ability to hear high-frequency sounds.

Nerve deafness is caused by damage to the inner ear or the auditory pathways. A common cause of nerve deafness is damage to the hair cells of the delicate organ of Corti by exposure to loud sounds such as jet engines, pneumatic drills, or highly amplified music. Consistent exposure to sounds above 85 decibels can damage hearing; this damage is cumulative and irreversible. Even using earphones can put you at risk for hearing loss because they generate high-pressure sound waves close to the tympanic membrane. Personal stereo earphones can reach 120 decibels, and people commonly use them at 100 decibels (equivalent to being at a rock concert).

Hair cells are sensitive to being bent

Hair cells are the mechanoreceptors in both the organs of hearing and equilibrium. Stereocilia project from the surface of each hair cell like a set of organ pipes (**Figure 46.10A**). The bending of stereocilia alters ion channels in the hair cell's plasma membrane. Bending in one direction causes the plasma membrane to depolarize, and bending in the other direction causes it to hyperpolarize (**Figure 46.10B**). Graded potentials in the hair cell plasma membrane control its release of neurotransmitter onto the sensory neuron associated with it, and the sensory neuron sends action potentials to the CNS.

How does the bending of the stereocilia open ion channels? The ion channels that are opened are at the ends of the stereocilia. This was discovered by exploring the areas around the stereocilia with microelectrodes, and seeing that local currents were created near the tips of the stereocilia when they were bent. Then, careful electron microscopic work revealed minute filaments that connected the tip of each stereocilium to its taller neighbor. It is hypothesized that these filaments are fine molecular attachments to the ion channels in the stereocilium plasma membrane, and they act like springs that open the channels. If the taller neighboring stereocilium is bent away, the spring tightens and the ion channel is opened. If the taller neighbor bends toward its shorter neighbor, the spring is relaxed and the channel closes (see Figure 46.10B).

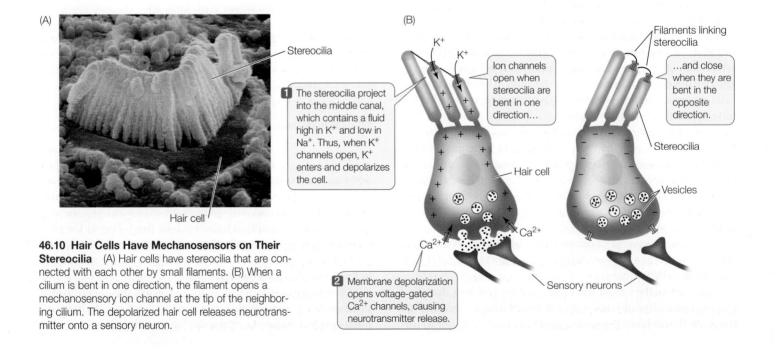

46.10 Hair Cells Have Mechanosensors on Their Stereocilia (A) Hair cells have stereocilia that are connected with each other by small filaments. (B) When a cilium is bent in one direction, the filament opens a mechanosensory ion channel at the tip of the neighboring cilium. The depolarized hair cell releases neurotransmitter onto a sensory neuron.

Stereocilia

Hair cell

1 The stereocilia project into the middle canal, which contains a fluid high in K^+ and low in Na^+. Thus, when K^+ channels open, K^+ enters and depolarizes the cell.

K^+

K^+

Ion channels open when stereocilia are bent in one direction…

Filaments linking stereocilia

…and close when they are bent in the opposite direction.

Stereocilia

Hair cell

Vesicles

Ca^{2+}

2 Membrane depolarization opens voltage-gated Ca^{2+} channels, causing neurotransmitter release.

Sensory neurons

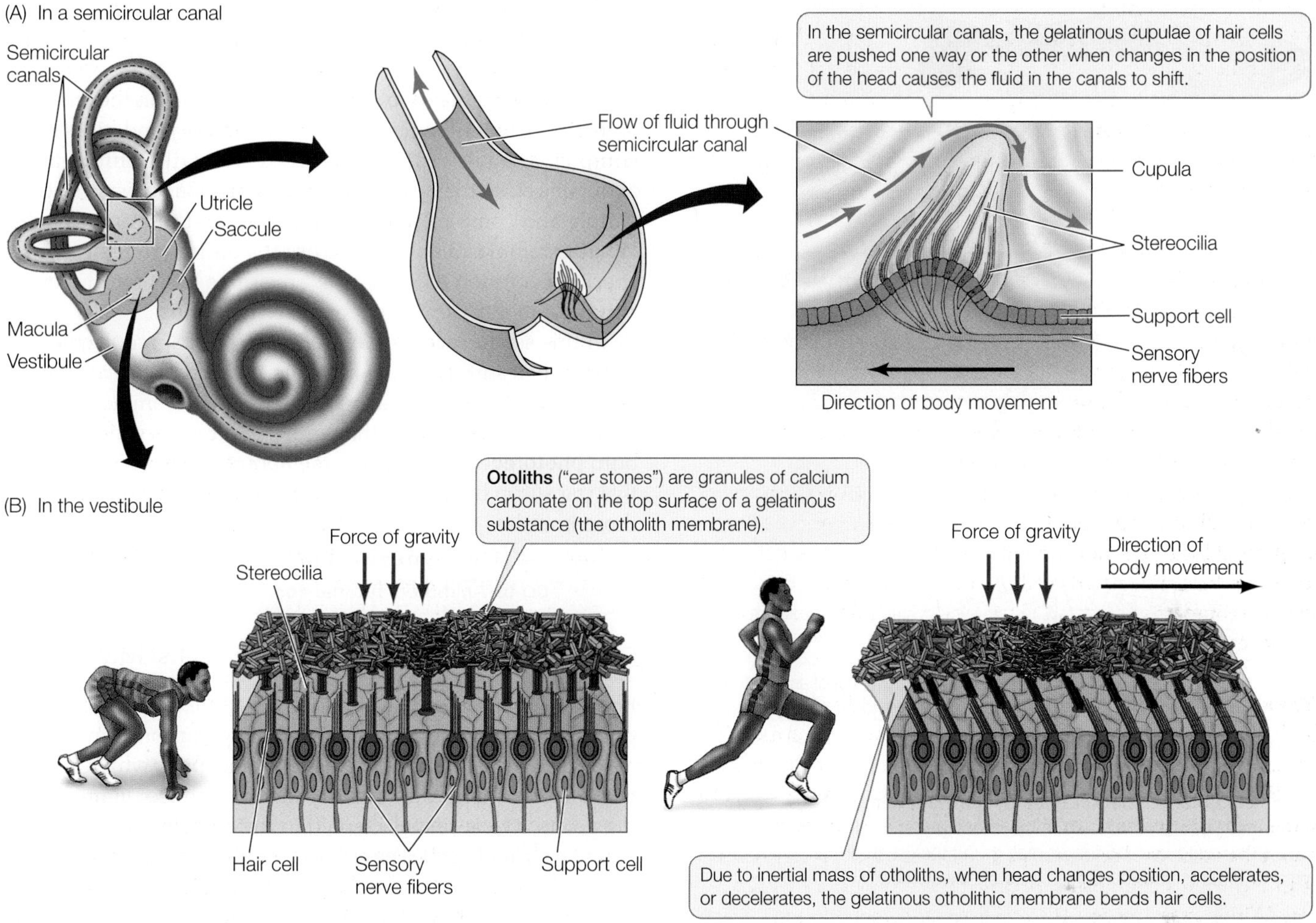

(A) In a semicircular canal

Semicircular canals

Utricle

Saccule

Macula

Vestibule

Flow of fluid through semicircular canal

In the semicircular canals, the gelatinous cupulae of hair cells are pushed one way or the other when changes in the position of the head causes the fluid in the canals to shift.

Cupula

Stereocilia

Support cell

Sensory nerve fibers

Direction of body movement

(B) In the vestibule

Force of gravity

Stereocilia

Otoliths ("ear stones") are granules of calcium carbonate on the top surface of a gelatinous substance (the otolith membrane).

Force of gravity

Direction of body movement

Hair cell

Sensory nerve fibers

Support cell

Due to inertial mass of otoliths, when head changes position, accelerates, or decelerates, the gelatinous otolithic membrane bends hair cells.

Hair cells detect forces of gravity and momentum

In the mammalian inner ear, the vestibular system consists of three bony **semicircular canals** and two bony chambers called the *saccule* and the *utricle*. Hair cells in the vestibular system detect the position and movement of the head—information that is essential for maintaining balance (equilibrium). The information from the vestibular system is also crucial for the control of eye movements. When you look at something, you can move your head while staying focused on the object because of your vestibulo-ocular reflex.

The entire vestibular system is filled with a fluid called endolymph. In the semicircular canals, the endolymph shifts when the head changes position. Since the three semicircular canals have different orientations, they respond differentially to the direction of movement. Projecting into the base of each canal is a gelatinous swelling called a *cupula* (plural *cupulae*) that encloses a cluster of hair cell stereocilia. When the shifting endolymph pushes on the cupulae, it bends the stereocilia and causes a graded potential in their hair cell plasma membranes (**Figure 46.11A**). The stereocilia in the saccule and utricle are bent in a different way. These stereocilia are embedded in *otoliths* (Latin for "ear stones"), gelatinous structures containing crystals of calcium carbonate. When the head changes position or when it accelerates or decelerates, gravitational forces are exerted on the otoliths and bend the stereocilia (**Figure 46.11B**).

46.11 Organs of Equilibrium The vestibular system consists of bony chambers and fluid-filled canals. (A) Each semicircular canal has a cupula containing stereocilia. When fluid moves against the cupula, the stereocilia bend. (B) In the saccule and utricle, stereocilia are bent by gravitational forces on the otoliths.

As in the cochlea, the hair cells of the vestibular system do not fire action potentials, but they release neurotransmitter at synapses with sensory neurons, which in turn fire action potentials.

Hair cells are evolutionarily conserved

An early evolutionary use of hair cells to measure movement can be seen in the **lateral line** sensory system of fishes. The lateral line is a canal just under the surface of the skin that runs down each side of the fish (**Figure 46.12**). Hair cells line the canal, and their stereocilia, protected by cupulae, project into the stream of water that flows through the lateral line canal when the fish moves through the water. Forward movement of the fish puts pressure on the cupulae, causing the stereocilia to bend in the direction that depolarizes the hair cells. Since water is incompressible, disturbances in the water around the fish are translated into pressure waves that can be picked up by the lateral line stereocilia. Thus, the lateral line system provides

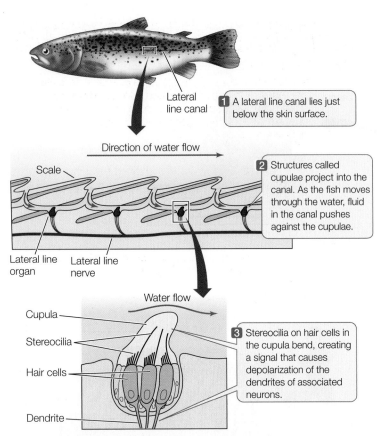

1 A lateral line canal lies just below the skin surface.

Direction of water flow

2 Structures called cupulae project into the canal. As the fish moves through the water, fluid in the canal pushes against the cupulae.

3 Stereocilia on hair cells in the cupula bend, creating a signal that causes depolarization of the dendrites of associated neurons.

46.12 The Lateral Line Acoustic System Contains Mechanosensors Hair cells in the lateral line of a fish detect movement of the water around the animal, giving the fish information about its own movements and the movements of objects nearby.

information about the fish's movements as well as about other moving objects, such as predators or prey.

46.3 RECAP

Sensations that derive from mechanoreceptors include touch, tickle, pressure, joint position, muscle load, hearing, and equilibrium.

- Describe some of the different mechanoreceptors in the skin and their properties. See pp. 970–971 and Figure 46.6

- How do different frequencies of sound result in action potentials being fired in different acoustic neurons? See pp. 972–973 and Figures 46.8 and 46.9

- How do hair cells transduce force into action potentials? See p. 974 and Figure 46.10

Chemoreception gave us good examples of metabotropic sensory receptors, and mechanoreception has given us good examples of ionotropic sensory receptors. Now let's turn to another example of metabotropic sensory reception—one in which light is the stimulus. We will see how light energy is converted into action potentials.

46.4 How Do Sensory Systems Detect Light?

Sensitivity to light—**photosensitivity**—confers on the simplest animals the ability to orient to the sun and sky and gives more complex animals rapid and extremely detailed information about objects in their environment. It is not surprising that both simple and complex animals can sense and respond to light. What is remarkable is that across the entire range of animal species, evolution has conserved the same basis for photosensitivity: a family of pigments called **rhodopsins**.

In this section we will learn how rhodopsin molecules respond when stimulated by light energy and how that response is transduced into neural signals. We will also examine the structures of eyes, the organs that gather light energy and focus it onto **photoreceptor cells**, the metabotropic sensory receptors that transform light energy into action potentials, and the routes those impulses travel to the brain.

— **yourBioPortal.com** —
GO TO Animated Tutorial 46.2 • Photosensitivity

Rhodopsins are responsible for photosensitivity

Photosensitivity depends on the ability of rhodopsins to absorb photons of light and to undergo a change in conformation. A rhodopsin molecule consists of a protein, **opsin** (which alone is not photosensitive), and an associated nonprotein light-absorbing group, **11-*cis*-retinal**, cradled in the center of the opsin and bound covalently to it. The entire rhodopsin molecule sits within the plasma membrane of a photoreceptor cell.

When 11-*cis*-retinal absorbs a photon of light energy, it changes into a different isomer of retinal, called all-*trans*-retinal. This change puts a strain on the bonds between retinal and opsin, changing the conformation of opsin. This change signals the detection of light. In vertebrate eyes, the retinal and the opsin eventually separate from each other—a process called *bleaching*, which causes the molecule to lose its photosensitivity. A series of enzymatic reactions is then required to return the all-*trans*-retinal to the 11-*cis* isomer, which then recombines with opsin so that it once again becomes the photosensitive pigment rhodopsin (**Figure 46.13**).

How does the conformational change of rhodopsin transduce light into a cellular response? After retinal is converted from the 11-*cis* to the all-*trans* form, its interactions with opsin pass through several unstable intermediate stages. One of these stages triggers a cascade of reactions involving a G protein signaling mechanism that results in the alteration of membrane potential that is the photoreceptor cell's response to light.

Rod cells respond to light

To get a better idea of how rhodopsin alters the membrane potential of a photoreceptor cell and how that photoreceptor cell signals that it has been stimulated by light, let's look at one type of vertebrate photoreceptor cell, the rod cell. The rod cell, named for its shape, is a modified neuron that does not produce action potentials. Rod cells release neurotransmitter from their bases

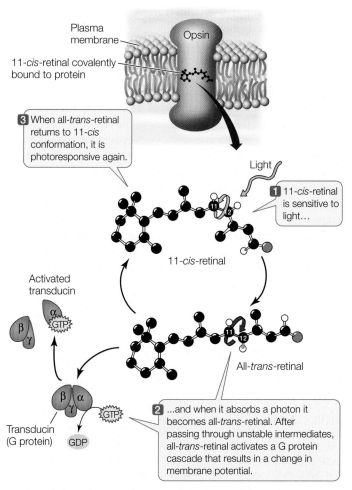

3 When all-*trans*-retinal returns to 11-*cis* conformation, it is photoresponsive again.

1 11-*cis*-retinal is sensitive to light...

11-*cis*-retinal

Activated transducin

All-*trans*-retinal

2 ...and when it absorbs a photon it becomes all-*trans*-retinal. After passing through unstable intermediates, all-*trans*-retinal activates a G protein cascade that results in a change in membrane potential.

Transducin (G protein)

46.13 Light Changes the Conformation of Rhodopsin The light-absorbing molecule 11-*cis*-retinal bonds with the protein opsin to form the pigment rhodopsin, the molecular agent of photosensitivity.

where they form synapses with the next neurons in the visual pathway (**Figure 46.14**). Each rod cell has an outer segment, an inner segment, and a synaptic terminal. The outer segment is highly specialized and contains a stack of discs of plasma membrane densely packed with rhodopsin. The function of the discs is to capture photons of light passing through the rod cell. The inner segment contains the cell nucleus, mitochondria, and other organelles. The synaptic terminal is where the rod cell communicates with other neurons.

To see how a rod cell responds to light, we can penetrate a single rod cell with an electrode and record its membrane potential in the dark and in the light. From what we have learned about other types of sensory receptors, we might expect stimulation of the rod cell by light would make its membrane potential less negative. But the opposite is true—it becomes more negative.

When a rod cell is kept in the dark, it has a relatively depolarized resting potential compared with other neurons. In fact, the plasma membrane of the rod cell is almost as permeable to Na^+ as to K^+. In the dark, Na^+ continually enters the outer segment of the cell—the dark current. When light is flashed on the dark-adapted rod cell, its membrane potential becomes more negative—it hyperpolarizes (see Figure 46.14). The rate of neu-

---

INVESTIGATING LIFE

46.14 A Rod Cell Responds to Light

The plasma membrane of a rod cell hyperpolarizes—becomes more negative—in response to a flash of light. Rod cells do not fire action potentials, but in response to the absorption of light energy, the neuron experiences a change in membrane potential.

HYPOTHESIS When a rod cell absorbs photons (light energy), its membrane potential changes in proportion to the strength of the light stimulus.

METHOD

1. Record membrane potentials from the inner segment of a rod cell.
2. Stimulate the rod cells with light flashes of varying intensity and record the results.

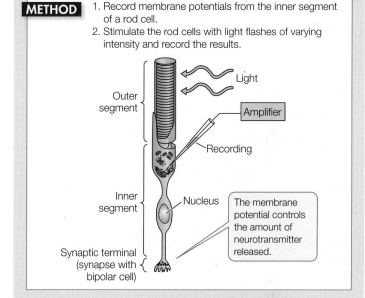

The membrane potential controls the amount of neurotransmitter released.

RESULTS

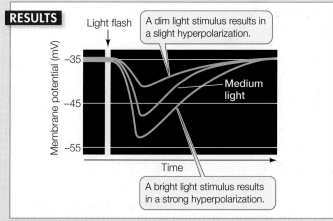

A dim light stimulus results in a slight hyperpolarization.

Medium light

A bright light stimulus results in a strong hyperpolarization.

CONCLUSION The membrane potential of rod cells is depolarized in the dark and hyperpolarizes (becomes more negative) in response to light.

FURTHER INVESTIGATION: How would you investigate the effect of background illumination on the rod cell's response to light?

rotransmitter release changes as membrane potential changes. As the rod cell hyperpolarizes, its release of neurotransmitter decreases.

How does the absorption of light by rhodopsin hyperpolarize the rod cell? When rhodopsin is excited by light, it initiates a cascade of events. The dark-adapted rod cell has open Na$^+$ channels, allowing a depolarizing dark current (**Figure 46.15A**). Light photoexcites rhodopsin, which activates a G protein called *transducin*. Activated transducin in turn activates a phosphodiesterase (PDE) (**Figure 46.15B**). Activated PDE converts cyclic GMP (cGMP) to GMP, which causes the Na$^+$ channels to close. Na$^+$ is pumped out and the cell hyperpolarizes (**Figure 46.15C**).

This mechanism may seem like a roundabout way of doing business, but its advantage is its enormous amplification ability. Each molecule of photoexcited rhodopsin can activate several hundred transducin molecules, thus activating a large number of PDE molecules. The catalytic capacity of PDE is great: one molecule can hydrolyze several hundred molecules of cGMP per second. The bottom line is that a single photon of light can cause a huge number of Na$^+$ channels to close.

Invertebrates have a variety of visual systems

Photoreceptors using rhodopsin are incorporated into a variety of visual systems, from simple to complex. Flatworms obtain directional information about light from photoreceptor cells that are organized into **eye cups**. The eye cups are paired bilateral structures, each partly shielded from light by a layer of pigmented cells lining the cup. The photoreceptors on the two sides of the animal are unequally stimulated unless the animal is facing directly toward or away from a light source. The flatworm generally uses directional information from the eye cups to move away from light.

Arthropods have **compound eyes** that provide them with information about patterns or images in the environment. These eyes are called compound because each eye consists of many optical units called **ommatidia** (singular *ommatidium*), each with its own narrow-angle lens (**Figure 46.16**). In contrast, a vertebrate eye consists of just one optical unit with a wide-angle lens. The number of ommatidia in a compound eye varies from only a few in some ants, to 800 in fruit flies, to 30,000 in some dragonflies.

Each ommatidium has a lens structure that directs light onto photoreceptor cells. Flies, for example, have eight elongated photoreceptors in each ommatidium. The inner borders of the photoreceptors are covered with microvilli that contain rhodopsin and trap light. Axons from the photoreceptors send the light information to the nervous system. Since each ommatidium of a compound eye is directed at a slightly different part of the visual world, only a low-resolution or pixillated image can be communicated from the compound eye to the CNS.

46.15 Light Absorption Closes Sodium Channels The absorption of light by rhodopsin initiates a signaling cascade that hyperpolarizes the rod cell. In the dark, Na$^+$ channels in the plasma membrane of the rod cell's outer segment are held open by cGMP, allowing positive charges to enter the cell (panel A, upper right). When the rod cell is stimulated by light, it activates transducin (lower portion of panel A). (B) Transducin activates a molecule of phosphodiesterase (PDE). (C) Activated PDE catalyzes the breakdown of cGMP to GMP. The depletion of cGMP results in closure of the Na$^+$ channels and hyperpolarization of the cell.

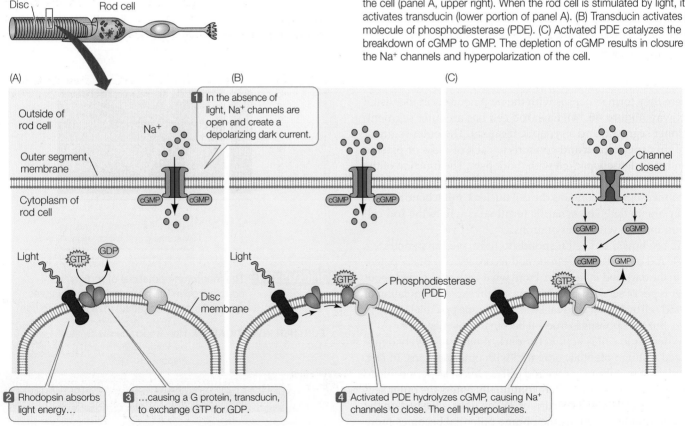

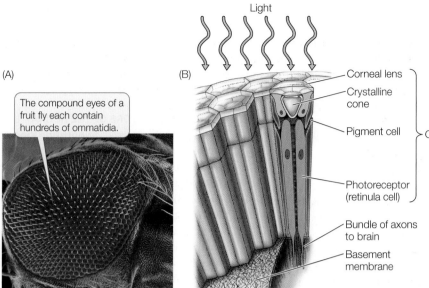

The compound eyes of a fruit fly each contain hundreds of ommatidia.

Light

Corneal lens
Crystalline cone
Pigment cell
Ommatidium
Photoreceptor (retinula cell)
Bundle of axons to brain
Basement membrane

46.16 Ommatidia: The Functional Units of Insect Eyes (A) The micrograph shows the compound eye of a fruit fly. (B) The rhodopsin-containing retinula cells are the photoreceptors in ommatidia.

iris controls the amount of light that reaches the photoreceptor cells at the back of the eye, just as the diaphragm of a camera controls the amount of light reaching the film. The central opening of the iris is the **pupil**. The iris is under neural control. In bright light, the iris constricts and the pupil is very small. As light levels fall, the iris relaxes and the pupil enlarges.

yourBioPortal.com

GO TO Web Activity 46.2 • Structure of the Human Eye

Image-forming eyes evolved independently in vertebrates and cephalopods

Both vertebrates and cephalopod mollusks have *image-forming eyes*—eyes with exceptional abilities to form detailed images of the visual world. Like cameras, both these eye types focus inverted images on an internal surface that is sensitive to light. Considering that they evolved independently, their degree of similarity is remarkable (**Figure 46.17**).

The vertebrate eye is a spherical, fluid-filled structure bounded by a tough connective tissue layer called the *sclera*. At the front of the eye, the sclera forms the transparent **cornea**, through which light passes to enter the eye. Just inside the cornea is the pigmented **iris**, which gives the eye its color. The

Behind the iris is the crystalline protein **lens**, which makes fine adjustments in the focus of images falling on the photosensitive layer—the **retina**—at the back of the eye. The cornea and the fluids within the eye bend light rays passing through them so that they are focused on the retina. The lens makes fine adjustments to the focus and allows the eye to *accommodate*—that is, to focus on objects at various locations in the near visual field. To focus a camera on objects close at hand, you adjust the distance between the lens and the internal surface sensitive to light. Fishes, amphibians, and reptiles accommodate in a similar manner, moving the lenses of their eyes closer to or farther from their

46.17 Convergent Evolution of Eyes The lenses of vertebrate (A) and cephalopod (B) eyes focus images on layers of photoreceptor cells.

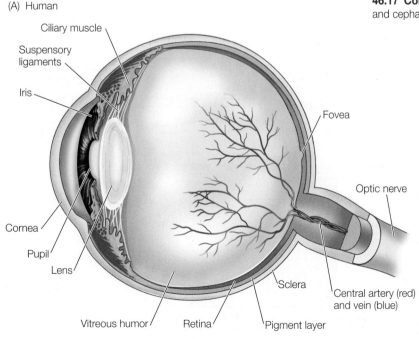

(A) Human

Ciliary muscle
Suspensory ligaments
Iris
Cornea
Pupil
Lens
Vitreous humor
Retina
Pigment layer
Fovea
Optic nerve
Central artery (red) and vein (blue)
Sclera

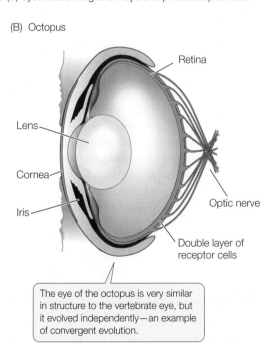

(B) Octopus

Retina
Lens
Cornea
Iris
Optic nerve
Double layer of receptor cells

The eye of the octopus is very similar in structure to the vertebrate eye, but it evolved independently—an example of convergent evolution.

46.18 Staying in Focus
Mammals and birds focus their eyes by changing the shape of the lens depending on the eye's distance from the object of focus.

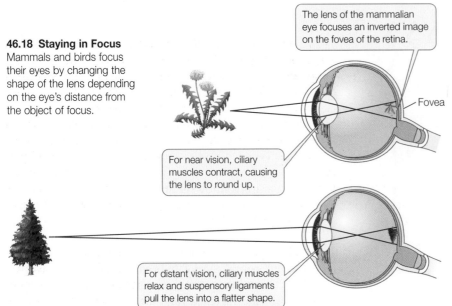

The lens of the mammalian eye focuses an inverted image on the fovea of the retina.

Fovea

For near vision, ciliary muscles contract, causing the lens to round up.

For distant vision, ciliary muscles relax and suspensory ligaments pull the lens into a flatter shape.

ated by the inner segments of those cells. The pigmented epithelial cells phagocytose the shed discs. Each outer segment is totally renewed about every two weeks.

THE PHOTORECEPTORS OF THE RETINA Until now we have referred to only one kind of photoreceptor—the rod cell. The other major class of vertebrate photoreceptor, the **cone cell**, is also named for its shape (**Figure 46.19**). Whereas rod cells are responsible for highly sensitive black-and-white vision, cone cells are responsible for the less sensitive color vision. A human retina has about 5 million cones and about 100 million rods, but their density is not the same across the entire retina.

In humans, light coming from the center of the visual field falls on the **fovea**, where the density of cone cells is highest. The human fovea has about 160,000 cones per mm². The fovea of a hawk has almost twice that number, making the hawk's vision much sharper than ours. The hawk also has *two* foveae in each eye. One receives light from straight ahead, the other from a more lateral field of vision. The forward-looking foveae make binocular vision possible, while the lateral-looking foveae provide high-acuity vision. Birds use both sets of foveae by frequently turning their heads slightly; they cannot move their eyes in the sockets as humans can.

Cones have low sensitivity to light and contribute little to night vision. Night vision depends mostly on rod cells, and therefore vision in dim light is mostly in shades of gray and acuity is low. You may have trouble seeing a small object such as a keyhole at night when you are looking straight at it—that is, when its image is falling on your fovea. If you look a little to the side, so that the image falls on a rod-rich area of your retina, you

retinas. Mammals and birds use a different method: they alter the shape of the lens.

The mammalian lens is contained in a connective tissue sheath that tends to keep it in a spherical shape, but it is attached to suspensory ligaments that pull it into a flatter shape. Circular *ciliary muscles* counteract the pull of the suspensory ligaments, permitting the lens to round up. When the ciliary muscles are at rest, the flatter lens has the correct optical properties to focus distant images on the retina. Contracting the ciliary muscles rounds up the lens, changing its light-bending properties to bring close images into focus (**Figure 46.18**).

Lenses become less elastic with age, so we lose the ability to focus on objects close at hand without the help of corrective lenses. Most people over the age of 45 need the assistance of reading glasses or bifocal lenses.

The vertebrate retina receives and processes visual information

During embryonic development, neural tissue grows out from the brain to form the retina. In addition to a layer of photoreceptor cells, the retina includes four additional layers of cells that process visual information from the photoreceptors. Light must pass through all the layers of retinal cells before being captured by rhodopsin. In humans and other day-active animals, the light that is not captured by rhodopsin is absorbed by a black pigment in a layer of epithelial cells behind the retina. In contrast, nocturnal animals such as deer and raccoons have a white reflective layer behind their retinas that maximize the capture of photons by reflecting them back onto the photoreceptors. Therefore, a deer in the headlights appears to have bright white eyes. Since we do not have the white reflective layer in our retinas, photographic flashes produce the red eye appearance on photos because the light is being reflected by the abundant blood vessels in the retina.

The pigmented epithelium also plays a role in the renewal of the photoreceptors. The photoreceptor cells are always shedding discs from their distal ends as new ones are being gener-

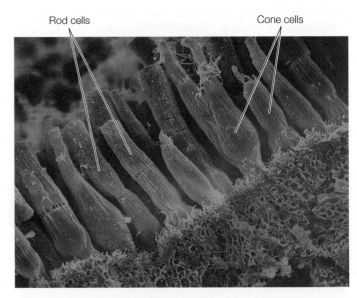

Rod cells   Cone cells

46.19 Rods and Cones This scanning electron micrograph of photoreceptors in the retina of a mud puppy (an amphibian) shows cylindrical rods and tapered cones.

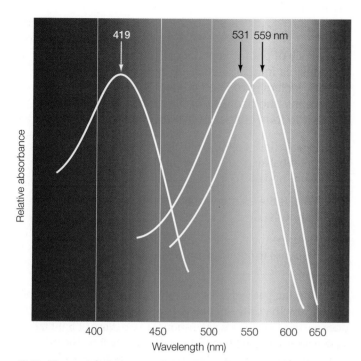

46.20 Absorption Spectra of Cone Cells The three kinds of cone cells contain slightly different opsin molecules, which absorb different wavelengths of light.

see the object better (astronomers looking for faint objects in the sky learned this trick a long time ago). The retinas of nocturnal animals, such as flying squirrels, contain a high percentage of rods. By contrast, some animals that are active only during the day (such as chipmunks) have mostly cones in their retinas.

COLOR VISION The human retina has three kinds of cone cells, each containing slightly different opsin molecules, which differ in the wavelengths of light they absorb best. Although the same 11-*cis*-retinal group is the light absorber in all three kinds of cones (see Figure 46.13), its molecular interactions with opsin determine the spectral sensitivity of the rhodopsin molecule

as a whole (**Figure 46.20**). Because different wavelengths of light are differentially absorbed by the different visual pigments, the brain can interpret the relative inputs from the different classes of cones as a full range of color. Some mammals have only one or two classes of cones, whereas birds have four.

Color blindness results from the absence or dysfunction of one or more classes of cone cells. By far the most common form is red–green color blindness, which occurs in about 10 percent of men of European descent. A genetic condition that affects a person's perception of red and green, its inheritance is sex-linked because the defective gene is on the X chromosome (see Figure 12.24).

INFORMATION FLOW IN THE RETINA The human retina is organized into five layers of neurons (including the photoreceptor cells) that receive visual information and process it before sending it to the brain (**Figure 46.21**). A first step in understanding how the retina tells the brain what it is seeing is to study how these layers are interconnected and how they influence one another. From our discussion of rod cells, we know that the photoreceptor cells at the back of the retina hyperpolarize in response to light and do not generate action potentials. The cells at the front of the retina (the cells closest to the lens) are **ganglion cells**. They do fire action potentials, and their axons form the optic nerve that travels to the brain. The layers of cells between the photoreceptors and the ganglion cells process information about the visual field.

The photoreceptors and ganglion cells are connected by *bipolar cells*. Changes in the membrane potential of rods and cones in response to light alter the rates at which the rods and cones release neurotransmitter at their synapses with the bipolar cells. In response to this neurotransmitter, the membrane potentials

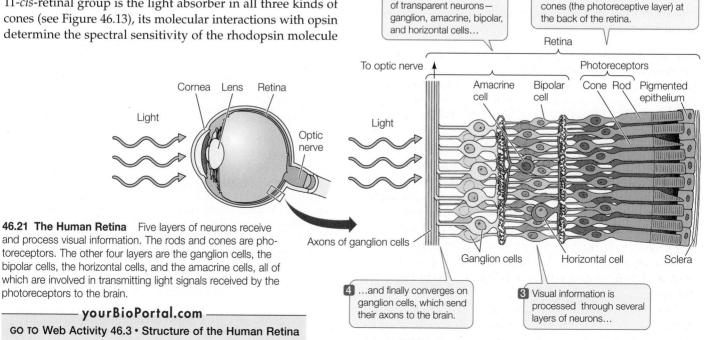

46.21 The Human Retina Five layers of neurons receive and process visual information. The rods and cones are photoreceptors. The other four layers are the ganglion cells, the bipolar cells, the horizontal cells, and the amacrine cells, all of which are involved in transmitting light signals received by the photoreceptors to the brain.

yourBioPortal.com

GO TO Web Activity 46.3 • Structure of the Human Retina

of the bipolar cells change, altering the rate at which they release neurotransmitter onto ganglion cells. The rate of neurotransmitter release from the bipolar cells determines the rate at which the ganglion cells fire action potentials. Thus, the direct flow of information in the retina is from photoreceptor to bipolar cell to ganglion cell. The ganglion cells send the information to the brain.

The other two cell layers, the horizontal cells and the amacrine cells, consist of interneurons that communicate laterally across the retina. *Horizontal cells* form synapses with neighboring photoreceptors and bipolar cells. Thus, light falling on one photoreceptor can influence the sensitivity of its neighbors to light. This lateral flow of information enables the retina to sharpen the perception of contrast between light and dark patterns. *Amacrine cells* form local interconnections between bipolar cells and ganglion cells. Some amacrine cell types are highly sensitive to changing illumination or to motion. Others assist in adjusting the sensitivity of the eyes according to the overall level of light falling on the retina. When background light levels change, amacrine cell connections to the ganglion cells help the ganglion cells remain sensitive to temporal changes in stimulation. Thus, even with large changes in background illumination, the eyes are sensitive to smaller, more rapid changes in the pattern of light falling on the retina.

46.4 RECAP

A family of photopigments called rhodopsins are responsible for light sensitivity in all animals. Receptor cells, including rod and cone cells in humans, transduce the photosensitivity of rhodopsins to light and use it to form images of the environment.

- Explain how a photon of light affects the membrane potential in a rod cell. See pp. 976–978 and Figures 46.14 and 46.15
- What is the mechanism of color vision? See p. 981 and Figure 46.20
- Describe the flow of signals that occurs in the eye in response to light. See pp. 981–982 and Figure 46.21

Knowing the path of information in the retina still does not tell us how that information is processed by the brain. What does the eye tell the brain in response to a pattern of light falling on the retina? In Chapter 47 we will describe how the brain reassembles that information into our view of the world.

CHAPTER SUMMARY

46.1 How Do Sensory Cells Convert Stimuli into Action Potentials?

- Sensory receptor cells, also known as sensors or receptors, transduce information about an animal's external and internal environment into action potentials. Some sensors do not fire action potentials, but respond with graded membrane potentials that control release of neurotransmitter onto sensory neurons that do fire action potentials.
- The interpretation of action potentials as particular sensations depends on which neurons in the CNS receive them.
- Sensory receptor cells have membrane **receptor proteins** that cause ion channels to open or close, affecting the resting potential of the cell. Metabotropic receptors act through signal transduction pathways to generate **receptor potentials**. Mechanoreceptors are ionotropic sensory receptors that open ion channels physically through forces such as pressure or stretch. **Review Figure 46.1**
- Receptor potentials initiated by a sensory cell can spread to regions of the cell's plasma membrane that generate action potentials, or they can release neurotransmitter in response to changes in membrane potential. **Review Figure 46.2**
- **Adaptation** enables the nervous system to ignore irrelevant or continuous stimuli while remaining responsive to relevant or new stimuli.

46.2 How Do Sensory Systems Detect Chemical Stimuli?

- **Chemoreceptors** are responsible for **olfaction**, **gustation**, and the sensing of **pheromones**.

- Mammalian olfactory sensors project directly to the olfactory bulb of the brain. Sensors for the same **odorant** project to the same area of the olfactory bulb.
- Each olfactory receptor cell expresses one receptor protein that can bind a specific molecule or ion. Binding causes a second messenger to open ion channels, which creates an action potential. **Review Figure 46.4**
- In vertebrates, **taste buds** in the mouth cavity are responsible for gustation. The five basic tastes are sweet, salty, sour, bitter, and umami. **Review Figure 46.5**

46.3 How Do Sensory Systems Detect Mechanical Forces?

- The skin contains a variety of ionotropic **mechanoreceptors** that respond to touch and pressure. The density of mechanoreceptors in any skin area determines the sensitivity of that area. **Review Figure 46.6**
- **Stretch receptors** in muscle spindles and in the Golgi tendon organ found in tendons and ligaments inform the CNS of the positions of and loads on parts of the body. **Review Figure 46.7**
- In mammalian **auditory systems**, ear pinnae collect and direct sound waves to the **tympanic membrane**, which vibrates in response to sound waves. The movements of the tympanic membrane are amplified through a chain of **ossicles** that conduct the vibrations to the **oval window**. Movements of the oval window create pressure waves in the fluid-filled **cochlea**. **Review Figure 46.8, WEB ACTIVITY 46.1**
- The **basilar membrane** running down the center of the cochlea is distorted by pressure waves at specific locations that depend on the frequency of the wave. These distortions cause the bending of hair cells in the **organ of Corti**. Receptor potentials

in hair cells cause them to release neurotransmitter, which creates action potentials in the **auditory nerve**. Review Figure 46.9, ANIMATED TUTORIAL 46.1

- **Hair cells** are also mechanoreceptors. The bending of their stereocilia alters receptor proteins and therefore their membrane potentials. Hair cells are found in the auditory organs and organs of equilibrium such as the **lateral line** system of fishes and the **semicircular canals** and **vestibular system** of mammals. Review Figures 46.10, 46.11, and 46.12

46.4 How Do Sensory Systems Detect Light?

- **Photosensitivity** depends on the absorption of photons of light by **rhodopsin**, a photoreceptor molecule that consists of a protein called **opsin** and a light-absorbing group called retinal. Absorption of light by retinal is the first step in a cascade of intracellular events leading to a change in the membrane potential of the **photoreceptor cell**. Review Figure 46.13, ANIMATED TUTORIAL 46.2

- When excited by light, vertebrate photoreceptor cells hyperpolarize and release less neurotransmitter onto the neurons with which they form synapses. They do not fire action potentials. Review Figures 46.14 and 46.15

- Visual systems range from the simple **eye cups** of flatworms, which sense the direction of a light source, to the **compound eyes** of arthropods, which detect shapes and patterns, to the **image-forming eyes** of vertebrates and cephalopods. Review Figures 46.16 and 46.17, WEB ACTIVITY 46.2

- Vertebrate and cephalopod eyes focus detailed images of the visual field onto dense arrays of photoreceptors that transduce the visual image into neural signals. Review Figure 46.18

- Vertebrates have two types of photoreceptors, **rod cells** and **cone cells**. In humans, the **fovea** contains almost exclusively cone cells, which are responsible for color vision but are not very sensitive in dim light. Color vision arises from three types of cone cells with different spectral absorption properties. Review Figures 46.19 and 46.20

- The vertebrate **retina** consists of five layers of neurons lining the back of the eye. The light-absorbing photoreceptor cells are at the back of the retina. Review Figure 46.21, WEB ACTIVITY 46.3

- The axons of the **ganglion cells**, in the innermost layer of the retina, are bundled together in the optic nerve. Between the photoreceptors and the ganglion cells are neurons that process information from the photoreceptors.

SELF-QUIZ

1. Which statement about sensory systems is *not* true?
   a. Sensory transduction involves the conversion (direct or indirect) of a physical or chemical stimulus into changes in membrane potentials.
   b. In general, a stimulus causes a change in the flow of ions across the plasma membrane of a sensory receptor cell.
   c. The term "adaptation" refers to the process by which a sensory system becomes insensitive to a continuing source of stimulation.
   d. The more intense a stimulus, the greater the magnitude of each action potential fired by a sensory neuron.
   e. Sensory adaptation plays a role in the ability of organisms to discriminate between important and unimportant information.

2. The female silkworm moth releases a chemical called bombykol from a gland at the tip of her abdomen. Bombykol is
   a. a sex hormone.
   b. detected by the male only when present in large quantities.
   c. not species-specific.
   d. detected by hairs on the antennae of male silkworm moths.
   e. a chemical basic to the taste process in arthropods.

3. Which statement about olfaction is *not* true?
   a. In general, mammals depend more on vision than on olfaction as their dominant sensory modality.
   b. Olfactory stimuli are recognized by the interaction between odorant molecules and receptor proteins on olfactory hairs.
   c. The more odorant molecules that bind to receptors, the more action potentials are generated.
   d. The greater the number of action potentials generated by an olfactory receptor, the greater the intensity of the perceived smell.
   e. The perception of different smells results from the activation of different combinations of olfactory receptors.

4. In general, the touch receptors located close to the surface of both hairy and nonhairy skin
   a. are relatively insensitive to light touch.
   b. adapt very quickly to stimuli.
   c. are uniformly distributed throughout the surface of the body.
   d. are called Pacinian corpuscles.
   e. adapt slowly and provide almost continuous information.

5. The membrane that is most directly responsible for the ability to discriminate different pitches of sound is the
   a. round window.
   b. oval window.
   c. tympanic membrane.
   d. tectorial membrane.
   e. basilar membrane.

6. Which statement is *not* true?
   a. The transmembrane potential of a rod cell becomes more negative when the rod cell is exposed to light.
   b. A photoreceptor releases the most neurotransmitter when in total darkness.
   c. Whereas in vision the intensity of a stimulus is encoded by the degree of hyperpolarization of photoreceptors, in hearing the intensity of a stimulus is encoded by changes in firing rates of sensory neurons.
   d. Stiffening of the ossicles in the middle ear can lead to deafness.
   e. The interaction among hammer (malleus), anvil (incus), and stirrup (stapes) conducts sound waves across the fluid-filled middle ear.

7. In humans, the region of the retina where the central part of the visual field falls is the
   a. central ganglion cell.
   b. fovea.
   c. optic nerve.
   d. cornea.
   e. pupil.

8. Which of the following statements about information flow in the vertebrate visual system is *true*?
   a. Action potentials in bipolar cells cause the release of neurotransmitter onto ganglion cells.
   b. Amacrine cells integrate the activity of neighboring rod and cone cells.
   c. When photons of light enter they eye, the first cells in the retina they encounter are ganglion cells.
   d. The highest density of rod cells in the human retina is centrally located in the fovea, resulting in high acuity dim light vision.
   e. Pigmented epithelial cells at the back of the retina provide information about the level of ambient light for contrast adjustments.

9. Which statement about the cone cells in a human eye is *not* true?
   a. They are responsible for our sharpest vision.
   b. They are responsible for color vision.
   c. They are more sensitive to light than rods are.
   d. They are fewer in number than rods.
   e. They exist in high numbers at the fovea.

10. The color in color vision results from the
   a. ability of each cone cell to absorb all wavelengths of light equally.
   b. lens of the eye acting like a prism and separating the different wavelengths of light.
   c. differential absorption of wavelengths of light by different kinds of rod cells.
   d. three different isomers of opsin in cone cells.
   e. absorption of different wavelengths of light by amacrine and horizontal cells.

FOR DISCUSSION

1. Compare and contrast the functioning of olfactory receptors and photoreceptors. How do these sensory cells enable the central nervous system to discriminate between an apple and an orange?

2. Amplification of signal is an important feature of sensory systems. Compare the mechanisms of amplification found in olfactory, visual, and auditory systems.

3. If you were blindfolded and sitting in a wheelchair, how would you know if you were being pushed forward or backward?

4. Animals can use visual, olfactory, tactile, and auditory signals to communicate. From what you know about these sensory systems, discuss the relative advantages and disadvantages of these systems for communication.

ADDITIONAL INVESTIGATION

A certain region of the brain contains neurons that are sensitive to the osmolarity of the blood. You could imagine that these cells are sensitive to the concentration of a particular solute such as NaCl or to tension in their plasma membranes if they take up or lose water osmotically. Using a slice of this brain region in a culture dish, and patch pipettes, how could you investigate the mechanism of signal transduction in these neurons?

WORKING WITH DATA (GO TO *yourBioPortal*.com)

Rod Cell Response In this exercise based on the experiments illustrated in Figure 46.14, you will analyze the data generated by electrode recordings from rod cells stimulated by light of varying intensities. You will then use the data to generate a maximum-response curve for the rod cell.

Drive a cab, grow your brain

Go to Google Maps and compare London with New York City at the same scale. In which city is it easier to drive a taxi? In the city with numbered avenues and streets at right angles to each other, or the city with a maze of named streets going in every which direction?

Eleanor Maguire at University College London was so impressed with the navigational abilities of London taxi drivers, she decided to see if there was anything "special" about their brains. Using the brain imaging technique known as MRI (*magnetic resonance imaging*), Maguire and her colleagues examined the brains of taxi drivers with varying years of experience and compared them with each other and with the brains of control subjects who were not taxi drivers. The studies revealed significant changes in the anatomy of the hippocampus among taxi drivers.

The hippocampus is an area of the brain involved in learning and memory. The posterior hippocampus in par-

ticular is implicated in the memory of spatial relationships among objects in the environment. Maguire found that the posterior hippocampi of taxi drivers was larger than that of the control subjects and that, among the cab drivers themselves, there was a positive correlation between the size of the posterior hippocampus and years of driving experience.

At the Massachusetts Institute of Technology, Matt Wilson and his colleagues can record the hippocampal activity of lab rats as the animals navigate a maze. By placing electrodes in the rats' hippocampi, Wilson has located specific neurons, sometimes referred to as "place cells," that fire only when the rat is at a particular location in the maze. In a sense he can even see what they are thinking, because when the rats are not moving through the maze, the neurons will occasionally fire in the same pattern as when the animal is running the maze—or they will fire in the reverse sequence, representing where the animal has just been. In a maze that had a choice point (go right or go left), the firing pattern seen when the animal was held at the start position predicted which direction it would turn. Similar firing patterns occur when the rats are sleeping. Are they dreaming about the maze, or are they transferring memory of today's experience into long-term memory?

Neural firing patterns that correlate with the hippocampal sequences are also seen in the visual cortex at the back of the brain

© Bluesky International Limited

A Mind-Expanding Maze The extraordinary ability of taxicab drivers in London to navigate its mazelike warren of streets and byways prompted a study that revealed London cabbies to have larger than normal posterior hippocampi—a brain region implicated in the memory of spatial relationships in the environment.

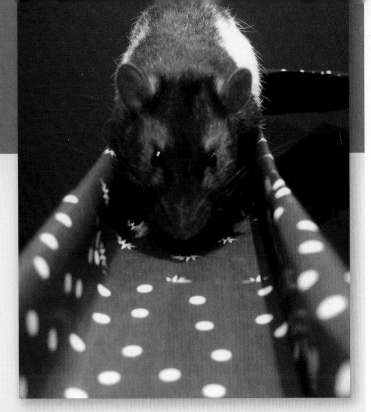

Seeing What a Rat Thinks By recording from many neurons in the hippocampus of a rat while it runs a maze, a sequence of "place cell" activities that correspond to different locations in the maze can be described. After the rat has learned the maze, a similar sequence of neural activity can be recorded from its visual cortex.

and in the prefrontal cortex at the front of the brain. In humans, the prefrontal cortex is involved in planning and decision-making.

The extensive ability of humans to remember, to learn, to communicate, and to feel emotions—the "higher functions" of the mammalian brain—is a fascinating subject that brings together the fields of neurobiology, psychology, and medicine. Are these functions simply the product of molecular changes in brain cells? To what extent can we reverse damage to the brain that interferes with higher functioning? These and countless other questions are only beginning to be articulated and studied.

IN THIS CHAPTER we will explore the cellular basis of important subsystems of the human nervous system. The human brain has about 100 billion neurons and probably a thousand times as many synapses, which account for its ability to handle vast amounts of information. But the ability of the brain to carry out specific functions—evaluating sensory input, controlling emotions, generating motor output, learning and remembering—arises from functional subsystems.

47.1 How Is the Mammalian Nervous System Organized?

We can describe the organization of the mammalian nervous system anatomically or functionally. In anatomical terms, all vertebrate nervous systems consist of three parts: a brain, a spinal cord, and a set of peripheral nerves that reach to all parts of the body. As discussed in Section 45.1, the brain and spinal cord are referred to as the *central nervous system*, or *CNS*, and the cranial and spinal nerves that connect the CNS to all the tissues of the body are referred to as the *peripheral nervous system*, or *PNS*. An additional division of the nervous system exists in the gut: the enteric nervous system, which we discuss in Chapter 51.

Recall from Section 45.1 that a *neuron* is an electrically excitable cell that communicates via an axon, and a *nerve* is a bundle of axons that carries information about many things simultaneously. Some axons in a nerve may be carrying information to the CNS, while other axons in the same nerve are carrying information from the CNS to the body's organs. In this chapter, we will further divide the anatomy of the brain, spinal cord, and PNS into smaller functional units.

A functional organization of the nervous system is based on flow and type of information

Figure 47.1 illustrates the major avenues of information flow through the nervous system. The *afferent* portion of the PNS carries sensory information to the CNS. We are conscious of much of this information (e.g., light, sound, skin temperature, the position of limbs), but we are usually not conscious of the information involved in physiological regulation (e.g., blood pressure, deep body temperature, blood oxygen levels).

The *efferent* portion of the PNS carries information from the CNS to the muscles and glands of the body. Efferent pathways can be divided into a voluntary division, which executes our conscious movements, and an involuntary, or *autonomic*, division, which controls physiological functions.

In addition to the neural information it receives from the PNS, the CNS receives chemical information from hormones circulating in the blood. In turn, *neurohormones* released by neurons enter the circulation and affect neurons and other cells distant from the site of release (see Figure 41.1).

The vertebrate CNS develops from the embryonic neural tube

Early in the development of a vertebrate embryo, a tube of neural tissue forms (see Section 44.3). At its anterior end, this *neural tube* forms three swellings that become the **hindbrain**, **midbrain**, and **forebrain**. The rest of the neural tube becomes the

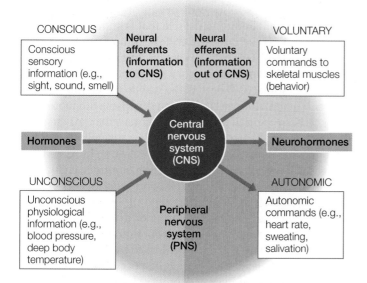

47.1 Organization of the Nervous System The peripheral nervous system (beige, green) carries information both to (afferent) and from (efferent) the central nervous system (purple). The CNS also receives hormonal inputs and produces hormonal outputs (lavender).

spinal cord. Peripheral nerves sprout from the midbrain and hindbrain (the cranial nerves) and from the spinal cord (the spinal nerves). From these early stages we see the linear axis of information flow in the nervous system. Although the developing brain will fold and become a complex structure, the information flow in the adult nervous system will follow paths that emerge from the simple linear neural tube.

Each of the three regions of the embryonic brain develops into several structures in the adult brain (**Figure 47.2**). From the embryonic midbrain come structures that process aspects of visual and auditory information. From the hindbrain come the **medulla**, the **pons**, and the **cerebellum**. The medulla is continuous with the spinal cord, the pons is in front of the medulla, and the cerebellum is a dorsal outgrowth of the pons. The medulla and pons contain distinct groups of neurons involved in controlling physiological functions such as breathing and circulation and basic motor patterns such as swallowing and vomiting. All information traveling between the spinal cord and higher brain areas must pass through the pons, the medulla, and the midbrain, which are collectively known as the **brainstem**.

The cerebellum is involved in coordinating muscle activity and maintaining balance. It is like the conductor of an orchestra; the cerebellum receives "copies" of the commands going to the muscles from higher brain areas, and it receives information coming up the spinal cord from the joints and muscles. It compares the motor "score" with the actual behavior of the muscles and refines the motor commands accordingly.

The embryonic forebrain develops a central region called the **diencephalon** and a surrounding structure called the **telencephalon**. The diencephalon is the core of the forebrain and consists of an upper structure called the **thalamus** and a lower structure called the *hypothalamus*. The thalamus is the final relay station for sensory information going to the telencephalon. The

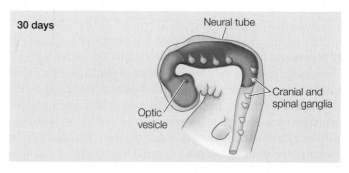

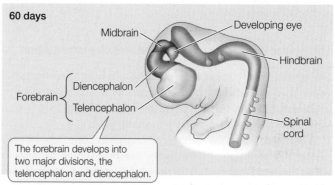

The forebrain develops into two major divisions, the telencephalon and diencephalon.

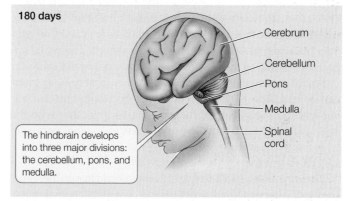

The hindbrain develops into three major divisions: the cerebellum, pons, and medulla.

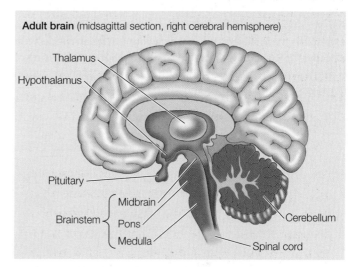

Adult brain (midsagittal section, right cerebral hemisphere)

47.2 Development of the Central Nervous System In vertebrate embryos, the anterior end of the hollow neural tube differentiates into forebrain, midbrain, and hindbrain. Each of these regions develops into several structures in the adult brain. The remainder of the neural tube becomes the spinal cord.

hypothalamus regulates many physiological functions and biological drives such as hunger and thirst; it receives a lot of physiological information of which we are not conscious.

The telencephalon—also called the **cerebrum**—consists of two **cerebral hemispheres**, left and right. The outer layer of the telencephalon is the **cerebral cortex**, a thin layer rich in cell bodies. If we compare the classes of vertebrates from fishes through amphibians, reptiles, birds, and mammals, the telencephalon increases in size, complexity, and importance—an evolutionary trend called *telencephalization* (see Figure 45.2). In humans, the telencephalon is by far the largest part of the brain and plays major roles in sensory perception, learning, memory, and conscious behavior.

The spinal cord transmits and processes information

The spinal cord conducts information in both directions between the brain and the body's organs. It also integrates much of the information coming from the PNS and responds to that information by issuing motor commands.

A cross section of the spinal cord reveals a central area of gray matter in the shape of a butterfly, surrounded by an area of white matter (**Figure 47.3**). In the nervous system, **gray matter** is rich in neural cell bodies, and **white matter** contains axons. The gray matter of the spinal cord contains the cell bodies of the spinal neurons; the white matter contains the axons that conduct information up and down the spinal cord. The white appearance is due to the myelin that wraps most of the axons.

Spinal nerves extend from the spinal cord at regular intervals on each side. Each spinal nerve has two roots, one connecting with the *dorsal horn* of the gray matter, the other with the *ventral horn*. The afferent (sensory) axons in a spinal nerve enter the spinal cord through the *dorsal root*, and the efferent (motor) axons leave through the *ventral root*.

The conversion of afferent to efferent information in the spinal cord without participation of the brain is called a **spinal reflex**. The simplest type of spinal reflex involves only two neurons and one synapse and is therefore called a **monosynaptic reflex**. An example is the knee-jerk reflex, which your physician checks with a mallet tap just below your knee. We can diagram the wiring of a monosynaptic reflex by following the flow of information through the spinal cord, as shown in Figure 47.3.

In the knee-jerk reflex, sensory information comes from stretch receptors in the leg muscle that is suddenly stretched when the mallet strikes the tendon running over the knee. Each stretch receptor initiates action potentials that are conducted by the axon of a sensory neuron through the dorsal horn of the spinal cord and all the way to the ventral horn. In the ventral horn, the sensory neuron synapses with motor neurons, causing them to fire action potentials that are then conducted back to the leg extensor muscle, causing it to contract. The function of this simple circuit is to sense an increased load on the limb and to increase the strength of muscle contraction to compensate for the added load and thereby keep muscle length constant.

Most spinal circuits are more complex than this monosynaptic reflex, as we can demonstrate by building on the circuit we just traced. Limb movement is controlled by *antagonistic* sets of muscles—muscles that work against each other. When one member of an antagonistic set of muscles contracts, it bends, or flexes, the limb; it is therefore called a *flexor*. The antagonist to this muscle, the *extensor*, straightens, or extends, the limb. For a limb to move, one muscle of the pair must relax while the other contracts. Thus, sensory input that activates the motor neuron of one muscle also inhibits its antagonist. This coordination is achieved by an *interneuron*, which makes an inhibitory synapse onto the motor neuron of the antagonistic

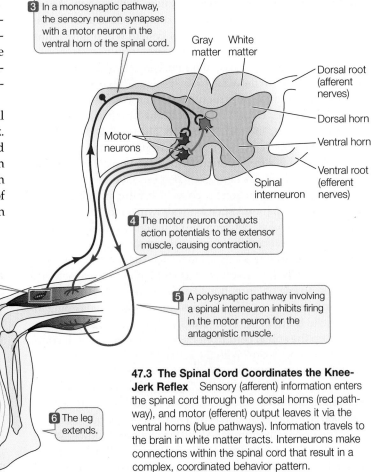

3 In a monosynaptic pathway, the sensory neuron synapses with a motor neuron in the ventral horn of the spinal cord.

Gray matter   White matter

Dorsal root (afferent nerves)

Dorsal horn

Motor neurons

Ventral horn

Spinal interneuron

Ventral root (efferent nerves)

4 The motor neuron conducts action potentials to the extensor muscle, causing contraction.

2 Stretch receptors fire action potentials.

5 A polysynaptic pathway involving a spinal interneuron inhibits firing in the motor neuron for the antagonistic muscle.

1 A hammer tap stretches the tendon in the knee, stretching receptors in the extensor muscle.

6 The leg extends.

47.3 The Spinal Cord Coordinates the Knee-Jerk Reflex Sensory (afferent) information enters the spinal cord through the dorsal horns (red pathway), and motor (efferent) output leaves it via the ventral horns (blue pathways). Information travels to the brain in white matter tracts. Interneurons make connections within the spinal cord that result in a complex, coordinated behavior pattern.

yourBioPortal.com

GO TO Animated Tutorial 47.1 • Information Processing in the Spinal Cord

muscle (see Figure 47.3). Thus the reciprocal inhibition of antagonistic muscles involves an interneuron between the sensory cell and the motor neuron of the inhibited muscle, and therefore at least two synapses.

The withdrawal reflex is an example of a polysynaptic spinal reflex that involves many interneurons. When you step on a tack, you immediately pull back your foot: the tack stimulates pain receptors in the foot, and the sensory neurons transmit action potentials into the dorsal horn of the spinal cord on the same side of the body. In the dorsal horn, these neurons synapse with interneurons that send information through their axons to the brain, resulting in the conscious sensation of pain. Before the brain is aware of the pain, however, synapses of the sensory neurons with other interneurons stimulate and inhibit a variety of different motor neurons in the spinal cord. Interneurons on the same side of the spinal cord coordinate the activity of the muscles that withdraw the foot and leg. To pull away, however, the other leg has to extend, and balance must be shifted. The coordination of these actions involves interneurons that make connections across the spinal cord to motor neurons on the opposite side. Thus a rather complex suite of movements is coordinated in the spinal cord without the brain's participation.

The reticular system alerts the forebrain

Sensory information ascending the spinal cord to final destinations in the forebrain passes through the brainstem. Many sensory axons give off branches in the brainstem that form synapses with a network of brainstem neurons called the **reticular system**. Within this highly complex network of axons and dendrites are many discrete groups of neurons that share a common characteristic such as the neurotransmitter they produce and release. Such an anatomically distinct group of neurons in the CNS is called a **nucleus** (not to be confused with the nucleus of a single cell).

As axons carrying sensory information ascend through the reticular formation, they make connections with nuclei that are involved in controlling many functions of the body. Information from joints and muscles, for example, is directed to nuclei in the pons and cerebellum that are involved in balance and coordination. Sensory information also goes to reticular formation nuclei that control sleep and wakefulness. High reticular formation activity produces waking; in the absence of such stimulation, sleep occurs. Therefore, the reticular core of the brainstem is called the *reticular activating system.*

The core of the forebrain controls physiological drives, instincts, and emotions

As mentioned above, the diencephalon consists of the thalamus and the hypothalamus. The thalamus communicates sensory information to the cerebral cortex; the hypothalamus receives information about physiological conditions in the body and regulates many homeostatic functions. Section 40.4 describes how the hypothalamus is involved in regulating body temperature, and Section 41.2 discusses the intimate association between the hypothalamus and the pituitary gland and its control of many homeostatic functions.

Surrounding the diencephalon of all vertebrates are phylogenetically older structures of the telencephalon called the **limbic system** (**Figure 47.4**). The limbic system is responsible for basic physiological drives such as hunger and thirst, instincts, long-term memory formation, and emotions such as fear. Within the limbic system are areas that, when stimulated with small electric currents, can cause intense sensations of pleasure, pain, or rage. If a rat is given the opportunity to stimulate its own pleasure centers by pressing a switch, it will ignore food, water, and even sex, pushing the switch until it is exhausted. Pleasure and pain centers in the limbic system are believed to play roles in learning and in physiological drives.

As described in the introduction to Chapter 45, one component of the limbic system—the **amygdala**—is involved in fear and fear memory. If a certain portion of the amygdala is damaged or chemically blocked, an animal cannot learn to be afraid of a stimulus or a situation that would normally induce a strong fear reaction.

Another part of the limbic system, the **hippocampus**, is necessary in humans for the transfer of short-term memory to long-term memory, as we'll discuss in Section 47.3.

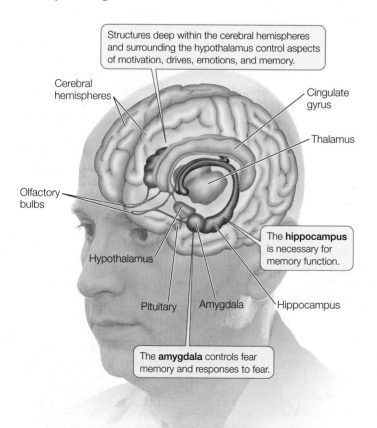

Structures deep within the cerebral hemispheres and surrounding the hypothalamus control aspects of motivation, drives, emotions, and memory.

Cerebral hemispheres

Cingulate gyrus

Thalamus

Olfactory bulbs

The **hippocampus** is necessary for memory function.

Hypothalamus

Pituitary   Amygdala   Hippocampus

The **amygdala** controls fear memory and responses to fear.

47.4 The Limbic System The evolutionarily primitive parts of the telencephalon are referred to as the limbic system. The hippocampus is involved in forming long-term memory. The amygdala triggers fear emotions and fear memories (see p. 943).

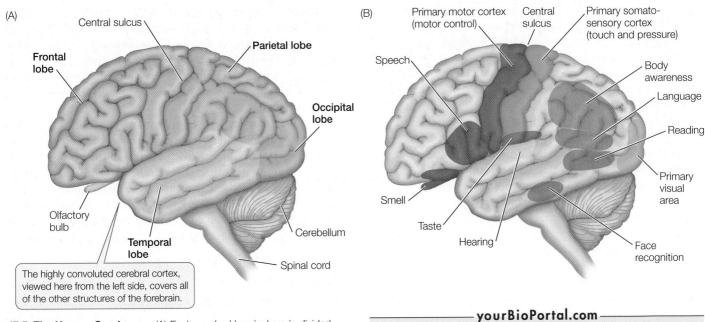

(A)

The highly convoluted cerebral cortex, viewed here from the left side, covers all of the other structures of the forebrain.

47.5 The Human Cerebrum   (A) Each cerebral hemisphere is divided into four lobes. (B) Different functions are localized in particular areas of the cerebral lobes.

yourBioPortal.com
GO TO Web Activity 47.1 • The Human Cerebrum

Regions of the telencephalon interact to produce consciousness and control behavior

The cerebrum is the dominant structure in the mammalian brain. In humans, it is so large that it covers all other parts of the brain except the cerebellum (**Figure 47.5A**). The cerebral cortex covering the cerebrum is only about 4 millimeters thick, but it covers a surface area larger than 1 m^2 because it is folded into ridges (*gyri;* singular *gyrus*) and valleys (*sulci;* singular *sulcus*). These foldings, or *convolutions,* allow the cortex to fit into the skull.

A curious feature of our nervous system is that the left side of the body is served (in both sensory and motor aspects) mostly by the right side of the brain, and the right side of the body is served mostly by the left side of the brain. Thus, sensory input from the right hand goes to the left cerebral hemisphere, and sensory input from the left hand goes to the right cerebral hemisphere. The exception is the head, where the left side is controlled by the left cerebral hemisphere and the right side by the right cerebral hemisphere. The two hemispheres are not symmetrical with respect to all functions. Language abilities, for example, reside predominantly in the left hemisphere.

Different regions of the cerebral cortex have specific functions (**Figure 47.5B**). Some of those functions are easily defined, such as receiving and processing sensory information or generating motor commands, but most of the cortex is involved in higher-order information processing that is less easy to define. These latter areas are given the general name of **association cortex**, so named because they integrate, or *associate,* information from different sensory modalities and from memory.

To understand the cerebral cortex, it helps to have an anatomical road map. Viewed from the left side, the left cerebral hemisphere looks like a boxing glove for the right hand with the fingers pointing forward, the thumb pointing out, and the wrist at

the rear. The "thumb" area is the **temporal lobe**, the fingers the **frontal lobe**, the back of the hand the **parietal lobe**, and the wrist the **occipital lobe** (see Figure 47.5A). The right cerebral hemisphere shows a mirror image of this arrangement. Let's look at each lobe separately.

As we explore the functions of the cerebral cortex and other parts of the brain, you will note frequent mention of persons whose brains were damaged by accidents or other unfortunate events. Until recently, the study of such individuals has been the main source of functional information about the human brain, but new imaging technologies such as positron emission tomography (PET) and magnetic resonance imaging (MRI) are providing a wealth of new information and opportunities to study the human brain.

THE TEMPORAL LOBE   The upper region of the temporal lobe receives and processes auditory information. The association areas of this lobe are involved in recognizing, identifying, and naming objects. Damage to the temporal lobe results in disorders called *agnosias,* in which the individual is aware of an object but cannot identify it.

Damage to a certain area of the temporal lobe results in the inability to recognize faces. Even old acquaintances cannot be identified by facial features, although they may be identified by other attributes such as voice, body features, and posture. Using monkeys, it has been possible to record the activity of neurons in this region that respond selectively to faces in general. These neurons do not respond to other stimuli in the visual field, and their responsiveness decreases if some of the facial features are missing or appear in inappropriate locations (**Figure 47.6**). Damage to other association areas of the temporal lobe causes deficits in understanding spoken language, although speaking, reading, and writing abilities may be intact.

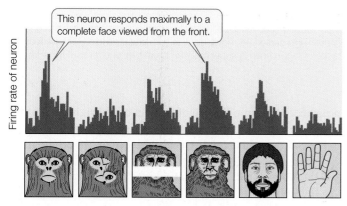

This neuron responds maximally to a complete face viewed from the front.

Firing rate of neuron

47.6 "Face Neurons" in One Region of the Temporal Lobe The electrode traces represent the firing rate of a neuron in the temporal lobe of a monkey in response to the pictures shown below them. Highest firing is stimulated by the appearance of a complete face.

THE FRONTAL LOBE The frontal and parietal lobes are separated by a deep valley called the *central sulcus*. A strip of the frontal lobe cortex just in front of the central sulcus is called the **primary motor cortex** (see Figure 47.5B). The neurons in this region control muscles in specific parts of the body; the parts of the body have been mapped onto the primary motor cortex largely during neurosurgical procedures. As part of these procedures, electrodes were used to stimulate small areas of cortex. In the area just anterior to the central sulcus, stimulation causes specific muscles to contract. Parts of the body with fine motor control, such as the face and hands, have disproportionate representation (**Figure 47.7A**). Stimulation of neurons in the primary motor cortex causes twitches of muscles, not coordinated movements.

The association functions of the frontal lobe are diverse and are best described as having to do with feeling and planning. They contribute significantly to personality. People with frontal lobe damage have drastic alterations of personality

and difficulty planning future events. A dramatic case of frontal lobe damage is the story of Phineas Gage, who in 1848 was an industrious and responsible young railroad construction foreman. Then a blasting accident shot a meter-long, 3-centimeter-wide iron tamping rod through his brain. The rod entered Gage's head below his left eye, passed through his frontal lobe, and exited the top of his head (**Figure 47.8**).

Remarkably, Gage survived, but he was a completely different person. He was quarrelsome, impatient, obstinate, and used profane language, which he had never done before. He lost his railroad job and spent his days as a drifter, earning money by telling his story and exhibiting his scars (and the tamping iron). He died of a seizure in 1860, at the age of 38. If you are in Boston, you can pay him a visit—his skull, death mask, and the tamping iron are on display in the Warren Anatomical Museum of Harvard Medical School.

THE PARIETAL LOBE The strip of parietal lobe cortex just behind the central sulcus is the **primary somatosensory cortex** (see Figure 47.5B). This area receives touch and pressure information relayed from the body through the thalamus.

The entire body surface can be mapped onto the primary somatosensory cortex (**Figure 47.7B**). Areas of the body that have a high density of tactile mechanoreceptors and are capable of making fine discriminations in touch (such as the lips and fingers) have disproportionately large representation. If a very small area of the primary somatosensory cortex is stimulated electrically, the subject reports feeling specific sensations, such as touch, in a localized part of the body.

47.7 The Body Is Represented in Primary Motor and Primary Somatosensory Cortex Neurons in the primary motor cortex (A) control muscles in specific parts of the body, while neurons in the primary somatosensory cortex (B) receive information from specific parts of the body. The locations of these neurons in the cortex correspond to "maps" on which regions of the body are represented in proportion to the amount of primary cortical area devoted to them.

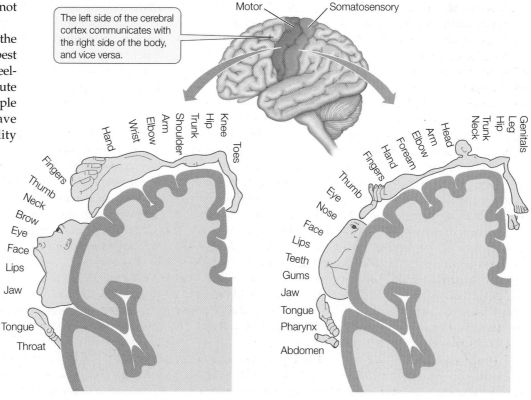

Motor / Somatosensory

The left side of the cerebral cortex communicates with the right side of the body, and vice versa.

Hand Wrist Elbow Arm Shoulder Trunk Hip Knee Toes
Fingers
Thumb
Neck
Brow
Eye
Face
Lips
Jaw
Tongue
Throat

(A) Motor cortex

Head Arm Elbow Foream Hand Trunk Hip Leg Genitals Neck
Fingers
Thumb
Eye
Nose
Face
Lips
Teeth
Gums
Jaw
Tongue
Pharynx
Abdomen

(B) Somatosensory cortex

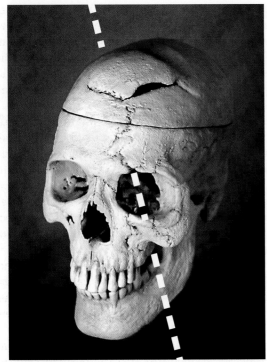

47.8 A Mind-Altering Experience *Phineas Gage miraculously survived a nine-teenth-century railroad construction accident in which an explosion blew an iron rod through his brain. His personality, however, was permanently altered. The path of the iron rod through Gage's brain is superimposed on this reconstruction of his skull.*

not cause the same degree of neglect of the right side of the body. We will see similar asymmetries in cortical function later in the chapter when we discuss language.

THE OCCIPITAL LOBE The occipital lobe receives and processes visual information. The association areas of the occipital cortex are essential for making sense of the visual world and translating visual experience into language. Some deficits resulting from damage to these areas are specific. In one case, a woman with limited damage was unable to see motion. Her vision was intact, but she could see a waterfall only as a still image, and an approaching car only as a series of a stationary objects at different distances.

The human brain is off the curve

Humans are sometimes called "big-brain primates," and that is an accurate characterization. Across vertebrate species there is a correlation between body size and brain size (**Figure 47.9**). Higher primates such as chimpanzees, baboons, and gorillas all fall above this regression line, but humans stand out because they are so far above the regression line. Gorillas are much larger than humans, but they have smaller brains. Elephants and whales have large brains, but they fall closer to the regression line. Dolphins and humans stand out as having bigger brains than would be predicted by their body sizes.

The correlation of brain size to body size does not tell the whole story of human brain evolution, however. In Figure 45.2, which compares the brains of four vertebrates, we see that the forebrain is larger than other brain regions, and in mammals this is seen as an elaboration of the cerebral cortex. If we look just at mammals, another feature is the degree of *convolution* of the cortex. Since the cortex is a layered, two-dimensional array of neurons, the area of cortex is increased by convolutions, which are greatest in humans. And finally, the percentage of the cortex that is association cortex (i.e., devoted to the integration of information) is by far the greatest in humans. It is these evolutionary changes, primarily in the cortex, that provide the resources for the great intellectual capacity of humans—a topic to which we will return at the end of the chapter.

A major association function of the parietal lobe is attending to complex stimuli. Damage to the right parietal lobe causes a condition called *contralateral neglect syndrome*, in which the individual tends to ignore stimuli from the left side of the body or the left visual field. Such individuals have difficulty performing complex tasks, such as dressing the left side of the body; an afflicted man may not be able to shave the left side of his face. When asked to copy simple drawings, a person who exhibits this syndrome can do well with the right side of the drawing but not the left.

The parietal cortex is not symmetrical with respect to its role in attention, however. Damage to the left parietal cortex does

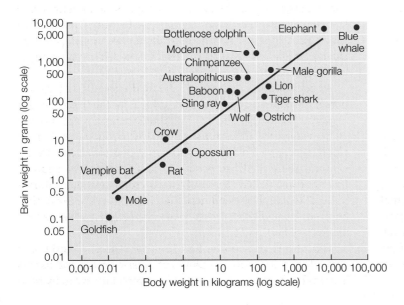

47.9 Evolution of the Human Brain *Brain size scales to body size across a wide range of vertebrates. The higher primates have larger brains than predicted by the correlation, and humans stand outside this relationship with much bigger brains. The increase in brain size in humans is mostly due to an increase in the cerebral cortex. The human brain is also highly convoluted, and more of it is devoted to associative functions.*

The central nervous system communicates with the rest of the body through the peripheral nervous system. We are conscious of some sensory information coming into the CNS, but are not conscious of other afferent information used in physiological regulation. Different regions of the brain have specific functions. Evolution of the human brain has resulted in a greatly increased cerebral cortex devoted to integration of information.

- Explain how the major functional divisions of the nervous system relate to their origins in the embryonic neural tube. See pp. 986–988 and Figure 46.2

- Trace the information flow that results in a spinal reflex. See pp. 988–989 and Figure 46.3

- Describe the spatial relations and functions of the major divisions of the telencephalon. See pp. 990–992 and Figure 47.5

- What distinguishes the human brain from the brains of other mammals? See p. 992 and Figure 47.9

Now that you have some knowledge of the structure and function of different regions of the nervous system, let's explore some examples of how information is processed in the neural circuitry in some specific regions.

47.2 How Is Information Processed by Neural Networks?

Specific functions are localized in specific parts of the nervous system and depend on the neural circuits, or networks, in those structures. A major focus of modern neuroscience is understanding how the various functions of the nervous system, ranging from simple reflexes to complex learning and memory, are accomplished by the interactions of neurons in circuits. Let's look at two examples of how neural networks process information: the autonomic nervous system (an output pathway) and the visual system (an input pathway).

The autonomic nervous system controls involuntary physiological functions

The **autonomic nervous system**, or **ANS**, comprises the output pathways of the CNS that control involuntary functions, such as heart rate, sweating, and some functions of the gut. Its control of diverse organs and tissues is crucial to homeostasis. The ANS has two divisions, **sympathetic** and **parasympathetic**, that work in opposition to each other in their effects on most organs: one division causes an increase in an activity and the other a decrease. The sympathetic and parasympathetic divisions of the ANS are easily distinguished by their anatomy, their neurotransmitters, and their actions (**Figure 47.10**).

The best-known functions of the ANS are those of the sympathetic division that produce the fight-or-flight response: increasing heart rate, blood pressure, and cardiac output and preparing the body for emergencies (see Figure 41.5). In contrast, the parasympathetic division slows the heart and lowers blood pressure; its actions have been characterized as "rest and digest." It is tempting to think of the sympathetic division as speeding things up and the parasympathetic division as slowing things down, but it is not that simple; for example, the sympathetic division slows down the digestive system whereas the parasympathetic division accelerates it.

Whether sympathetic or parasympathetic, every autonomic efferent pathway begins with a *cholinergic neuron* (one that uses acetylcholine as its neurotransmitter) that has its cell body in the brainstem or spinal cord. These cells are called *preganglionic neurons* because the second neuron in the pathway with which they synapse resides in a collection of neurons outside the CNS called a *ganglion*. The second neuron is called a *postganglionic neuron* because its axon extends out from the ganglion. The axon of the postganglionic neuron synapses with cells in the target organs (see Figure 47.10).

The postganglionic neurons of the sympathetic division are called *noradrenergic* because they use norepinephrine (also known as noradrenaline) as their neurotransmitter. In contrast, the postganglionic neurons of the parasympathetic division are mostly cholinergic. In organs that receive both sympathetic and parasympathetic input, the target cells respond in opposite ways to norepinephrine and to acetylcholine. This happens, for example, in a region of the heart called the *pacemaker*, which generates the heartbeat. Stimulating the sympathetic nerve to the heart or dripping norepinephrine onto the pacemaker cells increases their firing rate and causes the heart to beat faster. In contrast, stimulating the parasympathetic nerve to the heart or dripping acetylcholine onto the pacemaker cells decreases their firing rate and causes the heart to beat more slowly.

The sympathetic and parasympathetic divisions of the ANS can also be distinguished by anatomy. The preganglionic neurons of the parasympathetic division come from the brainstem and the *sacral* region; those of the sympathetic division come from the *thoracic* and *lumbar* regions (see Figure 47.10). Most of the ganglia of the sympathetic division are lined up in two chains, one on either side of the spinal cord. The parasympathetic ganglia are close to the target organs.

A specialization of the sympathetic division is its innervation of the adrenal gland, which is critical for the fight-or-flight response. The preganglionic sympathetic neuron sends its axon all the way out to the adrenal gland. The medulla (core) of the gland is composed of hormone-secreting cells that are really developmentally modified postganglionic sympathetic neurons that have lost their axons, and they secrete their "neurotransmitters" (epinephrine and norepinephrine) into the extracellular fluid, from which they enter the circulation and act as hormones (see Section 41.1).

The ANS is an important link between the CNS and many physiological functions. Its control of diverse organs and tissues is crucial to homeostasis. Despite its complexity, work by neu-

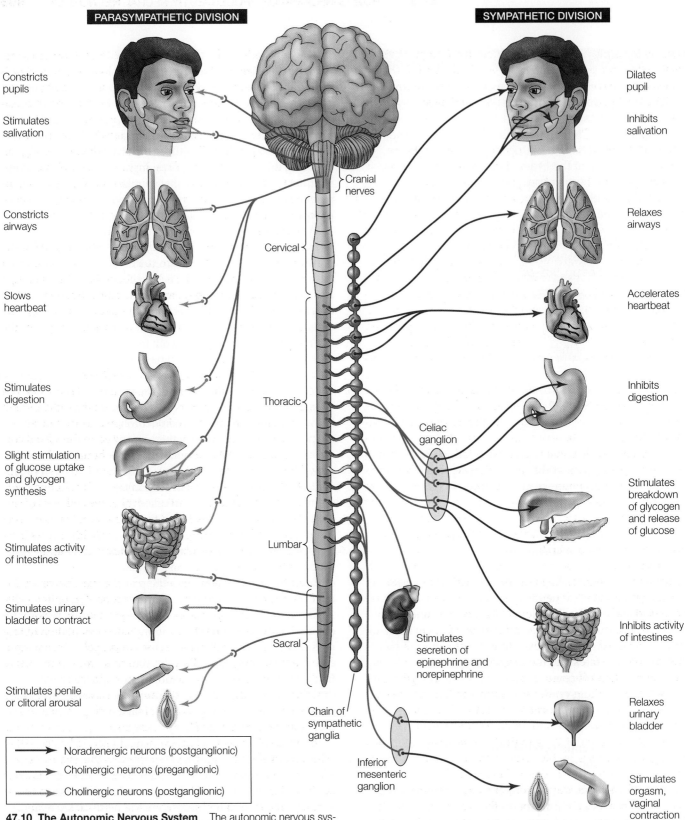

PARASYMPATHETIC DIVISION

Constricts pupils

Stimulates salivation

Constricts airways

Slows heartbeat

Stimulates digestion

Slight stimulation of glucose uptake and glycogen synthesis

Stimulates activity of intestines

Stimulates urinary bladder to contract

Stimulates penile or clitoral arousal

SYMPATHETIC DIVISION

Dilates pupil

Inhibits salivation

Relaxes airways

Accelerates heartbeat

Inhibits digestion

Stimulates breakdown of glycogen and release of glucose

Inhibits activity of intestines

Relaxes urinary bladder

Stimulates orgasm, vaginal contraction

Cranial nerves

Cervical

Thoracic

Lumbar

Sacral

Celiac ganglion

Stimulates secretion of epinephrine and norepinephrine

Chain of sympathetic ganglia

Inferior mesenteric ganglion

→ Noradrenergic neurons (postganglionic)
→ Cholinergic neurons (preganglionic)
→ Cholinergic neurons (postganglionic)

47.10 The Autonomic Nervous System The autonomic nervous system is divided into the sympathetic and parasympathetic divisions. The two divisions work in opposition to each other in their effects on most organs; one results in an increase and the other a decrease in activity.

robiologists and physiologists over many decades has made it possible to understand its functions in terms of neuronal properties and circuits.

Patterns of light falling on the retina are integrated by the visual cortex

The visual system is one of the most studied input pathways to the central nervous system. In Section 46.4 we described how light falling on the retina produces signals that are transmitted through the cellular circuits of the retina, resulting in action potentials in the optic nerve. But, how does the CNS use this information to reconstruct the visual world in the brain? The experiments that have investigated this question are some of the most famous experiments in neurobiology and have resulted in Nobel Prizes.

RETINAL RECEPTIVE FIELDS One aspect of the processing of visual information through the retina is the *convergence of information*. Each human retina contains over 100 million photoreceptors but only about 1 million *ganglion cells*. It is the axons of the ganglion cells that communicate information from the eyes to the brain. How is the information from so many photoreceptors integrated by the many fewer ganglion cells?

This question was addressed in classic experiments by Stephen Kuffler at Johns Hopkins University in 1953. He used electrodes to record the activity in the axons of single ganglion cells of cat eyes while stimulating the retinas with spots of light (**Figure 47.11**). These experiments were the starting point for understanding how the brain assembles information from single cells to create visual images—in other words, of how the brain sees.

Kuffler's experiments revealed that each ganglion cell has a well-defined **receptive field** composed of a group of photoreceptor cells that receive light from a small area of the entire visual field. Stimulating these photoreceptors with light activates the ganglion cell, which sends action potentials to the thalamus and on to the visual cortex (the area of the occipital lobe where visual information is processed; see Figure 47.5). Information from many photoreceptors is therefore communicated to the brain as a single message. However, individual photoreceptors may contribute to the receptive fields of multiple ganglion cells, so that receptive fields overlap.

The receptive fields of most ganglion cells are circular, but whether a spot of light falling on a receptive field excites or inhibits its ganglion cell depends both on the nature of the receptive field and on where the spot of light falls on it. Receptive fields have a *center* and a concentric *surround*, and can be either *on-center* or *off-center*. Light falling on the center of an on-center receptive field excites the ganglion cell, and light falling on the center of an off-center receptive field inhibits the ganglion cell. Light falling on the surround has the opposite effect: the surround for an on-center receptive field inhibits the ganglion cell, and the surround for an off-center field is excitatory. Thus the activity of the ganglion cell reflects how much of the light stimulus is on the center and how much is on the surround of its receptive field (see Figure 47.11).

Center effects are always stronger than surround effects. Thus a small dot of light directly on the center of a receptive field has the maximal effect, and a larger light stimulus illuminating the center and parts of the surround has a smaller effect. A uniform patch of light falling equally on the center and surround has very little effect on the firing rate of the ganglion cell for that receptive field.

---
yourBioPortal.com

GO TO Animated Tutorial 47.2 • Information Processing in the Retina
---

How are cells in the retina connected to each other to create receptive fields? The photoreceptors in a receptive field's center are connected to the associated ganglion cell by *bipolar cells*. The photoreceptors in the surround modify communication between the center photoreceptors and their bipolar cells through the lateral connections of *horizontal cells* and *amacrine cells* (see Figure 46.21). Thus, the receptive field of a ganglion cell results from a pattern of synapses between photoreceptors, horizontal cells, amacrine cells, and bipolar cells. A general lesson to learn from this seemingly confusing chain of events is that inhibition can be as important as excitation in neural circuits.

In summary, the neural circuitry of the retina results in the generation of signals in the axons of the optic nerve to the brain that communicate simple information about the patterns of light and dark falling on different parts of the retina. But once the action potentials in the optic nerve reach their destinations, how does the brain integrate them to construct visual images of the outside world?

RECEPTIVE FIELDS OF CELLS IN THE VISUAL CORTEX The axons of the optic nerves terminate in a region of the thalamus that is a relay station receiving information from both the right and left eyes. From the thalamus, the information encoded in the activity of axons in the optic nerves is relayed to the visual cortex in the occipital lobes at the back of the brain. In the 1960s, David Hubel and Torsten Wiesel of Harvard University studied the activity of neurons in the visual cortex by shining spots and bars of light on retinas while recording the activities of single cells in the cortex. They found that neurons in the visual cortex, like retinal ganglion cells, have receptive fields. For their pioneering work in visual neurophysiology, Hubel and Wiesel received the Nobel prize in 1981.

Receptive fields of neurons in the visual cortex, however, differ from the circular receptive fields of retinal ganglion cells. Neurons in the visual cortex called *simple cells* are maximally stimulated by bars of light with a particular orientation falling on a small region of the retina. These simple cells receive input from several ganglion cells whose circular retinal receptive fields are lined up in a row, creating a bar of sensitivity.

Other cells in the visual cortex are called *complex cells*. These cells responded maximally to bars of light with particular orientations, but the bars can fall anywhere on a large area of retina. Thus the receptive field of a complex cell appears to be due to that cell receiving inputs from several simple cells that share a certain stimulus orientation but have receptive fields in different locations on the retina (**Figure 47.12**). Some complex cells respond most strongly when the bar of light moves in a particular direction.

The concept that emerges from these experiments is that the brain assembles a mental image of the visual world by analyzing edges in patterns of light falling on the retina. Each retina sends a million axons to the brain, but there are *hundreds of millions* of neurons in the visual cortex. The action potentials from one retinal ganglion cell are received by hundreds of cortical neurons, each responsive to a different combination of orientation, position, color, and movement of contrasting lines in the patterns of light and dark falling on the retina.

INVESTIGATING LIFE

47.11 What Does the Eye Tell the Brain?

Stephen Kuffler's experiments recorded the activity in the axons of single ganglion cells in the eyes of cats. These groundbreaking experiments revealed the existence of a circular receptive field for each of the retina's ganglion cells. Signals from photoreceptor cells in a receptive field are either excitatory or inhibitory to the ganglion cell, which sends action potentials via the optic nerve to the brain.

HYPOTHESIS Retinal ganglion cells are excited or inhibited by light and dark stimuli falling on local areas of the retina.

METHOD

1. Place electrodes in the optic nerve of a cat.
2. Stimulate the retina with different combinations of light and dark stimuli and record the responses of ganglion cell axons.
3. Continue recording and move the stimuli around the retina to find the area of sensitivity—the receptive field—for a specific ganglion cell.

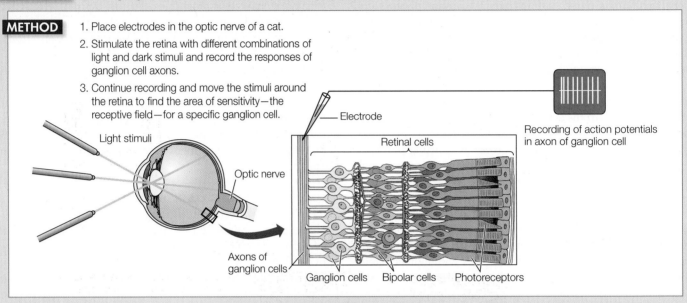

RESULTS

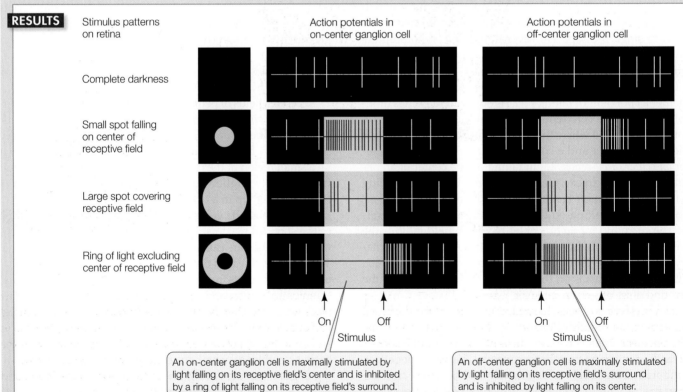

An on-center ganglion cell is maximally stimulated by light falling on its receptive field's center and is inhibited by a ring of light falling on its receptive field's surround.

An off-center ganglion cell is maximally stimulated by light falling on its receptive field's surround and is inhibited by light falling on its center.

CONCLUSION Ganglion cells use a center-surround dichotomy to encode patterns of contrast between light and dark.

Go to **yourBioPortal.com** for original citations, discussions, and relevant links for all INVESTIGATING LIFE figures.

INVESTIGATING LIFE

47.12 Cells in the Visual Cortex Respond to Specific Patterns of Light

Torsten and Wiesel showed that information from light falling on the circular receptive fields of ganglion cells (see Figure 47.11) converges on the visual cortex's simple cells, which have linear receptive fields. These simple cells transmit information to complex cells in the cortex, which in turn respond to linear stimuli falling on different areas of the retina.

HYPOTHESIS Signals from light falling on receptive fields of the ganglion cells elicit responses from cells in the visual cortex.

METHOD

1 The bar of light moves across the screen...

2 ...stimulating receptive fields in the retina.

3 As the cat views the screen, the electrode records activity in single cells in the occipital cortex...

4 ...and displays it on an oscilloscope.

On-center response of a cell in the occipital cortex

RESULTS Ganglion cells communicate with thalamic cells that relay information to the visual cortex. The area stimulated by each bar of light includes the receptive fields of several ganglion cells.

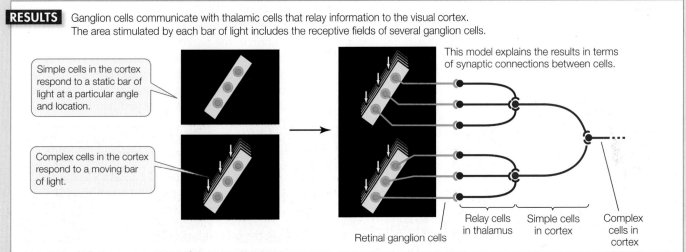

Simple cells in the cortex respond to a static bar of light at a particular angle and location.

Complex cells in the cortex respond to a moving bar of light.

This model explains the results in terms of synaptic connections between cells.

Retinal ganglion cells

Relay cells in thalamus

Simple cells in cortex

Complex cells in cortex

CONCLUSION Cells in the retina, thalamus, and cortex are connected in such a way as to respond to specific patterns of light.

Go to **yourBioPortal.com** for original citations, discussions, and relevant links for all INVESTIGATING LIFE figures.

Cortical cells receive input from both eyes

How do we see objects in three dimensions? The quick answer is that a person's two eyes, located at the front of the head, see overlapping, yet slightly different, visual fields—that is, humans have **binocular vision**. A person who is blind in one eye has great difficulty discriminating distances. Animals whose eyes are on the sides of the head have minimal overlap in their fields of vision and, as a result, poor depth vision; however, they can see predators creeping up from all sides.

The story of how the brain integrates information from two eyes begins with the paths of the optic nerves. The two optic nerves run along the underside of the brain, join just under the hypothalamus, and then separate again (**Figure 47.13A**). The place where they join is called the **optic chiasm**. Axons from the half of each retina closest to your nose cross in the optic chiasm and go to the opposite side of your brain. The axons from the outer half of each retina do not cross over at the optic chiasm; axons from the outer left retina go to left side of the brain, and vice versa for axons from the outer right retina.

The functional consequence of the optic chiasm is that all of the visual information from the left side of your field of vision when you are looking straight ahead goes to the right side of your brain, and all of the visual information from the right side

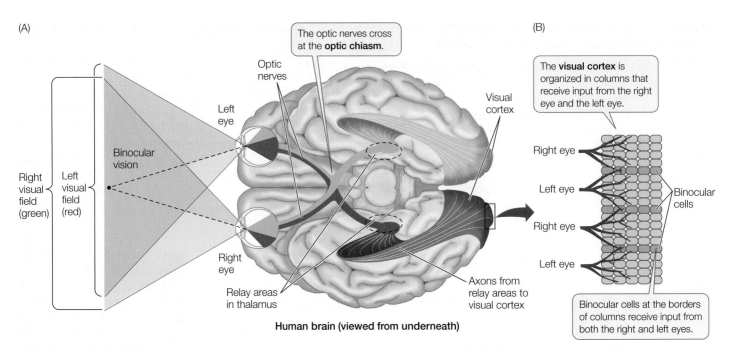

47.13 Anatomy of Binocular Vision (A) Each eye transmits information to both sides of the brain; however, the right side of the brain processes all information from the left visual field (red), and the left side of the brain processes all information from the right visual field (green). (B) The visual cortex sorts visual field information according to whether it comes from the right or left eye.

of your field of vision goes to the left side of your brain. These relationships are shown in red and green in Figure 47.13A.

Cells in the visual cortex are organized in stripes and columns. Stripes refer to the organization across the surface of the cortex, and columns to the organization through the depth of the cortex (**Figure 47.13B**). Stripes and columns alternate according to the source of their input: left eye, right eye, left eye, right eye, and so on. Cells closest to the border between two stripes or columns receive input from both eyes and are therefore called *binocular cells*. Binocular cells interpret distance by measuring the *disparity* between the points at which the same stimulus falls on the two retinas.

What is disparity? Hold your finger out in front of you and look at it, closing one eye and then the other. Your finger appears to jump back and forth, because its image falls on a different position on each retina. Repeat the exercise with a distant object. It doesn't jump back and forth as much, because there is less disparity in the positions of the image on the two retinas. Certain binocular cells respond optimally to a stimulus falling on both retinas with a particular disparity. Which set of binocular cells is stimulated depends on how far away the stimulus is.

When we look at something, we can detect its shape, color, depth, and movement. Where does all this information come together? Is there a single cell that fires only when a red sports car drives by? No. Specific visual experience comes from simultaneous activity in a large collection of cells. In addition, most visual experiences are enhanced by information from the other senses and from memory, which helps explain why about 75 percent of the cerebral cortex is association cortex.

47.2 RECAP

Information in the nervous system is processed by cellular interactions in neural networks. The opposing actions of the sympathetic and parasympathetic divisions of the ANS can be understood in terms of neural pathways consisting of just two neurons. Vision involves a more complex interaction of neurons, organized into receptive fields, to process patterns of light and dark falling on the retina.

- Describe the anatomical and functional differences between the sympathetic and parasympathetic divisions. See p. 993 and Figure 47.10

- Explain the cellular basis for the receptive fields of retinal ganglion cells. See p. 995 and Figure 47.11

- How do cells in the visual cortex get information about how far away an object is? See pp. 997–998 and Figure 47.13

By studying the neural circuitry of the visual system and the ANS, you have gained some understanding of how information reaches the CNS and how the CNS controls various functions of the body. But what about the higher functions of the mammalian CNS—the complex functions between input and output, such as language, learning, memory, and dreams?

47.3 Can Higher Functions Be Understood in Cellular Terms?

The higher brain functions discussed in the remaining pages of this chapter are undeniably complex. Nevertheless, neuroscientists, using a wide range of techniques, are making considerable progress in understanding some of the cellular and molecular mechanisms involved in those processes. The following discussion addresses several aspects of brain and behavior that present challenges to neuroscientists: sleep and dreaming, learning and memory, language use, and consciousness.

Sleep and dreaming are reflected in electrical patterns in the cerebral cortex

A dominant feature of behavior is the daily cycle of sleep and waking. All birds and mammals, probably all other vertebrates, and also many invertebrates, sleep. We humans spend one-third of our lives sleeping, yet we do not know why or how. We do know, however, that we need to sleep. Loss of sleep impairs alertness and performance. Many people in our society—certainly most college students—are chronically sleep-deprived. Every day, accidents and serious mistakes that endanger lives can be attributed to impaired alertness caused by lack of sleep. Insomnia (difficulty in falling or staying asleep) is one of the most common medical complaints.

THE ELECTROENCEPHALOGRAM A common tool of sleep researchers is the **electroencephalogram**, or **EEG** (**Figure 47.14A**). Rather than recording the activity of single neurons, the EEG characterizes activity in huge numbers of neurons. EEG electrodes are much larger than the very fine electrodes used to detect single cell activity. Placed at different locations on the scalp, EEG electrodes record changes in the electric potential differences between electrodes over time. These differences reflect the electrical activity of the neurons in the brain regions under the electrodes, primarily regions of the cerebral cortex. Usually the electrical activity of one or more skeletal muscles is also recorded on the chart; this record is called an *electromyogram*

(EMG). Movements of the eyes are recorded as an *electrooculogram* (EOG).

EEG, EMG, and EOG patterns reveal the transition from being awake to being asleep. They also reveal that there are different states of sleep. In mammals other than humans, two major sleep states are easily distinguished: **slow-wave sleep** and **rapid-eye-movement (REM) sleep**. Slow-wave sleep gets its name from the high-amplitude, slow-frequency waves in the EEG. REM sleep gets its name from jerky movements of the eyeballs that occur during this state. In humans, sleep states are characterized as non-REM sleep and REM sleep. Human non-REM sleep has four stages, of which only the two deepest stages are considered true slow-wave sleep.

When you fall asleep at night, the first sleep state entered is stage 1 non-REM sleep, which then progresses through stages 2, 3, and 4 (**Figure 47.14B**). Stages 3 and 4 are deep, restorative, slow-wave sleep. This first full cycle of non-REM sleep is followed by an episode of REM sleep. Throughout the night, you experience four or five cycles of non-REM and REM sleep (**Figure 47.14C**). About 80 percent of your sleep is non-REM sleep. Vivid dreams and nightmares occur during the 20 percent of sleep that is REM sleep.

CELLULAR CHANGES DURING SLEEP When we are awake, several nuclei in the brainstem reticular formation are continuously active. Axons from neurons in these nuclei extend to the thalamus and throughout the cerebral cortex, where they release depolarizing neurotransmitters (acetylcholine, norepinephrine, and serotonin). These broadly distributed neurotransmitters keep the resting potential of the neurons of the thalamus and cortex close to threshold and sensitive to synaptic inputs, thereby maintaining the responsiveness of the brain that characterizes being awake.

(A)

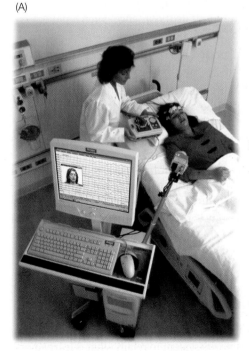

47.14 Stages of Sleep (A) Electrical activity in the cerebral cortex is detected by electrodes that are placed on the scalp and record changes in voltage between the electrodes through time. (B) The resulting record is an electroencephalogram, or EEG. (C) Humans cycle through different stages of sleep throughout the night.

(B)

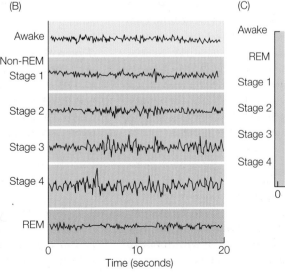

(C)

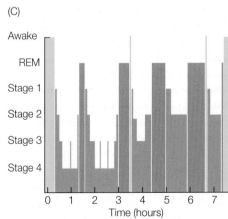

With the onset of sleep, activity in these brainstem nuclei decreases, and their axon terminals release less neurotransmitter. With the withdrawal of the depolarizing neurotransmitters, the resting potentials of the cells of the thalamus and cortex become more negative (hyperpolarized), and the cells are less sensitive to excitatory synaptic input. Their processing of information is inhibited, and consciousness is lost.

An interesting neural event happens as a result of the hyperpolarization with the onset of sleep: cells begin to fire action potentials in bursts. The synchronization of these bursts over broad areas of cerebral cortex results in the EEG slow-wave pattern that characterizes deep non-REM sleep. Studies of neurons of the thalamus and cortex have shown that their hyperpolarization during non-REM sleep is due to increased opening of K^+ channels, and that bursting is due to Ca^{2+} channels whose inactivation gates close rapidly and require hyperpolarization to be reopened. We can therefore explain the EEG pattern of non-REM sleep in terms of the properties of neurons and ion channels.

At the transition from non-REM to REM sleep, dramatic changes occur. Some of the brainstem nuclei that were inactive during non-REM sleep become active again, causing a general depolarization of cortical neurons. Thus in REM sleep the synchronized bursts of firing cease, and the EEG resembles that of the awake brain. Because the resting potentials of the neurons return to near threshold levels, the cortex can process information, and vivid dreams occur.

So, why don't we act out our dreams? During REM sleep the brain inhibits both afferent (sensory) and efferent (motor) pathways; we are paralyzed during REM sleep. Limb twitches and the jerky eye movements are motor signals breaking through the inhibition. The bizarre nature of dreams may be due to the lack of sensory feedback to the cortex from the body and the outside world. In other words, a functioning cortex is out of touch with reality. The function of muscle paralysis during REM sleep may be to prevent the acting out of dreams.

Knowing the cellular mechanisms of sleep has not yet led to an understanding of its function. Many questions remain. Why do we have two sleep states with very different neurophysiological characteristics? Why does non-REM sleep always occur first? Why do the two states cycle during the rest period? We know sleep is essential for life, but we don't know why. One set of hypotheses is that sleep is necessary for the maintenance and repair of neural connections and for the neural changes involved in learning and memory—and possibly forgetting. These hypotheses are supported by many experiments showing that performance of a learned task or recall of declarative information on the day following training is impaired if sleep is prevented, and is best following a good night's sleep.

Language abilities are localized in the left cerebral hemisphere

No aspect of brain function is as integrally related to human consciousness and intellect as language. Therefore, brain mechanisms that underlie the acquisition and use of language are extremely interesting to neuroscientists. A curious observation about language ability is that it resides in one cerebral hemisphere—which in 97 percent of people is the left hemisphere. This phenomenon is referred to as the *lateralization* of language functions.

Fascinating research on this subject was conducted by Roger Sperry and his colleagues at the California Institute of Technology. The two cerebral hemispheres are connected by a tract of white matter called the *corpus callosum*. In one severe form of epilepsy, bursts of action potentials travel between hemispheres via the corpus callosum. Cutting the tract eliminates the problem, and patients function well following surgery. However, these "split-brain" subjects display interesting deficits in language ability.

After the surgery, if an object is shown in the right visual field and the left eye is closed (see Figure 47.13), the patient can describe it verbally and in writing. If the object is shown in the left visual field and the right eye is closed, the patient cannot describe it verbally or in writing, but can use his left hand to point to a picture of it. Lacking the connecting tissue between the two hemispheres, knowledge or experience of the right hemisphere can no longer be expressed in language.

Individuals who have suffered damage to the left hemisphere are frequently left with some form of **aphasia**, a deficit in the ability to use or understand words. Studies of such individuals have identified several language areas in the left hemisphere.

- **Broca's area**, located in the frontal lobe just in front of the primary motor cortex, is essential for speech. Damage to Broca's area results in halting, slow, poorly articulated speech or even complete loss of speech, but the patient can still read and understand language.

- **Wernicke's area**, located in the temporal lobe close to its border with the occipital lobe, is more involved with sensory than with motor aspects of language. Damage to Wernicke's area can cause a person to lose the ability to speak sensibly while retaining the abilities to form the sounds of normal speech and to imitate its cadence. Such a patient cannot understand spoken or written language.

- The **angular gyrus**, located near Wernicke's area, is believed to be essential for integrating spoken and written language.

Normal language ability depends on the flow of information among various areas of the left cerebral cortex. Input from spoken language travels from the auditory cortex to Wernicke's area (**Figure 47.15A**). Input from written language travels from the visual cortex to the angular gyrus to Wernicke's area (**Figure 47.15B**). Commands to speak are formulated in Wernicke's area and travel to Broca's area and from there to the primary motor cortex. Damage to any one of those areas or the pathways between them can result in aphasia. Using modern methods of brain imaging, it is possible to see the metabolic activity in different brain areas when the brain is using language (**Figure 47.16**).

Some learning and memory can be localized to specific brain areas

Learning is the modification of behavior by experience. *Memory* is the ability of the nervous system to retain what is learned and

(A) Repeating a heard word

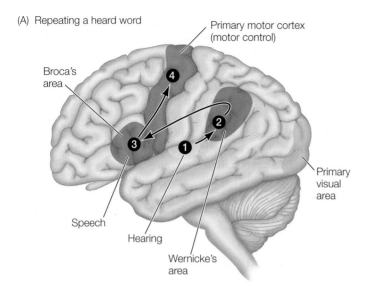

(B) Speaking a written word

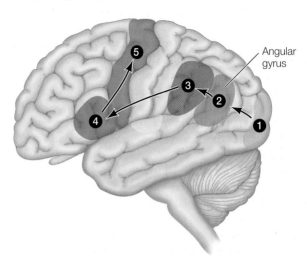

47.15 Language Areas of the Cortex Different regions of the left cerebral cortex participate in the processes of (A) repeating a word that is heard and (B) speaking a written word.

yourBioPortal.com
GO TO Web Activity 47.2 • Language Areas of the Cortex

experienced. Even very simple animals can learn and remember, but these two abilities are most highly developed in humans. Consider the amount of information associated with learning a language. The capacity of memory and the rate at which memories can be retrieved are remarkable features of the human nervous system.

LEARNING Learning that leads to long-term memory and modification of behavior must involve long-lasting synaptic changes. A phenomenon that may explain how long-term synaptic changes might arise is long-term potentiation, or LTP (see Figure 45.18). LTP results from high-frequency electrical

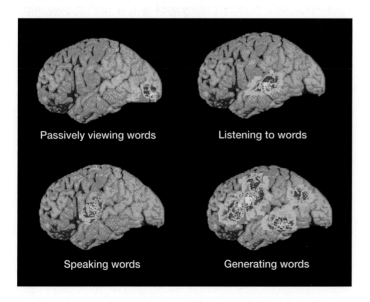

47.16 Imaging Techniques Reveal Active Parts of the Brain
Positron emission tomography (PET) scanning reveals the brain regions activated by different aspects of language use. Radioactively labeled glucose is given to the subject. Brain areas take up radioactivity in proportion to their metabolic use of glucose. The PET scan visualizes levels of radioactivity in specific brain regions when a particular activity is performed. The red and white areas are the most active.

stimulation of certain identifiable circuits that makes these circuits more sensitive to subsequent stimulation. In contrast, continuous, repetitive, low-level stimulation of these same circuits reduces their responsiveness, a phenomenon that has been called **long-term depression (LTD)**. LTP and LTD may be fundamental cellular or molecular mechanisms involved in learning and memory.

Several kinds of learning exist. A form that is widespread among animal species is **associative learning**, in which two unrelated stimuli become linked to the same response. The simplest example of associative learning is the **conditioned reflex**, described by the Russian physiologist Ivan Pavlov. Pavlov observed that a dog salivates at the sight or smell of food—a simple autonomic reflex. He discovered that if he rang a bell just before food was presented to the dog, after a few trials the dog would salivate at the sound of the bell, even if no food followed. The salivation reflex was conditioned to be associated with the sound of a bell, a stimulus that normally is unrelated to feeding and digestion. Experiments in the laboratory of Richard Thompson, now at University of Southern California, located for the first time the circuitry for a conditioned reflex, and it is in the cerebellum.

More complex forms of learning, often referred to as "observational learning" are the foundation of human intelligence. The general pattern of successful observational learning has three elements:

• We pay attention to another person's behavior.

• We retain a memory of what we have observed.

• We try to copy or use that information.

A key to this scheme of learning is the way in which we create and recall memories.

MEMORY Some of the first insights into memory processes came from clinical treatment of patients with severe epilepsy, a disorder characterized by uncontrollable, local increases in neural activity. The resulting *seizures* can endanger the afflicted individual. Serious cases of epilepsy are sometimes treated by destroying the part of the brain from which the surge of activity originates. To find the right area, the surgery is done under local anesthesia, with the patient remaining conscious. As different regions of the brain are electrically stimulated with electrodes, the patient reports the resulting sensations. Stimulation of some regions of the association cortex elicit recall of vivid memories. Such observations provided the first evidence that specific areas in the brain are associated with specific memories and that memory can be attributed to networks of neurons. Destroying a small area of the brain does not completely erase a memory, however, so it is postulated that memory is a function distributed over many brain regions and can be stimulated via many different routes.

You experience several forms of memory everyday. You have **immediate memory** for events that are happening now. Immediate memory is almost perfectly photographic, but it lasts only seconds. **Short-term memory** contains less information but lasts longer—on the order of 10 to 15 minutes. If you are introduced to a group of new people, you may remember most of their names for 5 or 10 minutes, but you will have forgotten them in an hour or so if you have not repeated them, written them down, or used them in a conversation. Repetition, use, or reinforcement by something that gets your attention (such as the title "president") facilitates the transfer of short-term memory to **long-term memory**, which can last for days, months, years, or a lifetime.

Knowledge about neural mechanisms for the transfer of short-term memory to long-term memory has come from observations of persons who have lost parts of the limbic system, notably the hippocampus. A famous case is that of the man identified as H.M., whose hippocampus on both sides of the brain was removed in 1953 in an effort to control severe epilepsy. After the surgery, H.M. was unable to transfer information to long-term memory. If someone was introduced to him, had a conversation with him, and then left the room for several minutes, when that person returned, H.M did not recognize him—it was as if the conversation had never taken place. Up until his death 55 years later, H.M. remembered events that happened before his surgery but could not remember postsurgery events for more than 10 or 15 minutes.

Memory of people, places, events, and things is called **declarative memory** because you can consciously recall and describe them. **Procedural memory** cannot be consciously recalled and described; it is the memory of how to perform a motor task. When you learn to ride a bicycle, ski, or use a computer keyboard, you form procedural memories. Although H.M. was incapable of forming declarative memories, he could form procedural memories. When taught a motor task day after day, he could not recall the lessons of the previous day, yet his performance steadily improved. Thus procedural learning and memory must involve mechanisms different from those used in declarative learning and memory.

Memories can have considerable emotional content. As mentioned earlier, the limbic system plays a major role in controlling emotions. The amygdala, a component of the limbic system, is necessary for the emotion of fear and the formation of fear memories. As we saw in the opening to Chapter 45, patients with a damaged amygdala do not associate fear reactions with their declarative memories. Memories can also have positive emotional content, and recalling those memories activates parts of the brain known to be associated with pleasurable sensations and reward, as revealed by brain imaging technologies.

We still cannot answer the question "What is consciousness?"

This chapter has only scratched the surface of the organization and functions of the human brain. Even with all of our knowledge of the human brain, and with all of the sophisticated new research tools, we still cannot answer the question "What is consciousness?"

The word "consciousness" is used in everyday language to refer to being awake in contrast to being asleep or in a coma. Here we are referring to the deeper meaning of being mentally aware of yourself, your environment, and events going on around you in such a way that you can plan for future events and make decisions based on experience, evidence, value systems, and predicted consequences. Speculations about consciousness have been the realm of philosophers, but we are getting closer to a neurobiological understanding.

The central requirement for conscious experience is a perception of self that can be integrated with information from the physical and social environment and information from past experience. The basis for a perception of self derives from the huge amount of somatosensory and visceral information that comes from all parts of the body. In the CNS of all vertebrates, this information is used for motor control and for homeostatic regulation. It enables animals to find food, seek mates, seek warmth, avoid cold, avoid danger, and so on. This afferent information goes to appropriate control and regulatory systems in the brainstem and forebrain.

In addition, some of this information goes to somatosensory areas of the cerebral cortex, so the animal is aware of certain information in the sense that it responds to it behaviorally. Visceral afferent information goes beyond its regulatory and control centers in the brainstem and hypothalamus to an area deep within the forebrain called the *insular cortex*, or *insula*. The insula appears to integrate physiological information from all over the body to create a sensation of how the body "feels." Thus, when an animal's actions restore homeostasis, it "feels" better, and this is motivation to do the right thing for well-being.

In humans and the great apes, the insula is greatly expanded and has even acquired new types of spindle-shaped neurons not seen in other animals. The circuitry involving the insula has also evolved to communicate with parts of the brain that are involved in planning and decision making. In imaging studies, the insula is seen to be active in a great diversity of situations that involve strong feelings, be they pleasure, disgust, humor, pain, lust, craving, humiliation, guilt, or empathy. Damage to the insula causes apathy, loss of ability to enjoy music, loss of sexual response, and even loss of the ability to distinguish good food from spoiled food. Humans and the few other species that have expanded insulas and the new spindle cells are the only species that can recognize themselves in a mirror. Could it be that this very discrete part of our brains and its circuitry are the neurobiological bases for self-awareness and conscious experience?

47.3 RECAP

Even complex functions of the nervous system are beginning to be understood in terms of the properties of neurons and neural networks.

- What events in the brain are associated with wakefulness and the stages of sleep? See p. 999

- Why do some neurobiologists think that the insular cortex might be involved in conscious experience? See p. 1002–1003

CHAPTER SUMMARY

47.1 How Is the Mammalian Nervous System Organized?

- The brain and spinal cord make up the central nervous system (CNS); the cranial and spinal nerves make up the peripheral nervous system (PNS).

- The nervous system can be modeled conceptually in terms of the direction of information flow and whether we are conscious of the information. The afferent component carries information from the PNS to the CNS, and the efferent component directs information from the CNS to the peripheral parts of the body. **Review Figure 47.1**

- The vertebrate nervous system develops from a hollow dorsal neural tube. The brain forms from three swellings at the anterior end of the neural tube, which become the **hindbrain**, the **midbrain**, and the **forebrain**. The forebrain develops into the **cerebral hemispheres** (the **telencephalon**, or **cerebrum**) and the underlying **thalamus** and hypothalamus (which together compose the **diencephalon**). The midbrain and hindbrain develop into the brainstem and the **cerebellum**. **Review Figure 47.2**

- The spinal cord communicates information between the brain and the rest of the body. It can issue some commands to the body without input from the brain. **Review Figure 47.3, ANIMATED TUTORIAL 47.1**

- The **reticular system** is a complex network that directs incoming information to appropriate brainstem **nuclei** that control autonomic functions, and transmits the information to the forebrain that results in conscious sensation. The reticular system controls the level of arousal of the nervous system, including sleep and wakefulness.

- The **limbic system** is an evolutionarily primitive part of the telencephalon that is involved in emotions, physiological drives (such as hunger or thirst), instincts, and memory. **Review Figure 47.4**

- The cerebral hemispheres are the dominant structures of the human brain. Their surfaces are layers of neurons called the **cerebral cortex**. The cerebral hemispheres can be divided into the **temporal**, **frontal**, **parietal**, and **occipital lobes**. Many motor functions are localized in parts of the frontal lobe. Information from many sensory receptors projects to a region of the parietal lobe. Visual information projects to the occipital lobe, and auditory information projects to a region of the temporal lobe. **Review Figures 47.5, 47.6, and 47.7, WEB ACTIVITY 47.1**

47.2 How Is Information Processed by Neural Networks?

- The **autonomic nervous system** (ANS) consists of efferent pathways that control the physiological function of organs and organ systems. Its **sympathetic** and **parasympathetic** divisions are characterized by their anatomy, neurotransmitters, and effects on target tissues. **Review Figure 47.10**

- The neural network of vision involves patterns of light falling on **receptive fields** in the retina. Receptive fields have a center and a surround, which have opposing effects on ganglion cell firing. **Review Figure 47.11, ANIMATED TUTORIAL 47.2**

- Information from retinal ganglion cells is communicated via the optic nerve to the thalamus and then to the visual cortex. The visual cortex seems to assemble an image of the visual world by analyzing edges of patterns of light. **Review Figure 47.12**

- **Binocular vision** is possible because information from both eyes is communicated to binocular cells in the visual cortex. These cells interpret distance by measuring the disparity between where the same stimulus falls on the two retinas. **Review Figure 47.13**

47.3 Can Higher Functions Be Understood in Cellular Terms?

- Humans have a daily cycle of sleep and waking. Sleep can be divided into **rapid-eye-movement (REM) sleep** and **non-REM sleep.** Deep non-REM sleep is known as slow-wave sleep because of its characteristic EEG patterns. **Review Figure 47.14**

- Language abilities are localized mostly in the left cerebral hemisphere, a phenomenon known as lateralization. Different areas of the left hemisphere—including **Broca's area**, **Wernicke's area**, and the **angular gyrus**—are responsible for different aspects of language. **Review Figure 47.16, WEB ACTIVITY 47.2**

- Some learning and memory processes have been localized to specific brain areas. Long-lasting changes in synaptic properties referred to as **long-term potentiation** (LTP) and **long-term depression** (LDP) may be involved in learning and memory.

- Complex memories can be elicited by stimulating small regions of association cortex. Damage to the hippocampus can destroy the ability to form long-term **declarative memory** but not **procedural memory**.

- A sense of the physiological state of the body may be created in the insular cortex from visceral afferent information. Evolution of this integrative function in higher primates and humans could be the basis for conscious experience.

SEE WEB ACTIVITY 47.3 for a concept review of this chapter.

SELF-QUIZ

1. Which of the following describes the route of sensory information from the foot to the brain?
   a. Ventral horn, spinal cord, medulla, cerebellum, midbrain, thalamus, parietal cortex
   b. Dorsal horn, spinal cord, medulla, pons, midbrain, hypothalamus, frontal cortex
   c. Dorsal horn, spinal cord, medulla, pons, midbrain, thalamus, parietal cortex
   d. Ventral horn, spinal cord, pons, cerebellum, midbrain, thalamus, parietal cortex
   e. Dorsal horn, spinal cord, medulla, pons, midbrain, thalamus, frontal cortex

2. Which statement about the reticular system is *not* true?
   a. Increased activity in the reticular system induces sleep.
   b. The reticular system is located in the brainstem.
   c. Damage to the reticular system in the midbrain can result in coma.
   d. Information from the spinal cord is routed to different nuclei in the reticular system and to the forebrain.
   e. There are groups of neurons called nuclei in the reticular system.

3. Which statement about afferent and efferent pathways is *not* true?
   a. Sympathetic and parasympathetic pathways carry only efferent information.
   b. Visceral afferents carry information about physiological functions of which we are not consciously aware.
   c. The voluntary division of the efferent portion of the peripheral nervous system executes conscious movements.
   d. The cranial nerves and spinal nerves are part of the peripheral nervous system.
   e. Afferent and efferent axons never travel in the same nerve.

4. Which statement about the limbic system is *not* true?
   a. Damage to one structure in the limbic system makes it impossible to form a fear memory.
   b. The limbic system is involved in basic physiological drives, instincts, and emotions.
   c. The limbic system consists of primitive forebrain structures.
   d. In humans, the limbic system is the largest part of the brain.
   e. In humans, a part of the limbic system is necessary for the transfer of short-term memory to long-term memory.

5. Which of the following represents the largest portion of the human cerebral cortex?
   a. The frontal lobes
   b. The primary somatosensory cortex
   c. The temporal cortex
   d. The association cortex
   e. The occipital cortex

6. Which statement about the autonomic nervous system is *true*?
   a. The sympathetic division is afferent, and the parasympathetic division is efferent.
   b. The transmitter norepinephrine is always excitatory, and acetylcholine is always inhibitory.
   c. Each pathway in the autonomic nervous system includes two neurons, and the neurotransmitter of the first neuron is acetylcholine.
   d. The cell bodies of many sympathetic preganglionic neurons are in the brainstem.
   e. The cell bodies of most parasympathetic postganglionic neurons are in or near the thoracic and lumbar spinal cord.

7. Which statement about cells in the visual cortex is *not* true?
   a. Many cortical cells receive inputs directly from single retinal ganglion cells.
   b. Many cortical cells respond most strongly to bars of light falling at specific locations on the retina.
   c. Some cortical cells respond most strongly to bars of light falling anywhere over large areas of the retina.
   d. Some cortical cells receive input from both eyes.
   e. Some cortical cells respond most strongly to an object when it is a certain distance from the eyes.

8. Which of the following characterizes non-REM sleep?
   a. Dreaming
   b. Paralysis of skeletal muscles
   c. EEG slow waves
   d. Rapid and jerky eye movements
   e. It makes up about 20 percent of total sleep time.

9. Which conclusion was supported by experiments on split-brain patients?
   a. Language abilities are localized mostly in the left cerebral hemisphere.
   b. Language abilities require both Wernicke's area and Broca's area.
   c. The ability to speak depends on Broca's area.
   d. The ability to read depends on Wernicke's area.
   e. The left hand is served by the left cerebral hemisphere.

10. In the withdrawal reflex,
    a. action potentials in a pain sensory neuron enter the spinal cord through the ventral root on the same side of the body.
    b. the axon from a pain sensory neuron makes inhibitory synapses with motor neurons on the same side of the spinal cord and inhibitory synapses with motor neurons on the other side of the spinal cord.
    c. coordinated escape reactions can be initiated before the brain registers the painful sensation.
    d. axons from the pain sensory neuron make synapses with sympathetic preganglionic neurons in the spinal cord.
    e. axons from motor neurons on the same side of the spinal cord that receives the input from the pain receptors cross the spinal cord to stimulate muscles on the other side of the body.

FOR DISCUSSION

1. A person receives a stab wound to the left side of his neck. Miraculously, blood vessels are spared. Afterward, however, the man's left pupil remains more constricted than his right pupil, and he drools out of the left side of his mouth. How can you explain these symptoms?

2. The stretch receptors in muscles are modified muscle cells, and they have their own motor neurons. What is the function of those motor neurons? In thinking about this question, remember that the function of the monosynaptic reflex is to adjust muscle tension to a change in load so that the position of the limb does not change.

3. A patient is unable to speak coherently. She can read and write and has no obvious loss of muscle function. Where would you expect to find an abnormality if you did brain scans of this patient?

4. How can you investigate how visual images falling on the retina are communicated to the brain? Start with the retina, and imagine how you would continue your investigation in visual areas of the brain.

5. We described the organization of the visual cortex as columns of cells that alternately receive input from the left and right eye. If a young kitten is allowed to see light out of only one eye for a day, more synapses are maintained in the cortical columns receiving input from that eye, while synapses decrease in the intervening columns. This redistribution of synapses does not occur, however, if the kitten is not allowed to sleep. What hypotheses could you propose on the basis of these results?

ADDITIONAL INVESTIGATION

The images in Figure 47.16 were made using technology that enables us to image regional activity in the brain. How would you use this imaging methodology to investigate a higher brain function that is of interest to you? First, formulate your hypothesis. Then design your investigation, keeping in mind that your subject has to lie motionless in a large machine during the imaging. You will have to plan how you will deliver the appropriate stimulus to elicit the response or activity you are interested in. Also, describe what controls you will use to draw clear conclusions about the stimulus–response relationships you expect to observe.

48 Musculoskeletal Systems

Champion jumpers

The Olympic record for the women's long jump is 7.4 meters, set in 1988 by Jackie Joyner-Kersee. Another world-record long jump that still stands was set two years earlier by Rosie the Ribeter, who jumped 6.5 meters. Rosie was a frog competing in the Calaveras County Jumping Frog Contest. In some ways, Rosie's jump is more impressive: while Jackie's jump was about 5 times her body length (i.e., her height), Rosie's was about *20 times* her body length.

Both jumps were powered by skeletal muscle. Muscle tissue responds to commands from the nervous system. The cellular mechanisms of muscle contraction are essentially the same in the frog and the human, so why is the frog's jump so much more impressive? The answer involves the concept of *leverage*, which depends on the muscles and skeletal elements working together.

Both frog and human jumping muscles pull on bones that are connected at joints to make levers. A lever makes it possible for the same force to move a large mass a small distance or a small mass a large distance. The ratio of a frog's leg length to its body mass is simply greater than that in a human. Thus the frog's legs are better at moving a small mass a long distance than are the human's legs.

Let's add a flea to our interspecies competition. The flea can jump more than 200 times its body length. This incredible performance is not due to feats of leverage, because no muscle can contract fast enough to explain the take-off velocity of the flea. A different mechanism evolved in the flea—a kind of slingshot action. At the base of the flea's jumping legs is an elastic material that is compressed by muscles while the flea is resting. When a trigger mechanism is released, the elastic material recoils and "fires" the flea into the air.

In a contest of jumping endurance, the uncontested champion would be the kangaroo. As a human runs faster, the number of strides and the energy expended per minute increase rapidly. Neither is true for the kangaroo. When moving at speeds from about 5 to 25 kilometers per hour, the kangaroo takes the same number of strides per minute and its metabolic rate does not increase. Why is this?

In kangaroos, as in frogs and humans, the muscles used to jump are attached to bones by tendons. Like the material at the base of the flea's

A Champion Jumper Jackie Joyner-Kersee set an Olympic record for the women's long jump at the Seoul Olympics in 1988.

Champion Jumpers   Relative to their size, many animals have more impressive jumping skills than humans. This leopard frog (*Rana pipiens*) can leap distances up to 20 times its body length.

legs, tendons can be elastic. The kangaroo's tendons stretch when it lands, and their recoil helps power the next jump—similar to the action of a pogo stick. To move faster, the kangaroo simply increases the length of its stride, thereby increasing both the stretch on its tendons each time it lands and the magnitude of the recoil at the initiation of each jump.

As discussed in Chapter 31, the ability to move is one of the things that distinguishes animals from the other multicellular organisms. It is our muscles and skeleton—the *musculoskeletal system*—that allow us to move.

IN THIS CHAPTER we will describe muscle structure and the molecular mechanisms of contraction. We will discuss the properties that enable different types of muscles to perform different kinds of tasks. To generate specific kinds of movement, muscles must have something to pull on; thus we will describe skeletal systems and their roles in generating movement.

48.1 How Do Muscles Contract?

Most behavior and many physiological actions, such as beating of the heart and moving of food through the digestive tract, depend on muscle contraction. Wherever tissues contract, muscle cells are responsible. As introduced in Section 40.1 and shown in Figure 40.4, there are three types of vertebrate muscle:

- **Skeletal muscle** is responsible for all voluntary movements, such as running or playing a piano. It is also involved in some involuntary actions, such as breathing, shivering and maintaining posture.
- **Cardiac muscle** is responsible for the beating action of the heart.
- **Smooth muscle** creates the movement in many hollow internal organs, such as the gut, bladder, and blood vessels, and is under the control of the autonomic (involuntary) nervous system.

All three muscle types use the same *sliding filament contractile mechanism*, and we begin our study of musculoskeletal movement by describing its underlying molecular mechanisms. We will use vertebrate skeletal muscle as our primary example. Later we will discuss the differences in cardiac and smooth muscle that adapt them to their particular functions.

Sliding filaments cause skeletal muscle to contract

Skeletal muscle is also called *striated muscle* because of its striped appearance (see Figure 40.4A). Skeletal muscle cells, called **muscle fibers**, are large and have many nuclei. These multinucleate cells form in development through the fusion of many individual embryonic muscle cells called *myoblasts*. A specific muscle such as your biceps (which bends your arm) is composed of hundreds or thousands of muscle fibers bundled together by connective tissue (**Figure 48.1**).

Muscle contraction is due to the interaction between the contractile proteins **actin** and **myosin**. Within muscle cells, actin and myosin molecules are organized into filaments consisting of many molecules. Actin filaments are also called *thin filaments*, and myosin filaments are *thick filaments*. The two kinds of filaments lie parallel to each other. When muscle contraction is triggered, the actin and myosin filaments slide past each other in a telescoping fashion.

What is the relationship between a skeletal muscle fiber and the actin and myosin filaments responsible for its contraction? Each muscle fiber is packed with **myofibrils**—bundles of thin actin and thick myosin filaments arranged in orderly fashion. In most regions of the myofibril, each thick myosin filament is surrounded by six thin actin filaments, and each thin actin filament sits within a triangle of three thick myosin filaments.

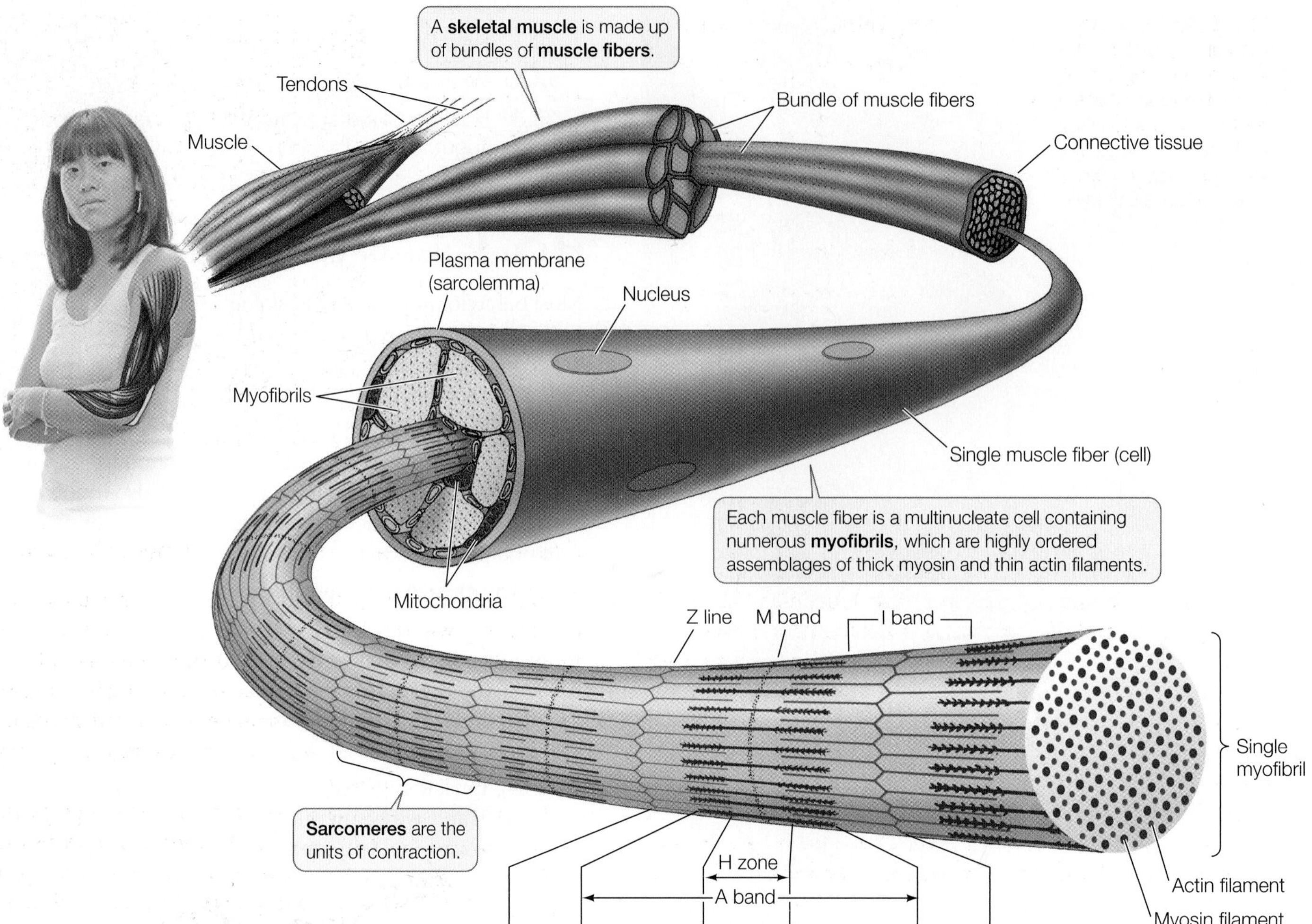

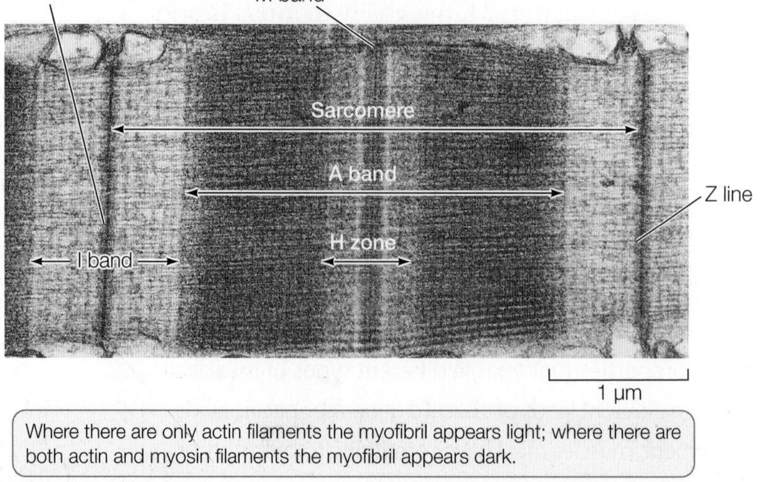

A **skeletal muscle** is made up of bundles of **muscle fibers**.

Tendons

Muscle

Bundle of muscle fibers

Connective tissue

Plasma membrane (sarcolemma)

Nucleus

Myofibrils

Single muscle fiber (cell)

Mitochondria

Each muscle fiber is a multinucleate cell containing numerous **myofibrils**, which are highly ordered assemblages of thick myosin and thin actin filaments.

Z line   M band   I band

Single myofibril

Sarcomeres are the units of contraction.

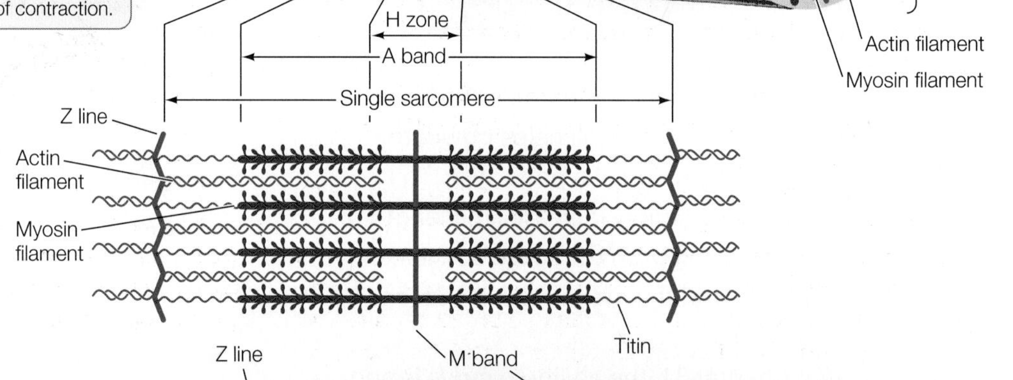

Actin filament

Myosin filament

H zone

A band

Single sarcomere

Z line

Actin filament

Myosin filament

48.1 The Structure of Skeletal Muscle
A skeletal muscle is made up of bundles of muscle fibers. Each muscle fiber is a multinucleate cell containing numerous myofibrils, which are highly ordered assemblages of thick myosin and thin actin filaments. The arrangement of the actin and myosin filaments gives skeletal muscle fibers their characteristic striated appearance.

Z line

M band

Titin

A longitudinal view of a myofibril reveals why skeletal muscle appears striated. The myofibril consists of repeating units called **sarcomeres**. Each sarcomere is made of overlapping filaments of actin and myosin, which create a distinct banding pattern (see Figure 48.1). Before the molecular nature of the muscle banding pattern was known, the bands were given names that are still used today. Each sarcomere is bounded by *Z lines*, which anchor the thin actin filaments. Centered in the sarcomere is the *A band*, which contains all the myosin filaments. The *H zone* and the *I band*, which appear light, are regions where actin and myosin filaments do not overlap in the relaxed muscle. The dark stripe within the H zone is called the *M band*; it contains proteins that help hold the myosin filaments in their regular arrangement.

Sarcomere

A band

Z line

I band

H zone

1 µm

Where there are only actin filaments the myofibril appears light; where there are both actin and myosin filaments the myofibril appears dark.

yourBioPortal.com
GO TO **Web Activity 48.1** • The Structure of a Sarcomere

48.2 Sliding Filaments
The banding pattern of the sarcomere changes as it shortens. Observations of electron micrographs such as those on the right led to the sliding filament hypothesis of muscle contraction.

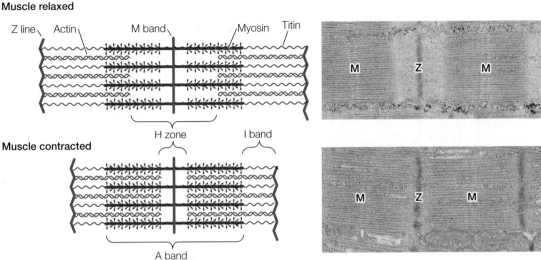

The bundles of myosin filaments are held in a centered position within the sarcomere by a protein called **titin**. Titin is the largest protein in the body; it runs the full length of the sarcomere from Z line to Z line. Each titin molecule runs right through a myosin bundle. Between the ends of the myosin bundles and the Z lines, titin molecules are very stretchable, like bungee cords. In a relaxed skeletal muscle, resistance to stretch is mostly due to the elasticity of the titin molecules.

As the muscle contracts, the sarcomeres shorten and the band pattern changes. The H zone and the I band become much narrower, and the Z lines move toward the A band as if the actin filaments were sliding into the H zone, the region occupied by the myosin filaments (**Figure 48.2**). In the mid 1950s, this observation independently led two teams of British biologists to propose the **sliding filament model** of muscle contraction.

It is not uncommon in science for critical breakthroughs to be made simultaneously in different laboratories, but in this case the coincidences are remarkable. The leaders of the two teams were named Hugh Huxley and Andrew Huxley—but they were not related. Working in separate Cambridge University labs, the two groups proposed the sliding filament model at the same time, and both papers were published in the same issue of the journal *Nature*.

yourBioPortal.com
GO TO Animated Tutorial 48.1 • Molecular Mechanisms of Muscle Contraction

Actin–myosin interactions cause filaments to slide

To understand how the sliding filament model explains muscle contraction, we must first examine the structures of actin and myosin (**Figure 48.3**). A myosin molecule consists of two long polypeptide chains coiled together, each ending in a large globular head. A myosin filament is made up of many myosin molecules arranged in parallel, with their heads projecting sideways at each end of the filament.

An actin filament consists of actin monomers polymerized into a long molecule that looks like two strands of pearls twisted together. Twisting around the actin chains is another protein, *tropomyosin*, and attached to tropomyosin at intervals are molecules of *troponin*. We'll discuss the latter two proteins in the next section.

The myosin heads can bind specific sites on actin, forming cross-bridges between the myosin and the actin filaments. Moreover, when a myosin head binds to an actin filament, the head's conformation changes, and it bends and exerts a tiny force that causes the actin filament to move 5 to 10 nanometers relative to the myosin filament. The myosin heads also have ATPase ac-

48.3 Actin and Myosin Filaments Overlap to Form Myofibrils
Myosin filaments are bundles of molecules with globular heads and polypeptide tails; the protein titin holds these filaments centered within the sarcomeres. Actin filaments consist of two chains of actin monomers twisted together. They are wrapped by chains of the polypeptide tropomyosin and are studded at intervals with another protein, troponin.

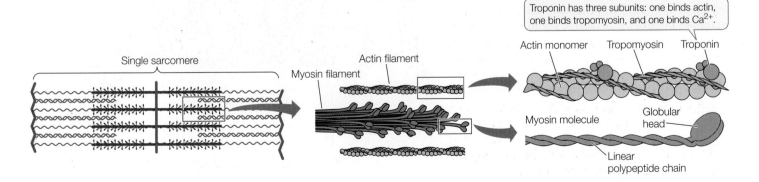

tivity; when they are bound to actin, they can bind and hydrolyze ATP. The energy released when this happens changes the conformation of the myosin head, causing it to release the actin and return to its extended position, from which it can bind to actin again.

Together, these details explain the cycle of events that cause the actin and myosin filaments to slide past each other and shorten the sarcomere. They also explain *rigor mortis*—the stiffening of muscles soon after death. ATP is needed to break the actin–myosin bonds, so when ATP production ceases with death, the actin–myosin bonds cannot be broken and the muscles stiffen. Eventually, however, the proteins begin to lose their integrity, and the muscles soften. The timing of these events helps a medical examiner estimate the time of death.

We have been discussing the cycle of contraction in terms of a single myosin head. Remember that each myosin filament has many myosin heads at both ends and is surrounded by six actin filaments; thus the contraction of the sarcomere involves a great many cycles of interaction between actin and myosin molecules. That is why when a single myosin head breaks its contact with actin, the actin filaments do not slip backward.

Actin–myosin interactions are controlled by calcium ions

Like neurons, muscle cells are *excitable*—that is, their plasma membranes can generate and conduct action potentials. In skeletal muscle fibers, action potentials are initiated by motor neurons arriving at a *neuromuscular junction*. The axon terminals of motor neurons are generally highly branched and form synapses with hundreds of muscle fibers (**Figure 48.4**). A motor neuron and all of the fibers with which forms synapses constitute a **motor unit**. The fibers contract simultaneously when its motor neuron fires. A muscle can consist of many motor units. Thus, there are two ways to increase a muscle's strength of contraction: increase the firing rate of an individual motor neuron, or recruit more motor neurons.

As described in Section 45.3, when an action potential arrives at a neuromuscular junction, the neurotransmitter acetylcholine is released from the motor neuron terminals, diffuses across the synaptic cleft, binds to receptors in the postsynaptic membrane, and causes ion channels in the motor end plate to open (see Figures 45.13 and 45.14). Most of the ions that flow through these channels are Na⁺, and therefore the motor end plate is depolarized. The depolarization spreads to the surrounding plasma membrane of the muscle fiber,

48.5 T Tubules Spread Action Potentials into the Fiber
An action potential at the neuromuscular junction spreads throughout the muscle fiber via a network of T tubules, triggering the release of Ca^{2+} from the sarcoplasmic reticulum.

yourBioPortal.com

GO TO Web Activity 48.2 • The Neuromuscular Junction

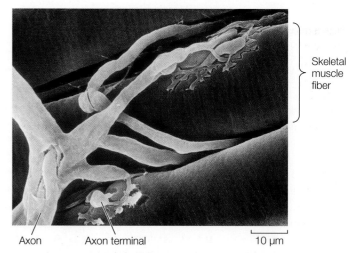

48.4 The Neuromuscular Junction Axons branching from a single motor neuron end in terminals that innervate multiple skeletal muscle fibers.

which contains voltage-gated sodium channels. When threshold is reached, the plasma membrane fires an action potential that is conducted rapidly to all points on the surface of the muscle fiber.

An action potential in a muscle fiber also travels deep within the cell. The plasma membrane is continuous with a distribution system of tubules that descend into the muscle fiber cytoplasm (also called the **sarcoplasm**). The action potential that spreads over the plasma membrane also spreads through this system of transverse tubules, or **T tubules** (**Figure 48.5**).

The T tubules come very close to the endoplasmic reticulum (ER) of the muscle cell. In muscle cells, the ER is called the **sar-**

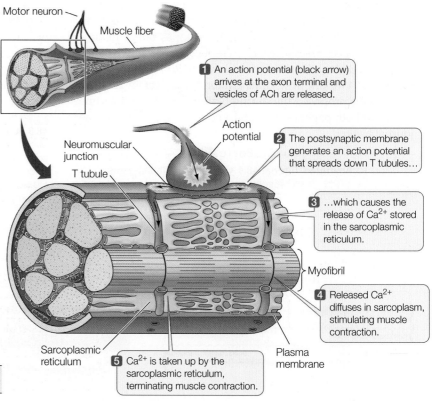

1 An action potential (black arrow) arrives at the axon terminal and vesicles of ACh are released.

2 The postsynaptic membrane generates an action potential that spreads down T tubules...

3 ...which causes the release of Ca^{2+} stored in the sarcoplasmic reticulum.

4 Released Ca^{2+} diffuses in sarcoplasm, stimulating muscle contraction.

5 Ca^{2+} is taken up by the sarcoplasmic reticulum, terminating muscle contraction.

coplasmic reticulum, and it is a closed compartment surrounding every myofibril. Calcium pumps in the sarcoplasmic reticulum take up Ca^{2+} ions from the sarcoplasm. Therefore, when the muscle fiber is at rest, there is a higher concentration of Ca^{2+} in the sarcoplasmic reticulum and a lower concentration in the sarcoplasm.

Spanning the space between the membranes of the T tubules and the membranes of the sarcoplasmic reticulum are two proteins. One protein, the *dihydropyridine (DHP) receptor*, is located in the T tubule membrane; it is voltage-sensitive and changes its conformation when an action potential reaches it. The other protein, the *ryanodine receptor*, is located in the sarcoplasmic reticulum membrane; it is a Ca^{2+} channel. These two proteins are physically connected. When the DHP receptor is activated by an action potential, it changes conformation, causing the ryanodine receptor to allow Ca^{2+} to leave the sarcoplasmic reticulum. Ca^{2+} ions diffuse into the sarcoplasm surrounding the actin and myosin filaments and trigger the interaction of actin and myosin and the sliding of the filaments. How do the Ca^{2+} ions do this?

An actin filament, as we have seen, is a helical arrangement of actin monomers. Twisted around the actin filament are two strands of the protein **tropomyosin (Figure 48.6**; see also Figure 48.3). At regular intervals, the filament also includes a globular protein, **troponin**. The troponin molecule has three subunits: one binds actin, one binds tropomyosin, and one binds Ca^{2+}.

When the muscle is at rest, the tropomyosin strands are positioned so that they block the sites on the actin filament where myosin heads can bind. When Ca^{2+} is released into the sarcoplasm, it binds to troponin, changing its conformation. Because the troponin is bound to the tropomyosin, this conformational change twists the tropomyosin enough to expose the actin–myosin binding sites. Thus the cycle of making and breaking actin–myosin bonds is initiated, the filaments are pulled past each other, and the muscle fiber contracts. When the calcium pumps remove the Ca^{2+} ions from the sarcoplasm, the conformation of the tropomyosin returns to the state in which it blocks the binding of myosin heads to actin, and the muscle fiber returns to its resting condition. Figure 48.6 summarizes this cycle.

48.6 Release of Ca^{2+} from the Sarcoplasmic Reticulum Triggers Muscle Contraction When Ca^{2+} binds to troponin, it exposes myosin-binding sites on the actin. As long as binding sites and ATP are available, the cycle of actin and myosin interactions continues and the filaments slide past each other.

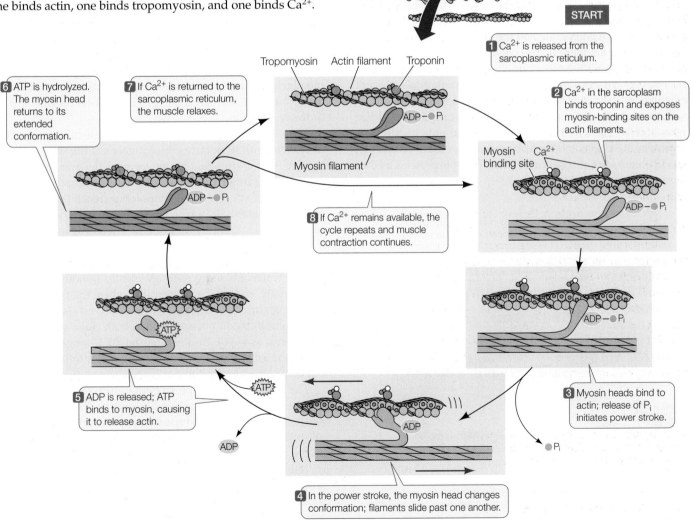

START

1 Ca^{2+} is released from the sarcoplasmic reticulum.

2 Ca^{2+} in the sarcoplasm binds troponin and exposes myosin-binding sites on the actin filaments.

6 ATP is hydrolyzed. The myosin head returns to its extended conformation.

7 If Ca^{2+} is returned to the sarcoplasmic reticulum, the muscle relaxes.

Tropomyosin   Actin filament   Troponin

ADP–P_i

Myosin filament

Myosin binding site Ca^{2+}

ADP–P_i

8 If Ca^{2+} remains available, the cycle repeats and muscle contraction continues.

ADP–P_i

3 Myosin heads bind to actin; release of P_i initiates power stroke.

P_i

5 ADP is released; ATP binds to myosin, causing it to release actin.

ATP

ATP

ADP

ADP

4 In the power stroke, the myosin head changes conformation; filaments slide past one another.

Cardiac muscle is similar to and different from skeletal muscle

Like skeletal muscle, cardiac muscle appears striated because of the regular arrangement of actin and myosin filaments into sarcomeres (**Figure 48.7**). The difference between cardiac and skeletal muscle is that cardiac muscle cells are much smaller and have only one nucleus each (*uninucleate*). Cardiac muscle cells branch, and the branches of adjoining cells interdigitate into a meshwork that is resistant to tearing. As a result, the heart walls can withstand high pressures while pumping blood, without the danger of developing leaks. Adding to the strength of cardiac muscle are *intercalated discs* that provide strong mechanical adhesions between adjacent cells. Gap junctions—protein structures that allow cytoplasmic continuity between cells—in the intercalated discs offer low-resistance pathways for ionic currents to flow between cells (see Figure 7.21A and Section 45.3). Therefore, cardiac muscle cells are electrically coupled. An action potential initiated at one point in the heart spreads rapidly through a large mass of cardiac muscle.

Certain cardiac muscle cells are specialized for generating and conducting electrical signals. These *pacemaker* and *conducting cells* have a low density of actin and myosin filaments, but they initiate and coordinate the rhythmic contractions of the heart. (The molecular basis for this pacemaking function is covered in Section 50.3.) Pacemaker cells make the vertebrate heartbeat *myogenic*, meaning it is generated by the heart muscle itself. A heart removed from a vertebrate can continue to beat with no input from the nervous system; although input from the autonomic nervous system modifies the *rate* of the pacemaker cells, it is not essential for their continued rhythmic function.

The mechanism of excitation–contraction coupling in cardiac muscle cells is different from that in skeletal muscle cells. The T tubules are larger, and the voltage-sensitive DHP proteins in the T tubules are Ca^{2+} channels. These T tubule proteins are not physically connected with the ryanodine receptors in the sarcoplasmic reticulum. Instead, the ryanodine receptors are ion-gated Ca^{2+} channels that are sensitive to Ca^{2+}. When an action potential spreads down the T tubules, it causes the voltage-gated channels to open, allowing extracellular Ca^{2+} to flow into the sarcoplasm. This slight rise in sarcoplasmic Ca^{2+} concentration opens the Ca^{2+} channels in the sarcoplasmic reticulum, which in turn causes a huge rise in sarcoplasmic Ca^{2+} concentration, resulting in fiber contraction. This mechanism is called *Ca^{2+}-induced Ca^{2+} release.*

Smooth muscle causes slow contractions of many internal organs

Smooth muscle provides the contractile force for most of our internal organs, which are under the control of the autonomic nervous system. Smooth muscle moves food through the digestive tract, controls the flow of blood through blood vessels, and empties the urinary bladder. Structurally, smooth muscle cells are the simplest muscle cells. They are smaller than skeletal muscle cells, usually long and spindle-shaped, and each has a single nucleus. They are "smooth" because the actin and myosin filaments are not as regularly arranged as they are in skeletal and cardiac muscle, and so do not produce the striated appearance.

Some smooth muscle tissue, such as that from the wall of the digestive tract, has interesting properties. The cells are arranged in sheets, and individual cells in a sheet are in electrical contact with one another through gap junctions, as they are in cardiac muscle. As a result, an action potential generated in the membrane of one smooth muscle cell can spread to all the cells in the sheet of tissue. Thus the cells in the sheet contract in a coordinated fashion.

The plasma membranes of smooth muscle cells are sensitive to stretch, with important consequences. If the wall of the digestive tract is stretched in one location (as by a mouthful of food passing down the esophagus to the stomach), the membranes of the stretched cells depolarize, reach threshold, and fire action potentials, which cause the cells to contract. Thus, smooth muscle contracts after being stretched, and the harder it is stretched, the stronger it contracts. This behavior of smooth muscle is important for moving food through the gut.

The walls of blood vessels are mostly smooth muscle. This is especially true on the arterial side where the blood is under higher pressure. The muscle tone in these vessels is responsive to changes in blood pressure and to both chemical and neural influences, as we discuss in Chapter 50. Changes in vascular smooth muscle tone are responsible for controlling the distribution of blood in the body.

The neural influences on smooth muscle come from the two divisions of the autonomic nervous system. The neurotransmitters of the sympathetic and parasympathetic postganglionic cells alter the membrane potential of smooth muscle cells. For example, in the digestive tract, acetylcholine causes smooth muscle cells to depolarize, making them more likely to fire action potentials and contract. Antagonistically, norepinephrine causes these muscle cells to hyperpolarize and thus be less likely to fire action potentials and contract (**Figure 48.8**). In contrast, norep-

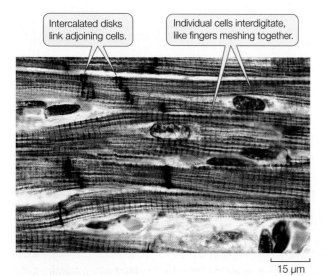

Intercalated disks link adjoining cells.

Individual cells interdigitate, like fingers meshing together.

15 μm

48.7 Cardiac Muscle Cells Form a Strong Meshwork Cardiac muscle cells branch and interdigitate, forming a tear-resistant mesh that can withstand the pressure of blood pumping through the heart.

INVESTIGATING LIFE

48.8 Neurotransmitters Alter the Membrane Potential of Smooth Muscle Cells

Earlier experiments showed that stretching the smooth muscles of the gut (as in the stretch applied by a full stomach) depolarizes the membranes, causing action potentials that activate the contractile mechanism. This follow-up experiment showed that the parasympathetic neurotransmitter acetylcholine and the sympathetic neurotransmitter norepinephrine act antagonistically to alter the membrane potential of smooth muscle.

HYPOTHESIS Stimulation from neurotransmitters of the autonomic nervous system induce contractions in the smooth muscles of the gut.

METHOD

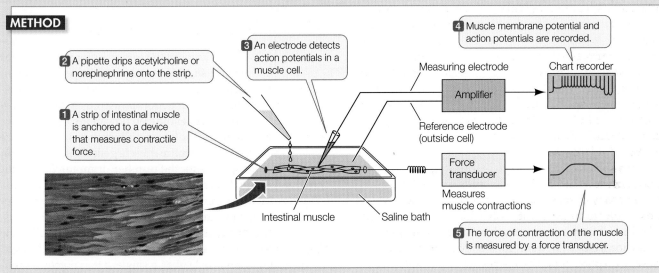

2 A pipette drips acetylcholine or norepinephrine onto the strip.

3 An electrode detects action potentials in a muscle cell.

4 Muscle membrane potential and action potentials are recorded.

Measuring electrode

Chart recorder

Amplifier

1 A strip of intestinal muscle is anchored to a device that measures contractile force.

Reference electrode (outside cell)

Force transducer

Measures muscle contractions

Intestinal muscle   Saline bath

5 The force of contraction of the muscle is measured by a force transducer.

RESULTS

When acetylcholine is dripped onto the muscle, the cells depolarize, fire action potentials more rapidly, and increase their force of contraction.

Norepinephrine, on the other hand, causes the cells to hyperpolarize, decreasing their rate of firing, and decreasing their force of contraction.

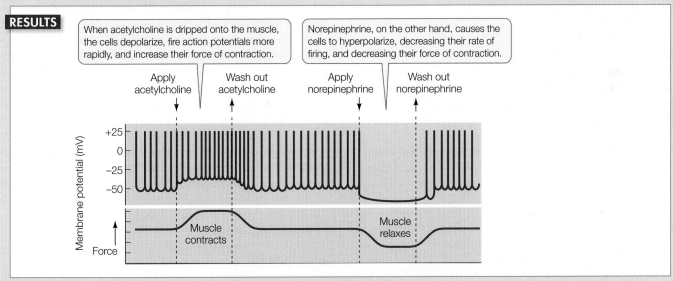

Apply acetylcholine   Wash out acetylcholine   Apply norepinephrine   Wash out norepinephrine

Membrane potential (mV)

+25
0
−25
−50

Muscle contracts

Muscle relaxes

Force

CONCLUSION ANS neurotransmitters can alter membrane resting potentials and affect the rate at which smooth muscle cells fire action potentials, thus controlling smooth muscle contraction.

Go to **yourBioPortal.com** for original citations, discussions, and relevant links for all INVESTIGATING LIFE figures.

inephrine acting through G protein-coupled receptors causes the smooth muscle in arteries serving the gut to contract. Remember that the action of the neurotransmitters depends on the receptors in the target tissues. Sympathetic activity is high in a fight-or-flight situation; in an emergency you don't need to digest your lunch, but you do need to send blood to the tissues critical for survival.

Although smooth muscle cell contraction is not controlled by the troponin–tropomyosin mechanism, calcium still plays a critical role. A Ca^{2+} influx into the sarcoplasm of a smooth muscle cell can be stimulated by action potentials, hormones, or stretching. The Ca^{2+} that enters the sarcoplasm combines with a protein called *calmodulin*. The calmodulin–Ca^{2+} complex activates an enzyme called myosin kinase, which can phosphory-

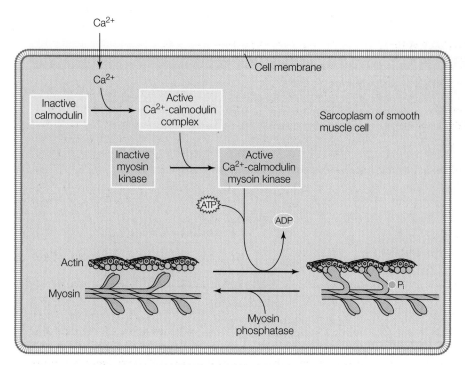

Ca²⁺

Ca²⁺

Cell membrane

Inactive calmodulin → Active Ca²⁺-calmodulin complex

Sarcoplasm of smooth muscle cell

Inactive myosin kinase → Active Ca²⁺-calmodulin mysoin kinase

ATP

ADP

Actin

Myosin

Pi

Myosin phosphatase

48.9 The Role of Ca²⁺ in Smooth Muscle Contraction   When a smooth muscle cell is stimulated by neurotransmitter, Ca²⁺ enters the sarcoplasm and binds to calmodulin, which in turn activates an enzyme that phosphorylates the myosin heads, causing them to bind to actin. As long as the myosin remains phosphorylated, actin and myosin go through cycles of binding and release. Thus in smooth muscle, the Ca²⁺-mediated change is on myosin, whereas in skeletal and cardiac muscle it is on the actin-tropomyosin filamant.

late myosin heads. When the myosin heads in smooth muscle are phosphorylated, they undergo cycles of binding and releasing actin, causing muscle contraction. As Ca²⁺ is removed from the sarcoplasm, it dissociates from calmodulin, and the activity of myosin kinase falls. An additional enzyme, myosin phosphatase, dephosphorylates the myosin to help reduce actin–myosin interactions (**Figure 48.9**).

─────── **yourBioPortal.com** ───────

GO TO Animated Tutorial 48.2 • Smooth Muscle Action

Single skeletal muscle twitches are summed into graded contractions

In skeletal muscle, the arrival of an action potential at a neuromuscular junction causes an action potential in a muscle fiber. The spread of that action potential through the muscle fiber's

T tubule system causes a minimum unit of contraction, called a **twitch**. A twitch can be measured in terms of the *tension*, or force, it generates (**Figure 48.10A**). A single action potential stimulates a single twitch, but the ultimate force generated by an action potential can vary enormously depending on how many muscle fibers are in the motor unit it innervates. The level of tension an entire muscle generates depends on two factors: the number of motor units activated, and the frequency at which the motor units fire

In muscles responsible for fine movements, such as those of the fingers, a motor neuron may innervate only one or a few muscle fibers, but in a muscle that produces large forces, such as the biceps, a motor neuron innervates a large number of muscle fibers.

At the level of a muscle fiber, a single action potential stimulates a single twitch. If action potentials reaching the muscle fiber are adequately separated in time, each twitch is a discrete, all-or-none phenomenon. If action potentials are fired more rapidly, however, new twitches are triggered before the myofibrils have a chance to return to their resting condition. As a result, the twitches sum, and the tension generated by the fiber increases and becomes more sustained. Thus an individual muscle fiber can show a graded response to increased levels of stimulation by its motor neuron.

Twitches sum at high levels of stimulation because the calcium pumps in the sarcoplasmic reticulum are not able to clear the Ca²⁺ ions from the sarcoplasm between action potentials.

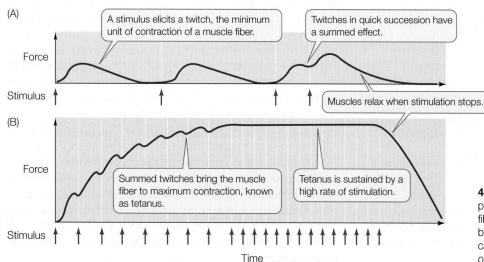

(A)

Force

A stimulus elicits a twitch, the minimum unit of contraction of a muscle fiber.

Twitches in quick succession have a summed effect.

Stimulus

Muscles relax when stimulation stops.

(B)

Force

Summed twitches bring the muscle fiber to maximum contraction, known as tetanus.

Tetanus is sustained by a high rate of stimulation.

Stimulus

Time

48.10 Twitches and Tetanus   (A) Action potentials from a motor neuron cause a muscle fiber to twitch. Twitches in quick succession can be summed. (B) Summation of many twitches can bring the muscle fiber to the maximum level of contraction, known as tetanus.

Eventually a stimulation frequency can be reached that results in continuous presence of Ca^{2+} in the sarcoplasm at high enough levels to cause continuous activation of the contractile machinery—a condition known as **tetanus** (**Figure 48.10B**). (Do not confuse this condition with the disease called tetanus, which is caused by a bacterial toxin and is characterized by spastic contractions of skeletal muscles.)

How long a muscle fiber can maintain a tetanic contraction depends on its supply of ATP. Eventually the fiber will become fatigued. It may seem paradoxical that the *lack* of ATP causes fatigue, since the action of ATP is to break actin–myosin bonds. But remember that the energy released from the hydrolysis of ATP "re-cocks" the myosin heads, allowing them to cycle through another power stroke. When a muscle is contracting against a load, the cycle of making and breaking actin–myosin bonds must continue, to prevent the load from stretching the muscle. The situation is like rowing a boat upstream: you cannot maintain your position relative to the stream bank by just holding the oars out against the current; you have to keep rowing. Likewise, actin–myosin bonds have to keep cycling to maintain tension in the muscle.

Many muscles of the body maintain a low level of tension even when the body is at rest. For example, the muscles of the neck, trunk, and limbs that maintain our posture against the pull of gravity are always working, even when we are standing or sitting still. *Muscle tone* comes from the activity of a small but changing number of motor units in a muscle; at any one time, some of the muscle's fibers are contracting and others are relaxed. The nervous system is constantly readjusting muscle tone.

48.1 RECAP

The contractile ability of muscle derives from interactions between actin and myosin filaments. The three types of muscle are skeletal, cardiac, and smooth. Contraction in all three depends on control of Ca^{2+} in the sarcoplasm. Tropomyosin and troponin are controlling elements in skeletal and cardiac muscle. Calmodulin is the controlling element in smooth muscle.

- Explain how the cellular and subcellular structure of skeletal muscle relate to the sliding filament theory of muscle contraction. See pp. 1007–1009 and Figures 48.1, 48.2, and 48.3

- What is the role of Ca^{2+} in the contractile mechanism of skeletal, cardiac, and smooth muscle? See pp. 1010–1011 and Figures 48.5 and 48.6

- What role does ATP play in the actin and myosin interactions that produce contraction? See Figure 48.6

Now that we understand how muscles generate force, we can look at what determines the characteristics of a muscle, and how individual muscles can change their characteristics with regular use and conditioning.

48.2 What Determines Muscle Performance?

The functions that different muscles perform place different demands on them. Some muscles, such as postural muscles, must sustain a load continuously over long periods of time. Other muscles, such as those that control your fingers, generally do not have to sustain long contractions, but they must be able to contract quickly. And what is "quick" for humans doesn't begin to compare with insect flight muscles that can contract as fast as 1,000 times per second. How are muscles adapted to specific functions and demands?

Muscle fiber types determine endurance and strength

Not all skeletal muscle fibers are alike, and a single muscle often contains more than one type of fiber. The two major types of skeletal muscle fibers express different genes for their myosin molecules, and these myosin variants have different rates of ATPase activity. Those with high ATPase activity can recycle their actin–myosin cross-bridges rapidly and are therefore called fast-twitch fibers. Slow-twitch fibers have lower ATPase activity; they develop tension more slowly but can maintain it longer.

Slow-twitch fibers are also called *oxidative* or *red muscle* because they contain the oxygen-binding protein *myoglobin*, have many mitochondria, and are well supplied with blood vessels. These characteristics both increase the fibers' capacity for oxidative metabolism and result in their red appearance. The maximum tension a slow-twitch fiber produces is low and develops slowly but is highly resistant to fatigue. Slow-twitch fibers have substantial reserves of fuel (glycogen and fat), so they can maintain steady, prolonged production of ATP as long as oxygen is available. Muscles with high proportions of slow-twitch fibers are good for long-term *aerobic* work (that is, work that requires oxygen). Long-distance runners, swimmers, cyclists, and other athletes whose activities require endurance have leg and arm muscles consisting mostly of slow-twitch fibers (**Figure 48.11**).

Some **fast-twitch fibers** are also called *glycolytic* or *white muscle* because, compared with slow-twitch fibers, they have few mitochondria, little or no myoglobin, and fewer blood vessels; thus they look pale. Fast-twitch glycolytic fibers can develop maximum tension more rapidly than slow-twitch fibers can, and that maximum tension is greater. However, fast-twitch fibers fatigue rapidly. The myosin of these fibers puts the energy of ATP to work very rapidly, but the fibers cannot replenish ATP quickly enough to sustain contraction for a long time. Fast-twitch fibers are especially good for short-term work that requires maximum strength. Weight lifters and sprinters have leg and arm muscles with high proportions of fast-twitch fibers.

The types of fibers that make up a muscle influence the performance properties of that muscle. A sprinter benefits from muscles that generate maximum force rapidly, but they do not need to sustain a particular load for a long time. By contrast, a marathon runner benefits from muscles with maximum endurance.

(A) Cross sections of leg muscles

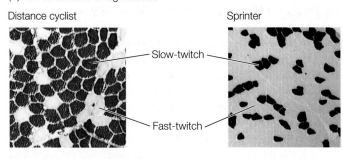

(B)

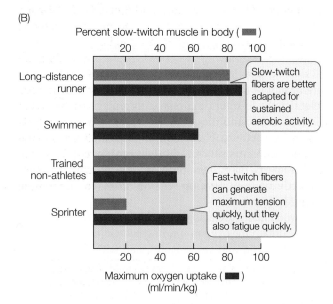

48.11 Slow- and Fast-Twitch Muscle Fibers The skeletal muscles in the micrographs were stained with a reagent that shows slow-twitch fibers as dark and fast-twitch muscle as light. Athletes in different sports have different distributions of muscle fiber types.

What can you do to optimize your performance in a particular activity? To a limited extent, you can alter the properties of your muscle fibers through training. There are fast-twitch fibers that are somewhat oxidative and therefore intermediate in their properties between slow-twitch and fast glycolytic fibers. These intermediate fibers can become more oxidative with endurance training and more glycolytic with strength training. However, the most important determinant of your muscle fiber types is your genetic heritage. There is some truth to the statement that champions are born, not made. A person born with a high proportion of fast-twitch fibers is unlikely to become a champion marathon runner, and one born with a high proportion of slow-twitch fibers is unlikely to become a champion sprinter.

A muscle has an optimal length for generating maximum tension

If you have ever done a pull-up, you know that two parts of this exercise are especially difficult. When you are hanging from the bar with your arms fully extended, it is hard to get

the pull-up started; and when your chin has just about reached the bar, pulling yourself up the last small distance is difficult. Why is this? Part of the explanation comes from the lever properties of the muscle–joint interaction that we discuss in the next section, and part comes from the structure of the sarcomere.

When a muscle is stretched and the sarcomeres are lengthened, there is less overlap between the actin and myosin filaments; therefore, fewer cross-bridges can form, and less force can be produced. In fact, if the sarcomeres are stretched too much, actin and myosin do not overlap and no force can be produced. How would a muscle recover from such a difficult situation? The bungee cord–like titin molecules create enough elastic recoil to pull the actin and myosin fibrils back into an overlapping arrangement.

When the muscle is fully contracted, the actin and myosin filaments overlap so much that the myosin bundles are pressed up against the Z lines. Because they have no place to go, additional shortening is difficult. You can see the relationship between the length of a muscle fiber and its ability to develop tension in **Figure 48.12**.

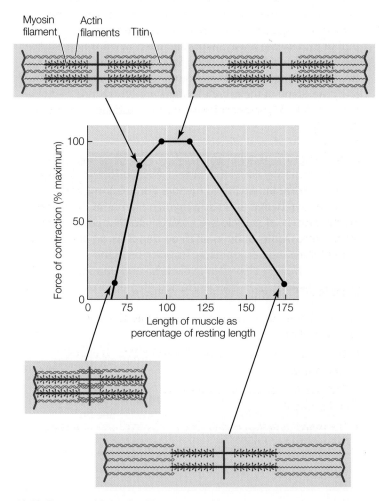

48.12 Force and Length The amount of force a sarcomere can generate depends on its resting length. When a muscle is stretched, the sarcomeres lengthen, there is less overlap between the actin and myosin filaments, and less force is produced. Overstretched sarcomeres produce no force because there is no overlap between the actin and myosin.

Exercise increases muscle strength and endurance

Different types of exercise produce different physical conditioning responses. In general, anaerobic activities, such as weight lifting, increase strength, and aerobic activities, such as jogging, increase endurance. Strength is the maximum force that a muscle can exert, and endurance is work capacity or how long a given workload can be sustained. What are the physiological bases for these differences?

Strength is a function of the cross-sectional area of muscles: the more actin and myosin filaments in a muscle fiber, and the more muscle fibers in a muscle, the more tension it can produce. When athletes undertake strength training, they use weights or exercises such as pull-ups to repeatedly contract specific muscles under heavy loads. Repetitions are usually done until the muscle is completely fatigued. Such stress on a muscle does minor tissue damage—hence the soreness the day after a hard workout—but it also induces the formation of new actin and myosin filaments in existing muscle fibers. The muscle fibers, and hence the muscles, get bigger and stronger. In extreme cases, and after serious muscle damage, new muscle fibers can also be produced from stem cells called *satellite cells* in the muscle. In general, however, the major effect of strength training is to produce bigger, rather than more, muscle fibers.

Aerobic exercise has a completely different effect on muscles: it enhances their oxidative capacity. This effect comes from increases in the number of mitochondria, in enzymes involved in energy use, and in the density of capillaries that deliver oxygen to the muscle. Myoglobin also increases in skeletal muscle cells. **Myoglobin** is an oxygen-binding protein similar to hemoglobin in red blood cells. However, myoglobin has a higher affinity for oxygen than does hemoglobin. Therefore, myoglobin accepts oxygen from the blood, facilitates the diffusion of oxygen throughout the muscle, and provides a store of oxygen for use when oxygen delivery by the blood is insufficient. By increasing the capacity of muscle to use oxygen to produce ATP, aerobic training increases the work load that muscles can sustain through time.

Muscle ATP supply limits performance

Muscles have three systems for supplying the ATP they need for contraction:

- The *immediate system* uses preformed ATP and creatine phosphate.
- The *glycolytic system* metabolizes carbohydrates to lactate and pyruvate.
- The *oxidative system* metabolizes carbohydrates or fats all the way to H_2O and CO_2.

The capacity of these three systems and the rates at which they can produce ATP determine both work capacity and endurance (**Figure 48.13**).

ATP is present in muscles in very small amounts. However, muscle fibers also contain a storage compound called *creatine phosphate* (CP). This molecule stores energy in a phosphate bond, which it can transfer to ADP. The total energy available in all the muscles of your body in the form of ATP and CP—the immediate energy system—is only about 10 kilocalories. When at rest, you metabolize a kilocalorie of energy in less than a minute. Even though the energy available from ATP and CP is limited, it is available immediately, and it enables fast-twitch fibers to generate a lot of force quickly. During burst activity, the immediate system is exhausted in seconds.

The glycolytic system activates within a few seconds to replace the ATP depleted at the onset of muscle activity. The glycolytic enzymes are located in the cytoplasm of the muscle fiber, and therefore the ATP they generate is rapidly available to the myosin filaments. However, as noted in Chapter 9, glycolysis alone is an inefficient way to produce ATP, and it leads to the accumulation of lactic acid, which slows the process. Thus, the glycolytic system and the immediate system together can provide most of the energy for active muscles for less than a minute (see Figure 48.13).

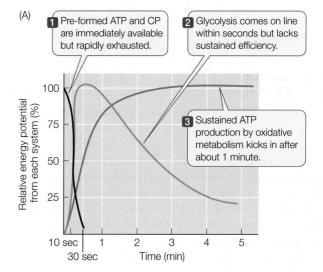

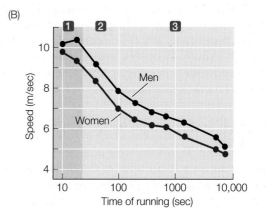

48.13 Supplying Fuel for High Performance (A) Muscles have three systems for obtaining the ATP they need for contraction during exertion such as running. (B) Looking at a plot of world-record times for running events of different durations, you can see that the performance of the athletes corresponds to the time courses of the three energy systems.

Oxidative metabolism becomes fully active in about a minute, producing relatively huge amounts of ATP because it can completely metabolize carbohydrates and fats. However, it requires many reactions (see Chapter 9), and it takes place in the mitochondria, so O_2 and substrate must diffuse into the mitochondria, and the formed ATP must diffuse from the mitochondria to the myosin filaments in the muscle. These processes are not instantaneous, so the rate at which oxidative metabolism can make ATP available to do work is slower than the rate at which the other two systems can supply ATP.

The fuel supply available to the muscles influences how long someone can sustain a high level of aerobic exercise. From the circulating blood, muscle receives glucose and free fatty acids, which it can metabolize to generate ATP. At high levels of aerobic exercise, however, most of the fuel used by muscles to produce ATP comes from the reserve of glycogen stored in the muscle itself. Depletion of muscle glycogen results in fatigue.

The rate at which muscle glycogen is replenished depends on diet: it is high with a high-carbohydrate diet, low with a high-fat diet, and intermediate with a mixed diet. This fact is the basis for a practice called "carbo-loading." For 3 to 5 days, athletes exercise at a level that depletes muscle glycogen. Then, 2 or 3 days before the event, they taper down their level of training and eat a diet rich in complex carbohydrates. The result can be glycogen supercompensation, in which the restoration of muscle glycogen stores "overshoots" and reaches above-normal levels.

Insect muscle has the greatest rate of cycling

Insect flight muscle can produce a wingbeat frequency of up to 1,000 cycles per second. Since neural action potentials last 1 to 3 milliseconds, that number of cycles per second would push the capacity of motor neurons, let alone the mechanism of cycling of the striated muscle contraction/relaxation. The wingbeat frequency of a hummingbird is less than 50 cycles per second, and that seems fast. How do insects do it?

The mechanism of excitation/contraction coupling is different in insect flight muscle. Vertebrate striated muscle and much of invertebrate striated muscle is called "synchronous" because the cycling of the contractile mechanism is linked to the firing of the motor neurons. This is not true of insect flight muscle, which is therefore called "asynchronous" muscle. The firing of action potentials in the insect flight motor neurons is not particularly fast, but it does cause depolarization of the muscle cell membrane, the spreading of an action potential throughout the membrane, and the release of Ca^{2+} from the sarcoplasmic reticulum. However, once the asynchronous muscle fiber is stimulated, its cycle of contraction/relaxation proceeds at its own characteristic frequency as long as the Ca^{2+} is available to bind to the troponin. The contraction of the muscle fiber deactivates the actin–myosin binding, which in turn permits a stretching of the muscle, which in turn activates the actin–myosin binding. Thus contractile cycling and the resulting wingbeat frequency are not tied to the firing rate of the flight motor neurons.

48.2 RECAP

Depending on the function a muscle serves, it may need to generate maximum force rapidly, sustain activity for a long period, or contract and relax at a very rapid rate. Properties of muscles can facilitate these types of activity.

- Describe the differences between slow-twitch and fast-twitch fibers. See p. 1015 and Figure 48.11
- How does exercise influence muscle strength and endurance? See p. 1017
- How do the different sources of ATP influence performance in different types of exercise? See pp. 1017–1018 and Figure 48.13
- Explain how insects can beat their wings 10 times faster than hummingbirds can. See p. 1018

Regardless of how much force a muscle can generate, how long it can sustain a work load, or how fast it can contract and relax, a muscle needs something to pull on; otherwise it would just be a lump of pulsating or quivering tissue. Let's look now at how skeletal systems help generate movement.

48.3 How Do Skeletal Systems and Muscles Work Together?

Muscles can contract and exert force, or they can relax. To create significant movement, they must have something to pull on and something that stretches the muscle back to a longer position. In some cases, muscles pull on each other, as in the trunk of an elephant or the arms of an octopus. In most cases, however, **skeletal systems** are the rigid supports against which muscles pull to create directed movement. In this section, we examine the three types of skeletal systems: hydrostatic skeletons, exoskeletons, and endoskeletons.

A hydrostatic skeleton consists of fluid in a muscular cavity

Cnidarians, annelids, and other soft-bodied invertebrates have **hydrostatic skeletons** consisting of a volume of fluid enclosed in a body cavity surrounded by muscle (see Section 31.2). When muscles oriented in one direction contract, the fluid-filled body cavity bulges out in the opposite direction.

An earthworm uses its hydrostatic skeleton to crawl. The earthworm's body cavity is divided into many separate segments, each of which contains a compartment filled with extracellular fluid. The body wall surrounding each segment has two muscle layers: a circular layer and a longitudinal layer. If the circular muscles in a segment contract, the compartment in that segment narrows and elongates. If the longitudinal muscles in a segment contract, the compartment shortens and bulges outward. Alternating contractions of the earthworm's circular and longitudinal muscles create waves of narrowing and widening,

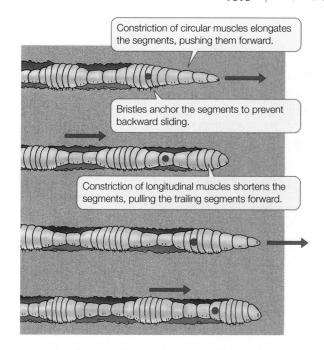

Constriction of circular muscles elongates the segments, pushing them forward.

Bristles anchor the segments to prevent backward sliding.

Constriction of longitudinal muscles shortens the segments, pulling the trailing segments forward.

48.14 A Hydrostatic Skeleton Alternating waves of muscle contraction move the earthworm through the soil. The red dot enables you to follow the changes in one segment as the worm moves forward.

lengthening and shortening, that travel down the body. Bulging, shortened segments serve as anchors as long, narrow segments project forward and longitudinal contractions pull other segments forward. Bristles help the widest parts of the body to hold firm against the substratum (**Figure 48.14**).

Exoskeletons are rigid outer structures

An *exoskeleton* is a hardened, rigid outer surface to which muscles can be attached. Contractions of the muscles cause jointed segments of the exoskeleton to move relative to each other. The simplest example of an exoskeleton is the shell of a mollusk. Some marine mollusks, such as clams, have shells composed of protein strengthened by crystals of calcium carbonate (a rock-hard material). These shells can be massive, affording significant protection against predators. The shells of land mollusks (snails) generally lack the hard mineral component and are much lighter (see Figure 32.15B,D).

The most complex exoskeletons are found among the arthropods. An exoskeleton, or *cuticle*, covers all the outer surfaces of the arthropod's body and all its appendages. It is made up of plates secreted by a layer of cells just below the exoskeleton. The cuticle contains stiffening materials everywhere except at the joints, where flexibility must be retained. Muscles attached to the inner surfaces of the arthropod exoskeleton move its parts around the joints (see Figure 32.4).

A drawback of the rigid arthropod exoskeleton is that it cannot expand. Therefore, if the animal is to become larger, it must *molt*, shedding its exoskeleton and forming a new, larger one. A molting animal is vulnerable because the new exoskeleton takes time to harden. The animal's body is temporarily unprotected, and without a firm exoskeleton against which its mus-

cles can exert maximum tension, it is unable to move rapidly. Soft-shelled crabs, a gourmet delicacy, are crabs caught when they are molting.

Vertebrate endoskeletons consist of cartilage and bone

The **endoskeleton** of vertebrates is an internal scaffolding. Muscles are attached to it and pull against it. Endoskeletons are composed of rodlike, platelike, and tubelike bones connected to one another at a variety of joints that allow a wide range of movements. An advantage of endoskeletons over the exoskeletons of arthropods is that bones in the body can grow without the animal shedding its skeleton.

The human skeleton consists of 206 bones, some of which are shown in **Figure 48.15**. It can be divided into an *axial skeleton*, which includes the skull, vertebral column, sternum, and ribs; and an *appendicular skeleton*, which includes the pectoral girdle, pelvic girdle, and bones of the arms, legs, hands, and feet.

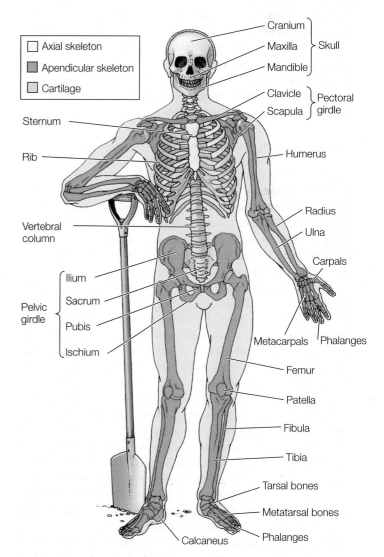

☐ Axial skeleton
☐ Apendicular skeleton
☐ Cartilage

Cranium
Maxilla } Skull
Mandible
Clavicle } Pectoral girdle
Scapula
Sternum
Rib
Humerus
Vertebral column
Radius
Ulna
Carpals
Ilium
Sacrum
Pelvic girdle
Pubis
Ischium
Metacarpals
Phalanges
Femur
Patella
Fibula
Tibia
Tarsal bones
Metatarsal bones
Calcaneus
Phalanges

48.15 The Human Endoskeleton Cartilage and bone make up the internal skeleton of a human being.

The vertebrate endoskeleton consists of two kinds of connective tissue, *cartilage* and *bone*, which are produced by two kinds of connective tissue cells. *Cartilage cells* produce an extracellular matrix that is a tough, rubbery mixture of polysaccharides and proteins—mainly fibrous collagen. Collagen fibers run in all directions like reinforcing cords through the gel-like matrix and give it the well-known strength and resiliency of "gristle." This matrix, called **cartilage**, is found in parts of the endoskeleton where both stiffness and resiliency are required, such as on the surfaces of joints where bones move against one another. Cartilage is also the supportive tissue in stiff but flexible structures such as the larynx (voice box), nose, and ear pinnae. Sharks and rays are called *cartilaginous fishes* because their skeletons are composed entirely of cartilage. In most other vertebrates, cartilage is the principal component of the embryonic skeleton, but during development most of it is gradually replaced by bone.

Bone also contains collagen fibers, but it gets its rigidity and hardness from an extracellular matrix of insoluble calcium phosphate crystals. Bone serves as a reservoir of calcium for the rest of the body and is in dynamic equilibrium with soluble calcium in the extracellular fluids of the body. This equilibrium is under the control of calcitonin and parathyroid hormone (see Figure 41.11). If too much calcium is taken from the skeleton, the bones are seriously weakened.

The living cells of bone—*osteoblasts, osteocytes,* and *osteoclasts*—are responsible for the constant dynamic remodeling of bone (**Figure 48.16**). **Osteoblasts** lay down new matrix material on bone surfaces. These cells gradually become surrounded by matrix and eventually become enclosed within the bone, at which point they cease laying down matrix but continue to exist within small lacunae (cavities) in the bone. In this state they are called **osteocytes**. Despite the vast amounts of matrix between them, osteocytes remain in contact with one another through long cellular extensions that run through tiny channels in the bone. Communication between osteocytes is important in controlling the activities of the cells that are laying down or removing bone.

The cells that resorb bone are the **osteoclasts**. They are derived from the same cell lineage that produces white blood cells. Osteoclasts erode bone, forming cavities and tunnels. Osteoblasts follow osteoclasts, depositing new bone. Thus the interplay of osteoblasts and osteoclasts constantly replaces and remodels the bones, allowing a bone to recover from damage and adjust to the forces placed on it.

How the activities of the bone cells are coordinated is not understood, but stress placed on bones somehow provides them with information. A remarkable finding in studies of astronauts who spent long periods in zero gravity was that their bones decalcified. Conversely, in athletes, certain bones thicken during training. Both thickening and thinning of bones are experienced by anyone who has had a leg in a cast for a long time: the bones of the uninjured leg carry the person's weight and thicken while the bones of the inactive leg in the cast thin.

Because of the positive effects of physical stress on bone deposition, weight-bearing exercise is effective in preventing and treating the loss of bone density (and hence strength) known as *osteoporosis*. More than 25 million people in the United States

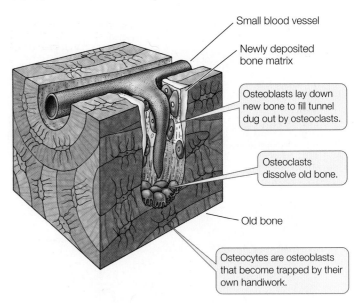

48.16 Bone Is Living Tissue Bones are constantly being remodeled by osteoblasts, which lay down bone, and osteoclasts, which resorb bone.

suffer from this debilitating condition. Although osteoporosis is most commonly a problem for postmenopausal women, it can occur in younger people as a result of malnutrition. For example, the condition known as *female athlete triad* includes eating disorders, cessation of menstrual cycling, and osteoporosis. These are interactive conditions in which the eating disorder and excessive training lead to malnutrition that can result in endocrine disruption and osteoporosis. Excessive training and malnutrition can lead to bone loss in males as well.

Bones develop from connective tissues

Bones are divided into two types on the basis of how they develop. **Membranous bone** forms on a scaffold of connective tissue membrane. **Cartilage bone** forms first as a cartilaginous structure resembling the future mature bone, then gradually hardens, or *ossifies*, to become bone. The outer bones of the skull are membranous bones; the bones of the limbs are cartilage bones.

Cartilage bones can grow throughout the ossification process. The long bones of the legs and arms, for example, ossify first at the centers and later at each end (**Figure 48.17**). Growth can continue until these areas of ossification join. The membranous bones forming the skull cap grow until their edges meet. The soft spot on the top of a baby's head (the fontanelle) is the point at which the skull bones have not yet joined.

The structure of bone may be **compact** (solid and hard) or **cancellous** (having numerous internal cavities that make it appear spongy, although it is rigid). The architecture of a specific bone depends on its position and function, but most bones have both compact and cancellous regions. The shafts of the long bones of the limbs, for example, are cylinders of compact bone surrounding central cavities that contain the bone marrow, where the cellular elements of the blood are made. The ends of the long bones are cancellous (see Figure 48.17). Cancellous bone is lightweight because of its numerous cavities, but it is also strong because its internal meshwork constitutes a support system. It can withstand considerable forces of compression. The

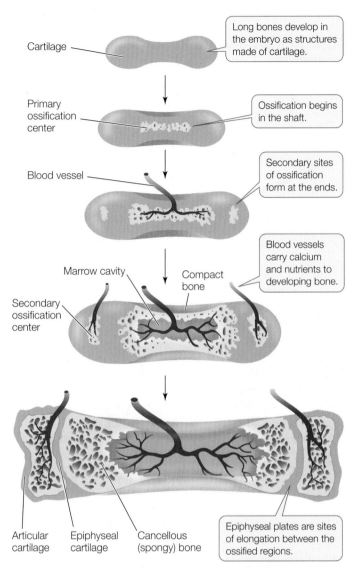

Cartilage — Long bones develop in the embryo as structures made of cartilage.

Primary ossification center — Ossification begins in the shaft.

Blood vessel — Secondary sites of ossification form at the ends.

Marrow cavity / Compact bone

Secondary ossification center

Blood vessels carry calcium and nutrients to developing bone.

Articular cartilage / Epiphyseal cartilage / Cancellous (spongy) bone

Epiphyseal plates are sites of elongation between the ossified regions.

48.17 The Growth of Long Bones In the long bones of human limbs, ossification occurs first at the centers and later at each end.

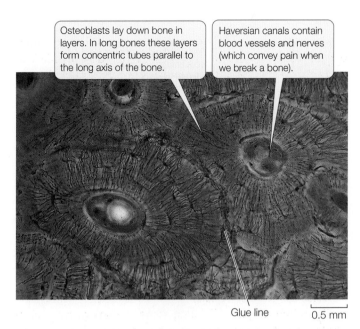

Osteoblasts lay down bone in layers. In long bones these layers form concentric tubes parallel to the long axis of the bone.

Haversian canals contain blood vessels and nerves (which convey pain when we break a bone).

Glue line

0.5 mm

48.18 Most Compact Bone Is Composed of Haversian Systems A micrograph of a section of a long bone shows Haversian systems with their central canals. Glue lines separate Haversian systems.

rigid, tubelike shaft of compact bone can withstand compression and bending forces. Architects and nature alike use hollow tubes as lightweight structural elements.

Most of the compact bone in mammals is called *Haversian bone* because it is composed of structural units called **Haversian systems** (**Figure 48.18**). Each Haversian system is a set of thin, concentric bony cylinders, between which are the osteocytes in their lacunae. Through the center of each Haversian system runs a narrow canal containing blood vessels and nerves. Adjacent Haversian systems are separated by boundaries called *glue lines*. Haversian bone is resistant to fracturing because cracks tend to stop at glue lines.

Bones that have a common joint can work as a lever

Muscles and bones work together around **joints**, where two or more bones come together. Different kinds of joints allow motion in different directions (**Figure 48.19**), but muscles can exert force in only one direction. Therefore, muscles create move-

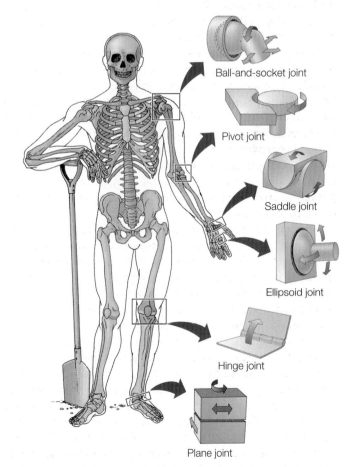

Ball-and-socket joint

Pivot joint

Saddle joint

Ellipsoid joint

Hinge joint

Plane joint

48.19 Types of Joints The designs of joints are similar to mechanical counterparts and enable a variety of movements.

── yourBioPortal.com ──
GO TO Web Activity 48.3 • Joints

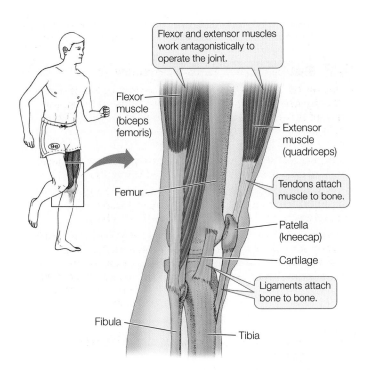

48.20 Joints, Ligaments, and Tendons A side view of the knee shows the interactions of muscle, bone, cartilage, ligaments, and tendons at this crucial and vulnerable human joint.

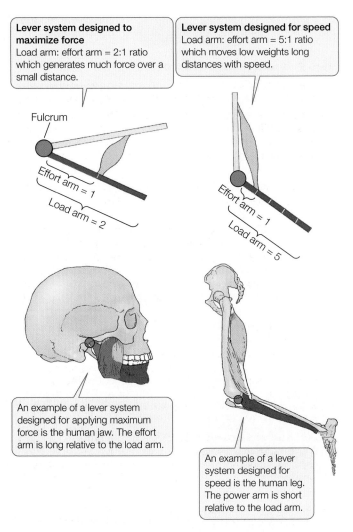

48.21 Bones and Joints Work Like Systems of Levers A lever system can be designed for maximizing either force or speed.

ment around joints by working in antagonistic pairs: when one muscle contracts, the other relaxes. When both contract, the joint becomes rigid (which is important for maintaining posture, for example).

With respect to a particular joint, such as the knee, we refer to the muscle that bends, or flexes, the joint as the **flexor**, and the muscle that straightens, or extends, the joint as the **extensor**. The bones that meet at the joint are held together by **ligaments**, which are flexible bands of connective tissue. Other straps of connective tissue, called **tendons**, attach the muscles to the bones (**Figure 48.20**). In many kinds of joints, only the tendon spans the joint, sometimes moving over the surfaces of the bones like a rope over a pulley. The tendon of the quadriceps muscle traveling over the knee joint is what is tapped to elicit the knee-jerk reflex (see Figure 47.3).

Bones constitute a system of levers that are moved around joints by the muscles. A lever has an *effort arm* and a *load arm* that work around a *fulcrum* (pivot). The length ratio of the two arms determines whether a particular lever can exert a lot of force over a short distance or is better at translating force into large or fast movements. Compare the jaw and knee joints, for example (**Figure 48.21**). The effort arm of the jaw is long relative to the load arm, allowing the jaw to apply great force over a small distance. Think of the powerful jaws of carnivores that can easily crack bones. The effort arm of the lower leg, by contrast, is short relative to the load arm, so you can run fast, jump high, and deliver swift kicks.

48.3 RECAP

Muscles can only contract and relax; to achieve organized movement, they must pull against rigid structures—other muscles, hydrostatic skeletons, exoskeletons, or endoskeletons.

- How do the muscles and fluid-filled body cavity of an earthworm interact to enable the animal to crawl? See pp. 1018–1019 and Figure 48.14

- Describe the differences between membranous and cartilaginous bone and between compact and cancellous bone. See p. 1020 and Figure 48.17

- In terms of levers, explain how specific joints can produce maximum force versus maximum speed. See pp. 1021–1022 and Figure 48.21

CHAPTER SUMMARY

48.1 How Do Muscles Contract?

- **Skeletal muscle** consists of bundles of **muscle fibers**. Each skeletal muscle fiber is a large cell containing multiple nuclei.

- Skeletal muscles contain numerous **myofibrils**, which are bundles of **actin** and **myosin** filaments. The regular, overlapping arrangement of the actin and myosin filaments into **sarcomeres** gives skeletal muscle its striated appearance. Review Figure 48.1, **WEB ACTIVITY 48.1**

- The changes in the banding patterns of sarcomeres led to the **sliding filament model** of muscle contraction. Review Figure 48.2, **ANIMATED TUTORIAL 48.1**

- The molecular mechanism of muscle contraction involves the binding of the globular heads of myosin molecules to actin. Upon binding, the myosin head changes its conformation, causing the two filaments to move past each other (the sliding filament mechanism). Release of the myosin heads from actin and their return to their original conformation requires ATP. Review Figures 48.3 and 48.6

- All the fibers activated by a single motor neuron constitute a **motor unit**. Each nerve ending of the motor neuron forms a synapse with the muscle cell membrane. When neurotransmitter is released at the synapse, the muscle cell membrane is depolarized and action potentials are generated. Action potentials spread across the plasma membrane and through the **T tubules**, causing Ca^{2+} to be released from the **sarcoplasmic reticulum**. Review Figure 48.5, **WEB ACTIVITY 48.2**

- Ca^{2+} binds to **troponin** and changes its conformation, pulling the **tropomyosin** strands away from the myosin-binding sites on the actin filament. The muscle fiber continues to contract until the Ca^{2+} is returned to the sarcoplasmic reticulum. Review Figure 48.6

- **Cardiac muscle** cells are striated, uninucleate, branching, and electrically connected by gap junctions, so that action potentials spread rapidly throughout sheets of cardiac muscle and cause coordinated contractions. Some cardiac muscle cells are pacemaker cells that generate and conduct electrical signals.

- **Smooth muscle** provides contractile force for internal organs. Smooth muscle cells respond to stretch and to neurotransmitters from the autonomic nervous system. Review Figure 48.8, **ANIMATED TUTORIAL 48.2**

- In skeletal muscle, a single action potential causes a minimum unit of contraction called a **twitch**. Twitches occurring in rapid succession can be summed, thus increasing the strength of contraction. Maximum sustained tension is known as **tetanus**. Review Figure 48.10

48.2 What Determines Muscle Performance?

- **Slow-twitch fibers** facilitate extended, aerobic work; **fast-twitch fibers** generate maximum forces for short periods of time. The ratio of slow-twitch to fast-twitch fibers in the muscles of an individual is largely genetically determined. Review Figure 48.11

- The force that a muscle fiber can produce depends on its initial state of extension or contraction. Review Figure 48.12

- Anaerobic exercise stimulates the enlargement of muscle fibers through production of new microfilaments. Aerobic exercise stimulates greater oxidative capacity of muscle fibers.

- Muscle performance depends on a supply of ATP. Available ATP and creatine phosphate can fuel maximum tension instantaneously but are exhausted within seconds. Glycolysis can regenerate ATP rapidly, but this process is inhibited by accumulation of lactic acid. Oxidative metabolism delivers ATP more slowly but can continue to do so for a long time. Review Figure 48.13

48.3 How Do Skeletal Systems and Muscles Work Together?

- **Skeletal systems** provide supports against which muscles can pull.

- **Hydrostatic skeletons** are fluid-filled body cavities that can be squeezed by muscles. Review Figure 48.14

- Exoskeletons are hardened outer surfaces to which internal muscles are attached.

- **Endoskeletons** are internal systems of rigid rodlike, platelike, and tubelike supports, consisting of **bone** and **cartilage** to which muscles are attached. Review Figure 48.15

- Bone is continually remodeled by **osteoblasts**, which lay down new bone, and **osteoclasts**, which erode bone. Review Figure 48.16

- Bones develop from connective tissue membranes (**membranous bone**) or from cartilage (**cartilage bone**) through ossification. Cartilage bone can grow until centers of ossification meet. Review Figure 48.17

- Bone can be **compact** (solid and hard), or **cancellous** (containing numerous internal spaces). Most of the compact bone of mammals is composed of **Haversian systems**. Review Figure 48.18

- **Joints** enable muscles to power movements in different directions. Muscles and bones work together around joints as systems of levers. Review Figures 48.19, 48.21, **WEB ACTIVITY 48.3**

- **Tendons** connect muscles to bones; **ligaments** connect bones to one another. Review Figure 48.20

SELF-QUIZ

1. Smooth muscle differs from both cardiac and skeletal muscle in that
   - *a.* it can act as a pacemaker for rhythmic contractions.
   - *b.* contractions of smooth muscle are not due to interactions between neighboring microfilaments.
   - *c.* neighboring cells are electrically connected by gap junctions.
   - *d.* neighboring cells are tightly coupled by intercalated discs.
   - *e.* the membranes of smooth muscle cells are depolarized by stretching.

2. Fast-twitch fibers differ from slow-twitch fibers in that
   - *a.* they are more common in the leg muscles of champion sprinters than marathon runners.
   - *b.* they have more mitochondria.
   - *c.* they fatigue less rapidly.
   - *d.* their abundance is more a product of training than of genetics.
   - *e.* they are more common in postural muscles than in finger muscles.

3. The role of Ca^{2+} in the control of muscle contraction is to
   a. cause depolarization of the T tubule system.
   b. change the conformation of troponin, thus exposing myosin-binding sites.
   c. change the conformation of myosin heads, thus causing microfilaments to slide past each other.
   d. bind to tropomyosin and break actin–myosin cross-bridges.
   e. block the ATP-binding site on myosin heads, enabling muscles to relax.

4. Fifteen minutes into a 10-k run, what is the major energy source of the leg muscles?
   a. Preformed ATP
   b. Glycolysis
   c. Oxidative metabolism
   d. Pyruvate and lactate
   e. High-protein drink consumed right before the race

5. Which statement about skeletal muscle contraction is *not* true?
   a. A single action potential at the neuromuscular junction is sufficient to cause a muscle to twitch.
   b. Once maximum muscle tension is achieved, no ATP is required to maintain that level of tension.
   c. An action potential in the muscle cell activates contraction by releasing Ca^{2+} into the sarcoplasm.
   d. Summation of twitches leads to a graded increase in the tension that can be generated by a single muscle fiber.
   e. The tension generated by a muscle can be varied by controlling how many of its motor units are active.

6. Which statement about the structure of skeletal muscle is *true*?
   a. The light bands of the sarcomere are the regions where actin and myosin filaments overlap.
   b. When a muscle contracts, the A bands of the sarcomere lengthen.
   c. The myosin filaments are anchored in the Z lines.
   d. When a muscle contracts, the H zone of the sarcomere shortens.
   e. The sarcoplasm of the muscle cell is contained within the sarcoplasmic reticulum.

7. Insects can beat their wings at exceptionally high frequencies because
   a. their wing muscles have mostly fast-twitch fibers.
   b. their motor neurons can fire action potentials at a very high frequency.
   c. their wings have exoskeletal supports.
   d. their wing muscles have extensive sarcoplasmic reticulum that cycles Ca^{2+} very fast.
   e. their wing muscles can generate a rapid oscillation of contraction asynchronous with motor neuron firing.

8. The long bones of our arms and legs are strong and can resist both compressional and bending forces because
   a. they are solid rods of compact bone.
   b. their extracellular matrix contains crystals of calcium carbonate.
   c. their extracellular matrix consists mostly of collagen and polysaccharides.
   d. they have a high density of osteoclasts.
   e. they consist of lightweight cancellous bone with an internal meshwork of supporting elements.

9. If we compare the jaw and knee joints as lever systems,
   a. the jaw joint can apply greater compressional forces.
   b. their ratios of power arm to load arm are about the same.
   c. the knee joint has greater rotational abilities.
   d. the knee joint has a greater ratio of power arm to load arm.
   e. only the jaw is a hinged joint.

10. Which statement about skeletons is *true*?
    a. They can consist of mostly cartilage.
    b. Hydrostatic skeletons cannot be used for locomotion.
    c. An advantage of exoskeletons is that they can continue to grow throughout the life of the animal.
    d. External skeletons must remain flexible, so they never include calcium carbonate crystals, as bones do.
    e. Internal skeletons consist of four different types of bone: compact, cancellous, membranous, and Haversian.

FOR DISCUSSION

1. You can see from the structure of a sarcomere that it can shorten only by a certain percentage of its resting length. Yet muscles can cause a wide variety of ranges of movement—compare the range of movement of a toe and a leg. What are two adaptive design features of muscles and skeletons that can maximize a muscle's ability to cause a greater range of movement of an appendage?

2. If athletes train up until a day before competition and then take a day of rest, performance in which types of events will be most affected by what they eat during the rest day?

3. Wombats are powerful digging animals, and kangaroos are powerful jumping animals. How do you think their leg structures compare in terms of their designs as lever systems?

4. Why are ducks better long-distance fliers than chickens?

5. If an adolescent breaks a leg bone close to the ankle joint, after the break heals, that leg may not grow as long as the other one. Why?

ADDITIONAL INVESTIGATION

For aerobic exercise endurance, the most important limiting factor is muscle glycogen reserve, but the muscles also use glucose from the blood. How would you design an experiment to test whether or not carbohydrate intake during exercise improves endurance, and if it does, what its effect on the sparing of muscle glycogen is?

Why do elephants have long noses?

Elephants use their trunks in a variety of ways. They pluck leaves from trees and pull up plants from the ground; they pick up objects ranging in size from trees to seeds; they suck in water and squirt it into their mouths; they spray water and dust over their bodies; they smell the air around them; they communicate by touch; and mothers can wallop their calves when they misbehave.

Of course, elephants also use their trunks to breathe. This function takes on special interest when elephants go into deep water—for example, to cross rivers. Elephants can swim or walk on the bottom. In either case, they use their trunks as snorkels. As long as the tip of the trunk is above water, an elephant can breathe.

You may be familiar with diving using a snorkel about a foot long. When you want to dive deeper, you hold your breath. Water is heavy, and as you swim deeper, water pressure increases and presses on your body. The air in your lungs is compressible, but the tissues of your body are not. So as you swim deeper, the volume of your lungs decreases.

Think what would happen if you tried to breathe through a long snorkel in deep water. Because the snorkel would be open to the air above, the pressure in your lungs would be at water surface level pressure, but the surrounding water would be pressing on your body. To expand your lungs, you would have to exert enough force to push against the surrounding water. Not so easy! Unlike elephants, we solve this problem with scuba equipment; by breathing air from a pressurized tank with a regulator valve, we keep the air in our lungs at the same pressure as the surrounding water.

The chest of a snorkeling elephant can be 3 meters below the surface—and for every 3 meters, the water pressure increases by 0.3 atmospheres. Even if an elephant's respiratory muscles were strong enough to work against this pressure, there would be a serious problem for the blood vessels lining the cavity in which the lungs are suspended. The pressure difference across the walls of those blood vessels is the difference between the pressure in the vessels (blood pressure plus water pressure) and the air pressure at the surface (where the trunk opens to the air). If the blood vessels couldn't withstand that pressure difference, they would rupture and fill the chest cavity with blood.

Proboscis as Snorkel Elephants can cross deep bodies of water either swimming or by walking on the bottom—whichever will allow it to keep its trunk above water.

Breathing Under Water Adapted to the aquatic life, fishes such as this whitetip shark have gills that extract respiratory gases from the water. The human cannot "breathe" water, so to enter this environment, he must carry a supply of respiratory gases in a scuba tank.

How does a snorkeling elephant avoid such damage? In all mammals except elephants, the surfaces of the lungs and the chest cavity move easily against each other. As the lungs inflate and deflate, their surface slips and slides against that of the chest cavity. In elephants, however, dense connective tissue attaches the surface of the lungs to the surface of the chest cavity. This connective tissue acts as reinforcement for the fragile blood vessels on the surface of the chest cavity and prevents them from rupturing and filling the cavity with blood, which would destroy the function of the lungs—exchange of oxygen and carbon dioxide.

IN THIS CHAPTER we will explore adaptations of the respiratory systems of both water and air breathers for exchanging oxygen and carbon dioxide with the environment. We will first discuss the physical factors that limit these gas exchange systems and identify those factors that natural selection has optimized. We will then examine the gas exchange systems of insects, fishes, birds, and humans and describe the adaptations of the blood for transporting respiratory gases. Finally, we will see how respiratory gas exchange systems are controlled and regulated.

49.1 What Physical Factors Govern Respiratory Gas Exchange?

Gas exchange systems are made up of gas exchange surfaces and the mechanisms that *ventilate* and *perfuse* those surfaces. The **respiratory gases** that organisms must exchange are oxygen (O_2) and carbon dioxide (CO_2). Cells need to obtain O_2 from the environment to produce an adequate supply of ATP by cellular respiration (see Chapter 9). CO_2 is an end product of cellular respiration, and it must be removed from the body to prevent toxic effects.

Diffusion is the only means by which respiratory gases are exchanged between an animal's internal body fluids and the outside medium (air or water). There are no active transport mechanisms to move respiratory gases across biological membranes. Because diffusion is a physical process, knowing what physical factors influence rates of diffusion helps us understand the diverse adaptations of gas exchange systems. (You should review the discussion about the physical nature of diffusion in Section 6.3.)

Diffusion is driven by concentration differences

Diffusion results from the random motion of molecules, and the net movement of a molecule is down its concentration gradient. Concentrations in solutions are easy to think about because they are simply the amount of solute per volume of solution. Concentrations of gases are more complicated because the total number of gas molecules in a specified volume depends on pressure; there are twice as many gas molecules in a liter of gas under 2 atmospheres of pressure as there are in a liter under 1 atmosphere of pressure. Biologists express the concentrations of different gases in a mixture by the *partial pressures* of those gases.

First we have to know what the total pressure is, and we measure that with an instrument called a barometer. There are many types of barometers, but the classical one is a glass tube closed at one end and filled with mercury. This is inverted over a pool of mercury with the open end of the tube under the surface of the mercury. At sea level, the pressure exerted by the atmosphere supports, and therefore equals, a column of mercury in the tube that is about 760 mm high (depending on the weather). Therefore, **barometric pressure** (atmospheric pressure) at sea level is 760 mm of mercury (mm Hg). Because dry air is 20.9 percent O_2, the **partial pressure of oxygen** (P_{O_2}) at sea level is 20.9 percent of 760 mm Hg, or about 159 mm Hg. If two gas mixtures are separated by a membrane permeable to O_2, O_2 will diffuse from the mixture where its partial pressure is higher to the mixture where its partial pressure is lower.

Describing the concentration of respiratory gases in a liquid such as water is more complicated because another factor is involved—the solubility of the gas in the liquid. Thus, the actual amount of a gas in a liquid depends on the partial pressure of that gas in the gas phase in contact with the liquid as well as on the solubility of that gas in that liquid. However, the diffusion of the gas between the gaseous phase and the liquid still depends on the partial pressures of the gas in the two phases. Whether in air or water, the diffusion rate of a substance depends on its concentration gradient and on other factors that are expressed in *Fick's law of diffusion.*

Fick's law applies to all systems of gas exchange

Diffusion is a physical phenomenon that can be described quantitatively with a simple equation called **Fick's law of diffusion**. All environmental variables that limit respiratory gas exchange and all adaptations that maximize respiratory gas exchange are reflected in one or more components of this equation. Fick's law is written as

$$Q = DA\frac{P_1 - P_2}{L}$$

where

- Q is the rate at which a gas such as O_2 diffuses between two locations.
- D is the *diffusion coefficient*, which is a characteristic of the diffusing substance, the medium, and the temperature. For

example, perfume has a higher D than motor oil vapor, and all substances diffuse faster at higher temperatures and faster in air than in water. Temperature is not expressed explicitly in Fick's law because the diffusion coefficient is usually determined at room temperature (about 20°C).

- A is the cross-sectional area through which the gas is diffusing.
- P_1 and P_2 are the partial pressures of the gas at the two locations.
- L is the path length, or distance, between the two locations.

Therefore, $(P_1 - P_2)/L$ is a *partial pressure gradient.*

Animals can maximize D for respiratory gases by using air rather than water as their gas exchange medium whenever possible. All other adaptations for maximizing respiratory gas exchange must influence the surface area (A) for gas exchange or the partial pressure gradient across that surface area.

Air is a better respiratory medium than water

The slow diffusion of O_2 molecules in water affects both air- and water-breathing animals. Eukaryotic cells carry out cellular respiration in their mitochondria, which are located in the cytoplasm—an aqueous medium. Cells are bathed in extracellular fluid—also an aqueous medium. In addition, all respiratory surfaces must be protected from drying out by a thin film of fluid through which O_2 must diffuse. Even in air-breathing animals, the slow rate of O_2 diffusion in water limits the efficiency of O_2 distribution from gas exchange surfaces to the sites of cellular respiration.

Diffusion of O_2 in water is so slow that even animal cells with low rates of metabolism can be no more than a few millimeters away from a good source of environmental O_2. Therefore, there are severe size and shape limits on the many species of invertebrates that lack internal systems for distributing O_2. Most of these species are very small, but some have grown larger by evolving a flat, thin body with a large external surface area (**Figure 49.1A**). Others have a very thin body built around a central cavity through which water can circulate (**Figure 49.1B**). A critical factor enabling larger, more complex animal bodies has been the evolution of specialized respiratory systems with large surface areas for enhancing gas exchange (**Figure 49.1C**).

O_2 can be obtained more easily from air than from water for several reasons:

- The O_2 content of air is much higher than the O_2 content of an equal volume of water. The maximum O_2 content of a bubbling stream in equilibrium with air is less than 10 ml of O_2 per liter of water. The O_2 content of the air over the stream is about 200 ml of O_2 per liter of air.

(A) *Pseudobiceros* sp.

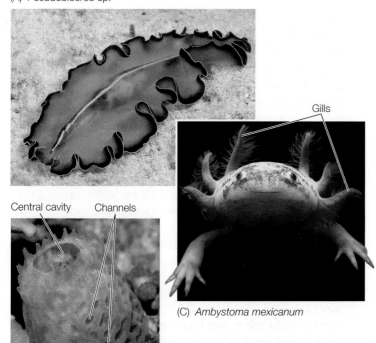

Gills

Central cavity   Channels

(C) *Ambystoma mexicanum*

(B) *Callyspongia plicifera*

49.1 Keeping in Touch with the Medium   (A) No cell in the leaflike body of this marine flatworm is more than a millimeter away from seawater. (B) Sponges have body walls perforated by many channels, which allow water to flow between the outside world and a central cavity. No cell in the sponge is more than a millimeter away from seawater. (C) A feathery fringe of gills on this aquatic larval salamander provides a large surface area for gas exchange. Blood circulating through the gills comes into close contact with the respiratory medium.

- O_2 diffuses about 8,000 times more rapidly in air than in water. That is why the O_2 content of a stagnant pond can be zero only a few millimeters below the surface.

- An animal has to do work to move water or air over its gas exchange surfaces. More energy is required to move water than air because water is 800 times denser than air and about 50 times more viscous. You can appreciate how important these facts were for the movement of life to the terrestrial environment—there were fewer constraints on the evolution of higher metabolic rates.

High temperatures create respiratory problems for aquatic animals

Animals that use water for their respiratory exchange medium are in a double bind when environmental temperatures rise. Most water-breathing animals are *ectotherms*—their body temperatures are closely tied to the temperature of the water around them. As the water temperature rises, an ectotherm's body temperature and metabolic rate rise (see Figure 40.8). Thus, water breathers need more O_2 as the water gets warmer. But warm water holds less dissolved gas than cold water does (think of the gases that escape when you open a warm bottle of soda). In addition, if an animal performs work to move water across its gas exchange surfaces (as fishes do), it must expend more energy to breath as water temperature rises. Therefore, as water temperature goes up, a water-breathing animal must extract more and more O_2 from an environment that is increasingly O_2 deficient, and a lower percentage of that O_2 is available to support activities other than breathing (**Figure 49.2**).

O_2 availability decreases with altitude

Just as a rise in water temperature reduces the supply of O_2 available to water-breathing animals, an increase in altitude reduces the O_2 supply for air breathers. At all altitudes, O_2 makes up 20.9 percent of the dry air; however, as you go up in altitude, the total amount of gas per unit volume decreases, as reflected in the barometric pressure. For example, at 5,800 m, barometric pressure is only half what it is at sea level, so the P_{O_2} at that altitude is only about 80 mm Hg. At the summit of Mount Everest (8,850 m), P_{O_2} is only about 50 mm Hg—roughly one-third what it is at sea level.

Because the movement of O_2 across respiratory gas exchange surfaces and into the body depends on diffusion, its rate of movement depends on the P_{O_2} difference between the air and the body fluids. Therefore, the drastically reduced P_{O_2} in the air at high altitudes constrains O_2 uptake. Because of this, mountain climbers attempting the highest peaks usually breathe O_2 from pressurized bottles.

CO_2 is lost by diffusion

Respiratory gas exchange is a two-way process: CO_2 diffuses out of the body as O_2 diffuses in. The direction and rate of diffusion of the respiratory gases across the exchange surfaces depend on the partial pressure gradients of the gases. The partial

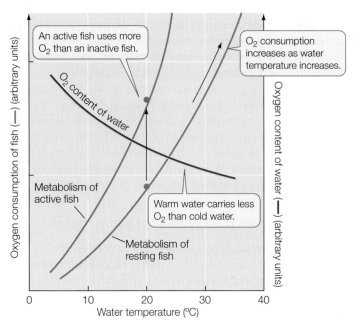

49.2 The Double Bind of Water Breathers Fishes need more O_2 when the water is warmer, but warm water carries less O_2 than cold water.

pressure gradients of O_2 and CO_2 across these exchange surfaces are quite different. The amount of CO_2 in the atmosphere is extremely low (0.03%), so for air-breathing animals there is always a large concentration gradient for diffusion of CO_2 from the body to the environment. Whereas the partial pressure gradient for O_2 decreases with increasing altitude, the gradient driving CO_2 out of the body hardly changes. The partial pressure of CO_2 in the atmosphere is close to zero both at sea level and atop Mount Everest.

In general, getting rid of CO_2 is not a problem for water-breathing animals because CO_2 is much more soluble in water than is O_2. Even in stagnant water, where the P_{CO_2} is higher than in moving water, the lack of O_2 becomes a problem for an animal long before CO_2 exchange difficulties arise.

49.1 RECAP

Respiratory gases are exchanged only by diffusion. Air is a better respiratory medium than water because a given volume of air has more O_2 than the same volume of water. O_2 diffuses faster in air than in water, and less work is required to move air over respiratory exchange surfaces.

- Describe how the variables in Fick's law of diffusion relate to respiratory systems. **See p. 1027**

- Why does a rise in water temperature create a double-bind situation for water-breathing animals? **See p. 1028 and Figure 49.2**

- Explain the concept of partial pressures of gases and how they relate to diffusion rates of O_2 and CO_2 at different altitudes. **See p. 1028**

Now that we have discussed the physical factors that influence diffusion rates of respiratory gases between animals and their environments, let's look at some of the adaptations that have evolved for maximizing respiratory gas exchange.

49.2 What Adaptations Maximize Respiratory Gas Exchange?

As you might expect from the components of Fick's law of diffusion, adaptations to maximize respiratory gas exchange can be categorized as those that:

- Increase the surface area for gas exchange
- Maximize the partial pressure difference driving diffusion
- Minimize the diffusion path length
- Minimize the diffusion that takes place in an aqueous medium

Respiratory organs have large surface areas

A variety of anatomical adaptations maximize the specialized body surface area (A) over which respiratory gases can diffuse. Water-breathing animals generally have *gills*, and air-breathing animals have *tracheae* or *lungs*. **External gills** are highly branched and folded extensions of the body surface that provide a large surface area for gas exchange with water (**Figure 49.3A**; see also Figure 49.1C). External gills are found in larval amphibians and in the larvae of many insects. Because they consist of thin, delicate tissues, external gills minimize the path length (L) traversed by diffusing molecules of O_2 and CO_2. External gills are vulnerable to damage, however, and are tempting morsels for predators, so in many animals protective body cavities for gills have evolved. Such **internal gills** are found in most mollusks and arthropods and in all fishes (**Figure 49.3B**).

   Lungs are internal cavities for respiratory gas exchange with air (**Figure 49.3C**). Their structure is quite different from that of gills. Lungs have a large surface area because they are highly divided; and because they are elastic, they can be inflated with air and deflated.

Insects have a respiratory gas exchange system consisting of a network of air-filled tubes called **tracheae** that branch through all tissues of the insect's body (**Figure 49.3D**). The terminal branches of these tubes are so numerous that they have an enormous surface area compared with the external surface area of the insect's body.

Transporting gases to and from exchange surfaces optimizes partial pressure gradients

Partial pressure gradients $[(P_1 - P_2)/L]$ drive diffusion across gas exchange surfaces; the larger the gradient, the greater the rate of gas exchange. These gradients can be maximized in several ways:

- *Minimization of path length*: Very thin tissues in gills and lungs reduce the diffusion path length (L).
- *Ventilation*: Actively moving the external medium over the gas exchange surfaces (i.e., breathing) regularly exposes those surfaces to fresh respiratory medium containing maximum O_2 and minimum CO_2 concentrations. This maximizes the concentration gradient.
- *Perfusion*: Actively moving the internal medium (e.g., blood) over the internal side of the exchange surfaces transports CO_2 to those surfaces and O_2 away from them, thus maximizing the concentration gradients driving diffusion.

This chapter describes four gas exchange systems. First we will look at the unique gas exchange system of insects. Then we will describe two highly efficient gas exchange systems, fish gills and bird lungs. Lastly we examine the structure and function of human lungs.

Insects have airways throughout their bodies

The tracheal system that enables insects to exchange respiratory gases extends to all tissues in the insect body. Thus, respiratory gases diffuse through air most of the way to and from every cell. The insect respiratory system communicates with the outside environment through gated openings called *spiracles* in the sides of the abdomen (**Figure 49.4A,B**). The spiracles open to allow gas exchange and then close to decrease water loss. They open into tubes called *tracheae* that branch into even finer tubes,

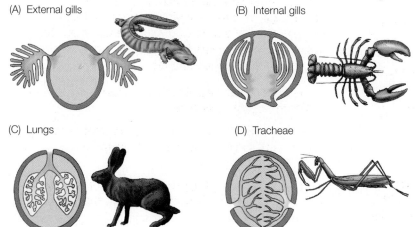

(A) External gills

(B) Internal gills

(C) Lungs

(D) Tracheae

49.3 Gas Exchange Systems Large surface areas (blue in these diagrams) for the diffusion of respiratory gases are common features of animals. External (A) and internal (B) gills are adaptations for gas exchange with water. Lungs (C) and tracheae (D) are organs for gas exchange with air.

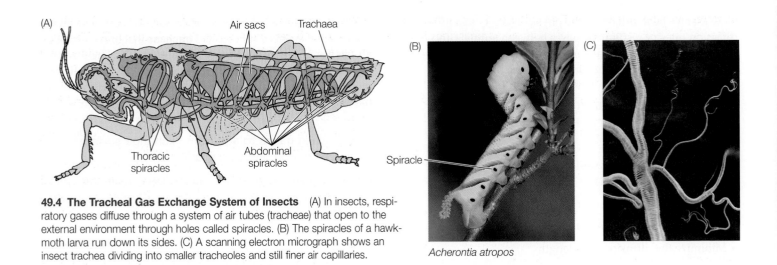

49.4 The Tracheal Gas Exchange System of Insects (A) In insects, respiratory gases diffuse through a system of air tubes (tracheae) that open to the external environment through holes called spiracles. (B) The spiracles of a hawk-moth larva run down its sides. (C) A scanning electron micrograph shows an insect trachea dividing into smaller tracheoles and still finer air capillaries.

Acherontia atropos

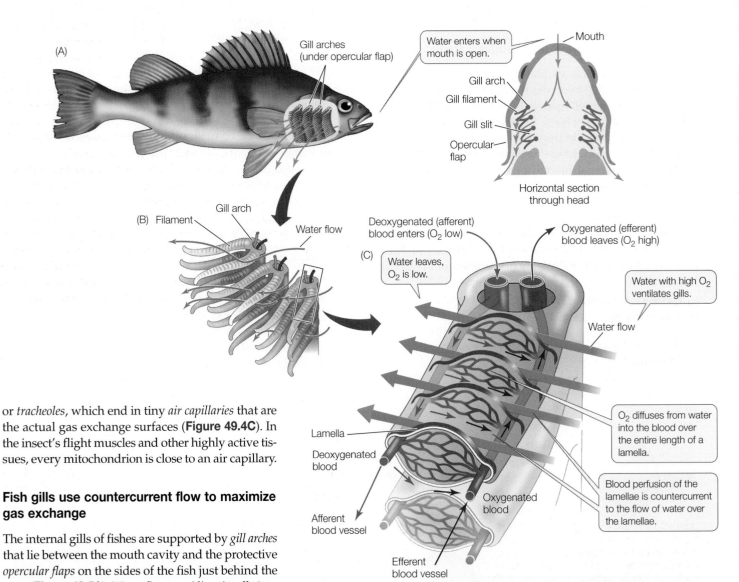

or *tracheoles*, which end in tiny *air capillaries* that are the actual gas exchange surfaces (**Figure 49.4C**). In the insect's flight muscles and other highly active tissues, every mitochondrion is close to an air capillary.

Fish gills use countercurrent flow to maximize gas exchange

The internal gills of fishes are supported by *gill arches* that lie between the mouth cavity and the protective *opercular flaps* on the sides of the fish just behind the eyes (**Figure 49.5A**). Water flows *unidirectionally* into the fish's mouth, over the gills, and out from under the opercular flaps. Thus, the gills are continuously bathed with fresh water. This constant, one-way flow

49.5 Fish Gills (A) Water flows unidirectionally over the gills of a fish. (B) Gill filaments have a large surface area and thin tissues. (C) Blood flows through the lamellae in the direction opposite (left to right, in this depiction; small blue and red arrows) to the flow of water (right to left; large blue arrows) over the lamellae.

of water over the gills maximizes the P_{O_2} on the external gill surfaces. On the internal side of the gill membranes, the circulation of blood minimizes the P_{O_2} by sweeping O_2 away as rapidly as it diffuses across.

Gills have an enormous surface area for gas exchange because they are so highly divided. Each gill consists of hundreds of ribbonlike *gill filaments* (**Figure 49.5B**). The upper and lower flat surfaces of each gill filament are covered with rows of evenly spaced folds, or *lamellae*. The lamellae are the actual gas exchange surfaces. Because they are exceedingly thin, the path length (*L*) for diffusion of gases between blood and water is minimized. The surfaces of the lamellae consist of highly flattened epithelial cells, so the water and the fish's red blood cells are separated by little more than 1 or 2 micrometers.

The flow of blood perfusing the inner surfaces of the lamellae, like the flow of water over the gills, is unidirectional. *Afferent* blood vessels bring deoxygenated blood to the gills, while *efferent* blood vessels take oxygenated blood away from the gills (**Figure 49.5C**). Blood flows through the lamellae in the direction opposite to the flow of water over the lamellae. This **countercurrent flow** optimizes the P_{O_2} gradient between water and blood, making gas exchange more efficient than it would be in a system using concurrent (parallel) flow (**Figure 49.6**).

Some fishes, including anchovies, tuna, and certain sharks, ventilate their gills by swimming almost constantly with their mouths open. Most fishes, however, ventilate their gills by means of a two-pump mechanism. The closing and contracting of the mouth cavity pushes water over the gills, and the expansion of the opercular cavity prior to opening of the opercular flaps pulls water over the gills.

These adaptations for maximizing the surface area (*A*) for diffusion, minimizing the path length (*L*) for diffusion, and maximizing the P_{O_2} gradient allow fishes to extract an adequate supply of O_2 from meager environmental sources.

Birds use unidirectional ventilation to maximize gas exchange

Birds are remarkable for their ability to sustain high levels of activity for a long time—for example, on long-distance flights—even at high altitudes where mammals cannot even survive. The first team to climb Mount Everest was sur-

prised to see birds flying over the mountain when they themselves could barely move without supplemental O_2. The highest recorded flight of a bird is from a vulture that collided with an airliner at 11,278 meters; a human could not exist at that altitude without supplemental O_2. Yet the lungs of a bird are smaller than the lungs of a similar-sized mammal, and bird lungs expand and contract less during a breathing cycle than do mammalian lungs. Furthermore, bird lungs *are compressed* during inhalation and *expand* during exhalation.

The structure of bird lungs allows air to flow unidirectionally through the lungs, rather than bidirectionally through all the same airways, as it does in mammals. Because mammalian lungs are never completely emptied of air during exhalation, there is always some lung volume that is not ventilated with fresh air. The air remaining in lungs and airways after exhalation is called **dead space**. Bird lungs, by contrast, have very little dead space, and the fresh incoming air is not mixed with stale air. In this way, a high P_{O_2} gradient is maintained.

In addition to lungs, birds have **air sacs** at several locations in their bodies (**Figure 49.7A**). The air sacs are interconnected with each other, with the lungs, and with air spaces in some of the bones. The air sacs receive inhaled air, but they are not gas exchange surfaces. As in other air-breathing vertebrates, air enters and leaves a bird's gas exchange system through the **trachea** (commonly known as the *windpipe*, and not to be confused with the air-conducting tracheae of insects), which divides into smaller airways called **bronchi** (singular *bronchus*).

The bronchi divide into tubelike **parabronchi** that run parallel to one another through the lungs (**Figure 49.7B**). Branching off the parabronchi are numerous tiny *air capillaries*. Air flows through the parabronchi and diffuses into the air capillaries, which are the gas exchange surfaces. They are so numerous that they provide an enormous surface area for gas exchange. The parabronchi coalesce into larger bronchi that take the air out of the lungs and back to the trachea. Thus the anatomy of a bird's airways allows air to flow unidirectionally and continuously through the lungs.

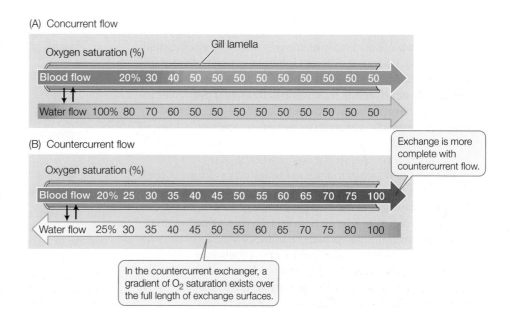

49.6 Countercurrent Exchange Is More Efficient In these models of concurrent and countercurrent gas exchange, the numbers represent the O_2 saturation percentages of blood and water. (A) In a concurrent exchanger, the saturation percentages of blood and water reach equilibrium halfway across the exchange surface. (B) A countercurrent exchanger allows more complete gas exchange because the water is always more O_2-saturated than the blood; thus a gradient of O_2 saturation is maintained.

(A) Concurrent flow

Oxygen saturation (%)   Gill lamella

Blood flow   20% 30 40 50 50 50 50 50 50 50 50 50

Water flow 100% 80 70 60 50 50 50 50 50 50 50 50

(B) Countercurrent flow

Oxygen saturation (%)

Blood flow 20% 25 30 35 40 45 50 55 60 65 70 75 100

Water flow 25% 30 35 40 45 50 55 60 65 70 75 80 100

Exchange is more complete with countercurrent flow.

In the countercurrent exchanger, a gradient of O_2 saturation exists over the full length of exchange surfaces.

(A) Avian air sacs and lungs

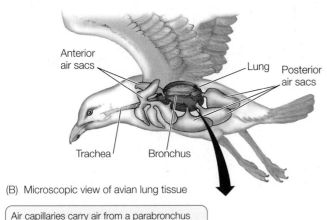

(B) Microscopic view of avian lung tissue

Air capillaries carry air from a parabronchus over blood capillaries, where O_2 is absorbed, and then out through the parabronchus.

Air

Parabronchus

Blood capillary

Air capillaries

49.7 The Respiratory System of a Bird (A) Air sacs and air spaces in the bones are unique to birds. (B) Air flows through bird lungs unidirectionally in parabronchi. Air capillaries, the site of gas exchange, branch off the parabronchi.

The puzzle of how birds breathe was solved by placing small O_2 electronic sensors at different locations in birds' air sacs and airways. The birds were then exposed to pure O_2 for just a single breath, which made it possible to track that particular inhalation. The experiment demonstrated that a single breath remains in a bird's gas exchange system for two cycles of inhalation and exhalation, and that the air sacs work like bellows; inhalation expands the sacs, and exhalation compresses them to maintain a continuous and unidirectional flow of fresh air through the lungs (**Figure 49.8**).

The advantages of the bird gas exchange system are similar to those of fish gills. The air sacs keep fresh air flowing unidirectionally over the gas exchange surfaces. Thus, a bird can supply its gas exchange surfaces with a continuous flow of fresh air that has a P_{O_2} close to that of the ambient air. Even when the P_{O_2} of the ambient air is only slightly above that of the blood, O_2 can diffuse from air to blood.

Tidal ventilation produces dead space that limits gas exchange efficiency

Lungs evolved in the first "air-gulping" vertebrates as outpocketings of the digestive tract. Although their structure has evolved considerably, lungs remain dead-end sacs in all air-breathing vertebrates except birds. Because of this, ventilation cannot be constant and unidirectional but must be **tidal**: air flows in and exhaled gases flow out by the same route. Since the lungs and airways can never be completely emptied of air, they always contain dead space. We can easily measure the volumes of air exchanged during breathing, but we have to use an indirect method to measure the dead space contained in the lungs and airways.

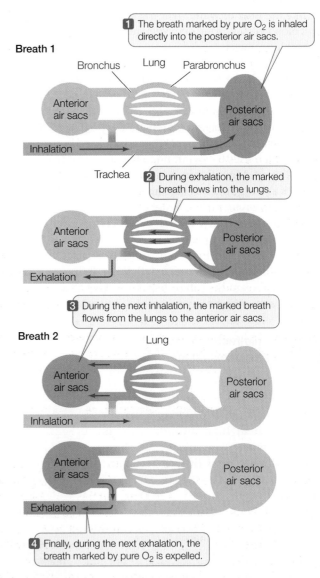

49.8 The Path of Air Flow through Bird Lungs The air a bird takes in by breathing (blue) travels through the lungs in one direction, from the posterior to the anterior air sacs. Each breath of air remains in the system for two breathing cycles.

yourBioPortal.com
GO TO Animated Tutorial 49.1 • Airflow in Birds

A *spirometer* is a device that measures the volume of air breathed in and out (**Figure 49.9**). Using a human as an example, the amount of air that moves in and out per breath when at rest is called the **tidal volume (TV)** (about 500 ml for an average human adult). When we breathe in as much as possible, the additional volume is the **inspiratory reserve volume (IRV)**. Conversely, if we forcefully exhale as much air as possible, the additional amount of air expelled is the **expiratory reserve volume (ERV)**. The maximum capacity for air exchange in one breath, or the **vital capacity (VC)**, is the sum of TV + IRV + ERV. The vital capacity of an athlete is generally greater than that of a non-athlete, and vital capacity decreases with age because of stiffening of the lung tissue.

Vital capacity is not the entire lung capacity because, as mentioned above, there is dead space, also called the **residual volume (RV)**. We can't measure RV directly with the spirometer, but we can measure it indirectly using the *helium dilution method*. Briefly, a person breathes from a reservoir with a known volume of air containing a known amount of helium (He). The helium is not absorbed from the lungs, so it becomes evenly distributed between the lungs and the reservoir as the subject inhales and exhales. Because the fixed amount of He becomes dispersed in a larger volume of air, its concentration decreases. That decrease in He concentration enables us to calculate the subject's **functional residual volume (FRV)**, which is the ERV + RV. Since we *can* measure the ERV with the spirometer, we can subtract it and obtain the RV.

───── **yourBioPortal.com** ─────
GO TO Working with Data • Calculating the Functional Residual Volume

Why is the RV important? Referring to Figure 49.9, you will see that for a normal person the ERV is about 1000 ml and the RV is 1000 ml. Thus the FRV is 2000 ml, but the tidal volume is only 500 ml. Thus, the air that reaches the alveoli with each breath consists of only 500 cc of fresh air diluted by 2000 cc of stale air. The maximum P_{O_2} in this mixed air is much below the P_{O_2} of the outside air, and because of the tidal ventilation pattern, the P_{O_2} in the alveoli is steadily dropping during the breathing cycle. The RV is important because it contributes to the FRC and to the dilution of the O_2 in the inhaled air. Any disease or condition that increases the RV (such as emphysema or pulmonary fibrosis) compromises a patients respiratory ability. Similarly, considering the mixing of fresh air with the FRV, you can understand why reductions in tidal volume can be a prob-

lem—and therefore why patients recovering from surgery are encouraged to breathe deeply, even if it hurts.

Offsetting the inefficiencies of tidal breathing, mammalian lungs have some design features to maximize the rate of gas exchange: an enormous surface area and a very short path length for diffusion.

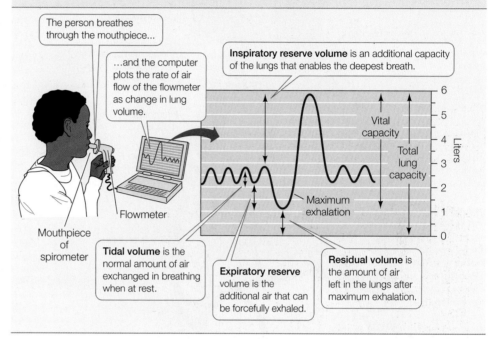

TOOLS FOR INVESTIGATING LIFE

49.9 Measuring Lung Ventilation
A spirometer is a device that measures the volume of air a person breathes through a mouthpiece. The combined tidal volume, inspiratory reserve volume, and expiratory reserve volume are the lungs' vital capacity.

The person breathes through the mouthpiece...

...and the computer plots the rate of air flow of the flowmeter as change in lung volume.

Inspiratory reserve volume is an additional capacity of the lungs that enables the deepest breath.

Vital capacity

Total lung capacity

Maximum exhalation

Mouthpiece of spirometer

Flowmeter

Tidal volume is the normal amount of air exchanged in breathing when at rest.

Expiratory reserve volume is the additional air that can be forcefully exhaled.

Residual volume is the amount of air left in the lungs after maximum exhalation.

49.2 RECAP

The major adaptations that increase animals' efficiency of respiratory gas exchange are a large surface area for exchange and a maximized concentration gradient across that surface.

- Describe three different ways that the concentration gradient for gas exchange is maximized across fish gills. See pp. 1030–1031 and Figures 49.5 and 49.6

- What respiratory adaptations enable birds to fly at extremely high altitudes? See pp. 1031–1032 and Figures 49.7 and 49.8

- Explain why residual volume limits the efficiency of tidal breathing. See pp. 1032–1033 and Figure 49.9

Despite their limitations, mammalian lungs serve the respiratory needs of mammals well. Let's look at the human respiratory system as an example.

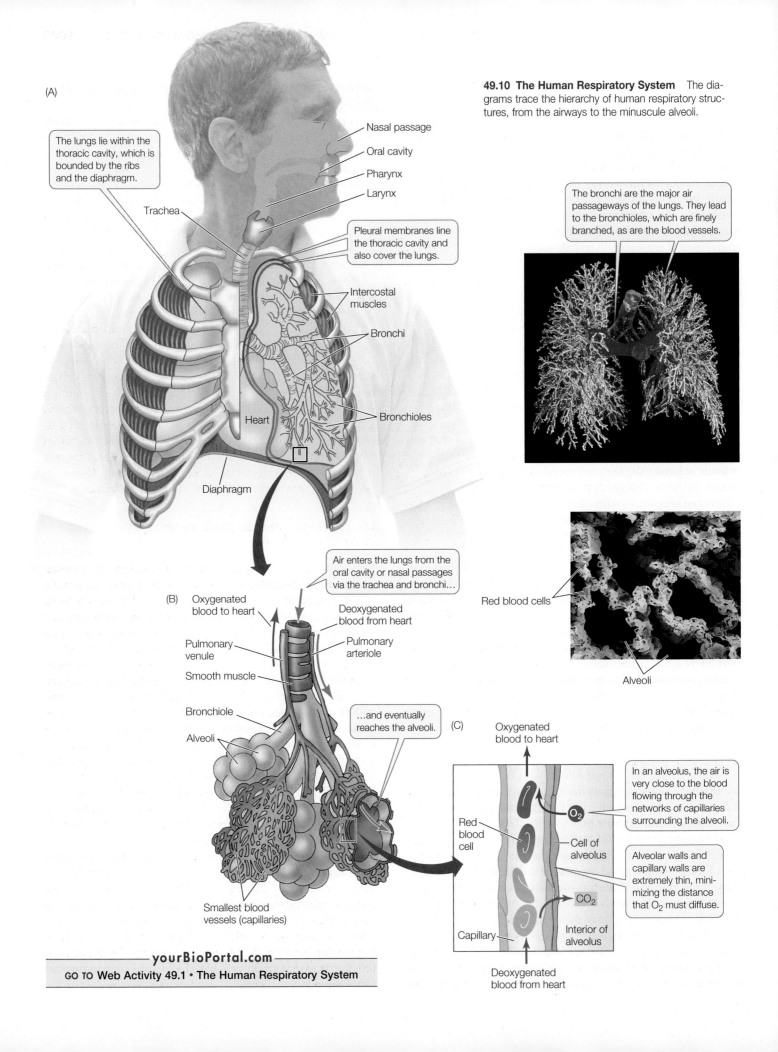

(A)

The lungs lie within the thoracic cavity, which is bounded by the ribs and the diaphragm.

Nasal passage

Oral cavity

Pharynx

Larynx

Trachea

Pleural membranes line the thoracic cavity and also cover the lungs.

Intercostal muscles

Bronchi

Bronchioles

Heart

Diaphragm

49.10 The Human Respiratory System The diagrams trace the hierarchy of human respiratory structures, from the airways to the minuscule alveoli.

The bronchi are the major air passageways of the lungs. They lead to the bronchioles, which are finely branched, as are the blood vessels.

Red blood cells

Alveoli

Air enters the lungs from the oral cavity or nasal passages via the trachea and bronchi…

(B)

Oxygenated blood to heart

Deoxygenated blood from heart

Pulmonary venule

Pulmonary arteriole

Smooth muscle

Bronchiole

Alveoli

…and eventually reaches the alveoli.

(C)

Oxygenated blood to heart

In an alveolus, the air is very close to the blood flowing through the networks of capillaries surrounding the alveoli.

Red blood cell

O_2

Cell of alveolus

Alveolar walls and capillary walls are extremely thin, minimizing the distance that O_2 must diffuse.

CO_2

Smallest blood vessels (capillaries)

Capillary

Interior of alveolus

Deoxygenated blood from heart

49.3 How Do Human Lungs Work?

In humans, air enters the lungs through the oral cavity or nasal passage, which join together in the *pharynx* (**Figure 49.10A**). Below the pharynx, the esophagus conducts food to the stomach, and the trachea leads to the lungs. At the beginning of this airway is the *larynx*, or voice box, which houses the vocal cords. The larynx is the "Adam's apple" that you can see or feel on the front of your neck. The trachea is about 2 cm in diameter. Its thin walls are prevented from collapsing by C-shaped bands of cartilage as air pressure changes during the breathing cycle. If you run your fingers down the front of your neck just below your larynx, you can feel a few of these bands of cartilage.

The trachea branches into two bronchi, one leading to each lung. The bronchi branch repeatedly to generate a treelike structure of progressively smaller airways extending to all regions of the lungs. After four branchings, the cartilage supports disappear, marking the transition to **bronchioles**. After about 16 branchings, the bronchioles are less than a millimeter in diameter, and tiny, thin-walled air sacs called **alveoli** begin to appear. Alveoli are the sites of gas exchange. After alveoli begin to appear, about six more branchings of the airways occur that end in clusters of alveoli (**Figure 49.10B**). Because the airways conduct air only to and from the alveoli and do not themselves participate in gas exchange, their volume is dead space.

Human lungs have about 300 million alveoli. Although each alveolus is very small, their combined surface area for diffusion of respiratory gases is about 70 square meters—about one-fourth the size of a basketball court. Each alveolus is made of very thin cells. Between and surrounding the alveoli are networks of capillaries whose walls are also made up of exceedingly thin cells. Where capillary meets alveolus, very little tissue separates them (**Figure 49.10C**), so the length of the diffusion path between air and blood is less than 2 micrometers.

Emphysema, a condition in which inflammation damages and eventually destroys the walls of the alveoli, is the fourth leading cause of death in the United States. As a result of this disease the lungs have fewer but larger alveoli, the RV increases, and the lungs loses elasticity. Although genetic factors can contribute to emphysema, the principal cause of the disease is smoking.

Respiratory tract secretions aid ventilation

Mammalian lungs produce two secretions that do not directly influence their gas exchange but do affect the process of ventilation: mucus and surfactant.

Many cells lining the airways produce sticky mucus that captures bits of dirt and microorganisms that are inhaled. Other cells lining the airways have cilia whose beating continually sweeps the mucus, with its trapped debris, up toward the pharynx, where it can be swallowed or spit out. This phenomenon, called the *mucus escalator*, can be adversely affected by inhaled pollutants. Smoking one cigarette can immobilize the cilia of the airways for hours. A smoker's cough results from the need to clear the obstructing mucus from the airways when the mucus escalator is out of order. The genetic disease *cystic fibrosis* causes

respiratory problems by affecting the respiratory mucus (see Figure 17.3B).

A **surfactant** is a substance that reduces the surface tension of a liquid. **Surface tension** gives the surface of a liquid such as water the properties of an elastic membrane, and it is why certain insects, such as water-striders, can walk on water. As discussed in Section 2.4, surface tension is the result of chemical forces of attraction between water molecules. The attractive forces working on the water molecules at the surface pull from below and from the sides but not from above. This imbalance of forces creates surface tension. The thin film of fluid covering the air-facing surfaces of the alveoli has surface tension that contributes to the lungs' elasticity. To inflate the lungs, enough force has to be generated to overcome both the elasticity of the lung tissue and the surface tension in the alveoli.

Lung surfactant is a fatty, detergent-like substance that is critical for reducing the work necessary to inflate the lungs. Certain cells in the alveoli release surfactant molecules when they are stretched. If a baby is born more than a month prematurely, these cells may not have developed the ability to produce surfactant. A baby with this condition, known as *respiratory distress syndrome*, will have great difficulty breathing and may die from exhaustion and lack of O_2. Common treatments for premature babies have been to put them on respirators to assist their breathing and to give them hormones to speed lung development. A new approach is to apply surfactant to the lungs via an aerosol.

Lungs are ventilated by pressure changes in the thoracic cavity

Human lungs are suspended in the **thoracic cavity**, a closed compartment bounded on the bottom by a sheet of muscle called the **diaphragm** (see Figure 49.10A). Each lung is covered by a continuous sheet of tissue called the **pleural membrane** that also lines the thoracic cavity adjacent to that lung. There is no real space between the pleural membranes of the lung and the thoracic cavity, but there is a thin film of fluid. This fluid lubricates the inner surfaces of the pleural membranes so they can slip and slide against each other during breathing movements. Just as we mentioned above in the explanation of surface tension, there are forces of attraction between the molecules of fluid in the pleural membranes. As a result, it is difficult to pull the pleural membranes apart. Think of two wet panes of glass or two wet microscope slides; you can slide them past each other, but it is difficult to separate them. While the inner surfaces of the pleural membranes are "stuck" to each other by surface tension, the outer surfaces are attached to the wall of the thoracic cavity and to the surface of the lung.

Inhalation and exhalation involve changes in the volume of the thoracic cavity (**Figure 49.11**). Because the pleural membranes are attached to the walls of the thoracic cavity, and because the pleural membranes covering the cavity wall and the lung surface are stuck to each other by surface tension, any attempt to increase the volume of the thoracic cavity increases the tension between the pleural membranes. Even between breaths, there is tension between the pleural membranes because the rib cage is pulling outward and the elasticity of the

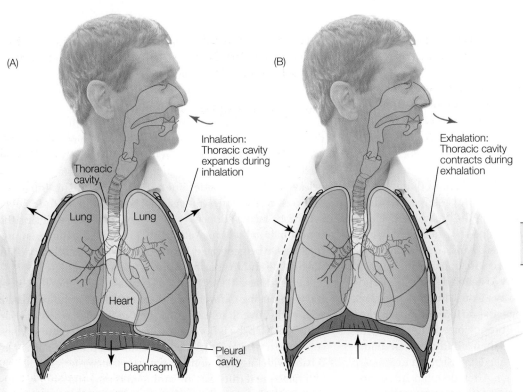

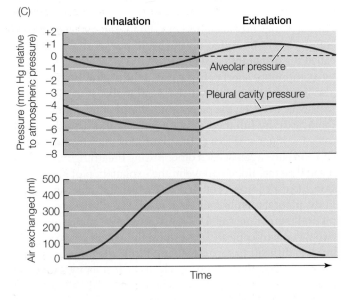

(A) Inhalation:
Thoracic cavity
expands during
inhalation

Thoracic
cavity

Lung   Lung

Heart

Pleural
cavity

Diaphragm

(B) Exhalation:
Thoracic cavity
contracts during
exhalation

**49.11 Into the Lungs and Out
Again** (A) Inhalation is an active
process spurred by contraction of
the diaphragm. (B) Exhalation generally is a passive process as the
diaphragm relaxes. (C) There is
always negative pressure in the
pleural cavity—the space between
the pleural membranes. Variations in
that negative pressure cause the
lungs to inflate and deflate during
the breathing cycle.

—yourBioPortal.com—
GO TO Animated Tutorial 49.2 •
Airflow in Mammals

During **inhalation**:
• Diaphragm contracts
• Thoracic cavity expands
• Intrapleural pressure becomes more negative
• Lungs expand
• Air rushes in

During **exhalation**:
• Diaphragm relaxes
• Thoracic cavity contracts
• Intrapleural pressure becomes less negative
• Lungs contract
• Gases in lungs are expelled

At rest, inhalation is initiated by contraction of the muscular diaphragm (see Figure 49.11A). As the domed diaphragm contracts, it pulls down, expanding the thoracic cavity and pulling on the pleural membranes. Since the pleural membranes cannot separate because of the surface tension of the thin film of fluid between them, they pull on the lungs. The lungs are not a closed cavity; they have an airway to the atmosphere, and they can expand. When the diaphragm contracts, air rushes in through the trachea from the outside and the lungs expand. Exhalation begins when contraction of the diaphragm ceases. As the diaphragm relaxes, the elastic recoil of the lung tissues pulls the diaphragm up and pushes air out through the airways (see Figure 49.11B). When a person is at rest, inhalation is an active process and exhalation is a passive process.

The diaphragm is not the only muscle that can change the volume of the thoracic cavity. Between the ribs are two sets of **intercostal muscles**. The *external intercostal muscles* expand the thoracic cavity by lifting the ribs up and outward. The *internal intercostal muscles* decrease the volume of the thoracic cavity by pulling the ribs down and inward. During strenuous exercise, the external intercostal muscles increase the volume of air inhaled, making use of the inspiratory reserve volume, and the internal intercostal muscles increase the amount of air exhaled, making use of the expiratory reserve volume. The abdominal muscles can also aid in breathing. When they contract, they cause the abdominal contents to push up on the diaphragm and thereby contribute to the expiratory reserve volume.

lung tissue is pulling inward. This slight negative pressure keeps the alveoli partly inflated even at the end of an exhalation. If the thoracic cavity is punctured—by a knife wound, for example—air can leak into the space between the pleural membranes and cause the lung to deflate. If the wound is not sealed, breathing movements pull air in between the pleural membranes rather than into the lung (a "sucking chest wound"), and there is no ventilation of the alveoli in that lung—a condition called "collapsed lung."

Remember that ventilation and perfusion work together to maximize the partial pressure gradients across the gas exchange surface. Ventilation delivers O_2 to the environmental side of the exchange surface, where it diffuses across and is swept away by the perfusing blood, which carries it to the tissues that need it. The reverse is true for the exchange of CO_2. Perfusion delivers CO_2 to the exchange surface, where it diffuses out and is swept away by ventilation.

49.3 RECAP

The mammalian respiratory system consists of a highly branching system of airways that lead to blind end sacs called alveoli which are the gas exchange surfaces. Respiratory muscles ventilate the alveoli by creating pressure differences between the lungs and the outside air. CO_2 and O_2 are exchanged across thin capillary and alveoli walls by diffusion.

- Describe the path that a breath of air takes from the nose to the gas exchange surfaces. **See p. 1035 and Figure 49.10**

- What roles do mucus and surfactant play in maintaining the function of the mammalian respiratory system? **See p. 1035**

- Explain the anatomical and functional relationships between the thoracic cavity, the pleural membranes, and the lungs. **See pp. 1035–1036 and Figure 49.11**

Now that we have discussed how respiratory gases get to and from the environmental side of the gas exchange membranes through ventilation, we can look at how these gases get to and from the internal side of the gas exchange membranes through perfusion.

49.4 How Does Blood Transport Respiratory Gases?

Perfusion of the lungs is one of the functions of the circulatory system. The circulatory system uses a pump (the heart) and a network of vessels to transport blood around the body. Circulatory systems are the subject of Chapter 50, so here we will discuss only one aspect of perfusion: how blood transports respiratory gases.

The liquid part of blood, the *blood plasma*, carries some O_2 in solution, but its ability to transport this nonpolar molecule is limited. The blood plasma of a human can contain in solution only about 0.3 ml of O_2 per 100 ml of plasma, which is inadequate to support even basal metabolism. However, the blood of most animals, vertebrate and invertebrate, contains molecules that reversibly bind O_2 and thus augment its transport capacity. These molecules pick up O_2 where P_{O_2} is high and release it where P_{O_2} is lower. There are many O_2 transport molecules in the animal kingdom, but in vertebrates this role is played by the protein hemoglobin con-

tained in red blood cells. Hemoglobin increases the capacity of blood to transport O_2 by about 60-fold, making high levels of metabolism possible.

Hemoglobin combines reversibly with O_2

Red blood cells contain enormous numbers of hemoglobin molecules. **Hemoglobin** is a protein consisting of four polypeptide subunits (see Figure 3.9), each of which surrounds a *heme group*—an iron-containing ring structure that can reversibly bind a molecule of O_2. Thus, each hemoglobin molecule can bind up to four O_2 molecules, enabling the blood to carry a large amount of O_2 to the body's tissues.

Hemoglobin's ability to pick up or release O_2 depends on the P_{O_2} in its environment. When the P_{O_2} of the blood plasma is high, as it usually is in the lung capillaries, each hemoglobin molecule can carry its maximum load of four O_2 molecules. As the blood circulates through the rest of the body, it releases some of the O_2 it is carrying when it encounters lower P_{O_2} values (**Figure 49.12**).

The relation between P_{O_2} and the amount of O_2 bound to hemoglobin is not linear but S-shaped (sigmoidal). The hemoglobin–oxygen binding curve in Figure 49.12 reflects interactions between the four subunits of the hemoglobin molecule. At low P_{O_2} values, only one subunit will bind an O_2 molecule. When it does so, the shape of that subunit changes, altering the quaternary structure of the whole hemoglobin molecule. That structural change makes it easier for the other subunits to bind an O_2 molecule; that is, their O_2 *affinity* is increased. Therefore, a smaller increase in P_{O_2} is necessary to get the hemoglobin molecules to bind a second O_2 molecule (that is, to become 50% saturated) than was necessary to get them to bind one O_2 molecule

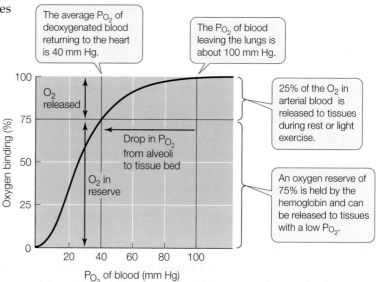

49.12 Binding of O_2 to Hemoglobin Depends on P_{O_2} Hemoglobin in blood leaving the lungs is 100 percent saturated (four O_2 molecules are bound to each hemoglobin molecule). Most hemoglobin molecules drop only one of their four O_2 molecules as they circulate through the body and are still 75 percent saturated when the blood returns to the lungs. The steep portion of this O_2-binding curve comes into play when tissue P_{O_2} falls below the normal 40 mm Hg, at which point hemoglobin "unloads" its O_2 reserves.

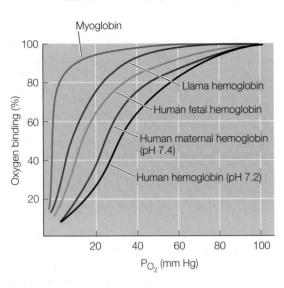

49.13 Oxygen-Binding Adaptations Myoglobin and the different hemoglobins have different O_2-binding properties adapted to different circumstances. The hemoglobin of South American camelids, for example, is adapted for binding O_2 at high altitudes, where P_{O_2} is low. Fetal hemoglobin has a higher affinity for O_2 than does maternal hemoglobin, facilitating O_2 transfer in the placenta. When high metabolism lowers the pH of the blood, hemoglobin releases more of its O_2.

Lama glama

── **yourBioPortal.com** ──
GO TO Web Activity 49.2 • Oxygen-Binding Curves

(to become 25% saturated). This influence of O_2 binding by one subunit on the O_2 affinity of the other subunits is called **positive cooperativity**.

Once the third O_2 molecule is bound, the relationship seems to change, as a larger increase in P_{O_2} is required for the hemoglobin to reach 100 percent saturation. This upper bend of the sigmoid curve is due to a probability phenomenon. The closer we get to having all subunits occupied, the less likely it is that any particular O_2 molecule will find a place to bind. Therefore, it takes a relatively greater P_{O_2} to achieve 100 percent saturation.

The O_2-binding/dissociation properties of hemoglobin help get O_2 to the tissues that need it most. In the lungs, where the P_{O_2} is about 100 mm Hg, hemoglobin is 100 percent saturated. The P_{O_2} in blood returning to the heart from the body is usually about 40 mm Hg. You can see in Figure 49.12 that at this P_{O_2}, the hemoglobin is still about 75 percent saturated. This means that as the blood circulates around the body, it releases only about one in four of the O_2 molecules it carries. This system seems inefficient, but it is really quite adaptive, because the hemoglobin keeps 75 percent of its O_2 in reserve to meet peak demands of highly active tissues.

When a tissue becomes starved of O_2 and its local P_{O_2} falls below 40 mm Hg, the hemoglobin flowing through that tissue is on the steep portion of its binding/dissociation curve. That means relatively small decreases in P_{O_2} below 40 mm Hg will result in the release of lots of O_2 to the tissue. Thus hemoglobin is very effective in making O_2 available to tissues precisely when and where it is needed most.

The O_2 transport function of hemoglobin can rapidly and tragically be disrupted by a common by-product of incomplete combustion: carbon monoxide (CO). If CO from a faulty heating system, engine exhaust, or burning charcoal accumulates in a closed space, the results can be deadly. Because CO binds to hemoglobin with a 240-fold higher affinity than O_2, it prevents hemoglobin from transporting O_2. In the United States, more than 5,000 people die each year from CO poisoning.

Myoglobin holds an O_2 reserve

Muscle cells have their own O_2-binding molecule, *myoglobin*. Myoglobin consists of just one polypeptide chain associated with an iron-containing ring structure that can bind one O_2 molecule. Myoglobin has a higher affinity for O_2 than hemoglobin does, so it picks up and holds O_2 at P_{O_2} values at which hemoglobin is releasing its bound O_2 (**Figure 49.13**).

Myoglobin facilitates the diffusion of O_2 in muscle cells and provides an O_2 reserve for times when metabolic demands are high and blood flow is interrupted. Interruption of blood flow in muscles is common because contracting muscles squeeze blood vessels. When tissue P_{O_2} values are low and hemoglobin can no longer supply more O_2, myoglobin releases its bound O_2. Diving mammals such as seals have high concentrations of myoglobin in their muscles, which is one reason they can stay under water for so long. (We discuss more adaptations for diving in Chapter 50.) Even in nondiving animals, muscles called on for extended periods of work frequently have more myoglobin than muscles that are used for short, intermittent periods, as noted in Section 48.2.

Hemoglobin's affinity for O_2 is variable

Various factors influence the O_2-binding/dissociation properties of hemoglobin, thereby influencing O_2 delivery to tissues. Three of those factors are the chemical composition of the hemoglobin, the blood pH, and the presence of 2,3-bisphosphoglyceric acid (BPG) in red blood cells.

HEMOGLOBIN COMPOSITION There is more than one type of hemoglobin, because the chemical composition of the polypeptide chains that form the hemoglobin molecule varies. The normal hemoglobin of adult humans has two each of two kinds of polypeptide chains—two α-globin chains and two β-globin chains—and the O_2-binding characteristics shown in Figure 49.13.

Before birth, the human fetus has a different form of hemoglobin, consisting of two α-globin and two γ-globin chains. The functional difference between fetal and adult hemoglobin is that fetal hemoglobin has a higher affinity for O_2. Therefore, the fetal hemoglobin–oxygen binding/dissociation curve is shifted to the left compared with that for adults (see Figure 49.13). You can see from these curves that if both types of hemoglobin are at the same P_{O_2} (as they are in the placenta), fetal hemoglobin will pick up O_2 that the maternal hemoglobin releases. This difference in O_2 affinities enables the efficient transfer of O_2 from the mother's blood to the fetus's blood.

South American camelids—llamas, alpacas, guanacos, and vicuñas—are native to the Andes Mountains. In the natural habitat of these mammals, more than 5,000 m above sea level, the P_{O_2} is below 85 mm Hg, and the P_{O_2} in their lungs is about 50 mm Hg. To increase their chances of survival in their low O_2 environment, the hemoglobins of these camelids have O_2-binding/dissociation curves to the left of those of most other mammals—in other words, their hemoglobin can become saturated with O_2 at lower P_{O_2} values than those of other mammals.

HEMOGLOBIN AND pH The O_2-binding properties of hemoglobin are also influenced by physiological conditions. The influence of pH on the function of hemoglobin is known as the **Bohr effect**. As blood passes through metabolically active tissue such as exercising muscle, it picks up acidic metabolites such as lactic acid, fatty acids, and CO_2. As a result, blood pH falls. The excess H^+ binds preferentially to deoxygenated hemoglobin and decreases its affinity for O_2 and the O_2-binding/dissociation curve of hemoglobin shifts to the right (see Figure 49.13). This shift means the hemoglobin will release more O_2 in tissues where pH is low—another way that O_2 is supplied where and when it is most needed.

2,3-BISPHOSPHOGLYCERIC ACID BPG is a metabolite of glycolysis (see Figure 7.5). Mammalian red blood cells respond to low P_{O_2} by increasing their rate of glycolysis and thus producing more BPG, which is an important regulator of hemoglobin function. BPG, like excess H^+, reversibly combines with deoxygenated hemoglobin and lowers its affinity for O_2. The result is that at any P_{O_2}, hemoglobin releases more of its bound O_2 than it otherwise would. In other words, BPG shifts the O_2-binding/dissociation curve of mammalian hemoglobin to the right.

When humans go to high altitudes, or when they cease being sedentary and begin to exercise, their red blood cells are exposed to a lower P_{O_2} and their level of BPG goes up, making it easier for hemoglobin to deliver more O_2 to tissues. The reason fetal hemoglobin has a left-shifted O_2-binding/dissociation curve is that its γ-globin chains have a lower affinity for BPG than do the β-globin chains of adult hemoglobin.

CO_2 is transported as bicarbonate ions in the blood

Delivering O_2 to tissues is only half the respiratory function of blood. Blood also must take CO_2, a metabolic waste product, away from tissues (**Figure 49.14**). CO_2 is highly soluble and readily diffuses through cell membranes, moving from its site of production in the tissues into the blood, where the partial pressure of CO_2 (P_{CO_2}) is lower. However, very little dissolved CO_2 is transported by the blood. Most CO_2 produced by the tissues is transported to the lungs in the form of **bicarbonate ions**, HCO_3^-. CO_2 is converted to HCO_3^-, transported to the lungs, and then converted back to CO_2 in several steps.

When CO_2 dissolves in water, some of it slowly reacts with the water molecules to form carbonic acid (H_2CO_3), some of which then dissociates into a proton (H^+) and a bicarbonate ion (HCO_3^-). This reversible reaction is expressed as follows:

$$CO_2 + H_2O \rightleftharpoons H_2CO_3 \rightleftharpoons H^+ + HCO_3^-$$

In the extracellular fluid, the reaction between CO_2 and H_2O proceeds slowly. But it is a different story in the endothelial cells of the capillaries and in the red blood cells, where the enzyme *carbonic anhydrase* speeds up the conversion of CO_2 to H_2CO_3.

49.14 Carbon Dioxide Is Transported as Bicarbonate Ions
Carbonic anhydrase in capillary endothelial cells and in red blood cells facilitates conversion of CO_2 produced by tissues into bicarbonate ions carried by the plasma. In the lungs, the process is reversed as CO_2 is exhaled.

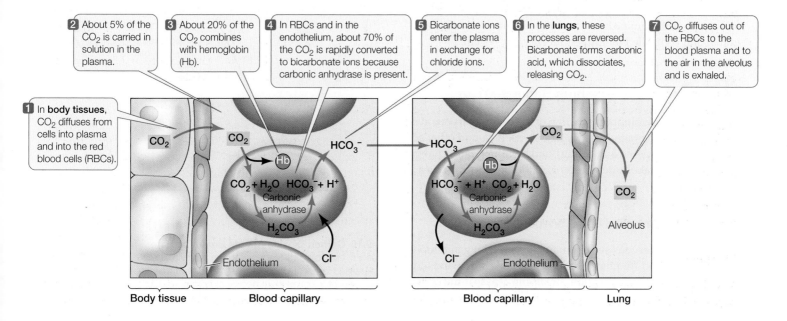

The newly formed H_2CO_3 dissociates, and the resulting bicarbonate ions enter the plasma in exchange for Cl^- (see Figure 49.14). By converting CO_2 to H_2CO_3, carbonic anhydrase reduces the P_{CO_2} in these cells and in the plasma, facilitating the diffusion of CO_2 from tissue cells to endothelial cells, plasma, and red blood cells. Some CO_2 is also carried in chemical combination with hemoglobin.

In the lungs, the reactions involving CO_2 and bicarbonate ions are reversed. Remember that an enzyme such as carbonic anhydrase only speeds up a reversible reaction; it does not determine its direction. The direction is determined by concentrations of reactants and products (see Section 6.1). Ventilation keeps the CO_2 concentration in the alveoli low, so CO_2 diffuses from the blood plasma into the alveoli, lowering the CO_2 concentration in the blood, which favors the conversion of HCO_3^- into CO_2.

49.4 RECAP

O_2 is transported from the lungs to the body's tissues in reversible combination with hemoglobin. Each hemoglobin molecule can reversibly combine with four O_2 molecules; the saturation of the binding sites is a function of the P_{O_2} in the hemoglobin's environment.

- Explain the advantage of having hemoglobin hold on to three O_2 molecules at the usual P_{O_2} of mixed venous blood. **See pp. 1037–1038 and Figure 49.12**

- How is the O_2-binding/dissociation curve of hemoglobin influenced by pH? By BPG? By development from fetus to newborn infant? **See p. 1039 and Figure 49.13**

- How is CO_2 transported in the blood? **See pp. 1039–1040 and Figure 49.14**

We must breathe every minute of our lives, but we don't usually worry about it, or even think about it very often. In the next section we will examine how the regular breathing cycle is generated and controlled by the central nervous system.

49.5 How Is Breathing Regulated?

Breathing is an involuntary function of the central nervous system. The breathing pattern easily adjusts itself around other activities (such as speech and eating), and breathing rates change to match the metabolic demands of our bodies. How is this accomplished?

Breathing is controlled in the brainstem

The autonomic nervous system maintains breathing and modifies its depth and frequency to meet the body's demands for O_2 supply and CO_2 elimination. Breathing ceases if the spinal cord is severed in the neck region, showing that breathing is

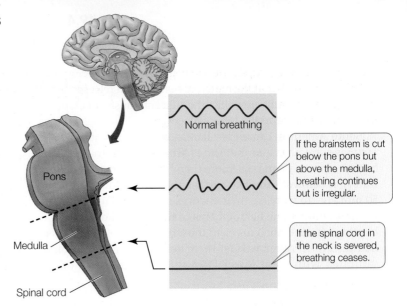

49.15 Breathing Is Controlled in the Brainstem Basic breathing rhythm is generated in the medulla and is modified by neurons in or above the pons.

generated in the brain. If the brainstem is cut just above the medulla (the segment of the brainstem just above the spinal cord), an irregular breathing pattern remains (**Figure 49.15**).

Groups of respiratory motor neurons in the medulla increase their firing rates just before an inhalation begins. As more and more of these neurons fire—and fire faster and faster—the diaphragm contracts. All of a sudden, the neurons stop firing, the

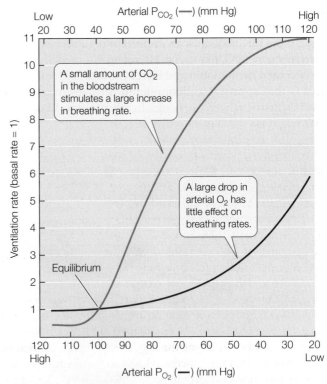

49.16 Carbon Dioxide Affects Breathing Rate The breathing mechanism, controlled by feedback to the medulla, is more sensitive to increased levels of CO_2 in arterial blood than to decreased amounts of O_2. The intersection point occurs at normal (equilibrium) ventilation rate and blood gas levels.

diaphragm relaxes, and exhalation begins. Exhalation is usually a passive process that depends on the elastic recoil of the lung tissues. When breathing demand is high, however, as during strenuous exercise, motor neurons for the intercostal muscles are recruited to a greater extent than during tidal breathing, increasing both the inhalation and the exhalation volumes. Brain areas above the medulla modify breathing to accommodate speech, ingestion of food, coughing, and emotional states.

Regulating breathing requires feedback information

When breathing or metabolism changes, it alters the P_{O_2} and P_{CO_2} in the blood. We should therefore expect the blood levels of one or both of these gases to provide feedback information to the breathing rhythm generator in the medulla. Experiments in which subjects breathe air with different P_{O_2} and P_{CO_2} concentrations lead us to conclude that humans (and other mammals) are remarkably insensitive to falling levels of O_2 in arterial blood but are extremely sensitive to increases in CO_2. That is, arterial P_{O_2} can deviate considerably from normal without causing much of an increase in ventilation rate, but even a small rise in arterial P_{CO_2} causes a large increase in ventilation (**Figure 49.16**). This relationship is reversed for water-breathing animals, in which O_2 is the primary feedback stimulus for gill ventilation.

We might ask whether it is an increase in the P_{CO_2} of the blood that stimulates increased breathing when we exercise. To answer this question, C. R. Bainton observed dogs running on treadmills at different speeds. As the speed of the treadmill increased, the dogs' respiratory gas exchange rate increased but their blood P_{CO_2} remained constant. Before concluding that blood P_{CO_2} does not control breathing rate, Bainton changed the experiment. Instead of increasing the speed of the treadmill, he gradually increased its slope so that the dogs were running at the same speed but were working harder because they were running uphill (**Figure 49.17**).

In this second experiment, the P_{CO_2} of the blood increased as the slope of the treadmill increased and as the respiratory gas exchange rate increased. Bainton concluded that the P_{CO_2} of the blood is the primary metabolic feedback information for breathing. However, when an animal starts to run or changes its running speed, additional feedback information from receptors in muscles and joints changes its sensitivity to CO_2—an example

of feedforward information. As noted in Section 40.1, feedforward information can change the sensitivity or the set point of a regulatory system.

Where are partial pressures of gases in the blood sensed? The major site of P_{CO_2} sensitivity is an area on the ventral surface of the medulla, not far from the groups of neurons that generate the breathing rhythm. The primary sensitivity of these chemosensitive cells is not to CO_2, however. Rather, they are stimulated by H^+ ions. The H^+ ion concentration, or pH, in the environment of these cells is a direct reflection of the P_{CO_2} of the

INVESTIGATING LIFE

49.17 Sensitivity of the Respiratory Control System Changes with Exercise

Experiments with dogs running on a treadmill show that sensitivity of the respiratory system to CO_2 changes when speed of running changes, but not when workload changes without a change in speed.

HYPOTHESIS Rising levels of blood CO_2 during exercise is the feedback signal that stimulates an increase in respiratory rate.

METHOD
1. Dogs are trained to run on a treadmill.
2. The trained dogs are equipped with instruments that measure their respiratory rate and with catheters that withdraw blood samples for measurement of CO_2 levels.
3. The dog runs with the treadmill set first at different speeds, and then with the slope of the treadmill elevated (which increases the workload).
4. Respiratory rate is plotted as a function of arterial CO_2 concentration.

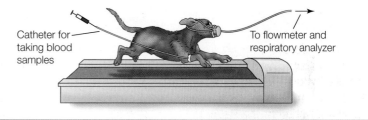

Catheter for taking blood samples

To flowmeter and respiratory analyzer

RESULTS

CHANGES IN SPEED
Respiratory rate changes with running speed but CO_2 levels do not. The hypothesis is not supported.

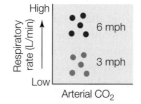

CHANGES IN SLOPE
Respiratory rate and CO_2 levels both rise as workload increases. The hypothesis is supported.

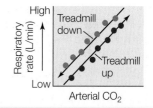

CONCLUSION Arterial CO_2 level is the metabolic feedback signal that regulates respiration in response to workload.

FURTHER INVESTIGATION: Exposure to cold is another variable that increases respiration. How would you test whether cold-induced respiration is driven by elevated blood CO_2 levels?

Go to **yourBioPortal.com** for original citations, discussions, and relevant links for all INVESTIGATING LIFE figures.

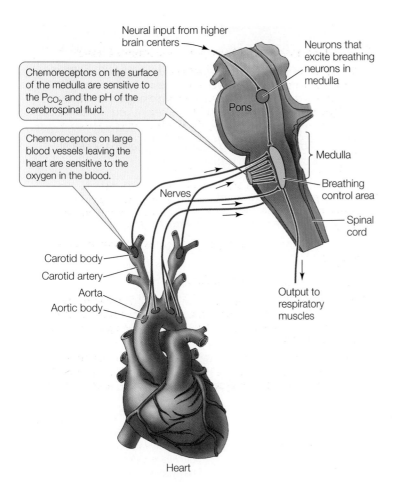

Neural input from higher brain centers

Chemoreceptors on the surface of the medulla are sensitive to the P_{CO_2} and the pH of the cerebrospinal fluid.

Chemoreceptors on large blood vessels leaving the heart are sensitive to the oxygen in the blood.

Pons

Neurons that excite breathing neurons in medulla

Medulla

Nerves

Breathing control area

Spinal cord

Carotid body
Carotid artery
Aorta
Aortic body

Output to respiratory muscles

Heart

49.18 Feedback Information Controls Breathing The body uses feedback information from chemosensors in the heart and brain to match breathing rate to metabolic demand.

blood. When the P_{CO_2} of the blood is higher than that of the extracellular fluid in this area, CO_2 diffuses out of the blood. That CO_2 interacts with H_2O to form carbonic acid (H_2CO_3), which dissociates into H^+ ions and HCO_3^- ions as shown in Figure 49.14. The H^+ ions that are produced stimulate the chemosensitive cells that increase respiratory gas exchange.

Thus, even though we measure blood P_{CO_2} as the stimulus for breathing, the real stimulus is pH.

Sensitivity to blood P_{O_2} resides in nodes of neural tissue on the large blood vessels leaving the heart: the aorta and the carotid arteries (**Figure 49.18**). These **carotid** and **aortic bodies** are chemosensors. If the blood supply to these structures decreases, or if the blood P_{O_2} falls dramatically, the chemosensors are activated and send nerve impulses to the breathing control center. Although we are not very sensitive to changes in blood P_{O_2}, the carotid and aortic bodies can stimulate increases in breathing during exposure to very high altitudes or when blood volume or blood pressure is very low. Also, there is a synergism between CO_2 and O_2 sensing. When blood P_{CO_2} increases, there is an increased sensitivity to low O_2 and vice versa.

49.5 RECAP

The rhythmic contractions of the respiratory muscles that drive breathing are generated by neurons in the brainstem.

- What is the primary chemical stimulus for controlling the respiratory rate, and where is it sensed? **See p. 1040 and Figures 49.16 and 49.18**

- Explain what feedforward information is. Can you give an example of feedforward information in the respiratory control system. **See Figure 49.17**

- What are the functions of the carotid and aortic bodies? **See p. 1042 and Figure 49.18**

CHAPTER SUMMARY

49.1 What Physical Factors Govern Respiratory Gas Exchange?

- Most cells require a constant supply of O_2 and continuous removal of CO_2. These **respiratory gases** are exchanged between an animal's body fluids and its environment by diffusion.

- **Fick's law of diffusion** shows how various physical factors influence the diffusion rate of gases. Adaptations to maximize respiratory gas exchange influence one or more variables of Fick's law.

- In water-breathing animals, gas exchange is limited by the low diffusion rate and low amount of O_2 in water. If water temperature rises, water-breathing animals face a double bind in that the amount of O_2 in water decreases, but their metabolism and the amount of work required to move water over the gas exchange surfaces increase. **Review Figure 49.2**

- In air, the **partial pressure of oxygen (P_{O_2})** decreases with altitude.

49.2 What Adaptations Maximize Respiratory Gas Exchange?

- Adaptations to maximize gas exchange include increasing the surface area for gas exchange, maximizing partial pressure gradients across those exchange surfaces, ventilating the outer surface with the respiratory medium, and perfusing the inner surface with blood.

- Insects distribute air throughout their bodies in a system of **tracheae, tracheoles,** and air capillaries. **Review Figure 49.4**

- The **gills** of fishes have large gas exchange surface areas that are ventilated continuously and unidirectionally with water. The **countercurrent flow** of blood helps increase the efficiency of gas exchange. **Review Figures 49.5 and 49.6**

- The gas exchange system of birds includes **air sacs** that communicate with the lungs but are not used for gas exchange. Air flows unidirectionally through bird lungs; gases are exchanged in air capillaries that run between **parabronchi**. Review Figure 49.7

- Each breath of air remains in a bird's respiratory system for two breathing cycles. The air sacs work as bellows to supply the air capillaries with a continuous unidirectional flow of fresh air. Review Figure 49.8, **ANIMATED TUTORIAL 49.1**

- In all air-breathing vertebrates except birds, breathing is **tidal**. This is a less efficient form of gas exchange than that of fishes and birds. Although the volume of air exchanged with each breath can vary considerably in tidal breathing, the inhaled air is always mixed with stale air. Review Figure 49.9

49.3 How Do Human Lungs Work?

- In mammalian lungs, the gas exchange surface area provided by the millions of **alveoli** is enormous, and the diffusion path length between the air and perfusing blood is short. **Surface tension** in the alveoli would make inflation of the lungs difficult if the alveoli did not produce **surfactant**. Review Figure 49.10, **WEB ACTIVITY 49.1**

- Inhalation occurs when contractions of the **diaphragm** pull on the **pleural membranes** and reduce the pressure in the **thoracic cavity**. Relaxation of the diaphragm increases pressure in the thoracic cavity and results in exhalation. Review Figure 49.11, **ANIMATED TUTORIAL 49.2**

- During periods of heavy metabolic demands such as strenuous exercise, the **intercostal muscles**, located between the ribs, increase the volume of air inhaled and exhaled.

49.4 How Does Blood Transport Respiratory Gases?

- O_2 is reversibly bound to **hemoglobin** in red blood cells. Each hemoglobin molecule can carry a maximum of four O_2 molecules. Because of **positive cooperativity**, hemoglobin's affinity for O_2 depends on the P_{O_2} to which the hemoglobin is exposed. Therefore, hemoglobin picks up O_2 as it flows through respiratory exchange structures and gives up O_2 in metabolically active tissues. Review Figure 49.12

- Myoglobin serves as an O_2 reserve in muscle.

- There is more than one type of hemoglobin. Fetal hemoglobin has a higher affinity for O_2 than does maternal hemoglobin, allowing fetal blood to pick up O_2 from the maternal blood in the placenta. Review Figure 49.13, **WEB ACTIVITY 49.2**

- CO_2 is transported in the blood principally as **bicarbonate ions** (HCO_3^-). Review Figure 49.14

49.5 How Is Breathing Regulated?

- The breathing rhythm is an autonomic function generated by neurons in the medulla and modulated by higher brain centers. The most important feedback stimulus for breathing is the level of CO_2 in the blood. Review Figures 49.16 and 14.17

- The breathing rhythm is sensitive to feedback from chemoreceptors on the ventral surface of the medulla and in the **carotid** and **aortic bodies** on the large vessels leaving the heart. Review Figure 49.18

SEE WEB ACTIVITY 49.3 for a concept review of this chapter.

SELF-QUIZ

1. Which of the following statements is *not* true?
   a. Respiratory gases are exchanged only by diffusion.
   b. O_2 has a lower rate of diffusion in water than in air.
   c. The O_2 content of water falls as the temperature of water rises.
   d. The amount of O_2 in the atmosphere decreases with increasing altitude.
   e. Birds have evolved active transport mechanisms to augment their respiratory gas exchange.

2. Which statement about the gas exchange system of birds is *not* true?
   a. Respiratory gases are not exchanged in the air sacs.
   b. A bird can achieve more complete exchange of O_2 from air to blood than humans can.
   c. Air passes through birds' lungs in only one direction.
   d. The gas exchange surfaces in bird lungs are the alveoli.
   e. A breath of air remains in the system for two breathing cycles.

3. Which statement about gas exchange in fish is *true*?
   a. Blood flows over the gas exchange surfaces in a direction opposite to the flow of water.
   b. Gases are exchanged across the gill arches.
   c. Ventilation of the gills is tidal in fast-swimming fishes.
   d. Less work is needed to ventilate gills in warm water than in cold water.
   e. The path length for diffusion of respiratory gases is determined by the length of the gill filaments.

4. In the human gas exchange system,
   a. the lungs and airways are completely collapsed after a forceful exhalation.
   b. the average P_{O_2} concentration of air inside the lungs is always lower than that in the air outside the lungs.
   c. the P_{O_2} of the blood leaving the lungs is greater than that of the exhaled air.
   d. the amount of air that is moved per breath during normal, at-rest breathing is termed the total lung capacity.
   e. O_2 and CO_2 are actively transported across the alveolar and capillary membranes.

5. Which statement about the human gas exchange system is *not* true?
   a. During inhalation, a negative pressure exists in the space between the lung and the thoracic wall.
   b. Smoking one cigarette can immobilize the cilia lining the airways for hours.
   c. The respiratory control center in the medulla responds more strongly to changes in arterial O_2 concentration than to changes in arterial CO_2 concentration.
   d. Without surfactant, the work of breathing is greatly increased.
   e. The diaphragm contracts during inhalation and relaxes during exhalation.

6. The hemoglobin of a human fetus
   a. is the same as that of an adult.
   b. has a higher affinity for O_2 than adult hemoglobin has.
   c. has only two protein subunits instead of four.
   d. is supplied by the mother's red blood cells.
   e. has a higher affinity for BPG than adult hemoglobin has.

7. The amount of O_2 carried by hemoglobin depends on the P_{O_2} in the blood. Hemoglobin in active muscles
   a. becomes saturated with O_2.
   b. takes up only a small amount of O_2.
   c. readily unloads O_2.
   d. tends to decrease the P_{O_2} in the muscle tissues.
   e. is denatured.

8. Most CO_2 in the blood is carried
   a. in the cytoplasm of red blood cells.
   b. as CO_2 dissolved in the plasma.
   c. in the plasma as bicarbonate ions.

   d. bound to plasma proteins.
   e. in red blood cells bound to hemoglobin.

9. Myoglobin
   a. binds O_2 at P_{O_2} values at which hemoglobin is releasing its bound O_2.
   b. has a lower affinity for O_2 than hemoglobin does.
   c. consists of four polypeptide chains, just as hemoglobin does.
   d. provides an immediate source of O_2 for muscle cells at the onset of activity.
   e. can bind four O_2 molecules at once.

10. When the level of CO_2 in the bloodstream increases,
    a. the rate of respiration decreases.
    b. the pH of the blood rises.
    c. the respiratory centers become dormant.
    d. the rate of respiration increases.
    e. the blood becomes more alkaline.

FOR DISCUSSION

1. The blood of a certain species of fish that lives in Antarctica has no hemoglobin. What anatomical and behavioral characteristics would you expect to find in this fish, and why is its distribution limited to the waters of Antarctica?

2. Blood banks store whole blood for a much shorter period than they store blood plasma. This is because when blood that has been stored for too long is infused into a patient, it can actually decrease the O_2 available to the patient's tissues. Why is this so? Explain in terms of the different physiological functions of BPG.

3. In 1875 three French physiologists went up in a hot air balloon called Zenith to see the effects of reduced air pressure on breathing. As they went higher and higher, their writing became more illegible and nonsensical, but they did not record any sense of danger. They continued to throw out ballast until they all were unconscious. When the balloon returned to the surface, only one of the scientists recovered; the other two were dead. From your knowledge of respira-

tory gas exchange and regulation of respiration, how would you explain this tragic episode? Why do you think the three men continued tossing out ballast to ascend higher and higher, without realizing they were in mortal danger of asphyxiation?

4. In patients who have emphysema, the fine structures of alveoli break down, resulting in the formation of larger air cavities in the lungs. Also, the lung tissue becomes less elastic. Give at least two explanations for why such patients have a low tolerance for exercise.

5. A condition called "the bends" affects scuba divers who surface too quickly after spending an extended period in deep water, where they have been breathing pressurized air. The cause of the bends is tiny bubbles of nitrogen coming out of solution in the blood plasma. Seals spend much more time under water and at deeper depths than scuba divers, yet they do not suffer the bends. Why?

ADDITIONAL INVESTIGATION

When you suddenly travel to a location at high altitude, you notice an unusual breathing pattern when you are resting. For a while you stop breathing completely; then suddenly you start breathing rapidly for a short time; then you stop breathing again.

This can go on and on in a cyclical pattern called Cheyne-Stokes breathing. It generally goes away after spending a few days at altitude. Can you hypothesize what causes Cheyne-Stokes breathing and design an experiment to test your hypothesis?

WORKING WITH DATA (GO TO yourBioPortal.com)

Describing Air Flow in Bird Lungs The path of avian air flow described in Figure 49.8 was established in experiments by Bretz and Schmidt-Nielsen at Duke University. This exercise presents their methods and data, along with questions to be considered in light of their results.

Calculating the Functional Residual Volume A spirometer is used to measure volumes of air exchanged during breathing (see Figure 49.9), but a spirometer cannot measure directly the

"dead space," or *residual volume*, of the lungs. The residual volume influences the degree to which incoming fresh air is diluted by the stale air not expelled by the lungs during exhalation. A technique called the *helium dilution method* is used to measure residual volume. In this exercise, you will use data from a helium dilution experiment to calculate the functional residual volume (FRV) of a subject. You will also determine the effect of an increase in FRV on the maximum P_{O_2} in the alveoli.

50 Circulatory Systems

You gotta have heart

On April 29, 1993, the Boston Celtics met the Charlotte Hornets in a National Basketball Association playoff game at the Boston Garden. The Celtics captain and star, 27-year-old Reggie Lewis, had just scored 10 points in 3 minutes when he suddenly slumped forward and fell to the floor. He was examined by the team doctor, who allowed him to return to the game. But Lewis's legs were "wobbly," and he played only briefly.

Lewis had been experiencing dizzy spells for about a month before the playoff incident. Afterward he underwent rigorous testing by cardiologists, who diagnosed him as having a dangerous arrhythmia (irregular heartbeat) caused by cardiomyopathy (diseased cardiac muscle). Accepting this diagnosis would mean the end of his professional athletic career, and Lewis sought a second opinion.

After another battery of tests, a second medical team concluded that Lewis had undergone a transient irregular heartbeat attributable to normal athletic stresses. The condition was thought to be due to Lewis's enlarged heart—a condition common in high-performing athletes. In July 1993, after an hour spent shooting baskets in a pick-up game, Reggie Lewis collapsed and died of heart failure.

Your heart is a muscular pump that, at rest, beats an average of 60 to 70 times per minute. With each beat, it circulates about 70 milliliters of blood through the body. Without taking work or exercise into account, that is 300 liters per hour, 7,200 liters per day, 2.6 million liters per year—no time-outs.

Heart failure is the leading cause of death in the United States, accounting for some one-fourth (about 600,000) of the deaths each year. Heart failure most commonly results from blockage of the vessels that supply the heart muscle with blood, and its risk increases with age. But heart failure is also the leading cause of death among young athletes.

In athletes, heart failure is usually due not to vessel blockage but to defects of the heart or blood vessels. The most common defect is hypertrophic cardiomyopathy (HCM), caused by a mutation that affects the contractile proteins of heart muscle. The heart is good at compensating for these mutant proteins. About 0.5 percent of the population has HCM, and most of them are unaware of it; they can live their entire lives without symptoms. With

An Athlete's Heart   Basketball star Reggie Lewis died of heart failure at the age of 27. The exact medical situation underlying his heart problems remains clouded in controversy.

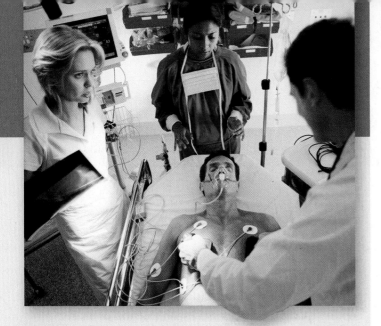

Clear! Hospital emergency rooms deal with heart attacks on a daily basis. Advances in heart surgery and emergency resuscitation techniques have saved many lives, but circulatory system failure remains the number one cause of death in Europe, Canada, and the United States.

continued heavy exercise, however, the heart compensates by getting larger. Eventually, this increase in size can disrupt the electrical impulses that coordinate contractions of the heart muscle. When heavy demand is placed on such an enlarged heart, muscle fiber contractions can suddenly become uncoordinated and render the heart incapable of pumping blood. The lack of prior symptoms is why the condition often goes undiagnosed.

Diagnosis is getting better. In December 2008, 33-year-old Cuttino Mobley, who had just joined the New York Knicks, announced his retirement from basketball after learning that his career could kill him. Like Reggie Lewis, Mobley has an enlarged heart. When a physical exam showed arrhythmia, Mobley's physician ordered an MRI that revealed the thickening of the heart walls characteristic of HCM. "Getting the MRI saved my life," Mobley said.

IN THIS CHAPTER we will learn about the adaptations of circulatory systems that enable them to match blood supply with demand in a variety of species. Taking the human circulatory system as a model, we will explore the mechanics of the beating heart and the characteristics of the arteries, capillaries, and veins of the vascular system. We will then describe the features of blood before ending with a discussion of the hormonal and neural regulation of the mammalian circulatory system.

50.1 Why Do Animals Need a Circulatory System?

A **circulatory system** consists of a muscular pump (the heart), a fluid (blood), and a series of conduits (blood vessels) through which the fluid can be pumped around the body. Heart, blood, and vessels are also known collectively as a **cardiovascular system** (Greek *kardia*, "heart," and Latin *vasculum*, "vessel"). The function of a circulatory system is to transport things around the body. Preceding chapters discuss how circulatory systems transport heat, hormones, respiratory gases, blood cells, platelets, and cells and molecules of the immune system. Succeeding chapters add nutrients and waste products to that list. In this section we will describe the general types of circulatory systems found in animals.

Some animals do not have a circulatory system

Single-celled organisms serve all of their needs through direct exchanges with the environment. Such organisms are found mostly in aquatic or very moist terrestrial environments. Similarly, many multicellular aquatic organisms are small or thin enough that all their cells are close to the external environment. Such species may not have a circulatory system because nutrients, respiratory gases, and wastes can diffuse directly between the cells of their bodies and the environment.

The cells of some larger aquatic multicellular animals without a circulatory system are served by highly branched central cavities called *gastrovascular systems* which essentially bring the external environment into the animal. All the cells of a sponge, for example, are in contact with, or very close to, the water that surrounds the animal and circulates through its central cavity (see Figure 49.1B). Very small animals without a circulatory system can maintain high levels of metabolism and activity, but larger animals without a circulatory system such as sponges, jellyfishes, and flatworms tend to be inactive, slow, or even sedentary. Large, active animals need a circulatory system.

Circulatory systems can be open or closed

The cells of large, mobile animals are supported by the extracellular fluid. All nutrients—oxygen, fuel, essential molecules—come from that fluid, and the waste products of cell metabolism go into it. Circulatory systems have muscular chambers, or *hearts*, that move the extracellular fluid around the body. In *open circulatory systems*, extracellular fluid is the same as the fluid in the circulatory system and is called *hemolymph*. This fluid leaves the vessels of the circulatory system, percolates between cells and through tissues, and then flows back into the heart or vessels of the circulatory system to be pumped out again. In contrast, *closed circulatory systems* completely contain the circulating fluid (*blood*) in a continuous system of vessels. Blood cells and large molecules stay within the system, but water and low-molecular-weight solutes leak out of the smallest vessels, the *capillaries*, which are highly permeable.

In animals with a closed circulatory system, *extracellular fluid* refers to both the fluid in the circulatory system and the fluid outside it. The fluid in the circulatory system is the *blood plasma*; the fluid around the cells is the *interstitial fluid* (see Figure 40.1). A 70-kilogram person has a total extracellular fluid volume of about 14 liters. Less than a quarter of it—about 3 liters—is the blood plasma.

Circulatory systems control the distribution of blood to the body's tissues and organs, with two reciprocal and complementary functions: to maintain the blood composition by picking up nutrients and eliminating wastes, and to supply nutrients to and remove wastes from the body's tissues.

Open circulatory systems move extracellular fluid

Open circulatory systems are found in arthropods, mollusks, and some other invertebrate groups. In these systems, a muscular pump, or heart, helps move the hemolymph through vessels leading to different regions of the body. The fluid leaves the vessels to trickle through the tissues before returning to the heart. In the generalized arthropod shown in **Figure 50.1A**, the fluid returns directly to the heart through openings called *ostia*. Ostia have valves that allow hemolymph to enter the relaxed heart but prevent it from flowing in the reverse direction when the heart contracts. In mollusks (**Figure 50.1B**), open vessels collect hemolymph from different regions of the body and return it to the heart.

Closed circulatory systems circulate blood through a system of blood vessels

In **closed circulatory systems**, a system of vessels keeps circulating blood separate from the interstitial fluid. Blood is pumped through this *vascular system* by one or more muscular hearts, and some components of the blood never leave the vessels. Closed circulatory systems characterize vertebrates and some invertebrate groups, among them annelids.

A simple example of a closed circulatory system is that of the earthworm (**Figure 50.1C**). One large ventral blood vessel carries blood from the worm's anterior end to its posterior end. Smaller vessels branch off and transport the blood to even smaller vessels serving the tissues in each body segment. In the smallest vessels, respiratory gases, nutrients, and metabolic wastes diffuse between the blood and interstitial fluid. The blood then flows from these vessels into larger vessels that lead into one large dorsal vessel, which carries the blood from the posterior to the anterior end of the body. Five pairs of muscular vessels connect the large dorsal and ventral vessels in the anterior end, thus completing the circuit. The dorsal vessel and

(A) Arthropod

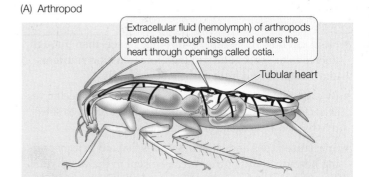

Extracellular fluid (hemolymph) of arthropods percolates through tissues and enters the heart through openings called ostia.

Tubular heart

(B) Mollusk

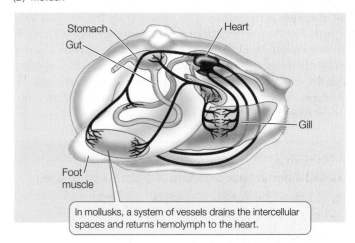

Stomach
Gut
Heart
Gill
Foot muscle

In mollusks, a system of vessels drains the intercellular spaces and returns hemolymph to the heart.

(C) Annelid worm

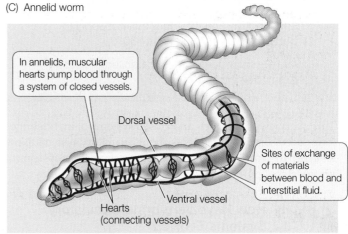

In annelids, muscular hearts pump blood through a system of closed vessels.

Dorsal vessel

Sites of exchange of materials between blood and interstitial fluid.

Hearts (connecting vessels)

Ventral vessel

50.1 Circulatory Systems Arthropods, illustrated here by an insect (A), and mollusks such as clams (B) have an open circulatory system. Hemolymph is pumped by a tubular heart and directed to different regions of the body through vessels that open into intercellular spaces. (C) Annelids such as earthworms have a closed circulatory system, in which the cellular and macromolecular elements of the blood are confined in a system of vessels and the blood is pumped through those vessels by one or more muscular hearts.

the five connecting vessels serve as hearts for the earthworm; their contractions keep the blood circulating. The direction of circulation is determined by one-way valves in the dorsal and connecting vessels.

Closed circulatory systems have several advantages over open systems:

● Fluid can flow more rapidly through vessels than through intercellular spaces and can therefore transport nutrients and wastes to and from tissues more rapidly.

● By changing the diameter (hence resistance) of specific vessels, closed systems can control the flow of blood to selective tissues and organs to match their needs.

● Specialized cells and large molecules that aid in transporting hormones and nutrients can be kept in the vessels but can drop their cargo in the tissues where it is needed.

It seems logical to accept the premise that in all but very small animals, closed circulatory systems can support higher levels of metabolic activity than open systems can. But we then have to ask: How do highly active insect species achieve high levels of metabolic output with an open circulatory system? The reason is that insects do not depend on their circulatory systems for respiratory gas exchange. As noted in Chapter 49, respiratory gas exchange in insects is through a separate system of air-filled tubes (see Figure 49.4).

50.1 RECAP

Circulatory systems consist of a pump and an open or closed set of vessels through which a fluid transports oxygen, nutrients, wastes, and a variety of other substances.

● Why are most animals with an open circulatory system rather inactive, and why does that not apply to insects? See p. 1047 and Figure 50.1

● What are some advantages of a closed circulatory system? See p. 1048

This overview of the open and closed systems of invertebrates introduced some basic concepts about circulatory systems. Now let's turn to describing the more complex circulatory systems of vertebrates.

50.2 How Have Vertebrate Circulatory Systems Evolved?

Vertebrates have a closed circulatory system and a heart with two or more chambers. When a heart chamber contracts, it squeezes the blood, putting it under pressure. Blood then flows out of the heart and into vessels, where pressure is lower. Valves prevent the backflow of blood as the heart cycles between contraction and relaxation.

As we explore the features of the circulatory systems of different classes of vertebrates, a general evolutionary theme will emerge: as circulatory systems become more complex, the blood that flows to the gas exchange organs (gills or lungs; see Chapter 49) becomes more completely separated from the blood that flows to the rest of the body.

In fishes, the phylogenetically oldest vertebrates, blood is pumped from the heart to the gills and then to the tissues of the body and back to the heart—a single circuit. In birds and mammals, blood is pumped from the heart to the lungs and back to the heart in a **pulmonary circuit**, and then from the heart to the rest of the body and back to the heart in a **systemic circuit**. In all other vertebrates we see various adaptations for separating the blood flow into pulmonary and systemic circuits.

Both pulmonary and systemic circuits begin with vessels called **arteries** that carry blood away from the heart. Arteries give rise to smaller vessels called **arterioles**, which feed blood into *capillary beds*. **Capillaries** are the tiny, thin-walled vessels where materials are exchanged between the blood and the tissue fluid. Small vessels called **venules** drain capillary beds. The venules join to form larger vessels called **veins**, which deliver blood back to the heart.

We can trace the evolutionary history of vertebrate circulatory systems by comparing the circulatory systems of fishes, lungfishes, amphibians, ectothermic reptiles, and birds and mammals.

Fishes have a two-chambered heart

The fish heart has two chambers. An **atrium** (plural *atria*) receives blood from the body and pumps it into a more muscular chamber, the **ventricle**. The ventricle pumps the blood to the gills, where gases are exchanged. Blood leaving the gills collects in a large dorsal artery, the **aorta**, which distributes blood to smaller arteries and arterioles leading to all the organs and tissues of the body. In the tissues, blood flows through beds of tiny capillaries, collects in venules and veins, and eventually returns to the atrium of the heart.

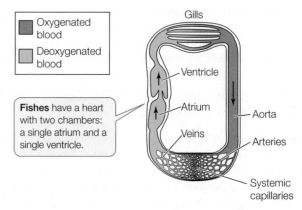

Most of the pressure imparted to the blood by the contraction of the ventricle is lost as a result of the blood entering the many narrow spaces in the gill lamellae. Therefore, blood leaving the gills and entering the aorta is under low pressure, limiting the maximum capacity of the fish circulatory system to supply the tissues with oxygen and nutrients. Yet this limitation on arterial blood pressure does not seem to hamper the performance of many rapidly swimming species, such as tuna and marlin.

The evolutionary transition from breathing water to breathing air had important consequences for the vertebrate circulatory system. An example of how the system changed to serve a primitive lung can be seen in the African lungfish.

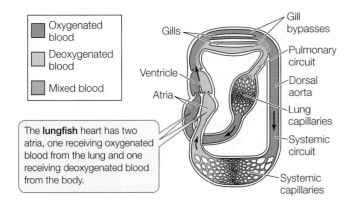

The **lungfish** heart has two atria, one receiving oxygenated blood from the lung and one receiving deoxygenated blood from the body.

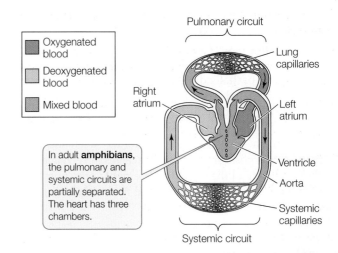

In adult **amphibians**, the pulmonary and systemic circuits are partially separated. The heart has three chambers.

Lungfish are periodically exposed to water with low oxygen content or to situations in which their aquatic environment dries up. The adaptation that deals with these conditions is an out-pocketing of the gut that serves as a lung. The lung contains many thin-walled blood vessels, so blood flowing through those vessels can pick up oxygen from air gulped into the lung.

How does the lungfish circulatory system take advantage of this new organ? In fishes, the gills are arranged on supportive gill arches (see Figure 49.5). Blood flows into the gill arch in an afferent arteriole and leaves in an efferent arteriole. In lungfishes, the blood vessels in the posterior pair of gill arteries have been modified to be a low-resistance conduit for blood to the lung, and a new vessel carries oxygenated blood from the lung back to the heart. In addition, two anterior gill arches have lost their gills, and their blood vessels deliver blood from the heart directly to the dorsal aorta. Because a few of the gill arches retain gills, the African lungfish can breathe either air or water.

The lungfish heart partially separates its flow of blood into pulmonary and systemic circuits; it has a partially divided atrium. The left side receives oxygenated blood from the lungs, and the right side receives deoxygenated blood from the other tissues. These two bloodstreams stay mostly separate as they flow through the ventricle and the large vessel leading to the gill arches. As a result, oxygenated blood goes mostly to the anterior gill arteries leading to the dorsal aorta, and deoxygenated blood goes mostly to the other gill arches that have functional gills as well as to the gill arteries that serve the lung.

We can conclude that the lungfish lung evolved as a means of supplementing oxygen uptake from the gills. When the water is fully oxygenated, the lungfish can depend on its gills; but in oxygen-depleted water, it can augment its oxygen intake by gulping air. Associated modifications of the lungfish vascular system set the stage for the evolution of separate pulmonary and systemic circulations in higher vertebrates.

Amphibians have a three-chambered heart

Pulmonary and systemic circulations are partially separated in adult amphibians. A single ventricle pumps blood to the lungs and the rest of the body. Two atria receive blood returning to the heart: one receives oxygenated blood from the lungs, while the other receives deoxygenated blood from the body.

Because both atria deliver blood to the same ventricle, the oxygenated and deoxygenated blood could mix, so that blood going to the tissues would not carry a full load of oxygen. Mixing is limited, however, because anatomical features of the ventricle direct the flow of deoxygenated blood from the right atrium to the pulmonary circuit and the flow of oxygenated blood from the left atrium to the aorta. Partial separation of pulmonary and systemic circulation has the advantage of allowing blood destined for the tissues to sidestep the large pressure drop that occurs in the gas exchange organ. Blood leaving the amphibian heart for the tissues moves directly to the aorta, and hence to the body, at a higher pressure than if it had first flowed through the lungs.

Amphibians have another adaptation for oxygenating their blood: they can pick up a considerable amount of oxygen in blood flowing through small blood vessels in their skin.

Reptiles have exquisite control of pulmonary and systemic circulation

Reptiles include turtles, snakes, lizards, crocodilians, and birds (see Figure 33.18). Crocodilians and birds have cardiovascular systems with two completely separated ventricles, creating a four-chambered heart. All other reptiles have a three-chambered heart because their ventricles are not completely separated into left and right chambers.

Consider the behavior, ecology, and physiology of *ectothermic* reptiles. Many are active, powerful, fast animals. But their activity comes in bursts that are interspersed with long periods of inactivity, during which these animals' metabolic rates are much lower than the resting metabolic rates of the *endothermic* birds and mammals. So enormous is the range of metabolic demand in ectothermic reptiles that they do not need to breathe continuously. Some species are accomplished divers and spend long periods under water, where they cannot breathe air.

When these animals are not breathing, it would be a waste of energy for them to pump blood through their lungs. Thus, they have evolved the capability to send blood to the lungs and the rest of the body when they are breathing, but when they are not breathing, they can bypass the pulmonary circuit and pump all the blood to the body. How do they do this?

Let's look first at the ectothermic reptiles with a three-chambered heart—the turtles, snakes, and lizards. The ventricle in these species is partially divided into left and right halves by a septum. Oxygenated blood from the lungs enters the left side of the ventricle through the left atrium. Deoxygenated blood from the body enters the right side of the ventricle through the right atrium. These species have two aortas—a left and a right. The left aorta is positioned so that it receives oxygenated blood from the left side of the ventricle. The right aorta, however, is positioned so that it can receive blood from either the right or left side of the ventricle.

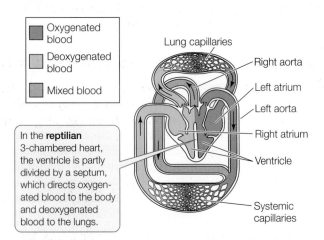

In the **reptilian** 3-chambered heart, the ventricle is partly divided by a septum, which directs oxygenated blood to the body and deoxygenated blood to the lungs.

When the animal is breathing air, the resistance in the pulmonary circuit is lower than the resistance in the systemic circuit, so blood from the right side of the ventricle tends to flow into the pulmonary artery rather than the right aorta. When the animal is not breathing, pulmonary vessels constrict, resistance in the pulmonary circuit goes up, and blood from the right side of the ventricle tends to flow into the right aorta. As a result, blood from both sides of the ventricle flows through both aortas to the systemic circuit.

Crocodilians, like birds, have two completely separated ventricles. Unlike birds, they have two aortas, one originating in each ventricle. There is a connection between the two aortas just

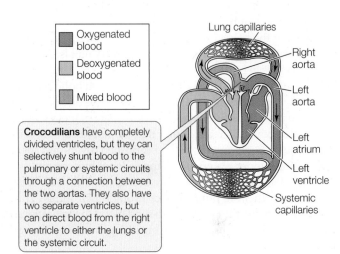

Crocodilians have completely divided ventricles, but they can selectively shunt blood to the pulmonary or systemic circuits through a connection between the two aortas. They also have two separate ventricles, but can direct blood from the right ventricle to either the lungs or the systemic circuit.

as they leave the heart, and this connection enables them to alter the proportions of blood going to their pulmonary and systemic circuits. When a crocodile or alligator is breathing and resistance in the pulmonary circuit is low, backpressure from the stronger left ventricle closes the valve between the right ventricle and the right aorta, forcing all of the blood from the right ventricle to flow into the pulmonary circuit. When the animal stops breathing, pulmonary vessels constrict, resistance in the pulmonary circuit rises, and blood from the right ventricle flows into the right aorta. This ability of all ectothermic reptiles to direct blood to their pulmonary or systemic circuits is highly adaptive for their lifestyle of intermittent breathing.

Birds and mammals have fully separated pulmonary and systemic circuits

The four-chambered hearts of birds and mammals have completely separate pulmonary and systemic circuits. Separate circuits have several advantages for these active animals with continuously high metabolic rates:

* Oxygenated and deoxygenated blood cannot mix; therefore, the systemic circuit always receives blood with the highest oxygen content.

* Respiratory gas exchange is maximized because the blood with the lowest oxygen content and highest CO_2 content is sent to the lungs.

* Separate systemic and pulmonary circuits can operate at different pressures.

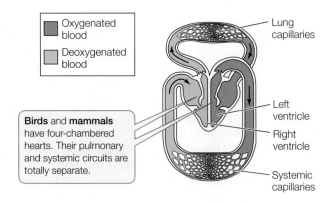

Birds and **mammals** have four-chambered hearts. Their pulmonary and systemic circuits are totally separate.

The tissues of birds and mammals have high nutrient demands and thus a very high density of blood vessels, requiring the heart to generate a high blood pressure to perfuse all the vessels of the systemic circuit. The pulmonary circuit of these animals receives a blood flow equal to that of the systemic circuit, but the lungs have far fewer blood vessels. Thus, the pulmonary circuit of birds and mammals can function at lower pressures, and the four-chambered heart makes that possible.

——— **yourBioPortal.com** ———
GO TO Web Activity 50.1 • Vertebrate Circulatory Systems

50.2 RECAP

The closed circulatory system of vertebrates has evolved from a two-chambered heart in fishes to a four-chambered heart in crocodilians, birds, and mammals.

- Explain why fishes cannot supply blood to their tissues at high pressure. **See p. 1048**

- By comparing lungfish and amphibian circulatory systems, explain how a three-chambered heart could have evolved. **See pp. 1048–1049**

- What are some of advantages of a four-chambered heart? **See p. 1050**

50.3 How Does the Mammalian Heart Function?

Humans have a typical mammalian heart, characterized by four chambers—a right and a left atrium and a right and a left ventricle (**Figure 50.2**). The right ventricle pumps blood through the pulmonary circuit, and the left ventricle pumps blood through the systemic circuit.

One-way valves between the atria and ventricles, the **atrioventricular (AV) valves**, prevent backflow of blood into the atria when the ventricles contract. The **pulmonary valve** and **aortic valve**, one-way valves between the ventricles and the major arteries, prevent backflow of blood into the ventricles when they relax.

In this section, we will describe the flow of blood through the heart and body and examine the unique electrical properties of cardiac muscle that result in the heartbeat.

Blood flows from right heart to lungs to left heart to body

The heart's right atrium receives deoxygenated blood from the **superior** (upper) **vena cava** and the **inferior** (lower) **vena cava** (see Figure 50.2), large veins that collect blood returning to the

yourBioPortal.com
GO TO Web Activity 50.2 • The Human Heart

50.2 The Human Heart and Circulation In the human heart, blood flows from right heart to lungs to left heart to body. The atrioventricular valves prevent blood from flowing back into the atria when the ventricles contract. The pulmonary and aortic valves prevent blood from flowing back into the ventricles from the arteries when the ventricles relax.

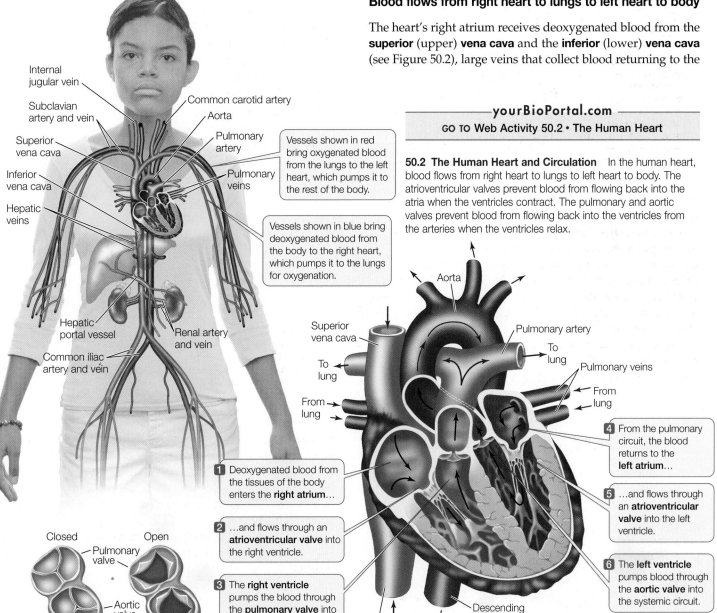

Internal jugular vein
Subclavian artery and vein
Superior vena cava
Inferior vena cava
Hepatic veins
Common carotid artery
Aorta
Pulmonary artery
Pulmonary veins
Hepatic portal vessel
Renal artery and vein
Common iliac artery and vein

Vessels shown in red bring oxygenated blood from the lungs to the left heart, which pumps it to the rest of the body.

Vessels shown in blue bring deoxygenated blood from the body to the right heart, which pumps it to the lungs for oxygenation.

Closed   Open
Pulmonary valve
Aortic valve

1 Deoxygenated blood from the tissues of the body enters the **right atrium**...

2 ...and flows through an **atrioventricular valve** into the right ventricle.

3 The **right ventricle** pumps the blood through the **pulmonary valve** into the pulmonary circuit.

Aorta
Superior vena cava
To lung
From lung
Inferior vena cava
Descending aorta
Pulmonary artery
To lung
Pulmonary veins
From lung

4 From the pulmonary circuit, the blood returns to the **left atrium**...

5 ...and flows through an **atrioventricular valve** into the left ventricle.

6 The **left ventricle** pumps blood through the **aortic valve** into the systemic circuit.

heart from the upper and lower body, respectively. The veins of the heart itself also drain into the right atrium. From the right atrium, the blood flows through an AV valve into the right ventricle. Most of the filling of the ventricle results from passive flow while the heart is relaxed between beats. Just at the end of this period of passive ventricular filling, the atrium contracts and adds a little more blood to the ventricular volume. The right ventricle then contracts, causing the AV valve to close and pumping the blood into the **pulmonary artery** leading to the lungs.

After gas exchange occurs in the lungs, **pulmonary veins** return oxygenated blood from the lungs to the left atrium, from which the blood enters the left ventricle through another AV valve. As on the right side of the heart, most left ventricular filling is passive, but ventricular filling is completed when the atria contract.

The walls of the left ventricle are powerful muscles that contract around the blood with a wringing motion starting from the bottom. When pressure in the left ventricle is high enough to push open the aortic valve, blood rushes into the aorta to begin its circulation throughout the body. In Figure 50.2, observe that the walls of the left ventricle are thicker than those of the right ventricle. The left ventricle has to propel blood through many more kilometers of blood vessels than does the right ven-

tricle, and must therefore push against more resistance, even though both ventricles pump the same volume of blood.

Both sides of the heart contract at the same time. Contraction of the two atria, followed by contraction of the two ventricles and then relaxation, is the **cardiac cycle**. The cardiac cycle is divided into two phases: **systole** (pronounced sís-toll-ee), when the ventricles contract, and **diastole** (die-ás-toll-ee), when the ventricles relax (**Figure 50.3**). Just at the end of diastole, the atria contract and top off the volume of blood in the ventricles.

The sounds of the cardiac cycle, the "lub-dup" heard through a stethoscope placed on the chest or the back, are created by the heart valves slamming shut. The closing and opening of these valves are simple mechanical events resulting from pressure differences on the two sides of the valves. As the ventricles begin to contract, the pressure in them rises above the pressure in the atria, so the AV valves close ("lub"). When the ventricles be-

---

yourBioPortal.com
GO TO Animated Tutorial 50.1 • The Cardiac Cycle

50.3 The Cardiac Cycle   The rhythmic contraction (systole) and relaxation (diastole) of the ventricles is called the cardiac cycle. The representation below shows pressure and volume changes during the cardiac cycle for the left ventricle only.

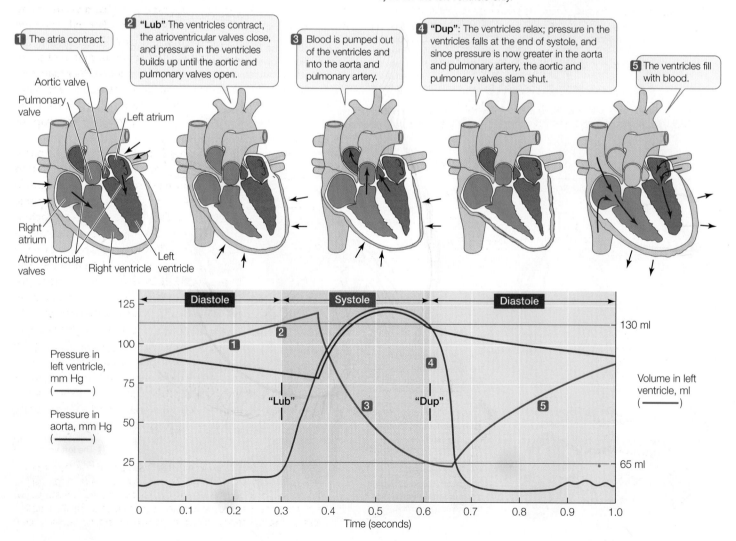

gin to relax, the high pressure in the aorta and pulmonary artery closes the aortic and pulmonary valves ("dup").

Defective valves that do not close completely produce turbulent blood flow and the sounds known as *heart murmurs*. For example, if an AV valve does not close completely, blood will flow back into the atrium with a "whoosh" following the "lub."

You can feel the rhythm of the cardiac cycle in the pulsation of arteries such as the one that supplies blood to your hand. You've probably taken your pulse by placing two fingers from one hand lightly over the wrist of your other hand just below the thumb. During systole, the pressure wave created by the contraction of the left ventricle surges through the arteries of your arms; this pressure wave is what you feel as the pulsing of the artery in your wrist.

Blood pressure changes associated with the cardiac cycle can be measured in the large artery in your arm by using an inflatable pressure cuff and a pressure gauge, together called a *sphygmomanometer*, and a stethoscope (**Figure 50.4**). This method measures the minimum pressure necessary to compress an artery so blood does not flow through it at all (the systolic value) and the minimum pressure that causes intermittent flow through the artery (the diastolic value). A conventional blood pressure reading is expressed as the systolic value placed over the diastolic value. Healthy values for a young adult might be 120 millimeters of mercury (mm Hg) during systole and 70 mm Hg during diastole, or 120/70.

The heartbeat originates in the cardiac muscle

Cardiac muscle (see Section 48.1) has unique adaptations that enable it to function as a pump. First, cardiac muscle cells are in electrical contact with one another through gap junctions, which enable *action potentials* (see Section 45.2) to spread rapidly from cell to cell. Because a spreading action potential stimulates contraction, large groups of cardiac muscle cells contract in unison. This coordinated contraction is essential for pumping blood effectively.

Second, some cardiac muscle cells are **pacemaker cells** that can initiate action potentials without stimulation from the nervous system. When they fire action potentials, they stimulate neighboring cells to contract. The primary pacemaker of the heart is a group of modified cardiac muscle cells, the **sinoatrial node**, located at the junction of the superior vena cava and right atrium (see Figure 50.6). The *resting membrane potentials* of these cells are less negative than those of other cardiac muscle cells and are not stable, but they gradually become even less negative until they reach threshold for initiating an action potential. The action potentials of pacemaker cells are very different from those of neurons and other muscle cells (see Figure 45.10). They are slower to rise; they are broader; and they are slower to return to resting potential. These properties of pacemaker cells are due to the ion channels in their membranes.

Pacemaker potentials involve Na^+, Ca^{2+}, and K^+ channels. As we discuss in Section 45.2, when Na^+ or Ca^{2+} channels open, positive charges flow into the cell and the membrane potential becomes less negative. When K^+ channels open, positive charges flow out of the cell and the membrane potential becomes more negative. Because the Na^+ channels of pacemaker cells are more open than are those of other cardiac muscle cells, the pacemaker resting potential is less negative. The action potential of pacemaker cells is due to voltage-gated Ca^{2+} channels rather than voltage-gated Na^+ channels as in neurons, skeletal muscle, and other cardiac muscle cells. These Ca^{2+} ion channels open and close more slowly than voltage-gated Na^+ channels, explaining the shape of pacemaker action potentials.

The unstable resting potential of pacemaker cells is due to the behavior of cation channels. As in neurons and skeletal muscle cells, there are voltage-gated K^+ channels that open on the rising phase of the action potential. The opening of these channels allows K^+ ions to leave the cell and restores the negative charge on the cell membrane. The hyperpolarization (negative membrane potential) causes the opening of a unique class of voltage-gated cation channels that mostly conduct Na^+. At the same time, the voltage-gated K^+ channels that opened during the action potential are slowly closing. The result is that there are more Na^+ ions coming into the cell than there are K^+ ions leaving, and the cell membrane potential gradually becomes less negative.

1 The cuff is inflated beyond the point that shuts off all blood flow.

2 Pressure in the cuff is gradually lowered until the sound of a pulsing flow of blood through the constriction in the artery is heard. At this time, pressure in the cuff is just below the peak **systolic pressure** in the artery.

3 Pressure is further lowered until the sound becomes continuous. At this time, the cuff is just below the **diastolic pressure** in the artery. This person's blood pressure is 120/70.

Pulsing sounds

Pulsing sound gives way to smooth "whoosh" of blood flow

50.4 Measuring Blood Pressure Blood pressure in the major artery of the arm can be measured with a device called a sphygmomanometer, which combines an inflatable cuff and a pressure gauge. A stethoscope is also used to detect sounds created by the blood vessels in response to changes in pressure during the cardiac cycle.

INVESTIGATING LIFE

50.5 The Autonomic Nervous System Controls Heart Rate

The membrane potentials of pacemaker cells spontaneously depolarize until action potential threshold is reached. Neurotransmitter signals from the two divisions of the autonomic nervous system speed up and slow down the rate at which the pacemaker membrane potential drifts upward, thereby controlling the rate at which pacemaker cells fire action potentials.

HYPOTHESIS The autonomic nervous system neurotransmitters norepinephrine and acetylcholine influence the membrane potentials of pacemaker cells.

METHOD
1. Culture living sinoatrial node tissue in a dish. Insert an intracellular recording electrode into pacemaker cells.
2. Measure the membrane potential of pacemaker cells during a resting heartbeat (the control) and after applications of the ANS neurotransmitters norepinephrine (sympathetic) and acetylcholine (parasympathetic).

RESULTS

Control recording shows that the membrane potential of pacemaker cells gradually depolarizes after an action potential is fired.

When norepinephrine is applied, the rate of depolarization of the membrane potential increases. Time between action potentials decreases and the heart rate increases.

When acetylcholine is applied, the membrane potential is more negative following an action potential and the rate of depolarization is slower. Time between action potentials increases and the heart rate slows down.

CONCLUSION The ANS neurotransmitters norepinephrine and ACh control heart rate by altering the properties of the pacemaker cell membrane potentials.

Go to **yourBioPortal.com** for original citations, discussions, and relevant links for all INVESTIGATING LIFE figures.

The gradual rise in membrane potential closes the channels that are allowing Na^+ to move into the cell, but as the membrane becomes less negative, some Ca^{2+} channels open, causing the membrane to continue its gradual rise. Eventually this rising membrane potential reaches threshold for the major voltage-gated Ca^{2+} channels, and another action potential is generated. The intricate interaction of these ion channels through their ef-

fects on membrane potential causes the rhythmic generation of action potentials that characterizes the pacemaker cells.

The nervous system controls the heartbeat (speeds it up or slows it down) by influencing the rate at which the pacemaker cell resting potentials drift upward (**Figure 50.5**). Norepinephrine released onto pacemaker cells by sympathetic nerves increases the permeability of the Na^+ channels and the Ca^{2+} channels. The result is that the resting potential of the pacemaker cells drifts up more rapidly, the interval between action potentials is decreased, and the heart beats faster. Conversely, the parasympathetic neurotransmitter acetylcholine has the opposite effect. ACh increases the permeability of K^+ channels so that the membrane potential becomes even more negative following an action potential and rises more slowly. ACh also decreases the permeability of the Ca^{2+} channels so that the rate of rise of the membrane potential slows, the interval between pacemaker action potentials lengthens, and the heart slows down.

A conduction system coordinates the contraction of heart muscle

A normal heartbeat begins with an action potential in the sinoatrial node (**Figure 50.6**). This action potential spreads rapidly throughout the electrically coupled cells of the atria, causing them to contract in unison. Because there are no gap junctions between the cells of the atria and those of the ventricles, the action potential does not spread directly to the ventricles. Therefore, the ventricles do not contract in unison with the atria.

How does the action potential move from the atria to the ventricles? Situated at the junction of the atria and the ventricles is a nodule of modified cardiac muscle cells—the **atrioventricular node**—which is stimulated by the depolarization of the atria. With a slight delay, it generates action potentials that are conducted to the ventricles via the **bundle of His**, which consists of modified cardiac muscle fibers that do not contract. These fibers divide into right and left *bundle branches* that run to the tips of the ventricles and then spread throughout the ventricular muscle mass as **Purkinje fibers**. These conducting fibers ensure that the cardiac action potential spreads rapidly and evenly throughout the ventricular muscle mass, starting at the very bottom of the ventricles. The short delay in the spread of the action potential imposed by the atrioventricular node ensures that the atria contract before the ventricles do, so that the blood passes progressively from the atria to the ventricles to the arteries.

Electrical properties of ventricular muscles sustain heart contraction

Electrical properties of ventricular muscle fibers allow them to contract for about 300 milliseconds—much longer than those of skeletal muscle fibers. As in neuronal and skeletal muscle

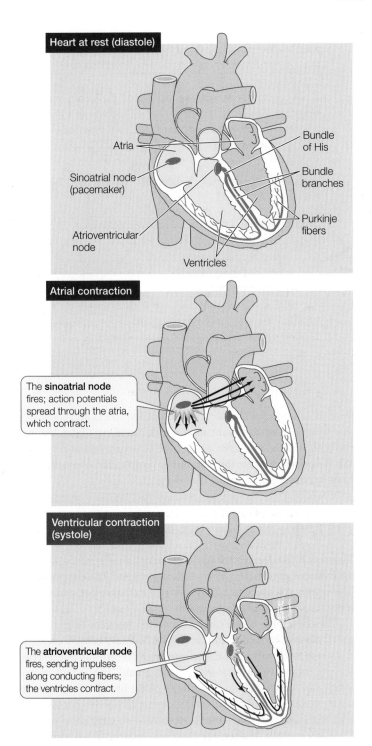

50.6 The Heartbeat Pacemaker cells in the sinoatrial node initiate the heartbeat by firing action potentials that spread through the electrically coupled atrial muscle. The atrial action potential eventually spreads to the atrioventricular node that, with a delay, conducts it through the Bundle of His and Purkinje fibers to the cells of the ventricles.

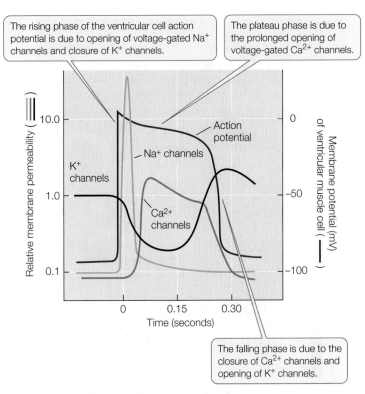

The rising phase of the ventricular cell action potential is due to opening of voltage-gated Na^+ channels and closure of K^+ channels.

The plateau phase is due to the prolonged opening of voltage-gated Ca^{2+} channels.

The falling phase is due to the closure of Ca^{2+} channels and opening of K^+ channels.

50.7 The Action Potential of Ventricular Muscle Fibers The rising phase of the action potential of ventricular muscle fibers (red) is due to the opening of voltage-gated Na^+ channels (gold). However, the membrane remains in a depolarized state for a prolonged time because of the opening of voltage-gated K^+ channels (black).

action potentials, the rising phase of the ventricular muscle cell action potentials is due to the opening of voltage-gated Na^+ channels. Unlike neurons and skeletal muscle fibers, however, ventricular muscle cells remain depolarized for a long time. This extended plateau of the action potential is due to sustained opening of voltage-gated Ca^{2+} channels (**Figure 50.7**). Like other muscle, cardiac muscle is stimulated to contract when Ca^{2+} is available to bind with troponin (see Figure 48.6). As long as Ca^{2+} remains in the sarcoplasm, the ventricular muscle cells continue to contract.

The drug digitalis has been used since the late 1700s to treat weakened hearts or hearts with irregular patterns of contraction. Digitalis strengthens and slows the heartbeat by slowing the reuptake of Ca^{2+} by the sarcoplasmic reticulum and thereby increasing the concentration of Ca^{2+} in the sarcoplasm. Before being introduced into the practice of medicine, digitalis prepared from the purple foxglove plant, *Digitalis purpurea,* was a folk remedy for heart problems.

The ECG records the electrical activity of the heart

Electrical events in the cardiac muscle during the cardiac cycle can be recorded by electrodes placed on the surface of the body. Such a recording is an **electrocardiogram**, or **ECG**. **EKG** is also used because German physicians who invented the method used the Greek word for heart and called it the *electrokardiogramm*. The ECG is an important tool for diagnosing heart problems (**Figure 50.8A**).

The action potentials that sweep through the muscles of the atria and ventricles before they contract are such massive, localized electrical events that they cause electric currents to flow throughout the body. Electrodes placed at different locations on

(A)

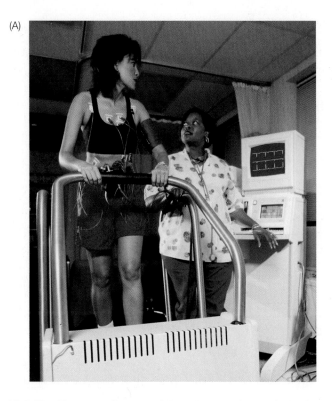

(B)

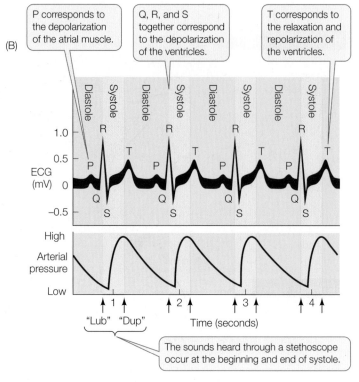

P corresponds to the depolarization of the atrial muscle.

Q, R, and S together correspond to the depolarization of the ventricles.

T corresponds to the relaxation and repolarization of the ventricles.

The sounds heard through a stethoscope occur at the beginning and end of systole.

50.8 The Electrocardiogram (A) An electrocardiogram (abbreviated as ECG or EKG) is used to monitor heart function. Electrodes attached to the person on the treadmill record an ECG that is amplified and displayed on a monitor. (B) Variations from the normal pattern shown here can be used to diagnose heart problems.

the skin detect those currents at different times and register a voltage difference between them. The appearance of the ECG depends on the placement of the electrodes. Electrodes placed on the right wrist and left ankle produced the normal ECG shown in **Figure 50.8B**. The wave patterns of the ECG are designated P, Q, R, S, and T, each letter representing a particular event in the cardiac muscle, as shown in the figure.

50.3 RECAP

The mammalian heart has two atria and two ventricles. Modified cardiac muscle tissue in the right atrium functions to spontaneously generate pacemaker action potentials. Other modified cardiac muscle tissue between the atria and ventricles and throughout the ventricles conducts those signals and coordinates the heart contraction.

- Trace the path of blood through both sides of the heart, naming the major blood vessels and heart valves. See pp. 1051–1052 and Figure 50.2

- Differentiate systole and diastole and describe the events of the cardiac cycle. See p. 1052 and Figure 50.3

- How do cells of the sinoatrial node generate the heartbeat? See pp. 1053–1054 and Figures 50.5 and 50.6

We'll now consider the composition of the blood and the characteristics of the vessels through which blood circulates around the body, illustrating once again how structure serves function. We will also consider the role of the lymphatic vessels that return interstitial fluid to the blood.

50.4 What Are the Properties of Blood and Blood Vessels?

Blood is a connective tissue. It consists of cells suspended in an extracellular matrix of complex, yet specific, composition. The unusual feature of blood is that the extracellular matrix is a liquid, so blood is a fluid tissue.

The cells of the blood can be separated from the fluid matrix, called **plasma**, by centrifugation (**Figure 50.9**). If a sample of blood is spun in a centrifuge, all the cells move to the bottom of the tube, leaving the clear, straw-colored plasma on top. The *packed-cell volume*, or *hematocrit*, is the percentage of the blood volume made up by cells. Normal hematocrit is about 42 percent for women and 46 percent for men, but these values can vary considerably. They are usually higher, for example, in people who live and work at high altitudes, because the low oxygen concentrations there stimulate the production of more red blood cells.

Here we will consider two elements in blood: the *red blood cells* and the *platelets*, which are pinched-off fragments of cells. *White blood cells*, or *leukocytes*, are the cells of the immune system, which we discussed in Chapter 42.

Red blood cells transport respiratory gases

Most of the cells in the blood are **erythrocytes**, or red blood cells. Mature red blood cells are biconcave, flexible discs packed with hemoglobin. Their function is to transport respiratory gases. Their shape gives them a large surface area for gas exchange,

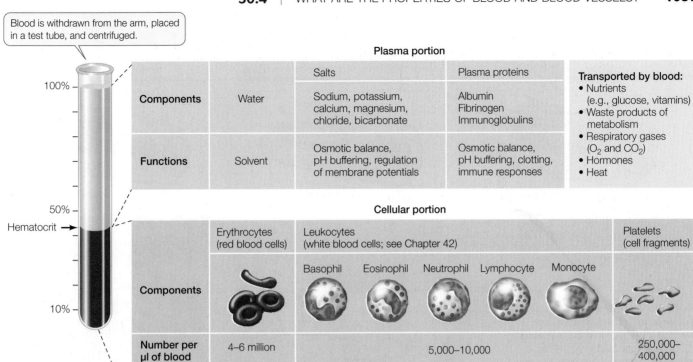

50.9 The Composition of Blood Blood consists of a complex aqueous solution (the plasma) and numerous cell types and cell fragments. The hematocrit (arrow) is a measure of the cellular portion as a percentage of total blood volume.

and their flexibility enables them to squeeze through narrow capillaries. Men have 4.5 to 6.0 million red blood cells per microliter of blood, and women have 3.5 to 5.0 million.

Red blood cells, as well as all the other cellular components of blood, are generated by stem cells in the bone marrow, particularly in the ribs, breastbone, pelvis, vertebrae, and the long bones of the limbs. Red blood cell production is controlled by a hormone, **erythropoietin**, which is released by cells in the kidneys in response to insufficient oxygen—**hypoxia**. Many tissues respond to hypoxia by expressing a transcription factor called *hypoxia-inducible factor 1 (HIF-1)*. When the kidneys become hypoxic and express HIF-1, one of the actions of the transcription factor is to activate the gene encoding erythropoietin. Increased circulating erythropoietin extends the lives of mature red blood cells and stimulates production of new red blood cells in the bone marrow.

Under normal conditions, your bone marrow produces about 2 million red blood cells every second. Developing, immature red blood cells divide many times while still in the bone marrow, and during this time they produce hemoglobin. When the hemoglobin content of a red blood cell approaches about 30 percent, the nucleus, endoplasmic reticulum, Golgi apparatus, and mitochondria of the cell begin to break down. This process is almost complete when the new red blood cell squeezes between the endothelial cells of blood vessels in the bone marrow and enters the circulation. Loss of nuclei from the red blood cells occurs in most mammalian species, but the red cells of some mammals and of all other vertebrates are nucleated.

Each red blood cell circulates for about 120 days. As it gets older, its membrane becomes less flexible and more fragile. Therefore, old red blood cells can rupture as they bend to fit through narrow capillaries. One place where they are really squeezed is in the **spleen**, an organ that sits near the stomach in the upper left side of the abdominal cavity. The spleen has many venous cavities, or sinuses, that serve as a reservoir for red blood cells, but to get into the sinuses, the red blood cells must squeeze between spleen cells. When old red blood cells are ruptured by this squeezing, their remnants are taken up and degraded by macrophages (a class of white blood cells that ingest debris and foreign materials).

Platelets are essential for blood clotting

Besides producing erythrocytes and leukocytes, the bone marrow stem cells described in Section 42.1 also produce cells called *megakaryocytes*. Megakaryocytes are large cells that remain in the bone marrow and release cell fragments called **platelets** into the circulation. A platelet is just a tiny fragment of a cell without cell organelles, but it is packed with enzymes and chemicals necessary for its function: sealing leaks in blood vessels and initiating **blood clotting** (**Figure 50.10**).

Damage to a blood vessel exposes collagen fibers. An encounter with collagen fibers activates a platelet. The platelet swells, becomes irregularly shaped and sticky, and releases chemicals that activate other platelets and initiate the clotting of blood. The sticky platelets also form a plug at the damaged site.

Blood clotting requires many steps and many clotting factors, most of which are circulating in the blood in an inactive form. The absence of any one of these proteins can impair clotting and cause excessive bleeding. Because the liver produces

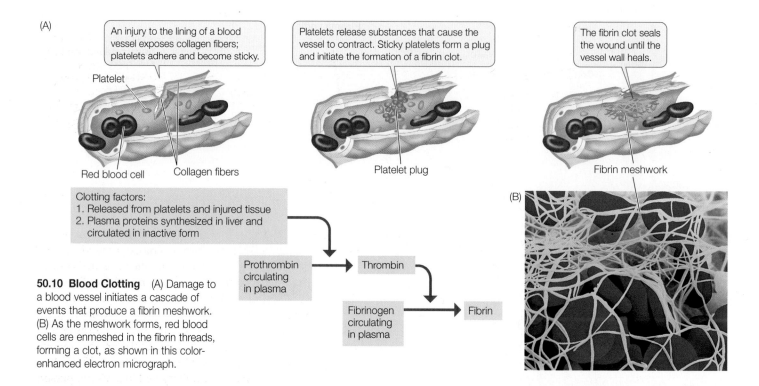

50.10 Blood Clotting (A) Damage to a blood vessel initiates a cascade of events that produce a fibrin meshwork. (B) As the meshwork forms, red blood cells are enmeshed in the fibrin threads, forming a clot, as shown in this color-enhanced electron micrograph.

most of the clotting factors, liver diseases such as hepatitis and cirrhosis can result in excessive bleeding. People with hemophilia experience uncontrolled bleeding because of a genetic inability to produce one of the clotting factors.

Blood clotting factors participate in a cascade of chemical activations of other substances circulating in the blood. The cascade begins with blood vessel and other tissue damage that exposes the blood to proteins such as collagen that are normally separated from the blood by endothelial cells lining the blood vessels. This exposure activates platelets and begins the clotting factor cascade. The end result of this cascade is to convert an inactive circulating enzyme, **prothrombin**, to its active form, **thrombin**. Thrombin cleaves molecules of **fibrinogen**, a plasma protein, forming insoluble threads of **fibrin**. The fibrin threads form the meshwork that binds platelets, seals the vessel, and provides a scaffold for the formation of scar tissue (see Figure 50.10).

Blood circulates throughout the body in a system of blood vessels

As we mentioned in Section 50.1, blood circulates through the vertebrate body in a system of closed vessels. Blood leaves the heart in arteries and is distributed throughout the body's tissues in arterioles, which feed capillary beds. Exchanges of nutrients, wastes, respiratory gases, and hormones occur in the capillaries. Blood leaving capillary beds collects in venules, which empty into veins that conduct the blood back to the heart.

The walls of the large arteries have many extracellular collagen and elastin fibers, which enable them to withstand the high blood pressures generated by the heart (**Figure 50.11A**). These elastic tissues have another important function: they are stretched during systole, and thereby store some of the energy imparted to the blood by the heart. Elastic recoil during dias-

tole returns this energy to the blood by squeezing it and pushing it forward. As a result, even though pressure in the arteries pulsates with the beating of the heart, the flow of blood is smoother than it would be through a system of rigid pipes.

Smooth muscle cells in the walls of the arteries and arterioles constrict or dilate those vessels. When the diameter of the vessels changes, their resistance to blood flow also changes, and the amount of blood flowing through them changes as a result. Neural and hormonal mechanisms act on smooth muscle cells in the walls of the arteries and arterioles, controlling the flow of blood through these vessels. The arterioles are referred to as *resistance vessels* because their resistance can vary to control the blood flow to specific tissues.

Materials are exchanged in capillary beds by filtration, osmosis, and diffusion

Beds of capillaries lie between arterioles and venules (**Figure 50.11B**). Few cells are more than a few cell diameters away from a capillary. (Notable exceptions include developing oocytes and the cells of the lens and cornea.) The cells' needs are served by the exchange of materials between blood and interstitial fluid across the capillary walls. This exchange is facilitated by the capillaries' thin, permeable walls, as well as by the slow flow of blood through the capillaries.

It may seem strange that blood flows through the large arteries rapidly at high pressures, but when it reaches the small capillaries, the pressure and rate of flow decrease (**Figure 50.11C**). When you restrict the diameter of a garden hose by placing your thumb over the opening, the pressure in the hose increases, which in turn increases the velocity of the water spraying out of the hose. But keep in mind that the arteries branch into many arterioles, which give rise to a huge number of capillaries. Even

(A)

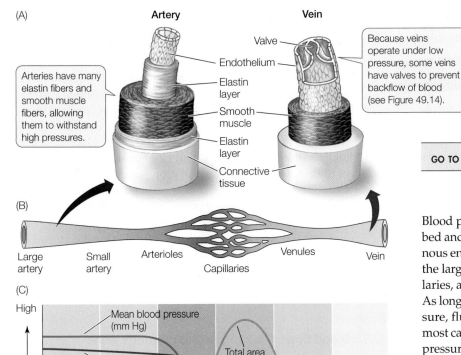

Artery

Vein

Arteries have many elastin fibers and smooth muscle fibers, allowing them to withstand high pressures.

Valve

Endothelium

Elastin layer

Smooth muscle

Elastin layer

Connective tissue

Because veins operate under low pressure, some veins have valves to prevent backflow of blood (see Figure 49.14).

50.11 Anatomy of Blood Vessels (A) The different anatomical characteristics of arteries and veins match their functions. (B) Blood from the arterial system feeds into capillary beds, where exchanges with the interstitial fluid occur. The venous system returns the blood to the heart. (C) The area encompassed by each vessel type is graphed along with the pressure and velocity of the blood within them.

──── **yourBioPortal.com** ────

GO TO **Web Activity 50.3 • Structure of a Blood Vessel**

(B)

Large artery

Small artery

Arterioles

Capillaries

Venules

Vein

(C)

High

Mean blood pressure (mm Hg)

Velocity (cm/sec)

Total area (cm²)

Total area (cm²)

Low

Large arteries

Small arteries

Arterioles

Capillaries

Venules

Veins

though each capillary has a diameter so small that red blood cells must pass through in single file, there are so many capillaries that their total cross-sectional area is much greater than that of any other class of vessels. As a result, all the capillaries together have a much greater capacity for blood than do the arterioles. Returning to our garden hose analogy, if we connected the hose to an increasing number of lawn sprinklers, eventually the pressure and flow in each sprinkler would be low.

Capillary walls consist of a single layer of thin endothelial cells (**Figure 50.12**). In most tissues of the body other than the brain, capillaries have tiny holes, or *fenestrations* (*fenestra*, "windows"). Capillaries are permeable to water, some ions, and some small molecules but not to large molecules such as proteins. At the arterial (high pressure) end, blood pressure squeezes water and some small solutes out of the capillaries and into the surrounding intercellular spaces. Why don't water and small-molecular-weight solutes collect in the intercellular spaces? How is the blood volume maintained if fluid is continuously leaking out of the capillaries?

An answer to this question was put forth more than 100 years ago by the physiologist E. H. Starling. Starling suggested that water movement across capillary walls is a result of two opposing forces, which are now known as **Starling's forces**:

• Blood pressure squeezes water and small solutes out of the capillaries.

• Osmotic pressure pulls water back into the capillaries.

Blood pressure is high at the arterial end of a capillary bed and steadily drops as blood moves towards the venous end (**Figure 50.13**). The osmotic pressure is due to the large protein molecules that cannot leave the capillaries, and it is relatively constant along the capillaries. As long as the blood pressure is above the osmotic pressure, fluid leaves the capillaries. At the venule end of most capillaries, blood pressure falls below the osmotic pressure, so fluid returns to the capillaries. The actual numbers for a normal capillary bed in a resting person suggest that there would be a *slight* net loss of fluid to the intercellular spaces. This loss, about 4 liters per day, percolates between cells as the interstitial fluid before it returns to the venous blood via the lymphatic system, which we will discuss later in this chapter.

Several observations supported Starling's model. In people with severe liver disease or protein starvation, a fall in blood protein concentration leads to an accumulation of fluid in the extracellular spaces, which results in tissue swelling, or *edema*. Edema is also characteris-

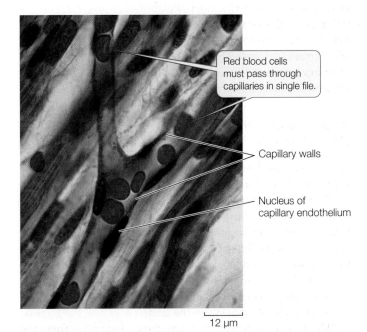

Red blood cells must pass through capillaries in single file.

Capillary walls

Nucleus of capillary endothelium

12 µm

50.12 A Narrow Lane Capillaries have a very small diameter, and blood flows through them slowly.

(A)

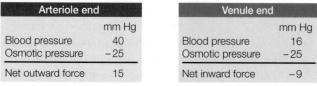

Fluid is squeezed out of the capillary by blood pressure.

Fluid is pulled back into the capillary by osmotic pressure.

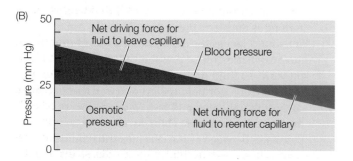

Blood pressure (40) Osmotic pressure (25) Blood pressure (16) Osmotic pressure (25)

Arteriole end	
	mm Hg
Blood pressure	40
Osmotic pressure	−25
Net outward force	15

Venule end	
	mm Hg
Blood pressure	16
Osmotic pressure	−25
Net inward force	−9

(B)

50.13 Starling's Forces Starling's model explains how blood volume is maintained in the capillary beds. (A) When blood pressure is greater than osmotic pressure, fluid leaves the capillaries; when blood pressure falls below osmotic pressure, fluid returns to the capillaries. (B) The balance of these two forces changes over the capillary bed as blood pressure falls.

tic of the inflammation response accompanying tissue damage or allergic responses (see Figure 42.3). *Histamine*, a mediator of inflammation released by certain white blood cells, increases capillary permeability and relaxes the smooth muscles of the arterioles, raising blood pressure in the capillaries and leading to fluid leakage into tissues.

A few situations are not explained by Starling's hypothesis. During strenuous exercise, the blood pressure in the arterioles serving the muscles rises substantially but does not result in edema. In birds, the blood pressure in arterioles is much higher than in mammals, and the osmotic pressure is lower. If edema is not a chronic problem in exercising muscles and in birds, what is missing from Starling's model?

Recent research suggests that bicarbonate ions (HCO_3^-) in the blood plasma contribute significantly to the osmotic attraction that draws water back into the capillaries. The CO_2 produced by cellular metabolism diffuses into the endothelial cells lining the capillaries, where it is converted into HCO_3^- and released into the plasma. When the subject is at rest, the increasing HCO_3^- concentration can cause the osmotic pressure of the blood at the venous end to be 30 mm Hg higher than at the arterial end, and during strenuous exercise this difference can be

much higher. Thus it appears that CO_2 and HCO_3^- are major factors that pull water back into the capillaries.

Lipid-soluble substances and many small solute molecules can easily pass through capillary walls from an area of higher concentration to one of lower concentration. The capillaries in different tissues, however, are differentially selective as to the sizes of molecules that can pass through them. You can imagine that this is an important issue in the design and delivery of drugs. All capillaries are permeable to O_2, CO_2, glucose, lactate, and small ions such as Na^+ and Cl^-. However, the capillaries of the brain do not have fenestrations, and therefore not much else can pass through them other than lipid-soluble substances, such as alcohol and anesthetics. This high selectivity of brain capillaries is known as the *blood–brain barrier*.

Much less selective capillaries are found in the digestive tract, where nutrients are absorbed, and in the kidneys, where wastes are filtered. Even in the brain there are specific regions where the capillaries are more permeable, enabling the brain to detect non-lipid soluble hormones.

Blood flows back to the heart through veins

The pressure of the blood flowing from capillaries to venules is extremely low and is insufficient to propel blood back to the heart. The walls of veins are more expandable than the walls of arteries, and blood tends to accumulate in veins. As much as 60 percent of your total blood volume may be in your veins when you are resting. Because of their high capacity to stretch and store blood, veins are called *capacitance vessels*.

Blood flow through veins that are above the level of the heart is assisted by gravity. Below the level of the heart, however, venous return is against gravity. The most important force propelling blood from these regions is the squeezing of the veins by the contractions of surrounding skeletal muscles. As muscles contract, the vessels are compressed and blood is squeezed through them. Blood flow may be temporarily obstructed during a prolonged muscle contraction, but when muscles relax, blood is free to move again. One-way valves in the veins of the extremities prevent backflow of blood. Thus, whenever a vein is squeezed, blood is propelled forward toward the heart (**Figure 50.14**).

In a resting person, gravity causes blood accumulation in the veins of the lower body and exerts backpressure on the capillary beds. This backpressure shifts the balance between blood pressure and osmotic pressure, causing increased loss of fluid to the intercellular spaces. That is why your feet swell during a long airline flight.

Because of the one-way valves in the veins of the legs, the contractions of leg muscles act as auxiliary vascular pumps when an animal walks or runs and facilitate the return of blood to the heart from the veins of the lower body. As a greater volume of blood is returned to the heart, the heart contracts more forcefully and its pumping action is enhanced. The heartbeat gets stronger because of a property of cardiac muscle cells described by the **Frank–Starling law**: if the cardiac muscle cells are stretched, as they are when the volume of returning blood increases, they contract more forcefully.

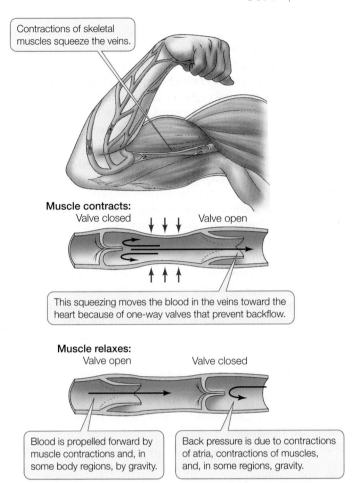

Contractions of skeletal muscles squeeze the veins.

Muscle contracts:
Valve closed   Valve open

This squeezing moves the blood in the veins toward the heart because of one-way valves that prevent backflow.

Muscle relaxes:
Valve open   Valve closed

Blood is propelled forward by muscle contractions and, in some body regions, by gravity.

Back pressure is due to contractions of atria, contractions of muscles, and, in some regions, gravity.

50.14 One-Way Flow Veins have valves that prevent blood from flowing backward, and contractions of skeletal muscle help move blood toward the heart.

The actions of breathing also help return venous blood to the heart. The muscles involved in inhalation create negative pressure that pulls air into the lungs (see Figure 49.11), and this negative pressure also pulls blood toward the chest, increasing venous return to the right atrium. In addition, some of the largest veins closest to the heart contain smooth muscle that contracts at the onset of exercise. Contraction of veins can rapidly increase venous return and stimulate the heart in accord with the Frank–Starling law, increasing cardiac output.

Lymphatic vessels return interstitial fluid to the blood

The interstitial fluid contains water and small molecules, but no red blood cells, and less protein than found in plasma. A separate system of vessels—the **lymphatic system**—returns interstitial fluid to the blood. Each capillary bed contains at least one blind-ended lymph capillary.

Once it enters the lymphatic vessels, the interstitial fluid is called **lymph**. Fine lymphatic capillaries merge into progressively larger vessels and ultimately into two lymphatic vessels—the **thoracic ducts**—that empty into large veins at the base of the neck (see Figure 42.1). The left thoracic duct carries most of the lymph from the lower part of the body and is much larger than the right thoracic duct. Lymph, like blood, is propelled to-

ward the heart by skeletal muscle contractions and breathing movements, and lymphatic vessels, like veins, have one-way valves that keep the lymph flowing toward the thoracic duct.

Mammals and birds have **lymph nodes** along the major lymphatic vessels. Lymph nodes are a major site of lymphocyte production and of the phagocytic action that removes microorganisms and other foreign materials from the circulation (see Section 42.1).

Vascular disease is a killer

As mentioned at the start of this chapter, cardiovascular disease is responsible for about one-fourth of all deaths each year in the United States, and the same is true for Europe. The immediate cause of most of these deaths is not a defect in heart muscle, as it is in athletes. Instead, the cause is mostly heart attack or stroke, which are usually the end result of a disease called **atherosclerosis** ("hardening of the arteries") that begins many years before symptoms are detected.

Healthy arteries have a smooth internal lining of endothelial cells (**Figure 50.15A**) that can be damaged by chronic high blood pressure, smoking, a high-fat diet, or microorganisms. Deposits called **plaque** begin to form at sites of endothelial damage. First, the damaged endothelial cells attract certain white blood cells to the site. These cells are then joined by smooth muscle cells migrating from the deeper layers of the arterial wall. Lipids, especially cholesterol, are deposited in these cells, so that the developing plaque becomes fatty. Fibrous connective tissue made by the invading smooth muscle cells in the plaque, along with

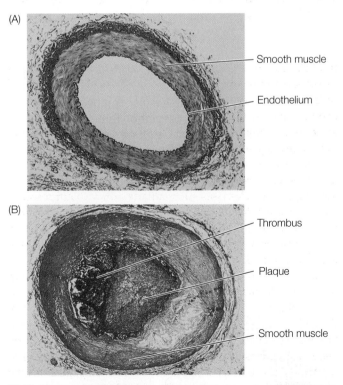

(A)

Smooth muscle

Endothelium

(B)

Thrombus

Plaque

Smooth muscle

50.15 Atherosclerotic Plaque (A) A healthy, clear artery. (B) An atherosclerotic artery, clogged with plaque and a thrombus.

deposits of calcium, make the artery wall less elastic—hence, "hardening of the arteries." The growing plaque deposit narrows the artery and causes turbulence in the blood flow. Blood platelets stick to the plaque (see Figure 50.10) and initiate formation of an intravascular blood clot, a **thrombus**, which can block the artery (**Figure 50.15B**).

The blood supply to the heart muscle flows through the **coronary arteries** which are highly susceptible to atherosclerosis. As these arteries narrow, blood flow to the heart muscle decreases, causing the symptoms of chest pain and shortness of breath during mild exertion. A person with atherosclerosis is at high risk of forming a thrombus in a coronary artery. This condition, called **coronary thrombosis**, can totally block the vessel, causing a *heart attack*, or **myocardial infarction**.

A piece of a thrombus that breaks loose, called an **embolus**, is likely to travel to and become lodged in a vessel of smaller diameter, blocking its flow (an **embolism**). Arteries already narrowed by plaque formation are likely places for an embolism. An embolism in an artery in the brain causes the cells fed by that artery to die. This event is a **stroke**. The specific damage resulting from a stroke, such as memory loss, speech impairment, or paralysis, depends on the location of the blocked artery.

Probably the most important determinants of whether you will get atherosclerosis are your genetic predisposition and your age. Environmental risk factors play a large role, however. If you do have a genetic predisposition to atherosclerosis, it is even more important to minimize environmental risk factors. These include high-fat and high-cholesterol diets, smoking, and a sedentary lifestyle. Certain untreated medical conditions such as *hypertension* (high blood pressure), obesity, and diabetes are also risk factors for atherosclerosis. Changes in diet and behavior and treatment of predisposing medical conditions can prevent and reverse early atherosclerosis and help fend off this silent killer.

50.4 RECAP

Blood is a fluid tissue with cellular components that play roles in transport of respiratory gases, immune system function, and blood clotting. Blood is distributed throughout the body in a system of vessels. Exchanges between the blood and interstitial fluids occur in the smallest of those vessels, the capillaries.

- How are the structural differences between the various classes of vessels related to their functions? **See pp. 1058–1061 and Figure 50.11**

- Why are arterioles called resistance vessels and veins called capacitance vessels? **See pp. 1058 and 1060**

- What factors control the movement of fluids between the vascular and extravascular spaces? **See pp. 1059–1060 and Figure 50.13**

- What propels blood from the lower part of the body back to the heart? **See pp. 1060–1061 and Figure 50.14**

Every tissue in the body requires an adequate flow of oxygen-saturated blood that carries essential nutrients and is relatively free of waste products. Blood flow depends on the maintenance of an appropriate blood pressure, and the distribution of blood flow throughout the body depends on control of the resistance in the blood vessels supplying different tissues.

50.5 How Is the Circulatory System Controlled and Regulated?

When we investigate how a physiological process is regulated, we start by identifying the critical components of that process, how they can be controlled, and the information used to govern that control. Because blood flow depends on pressure, we can identify the pressure in the aorta as a critical variable of the circulatory system. The pressure in the aorta oscillates between systole and diastole, so we define our variable as the *mean arterial pressure* (MAP). MAP is determined by the cardiac output (CO) and the resistance to flow in the blood vessels, or *total peripheral resistance* (TPR):

$$MAP = CO \times TPR$$

Since CO is a function of the heart rate (HR) and stroke volume (SV), however, the critical relationships we have to understand can be expressed as:

$$MAP = HR \times SV \times TPR$$

HR, SV, and TPR are controlled by neural and hormonal mechanisms at both the local and systemic levels. At the local level, each tissue controls its own blood flow through **autoregulatory mechanisms** that cause the arterioles supplying that tissue to constrict or dilate.

The collective autoregulatory actions in the capillary beds in all tissues of the body determine TPR and therefore MAP. If many arterioles suddenly dilate, TPR goes down and MAP falls. If many arterioles constrict, TPR goes up and MAP goes up. Changes in MAP provide information about changing needs of the body. As blood flows through capillary beds, its composition changes; its CO_2 content goes up and its O_2 content goes down. Thus, blood composition also provides information the body uses to regulate the circulatory system.

The nervous and endocrine systems respond to changes in MAP and blood composition by changing breathing rate, heart rate, stroke volume, and peripheral resistance to match the metabolic needs of the body. We will now see how these mechanisms work.

Autoregulation matches local blood flow to local need

The amount of blood that flows through a capillary bed is controlled by the smooth muscle of the arteries and arterioles feeding that bed. **Figure 50.16** illustrates the flow of blood in a typical capillary bed. Blood flows into the bed from an arteriole. Smooth muscle "cuffs," or **precapillary sphincters**, on the arteriole can shut off the supply of blood to the capillary bed. When the precapillary sphincters are relaxed and the arteriole is open, the arterial blood pressure pushes blood into the capillaries.

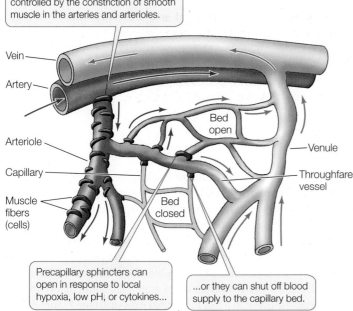

Blood flow through a capillary bed is controlled by the constriction of smooth muscle in the arteries and arterioles.

Vein

Artery

Arteriole

Capillary

Muscle fibers (cells)

Bed open

Venule

Throughfare vessel

Bed closed

Precapillary sphincters can open in response to local hypoxia, low pH, or cytokines...

...or they can shut off blood supply to the capillary bed.

50.16 Local Control of Blood Flow Low O_2 concentrations or high levels of metabolic by-products cause the smooth muscle of the arteries and arterioles to relax, thus increasing the supply of blood to the capillary bed.

Autoregulation depends on the sensitivity of the smooth muscle to its local chemical environment. Low O_2 concentrations and high CO_2 concentrations cause the smooth muscle to relax, thus increasing the supply of blood, which brings in more O_2 and carries away CO_2—a response known as **hyperemia**, which means "excess blood." Increases in other by-products of metabolism, such as lactic acid, hydrogen ions, potassium, and adenosine (all of which increase in exercising muscle), promote hyperemia through the same mechanism. Hence, activities that increase the metabolism of a tissue also induce hyperemia in that tissue.

Arterial pressure is regulated by hormonal and neural mechanisms

Control and regulation of the circulatory system begins with the local autoregulatory mechanisms that alter the resistance of arteries and arterioles feeding capillary beds. The demands of the capillary beds influence MAP and blood composition. Both of these provide information for the control of endocrine and neural responses that act to return blood pressure and composition to normal. Thus circulatory functions are matched to the regional and overall needs of the body.

Arteries and arterioles are innervated by the autonomic nervous system, particularly the sympathetic division. The sympathetic postganglionic neurotransmitter norepinephrine binds to receptors in smooth muscle in blood vessels in the gut and other tissues not essential for "fight or flight" and causes these

vessels to constrict, resulting in reduced blood flow through them and an elevation in MAP. In skeletal muscle, however, specialized sympathetic neurons release acetylcholine, causing the smooth muscle of these arterioles to relax and the vessels to dilate, increasing blood to flow to the muscle. As we discussed earlier in this chapter, increased sympathetic activity increases heart rate, and by increasing the strength of the cardiac muscle contraction, it also increases stroke volume.

Hormones also play a role in regulating arterial pressure. *Epinephrine* has actions similar to those of norepinephrine and is released from the adrenal medulla during massive sympathetic activation stimulated by a fall in arterial pressure or by activation of the fight-or-flight response to a dangerous threat. Another hormone, *angiotensin,* is produced when blood pressure to the kidneys falls (**Figure 50.17**). These hormones influence arterioles located for the most part in peripheral tissues (extremities) or in tissues whose functions need not be maintained continuously (such as the gut). By reducing blood flow in those arterioles, these hormones increase central blood pressure and blood flow to essential organs such as the heart, brain, and kidneys.

The autonomic nervous system activity that controls heart rate and constriction of blood vessels originates in a cardiovascular control center in the medulla. Many inputs converge on this central integrative network and influence the commands it issues via parasympathetic and sympathetic nerves (**Figure 50.18**). Of special importance is incoming information about changes in blood pressure and composition from both **baroreceptors** (stretch receptors) and *chemoreceptors* in the walls of the large arteries leading to the brain—the aorta and the carotid arteries.

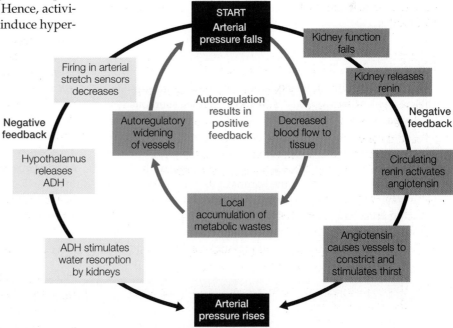

START
Arterial pressure falls

Kidney function fails

Kidney releases renin

Negative feedback

Circulating renin activates angiotensin

Angiotensin causes vessels to constrict and stimulates thirst

Arterial pressure rises

ADH stimulates water resorption by kidneys

Hypothalamus releases ADH

Negative feedback

Firing in arterial stretch sensors decreases

Autoregulatory widening of vessels

Autoregulation results in positive feedback

Decreased blood flow to tissue

Local accumulation of metabolic wastes

50.17 Control of Blood Pressure through Local and Systemic Mechanisms A drop in arterial pressure reduces blood flow to tissues, resulting in local accumulation of metabolic wastes. This change in the extracellular environment stimulates autoregulatory opening of the arteries. A fall in central blood pressure is prevented by negative feedback mechanisms that constrict arteries in less essential tissues and stimulate maintenance of blood volume and blood pressure.

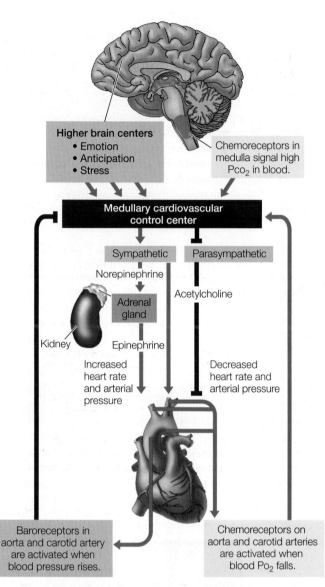

50.18 Regulating Cardiac Output   The autonomic nervous system controls heart rate in response to information about blood pressure and blood composition originating in baroreceptors and chemosensors shown at the bottom of the figure. Information from these sensors goes to the cardiovascular control center in the medulla, where it is integrated with other information. The medullary center generates responses in the sympathetic and parasympathetic nervous systems that control cardiac output.

Another hormone that helps stabilize blood pressure is antidiuretic hormone (ADH, also called vasopressin), which is secreted by the posterior pituitary in response to a fall in the activity of the baroreceptors, signaling a fall in arterial pressure. ADH causes the kidneys to resorb more water and thereby maintain blood volume and increase blood pressure. Increased activity of the baroreceptors inhibits the release of ADH, and as a result the kidneys excrete more water, reducing blood volume and contributing to a fall in arterial pressure (see Figure 50.17).

Other information that causes the cardiovascular control center to increase heart rate and blood pressure comes from chemoreceptors in the medulla, aorta, and the carotid arteries. As we discussed in Section 49.5, the medullary chemosensors are activated by increases in arterial CO_2 levels, and the carotid and aortic bodies are activated by falls in arterial O_2 levels. Chemosensors send signals to the cardiovascular regulatory center as well as to the respiratory regulatory center.

50.5 RECAP

The delivery of blood to tissues is controlled locally by autoregulatory mechanisms that dilate or constrict arterioles. These local actions are translated into alterations in central blood pressure and composition that are detected by neural and hormonal mechanisms, which then mediate corrective cardiovascular adjustments.

- How do autoregulatory changes in blood flow to capillary beds result in adjustments to MAP? **See pp. 1062–1063 and Figure 50.16**
- What are the roles of hormones in regulating blood pressure? **See p. 1063 and Figure 50.17**
- Describe the role of baroreceptors and chemoreceptors in regulating blood pressure. **See pp. 1063–1064 and Figure 50.18**

Increased activity in baroreceptors of the large arteries signals rising blood pressure and inhibits sympathetic nervous system signaling to arteries and arterioles while increasing parasympathetic signaling to the heart's pacemaker (see Figure 50.5). As a result, the heart slows and arterioles in peripheral tissues dilate, reducing blood pressure. If pressure in the large arteries falls, the activity of the baroreceptors decreases, stimulating sympathetic output to the arteries and arterioles while reducing parasympathetic output to the heart's pacemaker. As a result, the heart beats faster and the arterioles in peripheral tissues constrict, increasing blood pressure.

CHAPTER SUMMARY

50.1 Why Do Animals Need a Circulatory System?

- The metabolic needs of the cells of many small animals are met by direct exchange of materials with the external medium. The metabolic needs of the cells of larger animals are met by a **circulatory system** that transports nutrients, respiratory gases, and metabolic wastes throughout the body.

- In **open circulatory systems**, extracellular fluid leaves vessels and percolates through tissues. In **closed circulatory systems**, the blood is contained in a system of vessels. Closed circulatory systems have several advantages, including the ability to selectively direct blood, hormones, and nutrients to specific tissues. **Review Figure 50.1**

50.2 How Have Vertebrate Circulatory Systems Evolved?

SEE WEB ACTIVITY 50.1

- The circulatory system of vertebrates consists of a heart and a closed system of vessels containing blood that is separate from the interstitial fluid. **Arteries** and **arterioles** carry blood from the heart; **capillaries** are the site of exchange between blood and interstitial fluid; **venules** and **veins** carry blood back to the heart.

- The vertebrate heart evolved from two chambers in fishes to three in amphibians and some reptiles and four in crocodilians, birds, and mammals. This evolutionary progression has led to an increasing separation of blood that flows to the gas exchange organs and blood that flows to the rest of the body.

- The simplest (two-chambered heart) has an **atrium** that receives blood from the body and a **ventricle** that pumps blood out of the heart. An **aorta** distributes blood to arteries.

- In birds and mammals, blood circulates through two completely separate circuits: the **pulmonary**, which transports blood between the heart and lungs, and the **systemic**, which transports oxygen-rich blood between the heart and tissues.

50.3 How Does the Mammalian Heart Function?

- The human heart has four chambers. Valves in the heart prevent the backflow of blood. **Review Figure 50.2, WEB ACTIVITY 50.2**

- The **cardiac cycle** has two phases: **systole**, in which the ventricles contract, and **diastole**, in which the ventricles relax. The sequential heart sounds ("lub-dup") are made by the closing of the heart valves. **Review Figure 50.3, ANIMATED TUTORIAL 50.1**

- Blood pressure can be measured using a sphygmomanometer and a stethoscope. **Review Figure 50.4**

- The autonomic nervous system controls heart rate: sympathetic activity increases heart rate, and parasympathetic activity decreases it. Norepinephrine increases and acetylcholine decreases the rate of depolarization of the plasma membranes of **pacemaker cells**, affecting heart rate. **Review Figure 50.5**

- The **sinoatrial node** controls the cardiac cycle by initiating a wave of depolarization in the atria, which is conducted to the ventricles through a system consisting of the **atrioventricular node, bundle of His**, and **Purkinje fibers**. **Review Figure 50.6**

- Sustained contraction of ventricular muscle cells is due to long-duration action potentials that are generated by voltage-gated Na^+ and Ca^{2+} channels. **Review Figure 50.7**

- An **electrocardiogram** (**ECG**) records electrical events associated with the contraction and relaxation of the cardiac muscles. **Review Figure 50.8**

50.4 What Are the Properties of Blood and Blood Vessels?

- Blood can be divided into a **plasma** portion (water, salts, and proteins) and a cellular portion (**erythrocytes** or red blood cells, platelets, and white blood cells). All of the cellular components are produced from stem cells in the bone marrow. **Review Figure 50.9**

- Erythrocytes transport oxygen. Their production in the bone marrow is stimulated by **erythropoietin**, which is produced in response to **hypoxia** (low oxygen levels) in the tissues.

- **Platelets**, along with circulating proteins, are involved in **blood clotting**, which results in a meshwork of **fibrin** threads that help seal vessels. **Review Figure 50.10**

- Abundant smooth muscle cells allow vessels to change their diameter, altering their resistance and thus blood flow. Arteries and arterioles have many elastic fibers that enable them to withstand high pressures. **Review Figure 50.11, WEB ACTIVITY 50.3**

- Capillary beds are the site of exchange of materials between blood and tissue fluid.

- **Starling's forces** suggest that blood volume is maintained in the capillary beds by an exchange of fluids driven by both blood pressure and osmotic pressure. **Review Figure 50.13**

- An accumulation of fluid in the extracellular spaces leads to **edema**. Bicarbonate ions in the blood plasma contribute to the osmotic forces that draw water back into capillaries.

- The ability of a specific molecule to cross a capillary wall depends on the architecture of the capillary, the type of substance, and the concentration gradient between the blood and the tissue fluid.

- Veins have a high capacity for storing blood. Aided by gravity, by contractions of skeletal muscle, and by the actions of breathing, they return blood to the heart. **Review Figure 50.14**

- The **Frank–Starling law** describes forces that increase cardiac output, such as stretch of the cardiac muscles cells caused by increased venous return.

- The **lymphatic system** returns the interstitial fluid to the blood.

50.5 How Is the Circulatory System Controlled and Regulated?

- Blood flow through capillary beds is controlled by local **autoregulatory mechanisms**, hormones, and the autonomic nervous system. **Review Figure 50.16**

- Blood pressure is controlled in part by the hormones ADH and angiotensin, which stimulate contraction of blood vessels. **Review Figure 50.17**

- Heart rate is controlled by the autonomic nervous system, which responds to information about blood pressure and blood composition that is integrated by regulatory centers in the medulla. **Review Figure 50.18**

SELF-QUIZ

1. An open circulatory system is characterized by
   a. the absence of a heart.
   b. the absence of blood vessels.
   c. blood with a composition different from that of tissue fluid.
   d. the absence of capillaries.
   e. a higher-pressure circuit through gills than to other organs.

2. Which statement about vertebrate circulatory systems is *not* true?
   a. In fish, oxygenated blood from the gills returns to the heart through the left atrium.
   b. In mammals, deoxygenated blood leaves the heart through the pulmonary artery.

c. In amphibians, deoxygenated blood enters the heart through the right atrium.

d. In reptiles, the blood in the pulmonary artery has a lower oxygen content than the blood in the aorta.

e. In birds, the pressure in the aorta is higher than the pressure in the pulmonary artery.

3. Which statement about the human heart is *true*?
   a. The walls of the right ventricle are thicker than the walls of the left ventricle.
   b. Blood flowing through atrioventricular valves is always deoxygenated blood.
   c. The second heart sound is due to the closing of the aortic valve.
   d. Blood returns to the heart from the lungs in the vena cava.
   e. During systole, the aortic valve is open and the pulmonary valve is closed.

4. The pacemaker action potentials in the heart
   a. are due to opposing actions of norepinephrine and acetylcholine.
   b. are generated by the bundle of His.
   c. depend on the gap junctions between the cells that make up the atria and those that make up the ventricles.
   d. are due to spontaneous depolarization of the plasma membranes of modified cardiac muscle cells.
   e. result from hyperpolarization of cells in the sinoatrial node.

5. Blood velocity through capillaries is slow because
   a. much blood volume is lost from the capillaries.
   b. the pressure in venules is high.
   c. the total cross-sectional area of capillaries is larger than that of arterioles.
   d. the osmotic pressure in capillaries is very high.
   e. erythrocytes must pass through in single file.

6. How are lymphatic vessels like veins?
   a. Both have nodes where they join together into larger common vessels.
   b. Both carry blood under low pressure.
   c. Both are capacitance vessels.
   d. Both have valves.
   e. Both carry fluids rich in plasma proteins.

7. The production of erythrocytes
   a. ceases if the hematocrit falls below normal.
   b. is stimulated by erythropoietin.
   c. is about equal to the production of white blood cells.
   d. is inhibited by prothrombin.
   e. occurs in bone marrow before birth and in lymph nodes after birth.

8. Which of the following does *not* increase blood flow through a capillary bed?
   a. High concentration of CO_2
   b. High concentration of lactate and hydrogen ions
   c. Histamine
   d. ADH
   e. Increase in arterial pressure

9. Blood clotting
   a. is impaired in patients with hemophilia because they do not produce platelets.
   b. is initiated when platelets release fibrinogen.
   c. involves a cascade of factors produced in the liver.
   d. is initiated by leukocytes forming a meshwork.
   e. requires production of angiotensin.

10. Autoregulation of blood flow to a tissue is due to
    a. sympathetic innervation.
    b. the release of ADH by the hypothalamus.
    c. increased activity of baroreceptors.
    d. chemoreceptors in the aorta and the carotid arteries.
    e. the effect of the local chemical environment on arterioles.

FOR DISCUSSION

1. At the beginning of a race, cardiac output increases immediately before there is any change in blood O_2 or CO_2 concentrations. Explain two factors that contribute to this effect. Include the Frank–Starling law in your answer.

2. Explain how the hearts of crocodilians have the advantages of mammalian hearts during exercise but the efficiency of reptilian hearts during rest.

3. A sudden and massive loss of blood results in a decrease in blood pressure. Describe several mechanisms that help return blood pressure to normal.

4. You can describe the cycle of events in a ventricle of the heart by a graph that plots the pressure in the ventricle on the y axis and the volume of blood in the ventricle on the x axis. What would such a graph look like? Where would the heart sounds be on this graph? How would the graph differ for the left and the right ventricles?

5. If the major arteries become clogged with plaque and become less elastic because of atherosclerosis, the left ventricle must work harder and harder to pump an adequate supply of blood to the body. As a result, the left ventricle can become weakened and begin to fail, even though the right ventricle is healthy. A heart attack primarily affecting the left ventricle can have the same effect. This condition is known as congestive heart failure, and it commonly leads to fatal pulmonary edema. Explain how left ventricular failure can result in pulmonary edema, and why is it said that this condition creates a vicious circle that makes itself worse rapidly.

ADDITIONAL INVESTIGATION

1. There are strains of rats that are spontaneously hypertensive—they naturally have high blood pressure. Can you formulate a hypothesis about the basis for such inherited hypertension and explain experiments you would do to test your hypothesis?

2. If you were in training to compete in the Tour de France, a grueling 3-week-long bicycle race, a trainer might tell you to "train high, compete low." What do you think is the physiological basis for such advice, and how would you test whether or not that reasoning was correct?

Nutrition, Digestion, and Absorption

An obesity epidemic

For thousands of years, the Pima of southwestern North America were hunters and gatherers who supplemented their diet with subsistence agriculture. Their environment was arid, so they developed sophisticated irrigation systems; even so, they frequently encountered drought and subsequent starvation. Today most individuals of the ethnic Pima population in North America are clinically obese. In fact, as a population, they are one of the heaviest in the world.

With obesity come related health problems such as diabetes, high blood pressure, and heart disease. The incidence of diabetes among the Pima rose from an extremely high level of 45 percent of adults in 1965 to a staggering 80 percent in 1999. Moreover, diabetes is occurring in younger individuals than ever before. What has caused such a radical health change in an entire population? At least two interacting factors are involved: genetics and lifestyle.

Geneticists hypothesize that recurring episodes of starvation produce strong selective pressure for "thrifty genes"—particular alleles of the genes involved in digestion, absorption, and energy storage that result in greater-than-average efficiency in converting food into energy and into energy reserves, such as fat. Thrifty genes would give individuals a strong selective advantage when food is scarce. An example of a "thrifty" phenotype is seen among the Pima. They have a very low resting metabolic rate and convert food into fat readily. As we will see later in this chapter, the hormone insulin facilitates the conversion of dietary sugar into fat tissue. For many Pima, consuming a standard amount of glucose causes their insulin levels to rise three times higher than it does in Americans of European ancestry.

The other factor in the Pima obesity epidemic is an abrupt change in their traditional lifestyle. When food is plentiful and has high caloric content, thrifty genes con-

Efficiency Genes The Pima are an example of a human population that repeatedly experienced periods of severe food deprivation. These historic occurrences may have imposed selection for genes that improve the efficiency of managing the energy obtained from food. With modern diets and lifestyles, these "efficiency genes" can contribute to obesity.

The Great American Lunch High-fat fast foods have become prevalent in much of the developed world. A steady diet of such foods will mean weight gain for most people, and is a major reason for obesity in the United States.

tribute to obesity by maximizing fat storage. Today the Pima eat a modern Western diet that includes high-fat, high-calorie fast foods. In general, they also engage in less physical activity than their ancestors did.

A comparative study supports the hypotheses that have been put forward to explain the obesity epidemic in the Pima. Another population of Pima live in the Sierra Madre of northern Mexico. Genetically, they are the same as the Arizona population. However, they live a traditional lifestyle and eat traditional foods. Whereas the Arizona Pima engage in an average of only 2 hours of physical work per week, the Mexico Pima average 23 hours per week. Obesity and diabetes are not prevalent among the Mexico Pima.

A high-calorie diet and sedentary lifestyle affect not just the Pima but contribute to the overall increase in obesity throughout the U.S. population. Researchers are studying the Pima to learn more about the genetics of obesity and related pathologies.

IN THIS CHAPTER we will review the nutrients organisms require for energy, for molecular building blocks, and for specific biochemical functions. We will examine diverse adaptations for acquiring, ingesting, and digesting food and absorbing nutrients. Finally, we will learn how the body regulates its traffic in metabolic fuels and return to the question of control of body mass.

51.1 What Do Animals Require from Food?

Animals are **heterotrophs**—they derive their nutrition from eating other organisms. In contrast, **autotrophs** (most plants, some bacteria, some archaea, and some protists) can use solar energy or inorganic chemical energy to synthesize all of their components. Directly and indirectly, heterotrophs take advantage of—indeed, depend on—the organic synthesis carried out by autotrophs, and have evolved an enormous diversity of adaptations to exploit this resource (**Figure 51.1**). In this section we will discover how animals use food, be it plants or other animals, to obtain energy and building blocks of complex molecules. We will also consider the need for special mineral nutrients and organic molecules we call vitamins and the diseases that result when they are lacking in the diet.

Energy needs and expenditures can be measured

Energy is the capacity to do work, and it comes in different forms—electrical energy, heat energy, chemical energy, nuclear energy. As discussed in Chapter 8, a *calorie* (note the small *c*) is a unit of heat energy; specifically, it is the amount of heat necessary to raise the temperature of 1 gram of water 1°C. Since this is a tiny amount of energy, physiologists commonly use the **kilocalorie (kcal)** as a unit of measure (1 kcal = 1,000 calories). Nutritionists also use the kilocalorie as a standard unit of energy, but they traditionally refer to it as the **Calorie (Cal)** which is always capitalized to distinguish it from the single calorie.

Just about any food container you pick up in the U.S. carries a label, "Nutrition Facts," that includes the item "Calories." How do Calories (or kcal) relate to the discussion in Chapter 9 about how energy in the chemical bonds of food molecules is transferred to the high-energy phosphate bonds of adenosine triphosphate (ATP) and is then used to do cellular work? And why do we use heat energy as a measure of nutrition?

The reason is found in the laws of thermodynamics, which tell us that energy cannot be created or destroyed, but can be converted from one form to another (see Section 8.1). However, every energy conversion is inefficient and a large portion of the original energy always ends up as heat. Whether we are using the energy in glucose to make ATP or are using that ATP to power muscle contraction or ion transport, most of the available chemical energy is lost as heat. The bottom line is that if an animal is not growing, not doing work on its environment, and not changing its body temperature, the heat it loses to the environment is a measure of its total energy expenditure—its *metabolic rate*.

(A)

(B)

51.1 Heterotrophs Get Energy from Autotrophs (A) Herbivores get their energy directly from autotrophs. The large herbivores of the African grasslands must consume huge amounts of plant matter to fulfill their nutritional needs. (B) A carnivore's energy is indirectly obtained from autotrophs, since the energy stored in a prey animal was originally obtained from autotrophs.

The metabolic rate of an animal is a measure of its overall energy needs that must be met by the ingestion, digestion, and assimilation of food (see Section 40.3). The basal metabolic rate of a human is 1,300–1,500 Cal/day for an adult female and 1,600–1,800 Cal/day for an adult male. Physical activity adds to this basal energy requirement. For a person doing sedentary work, about 30 percent of the Calories expended are used for skeletal muscle activity; for a person doing heavy physical labor, more than 95 percent of total caloric expenditure is due to skeletal muscle activity.

The components of food that provide energy are fats, carbohydrates, and proteins. Fats yield 9.5 Cal/gram, carbohydrates 4.2 Cal/gram, and proteins about 4.1 Cal/gram. **Figure 51.2** shows some equivalencies of food, energy, and energy consumption.

Even though the units calorie, kilocalorie, and Calorie remain in popular use, most scientists now use the International System of Units. In this system, the basic unit of energy is the joule: 1 joule = 0.239 calories, and the measure of energy utilization is 1 joule/second = 1 watt. You are familiar with light bulb ratings, so think about that when you convert kcal/day into watts. The metabolic rate of the average man is 1,700 kcal/day or 82 watts (note that a watt includes the time dimension, so it is a rate of energy utilization).

Thus it is possible to quantify the caloric value of any food an animal eats. It is also possible to quantify the caloric expenditure of any activity or behavior an animal performs. By comparing calories consumed with calories expended, we can construct *energy budgets* that allow ecologists and evolutionary biologists to apply a *cost–benefit analysis* to feeding behavior, as we explain in Section 53.4.

51.2 Food Energy and How We Use It The energy contained in several common food items is shown at the left. The graphs indicate about how long it would take a person with a basal metabolic rate of about 1,800 Cal/day to utilize the equivalent amount of energy while resting, walking, or jogging.

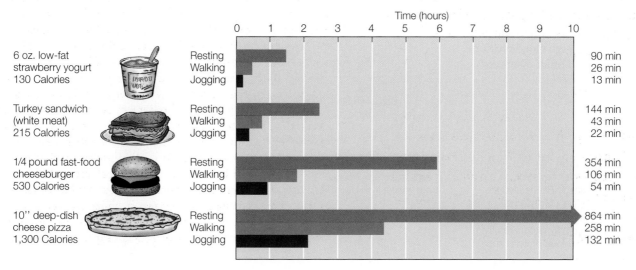

		Time (hours)	
6 oz. low-fat strawberry yogurt 130 Calories	Resting Walking Jogging		90 min 26 min 13 min
Turkey sandwich (white meat) 215 Calories	Resting Walking Jogging		144 min 43 min 22 min
1/4 pound fast-food cheeseburger 530 Calories	Resting Walking Jogging		354 min 106 min 54 min
10" deep-dish cheese pizza 1,300 Calories	Resting Walking Jogging		864 min 258 min 132 min

Sources of energy can be stored in the body

Although the cells of the body use energy continuously, most animals do not eat continuously. Therefore, animals must store fuel molecules that can be released as needed between meals. Carbohydrates are stored in liver and muscle cells as glycogen, but the total glycogen store represents only about a day's basal energy requirements (1,500–2,000 Cal). Fat is the most important form of stored energy in the bodies of animals. Not only does fat have more energy per gram than glycogen, but it can be stored with little associated water, making it more compact. Migrating birds store energy as fat to fuel their long flights; if they had to store the same amount of energy as glycogen, they would be too heavy to fly! Proteins are not used as energy storage compounds, although body protein can be metabolized as an energy source of last resort.

If an animal takes in too little food to meet its energy requirements, it is **undernourished** and must start metabolizing some of the molecules of its own body. This "self-consumption" begins with the energy storage compounds glycogen and fat. There is some protein loss due to normal protein turnover not being fully replaced because of a lack of amino acids for new protein synthesis. Once fat reserves are seriously depleted, the body increases its metabolism of proteins for energy (**Figure 51.3A**). The first proteins to be sacrificed are those of the blood plasma. The loss of plasma proteins decreases the osmotic concentration of the plasma, resulting in increased loss of fluid from the blood to the interstitial spaces (*edema*; see Section 50.4). Accumulation of fluid in the extremities and abdomen is the classic sign of *kwashiorkor*, a disease caused by chronic protein deficiency (**Figure 51.3B**). Continued protein loss damages the body's organs, leading eventually to death.

When an animal consistently takes in more food than it needs to meet its energy requirements, it is **overnourished**, and the excess nutrients are stored as increased body mass. First, glycogen reserves build up; then additional dietary carbohydrates, fats, and proteins are converted to body fat. In some species, such as hibernators, seasonal overnutrition is an important adaptation for surviving periods when food is not available. In humans, however, overnutrition can be a serious health hazard, increasing the risk of high blood pressure, heart attack, diabetes, and other disorders.

Food provides carbon skeletons for biosynthesis

Every animal requires certain basic organic molecules that it cannot synthesize for itself but needs as building blocks for its own complex organic molecules. The acetyl group ($CH_3CO—$) is one such required building block supplying the **carbon skeleton** of larger organic molecules (**Figure 51.4**). Animals cannot make acetyl groups from carbon, oxygen, and hydrogen molecules; they must obtain acetyl groups from food. Acetyl groups can be derived from the metabolism of almost any food, but they originate in plants. Acetyl groups are never in short supply for an adequately nourished animal. However, some groups supplying carbon skeletons can be deficient in an animal's diet even if caloric intake is adequate.

Amino acids, the building blocks of proteins, are a good example of carbon skeletons that can be in short supply. Animals can synthesize some of their own amino acids by utilizing carbon skeletons from acetyl or other groups and transferring to them amino groups ($—NH_2$) derived from other amino acids. Most animals cannot synthesize all the amino acids they need. Each species must obtain certain **essential amino acids** from food. Essential amino acids vary by species. In general, herbivores have fewer essential amino acids than carnivores. If an animal does not take in enough of even one of its essential amino acids, its protein synthesis is impaired and its capacity to maintain enzymatic and transport functions is challenged.

Most researchers agree that adult humans must obtain eight essential amino acids from their food: isoleucine, leucine, lysine, methionine, phenylalanine, threonine, tryptophan, and valine. All eight are available in milk, eggs, meat, and soybean products, but most plant foods do not contain adequate quantities

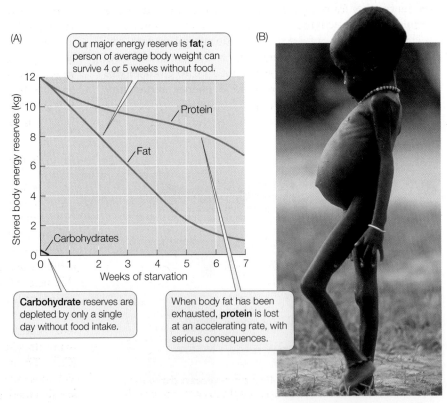

(A)

Our major energy reserve is **fat**; a person of average body weight can survive 4 or 5 weeks without food.

Protein

Fat

Carbohydrates

Stored body energy reserves (kg)

Weeks of starvation

(B)

Carbohydrate reserves are depleted by only a single day without food intake.

When body fat has been exhausted, **protein** is lost at an accelerating rate, with serious consequences.

51.3 The Course of Starvation (A) In a person subjected to undernutrition, the body's energy reserves are eventually depleted. (B) The swollen abdomen, face, hands, and feet of this Somali girl are due to edema. Along with her spindly limbs, these are symptoms of kwashiorkor, a syndrome resulting from the body breaking down blood proteins and muscle tissue to fuel metabolism.

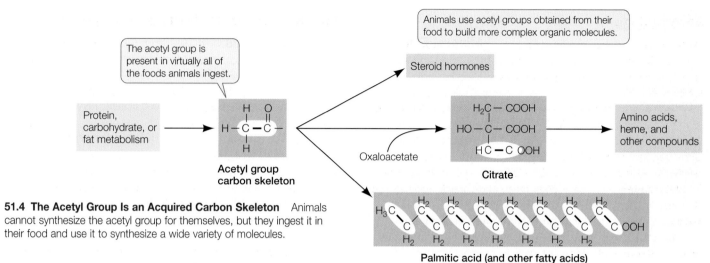

51.4 The Acetyl Group Is an Acquired Carbon Skeleton Animals cannot synthesize the acetyl group for themselves, but they ingest it in their food and use it to synthesize a wide variety of molecules.

of all eight, so a strict vegetarian diet carries a risk of protein malnutrition. A *complementary* diet of plant foods, however, supplies all eight essential amino acids (**Figure 51.5**). In general, grains (such as rice, wheat, and corn) are complemented by legumes (such as beans and peas). Long before the chemical basis for this complementarity was understood, societies with little access to meat developed complementary diets. Many Central and South American peoples traditionally eat beans with corn, and the native peoples of North America complemented their beans with squash.

Human infants are thought to require four additional amino acids in their diets: histidine, tyrosine, cysteine, and arginine. Also, some amino acids are *conditionally essential* for certain populations that cannot synthesize them in adequate amounts. For example, individuals with the genetic disease phenylketonuria lack the enzyme for converting phenylalanine to tyrosine (see Section 15.3) and thus must obtain tyrosine from their diets. They must keep their dietary intake of phenylalanine low to prevent its accumulation to toxic levels.

Why are dietary proteins completely digested to their constituent amino acids before being used by the body? Wouldn't it be more energy-efficient to reuse some dietary proteins directly? There are several reasons why ingested proteins are not used "as is":

- Macromolecules such as proteins are not readily absorbed by the cells of the gut, but their constituent monomers (such as amino acids) are readily absorbed.
- Protein structure and function are highly species-specific. A protein that functions optimally in one species might not function well in another.
- Foreign proteins entering the body directly from the gut would be recognized as invaders and would be attacked by the immune system.

Humans can synthesize almost all the lipids required by the body using acetyl groups obtained from food (see Figure 51.4), but we must have a dietary source of certain **essential fatty acids**—notably, linoleic acid—that we cannot synthesize. Linoleic acid is an unsaturated fatty acid needed by mammals to synthesize other unsaturated fatty acids, such as arachidonic acid, which is a component of several signaling molecules, including prostaglandins. Essential fatty acids are also necessary components of membrane phospholipids. A deficiency of linoleic acid can lead to problems such as infertility and impaired lactation, but since it is commonly present in vegetable oils, a deficiency is unlikely in an adequately nourished individual.

Animals need mineral elements for a variety of functions

Table 51.1 lists the principal mineral elements that animals require. Elements required in large amounts are called **macronutrients**; those required in only tiny amounts (generally less than 100 mg/day) are called **micronutrients**. Some micronutrients are required in such minute amounts that deficiencies are never observed, but they are nevertheless essential elements.

Calcium is an example of a macronutrient. It is the fifth most abundant element in the body; a 70-kg person contains about 1.2 kg of calcium. Calcium phosphate is the principal structural material in bones and teeth. Muscle contraction, neural function, and many other intracellular functions in animals require calcium ions (Ca^{2+}). The turnover of calcium in the extracellular fluid is high, as bones are constantly being remodeled and calcium is constantly entering and leaving cells. Calcium is lost from the body

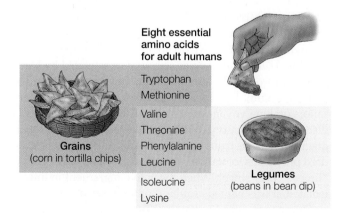

Eight essential amino acids for adult humans

Tryptophan
Methionine
Valine
Threonine
Phenylalanine
Leucine
Isoleucine
Lysine

Grains
(corn in tortilla chips)

Legumes
(beans in bean dip)

51.5 A Strategy for Vegetarians By combining cereal grains with legumes, an adult vegetarian can obtain all eight essential amino acids.

TABLE 51.1
Mineral Elements Required by Animals

ELEMENT	SOURCE IN HUMAN DIET	MAJOR FUNCTIONS
MACRONUTRIENTS		
Calcium (Ca)	Dairy foods, eggs, green leafy vegetables, whole grains, legumes, nuts, meat	Found in bones and teeth; blood clotting; nerve and muscle action; enzyme activation
Chlorine (Cl)	Table salt (NaCl), meat, eggs, vegetables, dairy foods	Water balance; digestion (as HCl); principal negative ion in extracellular fluid
Magnesium (Mg)	Green vegetables, meat, whole grains, nuts, milk, legumes	Required by many enzymes; found in bones and teeth
Phosphorus (P)	Dairy, eggs, meat, whole grains, legumes, nuts	Found in nucleic acids, ATP, and phospholipids; bone formation; buffers; metabolism of sugars
Potassium (K)	Meat, whole grains, fruits, vegetables	Nerve and muscle action; protein synthesis; principal positive ion in cells
Sodium (Na)	Table salt, dairy foods, meat, eggs	Nerve and muscle action; water balance; principal positive ion in extracellular fluid
Sulfur (S)	Meat, eggs, dairy foods, nuts, legumes	Found in proteins and coenzymes; detoxification of harmful substances
MICRONUTRIENTS		
Chromium (Cr)	Meat, dairy, whole grains, legumes, yeast	Glucose metabolism
Cobalt (Co)	Meat, tap water	Found in vitamin B_{12}; formation of red blood cells
Copper (Cu)	Liver, meat, fish, shellfish, legumes, whole grains, nuts	Found in active site of many redox enzymes and electron carriers; production of hemoglobin; bone formation
Fluorine (F)	Most water supplies	Found in teeth; helps prevent decay
Iodine (I)	Fish, shellfish, iodized salt	Found in thyroid hormones
Iron (Fe)	Liver, meat, green vegetables, eggs, whole grains, legumes, nuts	Found in active sites of many redox enzymes and electron carriers, hemoglobin, and myoglobin
Manganese (Mn)	Organ meats, whole grains, legumes, nuts, tea, coffee	Activates many enzymes
Molybdenum (Mo)	Organ meats, dairy, whole grains, green vegetables, legumes	Found in some enzymes
Selenium (Se)	Meat, seafood, whole grains, eggs, milk, garlic	Fat metabolism
Zinc (Zn)	Liver, fish, shellfish, and many other foods	Found in some enzymes and some transcription factors; insulin physiology

in urine, sweat, and feces, so it must be replaced regularly. Humans require 800–1,000 mg of calcium per day in their diet.

Iron is an example of a micronutrient. It is found throughout the body because it is the oxygen-binding atom in hemoglobin and myoglobin and is a component of enzymes in the electron transport chain. Nevertheless, the total amount of iron in a 70-kg person is only about 4 grams, and since iron is recycled efficiently in the body and is not lost in the urine, we require only about 15 mg per day in our food. Despite the small amount required, insufficient iron is the most common mineral nutrient deficiency in the world today, leading to *anemia*. Anemic individuals are weak and tired all the time.

---
yourBioPortal.com
GO TO Web Activity 51.1 • Mineral Elements
Required by Animals

Animals must obtain vitamins from food

Like essential amino acids and fatty acids, **vitamins** are carbon compounds that an animal requires for growth and metabolism

but cannot synthesize for itself. They are required in very small amounts compared with the essential amino acids and fatty acids, which are incorporated into large body structures. Most vitamins function as coenzymes or parts of coenzymes.

The list of vitamins varies from species to species. Most mammals, for example, can make their own ascorbic acid. Primates (including humans) cannot, so for primates, ascorbic acid is a vitamin—*vitamin C*. If we do not get vitamin C in our food, we develop a disease known as scurvy, characterized by bleeding gums, loss of teeth, subcutaneous hemorrhages, and slow wound healing. The disease was a frequently fatal problem for sailors on long voyages until a Scottish physician, James Lind, discovered that scurvy could be prevented if the sailors ate fresh greens and citrus fruit. The British Admiralty made limes standard provisions for its ships (and British sailors have been called "limeys" ever since). When the active ingredient in limes was isolated, it was named *ascorbic* ("without scurvy") *acid*.

Humans require 13 vitamins (**Table 51.2**). They are divided into two groups: water-soluble and fat-soluble. When water-soluble vitamins are ingested in excess of bodily needs, they are

TABLE 51.2
Vitamins in the Human Diet

VITAMIN	SOURCE	FUNCTION	DEFICIENCY SYMPTOMS
WATER-SOLUBLE			
B_1 (thiamin)	Liver, legumes, whole grains	Coenzyme in cellular respiration	Beriberi, loss of appetite, fatigue
B_2 (riboflavin)	Dairy, meat, eggs, green leafy vegetables	Coenzyme in FAD	Lesions in corners of mouth, eye irritation, skin disorders
Niacin	Meat, fowl, liver, yeast	Coenzyme in NAD and NADP	Pellagra, skin disorders, diarrhea, mental disorders
B_6 (pyridoxine)	Liver, whole grains, dairy foods	Coenzyme in amino acid metabolism	Anemia, slow growth, skin problems, convulsions
Pantothenic acid	Liver, eggs, yeast	Found in acetyl CoA	Adrenal problems, reproductive problems
Biotin	Liver, yeast, bacteria in gut	Found in coenzymes	Skin problems, loss of hair
B_{12} (cobalamin)	Liver, meat, dairy foods, eggs	Formation of nucleic acids, proteins, and red blood cells	Pernicious anemia
Folic acid	Vegetables, eggs, liver, whole grains	Coenzyme in formation of heme and nucleotides	Anemia
C (ascorbic acid)	Citrus fruits, tomatoes, potatoes	Formation of connective tissues; antioxidant	Scurvy, slow healing, poor bone growth
FAT-SOLUBLE			
A (retinol)	Fruits, vegetables, liver, dairy	Found in visual pigments	Night blindness
D (cholecalciferol)	Fortified milk, fish oils, sunshine	Absorption of calcium and phosphate	Rickets
E (tocopherol)	Meat, dairy foods, whole grains	Muscle maintenance, antioxidant	Anemia
K (menadione)	Intestinal bacteria, liver	Blood clotting	Blood-clotting problems

simply eliminated in the urine. (This is the fate of much of the excess vitamin C that people take.) Fat-soluble vitamins, however, can accumulate in body fat and may build up to toxic levels in the liver if taken in excess.

The fat-soluble vitamin D (cholecalciferol), which is essential for absorbing and metabolizing calcium, is a special case because the body can synthesize it. (As noted in Section 41.3, vitamin D is by definition a hormone.) Certain lipids present in the human body can be converted into vitamin D by the action of ultraviolet light on the skin. Thus vitamin D must be obtained in the diet by individuals with inadequate exposure to the sun.

The need for vitamin D may have been an important factor in the evolution of skin color. Human races that are adapted to equatorial and low latitudes have dark skin pigmentation as a protection against the damaging effects of ultraviolet radiation. These peoples generally expose extensive areas of skin to the sun on a regular basis, so their skin synthesizes adequate amounts of vitamin D. Most races that adapted to life in the higher latitudes lost this dark skin pigmentation, probably because lighter skin facilitates vitamin D production in the relatively small areas of skin exposed to sunlight during the short days of winter. The Inuit peoples of the Arctic seem to represent an exception to the correlation between latitude and skin pigmentation, but these dark-skinned people obtain ample vitamin D from the large amounts of animal fat (especially whale blubber) and fish oils in their diet.

—— **yourBioPortal**.com ——
GO TO **Web Activity 51.2 • Vitamins in the Human Diet**

Nutrient deficiencies result in diseases

The lack of any essential nutrient in the diet produces a state of deficiency called **malnutrition**, and chronic malnutrition leads to a characteristic **deficiency disease**. We have already discussed kwashiorkor (protein deficiency) and scurvy (vitamin C deficiency). A shortage of any of the vitamins results in specific deficiency symptoms (see Table 51.2). Another deficiency disease, *beriberi*, was directly involved in the discovery of vitamins.

Beriberi means "extreme weakness." It became prevalent in Asia in the nineteenth century after it became standard practice to mill rice to a high, white polish and discard the hulls present in brown rice. A critical observation was that birds—chickens and pigeons—developed beriberi-like symptoms when fed only polished rice. In 1912, Casimir Funk, a Polish scientist working in England, cured pigeons of beriberi by feeding them discarded rice hulls.

At the time of Funk's discovery, all diseases were thought to be either caused by microorganisms or inherited. Funk suggested the radical idea that beriberi and some other diseases are dietary in origin and result from deficiencies in specific substances. Funk coined the term "vitamines" because he mistakenly thought that all these substances vital for life were compounds with amino groups (vital amines). In 1926, thiamin (vitamin B_1)—the substance lost in the rice milling process—was the first vitamin to be isolated in pure form.

Deficiency diseases can also result from an inability to absorb or process an essential nutrient even if it is present in the diet. Vitamin B_{12} (cobalamin), for example, is present in all foods of animal origin. Since plants neither use nor produce vitamin B_{12}, a strictly vegetarian diet (not supplemented with dairy products or vitamin pills) can lead to a B_{12} deficiency disease called pernicious anemia, characterized by a failure of red blood cells to mature. The most common cause of pernicious anemia, however, is not a lack of vitamin B_{12} in the diet but an inability to absorb it. Normally, cells in the stomach lining secrete a peptide called *intrinsic factor*, which binds to vitamin B_{12} and makes it possible for it to be absorbed in the small intestine. Conditions that damage the stomach lining, such as alcoholism or gastritis, can thus lead to pernicious anemia.

Inadequate mineral nutrition can also lead to deficiency diseases. Examples are hypothyroidism and goiter resulting from iodine deficiency (see Section 41.3), and anemia resulting from iron deficiency. Iodine deficiency is almost unheard of today in the developed world because of the addition of iodide to salt. However, it is still a major health problem in large segments of the human population. Probably the single least expensive effective action to improve global health would be to provide iodized salt for everyone.

51.1 RECAP

As heterotrophs, animals must obtain the energy and molecular building blocks for biosynthesis from their food. Energy can come from the metabolism of carbohydrates, fats, and proteins. Molecular building blocks include carbon skeletons, vitamins, and minerals.

- Explain the different roles of food energy from proteins, carbohydrates, and fats in the mammalian body. **See p. 1070 and Figure 51.3**

- Give an example of an essential carbon skeleton, a micronutrient, and a macronutrient. **See pp. 1070–1073, Figure 51.4 and Table 51.1**

- Why should fat-soluble vitamins not be taken in excess? **See p. 1073**

We have surveyed the essential elements of nutrition in animals. Next we will look at various methods and adaptations by which animals obtain the food they need, and mechanisms by which the food is processed in the body to extract nutrients.

51.2 How Do Animals Ingest and Digest Food?

Heterotrophic organisms can be classified by how they acquire their nutrition. **Saprobes** (also called *saprotrophs* or *decomposers*) are organisms—mostly protists and fungi—that absorb nutrients from dead organic matter. **Detritivores**, such as earthworms and crabs, actively feed on dead organic material. Animals that feed on living organisms are **predators**: **herbivores** prey on plants, **carnivores** prey on animals, and **omnivores** prey on both. **Filter feeders**, such as clams and blue whales, prey on small organisms by filtering them from the aquatic environment. **Fluid feeders** include mosquitoes, aphids, leeches, and hummingbirds. The anatomical adaptations that enable a species to exploit a particular source of nutrition are usually obvious, but physiological and biochemical adaptations are also important.

The food of herbivores is often low in energy and hard to digest

Most vegetation is coarse and difficult to break down, but herbivores process large amounts of it because its energy content is low. Therefore, herbivores spend a great deal of time feeding. Many have striking adaptations for feeding, such as the trunk (a flexible, gripping nose) of the elephant or the huge bill of the fruit-eating toucan, which can be half as long as its body. Many types of grinding, rasping, cutting, and shredding mouthparts have evolved in invertebrates for ingesting plant material, and the teeth of herbivorous vertebrates have been shaped by selection to tear, crush, and grind coarse plant matter.

The digestive processes of herbivores can also be quite specialized. An example is the koala, which almost exclusively eats leaves of eucalyptus trees. These leaves are very fibrous, low in usable energy and protein, and high in toxic chemicals. Koalas actually smell like eucalyptol, a common ingredient in cough drops. The koala has strong jaws for grinding the leaves, a very long gut for fermenting them, enzymes in its liver for detoxifying chemicals in the leaves, and a low metabolic rate to compensate for low energy intake.

Carnivores must detect, capture, and kill prey

The predatory behaviors of many carnivores are legendary—the hunting skills of hawks, wolves, and tigers, for example. Carnivores have evolved stealth, speed, power, large jaws, sharp teeth, and strong gripping appendages. They also have evolved remarkable means of detecting prey. Bats use echolocation, pit vipers sense infrared radiation from the warm bodies of their prey, and certain fishes detect electric fields created in the water by their prey.

Adaptations for killing and ingesting prey are diverse and can be highly specialized. These adaptations are especially important when the prey can inflict damage on the predator. There are many fascinating examples of adaptations for capturing and immobilizing prey, such as the immobilizing venom in the bite of many snakes, the long sticky tongues of chameleons, and the webs of spiders. The long tentacles of certain jellyfish contain some of the most deadly toxins known and can kill humans. Specialized cells called *nematocysts* inject toxin into the victim (see Figure 31.9). Some jellyfish harbor enough nematocysts cells to kill 50 people. Some predators digest their prey externally. For example, a spider injects its insect prey with digestive enzymes and then sucks out the liquefied contents, leaving behind the empty exoskeletons frequently seen in old spider webs.

Vertebrate species have distinctive teeth

Teeth are adapted for the acquisition and initial processing of specific types of foods. Because they are among the hardest struc-

(A)

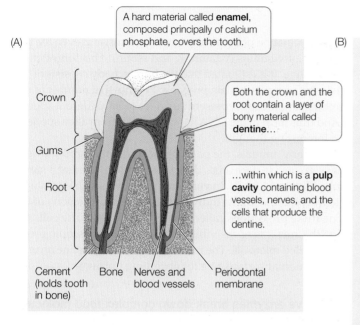

A hard material called **enamel**, composed principally of calcium phosphate, covers the tooth.

Crown

Gums

Root

Both the crown and the root contain a layer of bony material called **dentine**…

…within which is a **pulp cavity** containing blood vessels, nerves, and the cells that produce the dentine.

Cement (holds tooth in bone)

Bone

Nerves and blood vessels

Periodontal membrane

51.6 Mammalian Teeth (A) A mammalian tooth has three layers: enamel, dentine, and a pulp cavity. (B) The teeth of different mammalian species are specialized for different diets. This illustration depicts the teeth of the lower jaw, viewed from above.

— yourBioPortal.com —
GO TO Web Activity 51.3 • Mammalian Teeth

(B)

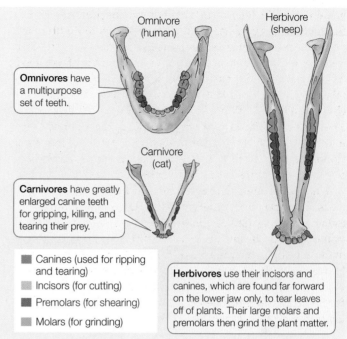

Omnivore (human)

Herbivore (sheep)

Omnivores have a multipurpose set of teeth.

Carnivore (cat)

Carnivores have greatly enlarged canine teeth for gripping, killing, and tearing their prey.

■ Canines (used for ripping and tearing)
□ Incisors (for cutting)
■ Premolars (for shearing)
□ Molars (for grinding)

Herbivores use their incisors and canines, which are found far forward on the lower jaw only, to tear leaves off of plants. Their large molars and premolars then grind the plant matter.

tures of the body, an animal's teeth remain in the environment long after it dies. Paleontologists use teeth to identify animals that lived in the distant past and to deduce their feeding behavior.

All mammalian teeth have the same general, three-layered structure (**Figure 51.6A**). An extremely hard material called **enamel**, composed principally of calcium phosphate, covers the crown of the tooth. Both the crown and root contain a layer of bony material called **dentine**, inside of which is a **pulp cavity** containing blood vessels, nerves, and the cells that produce the dentine.

There is a great deal of homology in the dentition of mammals, but the shapes and organization of mammalian teeth are adaptations to different diets (**Figure 51.6B**). In general, *incisors* are used for cutting, chopping, and gnawing; *canines* are used for stabbing, gripping, and ripping; and *molars* and *premolars* (the cheek teeth) are used for shearing, crushing, and grinding. The highly varied diet of humans is reflected in our multipurpose set of teeth, as is common among omnivores.

Digestion usually begins in a body cavity

Animals take food into a body cavity that is continuous with the outside environment. They secrete digestive enzymes into that cavity, and the enzymes break down the food into nutrient molecules that can be absorbed by the cells lining the cavity.

The simplest digestive systems are **gastrovascular cavities**, which connect to the outside world through a single opening. Cnidarians, such the jellyfish mentioned above, capture prey using stinging nematocysts and use their tentacles to cram the prey into their gastrovascular cavities. Enzymes in the gastrovascular

cavity partly digest the prey. Cells lining the cavity take in small food particles by endocytosis. The vesicles created by endocytosis then fuse with lysosomes containing digestive enzymes, and intracellular digestion completes the breakdown of the food. Nutrients are released to the cytoplasm as the vesicles break down.

Tubular guts have an opening at each end

The guts of most animals are tubular: a **mouth** takes in food; molecules are digested and absorbed throughout the length of the gut; and solid digestive wastes are eliminated through an **anus**. Different regions in the tubular gut are specialized for particular functions (**Figure 51.7**). These functions must be coordinated so they occur in the proper sequence and at rates that maximize the efficiency of digestion and absorption of nutrients.

At the anterior end of the gut is the mouth cavity where food can be fragmented, for example by teeth (in many vertebrates), by **radula** (in snails), or by **mandibles** (in many arthropods). In most birds, food is ground by small stones in an early, muscular, portion of the gut called the **gizzard**. Some animals, such as snakes, simply ingest whole prey with little or no fragmentation. **Stomachs** and **crops** are storage chambers that enable animals to ingest relatively large amounts of food when it is available, then digest it gradually. In these storage chambers, food may be further fragmented and mixed, and in most vertebrates it is an important site of digestion. Food delivered into the next section of the gut, the **intestine**, is in small particles, well mixed, and usually partially digested.

Most digestion occurs in the intestine, and nutrients, water, and ions are absorbed across its walls. Glands secrete digestive enzymes into the intestine, and other enzymes are produced by cells lining the intestine. The final segment of the intestine recovers water and ions and stores undigested wastes, or **feces**, so they can be released to the environment at an appropriate

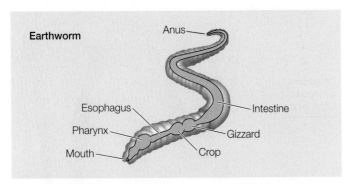

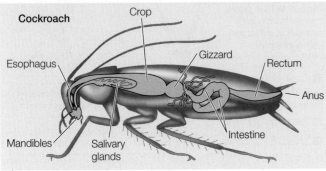

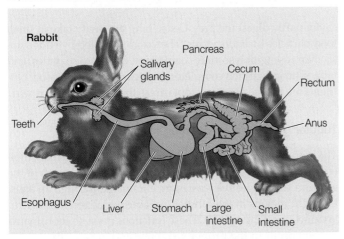

51.7 Compartments for Digestion and Absorption Most invertebrates and all vertebrates have a tubular gut that begins with a mouth, which takes in food, and ends in an anus, which eliminates wastes. Between these two structures are specialized regions for digestion and nutrient absorption; the structures in these regions are adapted to different diets and vary from species to species.

time or place. A muscular **rectum** near the anus assists in expelling feces.

Endosymbiotic bacteria colonize the intestines. These bacteria obtain their nutrition from the food passing through the host's gut while contributing to the host's digestive processes. Members of the leech genus *Hirudo*, for example, produce no enzymes that can digest the proteins in the blood they suck from vertebrates; instead, they depend on bacteria to perform this service. The resulting amino acids are subsequently used by both the leech and the bacteria.

The microorganisms in the human gut are called the "forgotten organ" because they provide important services in diges-

tion, elimination of harmful microorganisms, and even the production of vitamins (vitamin K and biotin). This forgotten organ is huge. It is estimated that the human body consists of 10^{13} cells, but our guts contain probably 10 times that number of unicellular organisms representing at least 500 different species. (And thus most of the DNA in our bodies is from alien species—an interesting thought.)

In many animals, the parts of the gut that absorb nutrients have greater surface areas than would be expected of a simple tube (**Figure 51.8A,B**). In vertebrates, the wall of the intestine is richly folded, with the individual folds bearing legions of tiny fingerlike projections called **villi** (**Figure 51.8C**). The cells that line the surfaces of the villi, in turn, have microscopic projections called **microvilli**. The microvilli give the intestine an enormous internal surface area for absorbing nutrients.

Digestive enzymes break down complex food molecules

Protein, carbohydrate, and fat macromolecules are broken down into their simplest monomeric units by hydrolytic enzymes produced at different locations in the digestive tract. Many are secreted into the lumen of the gut, and others remain associated with the membranes of the microvilli. All of these enzymes cleave the chemical bonds of macromolecules through *hydrolysis*, a reaction that adds a water molecule (see Figure 3.4B). Digestive enzymes are classified according to the substances they hydrolyze: *proteases* break the bonds between adjacent amino acids in proteins; *carbohydrases* hydrolyze carbohydrates; *peptidases*, peptides; *lipases*, fats; and *nucleases*, nucleic acids.

How can an organism produce enzymes that hydrolyze biological macromolecules without digesting itself? Many digestive enzymes are produced in an inactive form known as a **zymogen**, so that they cannot act on the cells that produce them. When secreted into the gut, a zymogen is generally activated by another enzyme. The cells lining the gut are not digested because they are protected by a covering of mucus.

51.2 RECAP

Heterotrophs have diverse adaptations for acquiring food. Once captured and/or ingested, food is digested extracellularly by secreted enzymes to release nutrients, which are absorbed into the animal's body, usually via a tubular gut.

- Why do herbivores typically spend a great deal of their time feeding? See p. 1074

- What is the primary purpose of the intestinal microvilli? See p. 1076 and Figure 51.8

- What is a zymogen, and why is it important? See p. 1076

Once ingested by an animal, food may be fragmented and moved into the gut for digestion by hydrolytic enzymes. The processes of digestion release the nutrients needed by the animal. Let's focus next on how those processes occur in vertebrates.

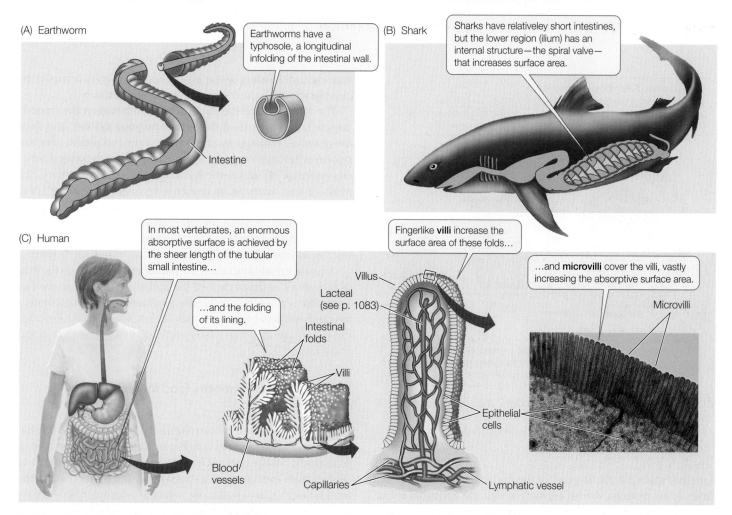

(A) Earthworm

Earthworms have a typhosole, a longitudinal infolding of the intestinal wall.

Intestine

(B) Shark

Sharks have relativeley short intestines, but the lower region (ilium) has an internal structure—the spiral valve—that increases surface area.

(C) Human

In most vertebrates, an enormous absorptive surface is achieved by the sheer length of the tubular small intestine...

...and the folding of its lining.

Fingerlike **villi** increase the surface area of these folds...

...and **microvilli** cover the villi, vastly increasing the absorptive surface area.

Microvilli

Villus

Lacteal (see p. 1083)

Intestinal folds

Villi

Epithelial cells

Blood vessels

Capillaries

Lymphatic vessel

51.8 Intestinal Surface Area and Nutrient Absorption Maximizing the surface area of the gut increases an animal's ability to absorb nutrients.

51.3 How Does the Vertebrate Gastrointestinal System Function?

Digestion in vertebrates occurs in the gastrointestinal system, which includes a tubular gut running from mouth to anus and several accessory structures that produce secretions that play important roles in digestion (**Figure 51.9**). In this section we will consider three important processes of this system: the movement of food through it, the sequential steps of digestion, and the absorption of nutrients. Our focus will be the human system.

The vertebrate gut consists of concentric tissue layers

The tissues of the vertebrate gut are arranged in concentric layers that have a similar organization throughout its

51.9 The Human Digestive System Different compartments in the long tubular gut specialize in digesting food, absorbing nutrients, and storing and expelling wastes. Accessory organs contribute secretions containing enzymes and other molecules.

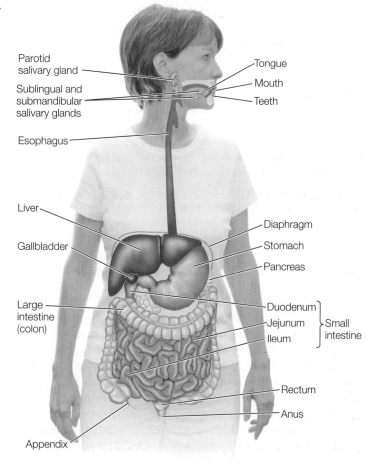

Parotid salivary gland

Sublingual and submandibular salivary glands

Esophagus

Liver

Gallbladder

Large intestine (colon)

Appendix

Tongue

Mouth

Teeth

Diaphragm

Stomach

Pancreas

Duodenum

Jejunum — Small intestine

Ileum

Rectum

Anus

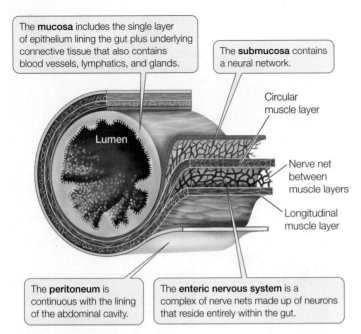

The **mucosa** includes the single layer of epithelium lining the gut plus underlying connective tissue that also contains blood vessels, lymphatics, and glands.

The **submucosa** contains a neural network.

Circular muscle layer

Lumen

Nerve net between muscle layers

Longitudinal muscle layer

The **peritoneum** is continuous with the lining of the abdominal cavity.

The **enteric nervous system** is a complex of nerve nets made up of neurons that reside entirely within the gut.

51.10 Tissue Layers of the Vertebrate Gut The organization of tissue layers is the same in all compartments of the gut, but specialized adaptations of specific tissues characterize different regions.

length (**Figure 51.10**). Starting in the cavity, or **lumen**, the first layer is the **mucosa**, which consists of delicate epithelial cells underlain by connective tissue. Some cells of this *mucosal epithelium* have secretory and absorptive functions. Some secrete mucus, which lubricates and protects the walls of the gut; others secrete digestive enzymes or hormones. Mucosal epithelial cells in the stomach secrete hydrochloric acid (HCl) and, as we noted in Section 51.1, some secrete intrinsic factor to aid the absorption of vitamin B_{12}. In some regions of the gut, nutrients are absorbed by mucosal epithelial cells. The apical plasma membranes of these absorptive cells have microvilli that increase the surface area over which absorption can take place (see Figure 51.8C).

At the base of the mucosa are some smooth muscle cells that can move the mucosa to improve contact with gut contents, and just under the mucosa is the submucosal tissue layer. Here we find the blood and lymph vessels that carry absorbed nutrients to the rest of the body. The **submucosa** also contains a network of nerves; the neurons in this network have sensory functions (responsible for stomach aches) and control various secretory functions of the gut.

External to the submucosa are two layers of smooth muscle responsible for the large movements of the gut. Innermost is the *circular muscle layer*, with its cells oriented around the gut. Outermost is the *longitudinal muscle layer*, with its cells oriented along the length of the gut (see Figure 51.10). The circular muscles constrict the gut, and the longitudinal muscles shorten it. Between the two layers of smooth muscle is another nerve network which controls and coordinates the movements of the gut. The coordinated activity of the two smooth muscle layers mixes the content of the gut and moves it continuously toward the rectum. Interestingly, the stomach has a third layer of smooth muscle that is closest to the lumen. The orientation of its fibers is oblique to

the longitudinal and circular layers. This third layer is important in generating the churning motions of the stomach.

The nerve nets in the submucosa and between the smooth muscle layers are called the **enteric nervous system**, and they are unusual. Whereas most neurons of the peripheral nervous system either receive synapses from neurons in the central nervous system (CNS) or contribute synapses to neurons in the CNS, most of the neurons in the enteric nervous system form synapses only with other neurons in their network. Thus, they are responsible for communication within the gut. The CNS can influence the activity in enteric nervous system and receive information from it, but truly, the gut has a "mind" of its own.

A tissue membrane called the **peritoneum** surrounds the gut, as it does all of the organs of the abdominal cavity as well as lining the wall of the cavity. The peritoneum includes connective and epithelial tissues, which secrete a fluid that lubricates the organs so they can easily slide against each other in the body cavity.

Mechanical activity moves food through the gut and aids digestion

In humans and most other mammals, food is chewed in the mouth and mixed with saliva. Periodically the *tongue* pushes a *bolus* (mass) of the chewed food toward the throat. By making contact with the *soft palate* at the back of the mouth cavity, the food bolus initiates *swallowing*, which is a complex series of reflexes. Swallowing propels the food through the pharynx (where the mouth cavity and nasal passages join) and into the **esophagus** (food tube). To prevent food from entering the trachea (windpipe), the *larynx* (voice box) closes, and a flap of tissue called the *epiglottis* covers the entrance to the larynx (**Figure 51.11A**).

Once a bolus of food enters the esophagus, it is moved toward the stomach both by the force of gravity and by waves of muscle contraction called **peristalsis** (**Figure 51.11B**). The muscle of the upper region of the esophagus is striated (i.e., skeletal muscle) and is controlled by the central nervous system; the muscles of the rest of the esophagus are smooth muscle controlled by the autonomic and enteric nervous systems.

The smooth muscles of the gut contract in response to being stretched. When a bolus of food reaches the smooth-muscle region of the esophagus and stretches it, the muscle responds by contracting, thus pushing the food toward the stomach. Why doesn't the contraction of the esophageal smooth muscle push the food back toward the mouth? The nerve net between the two smooth muscle layers coordinates the muscles so that contraction is always preceded by an *anticipatory wave of relaxation*. When a region of the gut smooth muscle contracts, the circular smooth muscle just beyond it relaxes while the longitudinal smooth muscle contracts, pushing the food into that area. The resulting stretch causes that circular smooth muscle to contract while the next region relaxes. In this way peristalsis moves food down the gut from the mouth to the anus.

The backward movement of food from the stomach into the esophagus is prevented by the *lower esophageal sphincter*, a thick ring of circular smooth muscle at the junction of the esopha-

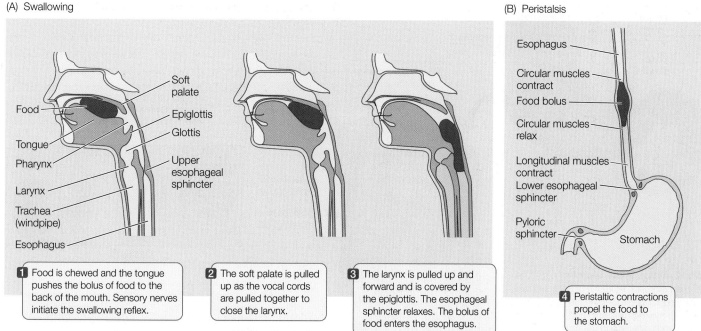

(A) Swallowing

- Food
- Soft palate
- Epiglottis
- Glottis
- Tongue
- Pharynx
- Larynx
- Upper esophageal sphincter
- Trachea (windpipe)
- Esophagus

1 Food is chewed and the tongue pushes the bolus of food to the back of the mouth. Sensory nerves initiate the swallowing reflex.

2 The soft palate is pulled up as the vocal cords are pulled together to close the larynx.

3 The larynx is pulled up and forward and is covered by the epiglottis. The esophageal sphincter relaxes. The bolus of food enters the esophagus.

(B) Peristalsis

- Esophagus
- Circular muscles contract
- Food bolus
- Circular muscles relax
- Longitudinal muscles contract
- Lower esophageal sphincter
- Pyloric sphincter
- Stomach

4 Peristaltic contractions propel the food to the stomach.

51.11 Swallowing and Peristalsis (A) Food pushed to the back of the mouth triggers the swallowing reflex. (B) Once a food bolus enters the esophagus, peristalsis propels it from mouth to anus by coordinated actions of the circular and longitudinal muscle layers of the gut.

gus and stomach. This sphincter is normally constricted, but waves of peristalsis cause it to relax enough to let food pass from the esophagus into the stomach. Sphincter muscles are found elsewhere in the digestive tract as well. The *pyloric sphincter* governs the passage of stomach contents into the intestine. Two other important sphincters also control movement of gut contents: the *ileocaecal sphincter* between the small and large intestine, and the internal *anal sphincter* which has to relax to allow defecation.

Chemical digestion begins in the mouth and the stomach

The main role of the stomach is to store food so that digestion can occur more slowly than ingestion. However, the enzyme **amylase** is secreted by the salivary glands and mixed with food as it is chewed. Amylase hydrolyzes the bonds between the glucose monomers that make up starch molecules. The action of amylase is what makes a chewed piece of bread or cracker taste slightly sweet if you hold it in your mouth long enough.

The stomach also has secretory functions. Deep infoldings in the walls of the stomach called **gastric pits** are lined with three types of secretory cells (**Figure 51.12A**). One type secretes a proteolytic enzyme, **pepsin**, that begins the digestion of protein. Another type secretes hydrochloric acid (HCl), which kills most ingested microorganisms. This nasty mix of substances could damage the stomach walls, but a third cell type secretes mucus that provides a protective coating for the walls of the gastric pits and stomach. This mucus keeps the pH at the surface of the mucosa near neutrality and also prevents pepsin from acting on the stomach cells.

The cells in the gastric pits that secrete pepsin are called *chief cells*, and they secrete it as an inactive zymogen called **pepsinogen**. The extremely low pH of the stomach juices initiates the conversion of pepsinogen to pepsin by cleaving away a sequence of amino acids that masks the active site of the enzyme. Newly activated pepsin can act on other pepsinogen molecules to activate them, creating a positive feedback process called **autocatalysis** (**Figure 51.12B**).

The cells in the gastric pits that secrete HCl are called *parietal cells*. They can secrete so much HCl—about 2 liters per day—that they can bring the pH of the stomach contents below 1, which is the same as battery acid and 10 times more acidic than pure lemon juice. This means that across their plasma membranes, gastric pits can create a H^+ concentration difference of 3 million-fold. Such a feat of transport is not seen anywhere else in the body. How do they do it? Enzymes and transporters are involved.

The enzyme carbonic anhydrase in these cells catalyzes the hydration of CO_2 to H_2CO_3, which dissociates into H^+ and bicarbonate ion (HCO_3^-). An *antiporter* transport protein (see Figure 6.15) exchanges HCO_3^- for Cl^- on the blood side of the gastric pits, and an antiporter on the gastric pit side exchanges H^+ for K^+ (**Figure 51.12C**). However, this K^+ can leak out again down its concentration gradient. Thus the inward transport of K^+ acts like an endless conveyer belt moving H^+ out into the stomach lumen. Cl^- also passively leaks out of the gastric lumen side of the parietal cells to maintain electrical neutrality.

Stomach ulcers can be caused by a bacterium

The secretions of the stomach are highly corrosive to living tissues and can result in *ulcers*—places where the mucosal lining of the stomach is damaged. Stomach ulcers can lead to maladies ranging from indigestion and heartburn to gastric bleeding and

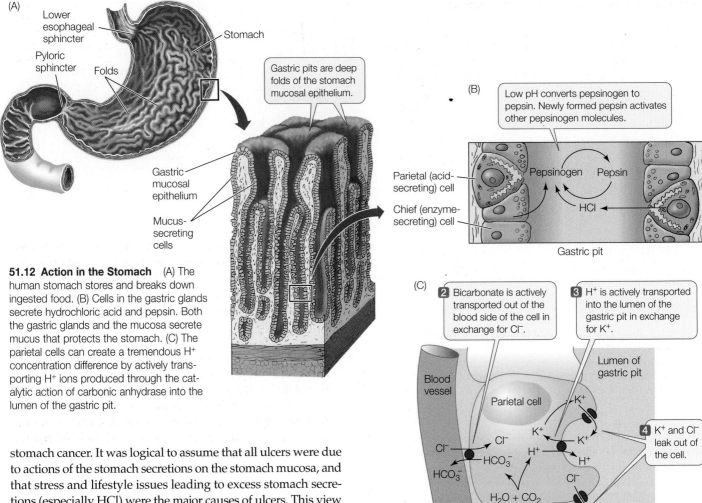

51.12 Action in the Stomach (A) The human stomach stores and breaks down ingested food. (B) Cells in the gastric glands secrete hydrochloric acid and pepsin. Both the gastric glands and the mucosa secrete mucus that protects the stomach. (C) The parietal cells can create a tremendous H^+ concentration difference by actively transporting H^+ ions produced through the catalytic action of carbonic anhydrase into the lumen of the gastric pit.

stomach cancer. It was logical to assume that all ulcers were due to actions of the stomach secretions on the stomach mucosa, and that stress and lifestyle issues leading to excess stomach secretions (especially HCl) were the major causes of ulcers. This view led the pharmaceutical industry to develop a plethora of drugs to decrease stomach acid production, which became "billion dollar drugs" because they were prescribed so widely. What seemed like the simplest fact in gastrointestinal medicine, however, was turned on its head by the work of two Australian researchers. Their work is a perfect example of the application of Koch's postulates (see Section 26.6) for the proof that a microorganism causes a disease (**Figure 51.13**).

In 1982, pathologist Robin Warren observed an unknown bacterium in biopsies from the stomachs of patients with ulcers. In a study of 100 patients, he and Barry Marshall of the University of Western Australia found that the bacterium was always present in the patients with ulcers. They isolated the bacterium, which they named *Helicobacter pylori*, and grew it in culture. Having thus satisfied Koch's first two postulates (microbe shown to be present in all instances of the disease, and isolating the microbe from diseased tissue), they turned to the last two, as described in Figure 51.13. Marshall actually drank a flask of the cultured bacteria (don't try this at home) and developed a pre-ulcerous condition that was subsequently cured with antibiotic. Their research showed not only that *H. pylori* causes stomach ulcers but that ulcer patients can be cured with antibiotics.

The medical profession was so certain that no microorganisms could live in the stomach and that stomach acid was the cause of ulcers that at first Warren and Marshall's findings were resisted

and even ridiculed. But in 2005 they received the Nobel Prize in Medicine for their important discovery, and antibiotic therapy is now a primary treatment for stomach ulcers worldwide.

Drugs that decrease stomach acid production are still important ulcer medications for several reasons. Once a person has an ulcer, stomach acid exacerbates it. In addition, in many individuals the lower esophageal sphincter muscle (see Figure 51.12A) is inadequate to prevent acid from entering the esophagus, resulting in irritations and lesions there.

The stomach gradually releases its contents to the small intestine

Contractions of the smooth muscles in the walls of the stomach churn its contents, thoroughly mixing them with the stomach secretions. The acidic, fluid mixture of gastric juice and partly digested food in the stomach is called **chyme**. A few substances can be absorbed across the stomach wall, including alcohol (hence its

Marshall and Warren set out to satisfy Koch's postulates:

Test 1

The microorganism must be present in every case of the disease.

Results: Biopsies from the stomachs of many patients revealed that the bacterium was always present if the stomach was inflamed or ulcerated.

Test 2

The microorganism must be cultured from a sick host.

Results: The bacterium was isolated from biopsy material and eventually grown in culture media in the laboratory.

Test 3

The isolated and cultured bacteria must be able to induce the disease.

Results: Marshall was examined and found to be free of bacteria and inflammation in his stomach. After drinking a pure culture of the bacterium, he developed stomach inflammation (gastritis).

Test 4

The bacteria must be recoverable from the infected volunteers.

Results: Biopsy of Marshall's stomach 2 weeks after he ingested the bacteria revealed the presence of the bacterium, now christened *Helicobacter pylori*, in the inflamed tissue.

Conclusion

Antibiotic treatment eliminated the bacteria and the inflammation in Marshall. The experiment was repeated on healthy volunteers, and many patients with gastric ulcers were cured with antibiotics. Thus Marshall and Warren demonstrated that the stomach inflammation leading to ulcers is caused by *H. pylori* infections in the stomach.

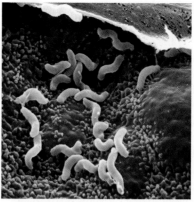

51.13 Satisfying Koch's Postulates Marshall and Warren showed that ulcers are caused not by the action of stomach acid but by infection with the bacterium *Helicobacter pylori*.

Helicobacter pylori

Digestion in the small intestine requires many specialized enzymes, as well as several other secretions. Two accessory organs that are not part of the digestive tract—the liver and the pancreas—produce many of these secretions and deliver them to the lumen of the intestine through ducts.

LIVER   The liver synthesizes **bile salts** from cholesterol and secretes them as **bile**. Bile includes bile salts and other substances, such as phospholipids and bilirubin (the breakdown product of hemoglobin). Bile flows from the liver through the *hepatic duct*. A side branch off the hepatic duct called the *cystic duct* goes to the **gallbladder**, where it is stored. Below this junction, the hepatic duct is called the *common bile duct*. Before it reaches the duodenum, it is joined by the *pancreatic duct* (**Figure 51.14**).

rapid effects), aspirin, and caffeine, but even these substances are absorbed in rather small quantities from the stomach.

Contractions of the stomach walls push the chyme toward the bottom of the stomach. These waves of contractions cause the pyloric sphincter to relax briefly so that little squirts of the chyme can enter the small intestine. In this manner, the human stomach empties itself gradually over a period of approximately 4 hours. This slow introduction of food into the small intestine enables it to work on a little material at a time.

Most chemical digestion occurs in the small intestine

In the **small intestine**, the digestion of carbohydrates and proteins continues, and the digestion of fats and absorption of nutrients begin. The small intestine takes its name from its diameter; it is in fact a very large organ, about 6 meters long in an adult human. Given its length, and because of the folds, villi, and microvilli of its lining, its inner surface area is roughly the size of a tennis court. Across this surface, the small intestine absorbs all the nutrient molecules derived from food.

The small intestine of humans has three sections. The initial section (about 25 cm long) is called the **duodenum** and is the site of most digestion; the **jejunum** and the **ileum** (together about 600 cm) carry out 90 percent of the absorption of nutrients (see Figure 51.9).

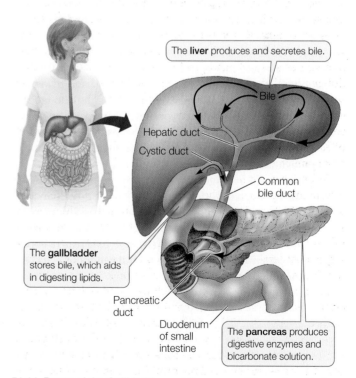

The **liver** produces and secretes bile.

Bile

Hepatic duct

Cystic duct

Common bile duct

The **gallbladder** stores bile, which aids in digesting lipids.

Pancreatic duct

Duodenum of small intestine

The **pancreas** produces digestive enzymes and bicarbonate solution.

51.14 Ducts of the Gallbladder and Pancreas   Bile produced in the liver leaves the liver via the hepatic duct. Branching off this duct is the gallbladder, which stores bile. Below the gallbladder, the hepatic duct is called the common bile duct and is joined by the pancreatic duct before entering the duodenum.

Fat entering the duodenum stimulates cells of the duodenal epithelium to release the hormone **cholecystokinin (CCK)**, which in turn stimulates the walls of the gallbladder to contract rhythmically. As a result, bile is squeezed out of the gallbladder and through the cystic duct to the common bile duct. A small sphincter at the junction of the common bile duct with the duodenum, the *sphincter of Oddi*, relaxes in response to waves of peristalsis and allows squirts of bile to enter the duodenal lumen.

To understand the role of bile in fat digestion, think of an oil and vinegar salad dressing. The oil, which is hydrophobic, tends to aggregate in large globules. For that reason, many salad dressings include an *emulsifier*—something that prevents oil droplets from aggregating. Mayonnaise, for example, is oil and vinegar with egg yolk added as an emulsifier. Bile contains salts that emulsify fats in the chyme and thereby greatly enlarge the surface area of the fats exposed to the **lipases**—the enzymes that digest fats. One end of each bile salt molecule is lipophilic (soluble in fat), and the other end is hydrophilic (soluble in water). The lipophilic ends of bile molecules merge with the fat droplets, leaving their hydrophilic ends sticking out. As a result, bile salts prevent the fat droplets from sticking together. The very small fat particles that result are called **micelles (Figure 51.15A)**.

PANCREAS The **pancreas** is a large gland that lies just behind and below the stomach (see Figures 51.9 and 51.14). It is both an endocrine gland (secreting hormones into the interstitial fluid; see Section 41.1) and an exocrine gland (secreting digestive juices through the pancreatic duct to the gut lumen). The exocrine tissues of the pancreas produce a host of digestive enzymes, including lipases, amylases, proteases, and nucleases (**Table 51.3**). As in the stomach, the protease enzymes are released as zymogens; otherwise, they would digest the pancreas and its ducts before they ever reached the duodenum. Once in the duodenum, one of these zymogens is activated by an enzyme called *enterokinase* (secreted by cells lining the duodenum) to produce the active protease **trypsin**. Active trypsin can cleave other zymogens to release more active trypsin, as well as activate other proteases as well. This activation process is similar to the activation of pepsinogen by low pH in the stomach.

The mixture of zymogens produced by the pancreas can be dangerous if the pancreatic duct is blocked or if the pancreas is injured by infection or physical trauma such as a blow to the abdomen. A few activated trypsin molecules can initiate a chain reaction of enzyme activity that digests the pancreas (a condition called *pancreatitis*), destroying both its endocrine and exocrine functions.

51.15 Digestion and Absorption of Fats (A) Dietary fats are broken up by bile into small micelles that present a large surface area to lipases. (B) The products of fat digestion are absorbed by intestinal mucosal cells, where they are resynthesized into triglycerides and exported to lymphatic vessels.

── **yourBioPortal.com** ──
GO TO Animated Tutorial 51.1 • The Digestion and Absorption of Fats

The pancreas also produces a secretion rich in bicarbonate ions (HCO_3^-). Bicarbonate ions are alkaline (basic) and neutralize the acidic pH of the chyme that enters the duodenum from the stomach. Intestinal enzymes function best at a neutral or slightly alkaline pH.

Nutrients are absorbed in the small intestine

The final step in digesting proteins and carbohydrates and absorbing their components occurs among the microvilli. Mucosal epithelial cells produce peptidases that cleave small peptides into absorbable amino acids. These epithelial cells also produce the enzymes maltase, lactase, and sucrase, which cleave the common disaccharides into absorbable monosaccharides—glucose, galactose, and fructose. There is also some lipase activity for fat digestion.

Many humans stop producing the enzyme lactase in childhood and thereafter have difficulty digesting lactose (the sugar in milk). Lactose is a disaccharide and cannot be absorbed without being cleaved into its constituents, glucose and galactose. Unabsorbed lactose is metabolized by bacteria in the large intestine, causing gas, diarrhea, and abdominal cramps.

The mechanisms by which cells of the intestinal epithelium absorb nutrients and inorganic ions are diverse and include dif-

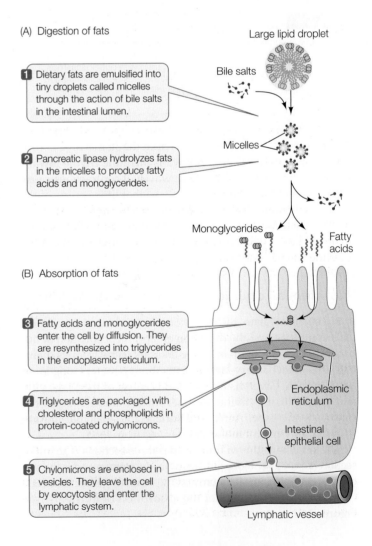

(A) Digestion of fats

Large lipid droplet

1 Dietary fats are emulsified into tiny droplets called micelles through the action of bile salts in the intestinal lumen.

Bile salts

Micelles

2 Pancreatic lipase hydrolyzes fats in the micelles to produce fatty acids and monoglycerides.

Monoglycerides

Fatty acids

(B) Absorption of fats

3 Fatty acids and monoglycerides enter the cell by diffusion. They are resynthesized into triglycerides in the endoplasmic reticulum.

4 Triglycerides are packaged with cholesterol and phospholipids in protein-coated chylomicrons.

Endoplasmic reticulum

Intestinal epithelial cell

5 Chylomicrons are enclosed in vesicles. They leave the cell by exocytosis and enter the lymphatic system.

Lymphatic vessel

TABLE 51.3

Major Digestive Enzymes of Humans

SOURCE/ENZYME	ACTION
SALIVARY GLANDS	
Salivary amylase	Starch → Maltose
STOMACH	
Pepsin	Proteins → Peptides; autocatalysis
PANCREAS	
Pancreatic amylase	Starch → Maltose
Lipase	Fats → Fatty acids and glycerol
Nuclease	Nucleic acids → Nucleotides
Trypsin	Proteins → Peptides; zymogen activation
Chymotrypsin	Proteins → Peptides
Carboxypeptidase	Peptides → Shorter peptides and amino acids
SMALL INTESTINE	
Aminopeptidase	Peptides → Shorter peptides and amino acids
Dipeptidase	Dipeptides → Amino acids
Enterokinase	Trypsinogen → Trypsin
Nuclease	Nucleic acids → Nucleotides
Maltase	Maltose → Glucose
Lactase	Lactose → Galactose and glucose
Sucrase	Sucrose → Fructose and glucose

fusion, facilitated diffusion, osmosis, active transport, and co-transport. Many inorganic ions such as sodium, calcium, and iron are actively transported by these cells. For example, active Na^+ transporters exist on the basal and lateral sides of the epithelial cells. They maintain a low concentration of Na^+ in those cells so that Na^+ can diffuse in from the chyme in the intestinal lumen. About 30 grams of Na^+ are transported this way every day, and Cl^- follows.

The transport of Na^+ and other ions is also important for water absorption because it creates an osmotic concentration gradient. At least 7 to 8 liters of water per day move through the spaces between the epithelial cells in response to this osmotic gradient. Because the water moves through *spaces* between the cells and not through the cells themselves, it can carry with it nutrients that are in solution—a transport mechanism called *solvent drag*.

Many different kinds of transport proteins exist in the epithelial cell membranes. Some, such as the transport protein for fructose, only facilitate diffusion, and that requires a concentration gradient. This mechanism works for fructose because once fructose enters the cell it is converted to glucose. Thus the concentration of fructose in the cell is always low and the concentration gradient is maintained. Transport proteins known as *symporters* (see Figure 6.15) exploit the concentration gradient of Na^+ between the inside and outside of the cell, which is maintained by the Na^+ / Ka^+ ATPase common to all cells. Symporters combine the transport of Na^+ and another molecule, such as glucose, galactose, or an amino acid. As Na^+ is pulled down its concentration gradient into the cell, the "hitchhiking" molecules are carried along with it.

The absorption of the products of fat digestion is relatively simple. Triglycerides are hydrolyzed to diglycerides, monoglycerides, and fatty acids, all of which are lipid-soluble and thus able to pass through the plasma membranes of the microvilli. In the intestinal epithelial cells, these molecules are resynthesized into triglycerides, combined with cholesterol and phospholipids, and coated with protein to form water-soluble **chylomicrons** (**Figure 51.15B**). Rather than enter the blood directly, chylomicrons pass into blind-ended lymph vessels called **lacteals** that are inside each villus (see Figure 51.8). They then flow through the lymphatic system, entering the bloodstream through the thoracic ducts at the base of the neck. After a meal rich in fats, chylomicrons can be so abundant in the blood that they give the plasma a milky appearance. Chylomicrons deliver their triglyceride and cholesterol cargo as they circulate through tissues.

The bile salts that emulsify fats are not absorbed along with the monoglycerides, diglycerides, and the fatty acids, but are shuttled back and forth between the gut contents and the microvilli. In the ileum, bile salts are actively reabsorbed and returned to the liver via the bloodstream.

Absorbed nutrients go to the liver

Blood leaving the digestive tract flows to the liver in the *hepatic portal vein*. This large vein delivers the blood to small spaces called sinusoids between groups of liver cells. These cells absorb the nutrients coming from the digestive tract and either store them or convert them to molecules the body needs. Glucose, sucrose, and fructose are used to synthesize glycogen. Amino acids are used to build proteins. Lipids from the chylomicrons are either stored as triglycerides or used to make *lipoproteins*, which are released by the liver and deliver the triglycerides and cholesterol to other tissues (see Section 51.4).

Water and ions are absorbed in the large intestine

The motility of the small intestine gradually pushes its contents into the *large intestine*, or **colon**. Most of the available nutrients have been removed from the chyme that enters the colon, but it contains a lot of water and inorganic ions.

The colon absorbs water and ions, producing semisolid feces from the chyme it receives from the small intestine. Feces are stored in the rectum of the colon until they are eliminated. Absorption of too much water from the colon can cause *constipation*. The opposite condition, *diarrhea*, results if too little water is absorbed; in this case, water in the colon is excreted with the feces. The excessive diarrhea caused by diseases such as cholera can produce such rapid loss of water and electrolytes that death can occur in hours.

Herbivores rely on microorganisms to digest cellulose

As the primary component of plant cell walls, cellulose is the principal component of the food of herbivores. Most herbivores, however, cannot produce *cellulases*—enzymes that hydrolyze

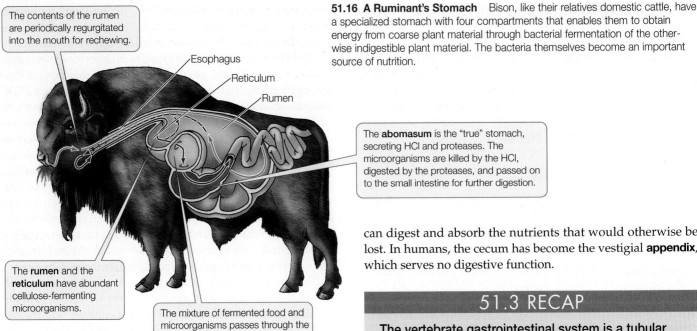

The contents of the rumen are periodically regurgitated into the mouth for rechewing.

Esophagus

Reticulum

Rumen

51.16 A Ruminant's Stomach Bison, like their relatives domestic cattle, have a specialized stomach with four compartments that enables them to obtain energy from coarse plant material through bacterial fermentation of the otherwise indigestible plant material. The bacteria themselves become an important source of nutrition.

The **abomasum** is the "true" stomach, secreting HCl and proteases. The microorganisms are killed by the HCl, digested by the proteases, and passed on to the small intestine for further digestion.

The **rumen** and the **reticulum** have abundant cellulose-fermenting microorganisms.

The mixture of fermented food and microorganisms passes through the **omasum**, where it is concentrated by water absorption.

cellulose. Exceptions include silverfish (insects that eat books and stored papers), earthworms, and shipworms. From termites to cattle, herbivores rely on microorganisms in their digestive tracts to digest cellulose.

The stomachs of **ruminants** (cud chewers) such as cattle are large, four-chambered organs that take advantage of their endosymbiotic microorganisms (**Figure 51.16**). The first two chambers, the **rumen** and the **reticulum**, are packed with microorganisms that break down cellulose by fermentation. The ruminant periodically regurgitates the contents of the rumen (the *cud*) into the mouth for rechewing. When swallowed again, the vegetal fibers present more surface area to the microorganisms. The microorganisms metabolize cellulose and other nutrients to simple fatty acids, which become nutrients for their host.

Enormous numbers of microorganisms leave the rumen along with the partially digested food. This mass is concentrated by water absorption in the **omasum** before it enters the true stomach, the **abomasum**, where the microorganisms are killed by secreted hydrochloric acid, digested by proteases, and passed on to the small intestine for further digestion and absorption. A cow derives more than 100 grams of protein per day from digestion of its endosymbiotic microorganisms. The rate of multiplication of microorganisms in the rumen offsets their loss, so a well-balanced, mutually beneficial relationship is maintained.

Some mammalian herbivores have a microbial fermentation chamber called a **cecum** extending from the large intestine. An example is the rabbit (see Figure 51.7). Since the cecum empties into the large intestine, absorption of some nutrients produced by the microorganisms is inefficient, because of the large intestine's limited surface area. Such species frequently produce two kinds of feces—ones that are pure waste and ones that contain cecal material. In a behavior known as **coprophagy**, these species reingest the cecal feces directly from the anus so they

can digest and absorb the nutrients that would otherwise be lost. In humans, the cecum has become the vestigial **appendix**, which serves no digestive function.

51.3 RECAP

The vertebrate gastrointestinal system is a tubular gut that is adapted to ingest food, fragment it, digest it, and absorb nutrients. Peristalsis moves food through the gut. Digestion and absorption of nutrients occur mostly in the small intestine; water and ions are absorbed in the large intestine.

- What digestive functions occur in the mouth and stomach? See p. 1079 and Figure 51.12

- How do bile salts assist in the digestion of fats? See p. 1082 and Figure 51.15

- Describe how symporters drive the absorption of nutrients. See p. 1083

The steps included in ingestion and digestion of food—from fragmentation in the mouth to the digestive processes in the gastrointestinal tract—make the nutrients in food available for absorption and ultimately for metabolism. Let's look at how the processes of digestion are controlled and how nutrients are handled by the body once food has been digested.

51.4 How Is the Flow of Nutrients Controlled and Regulated?

The vertebrate gut is an assembly line in reverse—a *dis*assembly line. As with a standard assembly line, the control and coordination of the sequential processes of digestion is critical. Both neuronal and hormonal controls govern these processes. Once the products of digestion are absorbed, their availability to the cells of the body must also be controlled.

You have certainly experienced salivation at the sight or smell of food. That response is an *unconscious reflex*, as is swallowing. Many such autonomic reflexes coordinate activity in different regions of the digestive tract. For example, the introduction of food into the stomach stimulates increased activity in the colon, which can lead to defecation.

As we already mentioned, the digestive tract has an intrinsic nervous system. Neuronal messages can travel from one region of the digestive tract to another without being processed by the CNS. One function of the gut's nervous system is coordinating the movement of food through the gut. Of course, this intrinsic nervous system communicates information to the CNS and receives input from the CNS, but its most important role is to coordinate actions throughout the digestive tract. In spite of this marvelous intrinsic nervous system, however, much of the control and regulation of the digestive system and nutrient management involves hormonal mechanisms.

Hormones control many digestive functions

Several hormones control the activities of the digestive tract and its accessory organs (**Figure 51.17**). The first hormone ever discovered came from the duodenum; it was called **secretin** because it causes the pancreas to secrete digestive juices. We now know that secretin is only one of several hormones that control pancreatic secretion; specifically, secretin stimulates the pancreas to secrete a solution rich in bicarbonate ions.

The stimulus that causes the duodenum to release secretin is low pH—the arrival of acidic chyme from the stomach. Similarly, the presence of fats and proteins in the chyme stimulates the release of cholecystokinin (CCK), the hormone mentioned earlier that stimulates the gallbladder to release bile. CCK also stimulates the pancreas to release digestive enzymes. Both CCK and secretin slow the movements of the stomach, which slows the delivery of chyme into the small intestine, allowing more complete digestion in the duodenum.

The presence of food in the stomach stimulates cells in the lower region of the stomach to secrete a hormone called **gastrin**. Gastrin returns to the stomach in the blood and stimulates the secretion of digestive juices and also the movements of the stomach. Gastrin release begins to be inhibited when the pH of the stomach contents falls below 3—an example of negative feedback.

Most animals do not eat continuously, so they can be either in an **absorptive state** (food in the gut) or in a **postabsorptive state** (no food in the gut). Nutrient requirements for energy metabolism and biosynthesis are continuous, however. Thus, nutrient traffic must be controlled so that reserves accumulate in the liver, muscle, and adipose (fat) tissue while the animal is in the absorptive state and are used appropriately during the postabsorptive state.

The liver directs the traffic of the molecules that fuel metabolism

When fuel molecules are abundant in the blood, the liver stores them in the form of glycogen and fats. The liver also synthesizes blood plasma proteins from circulating amino acids. When fuel molecule levels in the blood decline, the liver taps its reserves and delivers nutrients into the blood.

The liver has an enormous capacity to interconvert fuel molecules. Liver cells can convert monosaccharides into either glycogen or fats, and vice versa. The liver can also convert certain amino acids and some other molecules, such as pyruvate and lactate, into glucose—a process termed *gluconeogenesis* (see Section 9.5). The liver is also the major controller of fat metabolism through its production of lipoproteins. A **lipoprotein** is a particle made up of a core of hydrophobic fat and cholesterol with a covering of hydrophilic protein that allows it to be suspended in water.

LIPOPROTEINS: THE GOOD, THE BAD, AND THE UGLY Lipoproteins move fats, the most abundant fuel reserve in the body, from sites of storage to sites of use. We saw in the previous section how,

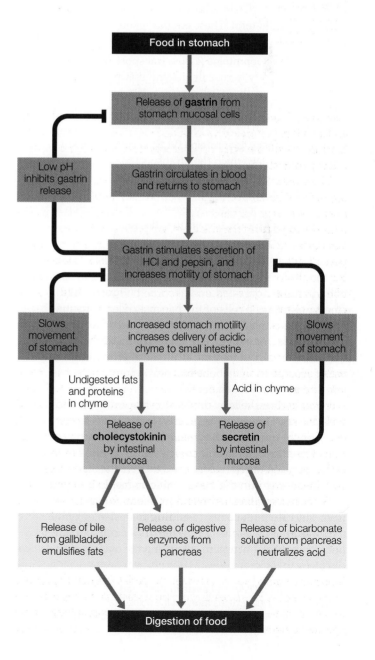

51.17 Hormones Control Digestion The hormones gastrin, cholecystokinin, and secretin are involved in feedback loops that control the sequential processing of food in the digestive tract. Red lines indicate inhibitory actions; green lines indicate stimulatory actions.

in the intestine, bile solves the problem of processing hydrophobic fats in an aqueous medium. The transport of fats in the circulatory system presents the same problem, and lipoproteins provide the solution.

The chylomicrons (see p. 1083) produced by the mucosal cells of the intestine are the largest lipoprotein particles in the blood. As the circulation carries chylomicrons through the liver and adipose tissue, lipoprotein lipases begin to break them down, and their triglyceride and cholesterol cargo is absorbed into the liver or fat cells.

Lipoproteins other than chylomicrons are synthesized in the liver. These lipoproteins can be classified according to their density. Fat has a low density (it floats in water) and protein has a high density, so the greater the fat-to-protein ratio in the lipoprotein, the lower its density.

- **High-density lipoproteins (HDLs)** remove cholesterol from tissues and carry it to the liver, where it can be used to synthesize bile. HDL consists of about 50% protein, 35% lipids, and 15% cholesterol. These are the "good" lipoproteins, and their levels are higher in people who exercise and are fit.

- **Low-density lipoproteins (LDLs)** transport cholesterol around the body for use in biosynthesis and for storage. LDL consists of about 25% protein, 25% lipids, and 50% cholesterol. These are the "bad" lipoproteins associated with a high risk for cardiovascular disease.

- **Very low-density lipoproteins (VLDLs)** contain mostly triglyceride fats, which they transport to fat cells in adipose tissues around the body. VLDL consists of about 2% protein, 94% lipids, and 3% cholesterol. These are the "ugly," as they are associated with excessive fat deposition as well as a high risk for cardiovascular disease.

INSULIN AND GLUCAGON CONTROL FUEL METABOLISM   During the absorptive state, blood glucose levels rise as carbohydrates are digested and absorbed (**Figure 51.18**). During this time, β cells of the pancreas release insulin, which plays a major role in directing glucose to where it will be used or stored. The actions of insulin vary in different tissues, but they are all aimed at promoting the use of glucose for metabolic fuel and getting the excess into storage as glycogen or fat.

Glucose enters cells by diffusion facilitated by glucose transporters. However, the glucose transporters in resting skeletal muscle and adipose tissues are normally sequestered in cytoplasmic vesicles until insulin binds its receptors on the cell surface and triggers the insertion of transporters into the plasma membrane. In adipose cells, insulin inhibits lipase and promotes fat synthesis from glucose. In the liver, insulin activates an enzyme that phosphorylates glucose as it enters the cell so it cannot diffuse back out again, enhancing the overall diffu-

sion of glucose into the cells. In liver cells, insulin also activates the enzymes that catalyze the synthesis of glycogen.

During the postabsorptive state, a fall in blood glucose decreases the release of insulin, and the uptake of glucose by most cells is curtailed (see Figure 51.18). To maintain blood glucose levels, liver cells break down their stored glycogen, which releases glucose into the blood. The liver and adipose tissues supply fatty acids to the blood, and most cells preferentially use fatty acids as their metabolic fuel.

One tissue that does not switch fuel sources when an animal is postabsorptive is the nervous system. The cells of the nervous system require a constant supply of glucose, and they can use other fuels to a very limited extent. Most neurons do not require insulin to absorb glucose from the blood, but they do need an adequate glucose concentration gradient to drive the facilitated diffusion of glucose across their plasma membranes. Therefore it is critical that blood glucose levels are maintained when an animal is postabsorptive. The overall dependence of neural tissues on glucose, and their requirement for constant blood glucose levels, are the reasons it is so important for other cells of the body to shift to fat metabolism during the postabsorptive state.

The metabolism of fuel molecules during the postabsorptive state is mostly controlled by the lack of insulin, but if blood glucose falls below a certain level, another pancreatic hormone, **glucagon**, is called into play. Glucagon's effect is opposite that of insulin: it stimulates liver cells to break down glycogen and to carry out gluconeogenesis. Thus, under the influence of glucagon, the liver produces glucose and releases it into the blood.

— yourBioPortal.com —

GO TO **Animated Tutorial 51.2** • Insulin and Glucose Regulation

---

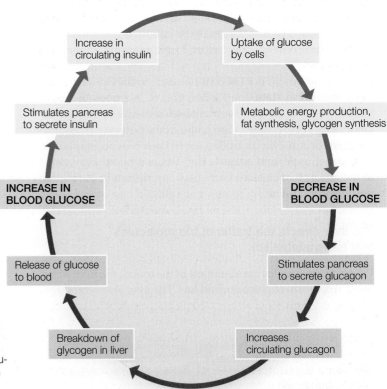

51.18 Regulating Glucose Levels in the Blood   Insulin (blue) and glucagon (brown) interactions maintain the homeostasis of circulating glucose. It is important for blood glucose to remain stable because it is the essential source fuel for the nervous system.

Regulating food intake is important

Obesity is a major health issue in the United States. A simple rule—take in fewer calories than your body burns while eating a balanced diet—should solve the problem, but it doesn't. Why? As we noted at the beginning of this chapter, lifestyle plays a major role in obesity, and genetic and regulatory factors "weigh in" as well.

The amount of food an animal eats is governed by its sensations of hunger and satiety. These sensations are influenced by the hypothalamus (see Section 41.2). If a region in the middle of the hypothalamus of rats (the *ventromedial hypothalamus*) is experimentally damaged, the animals will increase their food intake and become obese. If a different region, called the lateral hypothalamus, is damaged, they will decrease their food intake and become thin. In both cases, the rats eventually reach a new equilibrium body weight, which they maintain. Thus, regulation is maintained, but the set point is changed. Other brain regions have also been implicated in control of hunger and satiety.

In Section 40.1 we noted that regulation involves feedback information and a means of comparing that information with a set point. Criteria for a negative feedback signal for food intake include:

- Levels of the signal in the circulation reflect body energy reserves.

- The signal crosses the blood–brain barrier.

- Presence of the signal in the brain decreases food intake and body mass.

- Blocking the signal's actions in the brain results in increased food intake and gain of body mass.

Two molecules—insulin and leptin—satisfy these criteria. We have already discussed the roles of insulin in fuel management (see Figure 51.18); it may also have a role in signaling the hypothalamus about the fuel stores of the body.

The protein **leptin** (Greek *leptos*, "thin") was discovered in a strain of mice that eat enormous amounts of food and become obese. Experiments revealed this trait to be due to a recessive allele of a single gene, which they designate *ob*. It was subsequently learned that the wild-type *Ob* gene codes for leptin. The obese mice carried two copies of the recessive allele for the leptin gene and thus were designated *ob/ob*.

Leptin is a circulating hormone produced by fat cells. Receptors for leptin are found in the regions of the hypothalamus involved in controlling hunger and satiety. It seems that leptin provides feedback information to the brain about the status of the body's fat reserves. When leptin was injected into *ob/ob* mice, they ate less and lost body fat. However, a different strain of obese mice, *db/db*, did not respond to leptin injections. Further experiments revealed that *db/db* mice lacked the receptor for leptin (**Figure 51.19**).

Could leptin be used to treat human obesity? In the few obese people who do not produce leptin, injections curb appetite and facilitate weight loss. Most obese people, however, have higher than normal levels of leptin in their blood, so the likelihood is that it is their leptin receptors that are not completely functional.

INVESTIGATING LIFE

51.19 A Single-Gene Mutation Leads to Obesity in Mice

In mice the *Ob* gene codes for the protein leptin, a satiety factor that signals the brain when enough food has been consumed. The recessive *ob* allele is a loss-of-function allele, so *ob/ob* mice do not produce leptin; they do not experience satiety and become obese. The *Db* gene encodes the leptin receptor, so mice homozygous for the recessive loss-of-function allele *db*, even if they produce leptin, cannot use it and so become obese.

HYPOTHESIS Mice who cannot produce the satiety signal protein leptin will not become obese if they are able to obtain leptin from an outside source.

METHOD
1. Create two strains of genetically obese laboratory mice, one of which lacks functional leptin (genotype *ob/ob*) and one which lacks the receptor for leptin (genotype *db/db*).
2. Create parabiotic pairs by surgically joining the circulatory systems of a non-obese (wild-type) mouse with a partner from one of the obese strains.
3. Allow mice to feed at will.

Parabiotic pair

Wild-type mouse (*Ob/–* and *Db/–*)   Genetically obese mouse (either *ob/ob* or *db/db*)

RESULTS Parabiotic *ob/ob* mice obtain leptin from the wild-type partner and lose fat. Parabiotic *db/db* mice remain obese because they lack the leptin receptor and thus the leptin they obtain from their partner has no effect.

CONCLUSION The protein leptin is a satiety signal that acts to prevent overeating and resultant obesity.

Go to **yourBioPortal.com** for original citations, discussions, and relevant links for all INVESTIGATING LIFE figures.

Additional feedback signals are involved in regulating food intake. A hormone called *peptide YY* comes from the gut and reduces appetite in both rodents and humans. In contrast, *ghrelin*, a hormone produced and secreted by cells in the stomach, rises before a meal and falls after a meal. Fasting causes an increase in ghrelin levels. Ghrelin binding to its receptors in the hypothalamus stimulates appetite. Ghrelin also stimulates cells in the pituitary gland to release growth hormone.

Most interesting, however, is the discovery of the central role of the enzyme AMP-activated protein kinase (AMPK) as a signal of nutrient insufficiency. When most cells are nutrient deprived, they produce AMPK, which stimulates the oxidation of substrates to replenish ATP. However, in the hypothalamus, increased levels of AMPK stimulate food intake, and lowered levels of AMPK inhibit food intake. In addition, AMPK activity in the hypothalamus is inhibited by leptin and insulin but is stimulated by ghrelin. Thus, AMPK could be a final common pathway for various signals controlling food intake.

51.4 RECAP

The major controlling factors of gut function are an intrinsic nervous system and the hormones gastrin, secretin, and cholecystokinin. Insulin is the major hormonal controller of fuel metabolism. The hypothalamus controls food intake by generating sensations of hunger and satiety influenced by feedback from blood glucose and hormones, including leptin and ghrelin.

- What are the roles of the three different classes of lipoproteins? See pp. 1085–1086

- By what actions does insulin promote uptake and storage of energy during the absorptive state? See p. 1086 and Figure 51.18

- What evidence supports the hypothesis that leptin influences satiety? See p. 1087 and Figure 51.19

CHAPTER SUMMARY

51.1 What Do Animals Require from Food?

- Animals are **heterotrophs** that derive their energy and molecular building blocks, directly or indirectly, from **autotrophs**.

- Carbohydrates, fats, and proteins in food supply animals with metabolic energy. A measure of the energy content of food is the **kilocalorie (kcal)**. Excess caloric intake is stored as glycogen and fat. Review Figure 51.2

- For many animals, food provides essential **carbon skeletons** that they cannot synthesize themselves. Review Figure 51.4

- Most researchers consider 8 amino acids to be essential for adult humans; some believe that infants require as many as 12 **essential amino acids** in their diet. **Macronutrients** are mineral elements needed in large quantities, **micronutrients** are needed in small amounts. Review Figure 51.5, Table 51.1, **WEB ACTIVITY 51.1**

- **Vitamins** are organic molecules that must be obtained in food. Review Table 51.2, **WEB ACTIVITY 51.2**

- **Malnutrition** results when any essential nutrient is lacking from the diet. Chronic malnutrition causes **deficiency disease**.

51.2 How Do Animals Ingest and Digest Food?

- Animals can be characterized by how they acquire nutrients: **saprobes** and **detritivores** depend on dead organic matter, **filter feeders** strain the aquatic environment for small food items, **herbivores** eat plants, and **carnivores** eat animals. Behavioral and anatomical adaptations reflect these feeding strategies. **SEE WEB ACTIVITY 51.3**

- Digestion involves the breakdown of complex food molecules into monomers that can be absorbed and utilized by cells. In most animals, digestion takes place in a tubular gut. Review Figure 51.7

- Absorptive areas of the gut are characterized by a large surface area produced by extensive folding and numerous **villi** and **microvilli**. Review Figure 51.8

- Hydrolytic enzymes break down proteins, carbohydrates, and fats into their monomeric units. To prevent the organism itself from being digested, many of these enzymes are released as inactive **zymogens**, which become activated when secreted into the gut.

51.3 How Does the Vertebrate Gastrointestinal System Function?

- The vertebrate gut can be divided into several compartments with different functions. Review Figure 51.9, **WEB ACTIVITY 51.4**

- The cells and tissues of the vertebrate gut are organized in the same way throughout its length. The innermost tissue layer, the **mucosa**, is the secretory and absorptive surface. The **submucosa** contains blood and lymph vessels and a nerve network. External to the submucosa are two smooth muscle layers. Between the two muscle layers is another nerve network that controls the movements of the gut. Review Figure 51.10

- Swallowing is a reflex that pushes a bolus of food into the **esophagus**. **Peristalsis** and other movements of the gut move the bolus down the esophagus and through the entire length of the gut. Sphincters block the gut at certain locations, but they relax as a wave of peristalsis approaches. Review Figure 51.11

- Digestion begins in the **mouth**, where **amylase** is secreted with the saliva. Digestion of protein begins in the **stomach**, where parietal cells secrete HCl and chief cells secrete **pepsinogen**, a zymogen that becomes **pepsin** when activated by low pH and **autocatalysis**. The mucosa also secretes mucus, which protects the tissues of the gut. Review Figure 51.12

- In the **duodenum**, pancreatic enzymes carry out most of the digestion of food. Bile from the liver and **gallbladder** emulsify fats into micelles. Bicarbonate ions from the **pancreas** neutralize the pH of the **chyme** entering from the stomach to produce an environment conducive to the actions of pancreatic enzymes such as **trypsin**. Review Figure 51.14 and Table 51.3

- Final enzymatic cleavage of polypeptides and disaccharides occurs among the microvilli of the intestinal mucosa. Amino acids, monosaccharides, and inorganic ions are absorbed by the microvilli. Specific transporter proteins are sometimes involved. Symporters often power the active transport of nutrients.

- Fats broken down by **lipases** are absorbed mostly as monoglycerides and fatty acids and are resynthesized into triglycerides within cells. The triglycerides are combined with cholesterol and phospholipids and coated with protein to form **chylomicrons**, which pass out of the mucosal cells and into lymphatic vessels in the submucosa. Review Figure 51.15, **ANIMATED TUTORIAL 51.1**

- Water and ions are absorbed in the large intestine as waste matter and consolidated into **feces**, which are periodically eliminated.

- Microorganisms in some compartments of the gut digest materials that their host cannot. Review Figure 51.16

51.4 How Is the Flow of Nutrients Controlled and Regulated?

- Autonomic reflexes coordinate activity of the digestive tract, which has an intrinsic nervous system that can act independently of the CNS.

- The actions of the stomach and small intestine are largely controlled by the hormones **gastrin**, **secretin**, and **cholecystokinin**. Review Figure 51.17

- The liver plays a central role in directing the traffic of fuel molecules. In the **absorptive state**, the liver takes up and stores fats and carbohydrates, converting monosaccharides to glycogen or fats. The liver also takes up amino acids and uses them to produce blood plasma proteins, and can engage in gluconeogenesis.

- Fat and cholesterol are shipped out of the liver as **low-density lipoproteins**. **High-density lipoproteins** act as acceptors of cholesterol and are believed to bring fat and cholesterol back to the liver.

- Insulin largely controls fuel metabolism during the absorptive state and promotes glucose uptake as well as glycogen and fat synthesis. In the **postabsorptive state**, lack of insulin blocks the uptake and utilization of glucose by most cells of the body except neurons. If blood glucose levels fall, **glucagon** secretion increases, stimulating the liver to break down glycogen and release glucose to the blood. Review Figure 51.18, **ANIMATED TUTORIAL 51.2**

- Food intake is governed by sensations of hunger and satiety, which are determined by brain mechanisms responding to feedback signals such as insulin, **leptin**, and **ghrelin**. Review Figure 51.19

SELF-QUIZ

1. Most of the metabolic energy needed by a bird for a long-distance migratory flight is stored as
   a. glycogen.
   b. fat.
   c. protein.
   d. carbohydrates.
   e. ATP.

2. Which statement about essential amino acids is true?
   a. They are not found in vegetarian diets.
   b. They are stored by the body until they are needed.
   c. Without them, one is undernourished.
   d. All animals require the same ones.
   e. Humans can acquire all of theirs by eating milk, eggs, and meat.

3. Which statement about vitamins is true?
   a. They are essential inorganic nutrients.
   b. They are required in larger amounts than are essential amino acids.
   c. Many serve as coenzymes.
   d. Vitamin D can be acquired only by eating meat or dairy foods.
   e. When vitamin C is eaten in large quantities, the excess is stored in fat for later use.

4. The digestive enzymes of the small intestine
   a. do not function best at a low pH.
   b. are produced and released in response to circulating secretin.
   c. are produced and released under neural control.
   d. are all secreted by the pancreas.
   e. are all activated by an acidic environment.

5. Which statement about nutrient absorption by the intestinal mucosal cells is true?
   a. Carbohydrates are absorbed as disaccharides.
   b. Fats are absorbed as fatty acids and monoglycerides.
   c. Amino acids move across the plasma membrane only by diffusion.
   d. Bile transports fats across the plasma membrane.
   e. Most nutrients are absorbed in the duodenum.

6. Chylomicrons are like the tiny micelles of dietary fat in the lumen of the small intestine in that both
   a. are coated with bile.
   b. are lipid soluble.
   c. travel through the lymphatic system.
   d. contain triglycerides.
   e. are coated with lipoproteins.

7. Microbial fermentation in the gut of a cow
   a. produces fatty acids as a major nutrient for the cow.
   b. occurs in specialized regions of the small intestine.
   c. occurs in the cecum, from which food is regurgitated, chewed again, and swallowed into the true stomach.
   d. produces methane as a major nutrient.
   e. is possible because the stomach wall does not secrete hydrochloric acid.

8. Which of the following is stimulated by cholecystokinin?
   a. Stomach motility
   b. Release of bile
   c. Secretion of hydrochloric acid
   d. Secretion of bicarbonate ions
   e. Secretion of mucus

9. During the absorptive state,
   a. breakdown of glycogen supplies glucose to the blood.
   b. glucagon secretion is high.
   c. the number of circulating lipoproteins is low.
   d. glucose is the major metabolic fuel.
   e. the synthesis of fats and glycogen in muscle is inhibited.

10. During the postabsorptive state,
    a. glucose is the major metabolic fuel.
    b. glucagon stimulates the liver to produce glycogen.
    c. insulin facilitates the uptake of glucose by brain cells.
    d. fatty acids constitute the major metabolic fuel.
    e. liver functions slow down because of low insulin levels.

FOR DISCUSSION

1. Several currently popular diet books recommend high fat and protein intake and low carbohydrate intake as a means of losing body mass. What is the rationale for a high-fat and high-protein diet, and what health issues should be considered when someone considers going on such a diet?

2. Carnivores generally have more dietary vitamin requirements than herbivores do. Why?

3. It is said that the most important hormonal control of fuel metabolism in the postabsorptive state is the lack of insulin. Explain.

4. Why is obstruction of the common bile duct so serious? Consider in your answer the multiple functions of the pancreas and the way in which digestive enzymes are processed.

5. Trace the history of a fatty acid molecule from a slice of cheese pizza to a plaque on a coronary artery. Into what possible forms and structures might it have been converted as it passed through the body? Describe a direct and an indirect route it could have taken.

ADDITIONAL INVESTIGATION

You learned about brown fat in Chapter 40. This adipose tissue has lots of mitochondria and a rich blood supply and is found in many newborn mammals and cold-adapted mammals such as hibernators. The function of brown fat is to produce heat through uncoupling of oxidative phosphorylation and therefore the metabolism of lipids without the production of ATP. It has been thought for a long time that adult humans do not have brown fat, but recent studies have indicated that they do. It has further been hypothesized that the amount of brown fat an individual has is highly variable, and that this might explain why some individuals are more resistant to weight gain than others. Keeping in mind that brown fat is activated by the sympathetic nervous system, how would you investigate the possible role of brown fat in body weight control?

Blood, sweat, and tears

Blood, sweat, and tears taste salty because they have ionic concentrations similar to those of the interstitial fluids that bathe the cells of the body. The volume and composition of the interstitial fluids must remain within certain limits and be kept relatively free of wastes. Maintaining homeostasis of the interstitial fluids is the job of the excretory system, and it can be challenging. Sometimes it requires getting rid of excess fluids, and at other times it requires conserving fluids.

The nature of the challenge depends on the environment of the animal and its lifestyle. Some desert animals rarely if ever encounter free water; they must be able to live their entire lives without drinking. All animals derive water from the metabolism of food, but to survive on that amount of water, desert animals must conserve it to an exceptional degree. Accordingly, they excrete wastes that are extremely concentrated. Insects excrete semisolid wastes, and desert rodents excrete urine that is so concentrated it forms crystals of solute. Animals that live in fresh water have the opposite problem: water continuously enters their bodies by osmosis and with the food they eat, so they must constantly bail themselves out by producing copious amounts of dilute urine while at the same time conserving the solutes their bodies need.

The physiological adaptations animals have to maintain salt and water balance through excretion and conservation are remarkably flexible. Consider, for example, the vampire bat, a small tropical mammal that experiences extreme conditions within minutes of each other. The vampire bat feeds on the blood of other mammals, such as goats and cattle. Landing on an unsuspecting (usually sleeping) victim, the bat uses its sharp incisor teeth to make a small incision in a blood vessel and then laps up the blood that wells out.

Blood contains lots of nutritious protein, but it consists mostly of water. Blood meals may be few and far between, so the bat must quickly consume as much as it can—up to half its body mass. To maximize its protein intake and keep its weight low enough to fly, it rapidly eliminates the water from its meal. Within minutes of starting to feed, the bat is producing a lot of very dilute urine. The warm trickle down the neck of the victim is not blood!

Blood as a Fast Food The vampire bat *Desmodus rotundus* is able to adjust its excretory physiology from water-excreting to water-conserving, depending on whether it is ingesting or digesting its blood meal.

Living without Water Many kangaroo rat species (*Dipodomys*) inhabit arid deserts, and they rarely see free water.

Once feeding ends, this high rate of water loss must stop. Now the vampire bat is digesting and metabolizing protein and must excrete large amounts of nitrogenous wastes while conserving its body water. Within minutes, the bat's excretory system switches from producing abundant, dilute urine to producing a tiny amount of highly concentrated urine. In one feeding cycle, the vampire bat rapidly transitions from an excretory physiology typical of a mammal living in an environment with abundant fresh water to that of a desert mammal that never sees water. What an amazing case of physiological flexibility!

IN THIS CHAPTER we will describe adaptations for maintaining salt and water balance and for excreting nitrogenous wastes in environments that present animals with different challenges. Using some invertebrate examples, we will review the basic mechanisms common to all animal excretory systems. We will learn about the basic anatomical unit of the vertebrate excretory system—the nephron—and how it evolved. Finally, we will consider the mechanisms that control and regulate salt and water balance in mammals.

52.1 How Do Excretory Systems Maintain Homeostasis?

Controlling the volume, solute concentration, and composition of extracellular fluids are not huge problems for marine invertebrates because their extracellular fluids are very similar to seawater, and their nitrogenous wastes can simply diffuse into the surrounding seawater. The situation is different for marine vertebrates and for all freshwater and terrestrial animals because their extracellular fluids differ considerably from the external environment. These animals depend on **excretory systems** to maintain the volume, concentration, and composition of their extracellular fluids, and to excrete wastes.

Homeostasis of the extracellular fluids is critical because the concentration of solutes in them determines the water balance of the cells those fluids bathe, and the composition of the extracellular fluid influences the health and functions of all the cells of the body. Consider, for example, the importance of ion concentration gradients between the extracellular fluid and the cytoplasm of nerve and muscle cells (see Sections 45.2 and 48.1).

Water enters or leaves cells by osmosis

The volume of cells depends on whether they take up water from or lose water to the extracellular fluids. The movement of water across cell plasma membranes depends on differences in solute concentration on the two sides of the membrane and the permeability of the membrane. This is the process of osmosis, which we discuss in Sections 6.3 and 35.1. If the solute concentration of the extracellular fluid is less than that of the cytoplasm, water moves into the cells, causing them to swell and possibly burst. If the solute concentration of the extracellular fluid is greater than that of the cytoplasm, the cells lose water and shrink. Thus the solute concentration of the extracellular fluid affects both the volume and the solute concentration of the cells.

Animal physiologists use the term **osmolarity** in discussing osmosis. The osmolarity of a solution is the number of moles of osmotically active solutes per liter of solvent. Thus, a 1 molar solution of glucose is also a 1 osmolar (1 osmole per liter) solution, but a 1 molar solution of sodium chloride (NaCl) is a 2 osmolar solution, because each NaCl molecule dissociates into two osmotically active ions.

To achieve cellular water balance, animals must maintain the osmolarity of their extracellular fluid within an appropriate range. In addition, they must maintain an appropriate solute composition by saving some substances (reabsorption) and eliminating others (secretion). To accommodate these needs, most animals have excretory systems.

Excretory systems control extracellular fluid osmolarity and composition

Excretory systems control the osmolarity and composition of the extracellular fluids by excreting solutes that are present in excess (such as NaCl when we eat lots of salty food) and conserving solutes that are valuable or in short supply (such as glucose and amino acids). Excretory systems also eliminate the toxic waste products of protein metabolism. The output of the excretory system is called **urine**.

Three basic processes are common to a wide variety of animal excretory systems: filtration, secretion, and reabsorption. Extracellular fluid is filtered to produce a filtrate that contains no cells or large molecules, such as proteins. In animals with a closed circulatory system, the blood plasma is usually filtered from capillaries into associated tubules. The walls of the capillaries and of the tubules are the filter, and the filtration is driven by blood pressure. As the filtrate flows through the tubules, its composition and concentration are modified through processes of secretion and reabsorption to form the urine that leaves the body.

The processing of the filtrate into urine involves movement of water into and out of the tubules. There are no mechanisms for the active transport of water. The movement of water is due either to a pressure difference (filtration) or to a difference in solute concentration (osmosis). Water always flows down a pressure gradient or up a solute concentration gradient.

Animals can be osmoconformers or osmoregulators

Animals that live in marine, freshwater, or terrestrial environments face different salt and water balance problems. In the terrestrial environment, salts and water can be scarce and usually must be conserved by excretory systems. In the freshwater environment, water is plentiful but salts are scarce. Freshwater animals have to conserve salts and excrete the water that continuously invades their bodies through osmosis.

In contrast to fresh water, ocean water has a high osmolarity—over 1,000 milliosmoles/liter (mosm/l). Most marine invertebrates equilibrate their extracellular fluid osmolarity with the ocean water and are therefore called **osmoconformers**. Other marine animals maintain extracellular fluid osmolarities much lower than seawater and are therefore called **osmoregulators**. All marine vertebrates except for sharks and rays are osmoregulators. These animals maintain their extracellular fluids at about 300 mosm/l, like that of humans.

There are limits to osmoconformity. Even animals that can osmoconform over a wide range of osmolarities must osmoregulate in extreme environments. No animal could survive if its extracellular fluid had the osmolarity of fresh water. There would be too few solutes, including nutrients and ions essential for cell functions, and water would flow into the cells by osmosis, causing them to swell and burst. Nor could animals survive with internal osmolarities as high as those that may be

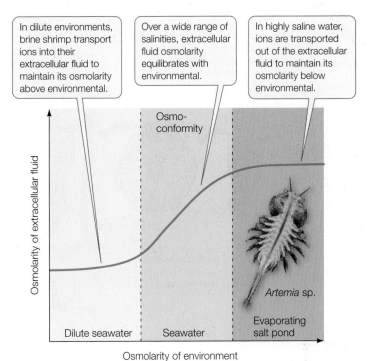

52.1 Osmoconformity Has Limits Animals that experience an extreme range of salt concentrations in their environment can be osmoconformers over much of that range but must osmoregulate at the extremes. They regulate by actively transporting ions across their gill membranes—in either direction, depending on the challenge.

reached in an evaporating tidal pool. High solute concentrations cause proteins to denature.

The brine shrimp *Artemia* is both an osmoconformer and an osmoregulator; this animal can live in environments of almost any osmolarity (**Figure 52.1**). *Artemia* are found in huge numbers in the most salty environments known, such as Utah's Great Salt Lake and in coastal evaporation ponds where salt is concentrated for commercial purposes (see Figure 26.23) and can reach an osmolarity of 2,500 mosm/l. At these high environmental osmolarities, *Artemia* maintains its tissue fluid osmolarity considerably below that of the environment by actively transporting NaCl from its extracellular fluid out across its gill membranes to the environment. *Artemia* cannot survive in fresh water, but it can live in dilute seawater by reversing the direction of transport of NaCl across its gill membranes to maintain its extracellular fluid above that of the environment.

Animals can be ionic conformers or ionic regulators

Osmoconformers can also be ionic conformers, allowing the ionic composition, as well as the osmolarity, of their extracellular fluid to match that of the environment. Most osmoconformers, however, are ionic regulators to some degree; they employ active transport mechanisms to excrete some ions and to maintain other ions in their extracellular fluid at optimal concentrations.

Terrestrial animals obtain their salts from food and regulate the ionic composition of their extracellular fluids by conserving some ions and excreting others. For example, herbivores have to conserve Na+ because the plants they eat have low concen-

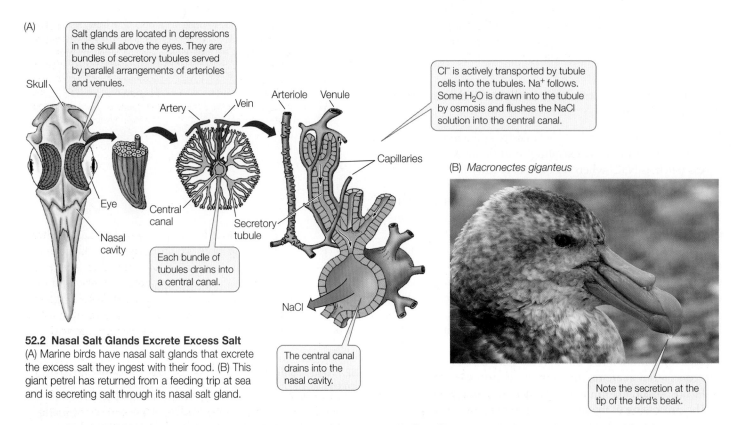

(A)

Salt glands are located in depressions in the skull above the eyes. They are bundles of secretory tubules served by parallel arrangements of arterioles and venules.

Cl⁻ is actively transported by tubule cells into the tubules. Na⁺ follows. Some H_2O is drawn into the tubule by osmosis and flushes the NaCl solution into the central canal.

Skull

Artery   Vein

Arteriole   Venule

Capillaries

Eye

Central canal

Secretory tubule

Nasal cavity

Each bundle of tubules drains into a central canal.

NaCl

(B) *Macronectes giganteus*

Note the secretion at the tip of the bird's beak.

52.2 Nasal Salt Glands Excrete Excess Salt
(A) Marine birds have nasal salt glands that excrete the excess salt they ingest with their food. (B) This giant petrel has returned from a feeding trip at sea and is secreting salt through its nasal salt gland.

The central canal drains into the nasal cavity.

trations of Na⁺. By contrast, birds that feed on marine animals must excrete the excess of sodium they ingest with their food. Their *nasal salt glands* excrete a concentrated solution of NaCl via a duct that empties into the nasal cavity. Birds, such as penguins and gulls, that have nasal salt glands can be seen frequently sneezing or shaking their heads to get rid of the salty droplets excreted from their nasal salt glands (**Figure 52.2**).

52.1 RECAP

Excretory systems control water and salt balance and the excretion of nitrogenous waste products through three mechanisms: filtration of body fluids to form urine, active secretion of substances into the urine, and active reabsorption of substances from the urine.

- Describe the two mechanisms used to move water across membranes. **See p. 1093**

- What different salt and water balance problems might animals encounter in marine, freshwater, and terrestrial environments? What are some of the ways they meet those challenges? **See pp. 1093–1094 and Figures 52.1 and 52.2**

In addition to maintaining salt and water balance, animals must eliminate the waste products of metabolism from their extracellular fluids. The major problem is nitrogen. When nitrogen-containing molecules are broken down by metabolism, the end product can be toxic.

52.2 How Do Animals Excrete Nitrogen?

The end products of the metabolism of carbohydrates and fats are water and carbon dioxide, which are not difficult to eliminate. Proteins and nucleic acids, however, contain nitrogen, so their metabolism produces *nitrogenous wastes* in addition to water and carbon dioxide.

Animals excrete nitrogen in a number of forms

The most common nitrogenous waste is **ammonia** (NH_3). Because it is highly toxic, ammonia is either excreted continuously to prevent its accumulation or is detoxified by conversion into **urea** or **uric acid** (**Figure 52.3**).

AMMONIA Ammonia is highly soluble in water and diffuses rapidly, so its continuous excretion is relatively simple for many aquatic animals that continuously lose ammonia from their blood to the environment by diffusion across their gill membranes. Animals that excrete ammonia, such as aquatic invertebrates and bony fishes, are **ammonotelic**.

If ammonia builds up in the extracellular fluids, it becomes toxic at rather low levels and is a dangerous metabolite for terrestrial animals and for those aquatic animals that cannot continuously excrete ammonia. These animals must convert ammonia into urea or uric acid.

UREA **Ureotelic** animals, such as mammals, amphibians, and cartilaginous fishes (sharks and rays), excrete urea as their principal nitrogenous waste product. Urea is quite soluble in water,

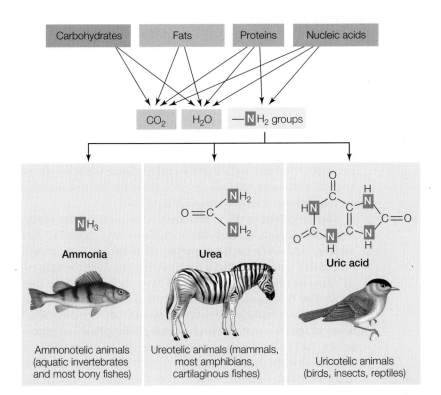

52.3 Waste Products of Metabolism The metabolism of proteins and nucleic acids produces nitrogenous wastes. Many aquatic animals, including most fishes, excrete nitrogenous wastes as ammonia, which is highly diffusible and soluble in an aqueous environment. Most terrestrial animals and some aquatic animals excrete either urea or uric acid. Urea is more soluble in water and is the major nitrogenous excretory product for mammals, amphibians, and some fishes. Uric acid is not very soluble in water and is the major nitrogenous excretory product for reptiles, birds, insects, and some amphibians.

Species that live in different habitats at different developmental stages may use more than one mechanism of nitrogen excretion. The tadpoles of frogs and toads, for example, excrete ammonia across their gill membranes, but adult frogs and toads generally excrete urea. Some adult amphibians that live in arid habitats excrete uric acid.

but its excretion can result in a large loss of water that many animals can ill afford. As we will see later in this chapter, mammals have evolved excretory systems that conserve water by producing concentrated urea solutions. The sharks and rays are another story. These marine species keep their extracellular fluids almost isosmotic (same osmotic concentration) to the marine environment by retaining high concentrations of urea and another nitrogen-containing compound, trimethylamine oxide, in their body fluids.

URIC ACID Animals that conserve water by excreting nitrogenous wastes mostly as uric acid are **uricotelic**. Insects, reptiles (including birds), and some amphibians are uricotelic. Uric acid is not very soluble in water, so it forms a colloidal suspension in the urine and is excreted as a semisolid (for example, the whitish material in bird droppings). A uricotelic animal loses very little water as it disposes of its nitrogenous wastes.

Most species produce more than one nitrogenous waste

Humans are ureotelic, but we also excrete uric acid. The uric acid in human urine comes largely from the metabolism of nucleic acids and caffeine. If uric acid levels in the extracellular fluids rise too high, uric acid crystals can precipitate in joints and cause the age-old malady called gout. Because solubility goes down with temperature, uric acid crystals usually precipitate first in the extremities, especially the big toe. Pain in the big toe is a telltale symptom of gout.

Humans can also excrete ammonia, which is an important mechanism for regulating the pH of the extracellular fluids. As we will see later in this chapter, excreted ammonia buffers the urine and enables the secretion of more hydrogen ions.

52.2 RECAP

Ammonia is a common metabolic waste product of nitrogen-containing molecules. Most aquatic animals excrete ammonia by diffusion into the water. Terrestrial animals and some aquatic animals detoxify it by conversion to urea or uric acid.

- Why might you expect a species from an arid habitat to use uric acid as its primary nitrogenous waste product? **See p. 1095 and Figure 52.3**

Animals exhibit a variety of adaptations for dealing with the challenges of salt and water balance in different environments. All of these adaptations, however, are based on two basic mechanisms—namely, filtration and tubular processing of the filtrate to conserve some solutes and excrete others.

52.3 How Do Invertebrate Excretory Systems Work?

Freshwater and terrestrial invertebrates have a wide variety of adaptations for maintaining salt and water balance and excreting nitrogen. In this section, we will explore three examples of invertebrate excretory systems: *protonephridia, metanephridia,* and *Malpighian tubules.* Each of these systems produces an extract of interstitial fluid lacking in large molecules. They then change the solute composition (ions and small molecules) of that fluid to form an excretory product.

The protonephridia of flatworms excrete water and conserve salts

Many flatworms, such as *Planaria*, live in fresh water. These animals excrete water through an elaborate network of tubules

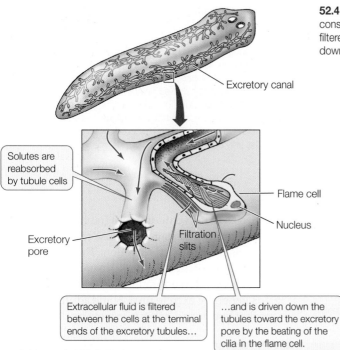

52.4 Protonephridia in Flatworms The protonephridia of the flatworm *Planaria* consist of tubules ending in flame cells. In the region of the flame cells, body fluid is filtered between the tubule cells. The composition of the filtrate is modified as it flows down the tubule.

Solutes are reabsorbed by tubule cells

Excretory pore

Filtration slits

Flame cell

Nucleus

Extracellular fluid is filtered between the cells at the terminal ends of the excretory tubules…

…and is driven down the tubules toward the excretory pore by the beating of the cilia in the flame cell.

Excretory canal

permeable capillary walls into the coelom. Some waste products, such as ammonia, diffuse directly from the tissues into the coelom. Where does this coelomic fluid go?

Each segment of the earthworm contains a pair of **meta-nephridia** (singular metanephridium; Greek *meta*, "akin to," and *nephros*, "kidney"). Each metanephridium begins as a ciliated, funnel-like opening called a *nephrostome*. The nephrostome resides in one segment and continues as a tubule in the next segment. The tubule ends in a pore, called a *nephridiopore*, that opens to the outside of the animal (**Figure 52.5**). Coelomic fluid is swept into the metanephridia through the ciliated nephrostomes. As the fluid passes through the tubules, their cells actively reabsorb certain molecules from it and actively secrete other molecules into it. What leaves the animal through the nephridiopores is a dilute urine containing nitrogenous wastes and other solutes.

running throughout their bodies. The tubules end in *flame cells*, so called because each cell has a tuft of cilia projecting into the tubule (**Figure 52.4**). The beating of the cilia gives the appearance of a flickering flame. A flame cell and a tubule together form a **protonephridium** (plural protonephridia; Greek *proto*, "before," and *nephros*, "kidney").

Extracellular fluid enters the tubules by filtration. The beating of the cilia causes a slight negative pressure in the tubule, and movements of the animal create positive pressure in the extracellular fluid. This pressure difference causes extracellular fluid to be filtered through tiny spaces between tubule cells. The filtrate flows toward the animal's excretory pore, and along the way the cells of the tubules modify the composition of the fluid by reabsorption and secretion of specific ions and molecules. Because more ions are reabsorbed than are secreted, the urine that leaves the flatworm's body is less concentrated than the extracellular fluid. Thus, the protonephridium conserves ions and excretes water and wastes.

The metanephridia of annelids process coelomic fluid

Filtration of body fluids and modification of urine by tubules are highly developed processes in annelid worms such as the earthworm. Annelids are segmented, and in each segment they have a fluid-filled body cavity called a *coelom* (see Figure 32.12). Annelids have a closed circulatory system through which blood is pumped under pressure. The pressure causes the blood to be filtered across the thin,

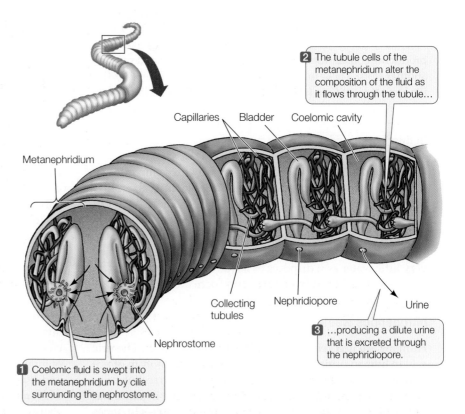

2 The tubule cells of the metanephridium alter the composition of the fluid as it flows through the tubule…

Capillaries   Bladder   Coelomic cavity

Metanephridium

Collecting tubules

Nephridiopore

Urine

3 …producing a dilute urine that is excreted through the nephridiopore.

Nephrostome

1 Coelomic fluid is swept into the metanephridium by cilia surrounding the nephrostome.

52.5 Metanephridia in Earthworms The metanephridia of annelids are arranged segmentally. The cross section at the left end shows a pair of metanephridia. Three longitudinal sections (right) show only one metanephridium of the two in each segment. Coelomic fluid enters the nephrostome and flows through tubules leading to the nephridiopore. A close association of the tubules and blood capillaries facilitates the active exchange of substances between the blood and the tubular fluid.

--- **yourBioPortal.com** ---

GO TO Web Activity 52.1 • Annelid Metanephridia

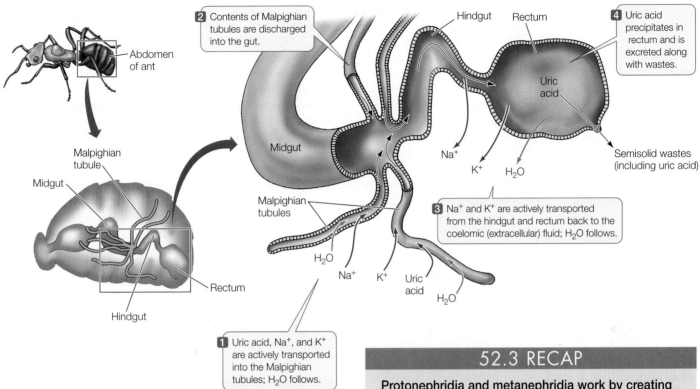

2 Contents of Malpighian tubules are discharged into the gut.

Abdomen of ant

Hindgut   Rectum

4 Uric acid precipitates in rectum and is excreted along with wastes.

Malpighian tubule

Midgut

Midgut

Uric acid

Malpighian tubules

Na^+

K^+ H_2O

Semisolid wastes (including uric acid)

3 Na^+ and K^+ are actively transported from the hindgut and rectum back to the coelomic (extracellular) fluid; H_2O follows.

Rectum

Hindgut

H_2O

Na^+ K^+ Uric acid H_2O

1 Uric acid, Na^+, and K^+ are actively transported into the Malpighian tubules; H_2O follows.

52.6 Malpighian Tubules in Insects The blind, thin-walled Malpighian tubules are attached to the junction of the insect's midgut and hindgut and project into the spaces containing extracellular fluid. This system makes it possible to excrete wastes with very little loss of water.

The Malpighian tubules of insects depend on active transport

Insects can excrete nitrogenous wastes with very little loss of water and can therefore live in the driest habitats on Earth. The insect excretory system consists of **Malpighian tubules**. An individual insect has from 2 to more than 100 of these blind-ended tubules that open into the gut between the midgut and hindgut (**Figure 52.6**).

Insects have an open circulatory system and therefore cannot use a pressure difference to filter extracellular fluids into the Malpighian tubules. Instead, the cells of the tubules actively transport uric acid, potassium ions, and sodium ions from the extracellular fluid into the tubules. The high concentration of solutes in the tubules causes water to follow osmotically, which flushes the tubule contents toward the gut.

The epithelial cells of the hindgut and rectum actively transport sodium and potassium ions from the gut contents back into the extracellular fluid. This local transport of salts creates an osmotic gradient that pulls water out of the rectal contents. As its concentration increases, the uric acid forms a colloidal suspension, freeing even more water to be reabsorbed. Remaining in the rectum is the uric acid mixed with other wastes; this semisolid matter is what the insect excretes. The Malpighian tubule system is a highly effective mechanism for excreting nitrogenous wastes and some salts without giving up much water.

52.3 RECAP

Protonephridia and metanephridia work by creating a filtrate of the body fluids that is modified by the secretion and reabsorption of specific substances before being excreted. Insect Malpighian tubules actively secrete uric acid and other solutes into closed tubules.

- Describe how an earthworm filters its blood and produces urine. See p. 1096 and Figure 52.5

- How do Malpighian tubules function, and how does this system make it possible for some insect species to survive in the desert? See p. 1097 and Figure 52.6

Having described how several invertebrate groups handle nitrogen excretion, we will next consider the nephron—the basic unit of the vertebrate excretory system—and how it evolved to be able to respond to a variety of salt and water balance challenges and maintain a relatively constant internal environment.

52.4 How Do Vertebrates Maintain Salt and Water Balance?

The main excretory organ of vertebrates is the **kidney**, and the functional unit of the kidney is the **nephron**, which has a blood vessel component and a tubule component. The vascular component begins with a knot of capillaries that are highly permeable and filter the blood into the tubule component. The blood vessels also carry substances to the tubules for secretion and carry away substances that the tubule cells reabsorb. Nephrons can filter large volumes of blood and achieve bulk reabsorption of salts and other valuable molecules such as glucose, making the vertebrate kidney well adapted for the excretion of excess water.

If the ancestors of vertebrates evolved in fresh water, as paleontologists propose, the excretory systems of vertebrates would have evolved to excrete excess water. How then have vertebrates adapted to environments where water must be conserved and salts excreted? The answer to this question differs among vertebrate groups. Even among the marine fishes, the excretory adaptations of the bony fishes are different from those of the cartilaginous fishes. Reptiles, birds, and mammals have excretory systems that conserve water. Reptiles and birds achieve this mainly by being uricotelic and producing a semisolid excretory product that contains little water. Mammals in contrast are ammonotelic, excrete a liquid waste product, but have evolved the ability to produce a highly concentrated urine.

Marine fishes must conserve water

Marine bony fishes osmoregulate their extracellular fluids to maintain them at one-third to one-half the osmolarity of seawater. Their only source of water is the sea around them, so they must conserve water and excrete excess solutes. Marine bony fishes cannot produce urine that is more concentrated than their extracellular fluids, so they minimize water loss by producing very little urine. In contrast, freshwater fishes produce lots of dilute urine.

If marine bony fishes cannot excrete excess solutes in their urine, how do they deal with the large salt loads they ingest with food? Marine bony fishes do not absorb from their guts some of the ions they take in, especially divalent ions such as Mg^{2+} or SO_4^{2-}. NaCl, the major salt ingested, is actively excreted across the gill membranes. As we mentioned earlier, bony fishes can lose their nitrogenous waste, ammonia, by diffusion across their gill membranes.

Cartilaginous fishes (sharks and rays) are osmoconformers but not ionic conformers. They raise the osmolarity of their body fluids in a unique way. Unlike boney fishes, they convert nitrogenous wastes to urea and trimethylamine oxide, and they retain large amounts of these compounds in their body fluids. As a result, their body fluids have an osmolarity close to that of seawater, so they do not lose body water to the environment by osmosis, and may actually gain water. These species have adapted to a concentration of urea in the body fluids that would be toxic to other vertebrates.

Sharks and rays still have the problem of excreting the large amounts of salts they take in with their food. They solve this problem by having a gland in the rectum that actively secretes NaCl by a mechanism similar to that of the nasal salt glands of seabirds.

Terrestrial amphibians and reptiles must avoid desiccation

Most amphibians live in or near fresh water, and they stay in humid habitats when they do venture from the water. Like freshwater fishes, most amphibians produce large amounts of dilute urine and conserve salts. Some amphibians, however, have adapted to habitats that require water conservation.

Amphibians living in dry terrestrial environments have reduced the permeability of their skin to water. Some secrete a waxy substance over the skin to waterproof it. Several species of frogs that live in arid regions of Australia burrow deep into the ground and remain there during long dry periods. They enter **estivation**, a state of very low metabolic activity and therefore low water turnover. When it rains, the frogs come out of estivation, feed, and reproduce. Their most interesting adaptation is an enormous urinary bladder. Before entering estivation, they fill the bladder with dilute urine, which can amount to one-third of their body weight. This dilute urine serves as a water reservoir that is gradually reabsorbed into the blood during the long period of estivation.

Reptiles occupy habitats ranging from aquatic to extremely hot and dry. In fact, snakes, lizards, and birds are among the most prominent members of many desert faunas. Three major adaptations have freed reptiles from the close association with water that is necessary for most amphibians (see Section 33.4). First, reptiles are *amniotes* that do not need fresh water to reproduce because they employ internal fertilization and lay eggs with shells that retard evaporative water loss. Second, they have a dry, epidermis (skin) that retards evaporative water loss. Third, they excrete nitrogenous wastes as uric acid semisolids, losing little water in the process.

Mammals can produce highly concentrated urine

Mammals occupy diverse habitats, many of which present special excretory system challenges. The most challenging environments are those in which water is severely limited. Mammals have a variety of adaptations to conserve water, but chief among them is the ability to produce a urine that is more concentrated than their extracellular fluids. They are able to concentrate their urine because of adaptations of their kidneys that we will explore in detail in Section 52.5. To understand how these adaptations work, however, we must first describe the structure and function of the vertebrate nephron.

The nephron is the functional unit of the vertebrate kidney

Urine formation in vertebrate nephrons involves three main processes (**Figure 52.7**):

- *Filtration.* Each nephron has a dense ball of capillaries called a **glomerulus** (plural glomeruli). The glomerulus is highly permeable to water, ions, and small molecules but impermeable to large molecules. Blood pressure drives the movement of water and small molecular weight solutes out of the glomerular capillaries.

- *Tubular reabsorption.* The filtrate from the glomerulus flows into the **renal tubule**. Cells in the renal tubule modify the filtrate by reabsorbing specific ions, nutrients, and water, returning these to the blood, and leaving behind and concentrating excess ions and waste products such as urea.

- *Tubular secretion.* The filtrate in the renal tubule is further modified by tubule cells transporting substances into the tubular contents. These are substances that the body needs to excrete.

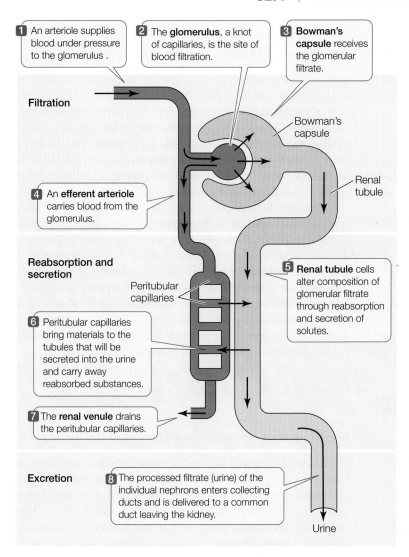

1 An arteriole supplies blood under pressure to the glomerulus.

2 The **glomerulus**, a knot of capillaries, is the site of blood filtration.

3 **Bowman's capsule** receives the glomerular filtrate.

Filtration

Bowman's capsule

4 An **efferent arteriole** carries blood from the glomerulus.

Renal tubule

Reabsorption and secretion

Peritubular capillaries

5 **Renal tubule** cells alter composition of glomerular filtrate through reabsorption and secretion of solutes.

6 Peritubular capillaries bring materials to the tubules that will be secreted into the urine and carry away reabsorbed substances.

7 The **renal venule** drains the peritubular capillaries.

Excretion

8 The processed filtrate (urine) of the individual nephrons enters collecting ducts and is delivered to a common duct leaving the kidney.

Urine

52.7 The Vertebrate Nephron The vertebrate nephron consists of a renal tubule closely associated with two capillary beds, the glomerulus and the peritubular capillaries.

yourBioPortal.com

GO TO **Web Activity 52.2 • The Vertebrate Nephron**

Blood is filtered into Bowman's capsule

The renal tubule begins with **Bowman's capsule** (see Figure 52.7), which encloses the glomerulus. The glomerulus appears to be pushed into Bowman's capsule much like a fist pushed into an inflated balloon. The cells of the capsule that are in direct contact with the glomerular capillaries are called **podocytes** (**Figure 52.8B**). These highly specialized cells have numerous armlike extensions, each with hundreds of fine, fingerlike processes. The podocytes wrap around the capillaries so that their processes interdigitate and intimately cover the capillaries.

The glomerulus filters the blood to produce a fluid (the renal filtrate) that lacks cells and large molecules. The walls of the capillaries, the basal lamina of the capillary endothelium, and the podocytes of Bowman's capsule all participate in filtration. *Fenestrations* in the walls of the capillaries (see Section 50.4) allow water and many solute molecules, but not red blood cells, to pass through. The meshwork of the basal lamina and the spaces between the processes of the podocytes are even finer and prevent large molecules from leaving the capillaries (**Figure 52.8C**). The arterial pressure of the blood entering the permeable capillaries causes the filtration of water and small molecules in the glomerulus. The glomerular filtration rate is high because blood

Blood enters the glomerular capillaries via an **afferent arteriole** and leaves the glomerulus in an **efferent arteriole**. This short vessel is called an arteriole because it feeds another capillary bed called the **peritubular capillaries** that intimately surround the renal tubules (**Figure 52.8A**). Peritubular capillaries deliver substances to the renal tubule cells that they will secrete into the urine and carry away substances these cells absorb from the urine.

52.8 A Tour of the Nephron Scanning electron micrographs illustrate the anatomical basis for blood filtration by the kidneys. (A) In a preparation showing only the blood vessels (tubular tissue has been digested away), the glomeruli appear as balls of capillaries served by arterioles. (B) Higher magnification of a glomerulus with the tubule cells intact shows the podocytes that wrap around the glomerular capillaries. (C) This cross section of an intact glomerulus shows the tubule cells that form Bowman's capsule.

(A) Afferent arterioles   Glomeruli

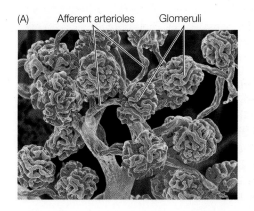

(B)   Podocytes

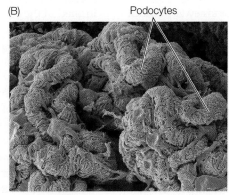

(C)   Proximal renal tubule

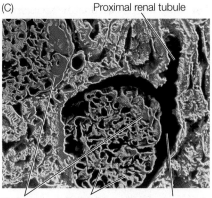

Podocytes   Capillaries   Bowman's capsule

pressure in the glomerular capillaries is unusually high, and because the capillaries of the glomerulus, along with their covering of podocytes, are more permeable to water than other capillary beds in the body.

The renal tubules convert glomerular filtrate to urine

The composition of the filtrate that enters the renal tubule is similar to that of the blood plasma, with the exception of high-molecular-weight solutes such as proteins. Reabsorption and secretion cause the composition of this fluid to change as it passes down the renal tubule. Cells of the tubule actively reabsorb certain molecules from the tubule fluid (which are returned to the blood flowing through the peritubular capillaries). For example, glucose and amino acids are reabsorbed. Most NaCl is reabsorbed. Other substances in the blood of the peritubular capillaries are actively secreted into the tubule fluid. An example is paraminohippuric acid (PAH), which is produced in the liver from benzoic acid, a common food preservative. Because of the actions of the renal tubules, the excreted urine is very different from the original filtrate.

52.4 RECAP

The kidney is the major excretory organ of vertebrates. Its functional unit is the nephron, which includes a glomerulus that filters blood and a renal tubule that secretes and reabsorbs solutes, modifying the filtrate to produce urine. The nephron is a mechanism for excreting excess water while conserving valuable solutes.

- Explain the difference in the osmoregulatory adaptations of marine bony and cartilaginous fishes. See p. 1098

- What are the functional relationships between the glomerular and peritubular capillaries? See p. 1099 and Figure 52.7

- Describe how blood is filtered by the glomerulus. See pp. 1099–1100

- How is the composition of the urine made different from the composition of the blood? See p. 1100

The adaptations that enable the mammalian kidney to produce urine more concentrated than extracellular fluids were important steps in vertebrate evolution, and they were largely achieved through changes in the structure and regional functions of the renal tubules.

52.5 How Does the Mammalian Kidney Produce Concentrated Urine?

Mammals have high body temperatures and high metabolic rates, and therefore have the potential for a high rate of water loss. Having an excretory system that minimizes water loss made it possible for these highly active species to occupy arid habitats.

Kidneys produce urine, and the bladder stores it

Mammalian excretory systems are similar, so we will use that of humans as our example. Humans have two kidneys at the back of the upper region of the abdominal cavity (**Figure 52.9A**). Each kidney filters blood, processes the filtrate into urine, and releases that urine into a duct called the **ureter**. The ureter of each kidney leads to the **urinary bladder**, where the urine is stored until it is excreted through the **urethra**, a short tube that opens to the outside of the body.

Two sphincter muscles surrounding the base of the urethra control urination. One of these sphincters is a smooth muscle and is controlled by the autonomic nervous system. As the bladder fills, stretch receptors in the walls of the bladder trigger a spinal reflex that relaxes this sphincter. This reflex is the only control of urination in infants, hence their frequent "accidents." The other sphincter is a skeletal muscle and is controlled by the voluntary nervous system. When the bladder is *very* full, only deliberate conscious effort prevents urination. Infant toilet training involves learning to control this sphincter.

yourBioPortal.com

GO TO Animated Tutorial 52.1 • The Mammalian Kidney

Nephrons have a regular arrangement in the kidney

The kidney is shaped like a kidney bean; when sliced along its long axis, its key anatomical features are revealed (**Figure 52.9B**). The ureter and the **renal artery** and **renal vein** enter the kidney on its concave (punched-in) side. The ureter extends into the kidney in several branches, the ends of which envelop kidney tissues called **renal pyramids**. The renal pyramids make up the internal core, or **medulla**, of the kidney. The medulla is covered by an outer layer, or **cortex**, that has a granular appearance. Between the cortex and the medulla, the renal artery divides into the many arterioles that serve the nephrons. In this same region, the renal vein collects blood from the many venules that drain the peritubular capillaries.

The organization of the nephrons within the kidney is very regular. All of the glomeruli with their Bowman's capsules are located in the cortex. The initial segment of a renal tubule is called the **proximal convoluted tubule**—"proximal" because it is closest to the glomerulus, and "convoluted" because it is twisted (**Figure 52.9C**). All of the proximal convoluted tubules are also located in the cortex.

At the point at which the renal tubule descends into the medulla, it straightens and descends directly down into the medulla. In the medulla the tubule makes a hairpin turn and ascends back to the cortex, forming what is called the **loop of Henle**. Some nephrons have longer loops of Henle than others. Some 20 to 30 percent of human nephrons that have glomeruli deep in the cortex (i.e., near the border with the medulla) have long loops of Henle that go deep into the medulla. Nephrons that have glomeruli farther up in the cortex generally have short loops of Henle that descend only a short distance into the medulla. As we will see, the long loops are the critical adaptation of the mammalian nephron that enables the kidney to concentrate the urine.

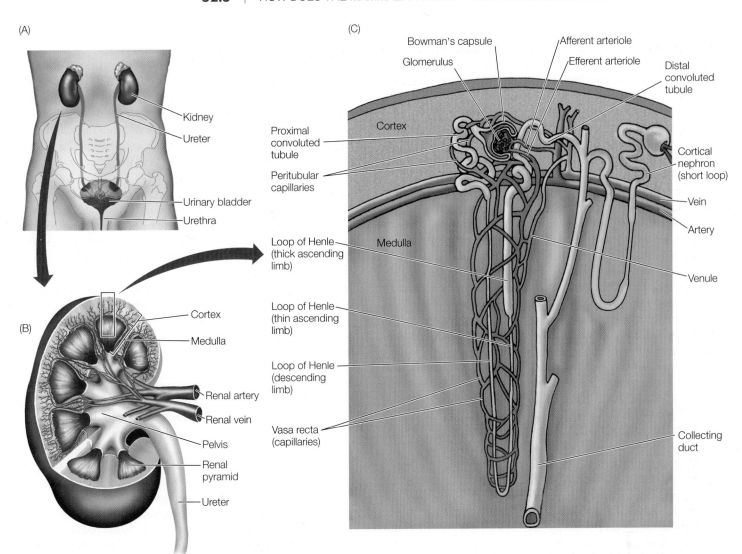

52.9 The Human Excretory System (A) The human kidneys lie against the back wall of the abdominal cavity, in the region of the middle back. (B) A highly organized internal structure is the basis for kidney function. Certain parts of the nephrons are in the organ's outer region, called the cortex; other parts are in the internal region, called the medulla. (C) The glomeruli and the proximal and distal convoluted tubules are located in the cortex of the kidney. The loops of Henle run in parallel as straight sections down into the renal medulla and back up to the cortex. Collecting ducts run from the cortex to the inner surface of the medulla, where they open into the ureter. The vasa recta are peritubular capillaries that parallel the loops of Henle.

──── **yourBioPortal.com** ────

GO TO **Web Activity 52.3** • The Human Excretory System

The *ascending limb* of the loop of Henle becomes the **distal convoluted tubule** when it reaches the cortex—"distal" because it is farther from the glomerulus. The distal convoluted tubules of many nephrons join a common **collecting duct** in the cortex. The collecting ducts descend back down through the renal pyramid, parallel to and past the tips of the loops of Henle, and empty into the funnel-shaped *pelvis*. Divisions of the pelvis that surround each renal pyramid join together to leave the kidney as the ureter (see Figure 52.9B).

The organization of the blood vessels of the kidney closely parallels the organization of the nephrons (see Figure 52.9C).

Smaller arteries branch from the renal artery and radiate into the cortex, forming the afferent arterioles that carry blood to each glomerulus. Each glomerulus is drained by an efferent arteriole that gives rise to the peritubular capillaries, most of which surround the cortical portions of the tubules. As we have seen, the intimate association between the glomerular and peritubular capillaries permits exchanges between the blood and the specialized regions of the nephron tubule.

Some of the peritubular capillaries run into the medulla in parallel with the loops of Henle and the collecting ducts, forming a vascular network called the **vasa recta**. All of the peritubular capillaries from a nephron join back together into a venule that joins with venules from other nephrons and eventually leads to the renal vein. As we will see, the concentrating ability of the mammalian kidney depends on water reabsorption in the renal medulla, and the vasa recta are the avenue by which that water gets out of the renal medulla and into the circulation.

Most of the glomerular filtrate is reabsorbed by the proximal convoluted tubule

Most of the water and solutes filtered by the glomerulus are reabsorbed and do not appear in the urine. We can reach this conclusion by comparing the rate of filtration by the glomeruli with the rate of urine production. The kidneys receive about 1 liter of blood per minute, or about 1,500 liters of blood per day. How much of this huge volume is filtered out of the glomeruli? The answer is about 12 percent. This is still a large volume—180 liters per day! We normally urinate 2 to 3 liters per day, so about 98 percent of the fluid volume that is filtered out of the glomerulus is returned to the blood. Where and how is this enormous fluid volume reabsorbed?

The proximal convoluted tubule (PCT) is responsible for most of the reabsorption of water and solutes from the glomerular filtrate. The cells of this section of the renal tubule have many microvilli that increase their apical (facing into the tubule) surface area for reabsorption, and they have many mitochondria—an indication that they are metabolically active. PCT cells actively transport Na^+ (with Cl^- following) and other solutes, such as glucose and amino acids, out of the tubule fluid.

Almost all glucose and amino acid molecules that are filtered from the blood are actively reabsorbed by PCT cells and transported into the extracellular fluid. The active transport of solutes from the proximal tubule into the interstitial fluid causes water to follow osmotically. The water and solutes moved into the interstitial fluid are taken up by the peritubular capillaries and returned to the venous blood. These processes accomplish the reabsorption of more than 75 percent of the fluid that initially enters the nephron.

Despite the bulk reabsorption of water and solutes by the PCT, the overall osmolarity of the fluid flowing through the PCT does not change. Thus, the process that is occurring in the PCT is called *isosmotic reabsorption*. The fluid that enters the loop of Henle has the same osmolarity as the blood plasma, although its composition is different. How then does the kidney produce urine that is more concentrated than the blood plasma?

The loop of Henle creates a concentration gradient in the renal medulla

Humans can produce urine that is four times more concentrated than their blood plasma. The vampire bat we encountered at the beginning of this chapter can produce urine that is twenty times more concentrated than its blood plasma. The concentrating ability of the mammalian kidney arises from a **countercurrent multiplier** mechanism made possible by the anatomical arrangement of the loops of Henle. The term "countercurrent" refers to the opposing directions in which the tubule fluid in the descending and ascending limbs flows. The term "multiplier" refers to the ability of this system to create a solute concentration gradient in the renal medulla.

The loops of Henle do not themselves produce concentrated urine; rather, they increase the osmolarity of the extracellular fluid in the medulla in a graduated way. In humans, for example, the extracellular fluid may be 300 mosm/l (the concentration of blood plasma) at the top of the medulla bordering the cortex, and increase to about 1,200 mosm/l near the hairpin turn of the loop of Henle (see Figure 52.10). How do the loops produce this effect?

The cells that make up the different segments of the loop of Henle differ anatomically and functionally. Cells of the descending limb and the initial cells of the ascending limb are thin, with no microvilli and few mitochondria. They are not specialized for transport. Partway up the ascending limb, the cells become specialized for active transport. These cells are thick and have many mitochondria. Accordingly, the segments of the loop of Henle are named the *thin descending limb*, the *thin ascending limb*, and the *thick ascending limb* (**Figure 52.10**).

The countercurrent multiplier mechanism may be more easily understood by first considering events occurring in the thick ascending limb (Figure 52.10, note 1). The cells of the thick ascending limb reabsorb Na^+ and Cl^- from the tubule fluid and move it into the interstitial fluid. (In the following discussion, we will distinguish between the two components of extracellular fluid—the blood plasma and the interstitial fluid.) The thick ascending limb is not permeable to water, so the reabsorption of Na^+ and Cl^- from the tubular fluid raises the concentration of those solutes in the surrounding interstitial fluid and decreases the concentration of the tubular fluid entering the distal convoluted tubule.

The thin descending limb, in contrast, is highly permeable to water but not very permeable to Na^+ and Cl^-. Since the local interstitial fluid has been made more concentrated by the Na^+ and Cl^- reabsorbed from the neighboring thick ascending limb, water is withdrawn osmotically from the tubule fluid in the descending limb. Therefore, the fluid in the descending limb becomes more concentrated as it flows toward the hairpin turn at the bottom of the renal medulla (Figure 52.10, note 2).

The thin ascending limb, like the thick ascending limb, is not permeable to water. It is, however, permeable to Na^+ and Cl^-. As the concentrated tubule fluid flows up the thin ascending limb, it is more concentrated than the surrounding interstitial fluid, so Na^+ and Cl^- diffuse out. When the tubule fluid reaches the thick ascending limb, active transport continues to move Na^+ and Cl^- from the tubule fluid to the interstitial fluid.

As a result of the processes described above, the tubule fluid reaching the distal convoluted tubule is less concentrated than the blood plasma (Figure 52.10, note 3), and the solutes that have been left behind in the renal medulla have created a concentration gradient in the interstitial fluid of the medulla (indicated by the background color gradient in Figure 52.10).

You may wonder why the blood flow through the medulla does not wash out the concentration gradient established by the loops of Henle. The parallel arrangement of the descending and ascending peritubular capillaries—the vasa recta—in the medulla helps preserve the concentration gradient in the medulla. These capillaries are permeable to both salt and water. Therefore, as blood flows down the descending limb of the vasa recta into the increasingly concentrated interstitial fluid of the medulla, it loses water and gains solutes. As blood flows up from the bottom of the medulla in the ascending limb of the vasa recta, the opposite happens (water is gained and solutes are lost) because now the blood is more concentrated than the surrounding interstitial fluid (Figure 52.10, notes 4–6). The dynamics of this countercurrent exchange of salts and water be-

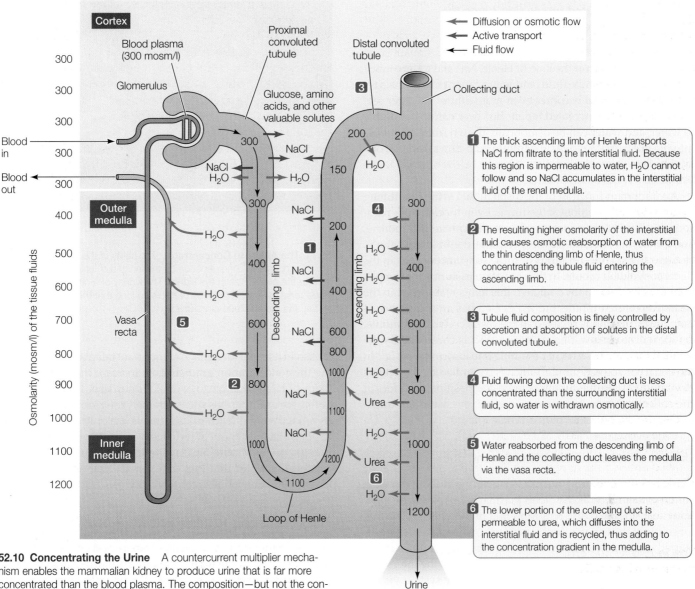

52.10 Concentrating the Urine A countercurrent multiplier mechanism enables the mammalian kidney to produce urine that is far more concentrated than the blood plasma. The composition—but not the concentration—of the filtrate is changed by the proximal convoluted tubule, which reabsorbs valuable molecules (including NaCl). Bulk reabsorption of water follows osmotically. The urine concentration process begins in the thick ascending limb of Henle, which reabsorbs NaCl but is impermeable to H_2O. Some of the reabsorbed NaCl enters the descending limb and is thereby trapped in the renal medulla, creating a concentration gradient in the interstitial fluid. As urine in the collecting duct passes through this concentration gradient, it can lose water osmotically and become highly concentrated.

The callouts in the figure read:

1. The thick ascending limb of Henle transports NaCl from filtrate to the interstitial fluid. Because this region is impermeable to water, H_2O cannot follow and so NaCl accumulates in the interstitial fluid of the renal medulla.

2. The resulting higher osmolarity of the interstitial fluid causes osmotic reabsorption of water from the thin descending limb of Henle, thus concentrating the tubule fluid entering the ascending limb.

3. Tubule fluid composition is finely controlled by secretion and absorption of solutes in the distal convoluted tubule.

4. Fluid flowing down the collecting duct is less concentrated than the surrounding interstitial fluid, so water is withdrawn osmotically.

5. Water reabsorbed from the descending limb of Henle and the collecting duct leaves the medulla via the vasa recta.

6. The lower portion of the collecting duct is permeable to urea, which diffuses into the interstitial fluid and is recycled, thus adding to the concentration gradient in the medulla.

tween the blood in the vasa recta and the interstitial fluids result in little net change in the composition of the interstitial fluid in the medulla.

Water permeability of kidney tubules depends on water channels

We have noted that some tubule regions, such as the PCT, are highly permeable to water while others, such as the thick ascending limb of the loop of Henle, are impermeable to water. How do differences in water permeability in different regions of the nephron arise? Regions of the nephron that are highly permeable to water have greater numbers of *aquaporins*, a class of membrane proteins that form water channels (see Section 6.3). Aquaporins are abundant in kidney PCT cells and in descending limbs of the loops of Henle, but not in the ascending limbs of the loop of Henle. The discovery of aquaporins resulted in Peter Agre of Johns Hopkins University receiving the Nobel Prize in Chemistry in 2003.

As an interesting evolutionary note, aquaporins are also important in maintaining water balance in amphibians. When not in an aqueous environment, many amphibians can gain water from a moist substrate because they have aquaporins in the epithelial cells of their belly skin. Thus, water can cross their skin into the interstitial fluid by osmosis.

The distal convoluted tubule fine-tunes the composition of the urine

The first portion of the distal convoluted tubule is similar to the thick ascending limb of the loop of Henle. Na+ and Cl− are transported out of the tubule fluid, and water cannot follow. As a result, the tubule fluid becomes even more dilute. The later sections of the distal convoluted tubule, however, can be permeable to water, and water can be osmotically drawn from the tubule into the interstitial fluid. As the tubule fluid flows from the distal tubule to the collecting duct, it can be below or equal to the osmolarity of the blood plasma.

An important function of the distal tubule is the fine-tuning of the ionic composition of the urine. Even though bulk reabsorption of substances such as calcium, phosphate, bicarbonate, and potassium occurs in the proximal convoluted tubule, changes in the concentrations of these substances occur in the distal convoluted tubule. In the case of potassium, for example, if a person is potassium depleted, this ion is reabsorbed in the distal convoluted tubule, but if a person has an abundance of potassium, this ion is secreted in the distal convoluted tubule. As we will see below, this exchange of K+ is controlled by the hormone *aldosterone*. Another example is reabsorption of Ca²⁺ in the distal convoluted tubule, which is controlled by the actions of vitamin D. The fine-tuning of urine composition continues in the collecting duct. As you can imagine, the list of ion transporters in the distal convoluted tubule is large.

Urine is concentrated in the collecting duct

The tubule fluid entering the collecting duct is at about the same solute *concentration* as the blood plasma, but its solute *composition* is considerably different from that of the plasma. The major solute in the tubular fluid is now urea, since salts were reabsorbed earlier in the nephron. As the tubule fluid flows down the collecting duct, it loses water osmotically to the interstitial fluid, and that water returns to the circulatory system via the vasa recta (see Figure 52.10).

The concentration gradient established in the renal medulla by the countercurrent multiplier actions of the loops of Henle creates the osmotic potential that withdraws water from the collecting ducts. The collecting ducts begin in the renal cortex and run through the renal medulla before emptying into the ureter at the tips of the renal pyramids. During this journey, the solute concentration of the surrounding interstitial fluid increases, and more and more water can be absorbed from the urine in the collecting duct. By the time it reaches the ureter, the urine can become greatly concentrated, with urea as a major solute.

As water is withdrawn from the collecting duct, some urea also leaks out into the medullary interstitial fluid, adding to its osmotic potential. This urea diffuses back into the loop of Henle and is returned to the collecting duct. The recycling of urea in the renal medulla contributes significantly to the concentration gradient and therefore the ability of the kidney to concentrate the urine in the collecting duct.

Overall, the ability of a mammal to concentrate its urine is determined by the maximum concentration gradient it can es-

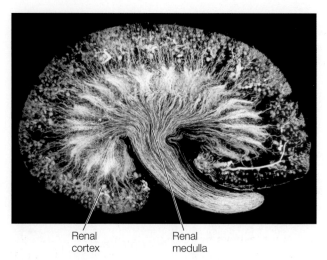

Renal   Renal
cortex   medulla

52.11 The Ability to Concentrate   The ability of the mammalian kidney to concentrate urine depends on the lengths of its loops of Henle relative to the overall size of the kidney. The kidney of a desert gerbil has a single renal pyramid with loops of Henle so long that the pyramid extends far into the ureter (ureter not shown).

tablish in its renal medulla. An important adaptation for increasing the concentration gradient is to increase the lengths of the loops of Henle relative to overall kidney size. The small desert gerbil, for example, has such extremely long loops of Henle that its renal pyramid (each of its kidneys has only one, in contrast to ours) extends far out of the concave surface of the kidney and into the ureter (**Figure 52.11**). These animals are so effective in conserving water that they can survive just on the water released by the metabolism of their food.

The kidneys help regulate acid–base balance

Besides regulating salt and water balance and excreting nitrogenous wastes, the kidneys have another important role: they regulate the hydrogen ion concentration (the pH) of the extracellular fluids. pH is a critical variable because it influences the structure and function of proteins.

One way to minimize pH changes in a chemical solution is to add a *buffer*—a substance that can either absorb or release hydrogen ions (see Section 2.4). The major buffers in the blood are bicarbonate ions (HCO_3^-; see Figure 49.14) that are formed from the dissociation of carbonic acid, which in turn is formed by the hydration of CO_2 according to the following equilibrium reaction:

$$CO_2 + H_2O \rightleftharpoons H_2CO_3 \rightleftharpoons H^+ + HCO_3^-$$

From this equation, you can see that if excess hydrogen ions are added to this reaction mixture, the reaction will move to the left and absorb the excess H+. If hydrogen ions are removed from the reaction mixture, however, the reaction will move to the right and supply more H+.

The HCO_3^- buffer system is important for controlling the pH of the blood, and therefore of the interstitial fluids as well, because the reaction can be pushed to the right and pulled to the left physiologically. The lungs control the levels of CO_2 in the blood, thus altering the acid portion of the reaction. CO_2 is considered the acid portion of the reaction because if you add additional CO_2, the reaction shifts to the right, producing more H^+ ions. The kidneys control the base portion of the reaction by removing H^+ from the blood and returning HCO_3^- to the blood. How are H^+ ions removed from the blood?

One mechanism for H^+ secretion and HCO_3^- reabsorption involves ammonia (NH_3) and ammonium ion (NH_4^+). The metabolism of glutamine in tubule cells produces NH_3 and HCO_3^-. The HCO_3^- is reabsorbed into the interstitial fluid. The NH_3 is transported into the tubule fluid by means of an NH_3 transporter that has been characterized only recently in an effort to identify novel proteins coming from the sequencing of the human and other genomes (**Figure 52.12**).

Another mechanism for H^+ secretion and HCO_3^- reabsorption is shown in **Figure 52.13**. The H^+ is transported into the tubule fluid in exchange for Na^+. The H^+ combines with HCO_3^- that has been filtered in the glomerulus, producing H_2CO_3 that

disassociates into H_2O and CO_2. The CO_2 diffuses into the tubule cell where in the presence of the enzyme carbonic anhydrase it produces HCO_3^- that is transported into the interstitial fluid and thence to the blood.

Kidney failure is treated with dialysis

Loss of kidney function, or *renal failure*, results in the retention of salts and water (hence high blood pressure), retention of urea (uremic poisoning), and a decreasing pH (acidosis). A person who suffers complete renal failure will die within 2 weeks if not treated. A drastic but highly successful treatment is kidney transplant, but it is usually necessary to sustain a patient for considerable time while waiting for a kidney to become available. Therefore, artificial kidneys, or renal dialysis machines, are essential modes of treatment.

In a dialysis machine, the patient's blood flows through many small channels made of semipermeable membranes

INVESTIGATING LIFE

52.12 An Ammonium Transporter in the Renal Tubules?

An important way the kidney excretes hydrogen ions and buffers the blood is to secrete ammonia (NH_3) into the renal tubules. It was thought that ammonia simply diffused into the tubules until the function of Rhcg, a protein in the Rh blood antigen family, was discovered. This experiment demonstrated that loss of the *Rhcg* gene in mice impairs their ability to buffer their blood pH by excreting excess H^+ (in the form of ammonium).

HYPOTHESIS The protein Rhcg is an ammonia transporter and is critical for the kidney's role in acid-base balance.

METHOD
1. Create a line of mice in which the gene for the protein Rhcg is knocked out (see Section 18.4).
2. Measure starting blood pH, plasma bicarbonate (HCO_3^-) levels, and urine ammonium (NH_4^+) levels in experimental and control (wild-type) mice.
3. For 6 days, administer drinking water containing NH_4Cl (ammonium chloride, a mild acid) to control mice and to Rhcg knockout mice.
4. Measure the three variables (see item 2 above) at days 2 and 6.

RESULTS

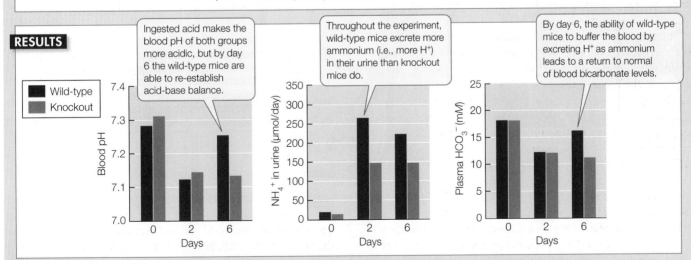

Ingested acid makes the blood pH of both groups more acidic, but by day 6 the wild-type mice are able to re-establish acid-base balance.

Throughout the experiment, wild-type mice excrete more ammonium (i.e., more H^+) in their urine than knockout mice do.

By day 6, the ability of wild-type mice to buffer the blood by excreting H^+ as ammonium leads to a return to normal of blood bicarbonate levels.

CONCLUSION Lack of functional Rhcg protein impairs a mouse's ability to secrete ammonium ions in its urine, thus limiting the capacity to regulate acid-base balance. This protein is probably an ammonia transporter in the renal tubules.

Go to **yourBioPortal.com** for original citations, discussions, and relevant links for all INVESTIGATING LIFE figures.

52.13 The Kidney Excretes Acids and Conserves Bases

Bicarbonate ions are filtered out of the blood at the glomerulus, and renal tubule cells secrete hydrogen ions into the tubule fluid. In the renal tubule, the filtered bicarbonate buffers the secreted hydrogen ions and keeps the urine from becoming too acidic. The CO_2 formed by the reaction of bicarbonate and hydrogen ions is converted back to bicarbonate by the renal tubule cells and transported back into the interstitial fluid.

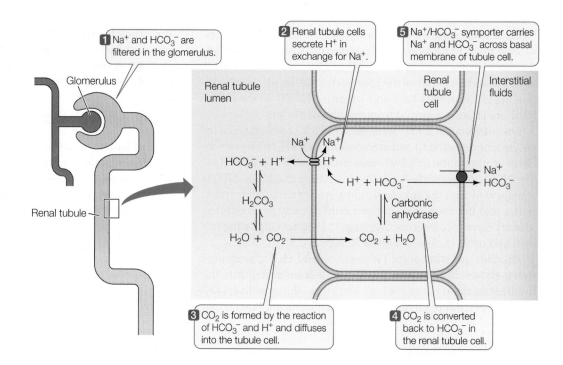

1 Na^+ and HCO_3^- are filtered in the glomerulus.

2 Renal tubule cells secrete H^+ in exchange for Na^+.

5 Na^+/HCO_3^- symporter carries Na^+ and HCO_3^- across basal membrane of tubule cell.

3 CO_2 is formed by the reaction of HCO_3^- and H^+ and diffuses into the tubule cell.

4 CO_2 is converted back to HCO_3^- in the renal tubule cell.

(**Figure 52.14**). A dialysis solution flows on the other side of these membranes, through which small molecules can diffuse. Molecules and ions diffuse from an area of higher concentration to an area of lower concentration, so the composition of the dialysis fluid is crucial. The concentrations of the molecules or ions that need to be conserved must be at the same concentra-

tion in the dialysis fluid as they are in the blood. The concentrations of molecules and ions that need to be removed from the blood are zero in the dialysis fluid. The total osmotic potential of the dialysis fluid must equal that of the plasma.

About 500 ml of the patient's blood is in the dialysis machine at any one time, and the unit processes several hundred milli-

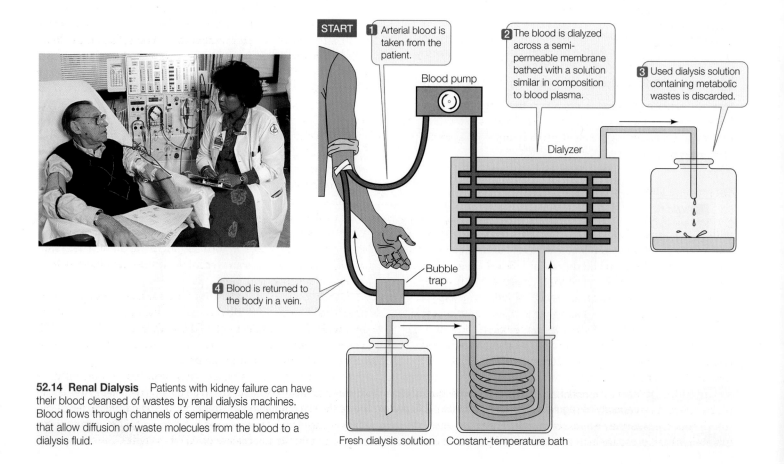

52.14 Renal Dialysis   Patients with kidney failure can have their blood cleansed of wastes by renal dialysis machines. Blood flows through channels of semipermeable membranes that allow diffusion of waste molecules from the blood to a dialysis fluid.

liters of blood per minute. A patient with no kidney function must be on the dialysis machine for 4 to 6 hours three times a week.

52.5 RECAP

The anatomical organization of the nephrons makes it possible for the mammalian kidney to produce a urine more concentrated than the blood, thereby conserving water to maintain extracellular fluid volume. Bulk reabsorption of salts, other valuable solutes, and water takes place in the proximal convoluted tubule. The loops of Henle act as a countercurrent multiplier, creating a concentration gradient of the interstitial fluids in the renal medulla. Collecting ducts run through the renal medulla and lose water osmotically to the surrounding interstitial fluids, concentrating the urine.

- Explain how the countercurrent multiplier mechanism of the nephron makes it possible for the kidney to form a concentrated urine. See p. 1102 and Figure 52.10

- How does the kidney contribute to acid–base balance? See pp. 1104–1105 and Figure 52.13

The kidney contributes to homeostasis in several ways, including regulating extracellular fluid volume, maintaining the osmotic concentration and ionic composition of the extracellular fluid, and regulating pH. As we will see next, the kidneys also play a major role in regulating blood pressure.

52.6 How Are Kidney Functions Regulated?

Several regulatory mechanisms act on the kidneys to maintain blood pressure, blood osmolarity, and blood composition. We will discuss these mechanisms separately, but keep in mind that they are always working together.

Glomerular filtration rate is regulated

If the kidneys stop filtering blood, they cannot accomplish any of their functions. The maintenance of a constant **glomerular filtration rate (GFR)** depends on an adequate blood supply to the kidneys at an adequate blood pressure. Renal arteries usually deliver blood to the kidneys at high pressure because they are early branches off the aorta. In addition, *autoregulatory mechanisms* ensure adequate blood supply and blood pressure for kidney function regardless of what is happening elsewhere in the body. The kidney's autoregulatory adjustments compensate for decreases in cardiac output or decreases in blood pressure so that the GFR remains constant.

One autoregulatory mechanism is the dilation (expansion) of the afferent renal arterioles when blood pressure falls. This dilation decreases the resistance in the arterioles and helps maintain blood pressure in the glomerulus. If arteriole dilation

does not keep the GFR from falling, the kidney releases an enzyme, **renin**, into the blood. Renin converts a circulating protein, angiotensinogen, into angiotensin I, which is then acted on by angiotensin converting enzyme (ACE) to form the active hormone angiotensin II, or simply, **angiotensin** (**Figure 52.15**). Angiotensin has several effects that help restore the GFR to normal:

- It constricts the efferent renal arterioles, raising the resistance for blood leaving the glomerulus. Like putting a finger over the end of a garden hose, this restriction of drainage elevates blood pressure in the glomerular capillaries.

- It constricts peripheral blood vessels all over the body, an action that elevates central blood pressure.

- It stimulates the adrenal cortex to release the hormone **aldosterone**. Aldosterone stimulates sodium reabsorption by the kidney, thereby making its reabsorption of water more effective. Enhanced water reabsorption helps maintain blood volume and therefore central blood pressure.

- It acts on the brain to stimulate thirst. Increased water intake in response to thirst increases blood volume and blood pressure.

Thus the renin-angiotensin-aldosterone system, or RAAS, coordinates many responses to maintain blood pressure and kidney function.

Blood osmolarity and blood pressure are regulated by ADH

Cells in the hypothalamus can stimulate the release from the posterior pituitary of a hormone called *antidiuretic hormone* (*ADH*, also called *vasopressin*) that can act on cells of the collecting duct to insert aquaporins (water channels) into their plasma membranes. The aquaporins increase the permeability of these membranes to water, and therefore more water is reabsorbed from the collecting duct fluid into the interstitial spaces of the renal medulla. The higher the circulating levels of ADH, the greater the number of aquaporins. Various factors can stimulate or inhibit the release of ADH. Of key importance to kidney function are osmoreceptors that monitor blood osmolarity and stretch receptors that monitor blood pressure (**Figure 52.16**).

Osmoreceptor neurons in the hypothalamus are activated by a rise in blood osmolarity, and they increase the release of ADH. ADH helps regulate blood osmolarity by controlling water reabsorption. The osmoreceptors also stimulate thirst. The resulting water retention and water intake dilute the blood as they expand blood volume.

Stretch receptors in the walls of the aorta and the carotid arteries (see Figure 50.17) that detect an increase in blood pressure will *inhibit* the release of ADH. With less circulating ADH, less water is reabsorbed, which decreases blood volume and hence acts to lower blood pressure.

If blood pressure falls, as when you lose blood volume through hemorrhage or excessive evaporative water loss, activity of the stretch receptors in the aorta and carotid arteries decreases. Input via cranial nerves to the hypothalamus from these receptors inhibits the release of ADH, so when the firing rates

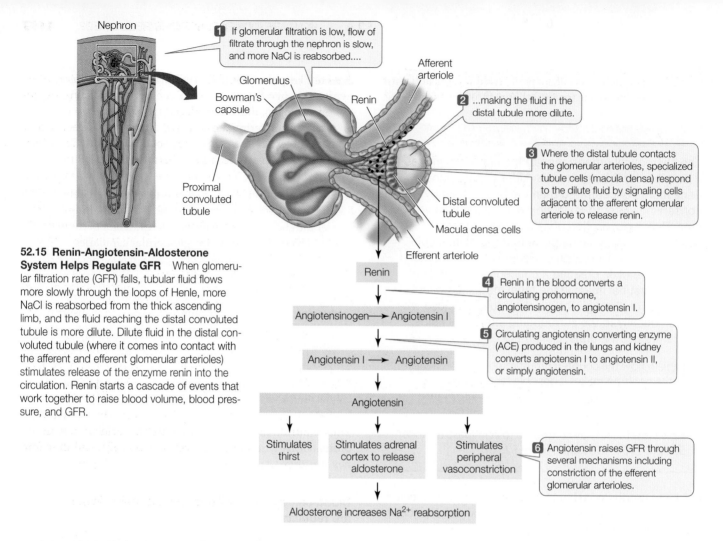

Nephron

1 If glomerular filtration is low, flow of filtrate through the nephron is slow, and more NaCl is reabsorbed....

Glomerulus

Bowman's capsule

Afferent arteriole

Renin

2 ...making the fluid in the distal tubule more dilute.

3 Where the distal tubule contacts the glomerular arterioles, specialized tubule cells (macula densa) respond to the dilute fluid by signaling cells adjacent to the afferent glomerular arteriole to release renin.

Proximal convoluted tubule

Distal convoluted tubule

Macula densa cells

Efferent arteriole

52.15 Renin-Angiotensin-Aldosterone System Helps Regulate GFR When glomerular filtration rate (GFR) falls, tubular fluid flows more slowly through the loops of Henle, more NaCl is reabsorbed from the thick ascending limb, and the fluid reaching the distal convoluted tubule is more dilute. Dilute fluid in the distal convoluted tubule (where it comes into contact with the afferent and efferent glomerular arterioles) stimulates release of the enzyme renin into the circulation. Renin starts a cascade of events that work together to raise blood volume, blood pressure, and GFR.

Renin

4 Renin in the blood converts a circulating prohormone, angiotensinogen, to angiotensin I.

Angiotensinogen ⟶ Angiotensin I

5 Circulating angiotensin converting enzyme (ACE) produced in the lungs and kidney converts angiotensin I to angiotensin II, or simply angiotensin.

Angiotensin I ⟶ Angiotensin

Angiotensin

Stimulates thirst

Stimulates adrenal cortex to release aldosterone

Stimulates peripheral vasoconstriction

6 Angiotensin raises GFR through several mechanisms including constriction of the efferent glomerular arterioles.

Aldosterone increases Na^{2+} reabsorption

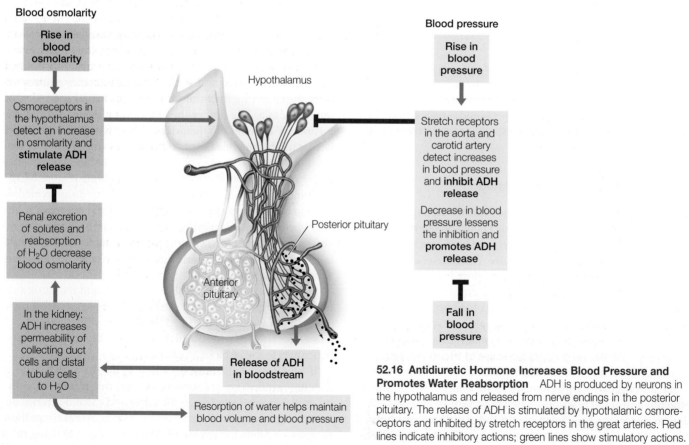

Blood osmolarity

Rise in blood osmolarity

Hypothalamus

Blood pressure

Rise in blood pressure

Osmoreceptors in the hypothalamus detect an increase in osmolarity and **stimulate ADH release**

Stretch receptors in the aorta and carotid artery detect increases in blood pressure and **inhibit ADH release**

Renal excretion of solutes and reabsorption of H_2O decrease blood osmolarity

Decrease in blood pressure lessens the inhibition and **promotes ADH release**

Posterior pituitary

In the kidney: ADH increases permeability of collecting duct cells and distal tubule cells to H_2O

Anterior pituitary

Release of ADH in bloodstream

Fall in blood pressure

Resorption of water helps maintain blood volume and blood pressure

52.16 Antidiuretic Hormone Increases Blood Pressure and Promotes Water Reabsorption ADH is produced by neurons in the hypothalamus and released from nerve endings in the posterior pituitary. The release of ADH is stimulated by hypothalamic osmoreceptors and inhibited by stretch receptors in the great arteries. Red lines indicate inhibitory actions; green lines show stimulatory actions.

INVESTIGATING LIFE

52.17 ADH Induces Insertion of Aquaporins into Plasma Membranes

Aquaporin proteins make different regions of renal tubules permeable to water. One of these, AQP-2, is responsible for the permeability of the collecting duct cells. How does antidiuretic hormone act on these proteins to control the level of permeability in renal cells?

HYPOTHESIS Antidiuretic hormone (ADH) controls the location of aquaporin proteins.

METHOD

1. Isolate collecting ducts from rat kidney.
2. Use immunochemical staining to localize the AQP-2 aquaporins in collecting duct cells both with and without the presence of ADH. Also localize the aquaporins after ADH is applied and then washed away.

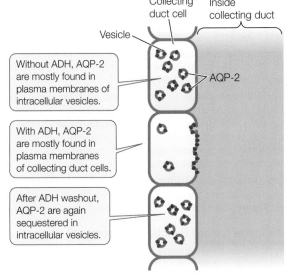

Without ADH, AQP-2 are mostly found in plasma membranes of intracellular vesicles.

With ADH, AQP-2 are mostly found in plasma membranes of collecting duct cells.

After ADH washout, AQP-2 are again sequestered in intracellular vesicles.

3. Perfuse collecting ducts and measure water permeability under the same three conditions.

RESULTS

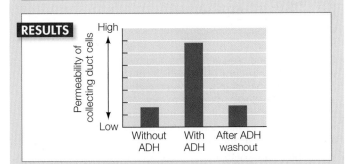

CONCLUSION In the absence of ADH, AQP-2 is sequestered intracellularly. When ADH is present, these water channels are inserted into the plasma membranes, making the cells more permeable to water.

Go to **yourBioPortal.com** for original citations, discussions, and relevant links for all INVESTIGATING LIFE figures.

of these stretch receptors fall, ADH release increases. More ADH results in more efficient water reabsorption and therefore a protection of blood volume and blood pressure.

Alcohol inhibits ADH release, explaining why excessive beer drinking leads to excessive urination and dehydration, which contributes to the symptoms of a hangover.

As mentioned earlier, the presence of aquaporins in plasma membranes determines their water permeability. Aquaporins play an important and unique role in the collecting duct. Several members of the aquaporin family of water channels are found in the plasma membranes of cells in the distal tubules. At least two aquaporins are localized in the basolateral membranes (facing the blood vessels). A different aquaporin, called AQP-2, is found in the apical plasma membranes (facing into the tubule). But the presence of AQP-2 in these membranes is controlled by ADH, as described in **Figure 52.17**.

Sometimes AQP-2 is sequestered in the membranes of intracellular vesicles, and when it is, the collecting duct permeability is low. ADH causes these vesicles to insert their AQP-2 channels into the apical plasma membrane, making that membrane (and therefore the cell) permeable to water. ADH also stimulates the synthesis of new AQP-2 proteins. Thus, circulating ADH controls the number of AQP-2 water channels in the plasma membranes of the collecting duct cells, and therefore their water permeability.

The heart produces a hormone that helps lower blood pressure

You may not think of the heart as an endocrine organ, but it is. When blood volume is high, blood pressure is high, and that puts strain on the heart. Under these conditions, the increased venous return stretches the atria of the heart. When the atrial muscle fibers are overly stretched, they release a peptide hormone called **atrial natriuretic peptide** (**ANP**). This peptide hormone enters the circulation, and in the kidney it decreases the reabsorption of sodium. If less sodium is reabsorbed, less water is reabsorbed, and more passes into the urine. Thus, ANP has the effect of lowering blood volume and therefore blood pressure.

52.6 RECAP

Glomerular filtration is essential for kidney function and is sustained by autoregulatory mechanisms. Sensors that monitor blood pressure and blood osmolarity may stimulate or inhibit the release of hormones that regulate kidney function.

- Explain how falling GFR results in an increase in circulating angiotensin and how angiotensin restores GFR. See p. 1107

- Explain how falling blood pressure or increasing blood osmolarity result in changes in permeability of the collecting ducts. See Figure 52.16

CHAPTER SUMMARY

52.1 How Do Excretory Systems Maintain Homeostasis?

- **Excretory systems** maintain the osmolarity and volume of the extracellular fluids and eliminate the waste products of nitrogen metabolism through the processes of filtration, reabsorption, and secretion. **Urine** is the output of excretory systems.

- There is no active transport of water, so water must be moved across membranes by a difference in either **osmolarity** or pressure.

- Water enters or leaves cells by osmosis. To achieve cellular water balance, animals must maintain the osmolarity of their extracellular fluids within an acceptable range.

- Marine animals can be **osmoconformers** or **osmoregulators**. Freshwater animals must be osmoregulators and must continually excrete water and conserve salts. Most animals are ionic regulators to some degree.

- Apart from regulating osmolarity of cells and extracellular fluids, most animals must also regulate their ionic composition by conserving some ions and secreting others. Salt glands are adaptations for secretion of NaCl. Review Figure 52.2

52.2 How Do Animals Excrete Nitrogen?

- Aquatic animals that breathe water can eliminate nitrogenous wastes such as **ammonia** by diffusion across their gill membranes. Terrestrial animals and some aquatic animals must detoxify ammonia by converting it to **urea** or **uric acid** before excretion. Review Figure 52.3

- Depending on the form in which they excrete their nitrogenous wastes, animals are classified as **ammonotelic**, **ureotelic**, or **uricotelic**.

52.3 How Do Invertebrate Excretory Systems Work?

- The **protonephridia** of flatworms consist of flame cells and excretory tubules. Extracellular fluid is filtered into the tubules, which process the filtrate to produce a dilute urine. Review Figure 52.4

- In annelid worms, blood pressure causes filtration of the blood across capillary walls. The filtrate enters the coelomic cavity, where it is taken up by **metanephridia**, which alter the composition of the filtrate by active transport mechanisms. Review Figure 52.5, WEB ACTIVITY 52.1

- The **Malpighian tubules** of insects receive ions and nitrogenous wastes by active transport across the tubule cells. Water follows by osmosis. Ions and water are reabsorbed from the rectum, so the insect excretes semisolid wastes. Review Figure 52.6

52.4 How Do Vertebrates Maintain Salt and Water Balance?

- Marine and terrestrial animals conserve water in various ways. Marine bony fishes produce little urine. Cartilaginous fishes retain urea so that the osmolarity of their body fluids remains close to that of seawater. Amphibians remain close to water. Reptiles have skin with low water permeability, lay shelled eggs, and excrete nitrogenous wastes as uric acid.

- Mammals produce urine more concentrated than their extracellular fluids.

- The **nephron**, the functional unit of the vertebrate **kidney**, consists of a **glomerulus**, in which blood is filtered, a **renal tubule**, which processes the filtrate into urine to be excreted, and a system of **peritubular capillaries**, which surround the tubule and serve as a site of secretion and reabsorption. Review Figure 52.7, WEB ACTIVITY 52.2

52.5 How Does the Mammalian Kidney Produce Concentrated Urine?

- The concentrating ability of the mammalian kidney is a function of its anatomy, which is ideal for countercurrent exchange. Review Figure 52.9, ANIMATED TUTORIAL 52.1

- The glomeruli and the **proximal** and **distal convoluted tubules** are located in the **cortex** of the kidney. Certain molecules are actively reabsorbed from the glomerular filtrate by the tubule cells, and other molecules are actively secreted. Straight sections of renal tubules called **loops of Henle** and **collecting ducts** are arranged in parallel in the **medulla** of the kidney. SEE WEB ACTIVITY 52.3

- Salts and water are reabsorbed in the proximal convoluted tubule without the renal filtrate becoming more concentrated, although its composition changes.

- The loops of Henle create a concentration gradient in the interstitial fluid of the renal medulla by a **countercurrent multiplier** mechanism. Urine flowing down the collecting ducts to the **ureter** is concentrated by the osmotic reabsorption of water caused by the concentration gradient in the surrounding interstitial fluid. Review Figure 52.10

- Hydrogen ions secreted by the renal tubules are buffered in the urine by bicarbonate and other chemical buffering systems. Review Figures 52.12 and 52.13

52.6 How Are Kidney Functions Regulated?

- Kidney function in mammals is controlled by autoregulatory mechanisms that maintain a constant high **glomerular filtration rate (GFR)** even if blood pressure varies.

- An important autoregulatory mechanism is the release of **renin** by the kidney when blood pressure falls. Renin activates **angiotensin**, which causes the constriction of efferent glomerular arterioles and peripheral blood vessels, causes the release of **aldosterone** (which enhances water reabsorption), and stimulates thirst.

- Changes in blood pressure and osmolarity influence the release of antidiuretic hormone (ADH), which controls the permeability of the collecting duct to water and therefore the amount of water that is reabsorbed from the urine. ADH stimulates the expression of and controls the intracellular location of aquaporins, which serve as water channels in the membranes of collecting duct cells. Review Figures 52.16 and 52.17

- When the volume of blood returning to the heart increases and stretches the atrial walls, **atrial natriuretic peptide (ANP)** is released, which causes increased excretion of salt and water.

SEE WEB ACTIVITY 52.4 for a review of the major human organ systems.

SELF-QUIZ

1. Which statement is *true*?
   a. Most marine invertebrates are osmoregulators.
   b. All freshwater invertebrates are osmoconformers.
   c. Marine bony and cartilaginous fishes have similar interstitial osmolarities.
   d. Freshwater fishes are ionic regulators.
   e. Marine mammals gain water osmotically.

2. The excretion of nitrogenous wastes
   a. by humans can be in the form of urea and uric acid.
   b. by mammals is never in the form of uric acid.
   c. by marine fishes is mostly in the form of urea.
   d. does not contribute to the osmolarity of the urine.
   e. requires more water if the waste product is the rather insoluble uric acid.

3. How are earthworm metanephridia like mammalian nephrons?
   a. Both process coelomic fluid.
   b. Both take in fluid through a ciliated opening.
   c. Both produce urine more concentrated than the blood.
   d. Both employ tubular secretion and reabsorption to control urine composition.
   e. Both involve a countercurrent multiplier effect.

4. What is the role of renal podocytes?
   a. They prevent red blood cells and large molecules from entering the renal tubules.
   b. They reabsorb most of the glucose that is filtered from the plasma.
   c. They control the glomerular filtration rate by changing the resistance of renal arterioles.
   d. They provide a large surface area for tubular secretion and reabsorption.
   e. They release renin when the glomerular filtration rate falls.

5. Which of the following are *not* found in a renal pyramid?
   a. Collecting ducts
   b. Vasa recta
   c. Peritubular capillaries
   d. Convoluted tubules
   e. Loops of Henle

6. Which part of the nephron is responsible for most of the difference in mammals between the glomerular filtration rate and the urine production rate?
   a. The glomerulus
   b. The proximal convoluted tubule
   c. The loops of Henle
   d. The distal convoluted tubule
   e. The collecting duct

7. For mammals of the same size, what feature of their excretory systems would give them the greatest ability to concentrate their urine?
   a. Higher glomerular filtration rate
   b. Longer convoluted tubules
   c. Increased number of nephrons
   d. More permeable collecting ducts
   e. Longer loops of Henle

8. Which of the following would *not* be a response stimulated by a large drop in blood pressure?
   a. Constriction of afferent renal arterioles
   b. Increased release of renin
   c. Increased release of antidiuretic hormone
   d. Increased thirst
   e. Constriction of efferent renal arterioles

9. Which statement about angiotensin is *true*?
   a. It is secreted by the kidney when the glomerular filtration rate falls.
   b. It is released by the posterior pituitary when blood pressure falls.
   c. It stimulates thirst.
   d. It increases the permeability of the collecting ducts to water.
   e. It decreases glomerular filtration rate when blood pressure rises.

10. Birds that feed on marine animals ingest a lot of salt, but they excrete most of it by means of
    a. Malpighian tubules.
    b. rectal salt glands.
    c. gill membranes.
    d. concentrated urine.
    e. nasal salt glands.

FOR DISCUSSION

1. Why is it said that the oceans are a physiological desert? For what animals would this apply?

2. Persons with uncontrolled diabetes mellitus can have very high levels of glucose in their blood. Why do such individuals have a high level of urine production?

3. Inulin is a molecule that is filtered out of the glomerulus but is not secreted or reabsorbed by the renal tubules. If you injected inulin into an animal and after a brief time measured the concentration of inulin in the animal's blood and urine, how could you determine the animal's glomerular filtration rate? Assume that the rate of urine production is 1 milliliter per minute.

4. After you did the inulin experiment to measure glomerular filtration rate, how could you use that information to determine whether another substance is secreted or reabsorbed by the renal tubules? Assume you can measure the concentration of that substance in the blood and urine. Urine production is still 1 milliliter per minute.

5. Explain what happens with respect to control and regulation of your salt and water balance when you eat a lot of very salty popcorn.

6. Review Figure 52.2, noting that the venules and the secretory tubules are arranged in parallel with blood flowing in one direction and the contents of the tubule flowing in the opposite direction. What is the advantage of this arrangement?

ADDITIONAL INVESTIGATION

We mention on p. 1107 that the glomerular filtration rate (GFR) is autoregulated, and this regulation is achieved by control over constriction/dilation of the afferent and efferent renal arterioles. What information could be used in this process? A clue comes from the anatomy of the nephron. When the ascending limb of Henle reaches the cortex, it makes direct contact with the affer-

ent and efferent arterioles of the glomerulus that produced the filtrate flowing in that particular nephron. What aspect of the tubular fluid in the early distal convoluted tubule would reflect changes in the volume of filtrate entering the nephron? How might the cells of the early distal convoluted tubule communicate with the smooth muscle cells of the arterioles?

WORKING WITH DATA (GO TO yourBioPortal.com)

What Kidney Characteristic Determines Urine Concentrating Ability?   Since the loops of Henle create the concentration gradient in the renal medulla, it is reasonable to expect that longer loops would make it possible to generate greater concentration gradients and therefore increase the ability to concentrate urine. However, mammals come in many different

sizes and so do their kidneys, so would you expect all big mammals to be able to concentrate their urine more than small mammals? In this exercise, we will explore anatomical data and urine concentration data from many species to see what characteristic of their kidneys relates to their urine concentrating abilities.

Monkey see, monkey do

Many years ago, scientists studying a troop of macaques (*Macaca fuscata*) on the Japanese island of Koshima lured the monkeys into the open by throwing pieces of sweet potato onto the beach from a passing boat. The monkeys tried to brush the sand off the potatoes, but the pieces remained gritty. One day, a young female named Imo took her sweet potatoes to the water and washed off the sand. Soon her siblings and other juveniles in her playgroup were imitating Imo's new behavior. Then their mothers began washing their potatoes. None of the adult males imitated this behavior, but young males learned the behavior from their mothers and their siblings and continued to wash their potatoes as they grew older.

The scientists were fascinated by the way Imo's creative, insightful behavior spread throughout the population, so they presented the monkeys with a new challenge: they threw wheat onto the beach. Picking wheat grains out of the sand was tedious and difficult. Imo, apparently a prodigy in the macaque world, came up with a solution: she carried handfuls of sand and wheat to the water and threw them in. The sand sank, but the wheat floated, enabling her to skim the grain off the surface and eat it. This efficient feeding behavior spread throughout the population just as washing sweet potatoes had—first to other juveniles, then to mothers, and then from mothers to their male and female offspring.

The macaques of that troop now routinely wash their food. They also play in the ocean, which they never did before, and they have added marine food items to their diet. Clearly, this population of monkeys has a *culture*: a set of invented behaviors shared by members of the population and transmitted through social learning.

Animals also display elaborate behaviors they do not learn. Web spinning by spiders, for example, requires no practice or prior experience. In many spider species, females die immediately after they lay their eggs; the hatching spiderlings never see their mother, her web, or webs spun by other spiders. Yet when they first construct their own webs,

Shared Learned Behaviors Become a Culture
In the space of a single generation, a population of Japanese macaques (*Macaca fuscata*) learned and transmitted a set of behaviors that included washing food, playing in the water, and eating marine food items—a new "culture" of water-related behaviors.

Spiders Are Born Web Designers Each spider performs a stereotypic sequence of movements that results in a species-specific web design. Spiders such as this banded garden spider (*Argiope trifasciata*) are born with this ability, with no need to learn from experience or to model their movements after those of a parent.

they do it perfectly, without the benefit of experience or a model to copy.

Web spinning requires thousands of movements performed in just the right sequence. Most of that sequence is *stereotypic*—performed the same way every time. Different spider species spin webs of different designs, using different sequences of movements. Yet every spider knows how to spin its species-specific web at birth; the behavior is part of its genetic inheritance.

Most animal behavior stems from an interaction of genetic inheritance and learning, and modern studies of animal behavior focus on the mechanisms of both. The evolution of behavior is also a topic of great interest, and research on that topic asks how particular behaviors are adaptive in specific environments. In short, behavioral biology is a highly integrative field that incorporates approaches from virtually all of the biological subdisciplines you have studied in this book.

IN THIS CHAPTER we will describe how the modern study of animal behavior emerged from earlier research focused on learning and on inherited behavior. We will give examples of how methods from genetics and molecular genetics, physiology, neurobiology, ecology, and evolution have contributed to our understanding of the behavior of animals—including ourselves.

53.1 What Are the Origins of Behavioral Biology?

Humans have studied animal behavior since prehistoric times. Understanding the habits of potential prey, as well as those of their predators, was of great value to hunters. Appreciation of behavioral traits led to the domestication of animal species. Accounts of dramatic animal behaviors such as seasonal appearances and disappearances, mating displays, aggression, prey capture, and even less dramatic behaviors such as parental care, communication, and sleep are found throughout recorded history. Yet the scientific study of animal behavior did not truly get under way until the early 1900s.

Conditioning was the focus of behaviorists

The discovery of the *conditioned reflex*, a simple form of learning, by the Russian physiologist Ivan Pavlov in the late 1800s was a seminal event. Pavlov was studying the neural control of digestive juice secretion, and he observed that a dog salivated when it smelled food. He also noticed that the dog salivated when the technician who normally fed the dog entered the room even when no food was present. Following up this observation, he substituted a sound stimulus for the technician. When the dog was fed, a metronome ticked. After a number of trials, the dog salivated when it heard the metronome even if no food was offered. Pavlov showed that other sounds worked as well (including a bell, which is commonly associated with his work).

Salivation in response to the sight, smell, or taste of food is a natural reflex response to a stimulus, but salivation in response to a sound was a learned response. The pairing of a sound with the experience of receiving food *conditioned* the dog's nervous system to generate a response, which Pavlov dubbed the **conditioned reflex**. The food was referred to as the *unconditioned stimulus* (US), and the sound as the *conditioned stimulus* (CS) (**Figure 53.1**). Thus Pavlov showed that a simple behavior controlled by the nervous system could be modified through experience.

Pavlov received a Nobel prize in 1904, and his work stimulated much new research because he had developed an experimental model of learning. All sorts of changes in the relationship between a CS and a US were investigated. A particularly interesting variation was developed by the psychologist B. F. Skinner, who showed that any random action of an animal could become a conditioned response to a stimulus if a reward was temporally associated with the action and the stimulus. For example, a rat could be conditioned to press a lever in response to a stimulus if it got a reward when it behaved as the experi-

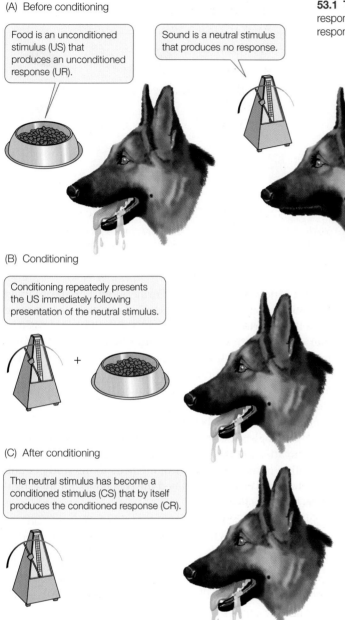

(A) Before conditioning

Food is an unconditioned stimulus (US) that produces an unconditioned response (UR).

Sound is a neutral stimulus that produces no response.

(B) Conditioning

Conditioning repeatedly presents the US immediately following presentation of the neutral stimulus.

+

(C) After conditioning

The neutral stimulus has become a conditioned stimulus (CS) that by itself produces the conditioned response (CR).

53.1 The Conditioned Reflex Ivan Pavlov discovered that when a normal response is paired with an artificial stimulus, an animal learns to produce the response even when only the artificial stimulus is presented.

- They focused on questions of learning and memory, largely to the exclusion of other types of behavior (e.g., mating, feeding, communication).

Thus defined, the field of animal behavior research became known as **behaviorism**, and it was largely the domain of psychologists.

Fixed action patterns were the focus of ethologists

An alternative approach to the study of animal behavior arose at the same time as behaviorism, but largely in Europe. Scientists there focused on describing the characteristics of animals in their natural environment, an approach that became known as **ethology** (Greek *ethos*, "character"; *logos*, "study"). In contrast to the behaviorists, the ethologists were interested in a wide variety of species, their evolutionary relationships, and the ways in which their behaviors were adapted to their environments. The leaders of the ethology movement were Karl von Frisch, who discovered the dance language of bees; Konrad Lorenz, who discovered that the strong bond between parent and offspring develops during a "critical period" following birth; and Niko Tinbergen, who studied inborn patterns of behavior commonly known as instincts. These three scientists shared the Nobel Prize in 1973 for "their discoveries concerning organization and elicitation of individual and social behavior patterns."

The instinctive behaviors that were the main interest of the ethologists were genetically determined patterns. Genetically determined behaviors

- are performed without learning
- are *stereotypic* (i.e., performed the same way each time)
- cannot be modified by learning

The ethologists called such behaviors **fixed action patterns**.

To demonstrate that a behavior was genetically determined, ethologists performed **deprivation experiments** in which an animal was raised in an environment devoid of opportunities to learn its species-specific behavior. An example of a natural deprivation experiment is the web spinning of the spider featured at the opening of this chapter. The parents of the young spider die before it hatches, and in a seasonal environment, it has no model webs to copy when it spins its first web, yet it creates a perfect, species-specific web the first and every time it spins a web.

Fixed action patterns are usually responses to specific stimuli. The ethologists carefully characterized such stimuli, which they called **releasers**. In general, releasers are very simple subsets of the information available in the environment. For example, Tinbergen studied the begging behavior of gull chicks. Adult gulls have a red dot on their bills. When a parent gull returns to the nest to feed its chicks, the chicks peck on the red

menter desired. Because the animal was conditioned to perform an operation on its environment, this experimental protocol was known as **operant conditioning** and was viewed as another model of learning.

The experimental approach to behavior initiated by Pavlov and Skinner had powerful effects on the nature of research on animal behavior. The focus of scientists using this approach was quite specific:

- They focused on laboratory environments rather than natural environments because in the laboratory, the variables in their conditioning experiments could be precisely controlled.
- They focused on only a few species (predominantly the albino rat) as model systems rather than studying diverse species from nature.

(A)

(B)

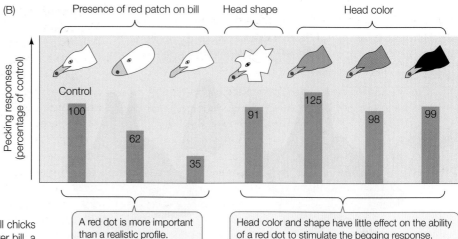

53.2 Releasing a Fixed Action Pattern (A) Gull chicks instinctively peck at the red dot on the parent's lower bill, a behavior that his induces the parent to regurgitate food into the chick's mouth. (B) Tinbergen's work showed that the red dot on the parent's beak was the critical component that released the pecking response.

dot, which stimulates the parent to regurgitate food (**Figure 53.2A**). Experimenters investigated what stimulated the chicks to peck their parents' bills. Models of gull heads of different shapes and colors were tested (**Figure 53.2B**), as were models of a beak without a head. The results showed that the red dot was necessary for the release of chick pecking behavior. In fact, a pencil with a red eraser elicited a more robust pecking response than an accurate model of a gull head without a red dot.

Ethologists probed the causes of behavior

The ethologists demonstrated the genetic basis for fixed action patterns by interbreeding closely related species. Konrad Lorenz studied the courtship behaviors of different species of dabbling ducks. Some of these species, such as mallards, teals, pintails, and gadwalls, are closely related and can interbreed, but they rarely do so in nature. Each male duck performs a courtship display consisting of a precise series of movements that is typical of his species (**Figure 53.3**). A female is not likely to accept him unless the entire display is successfully and correctly completed.

When Lorenz crossbred these duck species, the hybrid offspring expressed some elements of each parent's courtship dis-

play, but in novel combinations. Furthermore, Lorenz observed that hybrids sometimes exhibited display elements that were not in the repertoire of either parent species but were seen in other dabbling duck species. Lorenz's interbreeding studies demonstrated that the stereotypic motor patterns of the courtship displays are inherited. The observation that females were not interested in males performing hybrid displays was evidence that *sexual selection* (see Section 21.3) had shaped these genetically determined behaviors.

The ethologists recognized the importance of development and motivation in behavior, and they laid the foundation for the application of modern biological methods to the study of animal behavior. Tinbergen outlined the challenges for investigators as four questions:

- *Causation:* What is the stimulus for the behavior, and how has the relationship between stimulus and behavior been modified by learning?

- *Development:* How does the behavior change with age, and what experiences are necessary for it to be displayed?

53.3 Courtship Displays Are Highly Specific The courtship display of the male mallard (*Anas platyrhynchos*) contains the precise 10 elements illustrated here. The displays of closely related dabbling duck species contain some of the same elements, but have other elements not displayed by mallards.

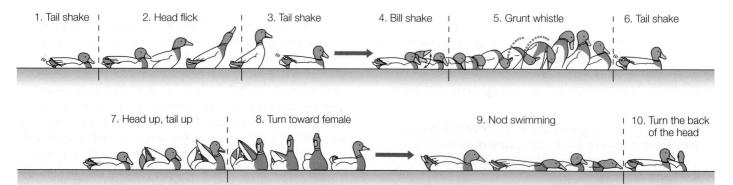

- *Function:* How does the behavior affect the animal's chances for survival and reproduction?
- *Evolution:* How does the behavior compare with similar behaviors in related species, and how might it have evolved?

The first two questions refer to the **proximate causes** of behavior: the immediate genetic, physiological, neurological, and developmental mechanisms that determine how an individual is behaving at a particular time. The third and fourth questions refer to the **ultimate causes** of behavior: the evolutionary processes that produced the animal's capacity and tendency to behave in particular ways. In the sections that follow, we will describe many experiments on animal behavior. For each one, ask yourself which of Tinbergen's four questions it addresses and whether it focuses on proximate or ultimate causes of behavior.

53.1 RECAP

Early scientific studies of animal behavior took two approaches. Behaviorists focused on the study of conditioned behavior in a few species of laboratory animals and asked questions about learning. Ethologists studied genetically determined behavior in many species in their natural environments and asked evolutionary questions.

- Describe the difference between conditioned reflexes and operant conditioning. **See pp. 1114–1115 and Figure 53.1**
- What is the relationship between a releaser and a fixed action pattern? **See p. 1115 and Figure 53.2**
- Explain the difference between proximate and ultimate causes of behavior and how Tinbergen's questions address these two types of causes. **See pp. 1116–1117**

The work of the ethologists left no doubt that behavior can be genetically determined, but how? Genes code for proteins. Behaviors are highly complex traits involving sensory input and intricate patterns of control over responses to that input. Is it reasonable to think that a single gene can have a specific effect on a behavior?

53.2 How Can Genes Influence Behavior?

Behaviors are complex traits that depend on many genes. Nevertheless, evidence from multiple approaches shows that alterations in single genes can result in discrete behavioral phenotypes on which natural selection can operate.

Breeding experiments can show whether behavioral phenotypes are genetically determined

Behavioral geneticists identify individuals with unusual behavioral phenotypes and conduct breeding experiments to see if those traits are inherited. If the inheritance of a trait follows a Mendelian inheritance pattern (see Section 12.1), then that trait is likely to be controlled by a small number of genes. A classic example is the genetics of hive cleaning by honey bees.

As we saw at the opening of Chapter 43, a honey bee colony consists of a reproductive queen bee and a huge number of nonreproductive workers. The worker bees have many tasks to perform. The queen deposits eggs in cells of the honeycomb structure that the workers construct. The eggs hatch into larvae, which are fed in their cells by workers. When the larvae are ready to pupate, the workers cap and seal their cells. If a pupa dies before emerging as an adult, workers will normally uncap the cell and drag the carcass outside the hive. This *hygienic* behavior increases the resistance of the hive to a bacterium that infects and kills bee larvae. Some hives do not show this behavior; these *nonhygienic* hives are more susceptible to the spread of this disease.

When a nonhygienic female bee was crossed with a hygienic male, the offspring were all nonhygienic, indicating that the genetic determinant of nonhygienic behavior is dominant. When these nonhygienic F_1 offspring were backcrossed to hygienic males, however, four phenotypes were produced. Two phenotypes were like those of the parents, either hygienic or nonhygienic, but the other two showed interesting deficits. One phenotype opened the cells of dead pupae, but did not remove them. The other phenotype did not open cells, but if a cell was already open, it would dispose of the dead pupa (**Figure 53.4**). Thus the complex hygienic behavior has at least two components that are controlled by separate genes. But the physical identity of these genes, and how they influence the behavior, could not be revealed by breeding experiments.

Knockout experiments can reveal the roles of specific genes

Some modern molecular genetic approaches start with identified genes and eliminate or silence them to see what effects their elimination has on a behavioral phenotype (see Section 18.4). As you might expect, knocking out genes involved in sensory pathways can have pronounced effects on behavior. One example is a gene for a specific olfactory receptor in mice.

As we saw in Section 46.2, mice have two olfactory organs: the nasal olfactory epithelium common to all mammals, and a small organ in the nasal passages, called the *vomeronasal organ*, or VNO (**Figure 53.5**). Catherine Dulac at Harvard University discovered that a large number of pheromone receptors were expressed in that organ. (As described in Section 46.2, *pheromones* are signaling molecules released into the environment.) Dulac hypothesized that when the receptors in the male's VNO bound to sex pheromones produced by female mice, they stimulated mating behavior. To test this hypothesis, Dulac created a genetically engineered male mouse in which the gene for an ion channel necessary for VNO receptor signaling was knocked out. Contrary to the prediction of the hypothesis, the knockout males in fact did pursue and mate with females placed in their cages. However, they also pursued and tried to mate with *males* placed in their cages. Normally a male mouse

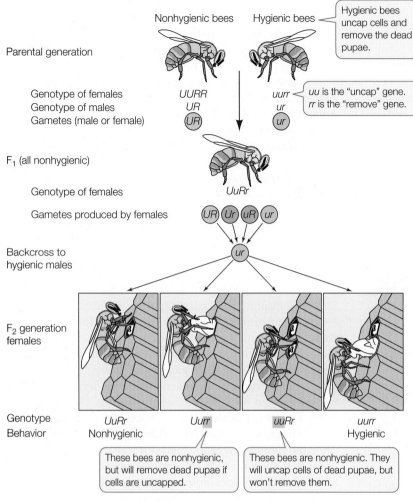

Parental generation

Nonhygienic bees   Hygienic bees

Hygienic bees uncap cells and remove the dead pupae.

Genotype of females
Genotype of males
Gametes (male or female)

UURR
UR
(UR)

uurr
ur
(ur)

uu is the "uncap" gene.
rr is the "remove" gene.

F₁ (all nonhygienic)

Genotype of females

UuRr

Gametes produced by females

(UR) (Ur) (uR) (ur)

Backcross to hygienic males

(ur)

F₂ generation females

Genotype
Behavior

UuRr
Nonhygienic

Uurr

uuRr

uurr
Hygienic

These bees are nonhygienic, but will remove dead pupae if cells are uncapped.

These bees are nonhygienic. They will uncap cells of dead pupae, but won't remove them.

53.4 Genes and Hygienic Behavior   Some worker honey bees remove carcasses of dead pupae from their hive. Two components of this behavior—cell uncapping and removal of the carcass—are under the control of separate recessive genes, designated *u* and *r*.

53.5 The Mouse Vomeronasal Organ Identifies Gender   (A) The mouse VNO is located adjacent to the nasal passages. It contains pheromone receptors whose input travels to a specific region of the olfactory bulb (the accessory olfactory bulb). (B) In mice, information from the VNO appears to be crucial in identifying gender and thus a potential sexual partner.

reacts aggressively to a strange male, but the knockout male could not discriminate between males and females placed in his cage. Thus properly functioning VNO receptors appear to be essential not for sexual attraction, but for gender identification. It is possible to imagine how selection working on this one gene could modify the intensity of male–male aggression and lead to changes in social behavior.

Behaviors are controlled by gene cascades

Male courtship behavior in *Drosophila melanogaster*, the laboratory fruit fly, is stereotypic, species-specific, and requires no learning—a classic fixed action pattern. When a male encounters a potential mate, he follows her, taps her body with his foreleg, extends and vibrates one wing, and licks her genitals (**Figure 53.6A**). The development of this complex male behavior is under the control of a single gene, called *fruitless* (*fru*), just as the development of male anatomy is under the control of another gene, called *doublesex* (*dsx*). In both males and females, these two genes are part of a gene expression cascade that results in different *dsx* and *fru* gene products in males and females (**Figure 53.6B**). The female version of the Dsx protein controls the development of female anatomy, and the expression of *fru* in the male nervous system results in the organization of the neural circuitry controlling male sexual behavior.

There are two take-home lessons from this example. First, genes that control aspects of behavior, like other genes, are generally embedded in gene cascades that offer multiple opportunities for simple genetic changes that will alter the phenotype of even complex behaviors. Second, certain genes, such as *dsx* and *fru*, influence a whole range of genes that contribute to complex behaviors. Modifications in any one of those genes or its ex-

(A)

Brain   Olfactory lobe of brain

Vomeronasal nerves

Nasal cavity

Nasal epithelium

Vomeronasal organ

Nostril

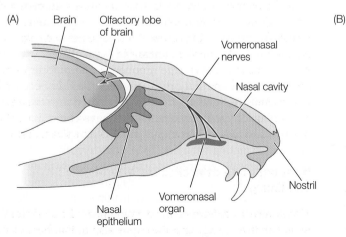

(B)

(A) Female fruit flies are XX... ...and males are XY.

Orienting — Tapping — Wing vibration — Licking — Attempted copulation — Copulation

(B) **1** Sex-determining pre-mRNAs are spliced in one specific way in female flies... **2** ...and another way in males.

Female-specific mRNA	Gene	Male-specific mRNA

Transcription and mRNA splicing — *sex-lethal (sxl)* — Transcription and mRNA splicing

sxl mRNA — Stop codon — Default splicing

Female Sxl protein — *transformer (tra)* — No functional Sxl protein

3 Female *sxl* and *tra* mRNAs make proteins that control splicing in the expression of genes in the female-specific hierarchy.

Female Tra protein — Stop codon — No functional Tra protein

4 Male *sxl* and *tra* mRNAs have stop codons that terminate translation.

Female Dsx protein — *doublesex (dsx)* — Male Dsx protein

5 The default splicing of *dsx* mRNAs controls male anatomy...

Female Fru protein — *fruitless (fru)* — Male Fru protein

Sequences colored black are introns (noncoding DNA).

6 ...and male-specific splicing of *fru*, which results in male courtship behavior.

53.6 The *fruitless* Gene (A) Male fruit flies display stereotypic, species-specific courtship behavior. (B) Sexual differentiation in *Drosophila* is controlled by a cascade of genes, including the *fru* gene, whose expression results in male sexual behavior.

pression can alter behavior. Thus, even though no behavior is coded for by a single gene, alterations in single genes can influence behavior in ways that affect an animal's fitness.

53.2 RECAP

Breeding experiments show that behavioral phenotypes can be inherited and modified by natural selection. Although most behaviors are controlled by complex cascades of genes, molecular genetic methods have shown that a single, identifiable gene in the cascade can control a complex behavior.

- How did breeding experiments on honey bees show the genetic basis for hygienic behavior? See p. 1117 and Figure 53.4

- What is the evidence that the vomeronasal organ in mice is responsible for gender identification? See pp. 1117–1118

- Describe some of the different ways a gene can control expression of a complex behavior. See p. 1118 and Figure 53.6

How can the genetic cascades that underlie complex behaviors be programmed to respond selectively to specific sets of stimuli? How can their expression be limited to appropriate times in an animal's life? The answers to these questions can by found by studying how behaviors develop over the life span.

53.3 How Does Behavior Develop?

The emergence of behavior as an animal develops and matures depends on the development of the nervous system as well as on the growth and maturation of other body systems. A bird cannot fly until its wings grow and its muscles and flight feathers mature. But even with anatomical and physiological competence, specific behaviors may not be expressed. Behaviors that are adaptive at one stage in an animal's life may not be adaptive at other stages. Behaviors typical of juvenile animals, such as begging for food, may disappear and new behavior patterns of a mature individual, such as courtship displays, appear.

Hormones can determine behavioral potential and timing

Hormones can determine the development of a behavioral potential at an early age and the expression of that behavior at a later age. An excellent example of this is sexual behavior in rats

(A) Female rats

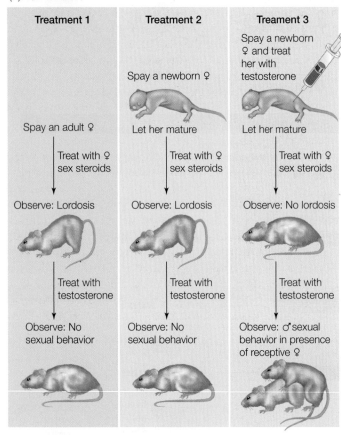

(B) Male rats

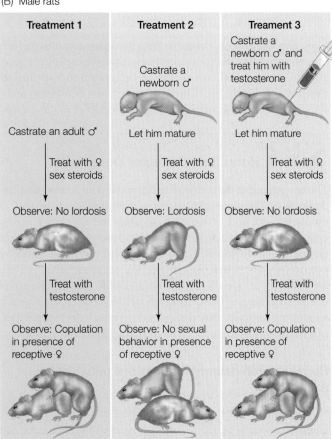

53.7 Hormonal Control of Sexual Behavior   Experimental hormonal treatments of rats demonstrated that the sex steroids present during early development determine what sexual behavior patterns develop, but the sex steroids present in adulthood control the expression of those patterns.

(**Figure 53.7**). Normally, adult male and female rats exhibit different patterns of sexual behavior: females adopt a sexually receptive posture, called *lordosis*, in the presence of males, and males copulate with receptive females. Neither sex, however, expresses these behaviors until the animals have reached adulthood. Experiments in which newborn and adult rats were neutered (to remove the influence of sex steroids naturally produced by their gonads) and artificially treated with hormones led to the following conclusions:

- Development of male sexual behavior requires that the brain of the newborn rat be exposed to testosterone, but development of female sexual behavior does not require exposure to estrogen.

- Testosterone masculinizes the nervous systems of both genetic males and genetic females.

- Exposure to sex steroids in adulthood is necessary for the expression of sexual behavior, but testosterone produces male sexual behavior only in adult rats whose brains were masculinized when they were newborns, and estrogen produces female sexual behavior only in adult rats whose brains were not masculinized when they were newborns.

Thus, the sex steroids that are present at birth determine which pattern of behavior develops, and the sex steroids that are present in adulthood determine when that pattern is expressed.

Some behaviors can be acquired only at certain times

Responsiveness to simple releasers is sufficient for certain behaviors such as begging behavior in gull chicks, but more complete information that cannot be genetically programmed is required for other behaviors. An example is parent–offspring recognition. When animals live in close proximity to other individuals, as in a herd or a nesting colony, it is important for a mother and her offspring to learn each other's identity soon after birth so they will be able to find each other in a crowded situation. In many such cases, a parent–offspring bond is formed by **imprinting**. What characterizes imprinting is that an animal learns a specific set of stimuli during a limited time called a **critical period** or **sensitive period**.

Konrad Lorenz demonstrated that young goslings imprint on their parents between 12 and 16 hours after hatching. By positioning himself to be present during this critical period, Lorenz succeeded in imprinting goslings on himself. The imprinted goslings followed him around as if he were their parent (**Figure 53.8A**).

Imprinting requires only a brief exposure, but its effects are strong and long-lasting. Emperor penguins reproduce during the coldest, darkest time of year in Antarctica. The parents walk up to 150 km inland to form a dense colony, where the female lays her egg. She then walks back to the ocean to feed while her

(A)

(B)

53.8 Imprinting Helps Parents and Offspring Recognize Each Other (A) Greylag geese that imprinted on Konrad Lorenz as hatchlings followed him everywhere he went. (B) Imprinting allows a male emperor penguin to find his own chick among many others.

mate incubates the egg. By the time she returns, the chick has hatched. She then takes over its care and feeding, and the father walks back to the ocean to feed. Generally, he is away so long that the mother must leave to find food as well to avoid starvation. Thus, after being away for weeks, the father must find his chick in a crowded, milling colony of chicks, all calling for their parents (**Figure 53.8B**). Yet he can unerringly locate his own offspring by recognizing its call, which he learned before he left to feed.

The critical or sensitive period for imprinting may be determined by a brief developmental or hormonal state. For example, if a mother goat does not nuzzle and lick her newborn within 10 minutes after its birth, she will not recognize it as her own offspring later. For goats, the sensitive period is associated with peaking levels of the hormone oxytocin in the mother's circulatory system at the time she gives birth and is sensing the olfactory cues emanating from her newborn kid. A female goat rendered incapable of smelling before giving birth is unable to differentiate between her own kid and other kids after giving birth.

Bird song learning involves genetics, imprinting, and hormonal timing

Male songbirds use species-specific song to claim and advertise a breeding territory, compete with other males, and declare

dominance. They also use song to attract females, who recognize the song of their species even though they do not sing it. For males of many species, such as the white-crowned sparrow, learning is an essential step in the acquisition of song, but *what* they can learn seems to be influenced by genes, and there is a limited *developmental time frame* for learning. A hatchling in nature hears his father and other white-crowned sparrows singing. He also hears the songs of many other bird species. But he does not sing until he approaches sexual maturity almost a year later, and when he does, he sings his father's type of song.

Studies of song learning in this species were initiated in the 1960s by Peter Marler at the University of California at Berkeley. Marler incubated eggs of white-crowned sparrows and hand-reared the hatchlings in the laboratory. He was therefore able to expose them to different tape-recorded songs at different times in their development. He discovered that male birds cannot produce their species-specific song as adults unless they hear it as nestlings in the first 2 months of their lives (**Figure 53.9**). If the nestlings hear tapes of white-crowned sparrow song during those first 2 months, they begin to sing, but poorly, as they approach sexual maturity. Through trial and error, the birds

(A)
Control or wild bird

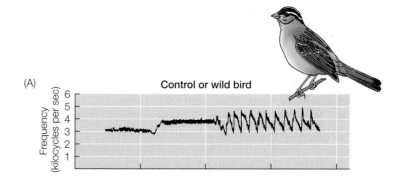

(B)
Isolated hand-reared bird

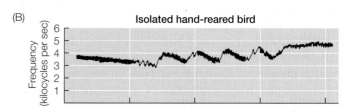

(C)
Deafened bird

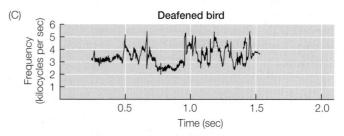

53.9 Sensitive Periods for Song Learning (A) Sonogram showing the species-specific song of an adult male white-crowned sparrow (*Zonotrichia leucophrys*). (B) Song of an adult male raised in isolation (never having heard the song as a nestling). (C) Song of an adult male who heard the song as a nestling but was deafened prior to ever singing himself. Marler's experiments showed the bird must first acquire a song memory by hearing the song as a nestling, and must then be able to hear himself as he attempts to match his singing to that song memory.

match their singing to their stored song memory, and from then on they sing their species-specific song. To reach this point, the young bird must be able to hear himself sing. If he is deafened just before he begins to sing, he will not be able to match his stored song memory. If he is deafened *after* he sings his correct species-specific song, however, he will continue to sing like a normal bird. We say that at this point the song behavior pattern is *crystallized.* Thus there are two sensitive periods for song learning: the first in the nestling stage, when a song memory is imprinted; the second as the bird approaches sexual maturity, when he learns to match that song memory.

In nature, nestling male white-crowned sparrows hear the songs of many species, so why do they learn only the song of their own species? Marler investigated this question in his isolation experiments by playing tape recordings of other songs to the hatchlings. The young male sparrows did not learn the songs of other species, even if they heard them many times, but hearing songs of their own species just a few times was sufficient for imprinting. Thus, although male sparrows must learn their song, they seem to have a genetic predisposition to learn their own song and not the songs of other species. Marler called this phenomenon "an instinct to learn." There is an important reason why male white-crowned sparrows should learn only their species-specific song: female white-crowned sparrows also listen to their fathers' songs while they are nestlings, and when they mature, they choose mates who sing like their fathers did.

Since Marler's pioneering work, more investigations have revealed additional complexity in white-crowned sparrow song learning capacity. First, it was demonstrated that each population of white-crowns has its own dialect, and that males from one population can learn the dialect of another population. Second, it was observed that in nature, white-crowns are occasionally heard singing the songs of other species. Could Marler's laboratory experiments have missed a critical natural variable?

Luis Baptista, a curator of birds at the California Academy of Sciences, took up the study of white-crowned sparrows to explore the effects of social interactions on song learning. He discovered that when a hand-reared white-crowned sparrow nestling was exposed to the sight and sound of a related species in an adjacent cage while tapes of his own species' song played in the background, the white-crowned sparrow sang the song of the other species when he matured. Thus social experience has a powerful effect on what the young bird can and will learn.

The timing and expression of bird song are under hormonal control

As we have seen, both male and female songbirds hear their species-specific song as nestlings, but only the males of most species sing as adults, and they do so only in spring. Hormones underlie both the difference in song expression between male and female songbirds and the timing of song expression.

To determine whether the absence of testosterone is the main reason female songbirds don't sing, investigators injected adult female songbirds with testosterone in spring. In response to these injections, the females sang their species-specific song just as the males did. Apparently females form a memory of their species-specific song when they are nestlings, and have the physical capacity to sing, but under normal circumstances they simply lack the hormonal stimulation.

How does testosterone cause a songbird to sing? A remarkable study revealed that each spring, an increase in circulating testosterone levels causes certain parts of the male's brain necessary for learning and developing song to grow larger. Individual neurons in those regions of the brain increase in size and grow longer extensions, and the number of neurons in those regions increases. Such research on the neurobiology of bird song revealed that, contrary to the once widely held belief that new neurons are not produced in the brains of adult vertebrates, hormones can control behavior by changing brain structure as well as brain function, both developmentally and in response to environmental cues.

What triggers the release of testosterone in response to the onset of spring? Takashi Yoshimura conducted a DNA microarray analysis of brain samples from another bird species, the Japanese quail, which is a model species for avian genetic analyses. Knowing that photoperiod was the environmental cue, he examined the responses of 38,000 genes to determine which genes were activated by changes in day length. Fourteen hours after dawn on the first day with the critical photoperiod that induces singing behavior, genes in the brain that code for thyroid-stimulating hormone were switched on. This hormone acts on the pituitary gland, which in turn regulates the production of other hormones that stimulate the growth of the testes and the production of testosterone (see Figure 43.10).

53.3 RECAP

Hormones can control what behavior patterns are programmed in the brain long before those behaviors are expressed. Learning and expression of behaviors can also be controlled by hormones, and therefore timed for particular stages in the life cycle.

- What hormonal conditions are necessary for the development of adult sexual behavior in male and female rats? See pp. 1119–1120 and Figure 53.7

- What is the adaptive value of imprinting? See pp. 1120–1121

- Describe the series of events necessary for a male white-crowned sparrow to sing its species-specific song in spring. See pp. 1121–1122 and Figure 53.9

Complex behaviors are the product of interactions of genetic, physiological, and environmental factors. Many genes are involved in shaping behavior, and therefore there are multiple opportunities for selection to favor behavioral modifications. Questions about how changes in behavior adapt animals to environmental conditions are the province of an evolution-based field called *behavioral ecology.*

53.4 How Does Behavior Evolve?

Environmental conditions are highly variable over both time and space. Animal behaviors are also variable within and between species. Behavioral ecologists strive to discover the relationships between behavior and environment with the intent of understanding the evolutionary mechanisms underlying behavior.

Animals must make many behavioral choices

Over an animal's lifetime, its behavior is largely a sequence of choices: where and when to move, where to build a nest, what to eat, when to fight and when to flee, with whom to associate, with whom to mate. Inability to make choices would make it difficult to survive, and making wrong choices reduces its fitness. Behavioral ecologists seek to discover what information animals use to make behavioral choices and how that information relates to aspects of the environment that influence their fitness.

The choice of a place to live, for example, is one that has many consequences for an animal. The environment in which an animal lives is its **habitat**. In most cases the habitat provides not only a protected nest site, but also food and access to mates. In many cases the environmental cues animals use to make their habitat choices are quite simple. For example, seabirds select cliffs or offshore rocks for nesting because those sites offer protection from predators. Animals with very specialized food requirements obviously select habitats where those foods are abundant. The general hypothesis that guides behavioral ecologists is that the cues animals use to select habitats are *reliable predictors of conditions suitable for future survival and reproduction.*

For many species, the presence of **conspecifics**—other members of the same species—is a reliable indicator of the suitability of a particular site for sustaining life. After all, you can't argue with success. Consider the planktonic larvae of marine organisms that drift freely in the water until they initiate a *settling response* that deposits them on the substratum to which they will attach themselves, usually for the rest of their lives. Molecules released into the water by conspecifics can aid larvae in detecting an appropriate substratum, and such molecules induce a settling response in many species.

Observing conspecifics can provide animals with information about the quality of a habitat in other ways. Collared flycatchers (*Ficedula albicollis*) on their breeding grounds in spring are nosy neighbors: they regularly visit the nests of conspecifics. Researchers hypothesized that this behavior allows the flycatchers to assess the quality of the habitat by seeing how well their neighbors are faring. To test this hypothesis, they created some areas with supersized broods—normally an indication of abundant food—by taking young birds from some nests and adding them to nests in another area. The next year, flycatchers preferentially settled in the areas where broods had been artificially enlarged (**Figure 53.10**).

Behaviors have costs and benefits

Behavioral ecologists often use a cost–benefit approach to investigate the relationship between behavior, environment, and

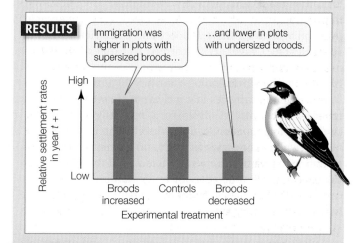

INVESTIGATING LIFE

53.10 Flycatchers Use Neighbors' Success to Assess Habitat Quality

Collared flycatchers (*Ficedula albicollis*) regularly "spy" on other flycatchers in their breeding range. Researchers hypothesized that the birds were gathering information on the reproductive success of already established pairs to help them decide where to settle and breed.

HYPOTHESIS Collared flycatchers will nest at higher densities in areas where their conspecifics appear to have reproduced successfully.

METHOD
1. Transfer nestlings from one territory into nests in another, thereby creating "supersized" broods (an indication of abundant food) in one area and undersized broods in the other. A third research territory is left unaltered to serve as a control.
2. A year later, observe settlement rates (i.e., number of pairs establishing nests) in each territory at the start of the breeding season.

RESULTS

Immigration was higher in plots with supersized broods…

…and lower in plots with undersized broods.

(Graph: y-axis "Relative settlement rates in year $t + 1$" from Low to High; x-axis "Experimental treatment" with bars for "Broods increased", "Controls", "Broods decreased")

CONCLUSION Collared flycatchers use the brood sizes of already settled conspecifics to assess how good a territory is for breeding.

Go to **yourBioPortal.com** for original citations, discussions, and relevant links for all INVESTIGATING LIFE figures.

fitness. A **cost–benefit approach** assumes that an animal has only a limited amount of time and energy, and therefore cannot afford to engage in behaviors that cost more to perform than they bring it in benefits. A cost–benefit approach provides a framework that behavioral ecologists can use to make observations, construct hypotheses, and design experiments to investigate why behavior patterns evolve as they do.

The benefits of a behavior can be measured in terms of the enhancement in fitness an animal accrues by performing the behavior. The cost of a behavior typically has three components:

- **Energetic cost** is the difference between the energy the animal expends performing the behavior and not performing it.

- **Risk cost** is the increased chance of being injured or killed as a result of performing the behavior.
- **Opportunity cost** is the benefit the animal forgoes by not being able to perform other behaviors during the same time interval.

Cost–benefit analysis has been used extensively in the study of **territorial behavior**: aggressive behavior used by an animal to actively deny other animals access to a habitat or resource. Optimal habitats and resources are frequently in short supply, so conspecifics must compete for them. Many animals—usually males—defend all-purpose territories that provide a nest site, food, and access to mates. The territory holder stakes out his boundaries by engaging in aggressive interactions with neighbors, and must then patrol those boundaries constantly and respond to trespassers. These aggressive interactions usually consist of highly stereotypic, species-specific displays such as bird song. Through territorial behavior, the male obtains the resources he needs for reproductive success, but he also pays a price.

Territorial displays require considerable expenditure of energy, they make a male more vulnerable to predation, and they detract from the time he has for feeding or engaging in parental behavior. Michael Moore and Catherine Marler at Arizona State University performed an experiment to estimate the costs incurred by male Yarrow's spiny lizards (*Sceloporus jarrovii*) when defending a territory (**Figure 53.11**). These lizards defend territories that include the habitats of several females. Their territorial behavior is normally most intense during September and October, when the circulating testosterone levels of the males are high and the females are most receptive to mating. The researchers varied the intensity of the lizards' territorial behavior by implanting testosterone capsules in some males in summer, when they are not normally highly territorial.

Testosterone-treated males spent more time patrolling their territories, performed more displays, and expended about one-third more energy than control males (an energetic cost). As a result, they had less time to feed (an opportunity cost), captured fewer insects, stored less energy, and had a higher death rate (a risk cost). In summer, when females are not normally receptive, these high costs of vigorous territorial defense outweigh the reproductive benefits of territoriality. Thus natural selection has favored seasonal variation in the level of the hormone controlling territorial behavior in this species.

yourBioPortal.com

GO TO **Animated Tutorial 53.1** • The Costs of Defending a Territory

The cost–benefit approach explains the diversity of territorial behaviors seen in different species. Even if a resource is absolutely essential to an animal, if it cannot be defended economically, the animal will not engage in territorial behavior. Food is essential for all animals, but if the food is widely distributed in space or fluctuating in availability, there is no benefit to balance the high costs of trying to defend it. For ex-

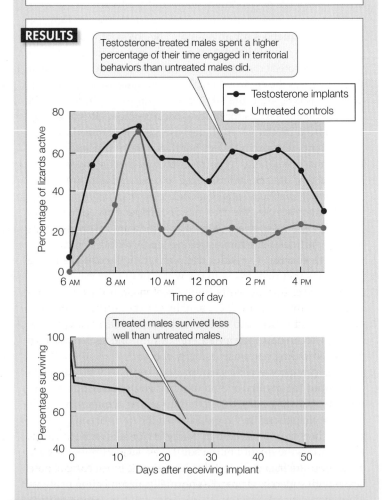

INVESTIGATING LIFE

53.11 The Costs of Defending a Territory

By using testosterone implants to increase territorial behavior, Moore and Marler measured the costs to male Yarrow's spiny lizards (*Sceloporus jarrovii*) of defending a territory during the summer, when they do not normally do so.

HYPOTHESIS Yarrow's spiny lizards do not defend a territory during summer because the energetic costs of territorial behavior in that season outweigh the benefits.

METHOD

1. During the summer, when female lizards are not sexually receptive, insert testosterone capsules under the skin of some males; leave other males untreated as controls.

2. Observe the patterns of territorial behavior and the survival rate of the two groups of males.

RESULTS

Testosterone-treated males spent a higher percentage of their time engaged in territorial behaviors than untreated males did.

Treated males survived less well than untreated males.

CONCLUSION For these lizards, the cost of defending territories during summer significantly reduces their survival rate without increasing their reproductive success.

Go to **yourBioPortal.com** for original citations, discussions, and relevant links for all INVESTIGATING LIFE figures.

(A) *Diomedea melanophris*

(B) *Mirounga angustirostris*

53.12 Animals Defend Territories of Different Sizes (A) The nesting territories of many seabirds consist of only as much space as they can defend without leaving the nest. (B) Male elephant seals fight vigorously to defend areas of beach where females haul out of the water to give birth to their pups. (C) Male sage grouse gather at a lek in Colorado to perform displays aimed at impressing females and winning the opportunity to mate.

(C) *Centrocercus urophasianus*

ample, the open ocean where seabirds feed cannot be defended. But safe nest sites on islands or rocky cliffs are in short supply, and they can be defended. Thus the territories of seabirds may be no larger than the distance the birds can reach while sitting on their nests (**Figure 53.12A**).

In some cases the resource that is defended is the female herself. Elephant seals spend most of their lives at sea, but females come to land at traditional beach sites to give birth to their pups. Male elephant seals arrive at these sites ahead of time and stake out territories through vigorous fighting (**Figure 53.12B**). When the females arrive on the beaches, they enter the territories of the males. As long as the male territory holder can fend off challengers, he will be able to mate with all the females using his piece of the beach.

The most unusual form of male territorial behavior arises in situations in which neither food, nest sites, nor females are defended. A **lek** is an area where males gather for the purpose of engaging in communal displays of their territorial prowess aimed at impressing females and winning the opportunity to mate. Even though space is not limited, each male defends a small piece of real estate on which he performs a display (**Figure 53.12C**). Those pieces of territory closest to the center of the lek are the prime sites, and males compete intensely for those locations. The females stroll into the lek, observe the males, and generally mate with the males holding the prime sites. The benefit of this system to the female is that she is inseminated by a successful competitor, and therefore her offspring will carry the

genes that contributed to his success. The costs of lekking to males are high, as they engage in continuous, intense territorial behavior that precludes eating, drinking, and sleeping until they are displaced. The benefit is the chance to maximize their fitness by mating with many females.

Cost–benefit analysis can be applied to foraging behavior

Cost–benefit analyses have also been used to investigate the food choices animals make. When an animal *forages* (searches for food), among the decisions it must make are how much time to spend in each location before giving up and moving on, what resources at each location are actually edible, and which of the different types of potential food should be eaten and which should be left alone. By applying cost–benefit approaches to feeding behavior, scientists have produced a body of knowledge known as **optimal foraging theory**, which helps them to identify the fitness value of feeding choices.

The primary benefit of foraging is the nutritional value of the food obtained: the energy, minerals, and vitamins it contains (see Section 51.1). The costs of foraging are similar to those of other behaviors: energy expended, time lost from other activities that could enhance fitness, and the risk of increased exposure to predators.

Animals frequently have to make choices among food items that may differ not only in terms of energy content, but also abundance or ease of acquisition and processing. Optimal foraging theory predicts that in such situations, animals will make choices that will maximize the rate at which they obtain energy. The more rapidly a foraging animal satisfies its energetic requirements, the lower the opportunity costs and risk costs of foraging.

Earl Werner and Donald Hall of Michigan State University performed laboratory experiments with bluegill sunfish to test this energy maximization hypothesis. In preparation for their experiments, they measured the energy content of water fleas (*Daphnia*) of different sizes (the different food types), how much time bluegill sunfish (the foragers) needed to capture and eat those different food types, the energy they spent pursuing and capturing the different food types, and the rates at which they encountered the different food types under different food densities. Werner and Hall then stocked experimental environments with different densities and proportions of large, medium, and small water fleas. They made two predictions from the energy maximization hypothesis: first, that in an environment with abundant large water fleas, the fish would ignore smaller water fleas; and second, that in an environment stocked with low densities of all three sizes of water fleas, the fish would eat every water flea they encountered. The proportions of large, medium, and small water fleas eaten by the fish under different conditions were close to those predicted by the hypothesis (**Figure 53.13**).

—— **yourBioPortal.com** ——
GO TO Animated Tutorial 53.2 • Foraging
Behavior

The energy maximization hypothesis considers food items in terms of the energy they provide, but animals have nutrient requirements in addition to energy that can play a role in shaping their foraging behavior. Essential minerals, for example, are in short supply in some animals' diets, and those animals may incur large energetic costs and risks to obtain them. Many herbivores and seed eaters, whose food contains low levels of mineral nutrients relative to their needs, obtain those nutrients by eating soil at particular sites where mineral-rich soil is exposed (**Figure 53.14**).

Foods may also be chosen for their medicinal value. Chimpanzees, for example, have been observed eating the pith of a plant called *Vernonia amygdalina*. The pith contains very small quantities of a secondary metabolite called vernonioside B1, which is toxic to chimps at high concentrations, but at low concentrations can kill their intestinal parasites. Chimps that consume this pith have fewer parasites.

INVESTIGATING LIFE

53.13 Bluegill Sunfish Are Energy Maximizers

Based on energy maximization calculations, Werner and Hall predicted (1) that in an environment with abundant large food items (i.e., the water flea *Daphnia*), bluegill sunfish (*Lepomis macrochirus*) would ignore smaller food items and feed preferentially on larger water fleas; and (2) that in an environment where all sizes of *Daphnia* were scarce, the fish would eat every one they encountered. Such a strategy is in keeping with the cost–benefit hypothesis of foraging behavior.

HYPOTHESIS Bluegill food item selection will match the energy maximization predictions of cost–benefit analysis.

METHOD
1. Measure the respective energy content of large, medium, and small water fleas (*Daphnia*).
2. Use cost–benefit energy maximization calculations to predict the rate at which bluegill sunfish will consume the different sized water fleas under different levels of food abundance (i.e., density of water fleas).
3. Provide bluegills with *Daphnia* of different sizes in varying proportions (represented by the different colored bars) and at different densities.

4. Note the proportions of small, medium, and large *Daphnia* actually eaten by the fish under the different conditions and compare these proportions to the predictions of energy maximization.

RESULTS The food choices made by bluegills match the predictions of the energy maximization hypothesis.

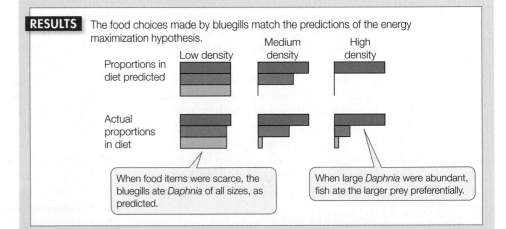

When food items were scarce, the bluegills ate *Daphnia* of all sizes, as predicted.

When large *Daphnia* were abundant, fish ate the larger prey preferentially.

CONCLUSION Bluegills select food items in accordance with the predictions of energy maximization calculations.

Go to **yourBioPortal.com** for original citations, discussions, and relevant links for all INVESTIGATING LIFE figures.

(A)

(B)

53.14 Herbivores Seek Out Unusual Sources of Minerals (A) Red-and-green macaws of the Amazon jungle obtain essential minerals by eating dried clay. (B) Pierid butterflies obtain needed salts by drinking secretions from the skin and nostrils of a caiman.

53.4 RECAP

Behavioral ecologists seek to explain relationships between variation in behavior and variation in environmental conditions. They seek to discover what information animals use to make behavioral choices and how that information relates to aspects of the environment that influence their fitness. Cost–benefit analysis has been applied to territorial and foraging behavior.

- What might the presence of conspecifics in a habitat tell an animal about that habitat? **See p. 1123 and Figure 53.10**

- Describe three types of territorial behavior and the costs and benefits of each. **See pp. 1124–1125 and Figure 53.11**

- How can cost–benefit analysis be applied to foraging behavior? **See pp. 1125–1126 and Figure 53.13**

Whereas behavioral ecologists are interested in understanding how the natural environment influences the fitness value of behavioral choices—the ultimate causes of those behaviors, in Tinbergen's terms—other behavioral biologists focus on the physiological mechanisms and principles that underlie behavior—the proximate causes.

53.5 What Physiological Mechanisms Underlie Behavior?

Control of behavior involves the nervous and endocrine systems. Execution of behavior involves the musculoskeletal system as well as other *effector mechanisms*, such as those that produce secretions, color changes, electrical impulses, sound, and even light. We have already considered many of the physiological systems that are involved in these processes, including endocrine mechanisms, reproductive systems, nervous systems, sensory systems, and feeding mechanisms. The field of behavioral physiology, which encompasses aspects of all of these systems, is enormous, so here we will dig deeper into just three different phenomena studied by behavioral physiologists: the timing of behavior, navigation, and communication.

Biological rhythms coordinate behavior with environmental cycles

Earth turns on its axis once every 24 hours, generating daily cycles of light and dark, temperature, humidity, and tides. In addition, Earth is tilted on its axis, so the light–dark cycle changes as it revolves around the sun. These daily and seasonal cycles profoundly influence the physiology and behavior of animals. Animals tend to be either day-active (diurnal) or night-active (nocturnal), and have appropriate sensory capabilities. Therefore it is adaptive to organize behavior on a cycle that corresponds with the environmental cycle of light and dark. Similarly, a behavior that is adaptive at one time of year may not be adaptive at other times. Thus it is important for animals to be able to time their behavior to appropriate times of the day or year and to be able to anticipate those times.

CIRCADIAN RHYTHMS Experimental animals kept in constant darkness and at a constant temperature, with food and water available all the time, still demonstrate daily cycles of activities such as sleeping, eating, drinking, and just about anything else that can be measured. The persistence of these daily cycles in the absence of environmental time cues suggests that animals have an internal clock. Because these daily cycles are not exactly 24 hours long, they are known as **circadian rhythms** (*circa*, "about," and *dies*, "day").

As described in Section 37.5, any biological rhythm can be viewed as a series of cycles, and the length of one of those cycles is the *period* of the rhythm. Any point in the cycle is a *phase* of that cycle. Hence, when two rhythms completely match, they are *in phase*, and if a rhythm is shifted (as in the resetting of a clock) it is *phase-advanced* or *phase-delayed*. Because the period of a circadian rhythm is not exactly 24 hours, it must be phase-advanced or phase-delayed each day to remain in phase with

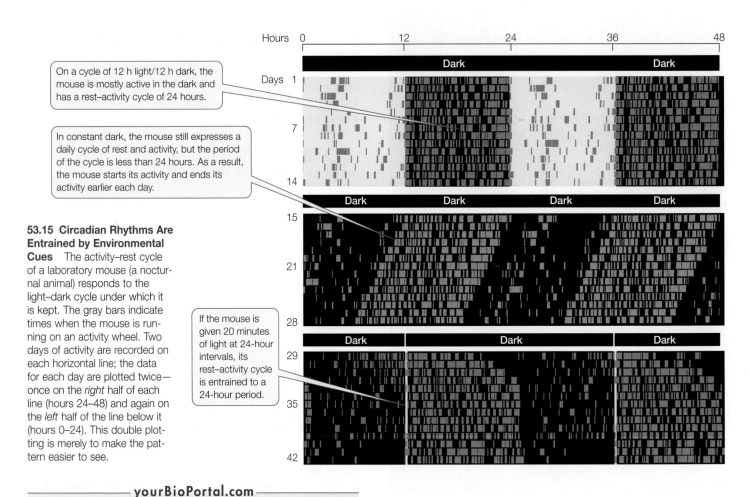

On a cycle of 12 h light/12 h dark, the mouse is mostly active in the dark and has a rest–activity cycle of 24 hours.

In constant dark, the mouse still expresses a daily cycle of rest and activity, but the period of the cycle is less than 24 hours. As a result, the mouse starts its activity and ends its activity earlier each day.

53.15 Circadian Rhythms Are Entrained by Environmental Cues The activity–rest cycle of a laboratory mouse (a nocturnal animal) responds to the light–dark cycle under which it is kept. The gray bars indicate times when the mouse is running on an activity wheel. Two days of activity are recorded on each horizontal line; the data for each day are plotted twice—once on the *right* half of each line (hours 24–48) and again on the *left* half of the line below it (hours 0–24). This double plotting is merely to make the pattern easier to see.

If the mouse is given 20 minutes of light at 24-hour intervals, its rest–activity cycle is entrained to a 24-hour period.

---

yourBioPortal.com

GO TO Animated Tutorial 53.3 • Circadian Rhythms

---

the daily cycle of the environment. In other words, the rhythm has to be **entrained** to the cycle of light and dark in the animal's environment.

An animal kept under constant conditions will not be entrained to the light–dark cycle of the environment, and its circadian clock will run according to its natural period—it will be *free-running*. If the period is less than 24 hours, the animal will begin its activity a little earlier each day (**Figure 53.15**). The period of the free-running circadian rhythm is under genetic control. Different species may have different average periods, and within a species, mutations can lead to different period lengths.

Under natural conditions, environmental time cues, such as the onset of light or dark, entrain the free-running rhythm to the light–dark cycle of the environment. In the laboratory, it is possible to entrain the circadian rhythms of free-running animals with short pulses of light or dark administered every 24 hours (see the bottom panel of Figure 53.15).

In mammals, the master circadian "clock" consists of two clusters of neurons just above the optic chiasm (the area of the brain where the optic nerves come together). These structures are called the **suprachiasmatic nuclei** (**SCN**). If they are destroyed, the animal becomes *arrhythmic*: it is just as likely to eat, drink, sleep, or wake at any time of day.

Martin Ralph (now at University of Toronto) and colleagues, who were then at the University of Virginia, used artificial selection to produce two strains of hamsters: one with a short circadian period and one with a long circadian period. When adult hamsters' SCN were destroyed, the animals became arrhythmic. After several weeks, the scientists transplanted SCN tissue from hamster fetuses into arrhythmic hamsters' brains. The long-period adult hamsters received tissue from short-period fetuses, and vice versa. This experiment produced two remarkable results. First, circadian rhythms were restored by the transplanted SCN tissue, demonstrating that the SCN is sufficient to generate circadian rhythms. This is the only case in which a behavior has been restored by a neural transplant. Second, the restored circadian rhythms had the period length of the donor strain, demonstrating that the specific phenotype of the behavior was a property of the donor neural tissue.

The molecular mechanism of the circadian clock involves negative feedback loops. Although there are a number of genes involved, we can generalize the mechanism by saying that when certain *clock genes* are expressed in SCN cells, the mRNA enters the cytoplasm, where it is translated. The resulting proteins combine and the dimer returns to the nucleus as a transcription factor that shuts off the expression of the clock genes. The period of this cycle is about a day. These findings show that it is possible to understand circadian rhythms of behavior at all levels, from the molecular rhythm generators to the environmental stimuli that entrain them to the daily cycle of light and dark.

CIRCANNUAL RHYTHMS Seasonal changes in the environment present challenges to many species. Most animals reproduce most successfully if they time their reproductive behavior to coincide with the most favorable time of year for the survival of their offspring. Many species require considerable advance preparation for reproduction. Migratory animals must arrive on their breeding grounds at the right time, and animals that have specialized structures used in mating displays, such as the antlers of deer, moose, and caribou, must grow their equipment before the breeding season arrives.

For many species, a change in day length—the *photoperiod*—is a reliable indicator of seasonal changes to come. For others, however, change in day length is not a reliable seasonal cue. Hibernators, for example, spend long months in dark burrows underground, but must be physiologically prepared to breed almost as soon as they emerge in the spring. A bird overwintering near the equator cannot use changes in photoperiod as a cue to time its migration to its temperate-zone breeding grounds. When held under constant laboratory conditions, such animals show endogenous **circannual rhythms**, built-in neural calendars that keep track of the time of year. Unlike circadian rythms, the neural basis for circannual rhythms is unknown.

Animals must find their way around their environment

To locate suitable habitats, find food and mates, and escape from predators and bad weather, an animal needs to be able to find its way around its environment. Within its local habitat, an animal can organize its behavior spatially by orienting to landmarks. But what if its destination is a considerable distance away?

PILOTING: ORIENTATION BY LANDMARKS Most animals find their way by knowing and remembering the structure of their environment. This form of navigation is called **piloting**. Gray whales, for example, migrate seasonally between the Bering Sea and the coastal lagoons of Mexico (**Figure 53.16**). They find their way in part by following the west coast of North America. Coastlines, mountain chains, rivers, water currents, and wind patterns can all serve as piloting cues for the whales. But some remarkable cases of long-distance orientation and movement cannot be explained by piloting.

HOMING: RETURN TO A SPECIFIC LOCATION The ability to return to a nest site, burrow, or other specific location is called **homing**. Homing can be accomplished by piloting in a known environment, but some animals that travel long distances through unfamiliar territory perform much more sophisticated homing. The ability of pigeons to return to their home loft even after being transported to remote sites is well known. How do they find their way home? Experiments have shown that pigeons use the sun as a compass, but they can still find their way home when the sun is not visible. Other experiments have shown that pigeons equipped with frosted contact lenses can find their way home, suggesting that visual cues are not essential. Most amazing has been the demonstration that pigeons can detect Earth's magnetic field and orient to it much as a human orients with a

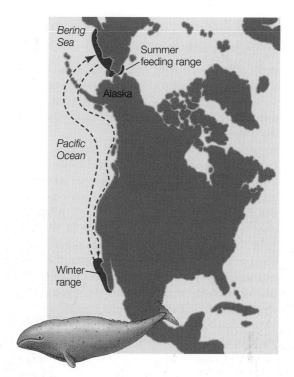

53.16 Piloting Gray whales (*Eschrichtius robustus*) migrate south in winter from the Bering Sea to the coast of Baja California by piloting, in part by following the western coast of North America.

compass. Taken together, the studies of homing by pigeons suggest that they can use multiple, redundant sources of directional information and can switch among those sources depending on the circumstances.

MIGRATION AND NAVIGATION OVER GREAT DISTANCES For as long as humans have inhabited temperate latitudes, they have been aware that whole populations of animals, especially birds, disappear and reappear seasonally. Not until the early nineteenth century, however, were patterns of migration traced by marking individual birds with identification bands around their legs. Only when individuals could be unmistakably identified was it possible to show that the same birds and their offspring returned to the same breeding grounds year after year, and that these same birds could be found during the nonbreeding season at locations hundreds or even thousands of kilometers from their breeding grounds.

Many homing and migrating species are able to take direct routes to their destinations through environments they have never experienced because they use mechanisms of navigation other than piloting. Humans use two major forms of navigation:

- *Distance–direction navigation* requires knowing in what direction and how far away the destination is. With a compass to determine direction and a means of measuring distance, humans can navigate.

- *Bicoordinate navigation*, also known as *true navigation*, requires knowing the latitude and longitude (the map coordinates) of both the current position and the destination, as well as a compass to determine direction.

Other animals, of course, do not carry compasses or GPS receivers, but many of them seem to have a *compass sense*, which allows them to use environmental cues to determine direction, and some appear to have a *map sense*, which allows them to determine their position.

The behavior of many animals suggests that they are capable of bicoordinate navigation. Gray-headed albatrosses, for example, breed on oceanic islands in the Southern Hemisphere. When a young gray-headed albatross first leaves its parents' nest, it flies widely over the southern oceans for 8 or 9 years before it reaches reproductive maturity (**Figure 53.17**). At that time, it flies back to the island where it was raised to select a mate and build a nest. How can it find a tiny island in an enormous ocean after years of wandering? A circadian clock probably gives the albatross information about the time of day and the additional information from the position of the sun might allow it to determine its map coordinates—much as sailors did in the days before global positioning satellites.

(A)

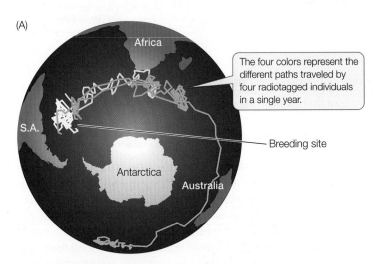

The four colors represent the different paths traveled by four radiotagged individuals in a single year.

 Breeding site

(B) *Thalassarche chrysostoma*

53.17 Coming Home (A) Gray-headed albatrosses are born on islands in the subantarctic oceans. Young birds roam widely over the southern oceans for 8 or 9 years. (B) Once they reach maturity, the birds return to the island where they were born to mate and raise their own young. A courting couple is shown here.

The ability to locate a position by calculating the angles between celestial objects such as the sun and stars and the horizon at specific times of day is called *celestial navigation*. During the day, the sun can serve as a compass for those species that can tell the time of day by means of their circadian clocks. This capacity has been demonstrated by "clock-shifting" experiments.

Researchers placed pigeons in a circular cage that enabled them to see the sun and sky, but no other visual cues. Food bins were arranged around the sides of the cage, and the birds were trained to expect food in the bin at one particular direction—south, for example. After training, no matter what time they were fed, and even if the cage was rotated between feedings, the birds always went to the bin at the southern end of the cage for food, even if that bin contained no food (**Figure 53.18**).

Next, the birds were placed in a room with a controlled light cycle in which the lights were turned on at midnight and off at noon. After 2 weeks of this treatment, the birds' circadian rhythms had been phase-advanced by 6 hours. Then the birds were returned to the circular cage under natural light conditions, with sunrise at 6:00 A.M. At sunrise, if the birds expected to find food in the south bin, then they should have looked for food 90 degrees to the right of the direction of the sun. But because their circadian clocks were telling them it was noon, they looked for food in the direction of the sun—in the east bin. The 6-hour shift in their circadian clocks resulted in a 90-degree error in their orientation, which showed that they were indeed using the sun as a compass. Similar experiments have shown that many animal species can orient by means of a *time-compensated solar compass*.

———— **yourBioPortal.com** ————
GO TO Animated Tutorial 53.4 • Time-Compensated Solar Compass

Many animals are normally nocturnal; in addition, many diurnal bird species migrate at night and thus cannot use the sun to determine direction. The stars offer two sources of information about direction: moving constellations and a fixed point. The positions of constellations (like that of the sun) change because Earth is rotating. With a star map and a clock, direction can be determined using any constellation. But one point that does not change position during the night is the point directly over the axis on which Earth turns. In the Northern Hemisphere, the star Polaris—the "North Star"—lies in that region of the sky and reliably indicates north.

Stephen Emlen at Cornell University showed that birds can learn to use the stars for orientation. As the time of year approaches when young birds would normally migrate to their winter range, young captive birds become more active and orient their activity in the direction they would fly. How do they know that direction? If these birds are raised in a planetarium with a natural star pattern, but one that does not rotate, the birds do not learn to orient, and their premigratory activity is random. However, if the planetarium sky rotates, and even if it rotates around a different point than the North Star, the birds orient their premigratory activity as if the fixed point in the sky was north.

INVESTIGATING LIFE

53.18 A Time-Compensated Solar Compass

Experiments show that pigeons use the sun to establish directions for navigation and finding food. "Clock-shifting" experiments demonstrate that the birds' circadian clocks factor into their ability to judge direction correctly based on the sun's position.

HYPOTHESIS Pigeons determine compass direction from the position of the sun with respect to their internal circadian clocks.

METHOD

1. Place a pigeon in a circular cage from which it can see the sun and sky, but not the horizon or any other visual cue.

2. Surround the cage with multiple food bins but place food only in the southernmost bin, thus training the bird to look for food in the south. (Rotating the cage but always placing the food in the southernmost bin confirms that the bird is navigating to find south.)

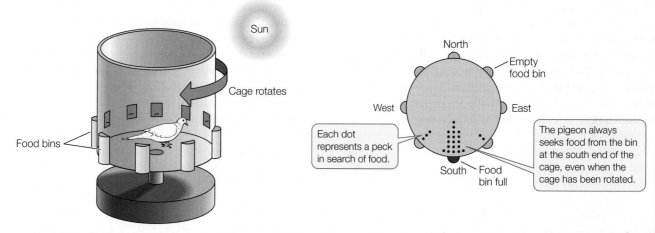

Sun

Cage rotates

Food bins

North

Empty food bin

West

East

Each dot represents a peck in search of food.

The pigeon always seeks food from the bin at the south end of the cage, even when the cage has been rotated.

South   Food bin full

3. Place the trained pigeon in a room with a controlled light cycle for 2 weeks. Turn the lights on at midnight and off at noon to phase-advance its circadian rhythm by 6 hours (i.e., 6 A.M. feels like noon to the bird).

4. Return the pigeon to the circular cage under natural light and observe its food-seeking behavior.

At sunrise, the phase-advanced pigeon seeks food in the bin at the east end of the cage—which would be south by the sun's position at noon.

RESULTS A 6-hour shift in the circadian clock results in a 90-degree error in the pigeon's orientation.

East

Food bin full

CONCLUSION Pigeons have the ability to determine direction using the position of the sun as a compass.

Go to **yourBioPortal.com** for original citations, discussions, and relevant links for all INVESTIGATING LIFE figures.

Animals use multiple modalities to communicate

As individual animals interact, they exchange information; therefore, animal behavior can evolve into systems of information exchange, or **communication**. The behaviors of individuals may become elaborated into communication signals, but only if the transmission of information benefits both the sender and the receiver. To understand why these conditions must be met, consider male courtship displays such as that of the mallard shown in Figure 53.3. This display will be favored if it increases the sender's probability of mating and passing on his

genes, and sexual selection will occur if the display conveys information to a female (the receiver) about his qualities as a potential father for her offspring.

Animals communicate using a variety of sensory modalities that vary in the nature of the signal produced, the specificity of the information conveyed, the speed and persistence of the signal, and its suitability in different environments. Behavioral physiologists interested in communication must take into consideration the sensory and motor characteristics of their study animals as well as the physics of the communication modalities they use and the environment in which the communication takes place.

CHEMICAL SIGNALS   Because of the diversity of their molecular structures, pheromones can communicate very specific, information-rich messages (see Section 46.2). Pheromones are effective day and night, and they can cover a broad range of transmission distances. Pheromones used in different types of communication vary in their volatility (ease of vaporization) and diffusibility; these chemical properties are functions of the nature and size of the pheromone molecule. Pheromones that act as alarm signals, for example, are highly volatile and diffusible, so their message spreads rapidly, but disappears rapidly. Territory-marking and trail-marking pheromones have low volatilities and diffusibilities and stay effective for a long time, so they can convey directional information. Sex pheromones, such as that of the gypsy moth (see Figure 46.3), are intermediate in these properties, so they can spread a long distance, but do not disappear rapidly.

Pheromones are an effective way to exchange species-specific information, and because the recipient must have the proper receptor molecule to detect the pheromone, it is not a signal that is easily intercepted by predators. Pheromonal signals cannot be changed rapidly, but they can convey static, complex information. Mammals that mark their territories with pheromones reveal a great deal of information about themselves: species, individual identity, reproductive status, size (indicated by the height of the marking), and how recently the animal has been in the area (indicated by the strength of the scent).

VISUAL SIGNALS   Visual signals offer the advantage of rapid delivery of information over considerable distances (depending on the environment and the visual acuity of the receiver); they also convey without ambiguity the position of the signaler. Signal content can be enhanced by movements (as in a courtship display) or by different postures. Effective visual signals, however, require sufficient light, and the receiver must be looking directly at the signaler. Thus, visual communication is not particularly useful at night or in environments that lack light, such as caves and ocean depths. Some species have overcome this constraint with light-emitting mechanisms. Fireflies, for example, use an enzymatic mechanism to create flashes of light. By emitting flashes in species-specific patterns, fireflies advertise for mates at night.

Another drawback of visual signals is that they can be intercepted by other species. There are predatory firefly species, for example, that mimic the flash pattern of females of other species. A male that approaches the mimicking "female," becomes a meal rather than a mate. Thus deception can be part of animal communication systems, just as it is part of human communication.

ACOUSTIC SIGNALS   Sound cannot convey complex information as rapidly as visual signals can. But acoustic signals, unlike visual signals, can be used at night and in dark environments. They are not hindered by objects that would interfere with visual signals, so they can be transmitted in complex environments such as forests. They are often better than visual signals

at getting the attention of a receiver because the receiver does not have to be looking at the signaler for the message to be received. Sounds are also useful for communicating over long distances. Even though the intensity of a sound decreases with distance from the source, loud sounds can transmit information over much longer distances than visual signals can. The complex songs of humpback whales, for example, when produced at ocean depths of about 1,000 meters, can be heard hundreds of kilometers away. In this way, humpback whales can locate one another across vast expanses of ocean.

The information content of acoustic signals can be increased by varying their frequency, as we can see in the sonograms of the species-specific songs of birds shown in Figure 53.9, and as we practice in our own speech. However, acoustic signals place the signaler at risk for detection by predators. This danger can be minimized by adjustments of frequency and signal structure that decrease the directional information the receiver can extract from the signal. Alarm calls tend to be pure tones (a single frequency) without much temporal structure (starts and stops). It is very difficult to localize such calls. On the other hand, territorial calls tend to cover a broad frequency range and have temporal structure. These calls are easy to localize. The frequencies and structures of acoustic signals are also adapted to specific habitats. Different vegetation types, for example, have different sound-absorbing properties: pure tones at lower frequencies carry better in forests, and more complex calls at higher frequencies carry well in open habitats.

MECHANOSENSORY SIGNALS   Animals in close contact with one another communicate by touch, especially under conditions in which visual communication is difficult. The best-studied use of mechanosensory communication is the dance of honey bees, first described by Karl von Frisch. Honey bees have a spectacular ability to navigate and can reliably communicate the location of food sources as far as 2.5 km away from their hive. When a forager bee finds food, she returns to the hive and communicates her discovery to her hive-mates by dancing in the dark hive on the vertical surface of the honeycomb. Other bees follow the dancer and receive her message.

If the food source is more than about 100 m away from the hive, the forager performs a *waggle dance* (**Figure 53.19**), which conveys information about both the distance and the direction of the food source. She repeatedly traces out a figure-eight pattern as she runs on the honeycomb. She alternates half-circles to the left and right with vigorous wagging of her abdomen in the short, straight run between turns. Bees use the sun as their compass, and the angle of the straight run indicates the direction of the food source relative to the position of the sun projected down to the horizon. Even under cloudy conditions, the forager can provide directions to the food source because she can see polarized light. The honey bee's internal clock allows the dancer to adjust her dance to take into account the sun's movement during her return flight. The speed of the dance indicates the distance to the food source: the farther away it is, the longer the duration of each waggle run.

If the food she has found is less than 80 m from the hive, the forager performs a *round dance*, running rapidly in a circle and

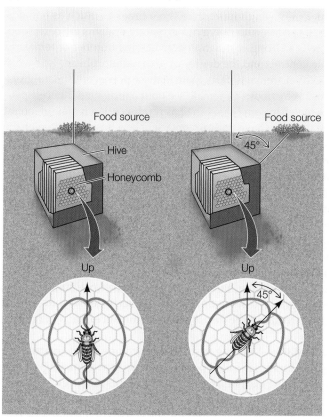

(A)   (B)

Food source   Food source

45°

Hive

Honeycomb

Up   Up

45°

Pattern of waggle dance   Pattern of waggle dance

53.19 The Waggle Dance of the Honey Bee (A) A honey bee runs straight up on the vertical surface of the honeycomb in the dark hive while wagging her abdomen to tell her hivemates that there is a food source in the direction of the sun. (B) When her waggle runs are at an angle from the vertical, the other bees know that the same angle separates the direction of the food source from the direction of the sun.

─── **yourBioPortal.com** ───
GO TO Web Activity 53.1 • Honey Bee Dance Communication

reversing her direction after each full circle. The odor on her body and the round dance combine tactile and chemical cues: the odor indicates the flower to be looked for, and the dance communicates the fact that the food source is close to the hive.

COMMUNICATION IN MULTIPLE SENSORY MODALITIES Avoiding ambiguity is a high priority in any signaling system. Signal specificity is enhanced if multiple sensory modalities are used. Courtship behavior in fruit flies, for example, involves tactile, chemical, visual, and acoustic signals (see Figure 53.6A). The male fruit fly orients toward the female and taps her body with his foreleg (touch). Upon detection of pheromones in her cuticle (chemical), the male begins to vibrate one wing, producing a species-specific "courtship song" (acoustic). The male then extends his mouthparts to taste the female's genitalia (chemical and tactile); if she is receptive, he initiates copulation. At any point, if any sensory feedback indicates to either the male or the female that their pairing is inappropriate, the courtship abruptly ends.

53.5 RECAP

Biological rhythms allow an animal to anticipate changes in its environment. In mammals, a circadian clock located in the suprachiasmatic nuclei generates a rhythm that is entrainable by environmental information. The navigational abilities of animals range from simple piloting by landmarks to distance-and-direction and bicoordinate navigation. Behaviors may evolve into communication signals if the transmission of information benefits both the sender and the receiver.

- What is meant by a "free-running" rhythm? Describe how it can be entrained to the 24-hour day. **See pp. 1127–1128 and Figure 53.15**

- Explain the difference between piloting, distance-and-direction navigation, and bicoordinate navigation. **See pp. 1129–1130**

- Give an advantage and a disadvantage for each of three modalities of communication. **See pp. 1131–1133**

We have just seen how a behavioral physiologist might look at the courtship behavior of a male fruit fly. A behavioral ecologist would probably approach this behavior in a different way, asking how it influences the fitness of both the male fly and his mate. Given that interactions between individual animals, especially those involved in mating and the care of offspring, are so directly tied to fitness, it is not surprising that they are an important focus of behavioral studies.

53.6 How Does Social Behavior Evolve?

The evolution of social behavior became a field of study in its own right in 1975, with the publication of E. O. Wilson's landmark book, *Sociobiology*. We begin our consideration of this enormous field with simple interactions that involve a single male and a single female, but already we see diversity. Species differ in their mating systems, which vary from monogamous to promiscuous; in the amount of parental care they give their young; and in the degree to which the male contributes to raising the young. Beyond these relatively simple mating systems, there are associations of larger numbers of reproductive individuals in *polygynous* mating systems, in which a male has more than one mate, or *polyandrous* mating systems, in which a female has more than one mate. Even more complex interactions exist in which extended families participate in raising young, and finally there are societies such as honey bee colonies in which large numbers of nonreproductive individuals assist a single reproductive individual. Sociobiology proposes that the evolution of all these variations can be understood by asking how social behavior contributes to the fitness of the *individuals* involved.

Mating systems maximize the fitness of both partners

At the start of Chapter 7, we learned about the mating behavior of two species of voles. Prairie voles (*Microtus ochrogaster*) are *monogamous*, forming strong pair bonds that can last for life, and both parents participate in rearing the young. In contrast, montane voles (*M. montanus*) are *promiscuous*: males mate with many females, and the young are raised by the females alone. Behavioral physiologists have explained the proximate mechanisms behind these stark behavioral differences in terms of the release of neurohormones and the distribution of the receptors for those hormones in the brains of the two species. The ultimate question—and the one asked here—is why two such different mating systems evolved in two species that are so closely related.

We begin with the premise that there is an asymmetry in the contributions of male and female animals to their offspring at the time of fertilization. As we learned in Section 44.1, females produce a limited number of eggs, and each egg is generously stocked with resources. Males produce an almost infinite number of sperm, which contain next to no resources. So the energetic and opportunity costs of reproduction are greater for the female than for the male. In mammals, this asymmetry increases throughout gestation as the female bears most of the costs. By the time of birth or hatching, the female's investment in the young is much greater than the male's investment, and the main way for the female to maximize her fitness is to make sure her young are healthy and survive to pass on her genes.

The male has different options for maximizing his fitness. He can simply move on after inseminating the female and seek additional mates as a means of maximizing his reproductive success—as in the case of the meadow vole. Or he can stay with the female he inseminated, protect her, and help care for their young—as in the case of the prairie vole. Which strategy maximizes his fitness depends on a number of factors that are influenced by the species' environment, such as the likelihood that a female and her offspring will survive without a male's help, and a male's likelihood of finding another fertile female. Thus sociobiologists seek to quantify these factors in nature as a means of explaining observed differences in mating systems.

POLYGYNY **Polygynous** mating systems, in which a male has more than one mate, involve a different asymmetry. In situations in which a male can sequester a group of females from other males, he can increase his fitness by increasing the number of females in his group. As we saw in Section 53.4, male elephant seals accomplish this by protecting an area of beach where females give birth. Male baboons do so by herding females. Male red-winged blackbirds may acquire more than one mate by defending high-quality nesting territories, building more than one nest in their territories, and attracting females to those nests. Since sex ratios in all these species are close to 50-50, a large differential in male fitness is established: some males have high reproductive success while many males have none. Thus selection favors males that are successful in competing with other males to obtain and protect access to many females. In general, bigger, stronger males are the winners, and sexual dimorphism in body size evolves. The elephant seal is an extreme example: males may weigh more than three times as much as females. When species with polygynous mating systems are compared, there is a strong correlation between the number of females a male controls and the degree of sexual dimorphism.

Why do females participate in these polygynous mating systems? Why doesn't a female seek out a nice, kind, noncompetitive male? In some cases she has no choice. If a female elephant seal wants to have her pup on a safe beach, she must enter the territory of a male. If a female red-winged blackbird wants to nest in an optimal territory, she will have to share the attentions of the territory's owner with other females. However, even if the female has a choice of mates, she is likely to maximize her fitness by mating with a male who is strong and dominant enough to control a number of females. Why? If her mate is a dominant male, her male offspring are likely to have their father's traits, become dominant males, and give her more grandchildren. The ultimate result of females selecting males for their prowess and dominance in male–male competition is the lek mating system (see Figure 53.12C), in which the *only* thing a male offers a female is the display of his dominance over other males.

POLYANDRY A mating system in which one female mates with multiple males is called **polyandry**. This type of mating system is relatively rare, but it is seen in some birds and a few mammal species in which paternal care for the young can have a large effect on fitness. That is the case the golden lion tamarins (*Leontopithecus rosalia*), primates native to the Brazilian tropical forests (**Figure 53.20**). In comparison to other primates, tamarins are tiny—under 1 kg—so they face high predation pressure. Females usually give birth to twins and thus newborns constitute a higher percentage of maternal weight than those of other primates. They also grow faster, so nursing costs are high. For all these reasons, young tamarins cared for by their mother alone are unlikely to survive.

Leontopithecus rosalia

53.20 Polyandry in a Small Primate The endangered golden lion tamarins of Brazil are small primates whose unique life history has given rise to polyandry in some groups, with males playing a major role in rearing the young.

What can a male tamarin do to help guarantee his reproductive success? Watching out for predators is one obvious contribution; gathering food for the female and her young is another. Like other primate parents, tamarins carry their young most of the time, but most other primates have single offspring. When tamarin mothers are carrying twins, they spend 92 percent of the time resting, in comparison to only 58 percent of the time when they are not carrying young. Resting is not compatible with foraging and filling the mother's high energy requirements. When a male is present, however, he carries the young about one-third of the time, so the mother has much more time for foraging and feeding.

If one male is helpful in protecting and raising young, then two should be even more helpful. Some females can attract a second mate by being sexually receptive to him. Neither male can be sure that any eventual offspring are his, so it is in the best interest of both to help in their rearing. Of the social groups observed in field studies, only 22 percent had one male and one female, whereas 61 percent had multiple males and one female.

Fitness can include more than producing offspring

As humans, we readily understand the concept of extended family—brothers, sisters, aunts, uncles, nieces, nephews. Extended families are a common form of social organization in other species as well, and members of these families may cooperate in territory defense, predator avoidance, foraging, and rearing of young. If behavior is favored when it increases the fitness of the individual performing it, then how can we explain the evolution of social behaviors that do not lead to the performer having more offspring and may even appear to be **altruistic**—benefiting another individual at a cost to the performer?

An individual's fitness is increased by having offspring because those offspring carry the parent's genes into the next generation. Fitness gained by producing offspring is referred to as **individual fitness**. However, an individual's genes are carried into the next generation by more than his or her own offspring. In diploid organisms, two offspring of the same parents share, on average, 50 percent of the same alleles, and an individual is

likely to share 25 percent of its alleles with its sibling's offspring. Therefore, by helping parents and other relatives raise their offspring, an individual increases the transmission of those shared alleles to the next generation. **Inclusive fitness** is the fitness derived from an individual's own reproductive success plus that derived through the reproductive success of its relatives.

The maximization of inclusive fitness is the mechanism behind **kin selection**: selection for behaviors that increase the reproductive success of relatives even when they come at a cost to the performer. An example is the phenomenon of *helping at the nest*, which was extensively studied in Florida scrub jays by Glen Woolfenden. Scrub jay pairs mate for life and establish large territories, which they defend aggressively. The mating pair may be assisted in rearing their young by 3 to 5 helpers (**Figure 53.21**). The helpers guard against predators, feed the young, clean the nest, and fly with fledglings. Why are these birds helping others rather than rearing their own young? Through a long-term study, Woolfenden was able to establish a number of important facts:

- The helpers are prior offspring of the mating pair and are usually 1 to 3 years old.
- Young birds that attempt to breed have almost zero reproductive success.
- Mating pairs with helpers have approximately 3 times the reproductive success of those without helpers.

These results support the conclusion that helper scrub jays are maximizing their inclusive fitness by helping their parents raise siblings until they are old and mature enough to have a reasonable probability of successfully raising their own offspring.

The concept of kin selection was formalized by W. D. Hamilton in what has become known as **Hamilton's rule**. He argued that, for an apparent altruistic behavior to be adaptive, the fitness benefit of that act to the recipient times the degree of relatedness between the performer and the recipient had to be greater than the cost to the performer. This relationship was clearly stated years before by the eminent geneticist J. B. S. Haldane, who said during an argument about altruism that he would not be willing to risk his life to save his brother—but for 2 brothers or 8 cousins, he would consider it.

53.21 Helpers at the Nest
Young Florida scrub jays (*Aphelocoma coerulescens*) often forego reproduction in their first few years of adulthood to help their parents raise their siblings. These young birds help their parents feed the nestlings, defend the territory, and protect the nest from predators.

Eusociality is the extreme result of kin selection

Species such as honey bees whose social groups include nonreproductive workers are called **eusocial** species. Prime examples are wasps, bees, and ants—members of the large insect order *Hymenoptera*. In a honey bee colony, the single reproductive female—the queen—occasionally produces a few male offspring, but most of the thousands of other individuals in the colony—her offspring—are sterile female workers. The key to understanding the evolution of eusociality in hymenopterans is their sex determination mechanism, called **haplodiploidy** because diploid individuals are female and haploid individuals are male. The queen carries a lifetime supply of sperm obtained during her single mating flight, and she controls whether her eggs are fertilized or not. An unfertilized egg develops into a male, a fertilized egg develops into a female.

If a queen copulates with only one male, all the sperm she receives are identical because a haploid male has only one set of chromosomes, all of which are transmitted to every sperm cell. Therefore, the queen's daughters share all of their father's genes and, on average, half of their mother's genes. As a result, the workers in the hive—all sisters—share, on average, 75 percent of their alleles. Were they to reproduce, they would share only 50 percent of their alleles with their own female offspring. Thus they can potentially increase their inclusive fitness more by caring for their sisters than by producing and caring for their own offspring.

Haplodiploidy is not essential for eusociality to evolve. This social system may also arise if it is costly or dangerous to establish new colonies. Nearly all eusocial animals construct elaborate nests or burrow systems within which their offspring are reared. Such a structure represents an enormous investment of resources. Naked mole-rats—the most eusocial mammals—live

Heterocephalus glaber

53.22 A Eusocial Mammal Naked mole-rats live in a large colony with only one reproductive female and a few reproductive males. They live in an elaborate tunnel system excavated by the colony over time.

in elaborate underground tunnel systems that can extend for as much as 5 kilometers in cumulative length (**Figure 53.22**). A colony includes 70 to 80 individuals but only one reproductive female and a few reproductive males. The other colony members are sterile workers that dig and maintain the tunnels, guard against intruders, harvest food (tubers), and use their feces to feed the queen and her offspring. Individuals attempting to found new colonies have a high risk of failing or being captured by predators. When chances of individual reproductive success are practically zero, an individual can best maximize its inclusive fitness by staying with and helping to maintaining the colony.

(A)

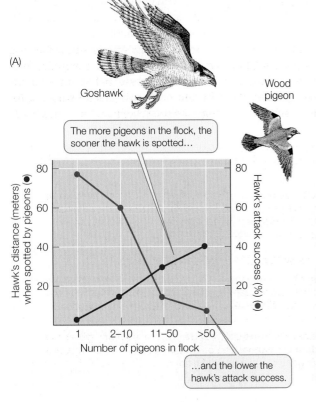

Goshawk

Wood pigeon

The more pigeons in the flock, the sooner the hawk is spotted…

...and the lower the hawk's attack success.

(B) *Spermophilus beldingi*

53.23 Group Living Provides Protection from Predators
Animals that live in groups can spread the cost of looking out for predators. (A) The larger the number of pigeons in a flock, the greater the chances that one of the pigeons will spot a hawk before it attacks, and the lower the chances that the hawk will capture one of the pigeons. (B) A young male Belding's ground squirrel gives an alarm call upon spotting a predator. Although this behavior increases his individual risk of becoming its prey, he increases the survival chances of many of his close relatives.

Group living has benefits and costs

Apart from their direct influences on reproductive success, social systems can contribute to survival in many ways, but they can also involve costs. Thus the cost–benefit approach of behavioral ecology is relevant to understanding the evolution of social behavior.

An obvious example of a benefit of group living is improved foraging efficiency. By hunting in packs, African wild dogs (see Figure 59.17) employ cooperative strategies that enable them to bring down larger prey than could a single dog. The larger the pack, the greater the hunting success rate. Once the prey is killed, the presence of conspecifics also reduces the risk that the wild dogs will lose their prey to larger scavengers, such as hyenas.

Living in a group can also reduce the risk of the members' becoming prey themselves. Many small birds forage in flocks. To test the hypothesis that flocking provides protection against predators, one investigator released a trained goshawk near wild wood pigeons in England. The hawk was most successful when it attacked solitary pigeons. Its success in capturing a pigeon in a flock decreased as the number of pigeons in the flock increased (**Figure 53.23A**). The larger the flock, the sooner some individual in the flock spotted the hawk and flew away. This escape behavior stimulated other individuals in the flock to take flight as well.

Alarm calling is another means of reducing predation risk, but the caller incurs a risk cost by calling attention to itself. Belding's ground squirrels live in large colonies in open meadows. When one squirrel announces the presence of a predator with loud, sharp barks, all the nearby squirrels dive into their burrows (**Figure 53.23B**). Paul Sherman showed that callers double their risk of being preyed on—so why do they do it? Sherman's and others' research has shown that this altruistic behavior is a product of kin selection. In this polygynous species, males establish large territories in the spring that include the territories of several females, whom they inseminate. The females then drive the males off. Female offspring settle near their mothers, so neighboring females in a colony tend to be sisters, and they defend each other's young. Sherman showed that males are less likely to give alarm calls than females, and that females are more likely to give alarm calls when related individuals are nearby.

Social behavior has many costs as well as benefits. Foraging in a group may reduce the amount of food available to each individual, and the foraging individuals may interfere with one another's foraging activities. Individuals living in groups may face more competition for mates, as well as for food, than solitary individuals would. A large group may actually attract the attention of predators. And living at high population densities can increase the risk of disease transmission. The study of disease transmission in wild animal populations is a relatively new field, but such studies have made it apparent that species living in social groups are more prone to outbreaks of disease than are solitary species.

53.6 RECAP

Social behavior can be understood by asking how it contributes to the fitness of the individuals involved. Asymmetry between the sexes in parental investment is a key factor in the evolution of mating systems. According to the theory of kin selection, an individual can increase its fitness by helping related individuals with whom it shares alleles. In extreme cases, kin selection has given rise to eusociality.

- What environmental conditions can lead to monogamy, polygamy, or polyandry? See pp. 1134–1135

- Explain how an individual can increase its fitness by helping its relatives. See p. 1135

- Why is eusociality so common among hymenopterans? See p. 1136

- What are some of the costs and benefits of group living? See p. 1137 and Figure 53.23

Knowledge of the behavior of particular species—how they use the environment, how they obtain food, how they organize their activities spatially and temporally—is essential for understanding how species interact in nature. These interactions are one focus of the science of ecology, the subject of Part Ten of this book.

CHAPTER SUMMARY

53.1 What Are the Origins of Behavioral Biology?

- Ivan Pavlov's discovery of **conditioned reflexes** and B. F. Skinner's research on **operant conditioning** as a model for learning led to an approach called **behaviorism** that mainly carried out laboratory experiments on rats and a few other animal models. Review Figure 53.1

- **Ethology** focuses on both the **proximate causes** of behavior—what is the immediate cause of the behavior, and how does the behavior develop?— and on the **ultimate causes**—how does the behavior affect the animal's fitness, and how does it compare with similar behaviors in related species?

- A major focus of the ethologists was **fixed action patterns** and their **releasers**. They performed **deprivation experiments** as well as breeding experiments to demonstrate that certain behaviors are genetically determined.

53.2 How Can Genes Influence Behavior?

- Breeding experiments can reveal whether a behavioral phenotype is inherited. Gene knockout experiments can reveal the roles of specific genes underlying a behavioral phenotype. Review Figures 53.4 and 53.5

- Most behaviors are complex traits involving many genes that function in cascades and offer many points for a change in a single gene to influence behavior. Review Figure 53.6

53.3 How Does Behavior Develop?

- Hormones can determine the pattern of behavior that develops and the timing of its expression. **Review Figure 53.7**

- **Imprinting** is a process by which an animal learns a specific set of stimuli during a limited **critical** or **sensitive period**. That critical period may be determined by hormones.

- The development and expression of song in white-crowned sparrows involves a genetic predisposition to learn the species-specific song, a critical period for imprinting of a song memory, and hormonally controlled timing of song expression. Social interactions may also play a role. **Review Figure 53.9**

53.4 How Does Behavior Evolve?

- An animal's behavior is a series of choices that influence its fitness. To make these choices, animals use environmental cues that are reliable predictors of the potential effects of their choices on their fitness. **Review Figure 53.10**

- The **cost–benefit approach** can be used to investigate the fitness value of specific behaviors. The cost of a behavior typically has three components: **energetic cost**, **risk cost**, and **opportunity cost**. **Review Figure 53.11, ANIMATED TUTORIAL 53.1**

- According to **optimal foraging theory**, animals should practice feeding behaviors that maximize their energetic gain at the least cost. **Review Figure 53.13, ANIMATED TUTORIAL 53.2**

53.5 What Physiological Mechanisms Underlie Behavior?

- **Circadian rhythms** control the daily cycle of behavior. Without environmental time cues, circadian rhythms free-run with a period that is genetically programmed. They are normally **entrained** to the light–dark cycle by environmental cues. **Review Figure 53.15, ANIMATED TUTORIAL 53.3**

- **Circannual rhythms** are endogenous yearly cycles whose neural basis is still unknown.

- Three forms of navigation used by animals to find their way in the environment are piloting (orienting to landmarks), distance–direction navigation, and bicoordinate navigation. Navigation mechanisms include celestial navigation and a time-compensated solar compass. **Review Figures 53.16–53.18, ANIMATED TUTORIAL 53.4**

- The behaviors of individuals may become communication signals if the transmission of information benefits both the sender and the receiver.

- Chemical communication signals (pheromones) can be highly specific and have different time courses depending on their volatility and diffusibility. Visual signals can convey complex messages rapidly, but the recipient must be looking at the sender. Acoustic signals travel well over distances, do not require a focused recipient, and can be modified to reveal or conceal directional information. Tactile signals are used by animals in close proximity and can convey complex messages. **Review Figure 53.19, WEB ACTIVITY 53.1**

53.6 How Does Social Behavior Evolve?

- The evolution of mating systems can be understood by looking at the fitness costs and benefits incurred by each partner in the species' environment. **Polygynous** mating systems, in which one male controls and mates with many females, can result in great variation in male reproductive success. **Polyandry**—a female mating with multiple males—can evolve in circumstances in which a male can make a substantial contribution to the survival of his offspring.

- The fitness an individual gains by producing offspring (**individual fitness**) plus the fitness it gains by increasing the reproductive success of relatives with whom it shares alleles is called **inclusive fitness**. **Kin selection** may favor **altruistic** behavior toward relatives, despite its cost to the performer, if it increases the performer's inclusive fitness.

- As a result of **haplodiploidy**, the sex determination mechanism of the Hymenoptera, nonreproductive workers share more alleles with one another than females share with their own offspring. Haplodiploidy has probably facilitated the evolution of **eusociality** in this group through kin selection. Eusociality has also risen in diploid species in which chances of individual reproductive success are extremely low.

- Group living confers benefits such as greater foraging efficiency, protection from predators, but it also has costs, such as increased competition for food and ease of transmission of diseases.

SEE WEB ACTIVITY 53.2 for a concept review of this chapter.

SELF-QUIZ

1. Which of the following is *not* true of a fixed action pattern?
   a. Its expression may depend on hormonal conditions.
   b. It is induced by complex, species-specific stimuli.
   c. It is highly stereotypic and species-specific.
   d. It can be expressed even if the animal has never seen it performed.
   e. Its genetic basis can be demonstrated by breeding experiments.

2. Which of the following is *not* a component of the cost of performing a behavior?
   a. Its energetic cost
   b. The risk of being injured
   c. Its opportunity cost
   d. The risk of being attacked by a predator
   e. Its information cost

3. Birds that migrate at night
   a. inherit a star map.
   b. determine direction by knowing the time and the position of a constellation in the sky.
   c. orient to a particular point in the sky.
   d. imprint on one or more key constellations.
   e. determine distance, but not direction, from the stars.

4. If a bird is trained to seek food on the western side of a cage open to the sky, and is then placed in a chamber with a controlled light cycle so that its circadian rhythm becomes phase-delayed by 6 hours (i.e., its circadian rhythm is 6 hours behind real time), when it is returned to the open cage at noon in real time, it will seek food in the
   a. north.
   b. south.
   c. east.
   d. west.

5. To be able to pilot, an animal must
   a. have a time-compensated solar compass.
   b. orient to a fixed point in the night sky.
   c. know the distance between two points.
   d. know landmarks.
   e. know its longitude and latitude.

6. Which of the following statements about communication is *true*?
   a. Complex information cannot be conveyed by pheromones.
   b. Visual signaling is advantageous in complex environments.
   c. Acoustic communication always reveals the location of the signaler.
   d. An advantage of pheromones is that the message can persist over time.
   e. The dance of honey bees is an example of visual signaling.

7. A cost commonly associated with group living is
   a. increased risk of predation.
   b. interference with foraging.
   c. higher exposure to diseases and parasites.
   d. poorer access to mates.
   e. All of the above

8. The choice of a mating partner may be based on
   a. the competitive ability of a potential mate.
   b. the territory held by a potential mate.
   c. both the competitive ability of a potential mate and the territory it holds.
   d. the courtship display of a potential mate.
   e. all of the above

9. Altruistic behavior
   a. can increase an individual's inclusive fitness.
   b. depends on haplodiploid sex determination.
   c. is most common between unrelated individuals.
   d. always causes a net decrease in the performer's fitness.
   e. characterizes a monogamous mating system.

10. A social group is said to be eusocial if
    a. group members interact very intensively.
    b. some group members produce many more offspring than others do.
    c. a dominance hierarchy exists among group members.
    d. young individuals remain in the group to help their parents rear other offspring.
    e. the group contains nonreproductive helper individuals.

FOR DISCUSSION

1. Cowbirds are nest parasites. A female cowbird lays an egg in the nest of another bird species, which then incubates the egg and raises the chick. What do you think would characterize the acquisition of song in cowbirds? In a given area, cowbirds tend to parasitize the nests of particular bird species. How do you think female cowbirds learn this behavior? How would you test your hypothesis?

2. Male dogs lift a hind leg when they urinate; female dogs squat. If a male puppy receives an injection of estrogen when it is a newborn, it will never lift its leg to urinate for the rest of its life; it will always squat. How might this result be explained?

3. The short-tailed shearwater is a bird that winters in Antarctica and summers in the Arctic. What problems would this species have in using either the sun or the stars for navigation? What is the most likely means it uses to find its way to its summer and its winter feeding grounds?

4. In most vertebrate species with helpers, the helpers are individuals capable of reproducing, and most of them later breed on their own. Among eusocial insects, sterile castes have evolved repeatedly. What differences between vertebrates and insects might explain the failure of sterile castes to evolve in the former?

ADDITIONAL INVESTIGATION

Experiments on animal migration show that animals can use a time-compensated solar compass or identification of a fixed point in the night sky as a basis for distance-and-direction navigation. However, the observations on the homing ability of seabirds such as the albatross indicate that animals might be capable of true or bicoordinate navigation. What experiments could you do to prove that animals have that capability, and what experiments could you do to test hypotheses about the mechanisms they might use? (*Hint*: At least two hypotheses might involve using the elevation of the sun or the angles of Earth's magnetic lines of force as a means of determining latitude.)

WORKING WITH DATA (GO TO yourBioPortal.com)

The Costs and Benefits of Foraging Behavior In this exercise based on Figure 53.13, you will analyze the techniques and results of Werner and Hall's experiments with bluegill sunfish and answer questions pertaining to experiments such as this one that seek to quantify behaviors.

Appendix A: The Tree of Life

Phylogeny is the organizing principle of modern biological taxonomy. A guiding principle of modern phylogeny is *monophyly*. A monophyletic group is considered to be one that contains an ancestral lineage and *all* of its descendants. Any such group can be extracted from a phylogenetic tree with a single cut.

The tree shown here provides a guide to the relationships among the major groups of extant (living) organisms in the tree of life as we have presented them throughout this book. The position of the branching "splits" indicates the relative branching order of the lineages of life, but the time scale is not meant to be uniform. In addition, the groups appearing at the branch tips do not necessarily carry equal phylogenetic "weight." For example, the ginkgo [63] is indeed at the apex of its lineage; this gymnosperm group consists of a single living species. In contrast, a phylogeny of the eudicots [55] could continue on from this point to fill many more trees the size of this one.

The glossary entries that follow are informal descriptions of some major features of the organisms described in Part Seven of this book. Each entry gives the group's common name, followed by the formal scientific name of the group (in parentheses). Numbers in square brackets reference the location of the respective groups on the tree.

It is sometimes convenient to use an informal name to refer to a collection of organisms that are not monophyletic but nonetheless all share (or all lack) some common attribute. We call these "convenience terms"; such groups are indicated in these entries by quotation marks, and we do not give them formal scientific names. Examples include "prokaryotes," "protists," and "algae." Note that these groups cannot be removed with a single cut; they represent a collection of distantly related groups that appear in different parts of the tree. We also use quotation marks here to designate two groups of fungi that are not believed to be monophyletic.

An interactive version of this tree, with links to much greater detail (such as photos, distribution maps, species lists, and identification keys), can be found at yourBioPortal.com.

– A –

acorn worms (*Enteropneusta*) Benthic marine hemichordates [120] with an acorn-shaped proboscis, a short collar (neck), and a long trunk.

"algae" A convenience term encompassing various distantly related groups of aquatic, photosynthetic chromalveolates [5] and certain members of the Plantae [8].

alveolates (*Alveolata*) [7] Unicellular eukaryotes with a layer of flattened vesicles (alveoli) supporting the plasma membrane. Major alveolate groups include the dinoflagellates [52], apicomplexans [53], and ciliates [54].

amborella (*Amborella*) [60] An understory shrub or small tree found in New Caledonia. Thought to be the sister-group of the remaining living angiosperms [14].

ambulacrarians (*Ambulacraria*) [30] The echinoderms [119] and hemichordates [120].

amniotes (*Amniota*) [37] Mammals, reptiles, and their extinct close relatives. Characterized by many adaptations to terrestrial life, including an amniotic egg (with a unique set of membranes—the amnion, chorion, and allantois), a water-repellant epidermis (with epidermal scales, hair, or feathers), and, in males, a penis that allows internal fertilization.

amoebozoans (*Amoebozoa*) [85] A group of eukaryotes [4] that use lobe-shaped pseudopods for locomotion and to engulf food. Major amoebozoan groups include the loboseans, plasmodial slime molds, and cellular slime molds.

amphibians (*Amphibia*) [129] Tetrapods [36] with glandular skin that lacks epidermal scales, feathers, or hair. Many amphibian species undergo a complete metamorphosis from an aquatic larval form to a terrestrial adult form, although direct development is also common. Major amphibian groups include frogs and toads (anurans), salamanders, and caecilians.

amphipods (*Amphipoda*) Small crustaceans [117] that are abundant in many marine and freshwater habitats. They are important herbivores, scavengers, and micropredators, and are an important food source for many aquatic organisms.

angiosperms (*Anthophyta* or *Magnoliophyta*) [14] The flowering plants. Major angiosperm groups include the monocots, eudicots, and magnoliids.

animals (*Animalia* or *Metazoa*) [21] Multicellular heterotrophic eukaryotes. The majority of animals are bilaterians [24]. Other groups of animals include the cnidarians [98], ctenophores [97], placozoans [96], and sponges [22]. The closest living relatives of the animals are the choanoflagellates [92].

annelids (*Annelida*) [105] Segmented worms, including earthworms, leeches, and polychaetes. One of the major groups of lophotrochozoans [26].

anthozoans (*Anthozoa*) One of the major groups of cnidarians [98]. Includes the sea anemones, sea pens, and corals.

anurans (*Anura*) Comprising the frogs and toads, this is the largest group of living amphibians [129]. They are tail-less, with a shortened vertebral column and elongate hind legs modified for jumping. Many species have an aquatic larval form known as a tadpole.

apicomplexans (*Apicomplexa*) [53] Parasitic alveolates [7] characterized by the possession of an apical complex at some stage in the life cycle.

arachnids (*Arachnida*) Chelicerates [115] with a body divided into two parts: a cephalothorax that bears six pairs of appendages (four pairs of which are usually used as legs) and an abdomen that bears the genital opening. Familiar arachnids include spiders, scorpions, mites and ticks, and harvestmen.

arbuscular mycorrhizal fungi (*Glomeromycota*) [88] A group of fungi [19] that associate with plant roots in a close symbiotic relationship.

archaeans (*Archaea*) [3] Unicellular organisms lacking a nucleus and lacking peptidoglycan in the cell wall. Once grouped with the bacteria, archaeans possess distinctive membrane lipids.

archosaurs (*Archosauria*) [39] A group of reptiles [38] that includes dinosaurs and crocodilians [134]. Most dinosaur groups became extinct at the end of the Cretaceous; birds [133] are the only surviving dinosaurs.

arrow worms (*Chaetognatha*) [107] Small planktonic or benthic predatory marine worms with fins and a pair of hooked, prey-grasping spines on each side of the head.

arthropods (*Arthropoda*) The largest group of ecdysozoans [27]. Arthropods are characterized by a stiff exoskeleton, segmented bodies, and jointed appendages. Includes the chelicerates [115], myriapods [116], crustaceans [117], and hexapods (insects and their relatives) [118].

ascidians (*Ascidiacea*) "Sea squirts"; the largest group of urochordates [122]. Also known as tunicates, they are sessile (as adults), marine, saclike filter feeders.

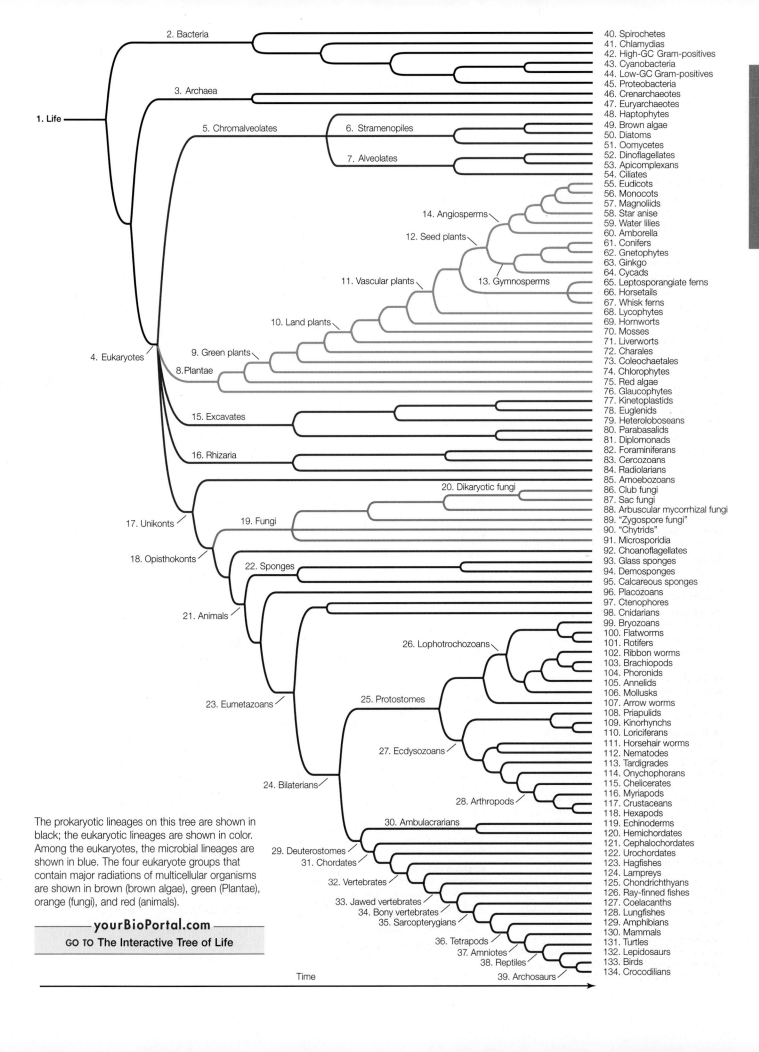

1. Life

2. Bacteria
- 40. Spirochetes
- 41. Chlamydias
- 42. High-GC Gram-positives
- 43. Cyanobacteria
- 44. Low-GC Gram-positives
- 45. Proteobacteria

3. Archaea
- 46. Crenarchaeotes
- 47. Euryarchaeotes

4. Eukaryotes

5. Chromalveolates
6. Stramenopiles
- 48. Haptophytes
- 49. Brown algae
- 50. Diatoms
- 51. Oomycetes
7. Alveolates
- 52. Dinoflagellates
- 53. Apicomplexans
- 54. Ciliates

14. Angiosperms
- 55. Eudicots
- 56. Monocots
- 57. Magnoliids
- 58. Star anise
- 59. Water lilies
- 60. Amborella
12. Seed plants
13. Gymnosperms
- 61. Conifers
- 62. Gnetophytes
- 63. Ginkgo
- 64. Cycads
11. Vascular plants
- 65. Leptosporangiate ferns
- 66. Horsetails
- 67. Whisk ferns
- 68. Lycophytes
10. Land plants
- 69. Hornworts
- 70. Mosses
- 71. Liverworts
9. Green plants
- 72. Charales
- 73. Coleochaetales
- 74. Chlorophytes
8. Plantae
- 75. Red algae
- 76. Glaucophytes

15. Excavates
- 77. Kinetoplastids
- 78. Euglenids
- 79. Heteroloboseans
- 80. Parabasalids
- 81. Diplomonads

16. Rhizaria
- 82. Foraminiferans
- 83. Cercozoans
- 84. Radiolarians

17. Unikonts
- 85. Amoebozoans
20. Dikaryotic fungi
- 86. Club fungi
- 87. Sac fungi
19. Fungi
- 88. Arbuscular mycorrhizal fungi
- 89. "Zygospore fungi"
- 90. "Chytrids"
- 91. Microsporidia
18. Opisthokonts
- 92. Choanoflagellates
22. Sponges
- 93. Glass sponges
- 94. Demosponges
- 95. Calcareous sponges
21. Animals
- 96. Placozoans
- 97. Ctenophores
- 98. Cnidarians
23. Eumetazoans

26. Lophotrochozoans
- 99. Bryozoans
- 100. Flatworms
- 101. Rotifers
- 102. Ribbon worms
- 103. Brachiopods
- 104. Phoronids
- 105. Annelids
- 106. Mollusks
25. Protostomes
- 107. Arrow worms
- 108. Priapulids
- 109. Kinorhynchs
- 110. Loriciferans
27. Ecdysozoans
- 111. Horsehair worms
- 112. Nematodes
- 113. Tardigrades
- 114. Onychophorans
- 115. Chelicerates
- 116. Myriapods
28. Arthropods
- 117. Crustaceans
- 118. Hexapods
24. Bilaterians

30. Ambulacrarians
- 119. Echinoderms
- 120. Hemichordates
29. Deuterostomes
31. Chordates
- 121. Cephalochordates
- 122. Urochordates
32. Vertebrates
- 123. Hagfishes
- 124. Lampreys
33. Jawed vertebrates
- 125. Chondrichthyans
- 126. Ray-finned fishes
34. Bony vertebrates
- 127. Coelacanths
35. Sarcopterygians
- 128. Lungfishes
- 129. Amphibians
36. Tetrapods
- 130. Mammals
37. Amniotes
- 131. Turtles
38. Reptiles
- 132. Lepidosaurs
- 133. Birds
39. Archosaurs
- 134. Crocodilians

The prokaryotic lineages on this tree are shown in black; the eukaryotic lineages are shown in color. Among the eukaryotes, the microbial lineages are shown in blue. The four eukaryote groups that contain major radiations of multicellular organisms are shown in brown (brown algae), green (Plantae), orange (fungi), and red (animals).

yourBioPortal.com
GO TO The Interactive Tree of Life

Time

– B –

bacteria (*Eubacteria*) [2]   Unicellular organisms lacking a nucleus, possessing distinctive ribosomes and initiator tRNA, and generally containing peptidoglycan in the cell wall. Different bacterial groups are distinguished primarily on nucleotide sequence data.

barnacles (*Cirripedia*)   Crustaceans [117] that undergo two metamorphoses—first from a feeding planktonic larva to a nonfeeding swimming larva, and then to a sessile adult that forms a "shell" composed of four to eight plates cemented to a hard substrate.

bilaterians (*Bilateria*) [24]   Those animal groups characterized by bilateral symmetry and three distinct tissue types (endoderm, ectoderm, and mesoderm). Includes the protostomes [25] and deuterostomes [29].

birds (*Aves*) [133]   Feathered, flying (or secondarily flightless) tetrapods [36].

bivalves (*Bivalvia*)   Major mollusk [106] group; clams and mussels. Bivalves typically have two similar hinged shells that are each asymmetrical across the midline.

bony vertebrates (*Osteichthyes*) [34]   Vertebrates [32] in which the skeleton is usually ossified to form bone. Includes the ray-finned fishes [126], coelacanths [127], lungfishes [128], and tetrapods [36].

brachiopods (*Brachiopoda*) [103]   Lophotrochozoans [26] with two similar hinged shells that are each symmetrical across the midline. Superficially resemble bivalve mollusks, except for the shell symmetry.

brittle stars (*Ophiuroidea*)   Echinoderms [119] with five long, whip-like arms radiating from a distinct central disk that contains the reproductive and digestive organs.

brown algae (*Phaeophyta*) [49]   Multicellular, almost exclusively marine stramenopiles [6] generally containing the pigment fucoxanthin as well as chlorophylls *a* and *c* in their chloroplasts.

bryozoans (*Ectoprocta* or *Bryozoa*) [99]   A group of marine and freshwater lophotrochozoans [26] that live in colonies attached to substrata; also known as ectoprocts or moss animals.

– C –

caecilians (*Gymnophiona*)   A group of burrowing or aquatic amphibians [129]. They are elongate, legless, with a short tail (or none at all), reduced eyes covered with skin or bone, and a pair of sensory tentacles on the head.

calcareous sponges (*Calcarea*) [95]   Filter-feeding marine sponges with spicules composed of calcium carbonate.

cellular slime molds (*Dictyostelida*)   Amoebozoans [85] in which individual amoebas aggregate under stress to form a multicellular pseudoplasmodium.

cephalochordates (*Cephalochordata*) [121]   A group of weakly swimming, eel-like benthic marine chordates [31]; also called lancelets.

cephalopods (*Cephalopoda*)   Active, predatory mollusks [106] in which the molluscan foot has been modified into muscular hydrostatic arms or tentacles. Includes octopuses, squids, and nautiluses.

cercozoans (*Cercozoa*) [83]   Unicellular eukaryotes [4] that feed by means of threadlike pseudopods. Group together with foraminiferans [82] and radiolarians [84] to comprise the rhizaria [16].

Charales [72]   Multicellular green algae with branching, apical growth and plasmodesmata between adjacent cells. The closest living relatives of the land plants [10], they retain the egg in the parent organism.

chelicerates (*Chelicerata*) [115]   A major group of arthropods [28] with pointed appendages (chelicerae) used to grasp food (as opposed to the chewing mandibles of most other arthropods). Includes the arachnids, horseshoe crabs, pycnogonids, and extinct sea scorpions.

chimaeras (*Holocephali*)   A group of bottom-dwelling, marine, scaleless chondrichthyan fishes [125] with large, permanent, grinding tooth plates (rather than the replaceable teeth found in other chondrichthyans).

chitons (*Polyplacophora*)   Flattened, slow-moving mollusks [106] with a dorsal protective calcareous covering made up of eight articulating plates.

chlamydias (*Chlamydiae*) [41]   A group of very small Gram-negative bacteria; they live as intracellular parasites of other organisms.

chlorophytes (*Chlorophyta*) [74]   The most abundant and diverse group of green algae, including freshwater, marine, and terrestrial forms; some are unicellular, others colonial, and still others multicellular. Chlorophytes use chlorophylls *a* and *c* in their photosynthesis.

choanoflagellates (*Choanozoa*) [92]   Unicellular eukaryotes [4] with a single flagellum surrounded by a collar. Most are sessile, some are colonial. The closest living relatives of the animals [21].

chondrichthyans (*Chondrichthyes*) [125]   One of the two main groups of jawed vertebrates [33]; includes sharks, rays, and chimaeras. They have cartilaginous skeletons and paired fins.

chordates (*Chordata*) [31]   One of the two major groups of deuterostomes [29], characterized by the presence (at some point in development) of a notochord, a hollow dorsal nerve cord, and a post-anal tail. Includes the cephalochordates [121], urochordates [122], and vertebrates [32].

chromalveolates (*Chromalveolata*) [5]   A contested group, said to have arisen from a common ancestor with chloroplasts derived from a red alga and supported by some molecular evidence. Major chromalveolate groups include the alveolates [7], stramenopiles [6], and haptophytes [48].

"chytrids" [90]   A convenience term used for a paraphyletic group of mostly aquatic, microscopic fungi [19] with flagellated gametes. Some exhibit alternation of generations.

ciliates (*Ciliophora*) [54]   Alveolates [7] with numerous cilia and two types of nuclei (micronuclei and macronuclei).

clitellates (*Clitellata*)   Annelids [105] with gonads contained in a swelling (called a clitellum) toward the head of the animal. Includes earthworms (oligochaetes) and leeches.

club fungi (*Basidiomycota*) [86]   Fungi [19] that, if multicellular, bear the products of meiosis on club-shaped basidia and possess a long-lasting dikaryotic stage. Some are unicellular.

club mosses (*Lycopodiophyta*) [68]   Vascular plants [11] characterized by microphylls. See lycophytes.

cnidarians (*Cnidaria*) [98]   Aquatic, mostly marine eumetazoans [23] with specialized stinging organelles (nematocysts) used for prey capture and defense, and a blind gastrovascular cavity. The closest living relatives of the ctenophores [97].

coelacanths (*Actinista*) [127]   A group of marine sarcopterygians [35] that was diverse from the Middle Devonian to the Cretaceous, but is now known from just two living species. The pectoral and anal fins are on fleshy stalks supported by skeletal elements, so they are also called lobe-finned fishes.

Coleochaetales [73]   Multicellular green algae characterized by flattened growth form composed of thin-walled cells. Thought to be the sister-group to the Charales [72] plus land plants [10].

conifers (*Pinophyta* or *Coniferophyta*) [61]   Cone-bearing, woody seed plants [12].

copepods (*Copepoda*)   Small, abundant crustaceans [117] found in marine, freshwater, or wet terrestrial habitats. They have a single eye, long antennae, and a body shaped like a teardrop.

craniates (*Craniata*)   Some biologist exclude the hagfishes [123] from the vertebrates [32], and use the term craniates to refer to the two groups combined.

crenarchaeotes (*Crenarchaeota*) [46]   A major and diverse group of archaeans [3], defined on the basis of rRNA base sequences. Many are extremophiles (inhabit extreme environments), but the group may also be the most abundant archaeans in the marine environment.

crinoids (*Crinoidea*)   Echinoderms [119] with a mouth surrounded by feeding arms, and a U-shaped gut with the mouth next to the anus. They attach to the substratum by a stalk or are free-swimming. Crinoids were abundant in the middle and late Paleozoic, but only a few hundred species have survived to the present. Includes the sea lilies and feather stars.

crocodilians (*Crocodylia*) [134]   A group of large, predatory, aquatic archosaurs [39]. The closest living relatives of birds [133]. Includes alligators, caimans, crocodiles, and gharials.

crustaceans (*Crustacea*) [117]   Major group of marine, freshwater, and terrestrial arthropods [28] with a head, thorax, and abdomen (although the head and thorax may be fused), covered with a thick exoskeleton, and with two-part appendages. Crustaceans undergo metamorphosis from a nauplius larva. Includes decapods, isopods, krill, barnacles, amphipods, copepods, and ostracods.

ctenophores (*Ctenophora*) [97]   Radially symmetrical, diploblastic marine animals [21], with a complete gut and eight rows of fused plates of cilia (called ctenes).

cyanobacteria (*Cyanobacteria*) [43]   A group of unicellular, colonial, or filamentous bacteria that conduct photosynthesis using chlorophyll *a*.

cycads (*Cycadophyta*) [64] Palmlike gymnosperms with large, compound leaves.

cyclostomes (*Cyclostomata*) This term refers to the possibly monophyletic group of lampreys [124] and hagfishes [123]. Molecular data support this group, but morphological data suggest that lampreys are more closely related to jawed vertebrates [33] than to hagfishes.

– D –

decapods (*Decapoda*) A group of marine, freshwater, and semiterrestrial crustaceans [117] in which five of the eight pairs of thoracic appendages function as legs (the other three pairs, called maxillipeds, function as mouthparts). Includes crabs, lobsters, crayfishes, and shrimps.

demosponges (*Demospongiae*) [94] The largest of the three groups of sponges [22], accounting for 90 percent of all sponge species. Demosponges have spicules made of silica, spongin fiber (a protein), or both.

deuterostomes (*Deuterostomia*) [29] One of the two major groups of bilaterians [24], in which the mouth forms at the opposite end of the embryo from the blastopore in early development (contrast with protostomes). Includes the ambulacrarians [30] and chordates [31].

diatoms (*Bacillariophyta*) [50] Unicellular, photosynthetic stramenopiles [6] with glassy cell walls in two parts.

dikaryotic fungi (*Dikarya*) [20] A group of fungi [19] in which two genetically different haploid nuclei coexist and divide within the same hypha; includes club fungi [86] and sac fungi [87].

dinoflagellates (*Dinoflagellata*) [52] A group of alveolates [7] usually possessing two flagella, one in an equatorial groove and the other in a longitudinal groove; many are photosynthetic.

diplomonads (*Diplomonadida*) [81] A group of eukaryotes [4] lacking mitochondria; most have two nuclei, each with four associated flagella.

– E –

ecdysozoans (*Ecdysozoa*) [27] One of the two major groups of protostomes [25], characterized by periodic molting of their exoskeletons. Nematodes [112] and arthropods [28] are the largest ecdysozoan groups.

echinoderms (*Echinodermata*) [119] A major group of marine deuterostomes [29] with five-fold radial symmetry (at some stage of life) and an endoskeleton made of calcified plates and spines. Includes sea stars, crinoids, sea urchins, sea cucumbers, and brittle stars.

elasmobranchs (*Elasmobranchii*) The largest group of chondrichthyan fishes [125]. Includes sharks, skates, and rays. In contrast to the other group of living chondrichthyans (the chimaeras), they have replaceable teeth.

embryophytes *See* land plants [10].

eudicots (*Eudicotyledones*)[55] A group of angiosperms [14] with pollen grains possessing three openings. Typically with two cotyledons, net-veined leaves, taproots, and floral organs typically in multiples of four or five.

euglenids (*Euglenida*) [78] Flagellate excavates characterized by a pellicle composed of spiraling strips of protein under the plasma membrane; the mitochondria have disk-shaped cristae. Some are photosynthetic.

eukaryotes (*Eukarya*) [4] Organisms made up of one or more complex cells in which the genetic material is contained in nuclei. Contrast with archaeans [3] and bacteria [2].

eumetazoans (*Eumetazoa*) [23] Those animals [21] characterized by body symmetry, a gut, a nervous system, specialized types of cell junctions, and well-organized tissues in distinct cell layers (although there have been secondary losses of some of these characteristics in some eumetazoans).

euphyllophytes (*Euphyllophyta*) The group of vascular plants [11] that is sister to the lycophytes [68] and which includes all plants with megaphylls.

euryarchaeotes (*Euryachaeota*) [47] A major group of archaeans [3], diagnosed on the basis of rRNA sequences. Includes many methanogens, extreme halophiles, and thermophiles.

eutherians (*Eutheria*) A group of viviparous mammals [130], eutherians are well developed at birth (contrast to prototherians and marsupials, the other two groups of mammals). Most familiar mammals outside the Australian and South American regions are eutherians (see Table 33.1).

excavates (*Excavata*) [15] Diverse group of unicellular, flagellate eukaryotes, many of which possess a feeding groove; some lack mitochondria.

– F –

"ferns" Vascular plants [11] usually possessing large, frondlike leaves that unfold from a "fiddlehead." Not a monophyletic group, although most fern species are encompassed in a monophyletic clade, the leptosporangiate ferns [65].

flatworms (*Platyhelminthes*) [100] A group of dorsoventrally flattened and generally elongate soft-bodied lophotrochozoans [26]. May be free-living or parasitic, found in marine, freshwater, or damp terrestrial environments. Major flatworm groups include the tapeworms, flukes, monogeneans, and turbellarians.

flowering plants *See* angiosperms [14].

flukes (*Trematoda*) A group of wormlike parasitic flatworms [100] with complex life cycles that involve several different host species. May be paraphyletic with respect to tapeworms.

foraminiferans (*Foraminifera*) [82] Amoeboid organisms with fine, branched pseudopods that form a food-trapping net. Most produce external shells of calcium carbonate.

fungi (*Fungi*) [19] Eukaryotic heterotrophs with absorptive nutrition based on extracellular digestion; cell walls contain chitin. Major fungal groups include the microsporidia [91], "chytrids" [90], "zygospore fungi" [89], arbuscular mycorrhizal fungi [88], sac fungi [87], and club fungi [86].

– G –

gastropods (*Gastropoda*) The largest group of mollusks [106]. Gastropods possess a well-defined head with two or four sensory tentacles (often terminating in eyes) and a ventral foot. Most species have a single coiled or spiraled shell. Common in marine, freshwater, and terrestrial environments.

ginkgo (*Ginkgophyta*) [63] A gymnosperm [13] group with only one living species. The ginkgo seed is surrounded by a fleshy tissue not derived from an ovary wall and hence not a fruit.

glass sponges (*Hexactinellida*) [93] Sponges [22] with a skeleton composed of four- and/or six-pointed spicules made of silica.

glaucophytes (*Glaucophyta*) [76] Unicellular freshwater algae with chloroplasts containing traces of peptidoglycan, the characteristic cell wall material of bacteria.

gnathostomes (*Gnathostomata*) *See* jawed vertebrates [33].

gnetophytes (*Gnetophyta*) [62] A gymnosperm [13] group with three very different lineages; all have wood with vessels, unlike other gymnosperms.

green plants (*Viridiplantae*) [9] Organisms with chlorophylls *a* and *b*, cellulose-containing cell walls, starch as a carbohydrate storage product, and chloroplasts surrounded by two membranes.

gymnosperms (*Gymnospermae*) [13] Seed plants [12] with seeds "naked" (i.e., not enclosed in carpels). Probably monophyletic, but status still in doubt. Includes the conifers [61], gnetophytes [62], ginkgo [63], and cycads [64].

– H –

hagfishes (*Myxini*) [123] Elongate, slimy-skinned vertebrates [32] with three small accessory hearts, a partial cranium, and no stomach or paired fins. *See also* craniata; cyclostomes.

haptophytes (*Haptophyta*) [48] Unicellular, photosynthetic chromalveolates [5] with two slightly unequal, smooth flagella. Abundant as phytoplankton, some form marine algal blooms.

hemichordates (*Hemichordata*) [120] One of the two primary groups of ambulacrarians [30]; marine wormlike organisms with a three-part body plan.

heteroloboseans (*Heterolobosea*) [79] Colorless excavates [15] that can transform among amoeboid, flagellate, and encysted stages.

hexapods (*Hexapoda*) [118] Major group of arthropods [28] characterized by a reduction (from the ancestral arthropod condition) to six walking appendages, and the consolidation of three body segments to form a thorax. Includes insects and their relatives (see Table 32.2).

high-GC Gram-positives (*Actinobacteria*) [42] Gram-positive bacteria with a relatively high (G+C)/(A+T) ratio of their DNA, with a filamentous growth habit.

hornworts (*Anthocerophyta*) [69] Nonvascular plants with sporophytes that grow from the base. Cells contain a single large, platelike chloroplast.

horsehair worms (*Nematomorpha*) [111] A group of very thin, elongate, wormlike freshwater ecdysozoans [27]. Largely nonfeeding as adults, they are parasites of insects and crayfish as larvae.

horseshoe crabs (*Xiphosura*) Marine chelicerates [115] with a large outer shell in three parts: a carapace, an abdomen, and a tail-like telson. There are only five living species, but many additional species are known from fossils.

horsetails (*Sphenophyta* or *Equisetophyta*) [66] Vascular plants [11] with reduced megaphylls in whorls.

hydrozoans (*Hydrozoa*) A group of cnidarians [98]. Most species go through both polyp and mesuda stages, although one stage or the other is eliminated in some species.

– I –

insects (*Insecta*) The largest group within the hexapods [118]. Insects are characterized by exposed mouthparts and one pair of antennae containing a sensory receptor called a Johnston's organ. Most have two pairs of wings as adults. There are more described species of insects than all other groups of life [1] combined, and many species remain to be discovered. The major insect groups are described in Table 32.2.

"invertebrates" A convenience term that encompasses any animal [21] that is not a vertebrate [32].

isopods (*Isopoda*) Crustaceans [117] characterized by a compact head, unstalked compound eyes, and mouthparts consisting of four pairs of appendages. Isopods are abundant and widespread in salt, fresh, and brackish water, although some species (the sow bugs) are terrestrial.

– J –

jawed vertebrates (*Gnathostomata*) [33] A major group of vertebrates [32] with jawed mouths. Includes chondrichthyans [125], ray-finned fishes [126], and sarcopterygians [35].

– K –

kinetoplastids (*Kinetoplastida*) [77] Unicellular, flagellate organisms characterized by the presence in their single mitochondrion of a kinetoplast (a structure containing multiple, circular DNA molecules).

kinorhynchs (*Kinorhyncha*) [109] Small (< 1 mm) marine ecdysozoans [27] with bodies in 13 segments and a retractable proboscis.

korarchaeotes (*Korarchaeota*) A group of archaeans [3] known only by evidence from nucleic acids derived from hot springs. Its phylogenetic relationships within the Archaea are unknown.

krill (*Euphausiacea*) A group of shrimplike marine crustaceans [117] that are important components of the zooplankton.

– L –

lampreys (*Petromyzontiformes*) [124] Elongate, eel-like vertebrates [32] that often have rasping and sucking disks for mouths.

lancelets (*Cephalochordata*) See cephalochordates [121].

land plants (*Embryophyta*) [10] Plants with embryos that develop within protective structures; also called embryophytes. Sporophytes and gametophytes are multicellular. Land plants possess a cuticle. Major groups are the liverworts [71], mosses [70], hornworts [69], and vascular plants [11].

larvaceans (*Larvacea*) Solitary, planktonic urochordates [122] that retain both notochords and nerve cords throughout their lives.

lepidosaurs (*Lepidosauria*) [132] Reptiles [38] with overlapping scales. Includes tuataras and

squamates (lizards, snakes, and amphisbaenians).

leptosporangiate ferns (*Pteridopsida* or *Polypodiopsida*) [65] Vascular plants [11] usually possessing large, frondlike leaves that unfold from a "fiddlehead," and possessing thin-walled sporangia.

life (*Life*) [1] The monophyletic group that includes all known living organisms. Characterized by a nucleic-acid based genetic system (DNA or RNA), metabolism, and cellular structure. Some parasitic forms, such as viruses, have secondarily lost some of these features and rely on the cellular environment of their host.

liverworts (*Hepatophyta*) [71] Nonvascular plants lacking stomata; stalk of sporophyte elongates along its entire length.

loboseans (*Lobosea*) A group of unicellular amoebozoans [85]; includes the most familiar amoebas (e.g., *Amoeba proteus*).

"lophophorates" Not a monophyletic group. A convenience term used to describe several groups of lophotrochozoans [26] that have a feeding structure called a lophophore (a circular or U-shaped ridge around the mouth that bears one or two rows of ciliated, hollow tentacles).

lophotrochozoans (*Lophotrochozoa*) [26] One of the two main groups of protostomes [25]. This group is morphologically diverse, and is supported primarily on information from gene sequences. Includes bryozoans [99], flatworms [100], rotifers [101], ribbon worms [102], brachiopods [103], phoronids [104], annelids [105], and mollusks [106].

loriciferans (*Loricifera*) [110] Small (< 1 mm) ecdysozoans [27] with bodies in four parts, covered with six plates.

low-GC Gram-positives (*Firmicutes*) [44] A diverse group of bacteria [2] with a relatively low (G+C)/(A+T) ratio of their DNA, often but not always Gram-positive, some producing endospores.

lungfishes (*Dipnoi*) [128] A group of aquatic sarcopterygians [35] that are the closest living relatives of the tetrapods [36]. They have a modified swim bladder used to absorb oxygen from air, so some species can survive the temporary drying of their habitat.

lycophytes (*Lycopodiophyta*) [68] Vascular plants [11] characterized by microphylls; includes club mosses, spike mosses, and quillworts.

– M –

magnoliids (*Magnoliidae*) [57] A major group of angiosperms [14] possessing two cotyledons and pollen grains with a single opening. The group is defined primarily by nucleotide sequence data; it is more closely related to the eudicots and monocots than to three other small angiosperm groups.

mammals (*Mammalia*) [130] A group of tetrapods [36] with hair covering all or part of their skin; females produce milk to feed their developing young. Includes the prototherians, marsupials, and eutherians.

marsupials (*Marsupialia*) Mammals [130] in which the female typically has a marsupium (a pouch for rearing young, which are born at an

extremely early stage in development). Includes such familiar mammals as opossums, koalas, and kangaroos.

metazoans (*Metazoa*) See animals [21].

microbial eukaryotes See "protists."

microsporidia (*Microsporidia*) [91] A group of parasitic unicellular fungi [19] that lack mitochondria and have walls that contain chitin.

mollusks (*Mollusca*) [106] One of the major groups of lophotrochozoans [26], mollusks have bodies composed of a foot, a mantle (which often secretes a hard, calcareous shell), and a visceral mass. Includes monoplacophorans, chitons, bivalves, gastropods, and cephalopods.

monilophytes (*Monilophyta*) A group of vascular plants [11], sister to the seed plants [12], characterized by overtopping and possession of megaphylls; includes the ferns [65], horsetails [66], and whisk ferns [67].

monocots (*Monocotyledones*) [56] Angiosperms [14] characterized by possession of a single cotyledon, usually parallel leaf veins, a fibrous root system, pollen grains with a single opening, and floral organs usually in multiples of three.

monogeneans (*Monogenea*) A group of ectoparasitic flatworms [100].

monoplacophorans (*Monoplacophora*) Mollusks [106] with segmented body parts and a single, thin, flat, rounded, bilateral shell.

mosses (*Bryophyta*) [70] Nonvascular plants with true stomata and erect, "leafy" gametophytes; sporophytes elongate by apical cell division.

moss animals See bryozoans [99].

myriapods (*Myriapoda*) [116] Arthropods [28] characterized by an elongate, segmented trunk with many legs. Includes centipedes and millipedes.

– N –

nanoarchaeotes (*Nanoarchaeota*) A group of extremely small, thermophilic archaeans [3] with a much-reduced genome. The only described example can survive only when attached to a host organism.

nematodes (*Nematoda*) [112] A very large group of elongate, unsegmented ecdysozoans [27] with thick, multilayer cuticles. They are among the most abundant and diverse animals, although most species have not yet been described. Include free-living predators and scavengers, as well as parasites of most species of land plants [10] and animals [21].

neognaths (*Neognathae*) The main group of birds [133], including all living species except the ratites (ostrich, emu, rheas, kiwis, cassowaries) and tinamous. See palaeognaths.

– O –

oligochaetes (*Oligochaeta*) An annelid [105] group whose members lack parapodia, eyes, and anterior tentacles, and have few setae. Earthworms are the most familiar oligochaetes.

onychophorans (*Onychophora*) [114] Elongate, segmented ecdysozoans [27] with many pairs of soft, unjointed, claw-bearing legs. Also known as velvet worms.

oomycetes (*Oomycota*) [51]   Water molds and relatives; absorptive heterotrophs with nutrient-absorbing, filamentous hyphae.

opisthokonts (*Opisthokonta*) [18]   A group of unikonts [17] in which the flagellum on motile cells, if present, is posterior. The opisthokonts include the fungi [19], animals [21], and choanoflagellates [92].

ostracods (*Ostracoda*)   Marine and freshwater crustaceans [117] that are laterally compressed and protected by two clamlike calcareous or chitinous shells.

– P –

palaeognaths (*Palaeognathae*)   A group of secondarily flightless or weakly flying birds [133]. Includes the flightless ratites (ostrich, emu, rheas, kiwis, cassowaries) and the weakly flying tinamous.

parabasalids (*Parabasalia*) [80]   A group of unicellular eukaryotes [4] that lack mitochondria; they possess flagella in clusters near the anterior of the cell.

phoronids (*Phoronida*) [104]   A small group of sessile, wormlike marine lophotrochozoans [26] that secrete chitinous tubes and feed using a lophophore.

placoderms (*Placodermi*)   An extinct group of jawed vertebrates [33] that lacked teeth. Placoderms were the dominant predators in Devonian oceans.

placozoans (*Placozoa*) [96]   A poorly known group of structurally simple, asymmetrical, flattened, transparent animals found in coastal marine tropical and subtropical seas. Most evidence suggests that placozoans are the sister-group of eumetazoans [23].

Plantae [8]   The most broadly defined plant group. In most parts of this book, we use the word "plant" as synonymous with "land plant" [10], a more restrictive definition.

plasmodial slime molds (*Myxogastrida*)   Amoebozoans [85] that in their feeding stage consist of a coenocyte called a plasmodium.

pogonophorans (*Pogonophora*)   Deep-sea annelids [105] that lack a mouth or digestive tract; they feed by taking up dissolved organic matter, facilitated by endosymbiotic bacteria in a specialized organ (the trophosome).

polychaetes (*Polychaeta*)   A group of mostly marine annelids [105] with one or more pairs of eyes and one or more pairs of feeding tentacles; parapodia and setae extend from most body segments. May be paraphyletic with respect to the clitellates.

priapulids (*Priapulida*) [108]   A small group of cylindrical, unsegmented, wormlike marine ecdysozoans [27] that takes its name from its phallic appearance.

"progymnosperms"   Paraphyletic group of extinct vascular plants [11] that flourished from the Devonian through the Mississippian periods. The first truly woody plants, and the first with vascular cambium that produced both secondary xylem and secondary phloem, they reproduced by spores rather than by seeds.

"prokaryotes"   Not a monophyletic group; as commonly used, includes the bacteria [2] and archaeans [3]. A term of convenience encompassing all cellular organisms that are not eukaryotes.

proteobacteria (*Proteobacteria*) [45]   A large and extremely diverse group of Gram-negative bacteria that includes many pathogens, nitrogen fixers, and photosynthesizers. Includes the alpha, beta, gamma, delta, and epsilon proteobacteria.

"protists"   This term of convenience does not describe a monophyletic group but is used to encompass a large number of distinct and distantly related groups of eukaryotes, many but far from all of which are microbial and unicellular. Essentially a "catch-all" term for any eukaryote group not contained within the land plants [10], fungi [19], or animals [21].

protostomes (*Protostomia*) [25]   One of the two major groups of bilaterians [24]. In protostomes, the mouth typically forms from the blastopore (if present) in early development (contrast with deuterostomes). The major protostome groups are the lophotrochozoans [26] and ecdysozoans [27].

prototherians (*Prototheria*)   A mostly extinct group of mammals [130], common during the Cretaceous and early Cenozoic. The five living species—the echidnas and the duck-billed platypus—are the only extant egg-laying mammals.

pterobranchs (*Pterobranchia*)   A small group of sedentary marine hemichordates [120] that live in tubes secreted by the proboscis. They have one to nine pairs of arms, each bearing long tentacles that capture prey and function in gas exchange.

pycnogonids (*Pycnogonida*)   Treated in this book as a group of chelicerates [115], but sometimes considered an independent group of arthropods [28]. Pycnogonids have reduced bodies and very long, slender legs. Also called sea spiders.

– R –

radiolarians (*Radiolaria*) [84]   Amoeboid organisms with needlelike pseudopods supported by microtubules. Most have glassy internal skeletons.

ray-finned fishes (*Actinopterygii*) [126]   A highly diverse group of freshwater and marine bony vertebrates [34]. They have reduced swim bladders that often function as hydrostatic organs and fins supported by soft rays (lepidotrichia). Includes most familiar fishes.

red algae (*Rhodophyta*) [75]   Mostly multicellular, marine and freshwater algae characterized by the presence of phycoerythrin in their chloroplasts.

reptiles (*Reptilia*) [38]   One of the two major groups of extant amniotes [37], supported on the basis of similar skull structure and gene sequences. The term "reptiles" traditionally excluded the birds [133], but the resulting group is then clearly paraphyletic. As used in this book, the reptiles include turtles [131], lepidosaurs [132], birds [133], and crocodilians [134].

rhizaria (*Rhizaria*) [16]   Mostly amoeboid unicellular eukaryotes with pseudopods, many with external or internal shells. Includes the foraminiferans [82], cercozoans [83], and radiolarians [84].

rhyniophytes (*Rhyniophyta*)   A group of early vascular plants [11] that appeared in the Silurian and became extinct in the Devonian. Possessed dichotomously branching stems with terminal sporangia but no true leaves or roots.

ribbon worms (*Nemertea*) [102]   A group of unsegmented lophotrochozoans [26] with an eversible proboscis used to capture prey. Mostly marine, but some species live in fresh water or on land.

rotifers (*Rotifera*) [101]   Tiny (< 0.5 mm) lophotrochozoans [26] with a pseudocoelomic body cavity that functions as a hydrostatic organ and a ciliated feeding organ called the corona that surrounds the head. They live in freshwater and wet terrestrial habitats.

roundworms (*Nematoda*) [112]   *See* nematodes.

– S –

sac fungi (*Ascomycota*) [87]   Fungi that bear the products of meiosis within sacs (asci) if the organism is multicellular. Some are unicellular.

salamanders (*Caudata*)   A group of amphibians [129] with distinct tails in both larvae and adults and limbs set at right angles to the body.

salps   *See* thaliaceans.

sarcopterygians (*Sarcopterygii*) [35]   One of the two major groups of bony vertebrates [34], characterized by jointed appendages (paired fins or limbs).

scyphozoans (*Scyphozoa*)   Marine cnidarians [98] in which the medusa stage dominates the life cycle. Commonly known as jellyfish.

sea cucumbers (*Holothuroidea*)   Echinoderms [119] with an elongate, cucumber-shaped body and leathery skin. They are scavengers on the ocean floor.

sea spiders   *See* pycnogonids.

sea squirts   *See* ascidians.

sea stars (*Asteroidea*)   Echinoderms [119] with five (or more) fleshy "arms" radiating from an indistinct central disk. Also called starfishes.

sea urchins (*Echinoidea*)   Echinoderms [119] with a test (shell) that is covered in spines. Most are globular in shape, although some groups (such as the sand dollars) are flattened.

"seed ferns"   A paraphyletic group of loosely related, extinct seed plants that flourished in the Devonian and Carboniferous. Characterized by large, frondlike leaves that bore seeds.

seed plants (*Spermatophyta*) [12]   Heterosporous vascular plants [11] that produce seeds; most produce wood; branching is axillary (not dichotomous). The major seed plant groups are gymnosperms [13] and angiosperms [14].

sow bugs   *See* isopods.

spirochetes (*Spirochaetes*) [40]   Motile, Gram-negative bacteria with a helically coiled structure and characterized by axial filaments.

sponges (*Porifera*) [22]   A group of relatively asymmetric, filter-feeding animals that lack a gut or nervous system and generally lack differentiated tissues. Includes glass sponges [93], demosponges [94], and calcareous sponges [95].

springtails (*Collembola*)   Wingless hexapods [118] with springing structures on the third and fourth segments of their bodies. Springtails are extremely abundant in some environments (especially in soil, leaf litter, and vegetation).

squamates (*Squamata*)   The major group of lepidosaurs [132], characterized by the possession of movable quadrate bones (which allow the upper jaw to move independently of the rest of the skull) and hemipenes (a paired set of eversible penises, or penes) in males. Includes the lizards (a paraphyletic group), snakes, and amphisbaenians.

star anise (*Austrobaileyales*) [58]   A group of woody angiosperms [14] thought to be the sister-group of the clade of flowering plants that includes eudicots [55], monocots [56], and magnoliids [57].

starfish (*Asteroidea*)   *See* sea stars.

stramenopiles (*Heterokonta* or *Stramenopila*) [6] Organisms having, at some stage in their life cycle, two unequal flagella, the longer possessing rows of tubular hairs. Chloroplasts, when present, surrounded by four membranes. Major stramenopile groups include the brown algae [49], diatoms [50], and oomycetes [51].

– T –

tapeworms (*Cestoda*)   Parasitic flatworms [100] that live in the digestive tracts of vertebrates as adults, and usually in various other species of animals as juveniles.

tardigrades (*Tardigrada*) [113]   Small (< 0.5 mm) ecdysozoans [27] with fleshy, unjointed legs and no circulatory or gas exchange organs. They live in marine sands, in temporary freshwater pools, and on the water films of plants. Also called water bears.

tetrapods (*Tetrapoda*) [36]   The major group of sarcopterygians [35]; includes the amphibians [129] and the amniotes [37]. Named for the presence of four jointed limbs (although limbs have been secondarily reduced or lost completely in several tetrapod groups).

thaliaceans (*Thaliacea*)   A group of solitary or colonial planktonic marine urochordates [122]. Also called salps.

therians (*Theria*)   Mammals [130] characterized by viviparity (live birth). Includes eutherians and marsupials.

theropods (*Theropoda*)   Archosaurs [39] with bipedal stance, hollow bones, a furcula ("wishbone"), elongated metatarsals with three-fingered feet, and a pelvis that points backwards. Includes many well-known extinct dinosaurs (such as *Tyrannosaurus rex*), as well as the living birds [133].

tracheophytes   *See* vascular plants [11].

trilobites (*Trilobita*)   An extinct group of arthropods [28] related to the chelicerates [115]. Trilobites flourished from the Cambrian through the Permian.

tuataras (*Rhyncocephalia*)   A group of lepidosaurs [132] known mostly from fossils; there are just two living tuatara species. The quadrate bone of the upper jaw is fixed firmly to the skull. Sister group of the squamates.

tunicates   *See* ascidians.

turbellarians (*Turbellaria*)   A group of free-living, generally carnivorous flatworms [100]. Their monophyly is questionable.

turtles (*Testudines*) [131]   A group of reptiles [38] with a bony carapace (upper shell) and plastron (lower shell) that encase the body.

– U –

unikonts (*Unikonta*) [17]   A group of eukaryotes [4] whose motile cells possess a single flagellum. Major unikont groups include the amoebozoans [85], fungi [19], and animals [21].

urochordates (*Urochordata*) [122]   A group of chordates [31] that are mostly saclike filter feeders as adults, with motile larvae stages that resemble a tadpole.

– V –

vascular plants (*Tracheophyta*) [11]   Plants with xylem and phloem. Major groups include the lycophytes [68] and euphyllophytes.

vertebrates (*Vertebrata*) [32]   The largest group of chordates [31], characterized by a rigid endoskeleton supported by the vertebral column and an anterior skull encasing a brain. Includes hagfishes [123], lampreys [124], and the jawed vertebrates [33], although some biologists exclude the hagfishes from this group. *See also* craniates.

– W –

water bears   *See* tardigrades.

water lilies (*Nymphaeaceae*) [59]   A group of aquatic, freshwater angiosperms [14] that are rooted in soil in shallow water, with round floating leaves and flowers that extend above the water's surface. They are the sister-group to most of the remaining flowering plants, with the exception of the genus *Amborella* [60].

whisk ferns (*Psilotophyta*) [67]   Vascular plants [11] lacking leaves and roots.

– Y –

"yeasts"   A convenience term for several distantly related groups of unicellular fungi [19].

– Z –

"zygospore fungi" (*Zygomycota*, if monophyletic) [89]   A convenience term for a probably paraphyletic group of fungi [19] in which hyphae of differing mating types conjugate to form a zygosporangium.

Appendix B: Some Measurements Used in Biology

MEASURES OF	UNIT	EQUIVALENTS	METRIC → ENGLISH CONVERSION
Length	meter (m)	base unit	1 m = 39.37 inches = 3.28 feet
	kilometer (km)	1 km = 1000 (10^3) m	1 km = 0.62 miles
	centimeter (cm)	1 cm = 0.01 (10^{-2}) m	1 cm = 0.39 inches
	millimeter (mm)	1 mm = 0.1 cm = 10^{-3} m	1 mm = 0.039 inches
	micrometer (μm)	1 μm = 0.001 mm = 10^{-6} m	
	nanometer (nm)	1 nm = 0.001 μm = 10^{-9} m	
Area	square meter (m^2)	base unit	1 m^2 = 1.196 square yards
	hectare (ha)	1 ha = 10,000 m^2	1 ha = 2.47 acres
Volume	liter (L)	base unit	1 L = 1.06 quarts
	milliliter (mL)	1 mL = 0.001 L = 10^{-3} L	1 mL = 0.034 fluid ounces
	microliter (μL)	1 μL = 0.001 mL = 10^{-6} L	
Mass	gram (g)	base unit	1 g = 0.035 ounces
	kilogram (kg)	1 kg = 1000 g	1 kg = 2.20 pounds
	metric ton (mt)	1 mt = 1000 kg	1 mt = 2,200 pounds = 1.10 ton
	milligram (mg)	1 mg = 0.001 g = 10^{-3} g	
	microgram (μg)	1 μg = 0.001 mg = 10^{-6} g	
Temperature	degree Celsius (°C)	base unit	°C = (°F − 32)/1.8
			0°C = 32°F (water freezes)
			100°C = 212°F (water boils)
			20°C = 68°F ("room temperature")
			37°C = 98.6°F (human internal body temperature)
	Kelvin (K)*	K = °C + 273	0 K = −460°F
Energy	joule (J)		1 J ≈ 0.24 calorie = 0.00024 kilocalorie[†]

*0 K (−273°C) is "absolute zero," a temperature at which molecular oscillations approach 0—that is, the point at which motion all but stops.

[†]A *calorie* is the amount of heat necessary to raise the temperature of 1 gram of water 1°C. The *kilocalorie*, or nutritionist's calorie, is what we commonly think of as a calorie in terms of food.

Answers to Self-Quizzes

Chapter 2

1.	b	6.	a
2.	d	7.	d
3.	c	8.	a
4.	c	9.	c
5.	d	10.	b

Chapter 3

1.	e	6.	a
2.	e	7.	c
3.	c	8.	e
4.	d	9.	b
5.	b	10.	d

Chapter 4

1.	c	6.	c
2.	c	7.	e
3.	c	8.	b
4.	d	9.	c
5.	e	10.	b

Chapter 5

1.	b	6.	e
2.	d	7.	a
3.	c	8.	d
4.	e	9.	b
5.	a	10.	d

Chapter 6

1.	e	6.	c
2.	c	7.	c
3.	a	8.	b
4.	d	9.	e
5.	c	10.	c

Chapter 7

1.	d	6.	a
2.	d	7.	e
3.	c	8.	d
4.	c	9.	c
5.	d	10.	a

Chapter 8

1.	c	6.	c
2.	e	7.	d
3.	b	8.	b
4.	c	9.	d
5.	c	10.	e

Chapter 9

1.	d	6.	d
2.	d	7.	e
3.	e	8.	d
4.	e	9.	a
5.	c	10.	e

Chapter 10

1.	e	6.	d
2.	b	7.	d
3.	d	8.	d
4.	b	9.	d
5.	e	10.	b

Chapter 11

1.	d	6.	d
2.	c	7.	e
3.	b	8.	d
4.	d	9.	d
5.	c	10.	c

Chapter 12*

1.	e	6.	d
2.	a	7.	b
3.	d	8.	b
4.	d	9.	b
5.	d	10.	b

Chapter 13

1.	c	6.	b
2.	a	7.	d
3.	c	8.	d
4.	b	9.	c
5.	e	10.	c

Chapter 14

1.	c	6.	d
2.	d	7.	b
3.	e	8.	d
4.	b	9.	d
5.	a	10.	c

Chapter 15

1.	a	6.	b
2.	c	7.	d
3.	b	8.	d
4.	b	9.	d
5.	d	10.	b

Chapter 16

1.	b	6.	c
2.	e	7.	c
3.	a	8.	d
4.	e	9.	b
5.	b	10.	a

Chapter 17

1.	c	6.	e
2.	c	7.	b
3.	a	8.	b
4.	e	9.	b
5.	e	10.	a

Chapter 18

1.	b	6.	b
2.	b	7.	c
3.	a	8.	a
4.	c	9.	e
5.	e	10.	d

Chapter 19

1.	c	6.	c
2.	b	7.	e
3.	a	8.	b
4.	b	9.	a
5.	d	10.	b

Chapter 20

1.	c	6.	b
2.	a	7.	a
3.	c	8.	e
4.	b	9.	d
5.	c	10.	b

Chapter 21

1.	d	6.	d
2.	e	7.	a
3.	d	8.	e
4.	c	9.	b
5.	d	10.	c

Chapter 22

1.	b	6.	b
2.	e	7.	e
3.	a	8.	a
4.	b	9.	e
5.	e	10.	d

Chapter 23

1.	c	6.	c
2.	e	7.	a
3.	d	8.	e
4.	c	9.	c
5.	a	10.	e

Chapter 24

1.	a	6.	e
2.	a	7.	e
3.	d	8.	e
4.	a	9.	b
5.	b	10.	e

Chapter 25

1.	d	6.	a
2.	b	7.	c
3.	e	8.	b
4.	c	9.	c
5.	a	10.	d

Chapter 26

1.	e	6.	b
2.	e	7.	d
3.	a	8.	d
4.	c	9.	b
5.	e	10.	d

*Answers to Chapter 12 Genetics Problems

1. $BB \times bb$; $bb \times bb$; $Bb \times bb$; $Bb \times Bb$

2. 1/32

3a. Autosomal dominant

3b. 1/4

4a. Males (XY) contain only one allele and will show only one color, black ($X^B Y$) or yellow ($X^b Y$). Females can be heterozygous ($X^B X^b$).

4b. $X^b Y$, yellow

5. The body color (G/g) and wing size (A/a) genes are linked; eye color (R/r) is unlinked to the other two genes. The map distance is 18.5 units.

6. Yellow, blue, and white in a 1:2:1 ratio

7. F_1 all wild-type, $PpSwsw$; F_2 will have phenotypes in the ratio 9:3:3:1. See Figure 12.7 (p. 244) for analogous genotypes.

8a. Ratio of phenotypes in F_2 is 3:1 (double dominant to double recessive).

8b. The F_1 are $PpByby$; they produce just two kinds of gametes (Pby and pBy). Combine them carefully and see the 1:2:1 phenotypic ratio fall out in the F_2.

8c. Pink-blistery

8d. See Figures 11.17 and 11.19 (pp. 224–226). Crossing over took place in the F_1 generation.

9. $Rraa$ and $RrAa$

10a. $w^+ > w^e > w$

10b. Parents are $w^e w$ and $w^+ Y$. Progeny are $w^+ w^e$, $w^+ w$, $w^e Y$, and wY.

11a. BX^a, BY, bX^a, bY

11b. Mother $bbX^A X^a$, father $BbX^a Y$, son $BbX^a Y$, daughter $bbX^a X^a$

12. 75 percent

13. Because the gene is carried on mitochondrial DNA, it is passed through the mother only. Thus if the woman does not have the disease but her husband does, their child will not be affected. On the other hand, if the woman has the disease but her husband does not, their child will have the disease.

Chapter 27
1. a 6. b
2. e 7. d
3. c 8. b
4. d 9. a
5. c 10. d

Chapter 28
1. d 6. e
2. c 7. c
3. e 8. b
4. b 9. b
5. b 10. d

Chapter 29
1. d 6. c
2. c 7. a
3. d 8. e
4. a 9. c
5. d 10. a

Chapter 30
1. b 6. a
2. d 7. e
3. e 8. a
4. c 9. e
5. d 10. c

Chapter 31
1. c 6. b
2. d 7. c
3. c 8. d
4. d 9. e
5. e 10. d

Chapter 32
1. b 6. e
2. e 7. b
3. c 8. d
4. d 9. d
5. a 10. e

Chapter 33
1. d 6. e
2. a 7. a
3. c 8. e
4. d 9. c
5. d 10. b

Chapter 34
1. c 6. b
2. b 7. b
3. e 8. c
4. e 9. a
5. a 10. d

Chapter 35
1. c 6. d
2. d 7. d
3. b 8. e
4. b 9. e
5. b 10. a

Chapter 36
1. d 6. c
2. d 7. e
3. c 8. d
4. a 9. d
5. c 10. e

Chapter 37
1. a 6. c
2. e 7. b
3. c 8. c
4. d 9. a
5. b 10. b

Chapter 38
1. d 6. e
2. b 7. a
3. e 8. c
4. b 9. c
5. d 10. d

Chapter 39
1. e 6. a
2. b 7. b
3. c 8. c
4. c 9. e
5. d 10. a

Chapter 40
1. c 6. b
2. c 7. e
3. a 8. a
4. d 9. e
5. b 10. b

Chapter 41
1. b 6. b
2. a 7. d
3. b 8. e
4. e 9. c
5. e 10. c

Chapter 42
1. a 6. a
2. b 7. d
3. e 8. d
4. e 9. a
5. c 10. d

Chapter 43
1. c 6. d
2. e 7. d
3. a 8. d
4. d 9. d
5. d 10. a

Chapter 44
1. a 6. c
2. c 7. b
3. e 8. b
4. a 9. b
5. d 10. c

Chapter 45
1. d 6. e
2. d 7. c
3. c 8. c
4. c 9. d
5. e 10. d

Chapter 46
1. d 6. e
2. d 7. b
3. a 8. c
4. e 9. c
5. e 10. d

Chapter 47
1. c 6. c
2. a 7. a
3. e 8. c
4. d 9. a
5. d 10. c

Chapter 48
1. e 6. d
2. a 7. e
3. b 8. a
4. c 9. a
5. b 10. e

Chapter 49
1. e 6. b
2. d 7. c
3. a 8. c
4. b 9. a
5. c 10. d

Chapter 50
1. d 6. d
2. a 7. b
3. c 8. d
4. d 9. c
5. c 10. e

Chapter 51
1. b 6. d
2. e 7. a
3. c 8. b
4. a 9. d
5. b 10. d

Chapter 52
1. d 6. b
2. a 7. e
3. d 8. a
4. a 9. c
5. d 10. e

Chapter 53
1. b 6. d
2. e 7. e
3. c 8. e
4. a 9. a
5. d 10. e

Chapter 54
1. d 6. c
2. a 7. c
3. a 8. c
4. d 9. d
5. a 10. b

Chapter 55
1. c 6. b
2. c 7. e
3. a 8. c
4. d 9. e
5. c 10. d

Chapter 56
1. a 6. a
2. c 7. b
3. b 8. d
4. c 9. e
5. d 10. b

Chapter 57
1. a 6. b
2. a 7. e
3. b 8. b
4. d 9. d
5. e 10. e

Chapter 58
1. e 6. e
2. d 7. a
3. b 8. e
4. c 9. d
5. c 10. b

Chapter 59
1. b 6. a
2. d 7. d
3. e 8. c
4. e 9. a
5. b 10. c

Glossary

- A -

abiotic (a' bye ah tick) [Gk. *a*: not + *bios*: life] Nonliving. (Contrast with biotic.)

abscisic acid (ABA) (ab sighs' ik) A plant growth substance with growth-inhibiting action. Causes stomata to close; involved in a plant's response to salt and drought stress.

abscission (ab sizh' un) [L. *abscissio*: break off] The process by which leaves, petals, and fruits separate from a plant.

absorption (1) Of light: complete retention, without reflection or transmission. (2) Of water or other molecules: soaking up (taking in through pores or by diffusion).

absorption spectrum A graph of light absorption versus wavelength of light; shows how much light is absorbed at each wavelength.

absorptive heterotroph An organism (usually a fungus) that obtains its food by secreting digestive enzymes into the environment to break down large food molecules, then absorbing the breakdown products.

abyssal plain (uh biss' ul) [Gk. *abyssos*: bottomless] The deep ocean floor.

accessory pigments Pigments that absorb light and transfer energy to chlorophylls for photosynthesis.

acetylcholine (ACh) A neurotransmitter that carries information across vertebrate neuromuscular junctions and some other synapses. It is then broken down by the enzyme acetylcholinesterase (AChE).

acetyl coenzyme A (acetyl CoA) A compound that reacts with oxaloacetate to produce citrate at the beginning of the citric acid cycle; a key metabolic intermediate in the formation of many compounds.

acid [L. *acidus*: sharp, sour] A substance that can release a proton in solution. (Contrast with base.)

acid growth hypothesis The hypothesis that auxin increases proton pumping, thereby lowering the pH of the cell wall and activating enzymes that loosen polysaccharides. Proposed to explain auxin-induced cell expansion in plants.

acid precipitation Precipitation that has a lower pH than normal as a result of acid-forming precursor molecules introduced into the atmosphere by human activities.

acidic Having a pH of less than 7.0 (a hydrogen ion concentration greater than 10^{-7} molar). (Contrast with basic.)

acoelomate An animal that does not have a coelom.

acrosome (a' krow soam) [Gk. *akros*: highest + *soma*: body] The structure at the forward tip of an animal sperm which is the first to fuse with the egg membrane and enter the egg cell.

ACTH *See* corticotropin.

actin [Gk. *aktis*: ray] A protein that makes up the cytoskeletal microfilaments in eukaryotic cells and is one of the two contractile proteins in muscle.

action potential An impulse in a neuron taking the form of a wave of depolarization or hyperpolarization.

action spectrum A graph of a biological process versus light wavelength; shows which wavelengths are involved in the process.

activation energy (E_a) The energy barrier that blocks the tendency for a chemical reaction to occur.

active site The region on the surface of an enzyme or ribozyme where the substrate binds, and where catalysis occurs.

active transport The energy-dependent transport of a substance across a biological membrane against a concentration gradient—that is, from a region of low concentration (of that substance) to one of high concentration. (*See also* primary active transport, secondary active transport; contrast with facilitated diffusion, passive transport.)

adaptation (a dap tay' shun) (1) In evolutionary biology, a particular structure, physiological process, or behavior that makes an organism better able to survive and reproduce. Also, the evolutionary process that leads to the development or persistence of such a trait. (2) In sensory neurophysiology, a sensory cell's loss of sensitivity as a result of repeated stimulation.

adaptive radiation An evolutionary radiation that results in an array of related species that live in a variety of environments and differ in the characteristics they use to exploit those environments.

adenine (A) (a' den een) A nitrogen-containing base found in nucleic acids, ATP, NAD, and other compounds.

adenosine triphosphate *See* ATP.

adrenal gland (a dree' nal) [L. *ad*: toward + *renes*: kidneys] An endocrine gland located near the kidneys of vertebrates, consisting of two glandular parts, the cortex and medulla.

adrenaline *See* epinephrine.

adrenocorticotropic hormone *See* corticotropin.

adsorption Binding of a gas or a solute to the surface of a solid.

adventitious roots (ad ven ti' shus) [L. *adventitius*: arriving from outside] Roots originating from the stem at ground level or below; typical of the fibrous root system of monocots.

aerenchyma In plants, parenchymal tissue containing air spaces.

aerobic (air oh' bic) [Gk. *aer*: air + *bios:* life] In the presence of oxygen; requiring oxygen. (Contrast with anaerobic.)

afferent (af' ur unt) [L. *ad*: toward + *ferre*: to carry] Carrying to, as in a neuron that carries impulses to the central nervous system (afferent neuron), or a blood vessel that carries blood to a structure. (Contrast with efferent.)

age structure The distribution of the individuals in a population across all age groups.

AIDS Acquired immune deficiency syndrome, a condition caused by human immunodeficiency virus (HIV) in which the body's T-helper cells are reduced, leaving the victim subject to opportunistic diseases.

air sacs Structures in the respiratory system of birds that receive inhaled air; they keep fresh air flowing unidirectionally through the lungs, but are not themselves gas exchange surfaces.

aldosterone (al dohs' ter own) A steroid hormone produced in the adrenal cortex of mammals. Promotes secretion of potassium and reabsorption of sodium in the kidney.

aleurone layer In some seeds, a tissue that lies beneath the seed coat and surrounds the endosperm. Secretes digestive enzymes that break down macromolecules stored in the endosperm.

allantoic membrane In animal development, an outgrowth of extraembryonic endoderm plus adjacent mesoderm that forms the allantois, a saclike structure that stores metabolic wastes produced by the embryo.

allantois (al lun twah') [Gk. *allant*: sausage] An extraembryonic membrane enclosing a sausage-shaped sac that stores the embryo's nitrogenous wastes.

allele (a leel') [Gk. *allos*: other] The alternate form of a genetic character found at a given locus on a chromosome.

allele frequency The relative proportion of a particular allele in a specific population.

allergic reaction [Ger. *allergie*: altered] An overreaction of the immune system to amounts of an antigen that do not affect most people; often involves IgE antibodies.

allometric growth A pattern of growth in which some parts of the body of an organism grow faster than others, resulting in a change in body proportions as the organism grows.

allopatric speciation (al' lo pat' rick) [Gk. *allos*: other + *patria*: homeland] The formation of two species from one when reproductive isolation occurs because of the interposition of (or crossing of) a physical geographic barrier such as a river. Also called geographic speciation. (Contrast with sympatric speciation.)

allopolyploidy The possession of more than two chromosome sets that are derived from more than one species.

allosteric regulation (al lo steer' ik) [Gk. *allos*: other + *stereos*: structure] Regulation of the activity of a protein (usually an enzyme) by the binding of an effector molecule to a site other than the active site.

alpha diversity Species diversity within a single community or habitat. (Contrast with beta diversity, gamma diversity.)

α (alpha) helix A prevalent type of secondary protein structure; a right-handed spiral.

alternation of generations The succession of multicellular haploid and diploid phases in some sexually reproducing organisms, notably plants.

alternative splicing A process for generating different mature mRNAs from a single gene by splicing together different sets of exons during RNA processing.

altruism Pertaining to behavior that benefits other individuals at a cost to the individual who performs it.

alveolus (al ve' o lus) (plural: alveoli) [L. *alveus*: cavity] A small, baglike cavity, especially the blind sacs of the lung.

amensalism (a men' sul ism) Interaction in which one animal is harmed and the other is unaffected. (Contrast with commensalism, mutualism.)

amine An organic compound containing an amino group (NH_2).

amino acid An organic compound containing both NH_2 and COOH groups. Proteins are polymers of amino acids.

amino acid replacement A change in the nucleotide sequence that results in one amino acid being replaced by another.

ammonotelic (am moan' o teel' ic) [Gk. *telos*: end] Pertaining to an organism in which the final product of breakdown of nitrogen-containing compounds (primarily proteins) is *ammonia*. (Contrast with ureotelic, uricotelic.)

amnion (am' nee on) The fluid-filled sac within which the embryos of reptiles (including birds) and mammals develop.

amniote egg A shelled egg surrounding four extraembryonic membranes and embryo-nourishing yolk. This evolutionary adaptation permitted mammals and reptiles to live and reproduce in drier environments than can most amphibians.

amphipathic (am' fi path' ic) [Gk. *amphi*: both + *pathos*: emotion] Of a molecule, having both hydrophilic and hydrophobic regions.

amplitude The magnitude of change over the course of a regular cycle.

amygdala A component of the limbic system that is involved in fear and fear memory.

amylase (am' ill ase) An enzyme that catalyzes the hydrolysis of starch, usually to maltose or glucose.

anabolic reaction (an uh bah' lik) [Gk. *ana*: upward + *ballein*: to throw] A synthetic reaction in which simple molecules are linked to form more complex ones; requires an input of energy and captures it in the chemical bonds that are formed. (Contrast with catabolic reaction.)

anaerobic (an ur row' bic) [Gk. *an*: not + *aer*: air + *bios*: life] Occurring without the use of molecular oxygen, O_2. (Contrast with aerobic.)

analogy (a nal' o jee) [Gk. *analogia*: resembling] A resemblance between two features that is due to convergent evolution rather than to common ancestry. The structures are said to be *analogous*, and each is an *analog* of the others. (Contrast with homology.)

anaphase (an' a phase) [Gk. *ana*: upward] The stage in cell nuclear division at which the first separation of sister chromatids (or, in the first meiotic division, of paired homologs) occurs.

ancestral trait The trait originally present in the ancestor of a given group; may be retained or changed in the descendants of that ancestor.

androgen (an' dro jen) Any of the several male sex steroids (most notably testosterone).

aneuploidy (an' you ploy dee) A condition in which one or more chromosomes or pieces of chromosomes are either lacking or present in excess.

angiotensin (an' jee oh ten' sin) A peptide hormone that raises blood pressure by causing peripheral vessels to constrict. Also maintains glomerular filtration by constricting efferent vessels and stimulates thirst and the release of aldosterone.

angular gyrus A part of the human brain believed to be essential for integrating spoken and written language.

animal hemisphere The metabolically active upper portion of some animal eggs, zygotes, and embryos; does not contain the dense nutrient yolk. (Contrast with vegetal hemisphere.)

anion (an' eye on) [Gk. *ana*: upward progress] A negatively charged ion. (Contrast with cation.)

anisogamous (an eye sog' a muss) [Gk. *aniso*: unequal + *gamos*: marriage] Having morphologically dissimilar male and female gametes. (Contrast with isogamous.)

annual A plant whose life cycle is completed in one growing season. (Contrast with biennial, perennial.)

antagonistic interactions Interactions between two species in which one species benefits and the other is harmed. Includes predation, herbivory, and parasitism.

antenna system In photosynthesis, a group of different molecules that cooperate to absorb light energy and transfer it to a reaction center.

anterior Toward or pertaining to the tip or headward region of the body axis. (Contrast with posterior.)

anterior pituitary The portion of the vertebrate pituitary gland that derives from gut epithelium and produces tropic hormones.

anther (an' thur) [Gk. *anthos*: flower] A pollen-bearing portion of the stamen of a flower.

antheridium (an' thur id' ee um) [Gk. *antheros*: blooming] The multicellular structure that produces the sperm in nonvascular land plants and ferns.

antibody One of the myriad proteins produced by the immune system that specifically binds to a foreign substance in blood or other tissue fluids and initiates its removal from the body.

anticodon The three nucleotides in transfer RNA that pair with a complementary triplet (a codon) in messenger RNA.

antidiuretic hormone (ADH) A hormone that promotes water reabsorption by the kidney. ADH is produced by neurons in the hypothalamus and released from nerve terminals in the posterior pituitary. Also called vasopressin.

antigen (an' ti jun) Any substance that stimulates the production of an antibody or antibodies in the body of a vertebrate.

antigenic determinant The specific region of an antigen that is recognized and bound by a specific antibody. Also called an epitope.

antiparallel Pertaining to molecular orientation in which a molecule or parts of a molecule have opposing directions.

antiporter A membrane transport protein that moves one substance in one direction and another in the opposite direction. (Contrast with symporter, uniporter.)

antisense RNA A single-stranded RNA molecule complementary to, and thus targeted against, an mRNA of interest to block its translation.

anus (a' nus) An opening through which solid digestive wastes are expelled, located at the posterior end of a tubular gut.

aorta (a or' tah) [Gk. *aorte*: aorta] The main trunk of the arteries leading to the systemic (as opposed to the pulmonary) circulation.

aortic body A chemosensor in the aorta that senses a decrease in blood supply or a dramatic decrease in partial pressure of oxygen in the blood.

aortic valve A one-way valve between the left ventricle of the heart and the aorta that prevents backflow of blood into the ventricle when it relaxes.

apex (a' pecks) The tip or highest point of a structure, as of a growing stem or root.

aphasia a deficit in the ability to use or understand words.

apical (a' pi kul) Pertaining to the *apex*, or tip, usually in reference to plants.

apical dominance In plants, inhibition by the apical bud of the growth of axillary buds.

apical hook A form taken by the stems of many eudicot seedlings that protects the delicate shoot apex while the stem grows through the soil.

apical meristem The meristem at the tip of a shoot or root; responsible for a plant's primary growth.

apomixis (ap oh mix' is) [Gk. *apo*: away from + *mixis*: sexual intercourse] The asexual production of seeds.

apoplast (ap' oh plast) In plants, the continuous meshwork of cell walls and extracellular spaces through which material can pass without crossing a plasma membrane. (Contrast with symplast.)

apoptosis (ap uh toh' sis) A series of genetically programmed events leading to cell death.

aposematism Warning coloration; bright colors or striking patterns of toxic or mimetic prey species that act as a warning to predators.

appendix In the human digestive system, the vestigial equivalent of the cecum, which serves no digestive function.

aquaporin A transport protein in plant and animal cell membranes through which water passes in osmosis.

aquatic (a kwa' tic) [L. *aqua*: water] Pertaining to or living in water. (Contrast with marine, terrestrial.)

aqueous (a' kwee us) Pertaining to water or a watery solution.

aquifer A large pool of groundwater.

archegonium (ar' ke go' nee um) The multicellular structure that produces eggs in nonvascular land plants, ferns, and gymnosperms.

archenteron (ark en' ter on) [Gk. *archos*: first + *enteron*: bowel] The earliest primordial animal digestive tract.

area phylogenies Phylogenies in which the names of the taxa are replaced with the names of the places where those taxa live or lived.

arms race A series of reciprocal adaptations between species involved in antagonistic interactions, in which adaptations that increase the fitness of a consumer species exert selection pressure on its resource species to counter the consumer's adaptation, and vice versa.

arteriole A small blood vessel arising from an artery that feeds blood into a capillary bed.

artery A muscular blood vessel carrying oxygenated blood away from the heart to other parts of the body. (Contrast with vein.)

artificial selection   The selection by plant and animal breeders of individuals with certain desirable traits.

ascus (ass' cus) (plural: asci) [Gk. *askos*: bladder] In sac fungi, the club-shaped sporangium within which spores (ascospores) are produced by meiosis.

asexual reproduction   Reproduction without sex.

assisted reproductive technologies (ARTs)   Any of several procedures that remove unfertilized eggs from the ovary, combine them with sperm outside the body, and then place fertilized eggs or egg–sperm mixtures in the appropriate location in a female's reproductive tract for development.

association cortex   In the vertebrate brain, the portion of the cortex involved in higher-order information processing, so named because it integrates, or associates, information from different sensory modalities and from memory.

associative learning   A form of learning in which two unrelated stimuli become linked to the same response.

astrocyte [Gk. *astron*: star]   A type of glial cell that contributes to the blood–brain barrier by surrounding the smallest, most permeable blood vessels in the brain.

atherosclerosis (ath' er oh sklair oh' sis) [Gk. *athero*: gruel, porridge + *skleros*: hard]   A disease of the arteries characterized by fatty, cholesterol-rich deposits in the walls of the arteries. When fibroblasts infiltrate these deposits and calcium precipitates in them, the disease become arteriosclerosis, or "hardening of the arteries."

atom [Gk. *atomos*: indivisible]   The smallest unit of a chemical element. Consists of a nucleus and one or more electrons.

atomic mass   *See* atomic weight.

atomic number   The number of protons in the nucleus of an atom; also equals the number of electrons around the neutral atom. Determines the chemical properties of the atom.

atomic weight   The average of the mass numbers of a representative sample of atoms of an element, with all the isotopes in their normally occurring proportions. Also called atomic mass.

ATP (adenosine triphosphate)   An energy-storage compound containing adenine, ribose, and three phosphate groups. When it is formed from ADP, useful energy is stored; when it is broken down (to ADP or AMP), energy is released to drive endergonic reactions.

ATP synthase   An integral membrane protein that couples the transport of protons with the formation of ATP.

atrial natriuretic peptide   A hormone released by the atrial muscle fibers of the heart when they are overly stretched, which decreases reabsorption of sodium by the kidney and thus blood volume.

atrioventricular node   A modified node of cardiac muscle that organizes the action potentials that control contraction of the ventricles.

atrium (a' tree um) [L. *atrium*: central hall]   An internal chamber. In the hearts of vertebrates, the thin-walled chamber(s) entered by blood on its way to the ventricle(s). Also, the outer ear.

auditory system   A sensory system that uses mechanoreceptors to convert pressure waves into receptor potentials; includes structures that gather sound waves, direct them to a sensory or-gan, and amplify their effect on the mechanoreceptors.

autocatalysis [Gk. *autos*: self + *kata*: to break down]   A positive feedback process in which an activated enzyme acts on other inactive molecules of the same enzyme to activate them.

autocrine   A chemical signal that binds to and affects the cell that makes it. (Contrast with paracrine.)

autoimmunity   An immune response by an organism to its own molecules or cells.

autonomic nervous system (ANS)   The portion of the peripheral nervous system that controls such involuntary functions as those of guts and glands. Also called the involuntary nervous system.

autopolyploidy   The possession of more than two entire chromosomes sets that are derived from a single species.

autosome   Any chromosome (in a eukaryote) other than a sex chromosome.

autotroph (au' tow trowf') [Gk. *autos*: self + *trophe*: food]   An organism that is capable of living exclusively on inorganic materials, water, and some energy source such as sunlight (photoautotrophs) or chemically reduced matter (see chemolithotrophs). (Contrast with heterotroph.)

auxin (awk' sin) [Gk. *auxein*: to grow]   In plants, a substance (the most common being indoleacetic acid) that regulates growth and various aspects of development.

avirulence (*Avr*) genes   Genes in a pathogen that may trigger defenses in plants. *See* gene-for-gene resistance.

Avogadro's number   The number of atoms or molecules in a mole (weighed out in grams) of a substance, calculated to be 6.022×10^{23}.

axillary bud   A bud that forms in the angle (axil) where a leaf meets a stem.

axon [Gk. axle]   The part of a neuron that conducts action potentials away from the cell body.

axon hillock   The junction between an axon and its cell body, where action potentials are generated.

axon terminals   The endings of an axon; they form synapses and release neurotransmitter.

- B -

B cell   A type of lymphocyte involved in the humoral immune response of vertebrates. Upon recognizing an antigenic determinant, a B cell develops into a plasma cell, which secretes an antibody. (Contrast with T cell.)

bacillus (bah sil' us) [L: little rod]   Any of various rod-shaped bacteria.

bacterial artificial chromosome (BAC)   A DNA cloning vector used in bacteria that can carry up to 150,000 base pairs of foreign DNA.

bacteriophage (bak teer' ee o fayj) [Gk. *bakterion*: little rod + *phagein*: to eat]   Any of a group of viruses that infect bacteria. Also called phage.

bacteroids   Nitrogen-fixing organelles that develop from endosymbiotic bacteria.

bark   All tissues external to the vascular cambium of a plant.

baroreceptor [Gk. *baros*: weight]   A pressure-sensing cell or organ. Sometimes called a stress receptor.

basal body   A centriole found at the base of a eukaryotic flagellum or cilium.

basal metabolic rate (BMR)   The minimum rate of energy turnover in an awake (but resting) bird or mammal that is not expending energy for thermoregulation.

base   (1) A substance that can accept a hydrogen ion in solution. (Contrast with acid.) (2) In nucleic acids, the purine or pyrimidine that is attached to each sugar in the sugar–phosphate backbone.

base pair (bp)   In double-stranded DNA, a pair of nucleotides formed by the *complementary base pairing* of a purine on one strand and a pyrimidine on the other.

basic   Having a pH greater than 7.0 (i.e., having a hydrogen ion concentration lower than 10^{-7} molar). (Contrast with acidic.)

basidiocarp   A fruiting structure produced by club fungi.

basidium (bass id' ee yum)   In club fungi, the characteristic sporangium in which four spores are formed by meiosis and then borne externally before being shed.

basilar membrane   A membrane in the human inner ear whose flexion in response to sound waves activates hair cells; flexes at different locations in response to different pitches of sound.

basophil   A type of phagocytic white blood cell that releases histamine and may promote T cell development.

Batesian mimicry   The convergence in appearance of an edible species (mimic) with an unpalatable species (model).

benefit   An improvement in survival and reproductive success resulting from performing a behavior or having a trait. (Contrast with cost.)

benthic zone [Gk. *benthos*: bottom]   The bottom of the ocean.

beta diversity   Between-habitat diversity; a measure of the change in species composition from one community or habitat to another. (Contrast with alpha diveristy, gamma diversity.)

β (beta) pleated sheet   A type of protein secondary structure; results from hydrogen bonding between polypeptide regions running antiparallel to each other.

biennial   A plant whose life cycle includes vegetative growth in the first year and flowering and senescence in the second year. (Contrast with annual, perennial.)

bilateral symmetry   The condition in which only the right and left sides of an organism, divided by a single plane through the midline, are mirror images of each other.

bilayer   A structure that is two layers in thickness. In biology, most often refers to the phospholipid bilayer of membranes. (*See* phospholipid bilayer.)

bile   A secretion of the liver made up of bile salts synthesized from cholesterol, various phospholipids, and bilirubin (the breakdown product of hemoglobin). Emulsifies fats in the small intestine.

binary fission   Reproduction of a prokaryote by division of a cell into two comparable progeny cells.

binocular vision   Overlapping visual fields of an animal's two eyes; allows the animal to see in three dimensions.

binomial nomenclature   A taxonomic naming system in which each species is given two names (a genus name followed by a species name).

biofilm A community of microorganisms embedded in a polysaccharide matrix, forming a highly resistant coating on almost any moist surface.

biogeochemical cycle Movement of inorganic elements such as nitrogen, photsphorus, and carbon through living organisms and the physical environment.

biogeographic region One of several defined, continental-scale regions of Earth, each of which has a biota distinct from that of the others. (Contrast with biome.)

biogeography The scientific study of the patterns of distribution of populations, species, and ecological communities across Earth.

bioinformatics The use of computers and/or mathematics to analyze complex biological information, such as DNA sequences.

biological species concept The definition of a species as a group of actually or potentially interbreeding natural populations that are reproductively isolated from other such groups. (Contrast with lineage species concept; morphological species concept.)

biology [Gk. *bios*: life + *logos*: study] The scientific study of living things.

bioluminescence The production of light by biochemical processes in an organism.

biomass The total weight of all the organisms, or some designated group of organisms, in a given area.

biome (bye' ome) A major division of the ecological communities of Earth, characterized primarily by distinctive vegetation. A given biogeographic region contains many different biomes.

bioremediation The use by humans of other organisms to remove contaminants from the environment.

biosphere (bye' oh sphere) All regions of Earth (terrestrial and aquatic) and Earth's atmosphere in which organisms can live.

biota (bye oh' tah) All of the organisms—animals, plants, fungi, and microorganisms—found in a given area. (Contrast with flora, fauna.)

biotechnology The use of cells or living organisms to produce materials useful to humans.

biotic (bye ah' tick) [Gk. *bios*: life] Alive. (Contrast with abiotic.)

biotic interchange The dispersal of species from two different biotas into the region they had not previously inhabited, as when two formerly separated land masses fuse.

blade The thin, flat portion of a leaf.

blastocoel (blass' toe seal) [Gk. *blastos*: sprout + *koilos*: hollow] The central, hollow cavity of a blastula.

blastocyst (blass' toe cist) An early embryo formed by the first divisions of the fertilized egg (zygote). In mammals, a hollow ball of cells.

blastodisc (blass' toe disk) An embryo that forms as a disk of cells on the surface of a large yolk mass; comparable to a blastula, but occurring in animals such as birds and reptiles, in which the massive yolk restricts incomplete cleavage.

blastomere Any of the cells produced by the early divisions of a fertilized animal egg.

blastula (blass' chu luh) An early stage of the animal embryo; in many species, a hollow sphere of cells surrounding a central cavity, the blastocoel. (Contrast with blastodisc.)

block to polyspermy Any of several responses to entry of a sperm into an egg that prevent more than one sperm from entering the egg.

blood A fluid tissue that is pumped around the body; a component of the circulatory system.

blood–brain barrier A property of blood vessels in the brain that prevents most chemicals from diffusing from the blood into the brain.

blue-light receptors Pigments in plants that absorb blue light (400–500 nm). These pigments mediate many plant responses including phototropism, stomatal movements, and expression of some genes.

body plan The general structure of an animal, the arrangement of its organ systems, and the integrated functioning of its parts.

bond *See* chemical bond.

bottleneck *See* population bottleneck.

bone A rigid component of vertebrate skeletal systems that contains an extracellular matrix of insoluble calcium phosphate crystals as well as collagen fibers.

Bowman's capsule An elaboration of the renal tubule, composed of podocytes, that surrounds and collects the filtrate from the glomerulus.

brain The centralized integrative center of a nervous system.

brainstem The portion of the vertebrate brain between the spinal cord and the forebrain, made up of the medulla, pons, and midbrain.

brassinosteroids Plant steroid hormones that mediate light effects promoting the elongation of stems and pollen tubes.

Broca's area A portion of the human brain essential for speech. Located in the frontal lobe just in front of the primary motor cortex.

bronchioles The smallest airways in a vertebrate lung, branching off the bronchi.

bronchus (plural: bronchi) The major airway(s) branching off the trachea into the vertebrate lung.

brown fat In mammals, fat tissue that is specialized to produce heat. It has many mitochondria and capillaries, and a protein that uncouples oxidative phosphorylation.

budding Asexual reproduction in which a more or less complete new organism grows from the body of the parent organism, eventually detaching itself.

buffer A substance that can transiently accept or release hydrogen ions and thereby resist changes in pH.

bulbourethral glands Secretory structures of the human male reproductive system that produce a small volume of an alkaline, mucoid secretion that helps neutralize acidity in the urethra and lubricate it to facilitate the passage of semen.

bulk flow The movement of a solution from a region of higher pressure potential to a region of lower pressure potential.

bundle of His Fibers of modified cardiac muscle that conduct action potentials from the atria to the ventricular muscle mass.

bundle sheath cell Part of a tissue that surrounds the veins of plants; contains chloroplasts in C_4 plants.

- C -

C3 plants Plants that produce 3PG as the first stable product of carbon fixation in photosynthesis and use ribulose bisphosphate as a CO_2 receptor.

C4 plants Plants that produce oxaloacetate as the first stable product of carbon fixation in photosynthesis and use phosphoenolpyruvate as CO_2 acceptor. C_4 plants also perform the reactions of C_3 photosynthesis.

calcitonin Hormone produced by the thyroid gland; lowers blood calcium and promotes bone formation. (Contrast with parathyroid hormone.)

calorie [L. *calor*: heat] The amount of heat required to raise the temperature of 1 gram of water by 1°C. Physiologists commonly use the kilocalorie (kcal) as a unit of measure (1 kcal = 1,000 calories). Nutritionists also use the kilocalorie, but refer to it as the *Calorie* (capital C).

Calvin cycle The stage of photosynthesis in which CO_2 reacts with RuBP to form 3PG, 3PG is reduced to a sugar, and RuBP is regenerated, while other products are released to the rest of the plant. Also known as the Calvin–Benson cycle.

calyx (kay' licks) [Gk. *kalyx*: cup] All of the sepals of a flower, collectively.

CAM *See* crassulacean acid metabolism.

Cambrian explosion The rapid diversification of multicellular life that took place during the Cambrian period.

cAMP (cyclic AMP) A compound formed from ATP that acts as a second messenger.

cancellous bone A type of bone with numerous internal cavities that make it appear spongy, although it is rigid. (Contrast with compact bone.)

canopy The leaf-bearing part of a tree. Collectively, the aggregate of the leaves and branches of the larger woody plants of an ecological community.

capillaries [L. *capillaris*: hair] Very small tubes, especially the smallest blood-carrying vessels of animals between the termination of the arteries and the beginnings of the veins. Capillaries are the site of exchange of materials between the blood and the interstitial fluid.

capsid The outer shell of a virus that encloses its nucleic acid.

carbohydrates Organic compounds containing carbon, hydrogen, and oxygen in the ratio 1:2:1 (i.e., with the general formula $C_nH_{2n}O_n$). Common examples are sugars, starch, and cellulose.

carbon skeleton The chains or rings of carbon atoms that form the structural basis of organic molecules. Other atoms or functional groups are attached to the carbon atoms.

carboxylase An enzyme that catalyzes the addition of carboxyl groups to a substrate.

cardiac (kar' dee ak) [Gk. *kardia*: heart] Pertaining to the heart and its functions.

cardiac cycle Contraction of the two atria of the heart, followed by contraction of the two ventricles and then relaxation.

cardiac muscle A type of muscle tissue that makes up, and is responsible for the beating of, the heart. Characterized by branching cells with single nuclei and a striated (striped) appearance. (Contrast with smooth muscle, skeletal muscle.)

cardiovascular system [Gk. *kardia*: heart + L. *vasculum*: small vessel] The heart, blood, and vessels are of a circulatory system.

carnivore [L. *carn*: flesh + *vorare*: to devour] An organism that eats animal tissues. (Contrast with detritivore, herbivore, omnivore.)

carotenoid (ka rah' tuh noid) A yellow, orange, or red lipid pigment commonly found as an accessory pigment in photosynthesis; also found in fungi.

carotid body   A chemosensor in the carotid artery that senses a decrease in blood supply or a dramatic decrease in partial pressure of oxygen in the blood.

carpel (kar′ pel) [Gk. *karpos*: fruit]   The organ of the flower that contains one or more ovules.

carrier   (1) In facilitated diffusion, a membrane protein that binds a specific molecule and transports it through the membrane. (2) In respiratory and photosynthetic electron transport, a participating substance such as NAD that exists in both oxidized and reduced forms. (3) In genetics, a person heterozygous for a recessive trait.

carrying capacity (K)   The number of individuals in a population that the resources of its environment can support.

cartilage   In vertebrates, a tough connective tissue found in joints, the outer ear, and elsewhere. Forms the entire skeleton in some animal groups.

cartilage bone   A type of bone that begins its development as a cartilaginous structure resembling the future mature bone, then gradually hardens into mature bone. (Contrast with membranous bone.)

Casparian strip   A band of cell wall containing suberin and lignin, found in the endodermis. Restricts the movement of water across the endodermis.

caspase   One of a group of proteases that catalyze cleavage of target proteins and are active in apoptosis.

catabolic reaction (kat uh bah′ lik) [Gk. *kata*: to break down + *ballein*: to throw]   A synthetic reaction in which complex molecules are broken down into simpler ones and energy is released. (Contrast with anabolic reaction.)

catabolite repression   In the presence of abundant glucose, the diminished synthesis of catabolic enzymes for other energy sources.

catalyst (kat′ a list) [Gk. *kata*: to break down]   A chemical substance that accelerates a reaction without itself being consumed in the overall course of the reaction. Catalysts lower the activation energy of a reaction. Enzymes are biological catalysts.

cation (cat′ eye on)   An ion with one or more positive charges. (Contrast with anion.)

caudal [L. *cauda*: tail]   Pertaining to the tail, or to the posterior part of the body.

cDNA   *See* complementary DNA.

cDNA library   A collection of complementary DNAs derived from mRNAs of a particular tissue at a particular time in the life cycle of an organism.

cecum (see′ cum) [L. blind]   A blind branch off the large intestine. In many nonruminant mammals, the cecum contains a colony of microorganisms that contribute to the digestion of food.

cell   The simplest structural unit of a living organism. In multicellular organisms, the building blocks of tissues and organs.

cell adhesion molecules   Molecules on animal cell surfaces that affect the selective association of cells into tissues during development of the embryo.

cell cycle   The stages through which a cell passes between one division and the next. Includes all stages of interphase and mitosis.

cell division   The reproduction of a cell to produce two new cells. In eukaryotes, this process involves nuclear division (mitosis) and cytoplasmic division (cytokinesis).

cell fate   The type of cell that an undifferentiated cell in an embryo will become in the adult.

cell junctions   Specialized structures associated with the plasma membranes of epithelial cells. Some contribute to cell adhesion, others to intercellular communication.

cell recognition   Binding of cells to one another mediated by membrane proteins or carbohydrates.

cell theory   States that cells are the basic structural and physiological units of all living organisms, and that all cells come from preexisting cells.

cell wall   A relatively rigid structure that encloses cells of plants, fungi, many protists, and most prokaryotes, and which gives these cells their shape and limits their expansion in hypotonic media.

cellular immune response   Immune system response mediated by T cells and directed against parasites, fungi, intracellular viruses, and foreign tissues (grafts). (Contrast with humoral immune response.)

cellular respiration   The catabolic pathways by which electrons are removed from various molecules and passed through intermediate electron carriers to O_2, generating H_2O and releasing energy.

cellulose (sell′ you lowss)   A straight-chain polymer of glucose molecules, used by plants as a structural supporting material.

central dogma   The premise that information flows from DNA to RNA to polypeptide.

central nervous system (CNS)   That portion of the nervous system that is the site of most information processing, storage, and retrieval; in vertebrates, the brain and spinal cord. (Contrast with peripheral nervous system.)

central vacuole   In plant cells, a large organelle that stores the waste products of metabolism and maintains turgor.

centrifuge [L. *centrum*: center + *fugere*: to flee]   A laboratory device in which a sample is spun around a central axis at high speed. Used to separate suspended materials of different densities.

centriole (sen′ tree ole)   A paired organelle that helps organize the microtubules in animal and protist cells during nuclear division.

centromere (sen′ tro meer) [Gk. *centron*: center + *meros*: part]   The region where sister chromatids join.

centrosome (sen′ tro soam)   The major microtubule organizing center of an animal cell.

cephalization (sef ah luh zay′ shun) [Gk. *kephale*: head]   The evolutionary trend toward increasing concentration of brain and sensory organs at the anterior end of the animal.

cerebellum (sair uh bell′ um) [L. diminutive of *cerebrum*, brain]   The brain region that controls muscular coordination; located at the anterior end of the hindbrain.

cerebral cortex   The thin layer of gray matter (neuronal cell bodies) that overlies the cerebrum.

cerebrum (su ree′ brum) [L. brain]   The dorsal anterior portion of the forebrain, making up the largest part of the brain of mammals; the chief coordination center of the nervous system; consists of two *cerebral hemispheres*.

cervix (sir′ vix) [L. neck]   The opening of the uterus into the vagina.

cGMP (cyclic guanosine monophosphate)   An intracellular messenger that is part of signal transmission pathways involving G proteins. (*See* G protein.)

channel protein   An integral membrane protein that forms an aqueous passageway across the membrane in which it is inserted through which specific solutes may pass.

chaperone   A protein that guards other proteins by counteracting molecular interactions that threaten their three-dimensional structure.

character   In genetics, an observable feature, such as eye color. (Contrast with trait.)

character displacement   An evolutionary phenomenon in which species that compete for the same resources within the same territory tend to diverge in morphology and/or behavior.

chemical bond   An attractive force stably linking two atoms.

chemical evolution   The theory that life originated through the chemical transformation of inanimate substances.

chemical reaction   The change in the composition or distribution of atoms of a substance with consequent alterations in properties.

chemical synapse   Neural junction at which neurotransmitter molecules released from a presynaptic cell induce changes in a postsynaptic cell. (Contrast with electrical synapse.)

chemically gated channel   A type of gated channel that opens or closes depending on the presence or absence of a specific molecule, which binds to the channel protein or to a separate receptor that in turn alters the three-dimensional shape of channel protein.

chemiosmosis   Formation of ATP in mitochondria and chloroplasts, resulting from a pumping of protons across a membrane (against a gradient of electrical charge and of pH), followed by the return of the protons through a protein channel with ATP synthase activity.

chemoautotroph   *See* chemolithotroph.

chemoheterotroph   An organism that must obtain both carbon and energy from organic substances. (Contrast with chemolithotroph, photoautotroph, photoheterotroph.)

chemolithotroph [Gk. *lithos*: stone, rock]   An organism that uses carbon dioxide as a carbon source and obtains energy by oxidizing inorganic substances from its environment; also called chemoautotroph. (Contrast with chemoheterotroph, photoautotroph, photoheterotroph.)

chemoreceptor   A sensory receptor cell that senses specific molecules (such as odorant molecules or pheromones) in the environment.

chiasma (kie az′ muh) (plural: chiasmata) [Gk. cross]   An X-shaped connection between paired homologous chromosomes in prophase I of meiosis. A chiasma is the visible manifestation of crossing over between homologous chromosomes.

chitin (kye′ tin) [Gk. *kiton*: tunic]   The characteristic tough but flexible organic component of the exoskeleton of arthropods, consisting of a complex, nitrogen-containing polysaccharide. Also found in cell walls of fungi.

chlorophyll (klor′ o fill) [Gk. *kloros*: green + *phyllon*: leaf]   Any of several green pigments associated with chloroplasts or with certain bacterial membranes; responsible for trapping light energy for photosynthesis.

chloroplast [Gk. *kloros*: green + *plast*: a particle] An organelle bounded by a double membrane containing the enzymes and pigments that perform photosynthesis. Chloroplasts occur only in eukaryotes.

choanocyte (ko' an uh site)   The collared, flagellated feeding cells of sponges.

cholecystokinin (ko' luh sis tuh kai' nin)   A hormone produced and released by the lining of the duodenum when it is stimulated by undigested fats and proteins. It stimulates the gallbladder to release bile and slows stomach activity.

chorion (kor' ee on) [Gk. *khorion*: afterbirth]   The outermost of the membranes protecting mammal, bird, and reptile embryos; in mammals it forms part of the placenta.

chromatid (kro' ma tid)   A newly replicated chromosome, from the time molecular duplication occurs until the time the centromeres separate (during anaphase of mitosis or of meiosis II).

chromatin   The nucleic acid–protein complex that makes up eukaryotic chromosomes.

chromosomal mutation   Loss of or changes in position/direction of a DNA segment on a chromosome.

chromosome (krome' o sowm) [Gk. *kroma*: color + *soma*: body]   In bacteria and viruses, the DNA molecule that contains most or all of the genetic information of the cell or virus. In eukaryotes, a structure composed of DNA and proteins that bears part of the genetic information of the cell.

chylomicron (ky low my' cron)   Particles of lipid coated with protein, produced in the gut from dietary fats and secreted into the extracellular fluids.

chyme (kime) [Gk. *kymus*: juice]   Created in the stomach; a mixture of ingested food with the digestive juices secreted by the salivary glands and the stomach lining.

cilium (sil' ee um) (plural: cilia) [L. eyelash]   Hairlike organelle used for locomotion by many unicellular organisms and for moving water and mucus by many multicellular organisms. Generally shorter than a flagellum.

circadian rhythm (sir kade' ee an) [L. *circa*: approximately + *dies*: day]   A rhythm of growth or activity that recurs about every 24 hours.

circannual rhythm [L. *circa*: + *annus*: year]   A rhythm of growth or activity that recurs on a yearly basis.

circulatory system   A system consisting of a muscular pump (heart), a fluid (blood or hemolymph), and a series of conduits (blood vessels) that transports materials around the body.

citric acid cycle   In cellular respiration, a set of chemical reactions whereby acetyl CoA is oxidized to carbon dioxide and hydrogen atoms are stored as NADH and FADH$_2$. Also called the Krebs cycle.

clade [Gk. *klados*: branch]   A monophyletic group made up of an ancestor and all of its descendants.

class I MHC molecules   Cell surface proteins that participate in the cellular immune response directed against virus-infected cells.

class II MHC molecules   Cell surface proteins that participate in the cell–cell interactions (of T-helper cells, macrophages, and B cells) of the humoral immune response.

cleavage   The first few cell divisions of an animal zygote. *See also* complete cleavage, incomplete cleavage.

climate   The long-term average atmospheric conditions (temperature, precipitation, humidity, wind direction and velocity) found in a region.

climax community   The final stage of succession; a community that is capable of perpetuating itself under local climatic and soil conditions and persists for a relatively long time.

clinal variation [Gk. *klinein*: to lean]   Gradual change in the phenotype of a species over a geographic gradient.

clonal deletion   Inactivation or destruction of lymphocyte clones that would produce immune reactions against the animal's own body.

clonal selection   Mechanism by which exposure to antigen results in the activation of selected T- or B-cell clones, resulting in an immune response.

clone [Gk. *klon*: twig, shoot]   (1) Genetically identical cells or organisms produced from a common ancestor by asexual means. (2) To produce many identical copies of a DNA sequence by its introduction into, and subsequent asexual reproduction of, a cell or organism.

closed circulatory system   Circulatory system in which the circulating fluid is contained within a continuous system of vessels. (Contrast with open circulatory system.)

coastal zone   The marine life zone that extends from the shoreline to the edge of the continental shelf. Characterized by relatively shallow, well-oxygenated water and relatively stable temperatures and salinities.

coccus (kock' us) (plural: cocci) [Gk. *kokkos*: berry, pit]   Any of various spherical or spheroidal bacteria.

cochlea (kock' lee uh) [Gk. *kokhlos*: snail]   A spiral tube in the inner ear of vertebrates; it contains the sensory cells involved in hearing.

codominance   A condition in which two alleles at a locus produce different phenotypic effects and both effects appear in heterozygotes.

codon   Three nucleotides in messenger RNA that direct the placement of a particular amino acid into a polypeptide chain. (Contrast with anticodon.)

coelom (see' loam) [Gk. *koiloma*: cavity]   An animal body cavity, enclosed by muscular mesoderm and lined with a mesodermal layer called peritoneum that also surrounds the internal organs.

coenocytic (seen' a sit ik) [Gk. *koinos*: common + *kytos*: container]   Referring to the condition, found in some fungal hyphae, of "cells" containing many nuclei but enclosed by a single plasma membrane. Results from nuclear division without cytokinesis.

coenzyme   A nonprotein organic molecule that plays a role in catalysis by an enzyme.

coevolution   Evolutionary processes in which an adaptation in one species leads to the evolution of an adaptation in a species with which it interacts; also known as reciprocal adaptation.

cofactor   An inorganic ion that is weakly bound to an enzyme and required for its activity.

cohesin   A protein involved in binding chromatids together.

cohesion   The tendency of molecules (or any substances) to stick together.

cohort (co' hort) [L. *cohors*: company of soldiers]   A group of similar-aged organisms.

coleoptile   A sheath that surrounds and protects the shoot apical meristem and young primary leaves of a grass seedling as they move through the soil.

collagen [Gk. *kolla*: glue]   A fibrous protein found extensively in bone and connective tissue.

collecting duct   In vertebrates, a tubule that receives urine produced in the nephrons of the kidney and delivers that fluid to the ureter for excretion.

collenchyma (cull eng' kyma) [Gk. *kolla*: glue + *enchyma*: infusion]   A type of plant cell, living at functional maturity, which lends flexible support by virtue of primary cell walls thickened at the corners. (Contrast with parenchyma, sclerenchyma.)

colon [Gk. *kolon*]   The large intestine.

commensalism [L. *com*: together + *mensa*: table]   A type of interaction between species in which one participant benefits while the other is unaffected.

communication   A signal from one organism (or cell) that alters the functioning or behavior of another organism (or cell).

community   Any ecologically integrated group of species of microorganisms, plants, and animals inhabiting a given area.

compact bone   A type of bone with a solid, hard structure. (Contrast with cancellous bone.)

companion cell   In angiosperms, a specialized cell found adjacent to a sieve tube element.

comparative experiment   Experimental design in which data from various unmanipulated samples or populations are compared, but in which variables are not controlled or even necessarily identified. (Contrast with controlled experiment.)

comparative genomics   Computer-aided comparison of DNA sequences between different organisms to reveal genes with related functions.

competition   In ecology, use of the same resource by two or more species when the resource is present in insufficient supply for the combined needs of the species.

competitive exclusion   A result of competition between species for a limiting resource in which one species completely eliminates the other.

competitive inhibitor   A nonsubstrate that binds to the active site of an enzyme and thereby inhibits binding of its substrate. (Contrast with noncompetitive inhibitor.)

complement system   A group of eleven proteins that play a role in some reactions of the immune system. The complement proteins are not immunoglobulins.

complementary base pairing   The AT (or AU), TA (or UA), CG, and GC pairing of bases in double-stranded DNA, in transcription, and between tRNA and mRNA.

complementary DNA (cDNA)   DNA formed by reverse transcriptase acting with an RNA template; essential intermediate in the reproduction of retroviruses; used as a tool in recombinant DNA technology; lacks introns.

complete cleavage   Pattern of cleavage that occurs in eggs that have little yolk. Early cleavage furrows divide the egg completely and the blastomeres are of similar size. (Contrast with incomplete cleavage.)

complete metamorphosis   A change of state during the life cycle of an organism in which the body is almost completely rebuilt to produce an individual with a very different body form. Characteristic of insects such as butterflies, moths, beetles, ants, wasps, and flies.

compound (1) A substance made up of atoms of more than one element. (2) Made up of many units, as in the *compound eyes* of arthropods.

concerted evolution The common evolution of a family of repeated genes, such that changes in one copy of the gene family are replicated in other copies of the gene family.

condensation reaction A chemical reaction in which two molecules become connected by a covalent bond and a molecule of water is released (AH + BOH → AB + H₂O.) (Contrast with hydrolysis reaction.)

conditional mutation A mutation that results in a characteristic phenotype only under certain environmental conditions.

conduction The transfer of heat from one object to another through direct contact.

cone (1) In conifers, a reproductive structure consisting of spore-bearing scales extending from a central axis. (Contrast with strobilus.) (2) In the vertebrate retina, a type of photoreceptor cell responsible for color vision.

conidium (ko nid' ee um) (plural: conidia) [Gk. *konis*: dust] A type of haploid fungal spore borne at the tips of hyphae, not enclosed in sporangia.

conjugation (kon ju gay' shun) [L. *conjugare*: yoke together] (1) A process by which DNA is passed from one cell to another through a *conjugation tube*, as in bacteria. (2) A nonreproductive sexual process by which *Paramecium* and other ciliates exchange genetic material.

connective tissue A type of tissue that connects or surrounds other tissues; its cells are embedded in a collagen-containing matrix. One of the four major tissue types in multicellular animals.

connexon In a gap junction, a protein channel linking adjacent animal cells.

consensus sequences Short stretches of DNA that appear, with little variation, in many different genes.

conservation biology An applied science that carries out investigations with the aim of maintaining the diversity of life on Earth.

conserved Pertaining to a gene or trait that has evolved very slowly and is similar or even identical in individuals of highly divergent groups.

conspecifics Individuals of the same species.

constant region The portion of an immunoglobulin molecule whose amino acid composition determines its class and does not vary among immunoglobulins in that class. (Contrast with variable region.)

constitutive Always present; produced continually at a constant rate. (Contrast with inducible.)

consumer An organism that eats the tissues of some other organism.

continental drift The gradual movements of the world's continents that have occurred over billions of years.

contractile vacuole (kon trak' tul) A specialized vacuole that collects excess water taken in by osmosis, then contracts to expel the water from the cell.

controlled experiment An experiment in which a sample is divided into groups whereby experimental groups are exposed to manipulations of an independent variable while one group serves as an untreated control. The data from the various groups are then compared to see if there are changes in a dependent variable as a result of the experimental manipulation. (Contrast with comparative experiment.)

convection The transfer of heat to or from a surface via a moving stream of air or fluid.

convergent evolution Independent evolution of similar features from different ancestral traits.

copulation Reproductive behavior that results in a male depositing sperm in the reproductive tract of a female.

cork cambium [L. *cambiare*: to exchange] In plants, a lateral meristem that produces secondary growth, mainly in the form of waxy-walled protective cells, including some of the cells that become bark.

cornea The clear, transparent tissue that covers the eye and allows light to pass through to the retina.

corolla (ko role' lah) [L. *corolla*: a small crown] All of the petals of a flower, collectively.

coronary artery (kor' oh nair ee) An artery that supplies blood to the heart muscle.

coronary thrombosis A fibrous clot that blocks a coronary artery.

corpus luteum (kor' pus loo' tee um) (plural: corpora lutea) [L. yellow body] A structure formed from a follicle after ovulation; produces hormones important to the maintenance of pregnancy.

corridor A connection between habitat patches through which organisms can disperse; plays a critical role in maintaining subpopulations.

cortex [L. *cortex*: covering, rind] (1) In plants, the tissue between the epidermis and the vascular tissue of a stem or root. (2) In animals, the outer tissue of certain organs, such as the adrenal gland (adrenal cortex) and the brain (cerebral cortex).

corticosteroids Steroid hormones produced and released by the cortex of the adrenal gland.

corticotropin A tropic hormone produced by the anterior pituitary hormone that stimulates cortisol release from the adrenal cortex. Also called adrenocorticotropic hormone (ACTH).

corticotropin-releasing hormone A releasing hormone produced by the hypothalamus that controls the release of cortisol from the anterior pituitary.

cortisol A corticosteroid that mediates stress responses.

cost–benefit analysis An approach to evolutionary studies that assumes an animal has a limited amount of time and energy to devote to each of its activities, and that each activity has fitness costs as well as benefits. (*See also* trade-off.)

cotyledon (kot' ul lee' dun) [Gk. *kotyledon*: hollow space] A "seed leaf." An embryonic organ that stores and digests reserve materials; may expand when seed germinates.

countercurrent flow An arrangement that promotes the maximum exchange of heat, or of a diffusible substance, between two fluids by having the fluids flow in opposite directions through parallel vessels close together.

countercurrent multiplier The mechanism that increases the concentration of the interstitial fluid in the mammalian kidney through countercurrent flow in the loops of Henle and selective permeability and active transport of ions by segments of the loops of Henle.

covalent bond Chemical bond based on the sharing of electrons between two atoms.

CpG islands DNA regions rich in C resides adjacent to G residues. Especially abundant in promoters, these regions are where methylation of cytosine usually occurs.

crassulacean acid metabolism (CAM) A metabolic pathway enabling the plants that possess it to store carbon dioxide at night and then perform photosynthesis during the day with stomata closed.

critical night length In the photoperiodic flowering response of short-day plants, the length of night above which flowering occurs and below which the plant remains vegetative. (The reverse applies in the case of long-day plants.)

critical period *See* sensitive period.

cross section A section taken perpendicular to the longest axis of a structure. Also called a transverse section.

crossing over The mechanism by which linked genes undergo recombination. In general, the term refers to the reciprocal exchange of corresponding segments between two homologous chromatids.

crypsis [Gk. *kryptos*: hidden] The resemblance of an organism to some part of its environment, which helps it to escape detection by enemies.

cryptochromes [Gk. *kryptos*: hidden + *kroma*: color] Photoreceptors mediating some blue-light effects in plants and animals.

ctene (teen) [Gk. *cteis*: comb] In ctenophores, a comblike row of cilia-bearing plates. Ctenophores move by beating the cilia on their eight ctenes.

culture (1) A laboratory association of organisms under controlled conditions. (2) The collection of knowledge, tools, values, and rules that characterize a human society.

cuticle (1) In plants, a waxy layer on the outer body surface that retards water loss. (2) In ecdysozoans, an outer body covering that provides protection and support and is periodically molted.

cyclic AMP *See* cAMP.

cyclic electron transport In photosynthetic light reactions, the flow of electrons that produces ATP but no NADPH or O₂.

cyclin A protein that activates a cyclin-dependent kinase, bringing about transitions in the cell cycle.

cyclin-dependent kinase (Cdk) A proetin kinase whose target proteins are involved in transitions in the cell cycle and which is active only when complexed with additional protein subunits, called cyclins.

cytokine A regulatory protein made by immune system cells that affects other target cells in the immune system.

cytokinesis (sy' toe kine ee' sis) [Gk. *kytos*: container + *kinein*: to move] The division of the cytoplasm of a dividing cell. (Contrast with mitosis.)

cytokinin (sy' toe kine' in) A member of a class of plant growth substances that plays roles in senescence, cell division, and other phenomena.

cytoplasm The contents of the cell, excluding the nucleus.

cytoplasmic determinants In animal development, gene products whose spatial distribution may determine such things as embryonic axes.

cytoplasmic segregation The asymmetrical distribution of cytoplasmic determinants in a developing animal embryo.

cytosine (C) (site' oh seen)   A nitrogen-containing base found in DNA and RNA.

cytoskeleton   The network of microtubules and microfilaments that gives a eukaryotic cell its shape and its capacity to arrange its organelles and to move.

cytosol   The fluid portion of the cytoplasm, excluding organelles and other solids.

cytotoxic T cells (T_C)   Cells of the cellular immune system that recognize and directly eliminate virus-infected cells. (Contrast with T-helper cells.)

- D -

DAG   See diacylglycerol.

daughter chromosomes   During mitosis, the separated chromatids from the beginning of anaphase onward.

dead space   The lung volume that fails to be ventilated with fresh air (because the lungs are never completely emptied during exhalation).

deciduous [L. *deciduus*: falling off]   Pertaining to a woody plant that sheds it leaves but does not die.

declarative memory   Memory of people, places, events, and things that can be consciously recalled and described. (Contrast with procedural memory.)

decomposer   An organism that metabolizes organic compounds in debris and dead organisms, releasing inorganic material; found among the bacteria, protists, and fungi. *See also* detritivore, saprobe.

defensin   A type of protein made by phagocytes that kills bacteria and enveloped viruses by insertion into their plasma membranes.

degeneracy   The situation in which a single amino acid may be represented by any of two or more different codons in messenger RNA. Most of the amino acids can be represented by more than one codon.

deletion   A mutation resulting from the loss of a continuous segment of a gene or chromosome. Such mutations almost never revert to wild type. (Contrast with duplication, point mutation.)

demethylase   An enzyme that catalyzes the removal of the methyl group from cytosine, reversing DNA methylation.

demography   The study of population structure and of the processes by which it changes.

denaturation   Loss of activity of an enzyme or nucleic acid molecule as a result of structural changes induced by heat or other means.

dendrite [Gk. *dendron*: tree]   A fiber of a neuron which often cannot carry action potentials. Usually much branched and relatively short compared with the axon, and commonly carries information to the cell body of the neuron.

denitrification   Metabolic activity by which nitrate and nitrite ions are reduced to form nitrogen gas; carried by certain soil bacteria.

denitrifiers   Bacteria that release nitrogen to the atmosphere as nitrogen gas (N_2).

density-dependent   Pertaining to a factor with an effect on population size that increases in proportion to population density.

density-independent   Pertaining to a factor with an effect on population size that acts independently of population density.

deoxyribonucleic acid   See DNA.

deoxyribose   A five-carbon sugar found in nucleotides and DNA.

depolarization   A change in the resting potential across a membrane so that the inside of the cell becomes less negative, or even positive, compared with the outside of the cell. (Contrast with hyperpolarization.)

derived trait   A trait that differs from the ancestral trait. (Contrast with shared derived trait.)

dermal tissue system   The outer covering of a plant, consisting of epidermis in the young plant and periderm in a plant with extensive secondary growth. (Contrast with ground tissue system and vascular tissue system.)

desmosome (dez' mo sowm) [Gk. *desmos*: bond + *soma*: body]   An adhering junction between animal cells.

desmotubule   A membrane extension connecting the endoplasmic retituclum of two plant cells that traverses the plasmodesma.

determination   In development, the process whereby the fate of an embryonic cell or group of cells (e.g., to become epidermal cells or neurons) is set.

determinate growth   A growth pattern in which the growth of an organism or organ ceases when an adult state is reached; characteristic of most animals and some plant organs. (Contrast with indeterminate growth.)

detritivore (di try' ti vore) [L. *detritus*: worn away + *vorare*: to devour]   An organism that obtains its energy from the dead bodies or waste products of other organisms.

developmental module   A functional entity in the embryo encompassing genes and signaling pathways that determine a physical structure independently of other such modules.

developmental plasticity   The capacity of an organism to alter its pattern of development in response to environmental conditions.

diacylglycerol (DAG)   In hormone action, the second messenger produced by hydrolytic removal of the head group of certain phospholipids.

diapause   A period of developmental or reproductive arrest, entered in response to day length, that enables an organism to better survive.

diaphragm (dye' uh fram) [Gk. *diaphrassein*: barricade]   (1) A sheet of muscle that separates the thoracic and abdominal cavities in mammals; responsible for breathing. (2) A method of birth control in which a sheet of rubber is fitted over the woman's cervix, blocking the entry of sperm.

diastole (dye ass' toll ee) [Gk. dilation]   The portion of the cardiac cycle when the heart muscle relaxes. (Contrast with systole.)

diencephalon   The portion of the vertebrate forebrain that develops into the thalamus and hypothalamus.

differential gene expression   The hypothesis that, given that all cells contain all genes, what makes one cell type different from another is the difference in transcription and translation of those genes.

differentiation   The process whereby originally similar cells follow different developmental pathways; the actual expression of determination.

diffuse coevolution   The evolution of similar traits in suites of species experiencing similar selection pressures imposed by other suites of species with which they interact.

diffusion   Random movement of molecules or other particles, resulting in even distribution of the particles when no barriers are present.

dihybrid cross   A mating in which the parents differ with respect to the alleles of two loci of interest.

dikaryon (di care' ee ahn) [Gk. *di*: two + *karyon*: kernel]   A cell or organism carrying two genetically distinguishable nuclei. Common in fungi.

dioecious (die eesh' us) [Gk. *di*: two + *oikos*: house]   Pertaining to organisms in which the two sexes are "housed" in two different individuals, so that eggs and sperm are not produced in the same individuals. Examples: humans, fruit flies, date palms. (Contrast with monoecious.)

diploblastic   Having two cell layers. (Contrast with triploblastic.)

diploid (dip' loid) [Gk. *diplos*: double]   Having a chromosome complement consisting of two copies (homologs) of each chromosome. Designated $2n$.

diplontic   A type of life cycle in which gametes are the only haploid cells and mitosis occurs only in diploid cells. (Contrast with haplontic.)

direct transduction   A cell signaling mechanism in which the receptor acts as the effector in the cellular response. (Contrast with indirect transduction.)

directional selection   Selection in which phenotypes at one extreme of the population distribution are favored. (Contrast with disruptive selection, stabilizing selection.)

disaccharide   A carbohydrate made up of two monosaccharides (simple sugars).

dispersal   Movement of organisms away from a parent organism or from an existing population.

dispersion   The distribution of individuals in space within a population.

disruptive selection   Selection in which phenotypes at both extremes of the population distribution are favored. (Contrast with directional selection; stabilizing selection.)

distal   Away from the point of attachment or other reference point. (Contrast with proximal.)

distal convoluted tubule   The portion of a renal tubule from where it reaches the renal cortex, just past the loop of Henle to where it joins a collecting duct. (Compare with proximal convoluted tubule.)

disturbance   A short-term event that disrupts populations, communities, or ecosystems by changing the environment.

disulfide bridge   The covalent bond between two sulfur atoms (–S—S–) linking two molecules or remote parts of the same molecule.

DNA (deoxyribonucleic acid)   The fundamental hereditary material of all living organisms. In eukaryotes, stored primarily in the cell nucleus. A nucleic acid using deoxyribose rather than ribose.

DNA fingerprint   An individual's unique pattern of allele sequences, commonly short tandem repeats and single nucleotide polymorphisms.

DNA helicase   An enzyme that functions to unwind the double helix.

DNA ligase   Enzyme that unites broken DNA strands during replication and recombination.

DNA methylation   The addition of methyl groups to to bases in DNA, usually cytosine or guanine.

DNA methyltransferase An enzyme that catalyzes the methylation of DNA.

DNA microarray A small glass or plastic square onto which thousands of single-stranded DNA sequences are fixed so that hybridization of cell-derived RNA or DNA to the target sequences can be performed.

DNA polymerase Any of a group of enzymes that catalyze the formation of DNA strands from a DNA template.

DNA topoisomerase An enzyme that unwinds and winds coils of DNA that form during replication and transcription.

docking protein A receptor protein that binds (docks) a ribosome to the membrane of the endoplasmic reticulum by binding the signal sequence attached to a new protein being made at the ribosome.

domain (1) An independent structural element within a protein. Encoded by recognizable nucleotide sequences, a domain often folds separately from the rest of the protein. Similar domains can appear in a variety of different proteins across phylogenetic groups (e.g., "homeobox domain"; "calcium-binding domain"). (2) In phylogenetics, the three monophyletic branches of life (Bacteria, Archaea, and Eukarya).

dominance In genetics, the ability of one allelic form of a gene to determine the phenotype of a heterozygous individual in which the homologous chromosomes carry both it and a different (recessive) allele. (Contrast with recessive.)

dormancy A condition in which normal activity is suspended, as in some spores, seeds, and buds.

dorsal [L. *dorsum*: back] Toward or pertaining to the back or upper surface. (Contrast with ventral.)

dorsal lip In amphibian embryos, the dorsal segment of the blastopore. Also called the "organizer," this region directs the development of nearby embryonic regions.

double fertilization In angiosperms, a process in which the nuclei of two sperm fertilize one egg. One sperm's nucleus combines with the egg nucleus to produce a zygote, while the other combines with the same egg's two polar nuclei to produce the first cell of the triploid endosperm (the tissue that will nourish the growing plant embryo).

double helix Refers to DNA and the (usually right-handed) coil configuration of two complementary, antiparallel strands.

downregulation A negative feedback process in which continuous high concentrations of a hormone can decrease the number of its receptors. (Contrast with upregulation.)

duodenum (do' uh dee' num) The beginning portion of the vertebrate small intestine. (Contrast with ileum, jejunum.)

duplication A mutation in which a segment of a chromosome is duplicated, often by the attachment of a segment lost from its homolog. (Contrast with deletion.)

- E -

ecdysone (eck die' sone) [Gk. *ek*: out of + *dyo*: to clothe] In insects, a hormone that induces molting.

ecological efficiency The overall transfer of energy from one trophic level to the next, expressed as the ratio of consumer production to producer production.

ecology [Gk. *oikos*: house] The scientific study of the interaction of organisms with their living (biotic) and nonliving (abiotic) environments.

ecosystem (eek' oh sis tum) The organisms of a particular habitat, such as a pond or forest, together with the physical environment in which they live.

ecosystem engineer An organism that builds structures that alter existing habitats or create new habitats.

ecosystem services Processes by which ecosystems maintain resources that benefit human society.

ectoderm [Gk. *ektos*: outside + *derma*: skin] The outermost of the three embryonic germ layers first delineated during gastrulation. Gives rise to the skin, sense organs, and nervous system.

ectotherm [Gk. *ektos*: outside + *thermos*: heat] An animal that is dependent on external heat sources for regulating its body temperature (Contrast with endotherm.)

edema (i dee' mah) [Gk. *oidema*: swelling] Tissue swelling caused by the accumulation of fluid.

edge effect The changes in ecological processes in a community caused by physical and biological factors originating in an adjacent community.

effector protein In cell signaling, a protein responsible for the cellular reponse to a signal transduction pathway.

efferent (ef' ur unt) [L. *ex*: out + *ferre*: to bear] Carrying outward or away from, as in a neuron that carries impulses outward from the central nervous system (efferent neuron), or a blood vessel that carries blood away from a structure. (Contrast with afferent.)

egg In all sexually reproducing organisms, the female gamete; in birds, reptiles, and some other vertebrates, a structure within which early embryonic development occurs. *See also* amniote egg, ovum.

electrical synapse A type of synapse at which action potentials spread directly from presynaptic cell to postsynaptic cell. (Contrast with chemical synapse.)

electrocardiogram (ECG or EKG) A graphic recording of electrical potentials from the heart.

electrochemical gradient The concentration gradient of an ion across a membrane plus the voltage difference across that membrane.

electroencephalogram (EEG) A graphic recording of electrical potentials from the brain.

electromagnetic radiation A self-propagating wave that travels though space and has both electrical and magnetic properties.

electron A subatomic particle outside the nucleus carrying a negative charge and very little mass.

electron shell The region surrounding the atomic nucleus at a fixed energy level in which electrons orbit.

electron transport The passage of electrons through a series of proteins with a release of energy which may be captured in a concentration gradient or chemical form such as NADH or ATP.

electronegativity The tendency of an atom to attract electrons when it occurs as part of a compound.

electrophoresis *See* gel electrophoresis.

element A substance that cannot be converted to simpler substances by ordinary chemical means.

elongation (1) In molecular biology, the addition of monomers to make a longer RNA or protein during transcription or translation. (2) Growth of a plant axis or cell primarily in the longitudinal direction.

embolus (em' buh lus) [Gk. *embolos*: stopper] A circulating blood clot. Blockage of a blood vessel by an embolus or by a bubble of gas is referred to as an *embolism*. (Contrast with thrombus.)

embryo [Gk. *en*: within + *bryein*: to grow] A young animal, or young plant sporophyte, while it is still contained within a protective structure such as a seed, egg, or uterus.

embryonic stem cell (ESC) A pluripotent cell in the blastocyst.

embryo sac In angiosperms, the female gametophyte. Found within the ovule, it consists of eight or fewer cells, membrane bounded, but without cellulose walls between them.

emergent property A property of a complex system that is not exhibited by its individual component parts.

emigration The deliberate and usually oriented departure of an organism from the habitat in which it has been living.

3′ end (3 prime) The end of a DNA or RNA strand that has a free hydroxyl group at the 3′ carbon of the sugar (deoxyribose or ribose).

5′ end (5 prime) The end of a DNA or RNA strand that has a free phosphate group at the 5′ carbon of the sugar (deoxyribose or ribose).

endemic (en dem' ik) [Gk. *endemos*: native] Confined to a particular region, thus often having a comparatively restricted distribution.

endergonic A chemical reaction in which the products have higher free energy than the reactants, thereby requiring free energy input to occur. (Contrast with exergonic.)

endocrine gland (en' doh krin) [Gk. *endo*: within + *krinein*: to separate] An aggregation of secretory cells that secretes hormones into the blood. The *endocrine system* consists of all *endocrine cells* and endocrine glands in the body that produce and release hormones. (Contrast with exocrine gland.)

endocytosis A process by which liquids or solid particles are taken up by a cell through invagination of the plasma membrane. (Contrast with exocytosis.)

endoderm [Gk. *endo*: within + *derma*: skin] The innermost of the three embryonic germ layers delineated during gastrulation. Gives rise to the digestive and respiratory tracts and structures associated with them.

endodermis In plants, a specialized cell layer marking the inside of the cortex in roots and some stems. Frequently a barrier to free diffusion of solutes.

endomembrane system A system of intracellular membranes that exchange material with one another, consisting of the Golgi apparatus, endoplasmic reticulum, and lysosomes when present.

endometrium The epithelial lining of the uterus.

endoplasmic reticulum (ER) [Gk. *endo*: within + L. *reticulum*: net] A system of membranous tubes and flattened sacs found in the cytoplasm of eukaryotes. Exists in two forms: rough ER, studded with ribosomes; and smooth ER, lacking ribosomes.

endorphins   Molecules in the mammalian brain act as neurotransmitters in pathways that control pain.

endoskeleton [Gk. *endo*: within + *skleros*: hard]   An internal skeleton covered by other, soft body tissues. (Contrast with exoskeleton.)

endosperm [Gk. *endo*: within + *sperma*: seed]   A specialized triploid seed tissue found only in angiosperms; contains stored nutrients for the developing embryo.

endospore   [Gk. *endo*: within + *spora*: to sow]   In some bacteria, a resting structure that can survive harsh environmental conditions.

endosymbiosis theory [Gk. *endo*: within + *sym*: together + *bios*: life]   The theory that the eukaryotic cell evolved via the engulfing of one prokaryotic cell by another.

endothelium   The single layer of epithelial cells lining the interior of a blood vessel.

endotherm [Gk. *endo*: within + *thermos*: heat]   An animal that can control its body temperature by the expenditure of its own metabolic energy. (Contrast with ectotherm.)

endotoxin   A lipopolysaccharide that forms part of the outer membrane of certain Gram-negative bacteria that is released when the bacteria grow or lyse. (Contrast with exotoxin.)

energetic cost   The difference between the energy an animal expends in performing a behavior and the energy it would have expended had it rested.

energy   The capacity to do work or move matter against an opposing force. The capacity to accomplish change in physical and chemical systems.

energy budget   A quantitative description of all paths of energy exchange between an animal and its environment.

enkephalins   Molecules in the mammalian brain act as neurotransmitters in pathways that control pain.

enthalpy (*H***)**   The total energy of a system.

entropy (*S***)** (en' tro pee) [Gk. *tropein*: to change]   A measure of the degree of disorder in any system. Spontaneous reactions in a closed system are always accompanied by an increase in entropy.

enveloped virus   A virus enclosed within a phospholipid membrane derived from its host cell.

environment   Whatever surrounds and interacts with or otherwise affects a population, organism, or cell. May be external or internal.

environmentalism   The use of ecological knowledge, along with economics, ethics, and many other considerations, to inform both personal decisions and public policy relating to stewardship of natural resources and ecosystems.

enzyme (en' zime) [Gk. *zyme*: to leaven (as in yeast bread)]   A catalytic protein that speeds up a biochemical reaction.

epi- [Gk. upon, over]   A prefix used to designate a structure located on top of another; for example, epidermis, epiphyte.

epiblast   The upper or overlying portion of the avian blastula which is joined to the hypoblast at the margins of the blastodisc.

epiboly   The movement of cells over the surface of the blastula toward the forming blastopore.

epitope   *See* antigenic determinant.

epidermis [Gk. *epi*: over + *derma*: skin]   In plants and animals, the outermost cell layers. (Only one cell layer thick in plants.)

epididymis (epuh did' uh mus) [Gk. *epi*: over + *didymos*: testicle]   Coiled tubules in the testes that store sperm and conduct sperm from the seminiferous tubules to the vas deferens.

epigenetics   The scientific study of changes in the expression of a gene or set of genes that occur without change in the DNA sequence.

epinephrine (ep i nef' rin) [Gk. *epi*: over + *nephros*: kidney]   The "fight or flight" hormone produced by the medulla of the adrenal gland; it also functions as a neurotransmitter. (Also known as adrenaline.)

epistasis   Interaction between genes in which the presence of a particular allele of one gene determines whether another gene will be expressed.

epithelium   A type of animal tissue made up of sheets of cells that lines or covers organs, makes up tubules, and covers the surface of the body; one of the four major tissue types in multicellular animals.

equilibrium   Any state of balanced opposing forces and no net change.

ER   *See* endoplasmic reticulum.

error signal   In regulatory systems, any difference between the set point of the system and its current condition.

erythrocyte (ur rith' row site) [Gk. *erythros*: red + *kytos*: container]   A red blood cell.

erythropoietin   A hormone produced by the kidney in response to lack of oxygen that stimulates the production of red blood cells.

esophagus (i soff' i gus) [Gk. *oisophagos*: gullet]   That part of the gut between the pharynx and the stomach.

essential acids   Amino acids or fatty acids that an animal cannot synthesize for itself and must obtain from its food.

essential element   A mineral nutrient required for normal growth and reproduction in plants and animals.

ester linkage   A condensation (water-releasing) reaction in which the carboxyl group of a fatty acid reacts with the hydroxyl group of an alcohol. Lipids are formed in this way.

estivation (ess tuh vay' shun) [L. *aestivalis*: summer]   A state of dormancy and hypometabolism that occurs during the summer; usually a means of surviving drought and/or intense heat. (Contrast with hibernation.)

estrogen   Any of several steroid sex hormones; produced chiefly by the ovaries in mammals.

estrus (es' trus) [L. *oestrus*: frenzy]   The period of heat, or maximum sexual receptivity, in some female mammals. Ordinarily, the estrus is also the time of release of eggs in the female.

ethology [Gk. *ethos*: character + *logos*: study]   An approach to the study of animal behavior that focuses on studying many species in natural environments and addresses questions about the evolution of behavior.

ethylene   One of the plant growth hormones, the gas $H_2C{=}CH_2$. Involved in fruit ripening and other growth and developmental responses.

eukaryotes (yew car' ree oats) [Gk. *eu*: true + *karyon*: kernel or nucleus]   Organisms whose cells contain their genetic material inside a nucleus. Includes all life other than the viruses, archaea, and bacteria.

eusocial   Pertaining to a social group that includes nonreproductive individuals, as in honey bees.

eutrophication (yoo trofe' ik ay' shun) [Gk. *eu*: truly + *trephein*: to flourish]   The addition of nutrient materials to a body of water, resulting in changes in ecological processes and species composition therein.

evaporation   The transition of water from the liquid to the gaseous phase.

evolution   Any gradual change. Most often refers to organic or Darwinian evolution, which is the genetic and resulting phenotypic change in populations of organisms from generation to generation. (*See* macroevolution, microevolution; contrast with speciation.)

evolutionary radiation   The proliferation of many species within a single evolutionary lineage.

evolutionary reversal   The reappearance of an ancestral trait in a group that had previously acquired a derived trait.

excision repair   A mechanism that removes damaged DNA and replaces it with the appropriate nucleotide.

excited state   The state of an atom or molecule when, after absorbing energy, it has more energy than in its normal, ground state. (Contrast with ground state.)

excretion   Release of metabolic wastes by an organism.

exergonic   A chemical reaction in which the products of the reaction have lower free energy than the reactants, resulting in a release of free energy. (Contrast with endergonic.)

exocrine gland (eks' oh krin) [Gk. *exo*: outside + *krinein*: to separate]   Any gland, such as a salivary gland, that secretes to the outside of the body or into the gut. (Contrast with endocrine gland.)

exocytosis   A process by which a vesicle within a cell fuses with the plasma membrane and releases its contents to the outside. (Contrast with endocytosis.)

exon   A portion of a DNA molecule, in eukaryotes, that codes for part of a polypeptide. (Contrast with intron.)

exoskeleton (eks' oh skel' e ton) [Gk. *exos*: outside + *skleros*: hard]   A hard covering on the outside of the body to which muscles are attached. (Contrast with endoskeleton.)

exotoxin   A highly toxic, usually soluble protein released by living, multiplying bacteria. (Contrast with endotoxin.)

expanding triplet repeat   A three-base-pair sequence in a human gene that is unstable and can be repeated a few to hundreds of times. Often, the more the repeats, the less the activity of the gene involved. Expanding triplet repeats occur in some human diseases such as Huntington's disease and fragile-X syndrome.

experiment   A testing process to support or disprove hypotheses and to answer questions. The basis of the scientific method. *See* comparative experiment, controlled experiment.

expiratory reserve volume   The amount of air that can be forcefully exhaled beyond the normal tidal expiration. (Contrast with inspiratory reserve volume, tidal volume, vital capacity.)

exploitation competition   Competition in which individuals reduce the quantities of their shared resources. (Contrast with interference competition.)

exponential growth   Growth, especially in the number of organisms in a population, which is a geometric function of the size of the growing entity: the larger the entity, the faster it grows. (Contrast with logistic growth.)

expression vector   A DNA vector, such as a plasmid, that carries a DNA sequence that includes the adjacent sequences for its expression into mRNA and protein in a host cell.

expressivity   The degree to which a genotype is expressed in the phenotype; may be affected by the environment.

extensor   A muscle that extends an appendage.

external fertilization   The release of gametes into the environment; typical of aquatic animals. Also called spawning. (Contrast with internal fertilization.)

external gills   Highly branched and folded extensions of the body surface that provide a large surface area for gas exchange with water; typical of larval amphibians and many larval insects.

extinction   The termination of a lineage of organisms.

extracellular matrix   A material of heterogeneous composition surrounding cells and performing many functions including adhesion of cells.

extraembryonic membranes   Four membranes that support but are not part of the developing embryos of reptiles, birds, and mammals, defining these groups phylogenetically as amniotes. (See amnion, allantois, chorion, and yolk sac.)

- F -

F$_1$ The first filial generation; the immediate progeny of a parental (P) mating.

F$_2$ The second filial generation; the immediate progeny of a mating between members of the F$_1$ generation.

facilitated diffusion   Passive movement through a membrane involving a specific carrier protein; does not proceed against a concentration gradient. (Contrast with active transport, diffusion.)

facilitation   In succession, modification of the environment by a colonizing species in a way that allows colonization by other species. (Contrast with inhibition.)

facultative anaerobe   A prokaryote that can shift its metabolism between anaerobic and aerobic operations modes on the presence or absence of O$_2$. (Alternatively, facultative aerobe.)

fast-twitch fibers   Skeletal muscle fibers that can generate high tension rapidly, but fatigue rapidly ("sprinter" fibers). Characterized by an abundance of enzymes of glycolysis.

fat   A triglyceride that is solid at room temperature. (Contrast with oil.)

fate map   A diagram of the blastula showing which cells (blastomeres) are "fated " to contribute to specific tissues and organs in the mature body.

fatty acid   A molecule made up of a long nonpolar hydrocarbon chain and a polar carboxyl group. Found in many lipids.

fauna (faw' nah)   All the animals found in a given area. (Contrast with flora.)

feces [L. *faeces*: dregs]   Waste excreted from the digestive system.

fecundity (m$_x$)   The average number of offspring produced by each female.

feedback information   In regulatory systems, information about the relationship between the set point of the system and its current state.

feedforward information   In regulatory systems, information that changes the set point of the system.

fermentation (fur men tay' shun) [L. *fermentum*: yeast]   The anaerobic degradation of a substance such as glucose to smaller molecules such as lactic acid or alcohol with the extraction of energy.

fertilization   Union of gametes. Also known as syngamy.

fetus   Medical and legal term for the stages of a developing human embryo from about the eighth week of pregnancy (the point at which all major organ systems have formed) to the moment of birth.

fiber   In angiosperms, an elongated, tapering sclerenchyma cell, usually with a thick cell wall, that serves as a support function in xylem. (See also muscle fiber.)

fibrin   A protein that polymerizes to form long threads that provide structure to a blood clot.

fibrinogen   A circulating protein that can be stimulated to fall out of solution and provide the structure for a blood clot.

fibrous root system   A root system typical of monocots composed of numerous thin adventitious roots that are all roughly equal in diameter. (Contrast with taproot system.)

Fick's law of diffusion   An equation that describes the factors that determine the rate of diffusion of a molecule from an area of higher concentration to an area of lower concentration.

fight-or-flight response   A rapid physiological response to a sudden threat mediated by the hormone epinephrine.

filter feeder   An organism that feeds on organisms much smaller than itself that are suspended in water or air by means of a straining device.

first law of thermodynamics   The principle that energy can be neither created nor destroyed.

fission   See binary fission.

fitness   The contribution of a genotype or phenotype to the genetic composition of subsequent generations, relative to the contribution of other genotypes or phenotypes. (See also inclusive fitness.)

flagellum (fla jell' um) (plural: flagella) [L. *flagellum*: whip]   Long, whiplike appendage that propels cells. Prokaryotic flagella differ sharply from those found in eukaryotes.

fixed action pattern   In ethology, a genetically determined behavior that is performed without learning, stereotypic (performed the same way each time), and not modifiable by learning.

flexor   A muscle that flexes an appendage.

flora (flore' ah)   All of the plants found in a given area. (Contrast with fauna.)

floral meristem   In angiosperms, a meristem that forms the floral organs (sepals, petals, stamens, and carpels).

floral organ identity genes   In angiosperms, genes that determine the fates of floral meristem cells; their expression is triggered by the products of meristem identity genes.

florigen   A plant hormone involved in the conversion of a vegetative shoot apex to a flower.

flower   The sexual structure of an angiosperm.

fluid feeder   An animal that feeds on fluids it extracts from the bodies of other organisms; examples include nectar-feeding birds and blood-sucking insects.

fluid mosaic model   A molecular model for the structure of biological membranes consisting of a fluid phospholipid bilayer in which suspended proteins are free to move in the plane of the bilayer.

follicle [L. *folliculus*: little bag]   In female mammals, an immature egg surrounded by nutritive cells.

follicle-stimulating hormone (FSH)   A gonadotropin produced by the anterior pituitary.

food chain   A portion of a food web, most commonly a simple sequence of prey species and the predators that consume them.

food vacuole   Membrane enclosed structure formed by phagocytosis in which engulfed food particles are digested by the action of lysosomal enzymes.

food web   The complete set of food links between species in a community; a diagram indicating which ones are the eaters and which are eaten.

forebrain   The region of the vertebrate brain that comprises the cerebrum, thalamus, and hypothalamus.

fossil   Any recognizable structure originating from an organism, or any impression from such a structure, that has been preserved over geological time.

fossil fuels   Fuels, including oil, natural gas, coal, and peat, formed over geologic time from organic material buried in anaerobic sediments.

founder effect   Random changes in allele frequencies resulting from establishment of a population by a very small number of individuals.

fovea [L. *fovea*: a small pit]   In the vertebrate retina, the area of most distinct vision.

frame-shift mutation   The addition or deletion of a single or two adjacent nucleotides in a gene's sequence. Results in the misreading of mRNA during translation and the production of a nonfunctional protein. (Contrast with missense mutation, nonsense mutation, silent mutation.)

Frank–Starling law   The stroke volume of the heart increases with increased return of blood to the heart.

free energy (G)   Energy that is available for doing useful work, after allowance has been made for the increase or decrease of disorder.

freeze-fracturing   Method of tissue preparation for transmission and scanning electron microscopy in which a tissue is frozen and a knife is then used to crack open the tissue; the fracture often occurs in the path of least resistance, within a membrane.

frequency-dependent selection   Selection that changes in intensity with the proportion of individuals in a population having the trait.

fruit   In angiosperms, a ripened and mature ovary (or group of ovaries) containing the seeds. Sometimes applied to reproductive structures of other groups of plants.

functional genomics   The assignment of functional roles to the proteins encoded by genes identified by sequencing entire genomes.

functional group   A characteristic combination of atoms that contribute specific properties when attached to larger molecules.

fundamental niche   A species' niche as defined by its physiological capabilities. (Contrast with realized niche.)

- G -

G cap   A chemically modified GTP added to the 5' end of mRNA; facilitates binding of mRNA to ribosome and prevents mRNA breakdown.

G1   In the cell cycle, the gap between the end of mitosis and the onset of the S phase.

G2   In the cell cycle, the gap between the S (synthesis) phase and the onset of mitosis.

G protein   A membrane protein involved in signal transduction; characterized by binding GDP or GTP.

gain of function mutation   A mutation that results in a protein with a new function. (Contrast with loss of function mutation.)

gallbladder   In the human digestive system, an organ in which bile is stored.

gametangium (gam uh tan' gee um) (plural: gametangia) [Gk. *gamos*: marriage + *angeion*: vessel]   Any plant or fungal structure within which a gamete is formed.

gamete (gam' eet) [Gk. *gamete/gametes*: wife, husband]   The mature sexual reproductive cell: the egg or the sperm.

gametogenesis (ga meet' oh jen' e sis)   The specialized series of cellular divisions that leads to the production of gametes. (*See also* oogenesis, spermatogenesis.)

gametophyte (ga meet' oh fyte)   In plants and photosynthetic protists with alternation of generations, the multicellular haploid phase that produces the gametes. (Contrast with sporophyte.)

gamma diversity   The regional diversity found over a range of communities or habitats in a geographic region. (Contrast with alpha diveristy, beta diversity.)

ganglion (gang' glee un) (plural: ganglia) [Gk. tumor]   A cluster of neurons that have similar characteristics or function.

ganglion cells   Cells at the front of the human retina that transmit information from the bipolar cells to the brain.

gap junction   A 2.7-nanometer gap between plasma membranes of two animal cells, spanned by protein channels. Gap junctions allow chemical substances or electrical signals to pass from cell to cell.

gastric pits   Deep infoldings in the walls of the stomach lined with secretory cells.

gastrin   A hormone secreted by cells in the lower region of the stomach that stimulates the secretion of digestive juices as well as movements of the stomach.

gastrovascular cavity   Serving for both digestion (gastro) and circulation (vascular); in particular, the central cavity of the body of jellyfish and other cnidarians.

gastrulation   Development of a blastula into a gastrula. In embryonic development, the process by which a blastula is transformed by massive movements of cells into a *gastrula*, an embryo with three germ layers and distinct body axes.

gated channel   A membrane protein that changes its three-dimensional shape, and therefore its ion conductance, in response to a stimulus. When open, it allows specific ions to move across the membrane.

gel electrophoresis (e lek' tro fo ree' sis) [L. *electrum*: amber + Gk. *phorein*: to bear]   A technique for separating molecules (such as DNA fragments) from one another on the basis of their electric charges and molecular weights by applying an electric field to a gel.

gene [Gk. *genes*: to produce]   A unit of heredity. Used here as the unit of genetic function which carries the information for a single polypeptide or RNA.

gene family   A set of similar genes derived from a single parent gene; need not be on the same chromosomes. The vertebrate globin genes constitute a classic example of a gene family.

gene flow   Exchange of genes between populations through migration of individuals or movements of gametes.

gene-for-gene resistance   In plants, a mechanism of resistance to pathogens in which resistance is triggered by the specific interaction of the products of a pathogen's *Avr* genes and a plant's *R* genes.

gene pool   All of the different alleles of all of the genes existing in all individuals of a population.

gene therapy   Treatment of a genetic disease by providing patients with cells containing functioning alleles of the genes that are nonfunctional in their bodies.

gene tree   A graphic representation of the evolutionary relationships of a single gene in different species or of the members of a gene family.

genetic code   The set of instructions, in the form of nucleotide triplets, that translate a linear sequence of nucleotides in mRNA into a linear sequence of amino acids in a protein.

genetic drift   Changes in gene frequencies from generation to generation as a result of random (chance) processes.

genetic map   The positions of genes along a chromosome as revealed by recombination frequencies.

genetic marker   (1) In gene cloning, a gene of identifiable phenotype that indicates the presence of another gene, DNA segment, or chromosome fragment. (2) In general, a DNA sequence such as a single nucleotide polymorphism whose presence is correlated with the presence of other linked genes on that chromosome.

genetic structure   The frequencies of different alleles at each locus and the frequencies of different genotypes in a Mendelian population.

genetic switches   Mechanisms that control how the genetic toolkit is used, such as promoters and the transcription factors that bind them. The signal cascades that converge on and operate these switches determine when and where genes will be turned on and off.

genetic toolkit   In evolutionary developmental biology, DNA sequences controlling developmental mechanisms that have been conserved over evolutionary time.

genetics   The scientific study of the structure, functioning, and inheritance of genes, the units of hereditary information.

genome (jee' nome)   The complete DNA sequence for a particular organism or individual.

genomic equivalence   The principle that no information is lost from the nuclei of cells as they pass through the early stages of embryonic development.

genomic imprinting   The form of a gene's expression is determined by parental source (i.e., whether the gene is inherited from the male or female parent).

genomic library   All of the cloned DNA fragments generated by the action of a restriction endonuclease on a genome.

genomics   The scientific study of entire sets of genes and their interactions.

genotype (jean' oh type) [Gk. *gen*: to produce + *typos*: impression]   An exact description of the genetic constitution of an individual, either with respect to a single trait or with respect to a larger set of traits. (Contrast with phenotype.)

genus (jean' us) (plural: genera) [Gk. *genos*: stock, kind]   A group of related, similar species recognized by taxonomists with a distinct name used in binomial nomenclature.

germ cell [L. *germen*: to beget]   A reproductive cell or gamete of a multicellular organism. (Contrast with somatic cell.)

germ layers   The three embryonic layers formed during gastrulation (ectoderm, mesoderm, and endoderm). Also called cell layers or tissue layers.

germ line mutation   Mutation in a cell that produces gametes (i.e., a germ line cell). (Contrast with somatic mutation.)

germination   Sprouting of a seed or spore.

gestation (jes tay' shun) [L. *gestare*: to bear]   The period during which the embryo of a mammal develops within the uterus. Also known as pregnancy.

ghrelin   A hormone produced and secreted by cells in the stomach that stimulates appetite.

gibberellin (jib er el' lin)   A class of plant growth hormones playing roles in stem elongation, seed germination, flowering of certain plants, etc.

gill   An organ specialized for gas exchange with water.

gizzard (giz' erd) [L. *gigeria*: cooked chicken parts]   A muscular part of the stomach of birds that grinds up food, sometimes with the aid of fragments of stone.

glia (glee' uh) [Gk. *glia*: glue]   Cells of the nervous system that do not conduct action potentials.

glomerular filtration rate (GFR)   The rate at which the blood is filtered in the glomeruli of the kidney.

glomerulus (glo mare' yew lus) [L. *glomus*: ball]   Sites in the kidney where blood filtration takes place. Each glomerulus consists of a knot of capillaries served by afferent and efferent arterioles.

glucagon   Hormone produced by alpha cells of the pancreatic islets of Langerhans. Glucagon stimulates the liver to break down glycogen and release glucose into the circulation.

gluconeogenesis   The biochemical synthesis of glucose from other substances, such as amino acids, lactate, and glycerol.

glucose [Gk. *gleukos*: sugar, sweet]   The most common monosaccharide; the monomer of the polysaccharides starch, glycogen, and cellulose.

glycerol (gliss' er ole)   A three-carbon alcohol with three hydroxyl groups; a component of phospholipids and triglycerides.

glycogen (gly' ko jen)   An energy storage polysaccharide found in animals and fungi; a branched-chain polymer of glucose, similar to starch.

glycolipid   A lipid to which sugars are attached.

glycolysis (gly kol' li sis) [Gk. *gleukos*: sugar + *lysis*: break apart]   The enzymatic breakdown of glucose to pyruvic acid.

glycoprotein   A protein to which sugars are attached.

glycosidic linkage   Bond between carbohydrate (sugar) molecules through an intervening oxygen atom (–O–).

glycosylation   The addition of carbohydrates to another type of molecule, such as a protein.

glyoxysome (gly ox' ee soam)   An organelle found in plants, in which stored lipids are converted to carbohydrates.

Golgi apparatus (goal' jee)   A system of concentrically folded membranes found in the cytoplasm of eukaryotic cells; functions in secretion from cell by exocytosis.

gonad (go' nad) [Gk. *gone*: seed]   An organ that produces gametes in animals: either an ovary (female gonad) or testis (male gonad).

gonadotropin   A type of trophic hormone that stimulates the gonads.

gonadotropin-releasing hormone (GnRH)   Hormone produced by the hypothalamus that stimulates the anterior pituitary to secrete ("release") gonadotropins.

Gondwana   The large southern land mass that existed from the Cambrian (540 mya) to the Jurassic (138 mya). Present-day remnants are South America, Africa, India, Australia, and Antarctica.

grafting   Artificial transplantation of tissue from one organism to another. In horticulture, the transfer of a bud or stem segment from one plant onto the root of another as a form of asexual reproduction.

Gram stain   A differential purple stain useful in characterizing bacteria. The peptidoglycan-rich cell walls of Gram-positive bacteria stain purple; cell walls of Gram-negative bacteria generally stain orange.

gravitropism [Gk. *tropos*: to turn]   A directed plant growth response to gravity.

gray matter   In the nervous system, tissue that is rich in neuronal cell bodies. (Contrast with white matter.)

greenhouse gases   Gases in the atmosphere, such as carbon dioxide and methane, that are transparent to sunlight, but trap heat radiating from Earth's surface, causing heat to build up at Earth's surface.

gross primary production   The amount of energy captured by the primary producers in a community.

gross primary productivity (GPP)   The rate at which the primary producers in a community turn solar energy into stored chemical energy via photosynthesis.

ground meristem   That part of an apical meristem that gives rise to the ground tissue system of the primary plant body.

ground tissue system   Those parts of the plant body not included in the dermal or vascular tissue systems. Ground tissues function in storage, photosynthesis, and support.

growth   An increase in the size of the body and its organs by cell division and cell expansion.

growth factor   A chemical signal that stimulates cells to divide.

growth hormone   A peptide hormone released by the anterior pituitary that stimulates many anabolic processes.

guanine (G) (gwan' een)   A nitrogen-containing base found in DNA, RNA, and GTP.

guard cells   In plants, specialized, paired epidermal cells that surround and control the opening of a stoma (pore). *See* stoma.

guild   In ecology, a group of species that exploit the same resource, but in slightly different ways.

gustation   The sense of taste.

gut   An animal's digestive tract.

- H -

habitat   The particular environment in which an organism lives. A *habitat patch* is an area of a particular habitat surrounded by other habitat types that may be less suitable for that organism.

hair cell   A type of mechanoreceptor in animals. Detects sound waves and other forms of motion in air or water.

half-life   The time required for half of a sample of a radioactive isotope to decay to its stable, nonradioactive form, or for a drug or other substance to reach half its initial dosage.

halophyte (hal' oh fyte) [Gk. *halos*: salt + *phyton*: plant]   A plant that grows in a saline (salty) environment.

Hamilton's rule   The principle that, for an apparent altruistic behavior to be adaptive, the fitness benefit of that act to the recipient times the degree of relatedness of the performer and the recipient must be greater than the cost to the performer.

haplodiploidy   A sex determination mechanism in which diploid individuals (which develop from fertilized eggs) are female and haploid individuals (which develop from unfertilized eggs) are male; typical of hymenopterans.

haploid (hap' loid) [Gk. *haploeides*: single]   Having a chromosome complement consisting of just one copy of each chromosome; designated $1n$ or n. (Contrast with diploid.)

haplontic   A type of life cycle in which the zygote is the only diploid cell and mitosis occurs only in haploid cells. (Contrast with diplontic.)

haplotype   Linked nucleotide sequences that are usually inherited as a unit (as a "sentence" rather than as individual "words").

Hardy–Weinberg equililbrium   In a sexually reproducing population, the allele frequency at a given locus that is not being acted on by agents of evolution; the conditions that would result in no evolution in a population.

haustorium (haw stor' ee um) (plural: haustoria)[L. *haustus*: draw up]   A specialized hypha or other structure by which fungi and some parasitic plants draw nutrients from a host plant.

Haversian systems   Units of organization in compact bone that reflect the action of intercommunicating osteoblasts.

heart   In circulatory systems, a muscular pump that moves extracellular fluid around the body.

heat of vaporization   The energy that must be supplied to convert a molecule from a liquid to a gas at its boiling point.

heat shock proteins   Chaperone proteins expressed in cells exposed to high or low temperatures or other forms of environmental stress.

helical   Shaped like a screw or spring; this shape occurs in DNA and proteins.

helper T cells   *See* T-helper cells.

hemiparasite   A parasitic plant that can photosynthesize, but derives water and mineral nutrients from the living body of another plant. (Contrast with holoparasite.)

hemizygous (hem' ee zie' gus) [Gk. *hemi*: half + *zygotos*: joined]   In a diploid organism, having only one allele for a given trait, typically the case for X-linked genes in male mammals and Z-linked genes in female birds. (Contrast with homozygous, heterozygous.)

hemoglobin (hee' mo glow bin) [Gk. *heaema*: blood + L. *globus*: globe]   Oxygen-transporting protein found in the red blood cells of vertebrates (and found in some invertebrates).

Hensen's node   In avian embryos, a structure at the anterior end of the primitive groove; determines the fates of cells passing over it during gastrulation.

hepatic (heh pat' ik) [Gk. *hepar*: liver]   Pertaining to the liver.

herbivore (ur' bi vore) [L. *herba*: plant + *vorare*: to devour]   An animal that eats plant tissues. (Contrast with carnivore, detritivore, omnivore.)

heritable trait   A trait that is at least partly determined by genes.

hermaphroditism (her maf' row dite ism)   The coexistence of both female and male sex organs in the same organism.

hetero- [Gk.: *heteros*: other, different]   A prefix indicating two or more different conditions, structures, or processes. (Contrast with homo-.)

heterochrony   Alteration in the timing of developmental events, leading to different results in the adult organism.

heterocyst   A large, thick-walled cell type in the filaments of certain cyanobacteria that performs nitrogen fixation.

heteromorphic (het' er oh more' fik) [Gk. *heteros*: different + *morphe*: form]   Having a different form or appearance, as two heteromorphic life stages of a plant. (Contrast with isomorphic.)

heterosporous (het' er os' por us)   Producing two types of spores, one of which gives rise to a female megaspore and the other to a male microspore. (Contrast with homosporous.)

heterosis   The superior fitness of heterozygous offspring as compared with that of their dissimilar homozygous parents. Also called hybrid vigor.

heterotherm   An animal that regulates its body temperature at a constant level at some times but not others, such as a hibernator.

heterotroph (het' er oh trof) [Gk. *heteros*: different + *trophe*: feed]   An organism that requires preformed organic molecules as food. (Contrast with autotroph.)

heterotrophic succession   Succession in detritus-based communities, which differs from other types of succession in taking place without the participation of plants.

heterotypic   Pertaining to adhesion of cells of different types. (Contrast with homotypic.)

heterozygous (het' er oh zie' gus) [Gk. *heteros*: different + *zygotos*: joined]   In diploid organisms, having different alleles of a given gene on the pair of homologs carrying that gene. (Contrast with homozygous.)

hexose [Gk. *hex*: six]   A sugar containing six carbon atoms.

hibernation [L. *hibernum*: winter] The state of inactivity of some animals during winter; marked by a drop in body temperature and metabolic rate.

hierarchical sequencing An approach to DNA sequencing in which genetic markers are mapped and DNA sequences are aligned by matching overlapping sites of known sequence. (Contrast with shotgun sequencing.)

high-density lipoproteins (HDLs) Lipoproteins that remove cholesterol from tissues and carry it to the liver; HDLs are the "good" lipoproteins associated with good cardiovascular health.

high-throughput sequencing Rapid DNA sequencing on a micro scale in which many fragments of DNA are sequenced in parallel.

highly repetitive sequences Short (less than 100 bp), nontranscribed DNA sequences, repeated thousands of times in tandem arrangements.

hindbrain The region of the developing vertebrate brain that gives rise to the medulla, pons, and cerebellum.

hippocampus [Gr. sea horse] A part of the forebrain that takes part in long-term memory formation.

histamine (hiss' tah meen) A substance released by damaged tissue, or by mast cells in response to allergens. Histamine increases vascular permeability, leading to edema (swelling).

histone Any one of a group of proteins forming the core of a nucleosome, the structural unit of a eukaryotic chromosome.

HIV Human immunodeficiency virus, the retrovirus that causes acquired immune deficiency syndrome (AIDS).

holoparasite A fully parasitic plant (i.e., one that does not perform photosynthesis).

homeobox 180-base-pair segment of DNA found in certain homeotic genes; regulates the expression of other genes and thus controls large-scale developmental processes.

homeostasis (home' ee o sta' sis) [Gk. *homos*: same + *stasis*: position] The maintenance of a steady state, such as a constant temperature or a stable social structure, by means of physiological or behavioral feedback responses.

homeotic genes Genes that act during development to determine the formation of an organ from a region of the embryo.

homeotic mutation Mutation in a homeotic gene that results in the formation of a different organ than that normally made by a region of the embryo.

homo- [Gk. *homos*: same] A prefix indicating two or more similar conditions, structures, or processes. (Contrast with hetero-.)

homolog (1) In cytogenetics, one of a pair (or larger set) of chromosomes having the same overall genetic composition and sequence. In diploid organisms, each chromosome inherited from one parent is matched by an identical (except for mutational changes) chromosome—its homolog—from the other parent. (2) In evolutionary biology, one of two or more features in different species that are similar by reason of descent from a common ancestor.

homology (ho mol' o jee) [Gk. *homologia*: of one mind; agreement] A similarity between two or more features that is due to inheritance from a common ancestor. The structures are said to be *homologous*, and each is a *homolog* of the others. (Contrast with analogy.)

homoplasy (home' uh play zee) [Gk. *homos*: same + *plastikos*: shape, mold] The presence in multiple groups of a trait that is not inherited from the common ancestor of those groups. Can result from convergent evolution, evolutionary reversal, or parallel evolution.

homosporous Producing a single type of spore that gives rise to a single type of gametophyte, bearing both female and male reproductive organs. (Contrast with heterosporous.)

homotypic Pertaining to adhesion of cells of the same type. (Contrast with heterotypic.)

homozygous (home' oh zie' gus) [Gk. *homos*: same + *zygotos*: joined] In diploid organisms, having identical alleles of a given gene on both homologous chromosomes. An individual may be a homozygote with respect to one gene and a heterozygote with respect to another. (Contrast with heterozygous.)

horizons The horizontal layers of a soil profile, including the topsoil (A horizon), subsoil (B horizon) and parent rock or bedrom (C horizon)

hormone (hore' mone) [Gk. *hormon*: to excite, stimulate] A chemical signal produced in minute amounts at one site in a multicellular organism and transported to another site where it acts on target cells.

host An organism that harbors a parasite or symbiont and provides it with nourishment.

Hox genes Conserved homeotic genes found in vertebrates, *Drosophila*, and other animal groups. Hox genes contain the homeobox domain and specify pattern and axis formation in these animals.

human chorionic gonadotropin (hCG) A hormone secreted by the placenta which sustains the corpus luteum and helps maintain pregnancy.

humoral immune response The response of the immune system mediated by B cells that produces circulating antibodies active against extracellular bacterial and viral infections. (Contrast with cellular immune response.)

humus (hew' mus) The partly decomposed remains of plants and animals on the surface of a soil.

hybrid (high' brid) [L. *hybrida*: mongrel] (1) The offspring of genetically dissimilar parents. (2) In molecular biology, a double helix formed of nucleic acids from different sources.

hybridize (1) In genetics, to combine the genetic material of two distinct species or of two distinguishable populations within a species. (2) In molecular biology, to form a double-stranded nucleic acid in which the two strands originate from different sources.

hybrid vigor *See* heterosis.

hybridoma A cell produced by the fusion of an antibody-producing cell with a myeloma (tumor) cell; it produces monoclonal antibodies.

hydrocarbon A compound containing only carbon and hydrogen atoms.

hydrogen bond A weak electrostatic bond which arises from the attraction between the slight positive charge on a hydrogen atom and a slight negative charge on a nearby oxygen or nitrogen atom.

hydrologic cycle The movement of water from the oceans to the atmosphere, to the soil, and back to the oceans.

hydrolysis reaction (high drol' uh sis) [Gk. *hydro*: water + *lysis*: break apart] A chemical reaction that breaks a bond by inserting the components of water ($AB + H_2O \rightarrow AH + BOH$). (Contrast with condensation reaction.)

hydrophilic (high dro fill' ik) [Gk. *hydro*: water + *philia*: love] Having an affinity for water. (Contrast with hydrophobic.)

hydrophobic (high dro foe' bik) [Gk. *hydro*: water + *phobia*: fear] Having no affinity for water. Uncharged and nonpolar groups of atoms are hydrophobic. (Contrast with hydrophilic.)

hydroponic Pertaining to a method of growing plants with their roots suspended in nutrient solutions instead of soil.

hydrostatic pressure Pressure generated by compression of liquid in a confined space. Generated in plants, fungi, and some protists with cell walls by the osmotic uptake of water. Generated in animals with closed circulatory systems by the beating of a heart.

hydrostatic skeleton A fluid-filled body cavity that transfers forces from one part of the body to another when acted on by surrounding muscles.

hydroxyl group The —OH group found on alcohols and sugars.

hyper- [Gk. *hyper*: above, over] Prefix indicating above, higher, more. (Contrast with hypo-.)

hyperpolarization A change in the resting potential across a membrane so that the inside of a cell becomes more negative compared with the outside of the cell. (Contrast with depolarization.)

hypersensitive response A defensive response of plants to microbial infection in which phytoalexins and pathogenesis-related proteins are produced and the infected tissue undergoes apoptosis to isolate the pathogen from the rest of the plant.

hypertonic Having a greater solute concentration. Said of one solution compared with another. (Contrast with hypotonic, isotonic.)

hypha (high' fuh) (plural: hyphae) [Gk. *hyphe*: web] In the fungi and oomycetes, any single filament.

hypo- [Gk. *hypo*: beneath, under] Prefix indicating underneath, below, less. (Contrast with hyper-.)

hypoblast The lower tissue portion of the avian blastula which is joined to the epiblast at the margins of the blastodisc.

hypothalamus The part of the brain lying below the thalamus; it coordinates water balance, reproduction, temperature regulation, and metabolism.

hypothermia Below-normal body temperature.

hypothesis A tentative answer to a question, from which testable predictions can be generated. (Contrast with theory.)

hypotonic Having a lesser solute concentration. Said of one solution in comparing it to another. (Contrast with hypertonic, isotonic.)

hypoxia A deficiency of oxygen.

- I -

ileum The final segment of the small intestine.

imbibition Water uptake by a seed; first step in germination.

immediate hypersensitivity A rapid, extensive overreaction of the immune system against an allergen, resuting in the release of large amounts of histamine. (Contrast with delayed hypersensitivity.)

immediate memory A form of memory for events happening in the present that is almost perfectly photographic, but lasts only seconds. (Contrast with short-term memory, long-term memory.)

immune system [L. *immunis*: exempt from] A system in an animal that recognizes and attempts to eliminate or neutralize foreign substances such as bacteria, viruses, and pollutants.

immunoassay The use of antibodies to measure the concentration of an antigen in a sample.

immunoglobulins A class of proteins containing a tetramer consisting of four polypeptide chains—two identical light chains and two identical heavy chains—held together by disulfide bonds; active as receptors and effectors in the immune system.

immunological memory The capacity to more rapidly and massively respond to a second exposure to an antigen than occurred on first exposure.

imperfect flower A flower lacking either functional stamens or functional carpels. (Contrast with perfect flower.)

implantation The process by which the early mammalian embryo becomes attached to and embedded in the lining of the uterus.

imprinting In animal behavior, a rapid form of learning in which an animal learns, during a brief critical period, to make a particular response, which is maintained for life, to some object or other organism. *See also* genomic imprinting.

in vitro [L. in glass] A biological process occurring outside of the organism, in the laboratory. (Contrast with in vivo.)

in vivo [L. alive] A biological process occurring within a living organism or cell. (Contrast with in vitro.)

inbreeding Breeding among close relatives.

inclusive fitness The sum of an individual's genetic contribution to subsequent generations both via production of its own offspring and via its influence on the survival of relatives who are not direct descendants.

incomplete cleavage A pattern of cleavage that occurs in many eggs that have a lot of yolk, in which the cleavage furrows do not penetrate all of it. (*See also* discoidal cleavage, superficial cleavage; contrast with complete cleavage.)

incomplete dominance Condition in which the heterozygous phenotype is intermediate between the two homozygous phenotypes.

incomplete metamorphosis Insect development in which changes between instars are gradual.

independent assortment During meiosis, the random separation of genes carried on nonhomologous chromosomes into gametes so that inheritance of these genes is random. This principle was articulated by Mendel as his second law.

indeterminate growth A open-ended growth pattern in which an organism or organ continues to grow as long as it lives; characteristic of some animals and of plant shoots and roots. (Contrast with determinate growth.)

indirect transduction Cell signaling mechanism in which a second messenger mediates the interaction between receptor binding and cellular response. (Contrast with direct transduction.)

individual fitness That component of inclusive fitness resulting from an organism producing its own offspring. (Contrast with kin selection.)

induced fit A change in the shape of an enzyme caused by binding to its substrate that exposes the active site of the enzyme.

induced mutation A mutation resulting from exposure to a mutagen from outside the cell. (Contrast with spontaneous mutation.)

induced pluripotent stem cells (iPS cells) Multipotent or pluripotent animal stem cells produced from differentiated cells in vitro by the addition of several genes that are expressed.

inducer (1) A compound that stimulates the synthesis of a protein. (2) In embryonic development, a substance that causes a group of target cells to differentiate in a particular way.

inducible Produced only in the presence of a particular compound or under particular circumstances. (Contrast with constitutive.)

induction In embryonic development, the process by which a factor produced and secreted by certain cells determines the fates other cells.

inflammation A nonspecific defense against pathogens; characterized by redness, swelling, pain, and increased temperature.

inflorescence A structure composed of several to many flowers.

inflorescence meristem A meristem that produces floral meristems as well as other small leafy structures (bracts).

ingroup In a phylogenetic study, the group of organisms of primary interest. (Contrast with outgroup.)

inhibitor A substance that blocks a biological process.

initials Cells that perpetuate plant meristems, comparable to animal stem cells. When an initial divides, one daughter cell develops into another initial, while the other differentiates into a more specialized cell.

initiation In molecular biology, the beginning of transcription or translation.

initiation complex In protein translation, a combination of a small ribosomal subunit, an mRNA molecule, and the tRNA charged with the first amino acid coded for by the mRNA; formed at the onset of translation.

initiation site The part of a promoter where transcription begins.

inner cell mass Derived from the mammalian blastula (bastocyst), the inner cell mass will give rise to the yolk sac (via hypoblast) and embryo (via epiblast).

inositol trisphosphate (IP₃) An intracellular second messenger derived from membrane phospholipids.

inspiratory reserve volume The amount of air that can be inhaled above the normal tidal inspiration. (Contrast with expiratory reserve volume, tidal volume, vital capacity.)

instar (in' star) An immature stage of an insect between molts.

insulin (in' su lin) [L. *insula*: island] A hormone synthesized in islet cells of the pancreas that promotes the conversion of glucose into the storage material, glycogen.

integrin In animals, a transmembrane protein that mediates the attachment of epithelial cells to the extracellular matrix.

integument [L. *integumentum*: covering] A protective surface structure. In gymnosperms and angiosperms, a layer of tissue around the ovule which will become the seed coat.

intercostal muscles Muscles between the ribs that can augment breathing movements by elevating and suppressing the rib cage.

interference competition Competition in which individuals actively interfere with one another's access to resources. (Contrast with exploitation competition.)

interference RNA (RNAi) *See* RNA interference.

interferon A glycoprotein produced by virus-infected animal cells; increases the resistance of neighboring cells to the virus.

intermediate filaments Components of the cytoskeleton whose diameters fall between those of the larger microtubules and those of the smaller microfilaments.

internal environment In multicelluar organisms, the extracellular fluid surrounding the cells.

internal fertilization The release of sperm into the female reproductive tract; typical of most terrestrial animals. (Contrast with external fertilization.)

internal gills Gills enclosed in protective body cavities; typical of mollusks, arthropods, and fishes.

interneuron A neuron that communicates information between two other neurons.

internode The region between two nodes of a plant stem.

interphase In the cell cycle, the period between successive nuclear divisions during which the chromosomes are diffuse and the nuclear envelope is intact. During interphase the cell is most active in transcribing and translating genetic information.

interspecific competition Competition between members of two or more species. (Contrast with intraspecific competition.)

interstitial fluid Extracellular fluid that is not contained in the vessels of a circulatory system.

intestine The portion of the gut following the stomach, in which most digestion and absorption occurs.

intraspecific competition Competition among members of the same species. (Contrast with interspecific competition.)

intrinsic rate of increase (r) The rate at which a population can grow when its density is low and environmental conditions are highly favorable.

intron Portion of a of a gene within the coding region that is transcribed into pre-mRNA but is spliced out prior to translation. (Contrast with exon.)

invasive species An exotic species that reproduces rapidly, spreads widely, and has negative effects on the native species of the region to which it has been introduced.

invasiveness The ability of a pathogen to multiply in a host's body. (Contrast with toxigenicity).

inversion A rare 180° reversal of the order of genes within a segment of a chromosome.

ion (eye' on) [Gk. *ion*: wanderer] An electrically charged particle that forms when an atom gains or loses one or more electrons.

ion channel An integral membrane protein that allows ions to diffuse across the membrane in which it is embedded.

ionic bond An electrostatic attraction between positively and negatively charged ions.

ionotropic receptors   A receptor that that directly alters membrane permeability to a type of ion when it combines with its ligand.

iris (eye' ris) [Gk. *iris*: rainbow]   The round, pigmented membrane that surrounds the pupil of the eye and adjusts its aperture to regulate the amount of light entering the eye.

island biogeography   A theory proposing that the number of species on an island (or in another geographically defined and isolated area) represents a balance, or equilibrium, between the rate at which species immigrate to the island and the rate at which resident species go extinct.

islets of Langerhans   Clusters of hormone-producing cells in the pancreas.

iso- [Gk. *iso*: equal]   Prefix used for two separate entities that share some element of identity.

isogamous   Having male and female gametes that are morphologically identical. (Contrast with anisogamous.)

isomers   Molecules consisting of the same numbers and kinds of atoms, but differing in the bonding patterns by which the atoms are held together.

isomorphic (eye so more' fik) [Gk. *isos*: equal + *morphe*: form]   Having the same form or appearance, as when the haploid and diploid life stages of an organism appear identical. (Contrast with heteromorphic.)

isotonic   Having the same solute concentration; said of two solutions. (Contrast with hypertonic, hypotonic.)

isotope (eye' so tope) [Gk. *isos*: equal + *topos*: place]   Isotopes of a given chemical element have the same number of protons in their nuclei (and thus are in the same position on the periodic table), but differ in the number of neutrons.

isozymes   Enzymes of an organism that have somewhat different amino acid sequences but catalyze the same reaction.

iteroparous [L. *itero*, to repeat + *pario*, to beget]   Reproducing multiple times in a lifetime. (Contrast with semelparous.)

- J -

jejunum (jih jew' num)   The middle division of the small intestine, where most absorption of nutrients occurs. (*See* duodenum, ileum.)

jelly coat   The outer protective layer of a sea urchin egg, which triggers an acrosomal reaction in sperm.

joint   In skeletal systems, a junction between two or more bones.

juvenile hormone   In insects, a hormone maintaining larval growth and preventing maturation or pupation.

- K -

K-strategist   A species whose life history strategy allows it to persist at or near the carrying capacity (*K*) of its environment. (Contrast with *r*-strategist.)

karyogamy   The fusion of nuclei of two cells. (Contrast with plasmogamy.)

karyotype   The number, forms, and types of chromosomes in a cell.

keystone species   Species that have a dominant influence on the composition of a community.

kidneys   A pair of excretory organs in vertebrates.

kin selection   That component of inclusive fitness resulting from helping the survival of relatives containing the same alleles by descent from a common ancestor. (Contrast with individual fitness.)

kinase   *See* protein kinase.

kinetic energy (kuh-net' ik) [Gk. *kinetos*: moving]   The energy associated with movement. (Contrast with potential energy.)

kinetochore (kuh net' oh core)   Specialized structure on a centromere to which microtubules attach.

knockout   A molecular genetic method in which a single gene of an organism is permanently inactivated.

Koch's postulates   A set of rules for establishing that a particular microorganism causes a particular disease.

Krebs cycle   *See* citric acid cycle.

- L -

lagging strand   In DNA replication, the daughter strand that is synthesized in discontinuous stretches. (*See* Okazaki fragments.)

larva (plural: larvae) [L. *lares*: guiding spirits]   An immature stage of any animal that differs dramatically in appearance from the adult.

lateral [L. *latus*: side]   Pertaining to the side.

lateral gene transfer   The transfer of genes from one species to another, common among bacteria and archaea.

lateral line   A sensory system in fishes consisting of a canal filled with water and hair cells running down each side under the surface of the skin, which senses disturbances in the surrounding water.

lateral meristem   Either of the two meristems, the vascular cambium and the cork cambium, that give rise to a plant's secondary growth.

lateral root   A root extending outward from the taproot in a taproot system; typical of eudicots.

laticifers (luh tiss' uh furs)   In some plants, elongated cells containing secondary plant products such as latex.

Laurasia   The northernmost of the two large continents produced by the breakup of Pangaea.

laws of thermodynamics [Gk. *thermos*: heat + *dynamis*: power]   Laws derived from studies of the physical properties of energy and the ways energy interacts with matter. (*See also* first law of thermodynamics, second law of thermodynamics.)

leaching   In soils, a process by which mineral nutrients in upper soil horizons are dissolved in water and carried to deeper horizons, where they are unavailable to plant roots.

leading strand   In DNA replication, the daughter strand that is synthesized continuously. (Contrast with lagging strand.)

leaf (plural: leaves)   In plants, the chief organ of photosynthesis.

leaf primordium (plural: primordia)   An outgrowth on the side of the shoot apical meristem that will eventually develop into a leaf.

leghemoglobin   In nitrogen-fixing plants, an oxygen-carrying protein in the cytoplasm of nodule cells that transports enough oxygen to the nitrogen-fixing bacteria to support their respiration, while keeping free oxygen concentrations low enough to protect nitrogenase.

lek   A display ground within which male animals compete for and defend small display areas as a means of demonstrating their territorial prowess and winning opportunities to mate.

lens   In the vertebrate eye, a crystalline protein structure that makes fine adjustments in the focus of images falling on the retina.

lenticel (len' ti sill)   In plants, a spongy region in the periderm that allows gas exchange.

leptin   A hormone produced by fat cells that is believed to provide feedback information to the brain about the status of the body's fat reserves.

leukocyte   *See* white blood cell.

lichen (lie' kun)   An organism resulting from the symbiotic association of a fungus and either a cyanobacterium or a unicellular alga.

life cycle   The entire span of the life of an organism from the moment of fertilization (or asexual generation) to the time it reproduces in turn.

life history strategy   The way in which an organism partitions its time and energy among growth, maintenance, and reproduction.

ligament   A band of connective tissue linking two bones in a joint.

ligand (lig' and)   Any molecule that binds to a receptor site of another (usually larger) molecule.

light reactions   The initial phase of photosynthesis, in which light energy is converted into chemical energy.

light-independent reactions   The phase of photosynthesis in which chemical energy captured in the light reactions is used to drive the reduction of CO_2 to form carbohydrates.

lignin   A complex, hydrophobic polyphenolic polymer in plant cell walls that crosslinks other wall polymers, strengthening the walls, especially in wood.

limbic system   A group of evolutionarily primitive structures in the vertebrate telencephalon that are involved in emotions, drives, instinctive behaviors, learning, and memory.

liming   Application of compounds such as calcium carbonate, calcium hydroxide, or magnesium carbonate—commonly known as *lime*— to soil to reverse its acidification and increase the availability of calcium to plants.

limiting resource   The required resource whose supply most strongly influences the size of a population.

lineage species concept   The definition of a species as a branch on the tree of life, which has a history that starts at a speciation event and ends either at extinction or at another speciation event. (Contrast with biological species concept; morphological species concept.)

linkage   Association between genes on the same chromosome such that they do not show random assortment and seldom recombine; the closer the genes, the lower the frequency of recombination.

lipase (lip' ase; lye' pase)   An enzyme that digests fats.

lipid (lip' id) [Gk. *lipos*: fat]   Nonpolar, hydrophobic molecules that include fats, oils, waxes, steroids, and the phospholipids that make up biological membranes.

lipid bilayer   *See* phospholipid bilayer.

liver   A large digestive gland. In vertebrates, it secretes bile and is involved in the formation of blood.

loam   A type of soil consisting of a mixture of sand, silt, clay, and organic matter. One of the best soil types for agriculture.

locus (low' kus) (plural: loci, low' sigh)   In genetics, a specific location on a chromosome. May be considered synonymous with *gene*.

logistic growth   Growth, especially in the size of an organism or in the number of organisms in a population, that slows steadily as the entity approaches its maximum size. (Contrast with exponential growth.)

long-day plant (LDP)   A plant that requires long days (actually, short nights) in order to flower.

long-term depression (LTD)   A long-lasting decrease in the responsiveness resulting from continuous, repetitive, low-level stimulation. (Contrast with long-term potentiation.)

long-term potentiation (LTP)   A long-lasting increase in the responsiveness of a neuron resulting from a period of intense stimulation. (Contrast with long-term depression.)

loop of Henle (hen' lee)   Long, hairpin loop of the mammalian renal tubule that runs from the cortex down into the medulla and back to the cortex; creates a concentration gradient in the interstitial fluids in the medulla.

lophophore   A U-shaped fold of the body wall with hollow, ciliated tentacles that encircles the mouth of animals in several different groups. Used for filtering prey from the surrounding water.

loss of function mutation   A mutation that results in the loss of a functional protein. (Contrast with gain of function mutation.)

low-density lipoproteins (LDLs)   Lipoproteins that transport cholesterol around the body for use in biosynthesis and for storage; LDLs are the "bad" lipoproteins associated with a high risk of cardiovascular disease.

lumen (loo' men) [L. *lumen*: light]   The open cavity inside any tubular organ or structure, such as the gut or a renal tubule.

lung   An internal organ specialized for respiratory gas exchange with air.

luteinizing hormone (LH)   A gonadotropin produced by the anterior pituitary that stimulates the gonads to produce sex hormones.

lymph [L. *lympha*: liquid]   A fluid derived from blood and other tissues that accumulates in intercellular spaces throughout the body and is returned to the blood by the lymphatic system.

lymph node   A specialized structure in the vessels of the lymphatic system. Lymph nodes contain lymphocytes,which encounter and respond to foreign cells and molecules in the lymph as it passes through the vessels.

lymphatic system   A system of vessels that returns interstitial fluid to the blood.

lymphocyte   One of the two major classes of white blood cells; includes T cells, B cells, and other cell types important in the immune system.

lymphoid tissues   Tissues of the immune system that are dispersed throughout the body, consisting of the thymus, spleen, bone marrow, and lymph nodes.

lysis (lie' sis) [Gk. *lysis*: break apart]   Bursting of a cell.

lysogeny   A form of viral replication in which the virus becomes incorporated into the host chromosome and remains inactive. Also called a lysogenic cycle. (Contrast with lytic cycle.)

lysosome (lie' so soam) [Gk. *lysis*: break away + *soma*: body]   A membrane-enclosed organelle originating from the Golgi apparatus and containing hydrolytic enzymes. (Contrast with secondary lysosome.)

lysozyme (lie' so zyme)   An enzyme in saliva, tears, and nasal secretions that hydrolyzes bacterial cell walls.

lytic cycle   A viral reproductive cycle in which the virus takes over a host cell's synthetic machinery to replicate itself, then bursts (lyses) the host cell, releasing the new viruses. (Contrast with lysogeny.)

- M -

M phase   The portion of the cell cycle in which mitosis takes place.

macroevolution [Gk. *makros*: large]   Evolutionary changes occurring over long time spans and usually involving changes in many traits. (Contrast with microevolution.)

macromolecule   A giant (molecular weight > 1,000) polymeric molecule. The macromolecules are the proteins, polysaccharides, and nucleic acids.

macronutrient   In plants, a mineral element required in concentrations of at least 1 milligram per gram of plant dry matter; in animals, a mineral element required in large amounts. (Contrast with micronutrient.)

macrophage (mac' roh faj)   Phagocyte that engulfs pathogens by endocytosis.

MADS box   DNA-binding domain in many plant transcription factors that is active in development.

maintenance methylase   An enzyme that catalyzes the methylation of the new DNA strand when DNA is replicated.

major histocompatibility complex (MHC)   A complex of linked genes, with multiple alleles, that control a number of cell surface antigens that identify self and can lead to graft rejection.

malignant   Pertaining to a tumor that can grow indefinitely and/or spread from the original site of growth to other locations in the body. (Contrast with benign.)

Malpighian tubule (mal pee' gy un)   A type of protonephridium found in insects.

map unit   The distance between two genes as calculated from genetic crosses; a recombination frequency.

marine [L. *mare*: sea, ocean]   Pertaining to or living in the ocean. (Contrast with aquatic, terrestrial.)

mark–recapture method   A method of estimating population sizes of mobile organisms by capturing, marking, and releasing a sample of individuals, then capturing another sample at a later time.

mass extinction   A period of evolutionary history during which rates of extinction are much higher than during intervening times.

mass number   The sum of the number of protons and neutrons in an atom's nucleus.

mast cells   Cells, typically found in connective tissue, that release histamine in response to tissue damage.

maternal effect genes   Genes coding for morphogens that determine the polarity of the egg and larva in fruit flies.

mating type   A particular strain of a species that is incapable of sexual reproduction with another member of the same strain but capable of sexual reproduction with members of other strains of the same species.

maximum likelihood   A statistical method of determining which of two or more hypotheses (such as phylogenetic trees) best fit the observed data, given an explicit model of how the data were generated.

mechanically gated channel   A molecular channel that opens or closes in response to mechanical force applied to the plasma membrane in which it is inserted.

mechanoreceptor   A cell that is sensitive to physical movement and generates action potentials in response.

medulla (meh dull' luh)   (1) The inner, core region of an organ, as in the adrenal medulla (adrenal gland) or the renal medulla (kidneys). (2) The portion of the brainstem that connects to the spinal cord.

medusa (plural: medusae)   In cnidarians, a free-swimming, sexual life cycle stage shaped like a bell or an umbrella.

megaphyll   The generally large leaf of a fern, horsetail, or seed plant, with several to many veins. (Contrast with microphyll.)

megaspore [Gk. *megas*: large + *spora*: to sow]   In plants, a haploid spore that produces a female gametophyte.

megastrobilus   In conifers, the female (seed-bearing) cone. (Contrast with microstrobilus.)

meiosis (my oh' sis) [Gk. *meiosis*: diminution]   Division of a diploid nucleus to produce four haploid daughter cells. The process consists of two successive nuclear divisions with only one cycle of chromosome replication. In *meiosis I*, homologous chromosomes separate but retain their chromatids. The second division *meiosis II*, is similar to mitosis, in which chromatids separate.

melatonin   A hormone released by the pineal gland. Involved in photoperiodicity and circadian rhythms.

membrane potential   The difference in electrical charge between the inside and the outside of a cell, caused by a difference in the distribution of ions.

membranous bone   A type of bone that develops by forming on a scaffold of connective tissue. (Contrast with cartilage bone.)

memory cells   Long-lived lymphocytes produced by exposure to antigen. They persist in the body and are able to mount a rapid response to subsequent exposures to the antigen.

Mendel's laws   *See* independent assortment; segregation.

meristem [Gk. *meristos*: divided]   Plant tissue made up of undifferentiated actively dividing cells.

meristem identity genes   In angiosperms, a group of genes whose expression initiates flower formation, probably by switching meristem cells from a vegetative to a reproductive fate.

mesenchyme (mez' en kyme) [Gk. *mesos*: middle + *enchyma*: infusion]   Embryonic or unspecialized cells derived from the mesoderm.

mesoderm [Gk. *mesos*: middle + *derma*: skin]   The middle of the three embryonic germ layers first delineated during gastrulation. Gives rise to the skeleton, circulatory system, muscles, excretory system, and most of the reproductive system.

mesoglea (mez' uh glee uh) [Gk. *mesos*: middle + *gloia*, glue]   A thick, gelatinous noncellular layer that separates the two cellular tissue layers of ctenophores, cnidarians, and scyphozoans.

mesophyll (mez' uh fill) [Gk. *mesos*: middle + *phyllon*: leaf]   Chloroplast-containing, photosynthetic cells in the interior of leaves.

messenger RNA (mRNA)   Transcript of a region of one of the strands of DNA; carries information (as a sequence of codons) for the synthesis of one or more proteins.

meta- [Gk.: between, along with, beyond]   Prefix denoting a change or a shift to a new form or level; for example, as used in metamorphosis.

metabolic pathway   A series of enzyme-catalyzed reactions so arranged that the product of one reaction is the substrate of the next.

metabolism (meh tab' a lizm) [Gk. *metabole*: change]   The sum total of the chemical reactions that occur in an organism, or some subset of that total (as in respiratory metabolism).

metabolome   The quantitative description of all the small molecules in a cell or organism.

metabotropic receptor   A receptor that that indirectly alters membrane permeability to a type of ion when it combines with its ligand.

metagenomics   The practice of analyzing DNA from environmental samples without isolating intact organisms.

metamorphosis (met' a mor' fo sis) [Gk. *meta*: between + *morphe*: form, shape]   A change occurring between one developmental stage and another, as for example from a tadpole to a frog. (*See* complete metamorphosis, incomplete metamorphosis.)

metanephridia   The paired excretory organs of annelids.

metaphase (met' a phase)   The stage in nuclear division at which the centromeres of the highly supercoiled chromosomes are all lying on a plane (the metaphase plane or plate) perpendicular to a line connecting the division poles.

metapopulation   A population divided into subpopulations, among which there are occasional exchanges of individuals.

metastasis (meh tass' tuh sis)   The spread of cancer cells from their original site to other parts of the body.

methylation   The addition of a methyl group (—CH_3) to a molecule.

MHC   *See* major histocompatibility complex.

micelle   A particle of lipid covered with bile salts that is produced in the duodenum and facilitates digestion and absorption of lipids.

microclimate   A subset of climatic conditions in a small specific area, which generally differ from those in the environment at large, as in an animal's underground burrow.

microevolution   Evolutionary changes below the species level, affecting allele frequencies. (Contrast with macroevolution.)

microfibril   Crosslinked cellulose polymers, forming strong aggregates in the plant cell wall.

microfilament   In eukaryotic cells, a fibrous structure made up of actin monomers. Microfilaments play roles in the cytoskeleton, in cell movement, and in muscle contraction.

microglia   Glial cells that act as macrophages and mediators of inflammatory responses in the central nervous system.

micronutrient   In plants, a mineral element required in concentrations of less than 100 micrograms per gram of plant dry matter; in animals, a mineral element required in concentrations of less than 100 micrograms per day. (Contrast with macronutrient.)

microphyll   A small leaf with a single vein, found in club mosses and their relatives. (Contrast with megaphyll.)

micropyle (mike' roh pile) [Gk. *mikros*: small + *pylon*: gate]   Opening in the integument(s) of a seed plant ovule through which pollen grows to reach the female gametophyte within.

microRNA   A small, noncoding RNA molecule, typically about 21 bases long, that binds to mRNA to inhibit its translation.

microspore [Gk. *mikros*: small + *spora*: to sow]   In plants, a haploid spore that produces a male gametophyte.

microstrobilus   In conifers, male pollen-bearing cone. (Contrast with megastrobilus.)

microtubules   Tubular structures found in centrioles, spindle apparatus, cilia, flagella, and cytoskeleton of eukaryotic cells. These tubules play roles in the motion and maintenance of shape of eukaryotic cells.

microvilli (sing.: microvillus)   Projections of epithelial cells, such as the cells lining the small intestine, that increase their surface area.

midbrain   One of the three regions of the vertebrate brain. Part of the brainstem, it serves as a relay station for sensory signals sent to the cerebral hemispheres.

middle lamella (la mell' ah) [L. *lamina*: thin sheet]   A layer of polysaccharides that separates plant cells; a shared middle lamella lies outside the primary walls of the two cells.

mineral nutrients   Inorganic ions required by organisms for normal growth and reproduction.

mismatch repair   A mechanism that scans DNA after it has been replicated and corrects any base-pairing mismatches.

missense mutation   A change in a gene's sequence that results in a change in the sequence of the amino acid specified by the corresponding codon. (Contrast with frame-shift mutation, nonsense mutation, silent mutation.)

mitochondrial matrix   The fluid interior of the mitochondrion, enclosed by the inner mitochondrial membrane.

mitochondrion (my' toe kon' dree un) (plural: mitochondria) [Gk. *mitos*: thread + *chondros*: grain]   An organelle in eukaryotic cells that contains the enzymes of the citric acid cycle, the respiratory chain, and oxidative phosphorylation.

mitosis (my toe' sis) [Gk. *mitos*: thread]   Nuclear division in eukaryotes leading to the formation of two daughter nuclei, each with a chromosome complement identical to that of the original nucleus.

model systems   Also known as model organisms, these include the small group of species that are the subject of extensive research. They are organisms that adapt well to laboratory situations and findings from experiments on them can apply across a broad range of species. Classic examples include white rats and the fruit fly *Drosophila*.

moderately repetitive sequences   DNA sequences repeated 10–1,000 times in the eukaryotic genome. They include the genes that code for rRNAs and tRNAs, as well as the DNA in telomeres.

Modern Synthesis   An understanding of evolutionary biology that emerged in the early twentieth century as the principles of evolution were integrated with the principles of modern genetics.

modularity   In evolutionary developmental biology, the principle that the molecular pathways that determine different developmental processes operate independently from one another. *See also* developmental module.

mole   A quantity of a compound whose weight in grams is numerically equal to its molecular weight expressed in atomic mass units. Avogadro's number of molecules: 6.023×10^{23} molecules.

molecular clock   The approximately constant rate of divergence of macromolecules from one another over evolutionary time; used to date past events in evolutionary history.

molecular evolution   The scientific study of the mechanisms and consequences of the evolution of macromolecules.

molecular tool kit   A set of developmental genes and proteins that is common to most animals and is hypothesized to be responsible for the evolution of their differing developmental pathways.

molecular weight   The sum of the atomic weights of the atoms in a molecule.

molecule   A chemical substance made up of two or more atoms joined by covalent bonds or ionic attractions.

molting   The process of shedding part or all of an outer covering, as the shedding of feathers by birds or of the entire exoskeleton by arthropods.

monoclonal antibody   Antibody produced in the laboratory from a clone of hybridoma cells, each of which produces the same specific antibody.

monoculture   In agriculture, a large-scale planting of a single species of domesticated crop plant.

monoecious (mo nee' shus) [Gk. *mono*: one + *oikos*: house]   Pertaining to organisms in which both sexes are "housed" in a single individual that produces both eggs and sperm. (In some plants, these are found in different flowers within the same plant.) Examples include corn, peas, earthworms, hydras. (Contrast with dioecious.)

monohybrid cross   A mating in which the parents differ with respect to the alleles of only one locus of interest.

monomer [Gk. *mono*: one + *meros*: unit]   A small molecule, two or more of which can be combined to form oligomers (consisting of a few monomers) or polymers (consisting of many monomers).

monophyletic (mon' oh fih leht' ik) [Gk. *mono*: one + *phylon*: tribe]   Pertaining to a group that consists of an ancestor and all of its descendants. (Contrast with paraphyletic, polyphyletic.)

monosaccharide   A simple sugar. Oligosaccharides and polysaccharides are made up of monosaccharides.

monosomic   Pertaining to an organism with one less than the normal diploid number of chromosomes.

monosynaptic reflex   A neural reflex that begins in a sensory neuron and makes a single synapse before activating a motor neuron.

morphogen   A diffusible substance whose concentration gradient determines a developmental pattern in animals and plants.

morphogenesis (more' fo jen' e sis) [Gk. *morphe*: form + *genesis*: origin] The development of form; the overall consequence of determination, differentiation, and growth.

morphological species concept The definition of a species as a group of individuals that look alike. (Contrast with biological species concept; lineage species concept.)

morphology (more fol' o jee) [Gk. *morphe*: form + *logos*: study, discourse] The scientific study of organic form, including both its development and function.

mosaic development Pattern of animal embryonic development in which each blastomere contributes a specific part of the adult body. (Contrast with regulative development.)

motif *See* structural motif.

motile (mo' tul) Able to move from one place to another. (Contrast with sessile.)

motor cortex The region of the cerebral cortex that contains motor neurons that directly stimulate specific muscle fibers to contract.

motor neuron A neuron carrying information from the central nervous system to a cell that produces movement.

motor proteins Specialized proteins that use energy to change shape and move cells or structures within cells. *See* dynein, kinesin.

motor unit A motor neuron and the muscle fibers it controls.

mouth An opening through which food is taken in, located at the anterior end of a tubular gut.

mRNA *See* messenger RNA.

mucosal epithelium An epithelial cell layer containing cells that secrete mucus; found in the digestive and respiratory tracts. Also called mucosa.

Müllerian mimicry Convergence in appearance of two or more unpalatable species.

multipotent Having the ability to differentiate into a limited number of cell types. (Contrast with pluripotent, totipotent.)

muscle fiber A single muscle cell. In the case of skeletal muscle, a syncitial, multinucleate cell.

muscle tissue Excitable tissue that can contract through the interactions of actin and myosin; one of the four major tissue types in multicellular animals. There are three types of muscle tissue: skeletal, smooth, and cardiac.

mutagen (mute' ah jen) [L. *mutare*: change + Gk. *genesis*: source] Any agent (e.g., chemicals, radiation) that increases the mutation rate.

mutation A change in the genetic material not caused by recombination.

mutualism A type of interaction between species that benefits both species.

mycelium (my seel' ee yum) [Gk. *mykes*: fungus] In the fungi, a mass of hyphae.

mycorrhiza (my' ko rye' za) (plural: mycorrhizae) [Gk. *mykes*: fungus + *rhiza*: root] An association of the root of a plant with the mycelium of a fungus.

myelin (my' a lin) Concentric layers of plasma membrane that form a sheath around some axons; myelin provides the axon with electrical insulation and increases the rate of transmission of action potentials.

myocardial infarction Blockage of an artery that carries blood to the heart muscle.

myofibril (my' oh fy' bril) [Gk. *mys*: muscle + L. *fibrilla*: small fiber] A polymeric unit of actin or myosin in a muscle.

myoglobin (my' oh globe' in) [Gk. *mys*: muscle + L. *globus*: sphere] An oxygen-binding molecule found in muscle. Consists of a heme unit and a single globin chain; carries less oxygen than hemoglobin.

myosin One of the two contractile proteins of muscle.

- N -

natural killer cell A type of lymphocyte that attacks virus-infected cells and some tumor cells as well as antibody-labeled target cells.

natural selection The differential contribution of offspring to the next generation by various genetic types belonging to the same population. The mechanism of evolution proposed by Charles Darwin.

nauplius (naw' plee us) [Gk. *nauplios*: shellfish] A bilaterally symmetrical larval form typical of crustaceans.

necrosis (nec roh' sis) [Gk. *nekros*: death] Premature cell death caused by external agents such as toxins.

negative feedback In regulatory systems, information that decreases a regulatory response, returning the system to the set point. (Contrast with positive feedback.)

negative regulation A type of gene regulation in which a gene is normally transcribed, and the binding of a repressor protein to the promoter prevents transcription. (Contrast with positive regulation.)

nematocyst (ne mat' o sist) [Gk. *nema*: thread + *kystis*: cell] An elaborate, threadlike structure produced by cells of jellyfishes and other cnidarians, used chiefly to paralyze and capture prey.

nephron (nef' ron) [Gk. *nephros*: kidney] The functional unit of the kidney, consisting of a structure for receiving a filtrate of blood and a tubule that reabsorbs selected parts of the filtrate.

Nernst equation A mathematical statement that calculates the potential across a membrane permeable to a single type of ion that differs in concentration on the two sides of the membrane.

nerve A structure consisting of many neuronal axons and connective tissue.

nervous tissue Tissue specialized for processing and communicating information; one of the four major tissue types in multicellular animals.

net primary productivity (NPP) The rate at which energy captured by photosynthesis is incorporated into the bodies of primary producers bodies through growth and reproduction.

net primary production The amount of primary producer biomass made available for consumption by heterotrophs.

neural network An organized group of neurons that contains three functional categories of neurons—afferent neurons, interneurons, and efferent neurons—and is capable of processing information.

neural tube An early stage in the development of the vertebrate nervous system consisting of a hollow tube created by two opposing folds of the dorsal ectoderm along the anterior–posterior body axis.

neurohormone A chemical signal produced and released by neurons that subsequently acts as a hormone.

neuromuscular junction Synapse (point of contact) where a motor neuron axon stimulates a muscle fiber cell.

neuron (noor' on) [Gk. *neuron*: nerve] A nervous system cell that can generate and conduct action potentials along an axon to a synapse with another cell.

neurotransmitter A substance produced in and released by a neuron (the presynaptic cell) that diffuses across a synapse and excites or inhibits another cell (the postsynaptic cell).

neurulation Stage in vertebrate development during which the nervous system begins to form.

neutral allele An allele that does not alter the functioning of the proteins for which it codes.

neutral theory A view of molecular evolution that postulates that most mutations do not affect the amino acid being coded for, and that such mutations accumulate in a population at rates driven by genetic drift and mutation rates.

neutron (new' tron) One of the three fundamental particles of matter (along with protons and electrons), with mass approximately 1 amu and no electrical charge.

niche (nitch) [L. *nidus*: nest] The set of physical and biological conditions a species requires to survive, grow, and reproduce. (*See also* fundamental niche, realized niche.)

nitrate reduction The process by which nitrate (NO_3^-) is reduced to ammonia (NH_3).

nitric oxide (NO) An unstable molecule (a gas) that serves as a second messenger causing smooth muscle to relax. In the nervous system it operates as a neurotransmitter.

nitrifiers Chemolithotrophic bacteria that oxidize ammonia to nitrate in soil and in seawater.

nitrogen fixation Conversion of atmospheric nitrogen gas (N_2) into a more reactive and biologically useful form (ammonia), which makes nitrogen available to living things. Carried out by nitrogen-fixing bacteria, some of them free-living and others living within plant roots.

nitrogenase An enzyme complex found in nitrogen-fixing bacteria that mediates the stepwise reduction of atmospheric N_2 to ammonia and which is strongly inhibited by oxygen.

node [L. *nodus*: knob, knot] In plants, a (sometimes enlarged) point on a stem where a leaf is or was attached.

node of Ranvier A gap in the myelin sheath covering an axon; the point where the axonal membrane can fire action potentials.

nodule A specialized structure in the roots of nitrogen-fixing plants that houses nitrogen-fixing bacteria, in which oxygen is maintained at a low level by leghemoglobin.

noncompetitive inhibitor A nonsubstrate that inhibits the activity of an enzyme by binding to a site other than its active site. (Contrast with competitive inhibitor.)

noncyclic electron transport In photosynthesis, the flow of electrons that forms ATP, NADPH, and O_2.

nondisjunction Failure of sister chromatids to separate in meiosis II or mitosis, or failure of homologous chromosomes to separate in meiosis I. Results in aneuploidy.

nonpolar Having electric charges that are evenly balanced from one end to the other. (Contrast with polar.)

nonrandom mating Selection of mates on the basis of a particular trait or group of traits.

nonsense mutation   Change in a gene's sequence that prematurely terminates translation by changing one of its codons to a stop codon.

nonsynonymous substitution   A change in a gene from one nucleotide to another that changes the amino acid specified by the corresponding codon (i.e., AGC →AGA, or serine → arginine). (Contrast with synonymous substitution.)

norepinephrine   A neurotransmitter found in the central nervous system and also at the postganglionic nerve endings of the sympathetic nervous system. Also called noradrenaline.

normal flora   Microorganisms that normally live and reproduce on or in the body without causing disease, and which form a nonspecific defense against pathogens by competing with them for space and nutrients.

notochord (no' tow kord) [Gk. *notos*: back + *chorde*: string]   A flexible rod of gelatinous material serving as a support in the embryos of all chordates and in the adults of tunicates and lancelets.

nucleic acid (new klay' ik)   A polymer made up of nucleotides, specialized for the storage, transmission, and expression of genetic information. DNA and RNA are nucleic acids.

nucleic acid hybridization   A technique in which a single-stranded nucleic acid probe is made that is complementary to, and binds to, a target sequence, either DNA or RNA. The resulting double-stranded molecule is a hybrid.

nucleoid (new' klee oid)   The region that harbors the chromosomes of a prokaryotic cell. Unlike the eukaryotic nucleus, it is not bounded by a membrane.

nucleolus (new klee' oh lus)   A small, generally spherical body found within the nucleus of eukaryotic cells. The site of synthesis of ribosomal RNA.

nucleoside   A nucleotide without the phosphate group; a nitrogenous base attached to a sugar.

nucleosome   A portion of a eukaryotic chromosome, consisting of part of the DNA molecule wrapped around a group of histone molecules, and held together by another type of histone molecule. The chromosome is made up of many nucleosomes.

nucleotide   The basic chemical unit in nucleic acids, consisting of a pentose sugar, a phosphate group, and a nitrogen-containing base.

nucleotide substitution   A change of one base pair to another in a DNA sequence.

nucleus (new' klee us) [L. *nux*: kernel or nut]   (1) In cells, the centrally located compartment of eukaryotic cells that is bounded by a double membrane and contains the chromosomes. (2) In the brain, an identifiable group of neurons that share common characteristics or functions.

nutrient   A food substance; or, in the case of mineral nutrients, an inorganic element required for completion of the life cycle of an organism.

- O -

obligate anaerobe   An anaerobic prokaryote that cannot survive exposure to O_2.

odorant   A molecule that can bind to an olfactory receptor.

oil   A triglyceride that is liquid at room temperature. (Contrast with fat.)

Okazaki fragments   Newly formed DNA making up the lagging strand in DNA replication. DNA ligase links Okazaki fragments together to give a continuous strand.

olfactory [L. *olfacere*: to smell]   Pertaining to the sense of smell (*olfaction*).

oligodendrocyte   A type of glial cell that myelinates axons in the central nervous system.

oligosaccharide   A polymer containing a small number of monosaccharides.

ommatidia [Gk. *omma*: eye]   The units that make up the compound eye of some arthropods.

omnivore [L. *omnis*: everything + *vorare*: to devour]   An organism that eats both animal and plant material. (Contrast with carnivore, detritivore, herbivore.)

oncogene [Gk. *onkos*: mass, tumor + *genes*: born]   A gene that codes for a protein product that stimulates cell proliferation. Mutations in oncogenes that result in excessive cell proliferation can give rise to cancer.

oocyte   *See* primary oocyte, secondary oocyte.

oogenesis (oh' eh jen e sis) [Gk. *oon*: egg + *genesis*: source]   Gametogenesis leading to production of an ovum.

oogonium (oh' eh go' nee um) (plural: oogonia)   (1) In some algae and fungi, a cell in which an egg is produced. (2) In animals, the diploid progeny of a germ cell in females.

operator   The region of an operon that acts as the binding site for the repressor.

open circulatory system   Circulatory system in which extracellular fluid leaves the vessels of the circulatory system, percolates between cells and through tissues, and then flows back into the circulatory system to be pumped out again. (Contrast with closed circulatory system.)

operon   A genetic unit of transcription, typically consisting of several structural genes that are transcribed together; the operon contains at least two control regions: the promoter and the operator.

opportunity cost   The sum of the benefits an animal forfeits by not being able to perform some other behavior during the time when it is performing a given behavior.

opsin (op' sin) [Gk. *opsis*: sight]   The protein portion of the visual pigment rhodopsin. (*See* rhodopsin.)

optic chiasm [Gk. *chiasma*: cross]   Structure on the lower surface of the vertebrate brain where the two optic nerves come together.

optical isomers   Two isomers that are mirror images of each other.

optimal foraging theory   The application of a cost–benefit approach to feeding behavior to identify the fitness value of feeding choices.

orbital   A region in space surrounding the atomic nucleus in which an electron is most likely to be found.

organ [Gk. *organon*: tool]   A body part, such as the heart, liver, brain, root, or leaf. Organs are composed of different tissues integrated to perform a distinct function. Organs, in turn, are integrated into organ systems.

organ identity genes   In angiosperms, genes that specify the different organs of the flower. (Compare with homeotic genes.)

organ of Corti   Structure in the inner ear that transforms mechanical forces produced from pressure waves ("sound waves") into action potentials that are sensed as sound.

organ system   An interrelated and integrated group of tissues and organs that work together in a physiological function.

organelle (or gan el')   Any of the membrane-enclosed structures within a eukaryotic cell. Examples include the nucleus, endoplasmic reticulum, and mitochondria.

organic   (1) Pertaining to any chemical compound that contains carbon. (2) Pertaining to any aspect of living matter, e.g., to its evolution, structure, or chemistry.

organism   Any living entity.

organizer   Region of the early amphibian embryo that directs early embryonic development. Also known as the primary embryonic organizer.

organogenesis   The formation of organs and organ systems during development.

origin of replication (ori)   DNA sequence at which helicase unwinds the DNA double helix and DNA polymerase binds to initiate DNA replication.

orthology (or thol' o jee)   Type of homology in which the divergence of homologous genes can be traced to speciation events. (Contrast with paralogy.)

osmoconformer   An aquatic animal that equilibrates the osmolarity of its extracellular fluid that is the same as with that of the external environment.

osmolarity   The concentration of osmotically active particles in a solution.

osmoregulation   Regulation of the chemical composition of the body fluids of an organism.

osmosis (oz mo' sis) [Gk. *osmos*: to push]   Movement of water across a differentially permeable membrane, from one region to another region where the water potential is more negative.

ossicle (oss' ick ul) [L. *os*: bone]   The calcified construction unit of echinoderm skeletons.

osteoblast (oss' tee oh blast) [Gk. *osteon*: bone + *blastos*: sprout]   A cell that lay down the protein matrix of bone.

osteoclast (oss' tee oh clast) [Gk. *osteon*: bone + *klastos*: broken]   A cell that dissolves bone.

osteocyte   An osteoblast that has become enclosed in lacunae within the bone it has built.

outgroup   In phylogenetics, a group of organisms used as a point of reference for comparison with the groups of primary interest (the ingroup).

oval window   The flexible membrane that, when moved by the bones of the middle ear, produces pressure waves in the inner ear.

ovarian cycle   In human females, the monthly cycle of events by which eggs and hormones are produced. (Contrast with uterine cycle).

ovary (oh' var ee) [L. *ovum*: egg]   Any female organ, in plants or animals, that produces an egg.

overtopping   Plant growth pattern in which one branch differentiates from and grows beyond the others.

oviduct   In mammals, the tube serving to transport eggs to the uterus or to outside of the body.

oviparity   Reproduction in which eggs are released by the female and development is external to the mother's body. (Contrast with viviparity.)

ovoviviparity   Pertaining to reproduction in which fertilized eggs develop and hatch within the mother's body but are not attached to the mother by means of a placenta.

ovulation   Release of an egg from an ovary.

ovule (oh' vule)   In plants, a structure comprising the megasporangium and the integument, which develops into a seed after fertilization.

ovum (oh' vum) (plural: ova) [L. egg]   The female gamete.

oxidation (ox i day' shun)   Relative loss of electrons in a chemical reaction; either outright removal to form an ion, or the sharing of electrons with substances having a greater affinity for them, such as oxygen. Most oxidations, including biological ones, are associated with the liberation of energy. (Contrast with reduction.)

oxidative phosphorylation   ATP formation in the mitochondrion, associated with flow of electrons through the respiratory chain.

oxygenase   An enzyme that catalyzes the addition of oxygen to a substrate from O_2.

oxytocin   A hormone released by the posterior pituitary that promotes social bonding.

- P -

pancreas (pan' cree us)   A gland located near the stomach of vertebrates that secretes digestive enzymes into the small intestine and releases insulin into the bloodstream.

Pangaea (pan jee' uh) [Gk. *pan*: all, every]   The single land mass formed when all the continents came together in the Permian period.

para- [Gk. *para*: akin to, beside]   Prefix indicating association in being along side or accessory to.

parabronchi   Passages in the lungs of birds through which air flows.

paracrine [Gk. *para*: near]   Pertaining to a chemical signal, such as a hormone, that acts locally, near the site of its secretion. (Contrast with autocrine.)

paralogy (par al' o jee)   Type of homology in which the divergence of homologous genes can be traced to gene duplication events. (Contrast with orthology.)

paraphyletic (par' a fih leht' ik) [Gk. *para*: beside + *phylon*: tribe]   Pertaining to a group that consists of an ancestor and some, but not all, of its descendants. (Contrast with monophyletic, polyphyletic.)

parasite   An organism that consumes parts of an organism much larger than itself (known as its host). Parasites sometimes, but not always, kill their host.

parasympathetic nervous system   The division of the autonomic nervous system that works in opposition to the sympathetic nervous system. (Contrast with sympathetic nervous system.)

parathyroid glands   Four glands on the posterior surface of the thyroid gland that produce and release parathyroid hormone.

parathyroid hormone (PTH)   A hormone secreted by the parathyroid glands that stimulates osteoclast activity and raises blood calcium levels. Also called parathormone.

parenchyma (pair eng' kyma)   A plant tissue composed of relatively unspecialized cells without secondary walls.

parent rock   The soil horizon consisting of the rock that is breaking down to form the soil. Also called bedrock, or the C horizon.

parental (P) generation   The individuals that mate in a genetic cross. Their offspring are the first filial (F_1) generation.

parsimony   Preferring the simplest among a set of plausible explanations of any phenomenon.

parthenocarpy   Formation of fruit from a flower without fertilization.

parthenogenesis [Gk. *parthenos*: virgin]   Production of an organism from an unfertilized egg.

particulate theory   In genetics, the theory that genes are physical entities that retain their identities after fertilization.

passive transport   Diffusion across a membrane; may or may not require a channel or carrier protein. (Contrast with active transport.)

patch clamping   A technique for isolating a tiny patch of membrane to allow the study of ion movement through a particular channel.

pathogen (path' o jen) [Gk. *pathos*: suffering + *genesis*: source]   An organism that causes disease.

pattern formation   In animal embryonic development, the organization of differentiated tissues into specific structures such as wings.

pedigree   The pattern of transmission of a genetic trait within a family.

pelagic zone [Gk. *pelagos*: sea]   The open ocean.

penetrance   The proportion of individuals with a particular genotype that show the expected phenotype.

penis   An accessory sex organ of male animals that enables the male to deposit sperm in the female's reproductive tract.

pentose [Gk. *penta*: five]   A sugar containing five carbon atoms.

PEP carboxylase   The enzyme that combines carbon dioxide with PEP to form a 4-carbon dicarboxylic acid at the start of C_4 photosynthesis or of crassulacean acid metabolism (CAM).

pepsin [Gk. *pepsis*: digestion]   An enzyme in gastric juice that digests protein.

pepsinogen   Inactive secretory product that is converted into pepsin by low pH or by enzymatic action.

peptide linkage   The bond between amino acids in a protein; formed between a carboxyl group and amino group (CO—NH⁻) with the loss of water molecules.

peptidoglycan   The cell wall material of many bacteria, consisting of a single enormous molecule that surrounds the entire cell.

perennial (per ren' ee al) [L. *per*: throughout + *annus*: year]   A plant that survives from year to year. (Contrast with annual, biennial.)

perfect flower   A flower with both stamens and carpels; a hermaphroditic flower. (Contrast with imperfect flower.)

pericycle [Gk. *peri*: around + *kyklos*: ring or circle]   In plant roots, tissue just within the endodermis, but outside of the root vascular tissue. Meristematic activity of pericycle cells produces lateral root primordia.

periderm   The outer tissue of the secondary plant body, consisting primarily of cork.

period   (1) A category in the geological time scale. (2) The duration of a single cycle in a cyclical event, such as a circadian rhythm.

peripheral nervous system (PNS)   The portion of the nervous system that transmits information to and from the central nervous system, consisting of neurons that extend or reside outside the brain or spinal cord and their supporting cells. (Contrast with central nervous system.)

peristalsis (pair' i stall' sis)   Wavelike muscular contractions proceeding along a tubular organ, propelling the contents along the tube.

peritoneum   The mesodermal lining of the body cavity in coelomate animals.

peroxisome   An organelle that houses reactions in which toxic peroxides are formed and then converted to water.

petal [Gk. *petalon*: spread out]   In an angiosperm flower, a sterile modified leaf, nonphotosynthetic, frequently brightly colored, and often serving to attract pollinating insects.

petiole (pet' ee ole) [L. *petiolus*: small foot]   The stalk of a leaf.

pH   The negative logarithm of the hydrogen ion concentration; a measure of the acidity of a solution. A solution with pH = 7 is said to be neutral; pH values higher than 7 characterize basic solutions, while acidic solutions have pH values less than 7.

phage (fayj)   *See* bacteriophage.

phagocyte [Gk. *phagein*: to eat + *kystos*: sac]   One of two major classes of white blood cells; one of the nonspecific defenses of animals; ingests invading microorganisms by phagocytosis.

phagocytosis   Endocytosis by a cell of another cell or large particle.

pharming   The use of genetically modified animals to produce medically useful products in their milk.

pharynx [Gk. throat]   The part of the gut between the mouth and the esophagus.

phenotype (fee' no type) [Gk. *phanein*: to show]   The observable properties of an individual resulting from both genetic and environmental factors. (Contrast with genotype.)

phenotypic plasticity   *See* developmental plasticity.

pheromone (feer' o mone) [Gk. *pheros*: carry + *hormon*: excite, arouse]   A chemical substance used in communication between organisms of the same species.

phloem (flo' um) [Gk. *phloos*: bark]   In vascular plants, the vascular tissue that transports sugars and other solutes from sources to sinks.

phosphate group   The functional group —OPO_3H_2.

phosphodiester linkage   The connection in a nucleic acid strand, formed by linking two nucleotides.

phospholipid   A lipid containing a phosphate group; an important constituent of cellular membranes. (*See* lipid.)

phospholipid bilayer   The basic structural unit of biological membranes; a sheet of phospholipids two molecules thick in which the phospholipids are lined up with their hydrophobic "tails" packed tightly together and their hydrophilic, phosphate-containing "heads" facing outward. Also called lipid bilayer.

phosphorylation   Addition of a phosphate group.

photoautotroph   An organism that obtains energy from light and carbon from carbon dioxide. (Contrast with chemolithotroph, chemoheterotroph, photoheterotroph.)

photoheterotroph   An organism that obtains energy from light but must obtain its carbon from organic compounds. (Contrast with chemolithotroph, chemoheterotroph, photoautotroph.)

photon (foe' ton) [Gk. *photos*: light]   A quantum of visible radiation; a "packet" of light energy.

photoperiodicity Control of an organism's physiological or behavioral responses by the length of the day or night.

photoreceptor (1) In plants, a pigment that triggers a physiological response when it absorbs a photon. (2) In animals, a sensory receptor cell that senses and responds to light energy.

photorespiration Light-driven uptake of oxygen and release of carbon dioxide, the carbon being derived from the early reactions of photosynthesis.

photosynthesis (foe tow sin' the sis) [literally, "synthesis from light"] Metabolic processes, carried out by green plants, by which visible light is trapped and the energy used to synthesize compounds such as ATP and glucose.

photosystem [Gk. *phos*: light + *systema*: assembly] A light-harvesting complex in the chloroplast thylakoid composed of pigments and proteins.

photosystem I In photosynthesis, the reactions that absorb light at 700 nm, passing electrons to ferrodoxin and thence to NADPH. Rich in chlorophyll *a*.

photosystem II In photosynthesis, the reactions that absorb light at 660 nm, passing electrons to the electron transport chain in the chloroplast. Rich in chlorophyll *b*.

phototropins A class of blue light receptors that mediate phototropism and other plant responses.

phototropism [Gk. *photos*: light + *trope*: turning] A directed plant growth response to light.

phycobilin Photosynthetic pigment that absorbs red, yellow, orange, and green light and is found in cyanobacteria and some red algae.

phylogenetic tree A graphic representation of lines of descent among organisms or their genes.

phylogeny (fy loj' e nee) [Gk. *phylon*: tribe, race + *genesis*: source] The evolutionary history of a particular group of organisms or their genes.

physiology (fiz' ee ol' o jee) [Gk. *physis*: natural form] The scientific study of the functions of living organisms and the individual organs, tissues, and cells of which they are composed.

phytoalexins Substances toxic to pathogens, produced by plants in response to fungal or bacterial infection.

phytochrome (fy' tow krome) [Gk. *phyton*: plant + *chroma*: color] A plant pigment regulating a large number of developmental and other phenomena in plants.

phytomers In plants, the repeating modules that compose a shoot, each consisting of one or more leaves, attached to the stem at a node; an internode; and one or more axillary buds.

phytoplankton Photosynthetic plankton.

phytoremediation A form of bioremediation that uses plants to clean up environmental pollution.

pigment A substance that absorbs visible light.

pineal gland Gland located between the cerebral hemispheres that secretes melatonin.

pinocytosis Endocytosis by a cell of liquid containing dissolved substances.

pistil [L. *pistillum*: pestle] The structure of an angiosperm flower within which the ovules are borne. May consist of a single carpel, or of several carpels fused into a single structure. Usually differentiated into ovary, style, and stigma.

pith In plants, relatively unspecialized tissue found within a cylinder of vascular tissue.

pituitary gland A small gland attached to the base of the brain in vertebrates. Its hormones control the activities of other glands. Also known as the hypophysis.

placenta (pla sen' ta) The organ in female mammals that provides for the nourishment of the fetus and elimination of the fetal waste products.

plankton Free-floating small aquatic organisms. Photosynthetic members of the plankton are referred to as phytoplankton.

planula (plan' yew la) [L. *planum*: flat] A free-swimming, ciliated larval form typical of the cnidarians.

plaque (plack) [Fr.: a metal plate or coin] (1) A circular clearing in a layer (lawn) of bacteria growing on the surface of a nutrient agar gel. (2) An accumulation of prokaryotic organisms on tooth enamel. Acids produced by these microorganisms cause tooth decay. (3) A region of arterial wall invaded by fibroblasts and fatty deposits. (*See* atherosclerosis.)

plasma (plaz' muh) The liquid portion of blood, in which blood cells and other particulates are suspended.

plasma cell An antibody-secreting cell that develops from a B cell; the effector cell of the humoral immune system.

plasma membrane The membrane that surrounds the cell, regulating the entry and exit of molecules and ions. Every cell has a plasma membrane.

plasmid A DNA molecule distinct from the chromosome(s); that is, an extrachromosomal element; found in many bacteria. May replicate independently of the chromosome.

plasmodesma (plural: plasmodesmata) [Gk. *plassein*: to mold + *desmos*: band] A cytoplasmic strand connecting two adjacent plant cells.

plasmogamy The fusion of the cytoplasm of two cells. (Contrast with karyogamy.)

plastid Any of the plant cell organelles that house biochemical pathways for photosynthesis.

plate tectonics [Gk. *tekton*: builder] The scientific study of the structure and movements of Earth's lithospheric plates, which are the cause of continental drift.

platelet A membrane-bounded body without a nucleus, arising as a fragment of a cell in the bone marrow of mammals. Important to blood-clotting action.

pleiotropy (plee' a tro pee) [Gk. *pleion*: more] The determination of more than one character by a single gene.

pleural membrane [Gk. *pleuras*: rib, side] The membrane lining the outside of the lungs and the walls of the thoracic cavity. Inflammation of these membranes is a condition known as pleurisy.

pluripotent [L. *pluri*: many + *potens*: powerful] Having the ability to form all of the cells in the body. (Contrast with multipotent, totipotent.)

podocytes Cells of Bowman's capsule of the nephron that cover the capillaries of the glomerulus, forming filtration slits.

point mutation A mutation that results from the gain, loss, or substitution of a single nucleotide.

polar Having separate and opposite electric charges at two ends, or poles. (Contrast with nonpolar.)

polar body A nonfunctional nucleus produced by meiosis during oogenesis.

polar nuclei In angiosperms, the two nuclei in the central cell of the megagametophyte; following fertilization they give rise to the endosperm.

polarity (1) In chemistry, the property of unequal electron sharing in a covalent bond that defines a polar molecule. (2) In development, the difference between one end of an organism or structure and the other.

pollen [L. *pollin*: fine flour] In seed plants, microscopic grains that contain the male gametophyte (microgametophyte) and gamete (microspore).

pollen tube A structure that develops from a pollen grain through which sperm are released into the megagametophyte.

pollination The process of transferring pollen from an anther to the stigma of a pistil in an angiosperm or from a strobilus to an ovule in a gymnosperm.

poly- [Gk. *poly*: many] A prefix denoting multiple entities.

poly A tail A long sequence of adenine nucleotides (50–250) added after transcription to the 3' end of most eukaryotic mRNAs.

polyandry Mating system in which one female mates with multiple males.

polygyny Mating system in which one male mates with multiple females.

polymer [Gk. *poly*: many + *meros*: unit] A large molecule made up of similar or identical subunits called monomers. (Contrast with monomer, oligomer.)

polymerase chain reaction (PCR) An enzymatic technique for the rapid production of millions of copies of a particular stretch of DNA where only a small amount of the parent molecule is available.

polymorphic (pol' lee mor' fik) [Gk. *poly*: many + *morphe*: form, shape] Coexistence in a population of two or more distinct traits.

polyp (pah' lip) [Gk. *poly*: many + *pous*: foot] In cnidarians, a sessile, asexual life cycle stage.

polypeptide A large molecule made up of many amino acids joined by peptide linkages. Large polypeptides are called proteins.

polyphyletic (pol' lee fih leht' ik) [Gk. *poly*: many + *phylon*: tribe] Pertaining to a group that consists of multiple distantly related organisms, and does not include the common ancestor of the group. (Contrast with monophyletic, paraphyletic.)

polyploidy (pol' lee ploid ee) The possession of more than two entire sets of chromosomes.

polyribosome (polysome) A complex consisting of a threadlike molecule of messenger RNA and several (or many) ribosomes. The ribosomes move along the mRNA, synthesizing polypeptide chains as they proceed.

polysaccharide A macromolecule composed of many monosaccharides (simple sugars). Common examples are cellulose and starch.

pons [L. *pons*: bridge] Region of the brainstem anterior to the medulla.

population Any group of organisms coexisting at the same time and in the same place and capable of interbreeding with one another.

population bottleneck A period during which only a few individuals of a normally large population survive.

population density The number of individuals of a population per unit of area or volume.

population dynamics   The patterns and processes of change in populations.

population genetics   The study of genetic variation and its causes within populations.

portal blood vessels   Blood vessels that begin and end in capillary beds.

positive cooperativity   Occurs when a molecule can bind several ligands and each one that binds alters the conformation of the molecule so that it can bind the next ligand more easily. The binding of four molecules of O_2 by hemoglobin is an example of positive cooperativity.

positive feedback   In regulatory systems, information that amplifies a regulatory response, increasing the deviation of the system from the set point. (Contrast with negative feedback.)

positive regulation   A form of gene regulation in which a regulatory macromolecule is needed to turn on the transcription of a structural gene; in its absence, transcription will not occur. (Contrast with negative regulation.)

post- [L. *postere*: behind, following after]   Prefix denoting something that comes after.

postabsorptive state   State in which no food remains in the gut and thus no nutrients are being absorbed. (Contrast with absorptive state.)

posterior   Toward or pertaining to the rear. (Contrast with anterior.)

postsynaptic cell   The cell that receives information from a neuron at a synapse. (Contrast with presynaptic neuron.)

postzygotic reproductive barriers   Barriers to the reproductive process that occur after the union of the nuclei of two gametes. (Contrast with prezygotic reproductive barriers.)

potential energy   Energy not doing work, such as the energy stored in chemical bonds. (Contrast with kinetic energy.)

precapillary sphincter   A cuff of smooth muscle that can shut off the blood flow to a capillary bed.

pre-mRNA (precursor mRNA)   Initial gene transcript before it is modified to produce functional mRNA. Also known as the primary transcript.

predator   An organism that kills and eats other organisms.

prereplication complex   In eukaryotes, a complex of proteins that binds to DNA at the initiation of DNA replication.

pressure flow model   An effective model for phloem transport in angiosperms. It holds that sieve element transport is driven by an osmotically driven pressure gradient between source and sink.

pressure potential   The hydrostatic pressure of an enclosed solution in excess of the surrounding atmospheric pressure. (Contrast with solute potential, water potential.)

presynaptic neuron   The neuron that transmits information to another cell at a synapse. (Contrast with postsynaptic cell.)

prey [L. *praeda*: booty]   An organism consumed by a predator as an energy source.

prezygotic reproductive barriers   Barriers to the reproductive process that occur before the union of the nuclei of two gametes (Contrast with postzygotic reproductive barriers.)

primary active transport   Active transport in which ATP is hydrolyzed, yielding the energy required to transport an ion or molecule against its concentration gradient. (Contrast with secondary active transport.)

primary cell wall   In plant cells, a structure that forms at the middle lamella after cytokinesis, made up of cellulose microfibrils, hemicelluloses, and pectins. (Contrast with secondary cell wall.)

primary consumer   An organism (herbivore) that eats plant tissues.

primary growth   In plants, growth that is characterized by the lengthening of roots and shoots and by the proliferation of new roots and shoots through branching. (Contrast with secondary growth.)

primary immune response   The first response of the immune system to an antigen, involving recognition by lymphocytes and the production of effector cells and memory cells. (Contrast with secondary immune response.)

primary lysosome   *See* lysosome.

primary meristem   Meristem that produces the tissues of the primary plant body.

primary oocyte   (oh' eh site) [Gk. *oon*: egg + *kytos*: container]   The diploid progeny of an oogonium. In many species, a primary oocyte enters prophase of the first meiotic division, then remains in developmental arrest for a long time before resuming meiosis to form a secondary oocyte and a polar body.

primary plant body   That part of a plant produced by primary growth. Consists of all the *nonwoody* parts of a plant; many herbaceous plants consist entirely of a primary plant body. (Contrast with secondary plant body.)

primary producer   A photosynthetic or chemosynthetic organism that synthesizes complex organic molecules from simple inorganic ones.

primary sex determination   Genetic determination of gametic sex, male or female. (Contrast with secondary sex determination.)

primary spermatocyte   The diploid progeny of a spermatogonium; undergoes the first meiotic division to form secondary spermatocytes.

primary succession   Succession that begins in an area initially devoid of life, such as on recently exposed glacial till or lava flows. (Contrast with secondary succession.)

primary structure   The specific sequence of amino acids in a protein.

primase   An enzyme that catalyzes the synthesis of a primer for DNA replication.

primer   Strand of nucleic acid, usually RNA, that is the necessary starting material for the synthesis of a new DNA strand, which is synthesized from the 3′ end of the primer.

primordium (plural: primordia) [L. origin]   The most rudimentary stage of an organ or other part.

prion   An infectious protein that can proliferate by converting the inactive form of a particular protein into an active protein.

pro- [L.: first, before, favoring]   A prefix often used in biology to denote a developmental stage that comes first or an evolutionary form that appeared earlier than another. For example, prokaryote, prophase.

probe   A segment of single stranded nucleic acid used to identify DNA molecules containing the complementary sequence.

procambium   Primary meristem that produces the vascular tissue.

procedural memory   Memory of motor tasks. Cannot be consciously recalled and described. (Contrast with declarative memory.)

processive   Pertaining to an enzyme that catalyzes many reactions each time it binds to a substrate, as DNA polymerase does during DNA replication.

progesterone [L. *pro*: favoring + *gestare*: to bear]   A female sex hormone that maintains pregnancy.

prolactin   A hormone released by the anterior pituitary, one of whose functions is the stimulation of milk production in female mammals.

proliferating cell nuclear antigen (PCNA)   A protein complex that ensures processivity of DNA replication in eukaryotes.

prometaphase   The phase of nuclear division that begins with the disintegration of the nuclear envelope.

promoter   A DNA sequence to which RNA polymerase binds to initiate transcription.

prophage (pro′ fayj)   The noninfectious units that are linked with the chromosomes of the host bacteria and multiply with them but do not cause dissolution of the cell. Prophage can later enter into the lytic phase to complete the virus life cycle.

prophase (pro′ phase)   The first stage of nuclear division, during which chromosomes condense from diffuse, threadlike material to discrete, compact bodies.

prostaglandin   Any one of a group of specialized lipids with hormone-like functions. It is not clear that they act at any considerable distance from the site of their production.

prostate gland   In male humans, surrounds the urethra at its junction with the vas deferens; supplies an acid-neutralizing fluid to the semen.

prosthetic group   Any nonprotein portion of an enzyme.

proteasome   In the eukaryotic cytoplasm, a huge protein structure that binds to and digests cellular proteins that have been tagged by ubiquitin.

protein (pro′ teen) [Gk. *protos*: first]   Long-chain polymer of amino acids with twenty different common side chains. Occurs with its polymer chain extended in fibrous proteins, or coiled into a compact macromolecule in enzymes and other globular proteins.

protein kinase (kye′ nase)   An enzyme that catalyzes the addition of a phosphate group from ATP to a target protein.

protein kinase cascade   A series of reactions in response to a molecular signal, in which a series of protein kinases activates one another in sequence, amplifying the signal at each step.

proteoglycan   A glycoprotein containing a protein core with attached long, linear carbohydrate chains.

proteolysis [protein + Gk. *lysis*: break apart]   An enzymatic digestion of a protein or polypeptide.

proteome   The set of proteins that can be made by an organism. Because of alternative splicing of pre-mRNA, the number of proteins that can be made is usually much larger than the number of protein-coding genes present in the organism's genome.

protoderm   Primary meristem that gives rise to the plant epidermis.

proton (pro′ ton) [Gk. *protos*: first, before]   (1) A subatomic particle with a single positive charge.

The number of protons in the nucleus of an atom determine its element. (2) A hydrogen ion, H⁺.

proton pump   An active transport system that uses ATP energy to move hydrogen ions across a membrane, generating an electric potential.

proton-motive force   Force generated across a membrane having two components: a chemical potential (difference in proton concentration) plus an electrical potential due to the electrostatic charge on the proton.

protonephridium   The excretory organ of flatworms, made up of a tubule and a flame cell.

protoplast   The living contents of a plant cell; the plasma membrane and everything contained within it.

provirus   Double-stranded DNA made by a virus that is integrated into the host's chromosome and contains promoters that are recognized by the host cell's transcription apparatus.

proximal   Near the point of attachment or other reference point. (Contrast with distal.)

proximal convoluted tubule   The initial segment of a renal tubule, closest to the glomerulus. (Compare with distal convoluted tubule.)

proximate cause   The immediate genetic, physiological, neurological, and developmental mechanisms responsible for a behavior or morphology. (Contrast with ultimate cause.)

pseudocoelomate (soo' do see' low mate) [Gk. *pseudes*: false + *koiloma*: cavity]   Having a body cavity, called a pseudocoel, consisting of a fluid-filled space in which many of the internal organs are suspended, but which is enclosed by mesoderm only on its outside.

pseudogene [Gk. *pseudes*: false]   A DNA segment that is homologous to a functional gene but is not expressed because of changes to its sequence or changes to its location in the genome.

pseudopod (soo' do pod) [Gk. *pseudes*: false + *podos*: foot]   A temporary, soft extension of the cell body that is used in location, attachment to surfaces, or engulfing particles.

pulmonary [L. *pulmo*: lung]   Pertaining to the lungs.

pulmonary circuit   The portion of the circulatory system by which blood is pumped from the heart to the lungs or gills for oxygenation and back to the heart for distribution. (Contrast with systemic circuit.)

pulmonary valve   A one-way valve between the right ventricle of the heart and the pulmonary artery that prevents backflow of blood into the ventricle when it relaxes.

Punnett square   Method of predicting the results of a genetic cross by arranging the gametes of each parent at the edges of a square.

pupa (pew' pa) [L. *pupa*: doll, puppet]   In certain insects (the Holometabola), the encased developmental stage between the larva and the adult.

pupil   The opening in the vertebrate eye through which light passes.

purine (pure' een)   One of the two types of nitrogenous bases in nucleic acids. Each of the purines—adenine and guanine—pairs with a specific pyrimidine.

Purkinje fibers   Specialized heart muscle cells that conduct excitation throughout the ventricular muscle.

pyrimidine (per im' a deen)   One of the two types of nitrogenous bases in nucleic acids. Each

of the pyrimidines—cytosine, thymine, and uracil—pairs with a specific purine.

pyrogen   Molecule that produces a rise in body temperature (fever); may be produced by an invading pathogen or by cells of the immune system in response to infection.

pyruvate   A three-carbon acid; the end product of glycolysis and the raw material for the citric acid cycle.

pyruvate oxidation   Conversion of pyruvate to acetyl CoA and CO_2 that occurs in the mitochondrial matrix in the presence of O_2.

- Q -

Q_{10}   A value that compares the rate of a biochemical process or reaction over 10°C temperature ranges. A process that is not temperature-sensitive has a Q_{10} of 1; values of 2 or 3 mean the reaction speeds up as temperature increases.

quantitative trait loci   A set of genes that determines a complex character that exhibits quantitative variation.

quaternary structure   The specific three-dimensional arrangement of protein subunits.

- R -

R group   The distinguishing group of atoms of a particular amino acid; also known as a side chain.

r-strategist   A species whose life history strategy allows for a high intrinsic rate of population increase (r). (Contrast with K-strategist.)

radial symmetry   The condition in which any two halves of a body are mirror images of each other, providing the cut passes through the center; a cylinder cut lengthwise down its center displays this form of symmetry.

radiation   The transfer of heat from warmer objects to cooler ones via the exchange of infrared radiation. *See also* electromagnetic radiation; evolutionary radiation.

radicle   An embryonic root.

radioisotope   A radioactive isotope of an element. Examples are carbon-14 (^{14}C) and hydrogen-3, or tritium (^{3}H).

reactant   A chemical substance that enters into a chemical reaction with another substance.

reaction center   A group of electron transfer proteins that receive energy from light-absorbing pigments and convert it to chemical energy by redox reactions.

realized niche   A species' niche as defined by its interactions with other species. (Contrast with fundamental niche.)

receptive field   The area of visual space that activates a particular cell in the visual system.

receptor   *See* receptor protein, sensory receptor cell.

receptor-mediated endocytosis   Endocytosis initiated by macromolecular binding to a specific membrane receptor.

receptor potential   The change in the resting potential of a sensory cell when it is stimulated.

receptor protein   A protein that can bind to a specific molecule, or detect a specific stimulus, within the cell or in the cell's external environment.

recessive   In genetics, an allele that does not determine phenotype in the presence of a dominant allele. (Contrast with dominance.)

reciprocal adaptation   *See* coevolution.

reciprocal crosses   A pair of matings in one of which a female of genotype A mates with a male of genotype B and in the other of which a female of genotype B mates with a male of genotype A.

recognition sequence   *See* restriction site.

recombinant   Pertaining to an individual, meiotic product, or chromosome in which genetic materials originally present in two individuals end up in the same haploid complement of genes.

recombinant DNA   A DNA molecule made in the laboratory that is derived from two or more genetic sources.

recombinant frequency   The proportion of offspring of a genetic cross that have phenotypes different from the parental phenotypes due to crossing over between linked genes during gamete formation.

reconciliation ecology   The practice of making exploited lands more biodiversity-friendly.

rectum   The terminal portion of the gut, ending at the anus.

redox reaction   A chemical reaction in which one reactant becomes oxidized and the other becomes reduced.

reduction   Gain of electrons by a chemical reactant; any reduction is accompanied by an oxidation. (Contrast with oxidation.)

refractory period   The time interval after an action potential during which another action potential cannot be elicited from an excitable membrane.

regeneration   The development of a complete individual from a fragment of an organism.

regulative development   A pattern of animal embryonic development in which the fates of the first blastomeres are not absolutely fixed. (Contrast with mosaic development.)

regulatory sequence   A DNA sequence to which the protein product of a regulatory gene binds.

regulatory gene   A gene that codes for a protein (or RNA) that in turn controls the expression of another gene.

regulatory system   A system that uses feedback information to maintain a physiological function or parameter at an optimal level. (Contrast with controlled system.)

regulatory T cells (T_{reg})   Class of T cells that mediates tolerance to self antigens.

reinforcement   The evolution of enhanced reproductive isolation between populations due to natural selection for greater isolation.

releaser   Sensory stimulus that triggers performance of a stereotyped behavior pattern.

REM (rapid-eye-movement) sleep   A sleep state characterized by vivid dreams, skeletal muscle relaxation, and rapid eye movements. (Contrast with slow-wave sleep.)

renal [L. *renes*: kidneys]   Relating to the kidneys.

renal tubule   A structural unit of the kidney that collects filtrate from the blood, reabsorbs specific ions, nutrients, and water and returns them to the blood, and concentrates excess ions and waste products such as urea for excretion from the body.

replication   The duplication of genetic material.

replication complex   The close association of several proteins operating in the replication of DNA.

replication fork A point at which a DNA molecule is replicating. The fork forms by the unwinding of the parent molecule.

replicon A region of DNA replicated from a single origin of replication.

reporter gene A genetic marker included in recombinant DNA to indicate the presence of the recombinant DNA in a host cell.

repressor A protein encoded by a regulatory gene that can bind to a specific operator and prevent transcription of the operon. (Contrast with activator.)

reproductive isolation Condition in which two divergent populations are no longer exchanging genes. Can lead to speciation.

rescue effect The process by which individuals moving between subpopulations of a metapopulation may prevent declining subpopulations from becoming extinct.

resistance (R) genes Plant genes that confer resistance to specific strains of pathogens.

resource Something in the environment required by an organism for its maintenance and growth that is consumed in the process of being used.

resource partitioning A situation in which selection pressures resulting from interspecific competition cause changes in the ways in which the competing species use the limiting resource, thereby allowing them to coexist.

respiration (res pi ra' shun) [L. *spirare*: to breathe] (1) Cellular respiration. (2) Breathing.

respiratory chain The terminal reactions of cellular respiration, in which electrons are passed from NAD or FAD, through a series of intermediate carriers, to molecular oxygen, with the concomitant production of ATP.

respiratory gases Oxygen (O_2) and carbon dioxide (CO_2); the gases that an animal must exchange between its internal body fluids and the outside medium (air or water).

resting potential The membrane potential of a living cell at rest. In cells at rest, the interior is negative to the exterior. (Contrast with action potential, electrotonic potential.)

restoration ecology The science and practice of restoring damaged or degraded ecosystems.

restriction enzyme Any of a type of enzyme that cleaves double-stranded DNA at specific sites; extensively used in recombinant DNA technology. Also called a restriction endonuclease.

restriction fragment length polymorphism *See* RFLP.

restriction point (R) The specific time during G1 of the cell cycle at which the cell becomes committed to undergo the rest of the cell cycle.

restriction site A specific DNA base sequence that is recognized and acted on by a restriction endonuclease.

reticular system A central region of the vertebrate brainstem that includes complex fiber tracts conveying neural signals between the forebrain and the spinal cord, with collateral fibers to a variety of nuclei that are involved in autonomic functions, including arousal from sleep.

retina (rett' in uh) [L. *rete*: net] The light-sensitive layer of cells in the vertebrate or cephalopod eye.

retinoblastoma protein A protein that inhibits an animal cell from passing through the restriction point; inactivation of this protein is necessary for the cell cycle to proceed.

retrovirus An RNA virus that contains reverse transcriptase. Its RNA serves as a template for cDNA production, and the cDNA is integrated into a chromosome of the host cell.

reverse genetics Method of genetic analysis in which a phenotype is first related to a DNA variation, then the protein involved is identified.

reverse transcriptase An enzyme that catalyzes the production of DNA (cDNA), using RNA as a template; essential to the reproduction of retroviruses.

RFLP Restriction fragment length polymorphism, the coexistence of two or more patterns of restriction fragments resulting from underlying differences in DNA sequence.

rhizoids (rye' zoids) [Gk. root] Hairlike extensions of cells in mosses, liverworts, and a few vascular plants that serve the same function as roots and root hairs in vascular plants. The term is also applied to branched, rootlike extensions of some fungi and algae.

rhizome (rye' zome) An underground stem (as opposed to a root) that runs horizontally beneath the ground.

rhodopsin A photopigment used in the visual process of transducing photons of light into changes in the membrane potential of photoreceptor cells.

ribonucleic acid *See* RNA.

ribose A five-carbon sugar in nucleotides and RNA.

ribosomal RNA (rRNA) Several species of RNA that are incorporated into the ribosome. Involved in peptide bond formation.

ribosome A small particle in the cell that is the site of protein synthesis.

ribozyme An RNA molecule with catalytic activity.

risk cost The increased chance of being injured or killed as a result of performing a behavior, compared to resting.

RNA (ribonucleic acid) An often single stranded nucleic acid whose nucleotides use ribose rather than deoxyribose and in which the base uracil replaces thymine found in DNA. Serves as genome from some viruses. (*See* rRNA, tRNA, mRMA, and ribozyme.)

RNA interference (RNAi) A mechanism for reducing mRNA translation whereby a double-stranded RNA, made by the cell or synthetically, is processed into a small, single-stranded RNA, whose binding to a target mRNA results in the latter's breakdown.

RNA polymerase An enzyme that catalyzes the formation of RNA from a DNA template.

RNA splicing The last stage of RNA processing in eukaryotes, in which the transcripts of introns are excised through the action of small nuclear ribonucleoprotein particles (snRNP).

rod cells Light-sensitive cell in the vertebrate retina; these sensory receptor cells are sensitive in extremely dim light and are responsible for dim light, black and white vision.

root The organ responsible for anchoring the plant in the soil, absorbing water and minerals, and producing certain hormones. Some roots are storage organs.

root cap A thimble-shaped mass of cells, produced by the root apical meristem, that protects the meristem; the organ that perceives the gravitational stimulus in root gravitropism.

root hair A long, thin process from a root epidermal cell that absorbs water and minerals from the soil solution.

root system The organ system that anchors a plant in place, absorbs water and dissolved minerals, and may store products of photosynthesis from the shoot system.

rough endoplasmic reticulum (RER) The portion of the endoplasmic reticulum whose outer surface has attached ribosomes. (Contrast with smooth endoplasmic reticulum.)

rRNA *See* ribosomal RNA.

rubisco Contraction of ribulose bisphosphate carboxylase/oxygenase, the enzyme that combines carbon dioxide or oxygen with ribulose bisphosphate to catalyze the first step of photosynthetic carbon fixation or photorespiration, respectively.

ruminant Herbivorous, cud-chewing mammals such as cows or sheep, characterized by a stomach that consists of four compartments: the rumen, reticulum, omasum, and abomasum.

- S -

S phase In the cell cycle, the stage of interphase during which DNA is replicated. (Contrast with G_1 phase, G_2 phase, M phase.)

saltatory conduction [L. *saltare*: to jump] The rapid conduction of action potentials in myelinated axons; so called because action potentials appear to "jump" between nodes of Ranvier along the axon.

saprobe [Gk. *sapros*: rotten] An organism (usually a bacterium or fungus) that obtains its carbon and energy by absorbing nutrients from dead organic matter.

sarcomere (sark' o meer) [Gk. *sark*: flesh + *meros*: unit] The contractile unit of a skeletal muscle.

sarcoplasm The cytoplasm of a muscle cell.

sarcoplasmic reticulum The endoplasmic reticulum of a muscle cell.

saturated fatty acid A fatty acid in which all the bonds between carbon atoms in the hydrocarbon chain are single bonds—that is, all the bonds are saturated with hydrogen atoms. (Contrast with unsaturated fatty acid.)

scientific method A means of gaining knowledge about the natural world by making observations, posing hypotheses, and conducting experiments to test those hypotheses.

Schwann cell A type of glial cell that myelinates axons in the peripheral nervous system.

scion In horticulture, the bud or stem from one plant that is grafted to a root or root-bearing stem of another plant (the stock).

sclerenchyma (skler eng' kyma) [Gk. *skleros*: hard + *kymus*: juice] A plant tissue composed of cells with heavily thickened cell walls. The cells are dead at functional maturity. The principal types of sclerenchyma cells are fibers and sclereids.

second law of thermodynamics The principle that when energy is converted from one form to another, some of that energy becomes unavailable for doing work.

second messenger A compound, such as cAMP, that is released within a target cell after a hormone (the first messenger) has bound to a surface receptor on a cell; the second messenger triggers further reactions within the cell.

secondary active transport A form of active transport that does not use ATP as an energy source; rather, transport is coupled to ion diffusion down a concentration gradient established by primary active transport.

secondary cell wall   A thick, cellulosic structure internal to the primary cell wall formed in some plant cells after cell expansion stops (Contrast with primary cell wall.)

secondary consumer   An organism that eat primary consumers.

secondary growth   In plants, growth that contributes to an increase in girth. (Contrast with primary growth.)

secondary immune response   A rapid and intense response to a second or subsequent exposure to an antigen, initiated by memory cells. (Contrast with primary immune response.)

secondary lysosome   Membrane-enclosed organelle formed by the fusion of a primary lysosome with a phagosome, in which macromolecules taken up by phagocytosis are hydrolyzed into their monomers. (Contrast with lysosome.)

secondary metabolite   A compound synthesized by a plant that is not needed for basic cellular metabolism. Typically has an antiherbivore or antiparasite function.

secondary plant body   That part of a plant produced by secondary growth; consists of woody tissues. (Contrast with primary plant body.)

secondary sex determination   Formation of secondary sexual characteristics (i.e., those other than gonads), such as external sex organs and body hair. (Contrast with primary sex determination.)

secondary spermatocyte   One of the products of the first meiotic division of a primary spermatocyte.

secondary structure   Of a protein, localized regularities of structure, such as the α helix and the β pleated sheet.

secondary succession   Succession after a disturbance that did not eliminate all the organisms originally living on the site. (Contrast with primary succession.)

secretin (si kreet' in)   A peptide hormone secreted by the upper region of the small intestine when acidic chyme is present. Stimulates the pancreatic duct to secrete bicarbonate ions.

sedimentary rock   Rock formed by the accumulation of sediment grains on the bottom of a body of water.

seed   A fertilized, ripened ovule of a gymnosperm or angiosperm. Consists of the embryo, nutritive tissue, and a seed coat.

segmentation   Division of an animal body into segments.

segmentation genes   Genes that determine the number and polarity of body segments.

segregation   In genetics, the separation of alleles, or of homologous chromosomes, from each other during meiosis so that each of the haploid daughter nuclei produced contains one or the other member of the pair found in the diploid parent cell, but never both. This principle was articulated by Mendel as his first law.

selective permeability   Allowing certain substances to pass through while other substances are excluded; a characteristic of membranes.

self-incompatability   In plants, the possession of mechanisms that prevent self-fertilization.

semelparous   [L. semel: once + pario: to beget] Reproducing only once in a lifetime. (Contrast with iteroparous.)

semen (see' men) [L. semin: seed]   The thick, whitish liquid produced by the male reproductive system in mammals, containing the sperm.

semiconservative replication   The way in which DNA is synthesized. Each of the two partner strands in a double helix acts as a template for a new partner strand. Hence, after replication, each double helix consists of one old and one new strand.

seminiferous tubules   The tubules within the testes within which sperm production occurs.

senescence [L. senescere: to grow old]   Aging; deteriorative changes with aging; the increased probability of dying with increasing age.

sensitive period   The life stage during which some particular type of learning must take place, or during which it occurs much more easily than at other times. Typical of song learning among birds.

sensor   See sensory receptor cell.

sensory neuron   A specialized neuron that transduces a particular type of sensory stimulus into action potentials.

sensory receptor cell   Cell that is responsive to a particular type of physical or chemical stimulation.

sensory transduction   The transformation of environmental stimuli or information into neural signals.

sepal (see' pul) [L. sepalum: covering]   One of the outermost structures of the flower, usually protective in function and enclosing the rest of the flower in the bud stage.

septum (plural: septa) [L. wall]   (1) A partition or cross-wall appearing in the hyphae of some fungi. (2) The bony structure dividing the nasal passages.

sequence alignment   A method of identifying homologous positions in DNA or amino acid sequences by pinpointing the locations of deletions and insertions that have occurred since two (or more) organisms diverged from a common ancestor.

Sertoli cells   Cells in the seminiferous tubules that nuture the developing sperm.

sessile (sess' ul) [L. sedere: to sit]   Permanently attached; not able to move from one place to another. (Contrast with motile.)

set point   In a regulatory system, the threshold sensitivity to the feedback stimulus.

sex chromosome   In organisms with a chromosomal mechanism of sex determination, one of the chromosomes involved in sex determination.

sex linkage   The pattern of inheritance characteristic of genes located on the sex chromosomes of organisms having a chromosomal mechanism for sex determination.

sexual reproduction   Reproduction involving the union of gametes.

sexual selection   Selection by one sex of characteristics in individuals of the opposite sex. Also, the favoring of characteristics in one sex as a result of competition among individuals of that sex for mates.

Shannon diversity index   A formula for quantifying diversity that takes both species richness and species evenness into account; based on a mathematical expression of the certainty with which the next item sampled in a series can be predicted.

shared derived trait   See synapomorphy.

shoot system   In plants, the organ system consisting of the leaves, stem(s), and flowers.

short tandem repeat (STR)   A short (1–5 base pairs), moderately repetitive sequence of DNA. The number of copies of an STR at a particular location varies between individuals and is inherited.

shotgun sequencing   A relatively rapid method of DNA sequencing in which a DNA molecule is broken up into overlapping fragments, each fragment is sequenced, and high-speed computers analyze and realign the fragments. (Contrast with hierarchical sequencing.)

side chain   See R group.

sieve tube element   The characteristic cell of the phloem in angiosperms, which contains cytoplasm but relatively few organelles, and whose end walls (sieve plates) contain pores that form connections with neighboring cells.

signal sequence   The sequence of a protein that directs the protein protein to a particular organelle.

signal transduction pathway   The series of biochemical steps whereby a stimulus to a cell (such as a hormone or neurotransmitter binding to a receptor) is translated into a response of the cell.

silent mutation   A change in a gene's sequence that has no effect on the amino acid sequence of a protein because it occurs in noncoding DNA or because it does not change the amino acid specified by the corresponding codon. (Contrast with frame-shift mutation, missense mutation, nonsense mutation.)

similarity matrix   A matrix used to compare the degree of divergence among pairs of objects. For molecular sequences, constructed by summing the number or percentage of nucleotides or amino acids that are identical in each pair of sequences.

single nucleotide polymorphisms (SNPs)   Inherited variations in a single nucleotide base in DNA that differ between individuals.

single-strand binding protein   In DNA replication, a protein that binds to single strands of DNA after they have been separated from each other, keeping the two strands separate for replication.

sink   In plants, any organ that imports the products of photosynthesis, such as roots, developing fruits, and immature leaves. (Contrast with source.)

sinoatrial node (sigh' no ay' tree al) [L. sinus: curve + atrium: chamber]   The pacemaker of the mammalian heart.

siRNAs (small interfering RNAs)   Short, double-stranded RNA molecules used in RNA interference.

sister chromatid   Each of a pair of newly replicated chromatids.

sister groups   Two phylogenetic groups that are each other's closest relatives.

skeletal muscle   A type of muscle tissue characterized by multinucleated cells containing highly ordered arrangements of actin and myosin microfilaments. Also called striated muscle. (Contrast with cardiac muscle, smooth muscle.)

skeletal systems   Organ systems that provide rigid supports against which muscles can pull to create directed movements.

sliding DNA clamp   Protein complex that keeps DNA polymerase bound to DNA during replication.

sliding filament theory   Mechanism of muscle contraction based on the formation and breaking of crossbridges between actin and myosin filaments, causing the filaments to slide together.

slow-wave sleep   A state of deep, restorative sleep characterized by high-amplitude slow waves in the EEG. (Contrast with REM sleep.)

small intestine   The portion of the gut between the stomach and the colon; consists of the duodenum, the jejunum, and the ileum.

small nuclear ribonucleoprotein particle (snRNP)   A complex of an enzyme and a small nuclear RNA molecule, functioning in RNA splicing.

smooth endoplasmic reticulum (SER)   Portion of the endoplasmic reticulum that lacks ribosomes and has a tubular appearance. (Contrast with rough endoplasmic reticulum.)

smooth muscle   Muscle tissue consisting of sheets of mononucleated cells innervated by the autonomic nervous system. (Contrast with cardiac muscle, skeletal muscle.)

sodium–potassium (Na⁺–K⁺) pump   Antiporter responsible for primary active transport; it pumps sodium ions out of the cell and potassium ions into the cell, both against their concentration gradients. Also called a sodium–potassium AT-Pase.

soil horizon   *See* horizon.

solute   A substance that is dissolved in a liquid (solvent) to form a solution.

solute potential   A property of any solution, resulting from its solute contents; it may be zero or have a negative value. The more negative the solute potential, the greater the tendency of the solution to take up water through a differentially permeable membrane. (Contrast with pressure potential, water potential.)

solution   A liquid (the solvent) and its dissolved solutes.

solvent   Liquid in which a substance (solute) is dissolved to form a solution.

somatic cell   [Gk. *soma*: body]   All the cells of the body that are not specialized for reproduction. (Contrast with germ cell.)

somatic mutation   Permanent genetic change in a somatic cell. These mutations affect the individual only; they are not passed on to offspring. (Contrast with germ line mutation.)

somatosensory cortex   The region of the cerebral cortex that receives input from mechanosensors distributed throughout the body.

somatostatin   Peptide hormone made in the hypothalamus that inhibits the release of other hormones from the pituitary and intestine.

somite (so' might)   One of the segments into which an embryo becomes divided longitudinally, leading to the eventual segmentation of the animal as illustrated by the spinal column, ribs, and associated muscles.

Sorenson's index   Mathematical formula that measures beta diversity.

source   In plants, any organ that exports the products of photosynthesis in excess of its own needs, such as a mature leaf or storage organ. (Contrast with sink.)

spatial summation   In the production or inhibition of action potentials in a postsynaptic cell, the interaction of depolarizations and hyperpolarizations produced at different sites on the postsynaptic cell. (Contrast with temporal summation.)

spawning   *See* external fertilization.

speciation (spee' see ay' shun)   The process of splitting one population into two populations that are reproductively isolated from one another.

species (spee' sees) [L. kind]   The base unit of taxonomic classification, consisting of an ancestor–descendant group of populations of evolutionarily closely related, similar organisms. The more narrowly defined "biological species" consists of individuals capable of interbreeding with each other but not with members of other species.

species–area relationship   The relationship between the size of an area and the numbers of species it supports.

species richness   The total number of species living in a region.

specific defenses   Defensive reactions of the vertebrate immune system that are based on the reaction of an antibody to a specific antigen. (Contrast with nonspecific defenses.)

specific heat   The amount of energy that must be absorbed by a gram of a substance to raise its temperature by one degree centigrade. By convention, water is assigned a specific heat of one.

sperm [Gk. *sperma*: seed]   The male gamete.

spermatid   One of the products of the second meiotic division of a primary spermatocyte; four haploid spermatids, which remain connected by cytoplasmic bridges, are produced for each primary spermatocyte that enters meiosis.

spermatogenesis (spur mat' oh jen' e sis) [Gk. *sperma*: seed + *genesis*: source]   Gametogenesis leading to the production of sperm.

spermatogonia   In animals, the diploid progeny of a germ cell in males.

spherical symmetry   The simplest form of symmetry, in which body parts radiate out from a central point such that an infinite number of planes passing through that central point can divide the organism into similar halves.

spicule [L. arrowhead]   A hard, calcareous skeletal element typical of sponges.

spinal reflex   The conversion of afferent to efferent information in the spinal cord without participation of the brain.

sphincter (sfink' ter) [Gk. *sphinkter*: something that binds tightly]   A ring of muscle that can close an orifice, for example, at the anus.

spindle   Array of microtubules emanating from both poles of a dividing cell during mitosis and playing a role in the movement of chromosomes at nuclear division. Named for its shape.

spleen   Organ that serves as a reservoir for venous blood and eliminates old, damaged red blood cells from the circulation.

spliceosome   RNA–protein complex that splices out introns from eukaryotic pre-mRNAs.

splicing   *See* RNA splicing.

spontaneous mutation   A genetic change caused by internal cellular mechanisms, such as an error in DNA replication. (Contrast with induced mutation.)

sporangiophore   A stalked reproductive structure produced by zygospore fungi that extends from a hypha and bears one or many sporangia.

sporangium (spor an' gee um) (plural: sporangia) [Gk. *spora*: seed + *angeion*: vessel or reservoir]   In plants and fungi, any specialized stucture within which one or more spores are formed.

spore [Gk. *spora*: seed]   (1) Any asexual reproductive cell capable of developing into an adult organism without gametic fusion. In plants, haploid spores develop into gametophytes, diploid spores into sporophytes. (2) In prokaryotes, a resistant cell capable of surviving unfavorable periods.

sporocyte   Specialized cells of the diploid sporophyte that will divide by meiosis to produce four haploid spores. Germination of these spores produces the haploid gametophyte.

sporophyte (spor' o fyte) [Gk. *spora*: seed + *phyton*: plant]   In plants and protists with alternation of generations, the diploid phase that produces the spores. (Contrast with gametophyte.)

stabilizing selection   Selection against the extreme phenotypes in a population, so that the intermediate types are favored. (Contrast with disruptive selection.)

stamen (stay' men) [L. *stamen*: thread]   A male (pollen-producing) unit of a flower, usually composed of an anther, which bears the pollen, and a filament, which is a stalk supporting the anther.

starch [O.E. *stearc*: stiff]   A polymer of glucose; used by plants to store energy.

Starling's forces   The two opposing forces responsible for water movement across capillary walls: blood pressure, which squeezes water and small solutes out of the capillaries, and osmotic pressure, which pulls water back into the capillaries.

start codon   The mRNA triplet (AUG) that acts as a signal for the beginning of translation at the ribosome. (Contrast with stop codon.)

stele (steel) [Gk. *stylos*: pillar]   The central cylinder of vascular tissue in a plant stem.

stem   In plants, the organ that holds leaves and/or flowers and transports and distributes materials among the other organs of the plant.

stem cell   In animals, an undifferentiated cell that is capable of continuous proliferation. A stem cell generates more stem cells and a large clone of differentiated progeny cells. (*See also* embryonic stem cell.)

steroid   Any of a family of lipids whose multiple rings share carbons. The steroid cholesterol is an important constituent of membranes; other steroids function as hormones.

sticky ends   On a piece of two-stranded DNA, short, complementary, one-stranded regions produced by the action of a restriction endonuclease. Sticky ends facilitate the joining of segments of DNA from different sources.

stigma [L. *stigma*: mark, brand]   The part of the pistil at the apex of the style that is receptive to pollen, and on which pollen germinates.

stimulus [L. *stimulare*: to goad]   Something causing a response; something in the environment detected by a receptor.

stock   In horticulture, the root or root-bearing stem to which a bud or piece of stem from another plant (the scion) is grafted.

stoma (plural: stomata) [Gk. *stoma*: mouth, opening]   Small opening in the plant epidermis that permits gas exchange; bounded by a pair of guard cells whose osmotic status regulates the size of the opening.

stomatal crypt   In plants, a sunken cavity below the leaf surface in which a stoma is sheltered from the drying effects of air currents.

stop codon   Any of the three mRNA codons that signal the end of protein translation at the ribosome: UAG, UGA, UAA.

stratosphere   The upper part of Earth's atmosphere, above the troposphere; extends from approximately 18 kilometers upward to approximately 50 kilometers above the surface.

stratum (plural strata) [L. *stratos*: layer]   A layer of sedimentary rock laid down at a particular time in the past.

stretch receptor   A modified muscle cell embedded in the connective tissue of a muscle that acts as a mechanoreceptor in response to stretching of that muscle.

striated muscle   *See* skeletal muscle.

strobilus (plural: strobili)   One of several cone-like structures in various groups of plants (including club mosses, horsetails, and conifers) associated with the production and dispersal of reproductive products.

stroma   The fluid contents of an organelle such as a chloroplast or mitochondrion.

structural gene   A gene that encodes the primary structure of a protein not involved in the regulation of gene expression.

structural isomers   Molecules made up of the same kinds and numbers of atoms, in which the atoms are bonded differently.

structural motif   A three-dimensional structural element that is part of a larger molecule. For example, there are four common motifs in DNA-binding proteins: helix-turn-helix, zinc finger, leucine zipper, and helix-loop-helix.

style [Gk. *stylos*: pillar or column]   In the angiosperm flower, a column of tissue extending from the tip of the ovary, and bearing the stigma or receptive surface for pollen at its apex.

sub- [L. under]   A prefix used to designate a structure that lies beneath another or is less than another. For example, subcutaneous (beneath the skin); subspecies.

suberin   A waxlike lipid that is a barrier to water and solute movement across the Casparian strip of the endodermis.

submucosa (sub mew koe' sah)   The tissue layer just under the epithelial lining of the lumen of the digestive tract.

subsoil   The soil horizon lying below the topsoil and above the parent rock (bedrock); the zone of infiltration and accumulation of materials leached from the topsoil. Also called the B horizon.

substrate (sub' strayte)   The molecule or molecules on which an enzyme exerts catalytic action.

substratum (plural: substrata)   The base material on which a sessile organism lives.

succession   The gradual, sequential series of changes in the species composition of a community following a disturbance.

succulence   In plants, possession of fleshy, water-storing leaves or stems; an adaptation to dry environments.

superficial cleavage   A variation of incomplete cleavage in which cycles of mitosis occur without cell division, producing a syncytium (a single cell with many nuclei).

suprachiasmatic nuclei (SCN)   In mammals, two clusters of neurons just above the optic chiasm that act as the master circadian clock.

surface area-to-volume ratio   For any cell, organism, or geometrical solid, the ratio of surface area to volume; this is an important factor in setting an upper limit on the size a cell or organism can attain.

surface tension   The attractive intermolecular forces at the surface of liquid; especially important in water.

surfactant   A substance that decreases the surface tension of a liquid. Lung surfactant, secreted by cells of the alveoli, is mostly phospholipid and decreases the amount of work necessary to inflate the lungs.

survivorship (l_x)   In life tables, the proportion of individuals in a cohort that are alive at age *x*. A graph of this data is a survivorship curve.

suspensor   In the embryos of seed plants, the stalk of cells that pushes the embryo into the endosperm and is a source of nutrient transport to the embryo.

sustainable   Pertaining to the use and management of ecosystems in such a way that humans benefit over the long term from specific ecosystem goods and services without compromising others.

symbiosis (sim' bee oh' sis) [Gk. *sym*: together + *bios*: living]   The living together of two or more species in a prolonged and intimate relationship.

symmetry   Pertaining to an attribute of an animal body in which at least one plane can divide the body into similar, mirror-image halves. (*See* bilateral symmetry, radial symmetry.)

sympathetic nervous system   The division of the autonomic nervous system that works in opposition to the parasympathetic nervous system. (Contrast with parasympathetic nervous system.)

sympatric speciation (sim pat' rik) [Gk. *sym*: same + *patria*: homeland]   Speciation due to reproductive isolation without any physical separation of the subpopulation. (Contrast with allopatric speciation.)

symplast   The continuous meshwork of the interiors of living cells in the plant body, resulting from the presence of plasmodesmata. (Contrast with apoplast.)

symporter   A membrane transport protein that carries two substances in the same direction. (Contrast with antiporter, uniporter.)

synapomorphy   A trait that arose in the ancestor of a phylogenetic group and is present (sometimes in modified form) in all of its members, thus helping to delimit and identify that group. Also called a shared derived trait.

synapse (sin' aps) [Gk. *syn*: together + *haptein*: to fasten]   A specialized type of junction where a neuron meets its target cell (which can be another neuron or some other type of cell) and information in the form of neurotransmitter molecules is exchanged across a synaptic cleft.

synapsis (sin ap' sis)   The highly specific parallel alignment (pairing) of homologous chromosomes during the first division of meiosis.

synergids [Gk. *syn*: together + *ergos*: work]   In angiosperms, the two cells accompanying the egg cell at one end of the megagametophyte.

syngamy   *See* fertilization.

synonymous (silent) substitution   A change of one nucleotide in a sequence to another when that change does not affect the amino acid specified (i.e., UUA → UUG, both specifying leucine). (Contrast with nonsynonymous substitution, missense substitution, nonsense substitution.)

systematics   The scientific study of the diversity and relationships among organisms.

systemic acquired resistance   A general resistance to many plant pathogens following infection by a single agent.

systemic circuit   Portion of the circulatory system by which oxygenated blood from the lungs or gills is distributed throughout the rest of the body and returned to the heart. (Contrast with pulmonary circuit.)

systems biology   The scientific study of an organism as an integrated and interacting system of genes, proteins, and biochemical reactions.

systole (sis' tuh lee) [Gk. *systole*: contraction]   Contraction of a chamber of the heart, driving blood forward in the circulatory system. (Contrast with diastole.)

- T -

T cell   A type of lymphocyte involved in the cellular immune response. The final stages of its development occur in the thymus gland. (Contrast with B cell; *see also* cytotoxic T cell, T-helper cell.)

T cell receptor   A protein on the surface of a T cell that recognizes the antigenic determinant for which the cell is specific.

T-helper cell (T_H)   Type of T cell that stimulates events in both the cellular and humoral immune responses by binding to the antigen on an antigen-presenting cell; target of the HIV-I virus, the agent of AIDS. (Contrast with cytotoxic T cells.)

T tubules   A system of tubules that runs throughout the cytoplasm of a muscle fiber, through which action potentials spread.

taproot system   A root system typical of eudicots consisting of a primary root (*taproot*) that extends downward by tip growth and outward by initiating lateral roots. (Contrast with fibrous root system.)

target cell   A cell with the appropriate receptors to bind and respond to a particular hormone or other chemical mediator.

taste bud   A structure in the epithelium of the tongue that includes a cluster of chemoreceptors innervated by sensory neurons.

TATA box   An eight-base-pair sequence, found about 25 base pairs before the starting point for transcription in many eukaryotic promoters, that binds a transcription factor and thus helps initiate transcription.

taxon (plural: taxa) [Gk. *taxis*: arrange, put in order]   A biological group (typically a species or a clade) that is given a name.

telencephalon   The outer, surrounding structure of the embryonic vertebrate forebrain, which develops into the cerebrum.

telomerase   An enzyme that catalyzes the addition of telomeric sequences lost from chromosomes during DNA replication.

telomeres (tee' lo merz) [Gk. *telos*: end + *meros*: units, segments]   Repeated DNA sequences at the ends of eukaryotic chromosomes.

telophase (tee' lo phase) [Gk. *telos*: end]   The final phase of mitosis or meiosis during which chromosomes became diffuse, nuclear envelopes reform, and nucleoli begin to reappear in the daughter nuclei.

template   A molecule or surface on which another molecule is synthesized in complementary fashion, as in the replication of DNA.

template strand   In double-stranded DNA, the strand that is transcribed to create an RNA transcript that will be processed into a protein. Also refers to a strand of RNA that is used to create a complementary RNA.

temporal summation   In the production or inhibition of action potentials in a postsynaptic cell, the interaction of depolarizations or hyperpolarizations produced by rapidly repeated stimulation of a single point on the postsynaptic cell. (Contrast with spatial summation.)

tendon   A collagen-containing band of tissue that connects a muscle with a bone.

tepal   A sterile, modified, nonphotosynthetic leaf of an angiosperm flower that cannot be distinguished as a petal or a sepal.

termination   In molecular biology, the end of transcription or translation.

terminator   A sequence at the 3′ end of mRNA that causes the RNA strand to be released from the transcription complex.

terrestrial (ter res′ tree al) [L. *terra*: earth]   Pertaining to or living on land. (Contrast with aquatic, marine.)

tertiary structure   In reference to a protein, the relative locations in three-dimensional space of all the atoms in the molecule. The overall shape of a protein. (Contrast with primary, secondary, and quaternary structures.)

test cross   Mating of a dominant-phenotype individual (who may be either heterozygous or homozygous) with a homozygous-recessive individual.

testis (tes′ tis) (plural: testes) [L. *testis*: witness]   The male gonad; the organ that produces the male gametes.

tetanus [Gk. *tetanos*: stretched]   (1) A state of sustained maximal muscular contraction caused by rapidly repeated stimulation. (2) In medicine, an often fatal disease ("lockjaw") caused by the bacterium *Clostridium tetani*.

tetrad [Gk. *tettares*: four]   During prophase I of meiosis, the association of a pair of homologous chromosomes or four chromatids.

thalamus [Gk. *thalamos*: chamber]   A region of the vertebrate forebrain; involved in integration of sensory input.

theory [Gk. *theoria*: analysis of facts]   A far-reaching explanation of observed facts that is supported by such a wide body of evidence, with no significant contradictory evidence, that it is scientifically accepted as a factual framework. Examples are Newton's theory of gravity and Darwin's theory of evolution. (Contrast with hypothesis.)

thermoneutral zone [Gk. *thermos*: temperature]   The range of temperatures over which an endotherm does not have to expend extra energy to thermoregulate.

thermophile (ther′ muh fyle)[Gk. *thermos*: temperature + *philos*: loving]   An organism that lives exclusively in hot environments.

thoracic cavity [Gk. *thorax*: breastplate]   The portion of the mammalian body cavity bounded by the ribs, shoulders, and diaphragm. Contains the heart and the lungs.

thoracic duct   The connection between the lymphatic system and the circulatory system.

threshold   The level of depolarization that causes an electrically excitable membrane to fire an action potential.

thrombus (throm′ bus) [Gk. *thrombos*: clot]   A blood clot that forms within a blood vessel and remains attached to the wall of the vessel. (Contrast with embolus.)

thylakoid (thigh la koid) [Gk. *thylakos*: sack or pouch]   A flattened sac within a chloroplast. Thylakoid membranes contain all of the chlorophyll in a plant, in addition to the electron carriers of photophosphorylation. Thylakoids stack to form grana.

thymine (T)   Nitrogen-containing base found in DNA.

thymus [Gk. *thymos*: warty]   A ductless, glandular lymphoid tissue, involved in development of the immune system of vertebrates. In humans, the thymus degenerates during puberty.

thyroid gland [Gk. *thyreos*: door-shaped]   A two-lobed gland in vertebrates. Produces the hormone thyroxin.

thyrotropin   Hormone produced by the anterior pituitary that stimulates the thyroid gland to produce and release thyroxine. Also called thyroid-stimulating hormone (TSH).

thyrotropin-releasing hormone (TRH)   Hormone produced by the hypothalamus that stimulates the anterior pituitary to release thyrotropin.

thyroxine   Hormone produced by the thyroid gland; controls many metabolic processes.

tidal volume   The amount of air that is exchanged during each breath when a person is at rest.

tight junction   A junction between epithelial cells in which there is no gap between adjacent cells.

tissue   A group of similar cells organized into a functional unit; usually integrated with other tissues to form part of an organ.

tissue system   In plants, any of three organized groups of tissues—dermal tissue, vascular tissue, and ground tissue—that are established during embryogenesis and have distinct functions.

titin   A protein that holds bundles of myosin filaments in a centered position within the sarcomeres of muscle cells. The largest protein in the human body.

tonoplast   The membrane of the plant central vacuole.

topsoil   The uppermost soil horizon; contains most of the organic matter of soil, but may be depleted of most mineral nutrients by leaching. Also called the A horizon.

totipotent [L. *toto*: whole, entire + *potens*: powerful]   Possessing all the genetic information and other capacities necessary to form an entire individual. (Contrast with multipotent, pluripotent.)

trachea (tray′ kee ah) [Gk. *trakhoia*: tube]   A tube that carries air to the bronchi of the lungs of vertebrates. When plural (*tracheae*), refers to the major airways of insects.

tracheary element   Either of two types of xylem cells—tracheids and vessel elements—that undergo apoptosis before assuming their transport function.

tracheid (tray′ kee id)   A type of tracheary element found in the xylem of nearly all vascular plants, characterized by tapering ends and walls that are pitted but not perforated. (Contrast with vessel element.)

trade-off   The relationship between the fitness benefits conferred by an adaptation and the fitness costs it imposes. For an adaptation to be favored by natural selection, the benefits must exceed the costs.

trait   In genetics, a specific form of a character: eye color is a character; brown eyes and blue eyes are traits. (Contrast with character.)

transcription   The synthesis of RNA using one strand of DNA as a template.

transcription factors   Proteins that assemble on a eukaryotic chromosome, allowing RNA polymerase II to perform transcription.

transduction   (1) Transfer of genes from one bacterium to another by a bacteriophage. (2) In sensory cells, the transformation of a stimulus (e.g., light energy, sound pressure waves, chemical or electrical stimulants) into action potentials.

transfection   Insertion of recombinant DNA into animal cells.

transfer RNA (tRNA)   A family of double-stranded RNA molecules. Each tRNA carries a specific amino acid and anticodon that will pair with the complementary codon in mRNA during translation.

transformation   (1) A mechanism for transfer of genetic information in bacteria in which pure DNA from a bacterium of one genotype is taken in through the cell surface of a bacterium of a different genotype and incorporated into the chromosome of the recipient cell. (2) Insertion of recombinant DNA into a host cell.

transgenic   Containing recombinant DNA incorporated into the genetic material.

translation   The synthesis of a protein (polypeptide). Takes place on ribosomes, using the information encoded in messenger RNA.

translocation   (1) In genetics, a rare mutational event that moves a portion of a chromosome to a new location, generally on a nonhomologous chromosome. (2) In vascular plants, movement of solutes in the phloem.

transmembrane protein   An integral membrane protein that spans the phospholipid bilayer.

transpiration [L. *spirare*: to breathe]   The evaporation of water from plant leaves and stem, driven by heat from the sun, and providing the motive force to raise water (plus mineral nutrients) from the roots.

transpiration–cohesion–tension mechanism   Theoretical basis for water movement in plants: evaporation of water from cells within leaves (transpiration) causes an increase ion surface tension, pulling water up through the xylem. Cohesion of water occurs because of hydrogen bonding.

transposable element   A segment of DNA that can move to, or give rise to copies at, another locus on the same or a different chromosome.

transposon   Mobile DNA segment that can insert into a chromosome and cause genetic change.

triglyceride   A simple lipid in which three fatty acids are combined with one molecule of glycerol.

triploblastic   Having three cell layers.

trisomic   Containing three rather than two members of a chromosome pair.

tRNA   *See* transfer RNA.

trochophore (troke′ o fore) [Gk. *trochos*: wheel + *phoreus*: bearer]   A radially symmetrical larval form typical of annelids and mollusks, distinguished by a wheel-like band of cilia around the middle.

trophic cascade   The progression over successively lower trophic levels of the indirect effects of a predator.

trophic level [Gk *trophes*: nourishment] A group of organisms united by obtaining their energy from the same part of the food web of a biological community.

trophoblast [Gk *trophes*: nourishment + *blastos*: sprout] At the 32-cell stage of mammalian development, the outer group of cells that will become part of the placenta and thus nourish the growing embryo. (Contrast with inner cell mass.)

tropic hormones Hormones produced by the anterior pituitary that control the secretion of hormones by other endocrine glands.

tropomyosin [troe poe my' oh sin] One of the three protein components of an actin filament; controls the interactions of actin and myosin necessary for muscle contraction.

troponin One of the three components of an actin filament; binds to actin, tropomyosin, and Ca^{2+}.

troposphere The lowest atmospheric zone, reaching upward from the Earth's surface approximately 10–17 km. Zone in which virtually all water vapor is located.

true-breeding A genetic cross in which the same result occurs every time with respect to the trait(s) under consideration, due to homozygous parents.

trypsin A protein-digesting enzyme. Secreted by the pancreas in its inactive form (trypsinogen), it becomes active in the duodenum of the small intestine.

tubulin A protein that polymerizes to form microtubules.

tumor [L. *tumor*: a swollen mass] A disorganized mass of cells. Malignant tumors spread to other parts of the body.

tumor necrosis factor A family of cytokines (growth factors) that causes cell death and is involved in inflammation.

tumor suppressor A gene that codes for a protein product that inhibits cell proliferation; inactive in cancer cells. (Contrast with oncogene.)

turgor pressure [L. *turgidus*: swollen] *See* pressure potential.

turnover In freshwater ecosystems, vertical movements of water that bring nutrients and dissolved CO_2 to the surface and O_2 to deeper water.

tympanic membrane [Gk. *tympanum*: drum] The eardrum.

- U -

ubiquinone (yoo bic' kwi known) [L. *ubique*: everywhere] A mobile electron carrier of the mitochondrial respiratory chain. Similar to plastoquinone found in chloroplasts.

ubiquitin A small protein that is covalently linked to other cellular proteins identified for breakdown by the proteosome.

ultimate cause In ethology, the evolutionary processes that produced an animal's capacity and tendency to behave in particular ways. (Contrast with proximate cause.)

uniporter [L. *unus*: one + *portal*: doorway] A membrane transport protein that carries a single substance in one direction. (Contrast with antiporter, symporter.)

unsaturated fatty acid A fatty acid whose hydrocarbon chain contains one or more double bonds. (Contrast with saturated fatty acid.)

upregulation A process by which the abundance of receptors for a hormone increases when hormone secretion is suppressed. (Contrast with downregulation.)

upwelling zones Areas of the ocean where cool, nutrient-rich water from deeper layers rises to the surface.

uracil (U) A pyrimidine base found in nucleotides of RNA.

urea A compound that is the main excreted form of nitrogen by many animals, including mammals.

ureotelic Pertaining to an organism in which the final product of the breakdown of nitrogen-containing compounds (primarily proteins) is urea. (Contrast with ammonotelic, uricotelic.)

ureter (your' uh tur) Long duct leading from the vertebrate kidney to the urinary bladder or the cloaca.

urethra (you ree' thra) In most mammals, the canal through which urine is discharged from the bladder and which serves as the genital duct in males.

uric acid A compound that serves as the main excreted form of nitrogen in some animals, particularly those which must conserve water, such as birds, insects, and reptiles.

uricotelic Pertaining to an organism in which the final product of the breakdown of nitrogen-containing compounds (primarily proteins) is uric acid. (Contrast with ammonotelic, ureotelic.)

urinary bladder A structure in which urine is stored until it can be excreted to the outside of the body.

urine (you' rin) In vertebrates, the fluid waste product containing the toxic nitrogenous byproducts of protein and amino acid metabolism.

uterine cycle In human females, the monthly cycle of events by which the endometrium is prepared for the arrival of a blastocyst. (Contrast with ovarian cycle).

uterus (yoo' ter us) [L. *utero*: womb] A specialized portion of the female reproductive tract in mammals that receives the fertilized egg and nurtures the embryo in its early development. Also called the womb.

- V -

vaccination Injection of virus or bacteria or their proteins into the body, to induce immunization. The injected material is usually attenuated (weakened) before injection.

vacuole (vac' yew ole) Membrane-enclosed organelle in plant cells that can function for storage, water concentration for turgor, or hydrolysis of stored macromolecules.

vagina (vuh jine' uh) [L. *sheath*] In female animals, the entry to the reproductive tract.

van der Waals forces Weak attractions between atoms resulting from the interaction of the electrons of one atom with the nucleus of another. This type of attraction is about one-fourth as strong as a hydrogen bond.

variable region The portion of an immunoglobulin molecule or T-cell receptor that includes the antigen-binding site and is responsible for its specificity. (Contrast with constant region.)

vas deferens (plural: vasa deferentia) Duct that transfers sperm from the epididymis to the urethra.

vasa recta Blood vessels that parallel the loops of Henle and the collecting ducts in the renal medulla of the kidney.

vascular (vas' kew lar) [L. *vasculum*: a small vessel] Pertaining to organs and tissues that conduct fluid, such as blood vessels in animals and xylem and phloem in plants.

vascular bundle In vascular plants, a strand of vascular tissue, including xylem and phloem as well as thick-walled fibers.

vascular cambium (kam' bee um) [L. *cambiare*: to exchange] In plants, a lateral meristem that gives rise to secondary xylem and phloem.

vascular tissue system The transport system of a vascular plant, consisting primarily of xylem and phloem.

vasopressin *See* antidiuretic hormone.

vector (1) An agent, such as an insect, that carries a pathogen affecting another species. (2) A plasmid or virus that carries an inserted piece of DNA into a bacterium for cloning purposes in recombinant DNA technology.

vegetal hemisphere The lower portion of some animal eggs, zygotes, and embryos, in which the dense nutrient yolk settles. The *vegetal pole* is to the very bottom of the egg or embyro. (Contrast with animal hemisphere.)

vegetative Nonreproductive, nonflowering, or asexual.

vegetative meristem An apical meristem that produces leaves.

vegetative reproduction Asexual reproduction through the modification of stems, leaves, or roots.

vein [L. *vena*: channel] A blood vessel that returns blood to the heart. (Contrast with artery.)

ventral [L. *venter*: belly, womb] Toward or pertaining to the belly or lower side. (Contrast with dorsal.)

ventricle A muscular heart chamber that pumps blood through the lungs or through the body.

venule A small blood vessel draining a capillary bed that joins others of its kind to form a vein. (Contrast with arteriole.)

vernalization [L. *vernalis*: spring] Events occurring during a required chilling period, leading eventually to flowering.

vertebral column [L. *vertere*: to turn] The jointed, dorsal column that is the primary support structure of vertebrates.

very low-density lipoproteins (VLDLs) Lipoproteins that consist mainly of triglyceride fats, which they transport to fat cells in adipose tissues throughout the body; associated with excessive fat deposition and high risk for cardiovascular disease.

vesicle Within the cytoplasm, a membrane-enclosed compartment that is associated with other organelles; the Golgi complex is one example.

vessel element A type of tracheary element with perforated end walls; found only in angiosperms. (Contrast with tracheid.)

vestibular system (ves tib' yew lar) [L. *vestibulum*: an enclosed passage] Structures within the inner ear that sense changes in position or momentum of the head, affecting balance and motor skills.

vicariant event (vye care' ee unt) [L. *vicus*: change] The splitting of a taxon's range by the imposition of some barrier to dispersal.

villus (vil' lus) (plural: villi) [L. *villus*: shaggy hair or beard] A hairlike projection from a membrane; for example, from many gut walls.

virion (veer' e on) The virus particle, the minimum unit capable of infecting a cell.

virulence [L. *virus*: poison, slimy liquid] The ability of a pathogen to cause disease and death.

virus Any of a group of ultramicroscopic particles constructed of nucleic acid and protein (and, sometimes, lipid) that require living cells in order to reproduce. Viruses evolved multiple times from different cellular species.

vital capacity The maximum capacity for air exchange in one breath; the sum of the tidal volume and the inspiratory and expiratory reserve volumes.

vitamin [L. *vita*: life] An organic compound that an organism cannot synthesize, but nevertheless requires in small quantities for normal growth and metabolism.

vitelline envelope The inner, proteinaceous protective layer of a sea urchin egg.

viviparity (vye vi par' uh tee) Reproduction in which fertilization of the egg and development of the embryo occur inside the mother's body. (Contrast with oviparity.)

vivipary Premature germination in plants.

voltage-gated channel A type of gated channel that opens or closes when a certain voltage exists across the membrane in which it is inserted.

vomeronasal organ (VNO) Chemosensory structure embedded in the nasal epithelium of amphibians, reptiles, and many mammals. Often specialized for detecting pheromones.

- W -

water potential In osmosis, the tendency for a system (a cell or solution) to take up water from pure water through a differentially permeable membrane. Water flows toward the system with a more negative water potential. (Contrast with solute potential, pressure potential.)

water vascular system In echinoderms, a network of water-filled canals that functions in gas exchange, locomotion, and feeding.

wavelength The distance between successive peaks of a wave train, such as electromagnetic radiation.

weathering The mechanical and chemical processes by which rocks are broken down into soil particles.

Wernicke's area A region in the temporal lobe of the human brain that is involved with the sensory aspects of language.

white blood cells Cells in the blood plasma that play defensive roles in the immune system. Also called leukocytes.

white matter In the central nervous system, tissue that is rich in axons. (Contrast with gray matter.)

wild-type Geneticists' term for standard or reference type. Deviants from this standard, even if the deviants are found in the wild, are usually referred to as mutant. (Note that this terminology is not usually applied to human genes.)

wood Secondary xylem tissue.

- X -

xerophyte (zee' row fyte) [Gk. *xerox*: dry + *phyton*: plant] A plant adapted to an environment with limited water supply.

xylem (zy' lum) [Gk. *xylon*: wood] In vascular plants, the tissue that conducts water and minerals; xylem consists, in various plants, of tracheids, vessel elements, fibers, and other highly specialized cells.

- Y -

yolk [M.E. *yolke*: yellow] The stored food material in animal eggs, rich in protein and lipids.

yolk sac In reptiles, birds, and mammals, the extraembryonic membrane that forms from the endoderm of the hypoblast; it encloses and digests the yolk.

- Z -

zeaxanthin A blue-light receptor involved in the opening of plant stomata.

zona pellucida A jellylike substance that surrounds the mammalian ovum when it is released from the ovary.

zoospore (zoe' o spore) [Gk. *zoon*: animal + *spora*: seed] In algae and fungi, any swimming spore. May be diploid or haploid.

zygote (zye' gote) [Gk. *zygotos*: yoked] The cell created by the union of two gametes, in which the gamete nuclei are also fused. The earliest stage of the diploid generation.

zymogen The inactive precursor of a digestive enzyme; secreted into the lumen of the gut, where a protease cleaves it to form the active enzyme.

Illustration Credits

Cover: © Dr. Merlin D. Tuttle/Photo Researchers, Inc.
Frontispiece: © Art Wolfe/www.artwolfe.com.

Photographs appearing behind chapter numbers:
Part 1 *HMS Beagle*: Painting by Ronald Dean, reproduced by permission of the artist and Richard Johnson, Esquire.
Part 2 *Plant cells*: © Ed Reschke/Peter Arnold Inc.
Part 3 *Stoma*: © Andrew Syred/SPL/Photo Researchers, Inc.
Part 4 *Anaphase*: © Nasser Rusan.
Part 5 *Sheep*: © Roddy Field, the Roslin Institute.
Part 6 *Trilobite*: Courtesy of the Amherst College Museum of Natural History and the Trustees of Amherst College.
Part 7 *Diatoms*: © Scenics & Science/Alamy.
Part 8 *Flowers*: © Ed Reschke/Peter Arnold Inc.
Part 9 *Leopard*: Courtesy of Andrew D. Sinauer.
Part 10 *Hummingbird*: © Yufeng Zhou/istockphoto.com.

Table of Contents

Chapter 1 *Frogs*: © Pete Oxford/Minden Pictures. *T. Hayes*: © Pamela S. Turner. 1.1A: © Eye of Science/SPL/Photo Researchers, Inc. 1.1B: Science Photo Library/Photolibrary.com. 1.1C: © Steve Gschmeissner/Photo Researchers, Inc. 1.1D: © Frans Lanting. 1.1E: © Glen Threlfo/Auscape/Minden Pictures. 1.1F: © Piotr Naskrecki/Minden Pictures. 1.1G: © Tui De Roy/Minden Pictures. 1.2A: From R. Hooke, 1664. *Micrographia*. 1.2B: © John Durham/SPL/Photo Researchers, Inc. 1.2C: © Biophoto Associates/Photo Researchers, Inc. 1.3 *Maple*: © Simon Colmer & Abby Rex/Alamy. 1.3 *Spruce*: David McIntyre. 1.3 *Lily pad*: © Pete Oxford/Naturepl.com. 1.3 *Cucumber*: David McIntyre. 1.3 *Pitcher plant*: © Nick Garbutt/Naturepl.com. 1.5A: © Frans Lanting. 1.5B: © Stefan Huwiler/Rolfnp/Alamy. 1.6 *Organism*: © Nico Smit/istockphoto.com. 1.6

Population: © blickwinkel/Alamy. 1.6 *Community*: © Georgie Holland/AGE Fotostock. 1.6 *Biosphere*: Courtesy of NASA. 1.7: © P&R Fotos/AGE Fotostock. 1.9: © Michael Abbey/Visuals Unlimited, Inc. 1.11: Courtesy of Wayne Whippen. 1.13: From T. Hayes et al., 2003. *Environ. Health Perspect.* 111: 568.

Chapter 2 *Hair*: © Steve Gschmeissner/Photo Researchers, Inc. *Barber*: © Digital Vision/Alamy. 2.3: From N. D. Volkow et al., 2001. *Am. J. Psychiatry* 158: 377. 2.14: © Pablo H Caridad/Shutterstock. 2.15: © Denis Miraniuk/Shutterstock.

Chapter 3 *T. Rex*: © The Natural History Museum, London. *Chicken*: David McIntyre. 3.8: Data from PDB 1IVM. T. Obita, T. Ueda, & T. Imoto, 2003. *Cell. Mol. Life Sci.* 60: 176. 3.10A: Data from PDB 2HHB. G. Fermi et al., 1984. *J. Mol. Biol.* 175: 159. 3.16C *left*: © Biophoto Associates/Photo Researchers, Inc. 3.16C *middle*: © Ken Wagner/Visuals Unlimited. 3.16C *right*: © CNRI/SPL/Photo Researchers, Inc. 3.17 *Cartilage*: © Robert Brons/Biological Photo Service. 3.17 *Beetle*: © Scott Bauer/USDA.

Chapter 4 *Phoenix*: Courtesy of NASA/JPL/UA/Lockheed Martin. *Ice*: Courtesy of NASA/JPL-Caltech/University of Arizona/Texas A&M University. 4.4: Data from S. Arnott & D. W. Hukins, 1972. *Biochem. Biophys. Res. Commun.* 47(6): 1504. 4.8: Courtesy of the Argonne National Laboratory. 4.13B: Courtesy of Janet Iwasa, Szostak group, MGH/Harvard. 4.14: © Stanley M. Awramik/Biological Photo Service. 4.14 *inset*: © Dennis Kunkel Microscopy, Inc.

Chapter 5 *Heart cell*: © Roger J. Bick & Brian J. Poindexter/UT-Houston Medical School/Photo Researchers, Inc. *Surgery*: © The Stock Asylum, LLC/Alamy. 5.1: After N. Campbell, 1990. *Biology*, 2nd Ed., Benjamin Cummings. 5.1 *Protein*: Data from PDB 1IVM. T. Obita, T. Ueda, & T. Imoto, 2003. *Cell. Mol. Life Sci.* 60: 176. 5.1 *T4*: © Dept. of Microbiology, Biozentrum/SPL/Photo Researchers, Inc. 5.1 *Bacterium*: © Jim Biddle/Centers for Disease Control. 5.1 *Plant cells*: © Michael Eichelberger/Visuals Unlimited, Inc. 5.1 *Frog egg*: David McIntyre. 5.1 *Bird*: © Steve Byland/Shutterstock. 5.1 *Baby*: Courtesy of Sebastian Grey Miller. 5.3 *Light microscope*: © Radu Razvan/Shutterstock. 5.3 *Bright-field*: Courtesy of the IST Cell Bank, Genoa. 5.3 *Phase-contrast*: © Michael W. Davidson, Florida State U. 5.3 *Stained*: © Richard J. Green/SPL/Photo Researchers, Inc. 5.3 *Confocal*: © Dr. Gopal Murti/SPL/Photo Researchers, Inc. 5.3 *Electron microscope*: © Sinclair Stammers/Photo Researchers, Inc. 5.3 *TEM*: © Dr. Gopal Murti/Visuals Unlimited. 5.3 *SEM*: © K. R. Porter/SPL/Photo Researchers, Inc. 5.3 *Freeze-fracture*: © D. W. Fawcett/Photo Researchers, Inc. 5.4: © J. J. Cardamone Jr. & B. K. Pugashetti/Biological Photo Service. 5.5A: © Dennis Kunkel Microscopy, Inc. 5.5B: Courtesy of David DeRosier, Brandeis U. 5.7 *Mitochondrion*:

© K. Porter, D. Fawcett/Visuals Unlimited. 5.7 *Cytoskeleton*: © Don Fawcett, John Heuser/Photo Researchers, Inc. 5.7 *Nucleolus*: © Richard Rodewald/Biological Photo Service. 5.7 *Peroxisome*: © E. H. Newcomb & S. E. Frederick/Biological Photo Service. 5.7 *Cell wall*: © Biophoto Associates/Photo Researchers, Inc. 5.7 *Ribosome*: From M. Boublik et al., 1990. *The Ribosome*, p. 177. Courtesy of American Society for Microbiology. 5.7 *Centrioles*: © Barry F. King/Biological Photo Service. 5.7 *Plasma membrane*: Courtesy of J. David Robertson, Duke U. Medical Center. 5.7 *Rough ER*: © Don Fawcett/Science Source/Photo Researchers, Inc. 5.7 *Smooth ER*: © Don Fawcett, D. Friend/Science Source/Photo Researchers, Inc. 5.7 *Chloroplast*: © W. P. Wergin, E. H. Newcomb/Biological Photo Service. 5.7 *Golgi apparatus*: Courtesy of L. Andrew Staehelin, U. Colorado. 5.8: © D. W. Fawcett/Photo Researchers, Inc. 5.9A: © Barry King, U. California, Davis/Biological Photo Service. 5.9B: © Biophoto Associates/Science Source/Photo Researchers, Inc. 5.10: © B. Bowers/Photo Researchers, Inc. 5.11: © Sanders/Biological Photo Service. 5.12: © K. Porter, D. Fawcett/Visuals Unlimited. 5.13 *left*: © W. P. Wergin, E. H. Newcomb/Biological Photo Service. 5.13 *right*: © W. P. Wergin/Biological Photo Service. 5.14A: © Michael Eichelberger/Visuals Unlimited, Inc. 5.14B: © Ed Reschke/Peter Arnold Inc. 5.14C: © Gerald & Buff Corsi/Visuals Unlimited. 5.15A: David McIntyre. 5.15A *inset*: © Richard Green/Photo Researchers, Inc. 5.15B: David McIntyre. 5.15B *inset*: Courtesy of R. R. Dute. 5.16: Courtesy of M. C. Ledbetter, Brookhaven National Laboratory. 5.17: Courtesy of Vic Small, Austrian Academy of Sciences, Salzburg, Austria. 5.19: Courtesy of N. Hirokawa. 5.20A *upper*: © SPL/Photo Researchers, Inc. 5.20A *lower*, 5.20B: © W. L. Dentler/Biological Photo Service. 5.22: From N. Pollock et al., 1999. *J. Cell Biol.* 147: 493. Courtesy of R. D. Vale. 5.23: © Michael Abbey/Visuals Unlimited. 5.24: © Biophoto Associates/Photo Researchers, Inc. 5.25 *left*: Courtesy of David Sadava. 5.25 *upper right*: From J. A. Buckwalter & L. Rosenberg, 1983. *Coll. Rel. Res.* 3: 489. Courtesy of L. Rosenberg. 5.25 *lower right*: © J. Gross, Biozentrum/SPL/Photo Researchers, Inc. 5.27: Courtesy of Noriko Okamoto and Isao Inouye.

Chapter 6 *Patient*: From Alzheimer, A. 1906. Über einen eigenartigen schweren Erkrankungsprozess der Hirnrinde. *Neurologisches Centralblatt* 23: 1129. *Plaques*: © G. W. Willis/Photolibrary.com. 6.2: After L. Stryer, 1981. *Biochemistry*, 2nd Ed., W. H. Freeman. 6.4: © D. W. Fawcett/Photo Researchers, Inc. 6.7A: Courtesy of D. S. Friend, U. California, San Francisco. 6.7B: Courtesy of Darcy E. Kelly, U. Washington. 6.7C: Courtesy of C. Peracchia. 6.10A *left*: © Stanley Flegler/Visuals Unlimited, Inc. 6.10A *center*: © David M. Phillips/Photo Researchers, Inc. 6.10B *left*: © Ed Reschke/Peter Arnold Inc. 6.13: From G. M. Preston et al., 1992. *Science* 256: 385. 6.19: From M. M. Perry, 1979. *J. Cell Sci.* 39: 26.

Audubon Society Collection/Photo Researchers, Inc.

Chapter 24 *Electric fish*: © Jane Burton/Naturepl.com. *Ray*: © David Fleetham/Alamy. 24.3: *Rice*: data from PDB 1CCR. H. Ochi et al., 1983. *J. Mol. Biol.* 166: 407. *Tuna*: data from PDB 5CYT. T. Takano, 1984. 24.4: From P. B. Rainey & M. Travisano, 1998. *Nature* 394: 69. © Macmillan Publishers Ltd. 24.7A *Langur*: © blickwinkel/Alamy. 24.7A *Longhorn*: Courtesy of David Hillis. 24.7B: © M. Graybill/J. Hodder/Biological Photo Service.

Chapter 25 *Meganeura*: © Graham Cripps/NHMPL. *Modern dragonfly*: © Natasha Litova/istockphoto.com. *Grand Canyon*: © Tim Fitzharris/Minden Pictures. 25.4A: Photos courtesy of P.F. Hoffman, Geological Survey of Canada. 25.4B: © Robin Smith/Photolibrary.com. 25.8: © Martin Bond/SPL/Photo Researchers, Inc. 25.9: David McIntyre. 25.11 *left*: © Ken Lucas/Visuals Unlimited. 25.11 *center*: © The Natural History Museum, London. 25.11 *right*: Courtesy of Martin Smith. 25.12 *Cambrian*: © John Sibbick/NHMPL. 25.12 *Marella*: Courtesy of the Amherst College Museum of Natural History, The Trustees of Amherst College. 25.12 *Ottoia*: © Alan Sirulnikoff/Photo Researchers, Inc. 25.12 *Anomalocaris*: © Kevin Schafer/Alamy. 25.12 *Devonian*: © Tom McHugh/Field Museum, Chicago/Photo Researchers, Inc. 25.12 *Codiacrinus*: Courtesy of the Amherst College Museum of Natural History, The Trustees of Amherst College. 25.12 *Phacops*: Courtesy of the Amherst College Museum of Natural History, The Trustees of Amherst College. 25.12 *Eridophyllum*: © Mark A. Schneider/Photo Researchers, Inc. 25.12 *Nautiloid*: © The Natural History Museum, London. 25.12 *Permian*: © John Sibbick/NHMPL. 25.12 *Cacops*: © Albert Copley/Visuals Unlimited, Inc. 25.12 *Walchia*: © The Natural History Museum, London. 25.12 *Triassic*: © OSF/Photolibrary.com. 25.12 *Ferns*: © Ken Lucas/Visuals Unlimited. 25.12 *Coelophysis*: © Ken Lucas/Visuals Unlimited. 25.12 *Cretaceous*: © OSF/Photolibrary.com. 25.12 *Gryposaurus*: Courtesy of the Amherst College Museum of Natural History, The Trustees of Amherst College. 25.12 *Magnolia*: © The Natural History Museum, London. 25.12 *Tertiary*: © Publiphoto/Photo Researchers, Inc. 25.12 *Hyracotherium*: Courtesy of the Amherst College Museum of Natural History, The Trustees of Amherst College. 25.12 *Plesiadapis*: © The Natural History Museum, London. 25.13: Courtesy of Conrad C. Labandeira, Department of Paleobiology, National Museum of Natural History, Smithsonian Institution.

Chapter 26 *Antarctica*: Courtesy of Emily Gercke. *Yellowstone*: Courtesy of Jim Peaco/National Park Service. 26.2: © Dennis Kunkel Microscopy, Inc. 26.3B: © Steve Gschmeissner/Photo Researchers, Inc. 26.4: From F. Balagaddé et al., 2005. *Science* 309: 137. Courtesy of Frederick Balagaddé. 26.5A: © David M. Phillips/Visuals Unlimited. 26.5B: Courtesy of the CDC. 26.6A: © J. A. Breznak & H. S. Pankratz/Biological Photo Service. 26.6B: © J. Robert Waaland/Biological Photo Service. 26.7: © USDA/Visuals Unlimited. 26.8: © Steven Haddock and Steven Miller. 26.12: Courtesy of David Cox/CDC. 26.13: Courtesy of Randall

C. Cutlip. 26.14: © David Phillips/Visuals Unlimited. 26.15A: © Paul W. Johnson/Biological Photo Service. 26.15B: © H. S. Pankratz/Biological Photo Service. 26.15C: © Bill Kamin/Visuals Unlimited. 26.16: © Dr. Kari Lounatmaa/Photo Researchers, Inc. 26.17: © Dr. Gary Gaugler/Visuals Unlimited. 26.18: © Michael Gabridge/Visuals Unlimited. 26.20: © Geoff Kidd/Photo Researchers, Inc. 26.21: From K. Kashefi & D. R. Lovley, 2003. *Science* 301: 934. Courtesy of Kazem Kashefi. 26.23: © David Sanger Photography/Alamy. 26.24: From H. Huber et al., 2002. *Nature* 417: 63. © Macmillan Publishers Ltd. Courtesy of Karl O. Stetter. 26.25A: © Science Photo Library RF/Photolibrary.com. 26.25B, C: © Russel Kightley/SPL/Photo Researchers, Inc. 26.25D: © Science Photo Library RF/Photolibrary.com. 26.25E: animate4.com ltd./Photo Researchers, Inc. 26.25F: © Russell Kightley/Photo Researchers, Inc. 26.26: © Nigel Cattlin/Alamy.

Chapter 27 *Bug*: © Martin Dohrn/Photo Researchers, Inc. *Leishmania*: © Dennis Kunkel Microscopy, Inc. 27.4: © Michael Abbey/Photo Researchers, Inc. 27.7A: © Wim van Egmond/Visuals Unlimited. 27.7B: © Science Photo Library RF/Photolibrary.com. 27.8A: © Georgette Douwma/Naturepl.com. 27.8B: © David Patterson, Linda Amaral Zettler, Mike Peglar, & Tom Nerad/micro*scope. 27.9: © London School of Hygiene/SPL/Photo Researchers, Inc. 27.10A: © Bill Bachman/Photo Researchers, Inc. 27.10B: © Markus Geisen/NHMPL. 27.11: © Science Photo Library RF/Photolibrary.com. 27.16: © Dennis Kunkel Microscopy, Inc. 27.17A: © SPL/Photo Researchers, Inc. 27.17B: © Dennis Kunkel Microscopy, Inc. 27.17C: © Paul W. Johnson/Biological Photo Service. 27.17D: © Steve Gschmeissner/Photo Researchers, Inc. 27.19: © Scenics & Science/Alamy. 27.20A: © Marevision/AGE Fotostock. 27.20B: © Larry Jon Friesen. 27.20C: © J. N. A. Lott/Biological Photo Service. 27.20D: © Gerald & Buff Corsi/Visuals Unlimited, Inc. 27.21: © James W. Richardson/Visuals Unlimited. 27.22A: © Wim van Egmond/Visuals Unlimited, Inc. 27.22B: © Doug Sokell/Visuals Unlimited, Inc. 27.23A: © Carolina Biological/Visuals Unlimited. 27.23B: © Marevision/AGE Fotostock. 27.24A: © J. Paulin/Visuals Unlimited. 27.24B: © Dr. David M. Phillips/Visuals Unlimited. 27.26: © Andrew Syred/SPL/Photo Researchers, Inc. 27.27A: © William Bourland/micro*scope. 27.27B: © David Patterson & Aimlee Laderman/micro*scope. 27.28A: © Eye of Science/Photo Researchers, Inc. 27.29: Courtesy of R. Blanton and M. Grimson.

Chapter 28 *Silurian*: © Richard Bizley/Photo Researchers, Inc. *Rainforest*: © Photo Resource Hawaii/Alamy. 28.2A: © Bob Gibbons/OSF/Photolibrary.com. 28.2B: © Larry Mellichamp/Visuals Unlimited. 28.4: © J. Robert Waaland/Biological Photo Service. 28.6: © U. Michigan Exhibit Museum. 28.8A: Courtesy of the Biology Department Greenhouses, U. Massachusetts, Amherst. 28.9: After C. P. Osborne et al., 2004. *PNAS* 101: 10360. 28.11A: David McIntyre. 28.11C: © Harold Taylor/Photolibrary.com. 28.12A: © mediacolor's/Alamy. 28.12B: © Ed Reschke/Peter Arnold, Inc. 28.13A: © Dr. John D. Cunningham/Visuals Unlimited, Inc. 28.13B: © Danilo Donadoni/AGE Fotostock. 28.14: © Daniel

Vega/AGE Fotostock. 28.15A: David McIntyre. 28.15B: © Carolina Biological/Visuals Unlimited. 28.16A: © Bjorn Svensson/SPL/Photo Researchers, Inc. 28.16B: © J. N. A. Lott/Biological Photo Service. 28.17: Courtesy of the Biology Department Greenhouses, U. Massachusetts, Amherst. 28.18A: © Michael & Patricia Fogden/Minden Pictures. 28.18B: © E.A. Janes/AGE Fotostock. 28.18C: Courtesy of the Talcott Greenhouse, Mount Holyoke College. 28.19 *inset*: David McIntyre.

Chapter 29 *Masada*: © Eddie Gerald/Alamy. *Coconut*: © Ben Osborne/Naturepl.com. 29.1 *Cycad*: David McIntyre. 29.1 *Ginkgo*: © hypnotype/Shutterstock. 29.1 *Conifer*: © Irina Tischenko/istockphoto.com. 29.1 *Magnolia*: © Dole/Shutterstock. 29.4: © Natural Visions/Alamy. 29.5A: © Susumu Nishinaga/Photo Researchers, Inc. 29.6A: © Victoria Field/Shutterstock. 29.6B: © Topic Photo Agency IN/AGE Fotostock. 29.6C: © Pichugin Dmitry/Shutterstock. 29.6D: © Frans Lanting. 29.7A *left*: David McIntyre. 29.7A *right*: © Stan W. Elems/Visuals Unlimited. 29.7B *left*: David McIntyre. 29.7B *right*: © Dr. John D. Cunningham/Visuals Unlimited. 29.9: Courtesy of Jim Peaco/National Park Service. 29.10A: David McIntyre. 29.10B: © koer/Shutterstock. 29.10C: © Marion Nickig/Picture Press/Photolibrary.com. 29.11A: © Plantography/Alamy. 29.11B: © Thomas Photography LLC/Alamy. 29.16A: © bamby/Shutterstock. 29.16B: © blickwinkel/Alamy. 29.16C: David McIntyre. 29.16D: © Michel de Nijs/istockphoto.com. 29.16E: © Yuri Vainshtein/istockphoto.com. 29.16F: © Denis Pogostin/istockphoto.com. 29.18A: Photo by David McIntyre, courtesy of the University of Massachusetts Biology Department Greenhouses. 29.18B: David McIntyre. 29.18C: © WILDLIFE GmbH/Alamy. 29.18D: © blickwinkel/Alamy. 29.18E: David McIntyre. 29.18F: © Mike Donenfeld/Shutterstock. 29.19A: © amit erez/istockphoto.com. 29.19B: © Tootles/Shutterstock. 29.19C: © Charlotte Erpenbeck/Shutterstock. 29.20A, B: Courtesy of David Hillis. 29.20C: © Susan Law Cain/istockphoto.com.

Chapter 30 *Fusarium*: © Dr. Gary Gaugler/Visuals Unlimited. *Ant*: © L. E. Gilbert/Biological Photo Service. 30.3: © David M. Phillips/Visuals Unlimited. 30.4B: © Dr. Jeremy Burgess/Photo Researchers, Inc. 30.6: © Richard Packwood/Photolibrary.com. 30.7A: © G. T. Cole/Biological Photo Service. 30.8: © N. Allin & G. L. Barron/Biological Photo Service. 30.9A: Courtesy of David Hillis. 30.9B: David McIntyre. 30.9C: Courtesy of David Hillis. 30.11A: © R. L. Peterson/Biological Photo Service. 30.11B: © Ken Wagner/Visuals Unlimited. 30.12A: © J. Robert Waaland/Biological Photo Service. 30.12B: © M. F. Brown/Visuals Unlimited. 30.12C: © Dr. John D. Cunningham/Visuals Unlimited. 30.12D: © Biophoto Associates/Photo Researchers, Inc. 30.13: © Eye of Science/Photo Researchers, Inc. 30.14: © John Taylor/Visuals Unlimited. 30.15: © G. L. Barron/Biological Photo Service. 30.16A: © Photoshot Holdings Ltd/Alamy. 30.16B: © Biosphoto/Le Moigne Jean-Louis/Peter Arnold Inc. 30.17: © Andrew Syred/SPL/Photo Researchers, Inc. 30.18A: © Jämsen Jorma/AGE Fotostock. 30.18B: © Matt Meadows/Peter Arnold Inc.

Chapter 31 *Placozoan:* © Ana Yuri Signorovitch. *Sponge:* © Fred Bavendam/Minden Pictures. *Ctenophore:* © Norbert Wu/Minden Pictures. 31.2A: Courtesy of J. B. Morrill. 31.2B: From G. N. Cherr et al., 1992. *Microsc. Res. Tech.* 22: 11. Courtesy of J. B. Morrill. 31.4A: © Ed Robinson/Photolibrary.com. 31.4B: © Steve Gschmeissner/Photo Researchers, Inc. 31.4C: © DEA/Christian Ricci/Photolibrary.com. 31.5A: © Jurgen Freund/Naturepl.com. 31.5B: © John Bell/istockphoto.com. 31.6A: © John A. Anderson/istockphoto.com. 31.6B: © Kevin Schafer/DigitalVision/Photolibrary.com. 31.6B inset: © Mike Rogal/Shutterstock. 31.7B: © David Patterson & Aimlee Laderman/micro*scope. 31.8A: © IntraClique LLC/Shutterstock. 31.8B: © Stockbyte/PictureQuest. 31.9: Adapted from F. M. Bayerand & H. B. Owre, 1968. *The Free-Living Lower Invertebrates,* Macmillan Publishing Co. 31.10A: © Cathy Keifer/Shutterstock. 31.10B: David McIntyre. 31.12A: © First Light/Alamy. 31.12B: © Charlie Bishop/istockphoto.com. 31.13A: © Dave Watts/Naturepl.com. 31.13B: © Larry Jon Friesen. 31.14 inset: © Scott Camazine/Phototake/Alamy. 31.15: © Larry Jon Friesen. 31.16A: © Jurgen Freund/Naturepl.com. 31.16B: David McIntyre. 31.16C: © David Wrobel/Visuals Unlimited. 31.18B: © Larry Jon Friesen. 31.19: Adapted from F. M. Bayerand & H. B. Owre, 1968. *The Free-Living Lower Invertebrates,* Macmillan Publishing Co. 31.20A: © Larry Jon Friesen. 31.20B: © Georgette Douwma/Naturepl.com. 31.20C: © Larry Jon Friesen. 31.21A: © Jurgen Freund/Naturepl.com. 31.21B: © Stephan Kerkhofs/Shutterstock. 31.22: Adapted from F. M. Bayerand & H. B. Owre, 1968. *The Free-Living Lower Invertebrates,* Macmillan Publishing Co.

Chapter 32 *Strepsipterans:* Courtesy of Dr. Hans Pohl. *Wasp:* © Sean McCann. 32.2: © blickwinkel/Alamy. 32.3A: From D. C. García-Bellido & D. H. Collins, 2004. *Nature* 429: 40. Courtesy of Diego García-Bellido Capdevila. 32.3B: © Piotr Naskrecki/Minden Pictures. 32.6: © Alexis Rosenfeld/Photo Researchers, Inc. 32.7A: © Larry Jon Friesen. 32.8B: © Robert Brons/Biological Photo Service. 32.9B: © Larry Jon Friesen. 32.10A: © Fred Bavendam/Minden Pictures. 32.11: © David Wrobel/Visuals Unlimited. 32.13A: © Larry Jon Friesen. 32.13B: Courtesy of Cindy Lee Van Dover. 32.13C: © Pakhnyushcha/Shutterstock. 32.13D: © Larry Jon Friesen. 32.15A: © Larry Jon Friesen. 32.15B: © Dave Fleetham/Photolibrary.com. 32.15C: © Larry Jon Friesen. 32.15F: © Reinhard Dirscherl/Alamy. 32.16A: Courtesy of Jen Grenier and Sean Carroll, U.Wisconsin. 32.16B: Courtesy of Graham Budd. 32.16C: Courtesy of Reinhardt Møbjerg Kristensen. 32.17B: © Grave/Photo Researchers, Inc. 32.17C: © Steve Gschmeissner/Photo Researchers, Inc. 32.18: © Pascal Goetgheluck/Photo Researchers, Inc. 32.19A: © Michael Fogden/Photolibrary.com. 32.19B: © Steve Gschmeissner/Photo Researchers, Inc. 32.20: © Kevin Schafer/Alamy. 32.21A: © David M Dennis/OSF/Photolibrary.com. 32.21B: © John R. MacGregor/Peter Arnold, Inc. 32.22A: © David Shale/Naturepl.com. 32.22B: © Frans Lanting. 32.23A: © Kelly Swift, www.swiftinverts.com. 32.23B: © Larry Jon Friesen. 32.23C: © Nigel Cattlin/Alamy. 32.23D: Photo by Eric Erbe; colorization by

Chris Pooley/USDA ARS. 32.24A, B: © Larry Jon Friesen. 32.24C: © Solvin Zankl/Naturepl.com. 32.24D: © Larry Jon Friesen. 32.24E: © Norbert Wu/Minden Pictures. 32.27: © Oxford Scientific Films/Photolibrary.com. 32.28: © Mark Moffett/Minden Pictures. 32.29A: © John C. Abbott/Abbott Nature Photography. 32.29B: © Dr. Torsten Heydenreich/imagebroker/Alamy. 32.29C: © Larry Jon Friesen. 32.29D: David McIntyre. 32.29E: © Papilio/Alamy. 32.29F: © CorbisRF/Photolibrary.com. 32.29G: © Larry Jon Friesen. 32.29H: © Juniors Bildarchiv/Alamy.

Chapter 33 *Gastric-brooding frog:* © Michael Tyler/ANTPhoto.com. *Marsupial frog:* © Michael Fogden/OSF/Photolibrary.com. 33.2: From S. Bengtson, 2000. Teasing fossils out of shales with cameras and computers. *Palaeontologia Electronica* 3(1). 33.4A: © Hal Beral/Visuals Unlimited. 33.4B: © Jose B. Ruiz/Naturepl.com. 33.4C: © WaterFrame/Alamy. 33.4D: © Peter Scoones/Photo Researchers, Inc. 33.4E: © Larry Jon Friesen. 33.5A: © C. R. Wyttenbach/Biological Photo Service. 33.6A: © Stan Elems/Visuals Unlimited, Inc. 33.6B: © Larry Jon Friesen. 33.7A: © Marevision/AGE Fotostock. 33.7B: © David Wrobel/Visuals Unlimited. 33.9A: © Brandon Cole Marine Photography/Alamy. 33.9B left: © Marevision/AGE Fotostock. 33.9B right: © anne de Haas/istockphoto.com. 33.11B: © Roger Klocek/Visuals Unlimited. 33.12A: © David B. Fleetham/OSF/Photolibrary.com. 33.12B: © Kelvin AitkenAGE Fotostock. 33.12C: © Norbert Wu/Minden Pictures. 33.13A: © Peter Pinnock/ImageState/Alamy. 33.13B, C: © Larry Jon Friesen. 33.13D: © Norbert Wu/Minden Pictures. 33.14A: © Peter Scoones/Photo Researchers, Inc. 33.14B: © Tom McHugh/Photo Researchers, Inc. 33.14C: © Ted Daeschler/Academy of Natural Sciences/VIREO. 33.16A: © Morley Read/Naturepl.com. 33.16B: © Michael & Patricia Fogden/Minden Pictures. 33.16C: © Jack Goldfarb/Design Pics, Inc./Photolibrary.com. 33.16D: Courtesy of David Hillis. 33.19A: © Dave B. Fleetham/OSF/Photolibrary.com. 33.19B: © C. Alan Morgan/Peter Arnold, Inc. 33.19C: © Cathy Keifer/Shutterstock. 33.19D: © Larry Jon Friesen. 33.20A: © Susan Flashman/istockphoto.com. 33.20B: © Gerry Ellis, DigitalVision/PictureQuest. 33.21A: From X. Xu et al., 2003. *Nature* 421: 335. © Macmillan Publishers Ltd. 33.21B: © Tom & Therisa Stack/Painet, Inc. 33.22: © Melinda Fawver/istockphoto.com. 33.23A: © Tim Zurowski/All Canada Photos/Photolibrary.com. 33.23B: © Roger Wilmhurst/Foto Natura/Minden Pictures. 33.23C: Courtesy of Andrew D. Sinauer. 33.23D: © Tom Vezo/Minden Pictures. 33.24A: © imagebroker.net/Photolibrary.com. 33.24B: © Dave Watts/Alamy. 33.25A: © Ingo Arndt/Naturepl.com. 33.25B: © JTB Photo Communications/Photolibrary.com. 33.25C: © Michael & Patricia Fogden/Minden Pictures. 33.26A: © Design Pics Inc/Photolibrary.com. 33.26B: © Barry Mansell/Naturepl.com. 33.26C: © Michael S. Nolan/AGE Fotostock. 33.26D: © John E Marriott/All Canada Photos/AGE Fotostock. 33.28: © Pete Oxford/Minden Pictures. 33.29A: © mike lane/Alamy. 33.29B: © Eric Isselée/Shutterstock. 33.30A: © Steve Bloom Images/Alamy. 33.30B: © Anup Shah/AGE Fotostock. 33.30C: © Lars Christensen/

istockphoto.com. 33.30D: © Anup Shah/Minden Pictures.

Chapter 34 *Svalbard:* Courtesy of the Global Crop Diversity Trust. *Seed:* © Dr. Richard Kessel & Dr. Gene Shih/Visuals Unlimited, Inc. 34.2A: David McIntyre. 34.2C: © Biosphoto/Thiriet Claudius/Peter Arnold Inc. 34.3A: David McIntyre. 34.3B: © Steven Wooster/Garden Picture Library/Photolibrary.com. 34.3C: © Renee Lynn/Photo Researchers, Inc. 34.6: © Biophoto Associates/Photo Researchers, Inc. 34.9A: © Biodisc/Visuals Unlimited. 34.9B: © P. Gates/Biological Photo Service. 34.9C: © Biophoto Associates/Photo Researchers, Inc. 34.9D: © Jack M. Bostrack/Visuals Unlimited. 34.9E: © John D. Cunningham/Visuals Unlimited. 34.9F: © J. Robert Waaland/Biological Photo Service. 34.9G: © Randy Moore/Visuals Unlimited. 34.10 upper: © Larry Jon Friesen. 34.10 lower: © Biodisc/Visuals Unlimited. 34.11B: © Microfield Scientific LTD/Photo Researchers, Inc. 34.13A: © Larry Jon Friesen. 34.13B: © Ed Reschke/Peter Arnold, Inc. 34.13C: © Dr. James W. Richardson/Visuals Unlimited. 34.14A left: © Ed Reschke/Peter Arnold, Inc. 34.14A right: © Biodisc/Visuals Unlimited. 34.14B: © Ed Reschke/Peter Arnold, Inc. 34.15B: Courtesy of Thomas Eisner, Cornell U. 34.15C: © Susumu Nishinaga/Photo Researchers, Inc. 34.18: © Biodisc/Visuals Unlimited. 34.19: © David M Dennis/OSF/Photolibrary.com.

Chapter 35 *Planting:* © Robert Harding Picture Library Ltd/Alamy. *Drought:* Courtesy of the International Rice Research Institute/IRRI file photo. 35.4: © Nigel Cattlin/Alamy. 35.9A: © David M. Phillips/Visuals Unlimited. 35.10: After G. D. Humble & K. Raschke, 1971. *Plant Physiology* 48: 447. 35.12: © R. Kessel & G. Shih/Visuals Unlimited. 35.13: © M. H. Zimmermann.

Chapter 36 *Family:* Dorothea Lange/Library of Congress, Prints & Photographs Division, FSA/OWI Collection. *Haiti:* Courtesy of the NASA/Goddard Space Flight Center Scientific Visualization Studio. 36.1: David McIntyre. 36.5: © Kathleen Blanchard/Visuals Unlimited. 36.9 left: © E. H. Newcomb & S. R. Tandon/Biological Photo Service. 36.9 right: © Dr. Jeremy Burgess/SPL/Photo Researchers, Inc. 36.12A: © J. H. Robinson/The National Audubon Society Collection/Photo Researchers, Inc. 36.12B: © Kim Taylor/Naturepl.com. 36.13: Courtesy of Susan and Edwin McGlew.

Chapter 37 *N. Borlaug:* © Micheline Pelletier/Sygma/Corbis. *Rice:* Courtesy of Drs. Matsuoka and Ashikari. 37.2: © Visions of America/Joe Sohm/Photolibrary.com. 37.3: From J. M. Alonso and J. R. Ecker, 2006. *Nature Reviews Genetics* 7: 524. 37.4: Courtesy of J. A. D. Zeevaart, Michigan State U. 37.5: © Sylvan Wittwer/Visuals Unlimited, Inc. 37.12: © Ed Reschke/Peter Arnold, Inc. 37.16: David McIntyre.

Chapter 38 *Girl:* © Image Source/ZUMA Press. *Postcards:* © Amoret Tanner/Alamy. 38.1A: David McIntyre. 38.1B1: © Tish1/Shutterstock. 38.1B2: © Pierre BRYE/Alamy. 38.1C: David McIntyre. 38.3A: © Biosphoto/Hazan Muriel/Peter Arnold Inc.

38.3B: © Rolf Nussbaumer/Imagebroker/ Photolibrary.com. 38.5: © Biosphoto/Gautier Christian/Peter Arnold Inc. 38.9A: David McIntyre. 38.11: Courtesy of Richard Amasino. 38.17: Courtesy of Richard Amasino and Colleen Bizzell. 38.18A: © Nigel Cattlin, Holt Studios International/Photo Researchers, Inc. 38.18B: © Maurice Nimmo/SPL/Photo Researchers, Inc. 38.19: © Russ Munn/ Agstockusa/AGE Fotostock.

Chapter 39 *Engraving:* © Mary Evans Picture Library/Alamy. *Wormwood:* © Andy Crump, TDR, WHO/SPL/Photo Researchers, Inc. 39.1: Courtesy of Robert Bowden/Kansas State University. 39.1 *inset:* © Nigel Cattlin/Alamy. 39.4: © Holt Studios International Ltd/Alamy. 39.5A: © Kim Taylor/Naturepl.com. 39.5B: © Daniel Borzynski/Alamy. 39.7: After A. Steppuhn et al., 2004. *PLoS Biology* 2: 1074. 39.9: Courtesy of Thomas Eisner, Cornell U. 39.10: © Carr Clifton/Minden Pictures. 39.11: © Dr. Jack Bostrack/Visuals Unlimited, Inc. 39.12: © Martin Harvey/Peter Arnold Inc. 39.13: © Simon Fraser/SPL/Photo Researchers, Inc. 39.14: © imagebroker/Alamy. 39.15: © John N. A. Lott/Biological Photo Service. 39.18: © Jurgen Freund/Naturepl.com. 39.19: Courtesy of Ryan Somma.

Chapter 40 *P. Radcliffe:* © PCN Black/Alamy. *Soldier:* Courtesy of Major Bryan "Scott" Robison. 40.3A: © Gladden Willis/Visuals Unlimited. 40.3B: From Ross, Pawlina, and Barnash, 2009. *Atlas of Descriptive Histology.* Sinauer Associates: Sunderland, MA. 40.3C: © Ed Reschke/Peter Arnold, Inc. 40.4A: From Ross, Pawlina, and Barnash, 2009. *Atlas of Descriptive Histology.* Sinauer Associates: Sunderland, MA. 40.4B: © Manfred Kage/ Peter Arnold Inc. 40.4C: © SPL/Photo Researchers, Inc. 40.5A: © Ed Reschke/Peter Arnold, Inc. 40.5B: From Ross, Pawlina, and Barnash, 2009. *Atlas of Descriptive Histology.* Sinauer Associates: Sunderland, MA. 40.5C: © Dennis Kunkel Microscopy, Inc. 40.6A: © James Cavallini/Photo Researchers, Inc. 40.6B: © Innerspace Imaging/SPL/Photo Researchers, Inc. 40.10B: © Frans Lanting. 40.10C: © Akira Kaede/DigitalVision/ Photolibrary.com. 40.11: © Greg Epperson/ istockphoto.com. 40.12: © Gerry Ellis/ DigitalVision. 40.14: Courtesy of Anton Stabentheiner. 40.17: From Ross, Pawlina, and Barnash, 2009. *Atlas of Descriptive Histology.* Sinauer Associates: Sunderland, MA. 40.18A: © Robert Shantz/Alamy. 40.18B: © Jim Brandenburg/Minden Pictures.

Chapter 41 *J. Canseco:* © Ezra O. Shaw/ Allsport/Getty Images. *Body builder:* © David Reed/Alamy. 41.2 *Prolactin:* Data from PDB 1RW5. K. Teilum et al., 2005. *J. Mol. Biol.* 351: 810. 41.2 *Fish:* © Alaska Stock LLC/Alamy. 41.2 *Amphibian:* © Gustav W. Verderber/Visuals Unlimited, Inc. 41.2 *Birds:* © Peter Bisset/Stock Connection Distribution/Alamy. 41.2 *Mammals:* © Ale Ventura/PhotoAlto/ Photolibrary.com. 41.10A: © Ed Reschke/Peter Arnold, Inc. 41.10C: © Scott Camazine/Photo Researchers, Inc. 41.16: Courtesy of Gerhard Heldmaier, Philipps U.

Chapter 42 *Painting:* © The Bridgeman Art Library/Getty Images. 42.8: © Steve Gschmeissner/Photo Researchers, Inc.

Chapter 43 *Queen bee:* David McIntyre. *Swarm:* © Stephen Dalton/Minden Pictures. 43.1A: © P&R Photos/AGE Fotostock. 43.1B: © Constantinos Petrinos/Naturepl.com. 43.2A: © Patricia J. Wynne. 43.5: © SIU/Peter Arnold, Inc. 43.6: © ullstein - Ibis Bildagentur/Peter Arnold, Inc. 43.7A: © Morales/AGE Fotostock. 43.7B: © Dave Watts/Naturepl.com. 43.9B: © Ed Reschke/Peter Arnold, Inc. 43.12B: © P. Bagavandoss/Photo Researchers, Inc. 43.15C: © S. I. U. School of Med./Photo Researchers, Inc. 43.17: Courtesy of The Institute for Reproductive Medicine and Science of Saint Barnabas, New Jersey.

Chapter 44 *Organs:* © Mads Abildgaard/ istockphoto.com. *Egg and sperm:* © Dr. David M. Phillips/Visuals Unlimited. 44.1: Courtesy of Richard Elinson, U. Toronto. 44.3A *left:* From H. W. Beams and R. G. Kessel, 1976. *American Scientist* 64: 279. 44.3A *center, right:* © Dr. Lloyd M. Beidler/Photo Researchers, Inc. 44.3B: From H. W. Beams and R. G. Kessel, 1976. *American Scientist* 64: 279. 44.3C: Courtesy of W. Baker and G. Schubiger. 44.4: © Dr. Y. Nikas/ Phototake/Alamy. 44.4: From J. G. Mulnard, 1967. *Arch. Biol.* (Liege) 78: 107. Courtesy of J. G. Mulnard. 44.14D: Courtesy of K. W. Tosney and G. Schoenwolf. 44.15B: Courtesy of K. W. Tosney. 44.19A: © CNRI/SPL/Photo Researchers, Inc. 44.19B: © Dr. G. Moscoso/ SPL/Photo Researchers, Inc. 44.19C: © Tissuepix/SPL/Photo Researchers, Inc. 44.19D: © Petit Format/Photo Researchers, Inc.

Chapter 45 *Dentist:* © Image Source/ZUMA Press. *Brain scan:* Courtesy of Dr. Kevin LaBar, Duke U. 45.4B: © C. Raines/Visuals Unlimited. 45.7: From A. L. Hodgkin & R. D. Keynes, 1956. *J. Physiol.* 148: 127.

Chapter 46 *Snake:* © Casey K. Bishop/ Shutterstock. *Bat:* © Stephen Dalton/Photo Researchers, Inc. 46.3A: © Hans Pfletschinger/ Peter Arnold, Inc. 46.3B: David McIntyre. 46.10A: © P. Motta/Photo Researchers, Inc. 46.16A: © Dennis Kunkel Microscopy, Inc. 46.19: © Omikron/Science Source/Photo Researchers, Inc.

Chapter 47 *London:* © Bluesky International Limited. *Rat:* Courtesy of Daoyun Ji and Matthew Wilson. 47.8: Photo from "Brain: The World Inside Your Head," © Evergreen Exhibitions. 47.14A: © David Joel Photography, Inc. 47.16: © Wellcome Dept. of Cognitive Neurology/SPL/Photo Researchers, Inc.

Chapter 48 *J. Joyner-Kersee:* © AFP/Getty Images. *Frog:* © OSF/Photolibrary.com. 48.1 *Micrograph:* © Frank A. Pepe/Biological Photo Service. 48.2: © Tom Deerinck/Visuals Unlimited. 48.4: © Don W. Fawcett/Photo Researchers, Inc. 48.7: © Manfred Kage/Peter Arnold Inc. 48.8: © SPL/Photo Researchers, Inc. 48.11: Courtesy of Jesper L. Andersen. 48.18: © Robert Brons/Biological Photo Service.

Chapter 49 *Elephant:* © Steve Bloom, stevebloom.com. *Shark:* © Thomas Aichinger/ VW Pics/ZUMA Press. 49.1A: © Ross Armstrong/AGE Fotostock. 49.1B: © WaterFrame/Alamy. 49.1C: © Photoshot Holdings Ltd/Alamy. 49.4B: © Andrew Darrington/Alamy. 49.4C: Courtesy of Thomas Eisner, Cornell U. 49.10 *Bronchi:* © SPL/Photo

Researchers, Inc. 49.10 *Alveoli:* © P. Motta/Photo Researchers, Inc. 49.13: © Ross Couper-Johnston/Naturepl.com. 49.17: After C. R. Bainton, 1972. *J. Appl. Physiol.* 33: 775.

Chapter 50 *R. Lewis:* © NBAE/Getty Images. *Emergency room:* © BananaStock/Alamy. 50.8A: © Brand X Pictures/Alamy. 50.9: After N. Campbell, 1990. *Biology,* 2nd Ed., Benjamin Cummings. 50.10B: © CNRI/Photo Researchers, Inc. 50.12: © Science Source/ Photo Researchers, Inc. 50.15A: © Chuck Brown/Science Source/Photo Researchers, Inc. 50.15B: © Biophoto Associates/Science Source/Photo Researchers, Inc.

Chapter 51 *Pima:* © Marilyn "Angel" Wynn/ Nativestock.com. *Burger:* © matt/Shutterstock. 51.1: Courtesy of Andrew D. Sinauer. 51.3: © Hartmut Schwarzbach/Peter Arnold Inc. 51.8C *Microvilli:* © Dennis Kunkel Microscopy, Inc. 51.13: © Eye of Science/SPL/Photo Researchers, Inc. 51.19: © Science VU/Jackson/Visuals Unlimited.

Chapter 52 *Bat:* © Michael Fogden/OSF/ Photolibrary.com. *Kangaroo rat:* © Mary McDonald/Naturepl.com. 52.1 *inset:* © Kim Taylor/Naturepl.com. 52.2B: © Morales/AGE Fotostock. 52.8A: © Susumu Nishinaga/Photo Researchers, Inc. 52.8B: © Science Photo Library/Photolibrary.com. 52.8C: © CNRI/SPL/ Photo Researchers, Inc. 52.11: From L. Bankir & C. de Rouffignac, 1985. *Am. J. Physiol.* 249: R643-R666. Courtesy of Lise Bankir, INSERM Unit, Hôpital Necker, Paris. 52.14: © Hank Morgan/Photo Researchers, Inc.

Chapter 53 *Monkey:* © Kennosuke Tsuda/ Minden Pictures. *Spider:* © Don Johnston/All Canada Photos/Photolibrary.com. 53.2A: © Arco Images GmbH/Alamy. 53.5B: © Maximilian Weinzierl/Alamy. 53.8A: © Nina Leen/Time Life Pictures/Getty Images. 53.8B: © Frans Lanting. 53.12A: © Interfoto/Alamy. 53.12B: © Shaun Cunningham/Alamy. 53.12C: © Shattil & Rozinski/Naturepl.com. 53.14A: © Jeremy Woodhouse/Photolibrary.com. 53.14B: © Elvele Images Ltd/Alamy. 53.17B: © Tui De Roy/Minden Pictures. 53.20: © Tui De Roy/Minden Pictures. 53.22: © J. Jarvis/ Visuals Unlimited, Inc. 53.23: © GeorgeLepp.com.

Chapter 54 *Beetle:* David McIntyre. *Cattle:* © Robert Harding Picture Library Ltd/Alamy. 54.4A: Courtesy of Jan M. Storey. 54.4B *upper:* © jack thomas/Alamy. 54.4B *lower:* © Lee & Marleigh Freyenhagen/Shutterstock. *Tundra, left:* © John Schwieder/Alamy. *Tundra, right:* © Mike Harding/Imagestate/Photolibrary.com. *Ptarmigan:* © Rolf Hicker Photography/Alamy. *Boreal, left:* © John E Marriott/All Canada Photos/AGE Photostock. *Boreal, right:* © Robert Harding Picture Library Ltd/Alamy. *Boreal owl:* © John Cancalosi/Alamy. *Temperate deciduous, left:* © Jack Milchanowski/AGE Fotostock. *Temperate deciduous, right:* © Carr Clifton/Minden Pictures. *Butterfly:* © Rick & Nora Bowers/Alamy. *Temperate grasslands, left:* © Tim Fitzharris/Minden Pictures. *Temperate grasslands, right:* © Mikhail Yurenkov/Alamy. *Rhea:* © Biosphoto/Muñoz Juan-Carlos/Peter Arnold Inc. *Hot desert, left:* © Peter Lilja/AGE Fotostock. *Hot desert, right:* © David M. Schrader/Shutterstock. *Beetle:* © Michael

Index